Polarimetric Microwave Imaging

Ruliang Yang · Bowei Dai · Lulu Tan ·
Xiuqing Liu · Zhen Yang · Haiying Li

Polarimetric Microwave Imaging

国防工業出版社
National Defense Industry Press

Springer

Ruliang Yang
Institute of Electronics
Chinese Academy of Sciences
Beijing, China

Bowei Dai
Institute of Microelectronics
Chinese Academy of Sciences
Beijing, China

Lulu Tan
East China Research Institute of Electronics
Engineering
Hefei, China

Xiuqing Liu
Institute of Electronics
Chinese Academy of Sciences
Beijing, China

Zhen Yang
Institute of Electronics
Chinese Academy of Sciences
Beijing, China

Haiying Li
National Astronomical Observatories
Chinese Academy of Sciences
Beijing, China

ISBN 978-981-15-8896-9 ISBN 978-981-15-8897-6 (eBook)
https://doi.org/10.1007/978-981-15-8897-6

Jointly published with National Defense Industry Press
The print edition is not for sale in China (Mainland). Customers from China (Mainland) please order the print book from: National Defense Industry Press.
ISBN of the Co-Publisher's edition: 978-7-118-10563-6

This Springer imprint is published by the registered company Springer Nature Singapore Pte Ltd.
The registered company address is: 152 Beach Road, #21-01/04 Gateway East, Singapore 189721, Singapore

Foreword

The basic principle of remote sensing is the interaction process between electromagnetic waves and the scattering radiation characteristics of different substances in nature. Due to its all-weather day and night detection capability, microwave remote sensing is different from other electromagnetic spectrum segments in the scattering and radiation characteristics of the target macrostructure and components properties and plays an important role in the fields of Earth observation, deep space exploration, and military. Such a technology has become the frontier of space remote sensing.

The characteristic information of a target is mainly manifested in the four kinds of information of the target electromagnetic wave, including the changes of the main electromagnetic wave elements such as amplitude (intensity), phase change, frequency, and polarization characteristics. Microwave remote sensing technology and its wide application have developed more than half a century. General speaking, scientists and users mainly focus on how to extract the basic information that reflects the authenticity of the observation target from the four elements of the target electromagnetic wave. The amplitude, phase, and frequency response information are proposed relatively earlier and have been successfully applied to remote sensing. Due to the limited technology and theory development, the extraction of full polarization information was just achieved in the past 10 years, which has made a significant improvement of remote sensing theory and technology. Remote sensing becomes a full electromagnetic wave information technology. Polarimetric microwave imaging technology has become one of the most important development directions in the field of microwave remote sensing.

The polarimetric microwave imaging radar transmits and receives electromagnetic waves of different polarization modes through multiple channels to form a complete polarization base, obtains a polarization scattering matrix, comprehensively obtains the scattering characteristics of the target in the observation direction, and greatly improves the ability to acquire the target scattering information. The polarization scattering matrix contains more abundant information, which enables us to deeply analyze and extract the physical properties (direction, shape, roughness, dielectric constant, etc.) of the target, and facilitate the research of parameter inversion. Polarimetric microwave imaging radar obtains completed scattering information of the target and extracts the polarization characteristic parameters, which provides more information for large-area object classification, target detection, and recognition.

In the past decade, polarimetric microwave imaging radar technology has been booming, and a series of spaceborne and airborne polarimetric microwave imaging radar systems have been developed. A large amount of polarimetric microwave imaging data has been acquired by these systems, which provide rich material for people to study polarization information. Scholars from various countries have conducted many studies on the analysis and application of polarimetric microwave imaging radar data. Many research results have been obtained in the aspects of scattering matrix decomposition, target polarization analysis, target classification, surface parameter inversion, and polarization image filtering. At present, polarimetric microwave imaging radar has been used in various areas, such as crop yield estimation, forest survey, biomass estimation, oceanography, geology, hydrology, resources, environment, disaster monitoring, military, and other fields.

In recent years, China's polarimetric microwave imaging radar technology has made great progress. Spaceborne, airborne PolSAR, PolInSAR, polarized microwave scatterometer, and radiometer have been put into operation. They greatly facilitate the development of China's polarimetric microwave imaging radar technology and narrow down the gap with the international advanced level. Professor Yang Ruliang has been engaging in the research of Synthetic Aperture Radar (SAR) theory and technology for a long time. As the chief designer of the main payload of China's environmental satellite "HJ-1C," he successfully developed and launched the spaceborne SAR, and obtained a large number of high-quality images. Yang Ruliang and his team are one of the pioneer teams in China to develop polarimetric synthetic aperture radar (PolSAR) imaging technology and have gained abundant results in polarimetric microwave imaging radar system technology, information processing, and its applications.

The main chapters, embedded with a large number of simulations and practical examples of *Polarimetric Microwave Imaging*, are valuable theoretical sublimation and systematic in-depth summary of the authors' main research results, which reflect a large number of new theories, new methods, and new achievements in polarimetric microwave imaging radar systems and polarization information processing fields. Therefore, it is an excellent work with the integration of theories and practices.

It is expected that the publication of *Polarimetric Microwave Imaging* will facilitate the development of polarimetric microwave imaging radar technology in China, and bring help to scientific researchers and students in related fields who are engaged in microwave, remote sensing, space technology, and their applications.

Beijing, China

Prof. Jingshan Jiang
Academician of Chinese Academy of Engineering

Preface

Polarimetric microwave imaging technology is an important development direction in the field of microwave remote sensing.

As a typical representative of microwave imaging technology, SAR is a high-resolution microwave imaging radar. It can not only image ground stationary targets and moving targets in all-weather day and night but also penetrate natural vegetation, artificial camouflage to discover important military targets. SAR has become an important means of Earth observation and has attracted wide attention from all countries. Rapid development of microwave imaging radar system has been driven by extensive military and civilian demands. It is an urgent task to further develop a new type of microwave imaging radar system and improve the performance of the earth observation system. Polarimetric microwave imaging technology is embedded at the core of these important developments.

Polarimetric microwave imaging radar is a discipline that acquires, processes, and analyzes the polarization state of a dielectric scattering field, and deals with the complete vector characteristics of scattered electromagnetic waves. Conventional SAR uses a single polarization channel to transmit and receive electromagnetic waves, which only obtains the scattering characteristics of the ground object to a single electromagnetic wave. PolSAR transmits and receives electromagnetic waves of different polarization modes through multiple channels to form a complete polarization base, obtaining a polarization scattering matrix and the scattering characteristics of the target in the observation direction, in a comprehensive manner. Therefore, the ability to acquire the target scattering information is greatly improved. The polarization scattering matrix contains more abundant information, which enables people to deeply analyze and extract the physical properties of the target, such as direction, shape, roughness, dielectric constant, etc., and facilitate the research of parameter inversion. By obtaining the complete scattering information of the target and extracts the polarization characteristic parameters, PolSAR provides more useful information for large-area object classification, target detection, and recognition. At present, polarimetric SAR has been applied in crop classification and estimation, forest survey, biomass estimation, geology, hydrology, resources, environment, disaster monitoring, ocean, military, and other fields.

The research group has completed research projects such as the National "Airborne Multi-Polarization Synthetic Aperture Radar System Research" and "Spatial Dual-Frequency Multi-Polarization Synthetic Aperture Radar System Technology Research," and carried out international cooperation and accumulated of the research experiences in polarimetric microwave imaging technology. At the same time, a group of doctoral students were organized to study the polarization microwave imaging and polarization information processing technology systematically. This book is a research summary of these doctoral dissertations. As a companion piece to the book "High-Resolution Microwave Imaging," "Polarimetric Microwave Imaging" attempts to systematically discuss polarimetric microwave imaging radar systems and polarization information processing techniques, linking theory with practice and reflecting as much as possible in recent years the new theories, new methods, and new achievements in this field.

The book was planned and designed by Prof. Ruliang Yang and written by the following authors: Prof. RuliangYang (Chaps. 1, 3), Dr. HaiyingLi (Chaps. 1, 3), Dr. Bowei Dai (Chaps. 2, 3, 7–10, 13), Dr. Lulu Tan (Chaps. 2, 4, 5, 11, 12, 15, 16), Dr. Xiuqing Liu (Chaps. 6, 9, 10, 13, 14, 17), Dr. Zhen Yang (Chaps. 6, 13), Yuxiao Xi (Chap. 6). Prof. Ruliang Yang conducted coordination and finalization.

This book studies polarimetric microwave imaging and polarization information processing techniques, including three parts: polarization SAR system technology, polarization SAR information processing technology, and applied technology of polarized SAR image.

The part of polarization SAR system technology discusses the theoretical basis of polarization SAR, and studies the polarization SAR, polarization interferometric SAR, compact polarization SAR, and polarization SAR calibration technology (Fig. 1).

The part of polarization SAR information processing technology mainly discusses the understanding of polarization SAR image, the statistical characteristics of polarization SAR, the enhancement processing of polarization SAR image, the speckle reduction technique of polarization SAR image, target decomposition theory of polarization SAR, classification of polarization SAR image, and target detection technology.

The part of applied technology of polarized SAR image discusses the studies of the surface parameter inversion of PolSAR data, the processing of PolInSAR image, the processing of compact polarization interferometric SAR image. The impact and compensation of tilting of the azimuth for SAR data are discussed as well.

The structure of the book with 17 chapters is presented in the configuration of polarimetric microwave imaging above.

This book systematically and deeply studies the polarimetric microwave imaging and polarization information processing technology, which can be used by researchers, engineers, and technicians of electronic systems, especially microwave remote sensing and microwave imaging radars areas. The book can also be used as a reference book for research projects or courses by college and university teachers and postgraduates majoring in related fields, such as communication and information systems, electronic technology, and applied computer.

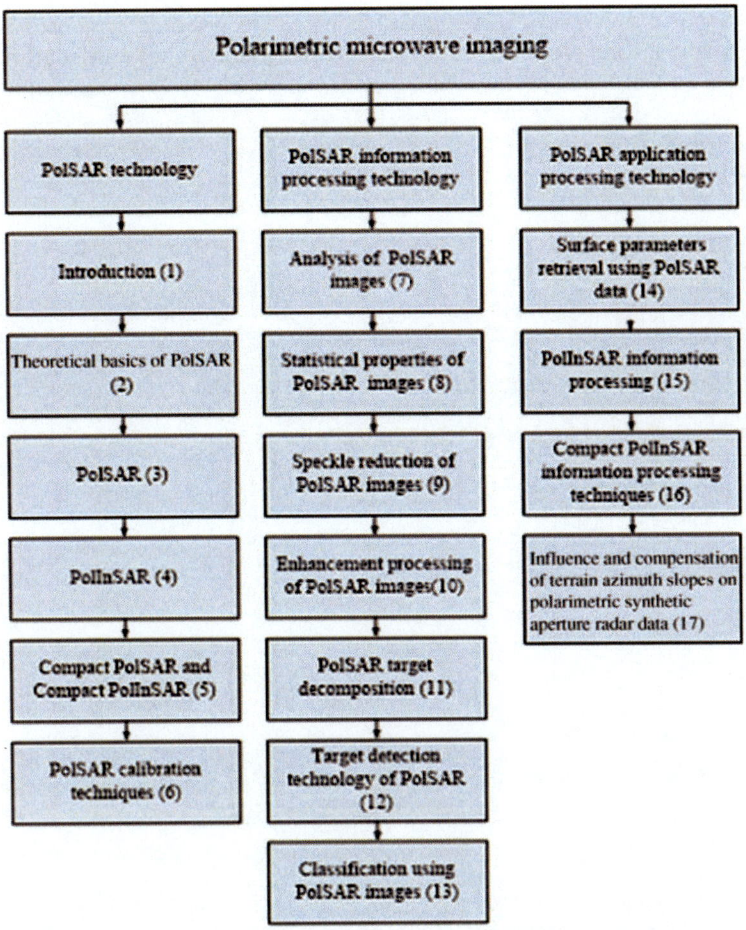

Fig. 1 Book structure table, The configuration of Polarimetric Microwave Imaging

The author would like to express sincere thanks to Academician Prof. Jingshan Jiang from the Chinese Academy of Sciences, who prefaces the book. Professor Yingning Peng, Prof. Jian Yang from Tsinghua University, and Prof. Shiyi Mao from Beijing University of Aeronautics and Astronautics are highly appreciated for their valuable opinions on this book.

The author would like to thank Dr. Xiaolei Han who translated Chaps. 1 and 6 from Chinese to English in this book, Dr. Ping Zhang who translated Chaps. 4, 7, 8, and 12. The author would like to thank the people who are concerned with and helped in this book's publication in the Institute of Electronics Chinese Academy of Sciences.

Last but not the least, I would like to thank to the National Defense Industry Press for its support of the publication of this book.

Polarimetric microwave imaging and polarization information processing are rapidly evolving technologies. Due to author's limited ability and knowledge, reader's comments on omissions, inaccuracies, and errors, are highly appreciated.

Beijing, China
May 2019

Prof. Ruliang Yang

Brief Introduction

This book introduces the radar system, information processing, and image application of the PolSAR.

The book contains 17 chapters that cover the three main parts. The research scope of the PolSAR includes the theory and system design of the PolSAR, PolInSAR, Compact PolSAR, and Calibration of the PolSAR.

The information processing of PolSAR includes the new theories and new methods of the PolSAR, such as statistical properties analysis of PolSAR images, speckle reduction, images enhancement, polarimetric target decomposition, and classification of PolSAR target detection. The polarimetric microwave imaging radar system includes the technical basics, PolSAR system, PolInSAR, Compact Polarimetric SAR, and Calibration techniques of PolSAR.

Application technology of PolSAR includes geophysical parameters retrieval of PolSAR data, PolInSAR information processing, compact PolInSAR information processing, and effects of terrain tilt in azimuth direction on PolSAR data.

The book offers a close and systematic framework. Theories are integrated with practices, which possess strong application background. It reflects new theories, methods, and achievements in the polarimatric microwave imaging fields in recent years. The readers of this book could be the researches, engineers, and scientists of electronics engineering, especially in the remote sensing and radar imaging. And it can also be used as a reference book for the faculty and postgraduate students of the engineering such as communication and information system and computer science, in colleges and universities.

Contents

About the Authors

Prof. Ruliang Yang supervisor, graduated from the Radar System Engineering Department of the University of Electronics Science and Technology of China in 1965. He was supported by K. C. Wang Foundation, Hong Kong and is a senior visiting scholar in the Engineering Department of Aberdeen University UK.

Prof. Ruliang Yang has been engaging in research of airborne and spaceborne electronics system engineering for the 50 years.

From 1965 to 1980, he was engaged in research for satellite-borne and missile-borne electronics equipment of the ballistic trajectory measurement and security system in satellites and missiles. He has taken part in satellite assembly and satellite launched for the Satellites "Orient Red" No. 1 and "Practice" No. 1 in China.

After joining the Institute of Electronics, Chinese Academy of Sciences at 1980, he was engaged in research of system collectivity engineering for satellite-borne and airborne (SAR systems. As a chief general designer, he developed the satellite-borne SAR and airborne SAR systems. Successively, he was engaged in researches of new technology and new systems in microwave imaging radars, such as the high-resolution microwave imaging radar system, the polarization SAR, the Inteferematric microwave imaging, the super-resolution information processing, the radar target identify technology, GMTI, the chaos signal SAR, and SAR radar control and guide technology, etc.

He was awarded the special level prize once and the second level prize once by the Progress of Chinese Academy of Sciences, the first level prize once and the second level prize once by the Department of State, and the National Science and Technology congress prize once. He came out with six item invent patents. He has published 240 papers, wrote 4 books, and translated 1 book. He has supervised more than 50 students, who have received their Ph.D. and master's degrees.

He was the Deputy Director of the Institute of Electronics Chinese Academy of Sciences (IECAS) and the Director of the Department of Imaging Radar System IECAS. He is the Director Member of the Unit Remote Sensing Centre of Chinese Academy of Sciences, the Commissioner of Radio Communication Committee of Chinese Communication Academy, the Commissioner of Remote Sensing Committee of Chinese Space Sciences Academy, the Committeeman of Radar and Navigation Committee of Chinese Aviation Academy, the Vice-Chief Editor of "Remote Sensing Technology and Application" journal, the Committeeman of "Journal of Radars" editor committee. He was awarded the government special allowance from the State Council.

Prof. Bowei Dai who was born in May of 1973, is the Deputy General Director of the Institute of Microelectronics, Chinese Academy of Science. His primary research focused on the PolSAR system design and information processing. He participated in multiple R&D programs of airborne and spaceborne SAR. Recently, his research areas mainly focused on the development strategy research of new-generation information technology, the advanced scientific research network construction and application, the future network technology research, and also major projects organization and management in strategic high-tech fields including IT, materials, and energy. He graduated from the Department of Electronic Engineering, Tsinghua University in July 1996. He obtained his Ph.D. in Communication and Information System from the Institute of Electronics, Chinese Academy of Sciences in July 2000. He has been working for the Chinese Academy of Science till now after his graduation. He was the Deputy General Director of the Computer Network Information Center and the Director of China Science and Technology Network Center from 2010 to 2012. He was the Deputy General Director of the Bureau of Major R&D Programs, Chinese Academy of Sciences from 2012 to 2018.

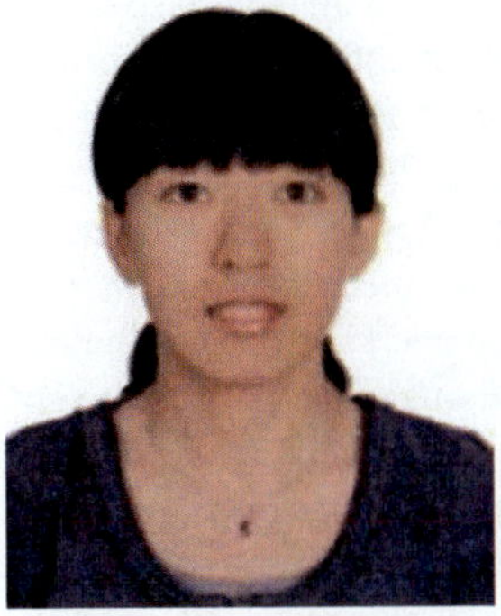

Dr. Lulu Tan Ph.D. senior engineer was born in Anqing, Anhui Province. She received the B.S. degree in electrical engineering from the Dalian University of Technology in 2004, and Ph.D. degree in communication and information system from Institute of Electronics, Chinese Academy of Sciences (IECAS) in 2010. From 2006 to 2010, she focused on the study of polarimetric SAR Interferometry. From 2010 till now, she is with East China Research Institute of Electronic Engineering (ECRIEE), where her research activity has been focusing on the design of space borne SAR system.

Dr. Xiuqing Liu Ph.D. an Associate Professor, graduated from the Institute of Electronics, Chinese Academy of Sciences and received a Ph.D. in Communication and Information Systems in January 2004. Her research direction is PolSAR information processing and application. She is mainly engaged in research on PolSAR image processing and information extraction. She has hosted one National Natural Science Foundation project; participated as an important member of one National 863 Key Project and four National Major Projects. She has conducted systematic research on information extraction such as SAR, PolSAR image and data filtering, data compensation, image enhancement, clutter suppression, image classification, inversion, and PolInSAR data processing. She has published more than ten papers and applied for three national invention patents.

Dr. Zhen Yang Ph.D. is an Associate Professor. She received her Ph.D. in Communication and Information System from the Institute of Electronics, Chinese Academy of Sciences in June 2003. She has been engaged in research of the SAR satellite missions. Her research interest was spaceborne inteferomatric SAR system, inner calibration of spaceborne SAR system. She has published some papers and came out with some item invent patents.

Dr. Haiying Li Ph.D. is an Associate Professor. She received her Ph.D. in Communication and Information System from the Institute of Electronics, Chinese Academy of Sciences in July 2002. She has been engaged in research of the SAR satellite missions in IECAS from 2002 to 2013. Her research interest was spaceborne SAR system design and simulation. She has published ten papers and came out with one item invent patents. Presently, she works at the National Astronomical Observatories, Chinese Academy of Sciences (NAOC). Her research interests are radar technology and applications on the moon and deep space exploration.

Acronyms

AASR	Azimuth Ambiguity to Signal Ratio
ANN	Artificial Neural Network
BSA	Back Scatter Alignment
CFAR	Constant false alarm ratio
CO-Null	Common Null of target
CO-POL	Common Polarization
CO-POL N	CO-Polarization Null
CP	Compact Polarization
C-PolInSAR	Compact Polarimetric Synthetic Aperture Radar Interferometry
CR	Coherence Region
CTLR	Circular Transmitting and Linear Receiving
CV	Coefficient of Variation
DBS	Double-bounce Scattering
DCP	Dual Circular Polarization
DEM	Digital Elevation Model
DOA	Direct of Arrival
DOP	Degree of Polarization
ESPRIT	Estimation of Signal Parameters via Rotational Invariance Techniques
FSA	Forward Scattering Alignment
GLRT	Generalized likelihood ration
GOPCE	Generalized Optimal Polarization Contrast enhancement
InSAR	Interferometric SAR
ISLR	Integrated Side-Lobe Ratio
LCP	Left Circular Polarization
LSE	Least Squares Estimate
ML CFAR	Maximum likelihood Constant false alarm ratio
ML	Maximum likelihood estimation
MUSIC	Multiple Signal Classification
MVM	Minimum Variance Method
NEσ^0	Noise Equivalent Sigma 0
OPCE	Optimal Polarization Contrast Enhancement

OPD	Optimal Polarization Detector
OS CFAR	Ordered Statistic Constant false alarm ratio
OVoG	Oriented Volume over Ground
PCD	Polarization Code Division
PCT	Polarization Coherence Tomography
PFD	Polarization Frequency Division
PMD	Polarization Match Detector
PMS	Power Maximization Synthesis
POA	Polarization Orientation Angle
PolInSAR	Polarimetric Synthetic Aperture Radar Interferometry
PolSAR	Polarimetric Synthetic Aperture Radar
PR	Power ring
PRF	Pulse Repetition Frequency
PRT	Pulse Repetition Time
Pseudo quad-pol	Pseudo quad Polarization
PSLR	Peak Side-Lobe Ratio
PSM	Polarization Subspace Method
PTD	Polarization Time Division
PWF	Polarization Whitening Filter
RASR	Range Ambiguity to Signal Ratio
RCP	Right Circular Polarization
rms	Root mean square
RVoG	Random Volume over Ground
SAR	Synthetic Aperture Radar
SLC	Single Look Complex
SNR	Signal to Noise Ratio
Span	Span
SPM	Small Perturbation Model
SRTM	Shuttle Radar Topography Mapper
SS	Surface Scattering
SSLR	Second Side-Lobe Ratio
SUMT	Sequential Unconstrained Minimization Technique
TLS-ESPRIT	Total Least Square ESPRIT
VS	Volume Scattering
X-POL	Cross Polarization
X-POL N	X-Polarization Null

Chapter 1
Introduction

Synthetic Aperture Radar (SAR) is a high-resolution active microwave remote sensing system for all-weather, day and night, and wide-coverage imaging capability. Since its first appearance in the 1950s, SAR has gradually developed into an important remote sensing tool. In recent years, SAR technology has made great progress, and Polarization SAR (PolSAR) has become an important trend in the development of microwave imaging remote sensing.

1.1 Polarization Synthetic Aperture Radar System

1.1.1 Development of Polarization SAR System [1, 2]

The propagation and scattering of electromagnetic waves are vector phenomena, and polarization is the vector feature used to study electromagnetic waves. Polarization Synthetic Aperture Radar (PolSAR) measures the polarization scattering characteristics of a ground object under different combinations of transmitting and receiving polarization and is expressed in the form of a polarization scattering matrix. Since the polarization of electromagnetic waves is sensitive to the surface roughness, physical properties, dielectric constant, geometry, and orientation of the target, polarization measurement can greatly improve the ability of the imaging radar to acquire target information.

Conventional single-polarization SAR uses a single fixed-polarized antenna to transmit and receive electromagnetic waves. It can only measure one component of the scattered wave vector and will lose the other components of the polarization scattering characteristics. In order to preserve all the information of the backscattered echo, a vector measurement process must be used to measure the polarization characteristics of the scattered wave. The PolSAR describes the amplitude and phase characteristics of the target scattered echo by measuring the full polarization scattering matrix in each resolution unit of the ground object to obtain the response of

R. Yang et al., *Polarimetric Microwave Imaging*,
https://doi.org/10.1007/978-981-15-8897-6_1

the ground object to different electromagnetic waves. The difference between multi-polarization SAR and single-polarization SAR is that the former is so sensitive to the polarization state of the transmitting and receiving signals that it can measure the amplitude and relative phase of the echo under different polarization states, thus PolSAR will improve the ability of the radar to acquire target information and provide the basis for further studying the target scattering characteristics.

Although the concept of polarization had a long history, it was not until the 1950s that people began to study the application of polarization in radar and made some progress [3]. In 1950, G. W. Sinclair introduced the concept of scattering matrix to describe the radar cross-sectional area of coherent scatters in his study on the transmitting and receiving of elliptical polarization wave [3]. In 1951, E. M. Kennaugh used the scattering matrix to study the polarization characteristics of radar echoes in the microwave remote sensing field and gave the concept of the optimal polarization state. After Kennaugh, due to the imperfect development of polarization theory, the research on the application of polarization in radar was still going on, but the results were few.

It was not until the publication of Dr. Huynen's thesis in 1970 [4] that a new boom of theoretical and experimental research on polarization has begun, and a lot of achievements have been obtained. In his Ph.D. thesis, Dr. Huynen deduced the theory of radar target phenomenology based on Kennaugh optimal polarization state and put forward the concept of random medium scattering decomposition, so he promoted the theoretical research of radar polarization to a new level. In 1981, Poleman proposed the concept of polarization synthesis and made an important contribution to the practical application of polarization. W. M. Boerner further completed the study of the polarization characteristics for scattered electromagnetic waves of terrain targets [5], extended Kennaugh's optimal polarization theory, and applied polarization analysis to target recognition. Ulaby and Elachi demonstrated the application of polarization in the geophysical parameter estimation [6], Valenzuela, Plant, and Alpers described the application of multi-polarization SAR and polarimetric radar imaging in ocean wave and ocean current remote sensing [7].

In 1985, NASA Jet Propulsion Laboratory (JPL) successfully developed the first L-band fully polarization Airborne Synthetic Aperture Radar (AIRSAR), which was loaded on the CV990 aircraft and pioneered the practical application of polarimetric microwave imaging radar. The radar used two orthogonal linear polarization antennas to transmit and receive electromagnetic waves. For the first time, the fully polarimetric measurement was achieved, the backscattered power and relative phase of all combinations of transmitting and receiving polarization states were obtained. Since then, NASA/JPL has extended the AIRSAR system from single-band to full polarization in C-, L-, and P-bands, and achieved the first successful flight in 1988.

With the rapid development of polarization measurement technology, a series of multi-frequency and multi-polarization AIRSAR systems have been developed one after another, such as ERIM SAR (X-, C-, and L-band full polarization) of the University of Michigan Environmental Research Institute (ERIM), SAR 580 (X- and C-band full polarization) of the Canadian Remote Sensing Centre (CCRS), E-SAR (S-, L-, and P-band full polarization) and its improved F-SAR (X-, C-, Land P-band full polarization) of German Aerospace Center (DLR), EMI-SAR (C- and L-band full polarization) of the Danish Center for Remote Sensing (DCRS), RAMSES multi-band radar (P-, L-, S-, C-, X- and Ku-band full polarization) from French Aerospace Research Agency (ONERA), and PiSAR (X- and L-band full polarization) jointly developed by the Japan National Space Development Agency (NASDA) and the Communication Research Laboratory (CRL). The above AIRSAR systems have acquired a large number of high-resolution and accurately calibrated fully polarimetric data in many flights, which greatly improved the ability of radar target detection, identification, classification, and parameter quantitative inversion.

The spaceborne PolSAR system began in 1994, the USA cooperated with Germany and Italy, SIR-C/X-SAR. The SIR-C/X-SAR loaded on the space shuttle successfully carried out two flight tests in April and October 1994 and obtained C- and L-band full-polarization and X-band single-polarization SAR images. On this basis, the research on vegetation distribution, soil moisture, and ocean dynamics was carried out. The SIR-C/X-SAR with the superiority of all-weather, day and night, multi-frequency, multi-polarization, high-resolution, and fast large area imaging ability leads the development direction of spaceborne SAR toward the multi-frequency and multi-polarization, and its advanced technologies have been inherited and developed by the subsequent spaceborne SAR systems.

At present, a number of PolSAR satellites have been successfully launched in the world, as listed in Table 1.1.

The Envisat satellite, launched by ESA in 2002, carries the C-band multi-polarization Advanced Synthetic Aperture Radar (ASAR). The Alternating Polarization (AP) mode of ASAR allows near-simultaneous collection of data from two linear polarization combinations (HH/VV or HH/HV or VH/VV) for SAR image.

The ALOS satellite, launched by Japan in 2006, is equipped with an L-band Polarimetric SAR (PALSAR). PALSAR can be operated in fine-beam mode with single polarization (HH or VV), dual polarization (HH/HV or VV/VH), or full polarization (HH/HV/VH/VV).

The TerraSAR-X satellite was launched by Germany in 2007. The experimental DRA (Dual Receive Antenna) Mode allows splitting of the receiving aperture into two halves to enable quad-polarization measurements.

COSMO-SkyMed (Constellation of Small Satellites for Mediterranean Basin Observation) is the largest Italian investment in Space Systems for Earth Observation. This mission consists of a constellation of four Low Earth Orbit mid-sized satellites, each equipped with a multi-mode high-resolution X-band SAR. In Stripmap mode, SAR is operated mainly with a fixed antenna beam in either single-polarization or multi-polarization models.

Radarsat-2, launched by Canada in December 2007, provides multiple modes of dual polarization (HH/HV or VH/VV) and full polarization in the C-band for meeting the requirements of marine and land imaging.

The RISAT-1 satellite, launched by India in 2012, carries a C-band SAR, which can obtain the multi-polarization (HH or VV, HV or VH) and fully polarimetric image.

Table 1.1 Spaceborne multi-polarization SAR system launched at abroad

Name	SIR-C\X-SAR [8]	ASAR [9, 10]	ALOS\PALSAR [11]	RadarSat-2 [12]	TerraSAR-X [13, 14]	Cosmo-Skymed [15, 2]	RiSAT [16, 17]
Country	USA	ESA	Japan	Canada	Germany	Italy	India
Launch time	1994.4 1994.10	2002.3	2006.1.24	2007.12	2007.6.15	1#, 2007.6.8 2#, 2007.12.9 3#, 2008.10.25 4#, 2010.11. 5	2012.4.26
Altitude (km)	225	768	691.65	798	514	619	536
Frequency (GHz)	L:1.28; C:5.3; X:9.6	5.331	1.27	5.405	9.65	9.65	5.35
Incident angle (°)	20–55°	15–45°	8–60°	10–60°	20–55°	20–60°	12–55°
Antenna dimensions (m × m)	L:12 × 2.95 C:12 × 0.75	10 × 1.3	8.9 × 2.9	15 × 1.5	4.8 × 0.7	6 × 1.2	6 × 2
Antenna type	Phased array	Phased array	Phased array	Phased array	Phased array	Phased array	Phased array
T/R modules	Dual-polarization T/R	Single-polarization T/R	Dual-polarization T/R	Dual-polarization T/R	Single-polarization T/R	Single-polarization	Dual-polarization T/R
Polarization operating mode	Polarization time division (H/V Tx, H and V Rx)	Polarization space division (H/V twin-bursts Tx in azimuth, H/V alternate Rx)	Polarization time division (H/V Tx, H and V Rx)	Polarization time division (H/V Tx with orthogonal LFM slope, H and V Rx) In Chap. 2	Polarization space division in azimuth (full aperture H/VTx, two sub-aperture H and V Rx)	Polarization time division (H/V Tx with orthogonal LFM slope, H and V Rx)	Compact polarization (Circular polarization Tx, H&V Rx)

(continued)

Table 1.1 (continued)

Name	SIR-C\X-SAR [8]	ASAR [9, 10]	ALOS\PALSAR [11]	RadarSat-2 [12]	TerraSAR-X [13, 14]	Cosmo-Skymed [15, 2]	RiSAT [16, 17]
ImagingMode (resolution/swath width/Polarizarion)	Stripmap: 30 m/15–90 km (SS[a]) Stripmap: 150 km (Co[b]) ScanSAR: 300–500 km	Stripmap: 30 m/100 km (Co) AP: 30 m/100 km (HH/VV or Dual[c]) ScanSAR: 150 m/400 km (Co) Global: 1000 m/400 km (Co) Wave: 10 m/5 km × 5 km(Co)	Fine: 10 m (2 looks)/20 m (4 looks), 70 km (Co or Dual) ScanSAR: 100 m (multi-looks)/250–350 km (Co) QuadPol-Stripmap: 30 m/30 km (Quad[d])	UltraFine: 3 m/20 km (SS) Multi look fine: 8 m/50 km (SS) Fine: 8 m/50 km (SS or Dual) Standard: 25 m/100 km (SS or Dual) Wide: 30 m/150 km (SS or Dual) Fine QuadPol: 8 m/25 km (Quad) Standard QuadPol: 25 m/25 km (Quad) ScanSAR narrow: 50 m/300 km (SS or Dual) ScanSAR wide: 100 m/500 km (SS or Dual)	Stripmap: 3 m/30 km (Co) 6 m/5–30 km (HH/VV or Dual) Spotlight: 1 m/5 km V or) m (S 2 m/5 km × 10 km (HH/VV) 2 m/10 km × 10 km (Co) 4 m/10 km × 10 km (HH/VV) ScanSAR: 16 m/100 km (Co) DRA: QuadPol	Spotlight: 1 m/10 km/Co Stripmap (Himage): 3 m/40 km/SS Stripmap (PingPong): 15 m/30 km/Co or Dual ScanSAR (Wide Region): 30 m/100 km /SS ScanSAR (Huge Region): 100 m/200 km /SS	Spotlight: 2 m/10 km × 10 km(Co and/or Cross[e]) Stripmap1: 3 m/25 km (Co and/or Cross) Stripmap2: 9 m/25 km (Quad) ScanSAR1: 25 m/115 km (Co and/or Cross) ScanSAR2: 50 m/223 km (Co and/or Cross)

[a]SS (HH or VV or HV or VH)
[b]Co (HH or VV)
[c]Dual (HH/HV or VH/VV)
[d]Quad (HH/HV/VH/VV)
[e]Cross (HV or VH)

The above-mentioned multi-polarization SAR satellite acquires multi-polarization data of the imaging region in different frequency bands and with different resolutions, which greatly promotes the on-board PolSAR system technology, the PolSAR data processing and calibration technology, the PolSAR information extraction and application technology development.

Up till now many airborne and spaceborne PolSAR systems at home and abroad have acquired a large number of data for polarization microwave imaging, which provides rich materials for people to study polarization. On this basis, scholars from various countries have carried out many researches on the analysis and application of PolSAR data such as scattering matrix decomposition [6], target polarization characteristics analysis [6, 18–28], target classification [1, 29–37], surface parameters inversion [38–46], and polarization image filtering [31, 46–57], etc., and a lot of results have been achieved. At present, PolSAR has developed into a relatively mature technology, which has been widely applied in agriculture (crop identification), forest (vegetation height and biomass estimation, species identification), geology (geologic structure survey), hydrology (surface roughness, soil and snow moisture, sea ice age and thickness estimation), sea ice monitoring (ice age and thickness estimation), oceanography (ocean wave property estimation, heat flow, and wavefront detection), military field, and other fields.

1.1.2 Polarization SAR Hierarchy and Operating Architecture [37, 16]

PolSAR acquires the polarization scattering matrix of target by transmitting and receiving electromagnetic waves of different polarization, which are used to completely describe the amplitude and phase information of the echo. The polarization scattering matrix contains abundant target information, which greatly improves the ability of SAR to acquire the target scattering information. According to the different polarization scattering matrix, polarization hierarchy of PolSAR can be divided into single polarization, dual polarization, full polarization, and compact polarization, etc.

The PolSAR system parameters are composed of technical parameters and polarization characteristic parameters. The technical parameters mainly include polarization hierarchy, resolution, Peak-to-Side-lobe Ratio (PSLR), Integrated Side-lobe Ratio (ISLR), system sensitivity, ambiguity ratio, polarization isolation, amplitude and phase unbalance between the polarization channels. The polarization characteristic parameters extracted from fully PolSAR data, include the span of a scattering matrix, polarization entropy, alpha angle, and similarity parameters, which represent the polarization characteristics of different targets to some extent.

In order to obtain the full-polarization scattering matrix, it is necessary that PolSAR has the ability to receive quad-polarization (HH/VH/HV/VV) echoes. For simultaneously obtained quad-polarization, the transmitting dual channels and receiving quad-channels are required, but the complexity of the instrument will be increased greatly. Therefore, for multi-polarization systems, the dual-polarization antennas, single transmit channels, single- or dual-receive channels usually adopted, and combinations of transmit and receive channels are used to obtain quasi-simultaneous four-polarization echoes. The combination of the transmitting and receiving channels can be implemented by time-division multiplexing polarization pulses, frequency-division multiplexing polarization pulses, code-division multiplexing polarization pulses, space-division Twin-burst polarization pulses in azimuth and compact polarization.

This book discusses polarization hierarchy, system parameters, and operating architecture of PolSAR in detail, and compares the advantages and disadvantages of different polarization hierarchies and operating architectures.

1.1.3 Polarization Synthetic Aperture Radar Interferometry (PolInSAR) [58]

The Polarization Synthetic Aperture Radar Interferometry (PolInSAR) effectively combines the polarization information and interference information and obtains the spatial three-dimensional structural feature information and polarization scattering information of the observed object, which can provide quantitative microwave remote sensing, high-precision digital elevation information and fine deformation of the observed object information [11, 58–73].

(1) **Interferometric Synthetic Aperture Radar (InSAR)**

The Interferometric Synthetic Aperture Radar (InSAR) uses the antennas that spatially separate the two locations to image the same area, estimates the interference phase difference between the corresponding pixels in the two complex images, and obtains the spatial vertical structure information of the features. Two important parameters of interference phase and coherence coefficient can be obtained by using InSAR technology.

The initial report of InSAR can be traced back to the study of surface observations of Venus by Rogers and Ingalls in 1969 [65].

In 1974, L. C. Graham obtained the first topographic contour map by optical processing of the data of Goodyear aerospace's airborne dual-antenna X-band SAR. It is the first time for InSAR to show the capability in topographic elevation imaging [66].

In the 1980s, R. M. Goldstein and H. A. Zebker of JPL improved InSAR technology in theory and practice, converted AIRSAR into a dual-antenna single-track InSAR system, and successfully implemented interferometry [67].

Since the launch of ERS-1 in 1991, the research of spaceborne InSAR has developed rapidly. In 1995, ESA launched the ERS-2 satellite again, which together with ERS-1 formed the TANDEM mission, enabling repeated observations of the same area at one-day intervals. Japan's JERS (L-band) and Canada's RADARSAT (C-band) can also provide a large amount of high-resolution InSAR data [68–72].

Shuttle Radar Topography Mission (SRTM) is the first spaceborne dual-antenna InSAR system, which successfully implemented the global topographic mapping in February 2000. The space shuttle carries two sets of SIR-C/X-SAR radars, their antenna are located at the two ends of a 60 m long telescopic arm. During 11 days of flight, SRTM observed the terrain between latitude 60° north and 56° south, covering 80% of the earth's surface, and repeated the observation in some areas. The multi-frequency and multi-polarization InSAR data of single track were obtained in the experiment by dual-antennas, which provided rich experimental data for studying the role of polarimetry in interferometry.

However, some factors limit the development of SAR interferometry. The topographic altitude estimation of terrain by InSAR usually leads to the offset of a vegetation layer, which will result in the deviation of several meters or even more than 10 m.

It is a feasible solution to introduce polarization technology into interferometry application, which can combine the fine structural characteristics of the target (polarization is sensitive) with the spatial distribution characteristics (interferometry is sensitive), and extract the relationship between them. Therefore, polarization will greatly improve the performance of SAR interferometry. PolSAR interferometry technology uses the fully PolSAR system for interferometric measurement, and performs interferometry analysis on the polarization scattering matrix, combining the characteristics of interferometry and polarization measurement. Not only can improve the accuracy of interferometry but also explain the polarimetric scattering mechanism of the target well.

(2) **Polarimetric SAR Interferometry (PolInSAR)**

Polarimetric SAR interferometry (PolInSAR) is a kind of earth observation technique based on the fusion of the theory, technology, and method of polarization and interferometry, which are fully developed and accumulated.

The key problem to be solved by the PolInSAR is that, under any combination of polarization states, the interferometry coherence is the highest or optimal. The theoretical basis of PolInSAR is polarization interferometry coherent optimal process and coherent optimal target decomposition theory.

The coherent optimal polarimetric interferometry process is to seek an optimal polarization state combination in the space of the full-polarization state combinations for optimal coherent interferometry. The optimal coherent interferometry includes the optimal coherent coefficient and the most accurate phase estimation, it can be understood mathematically as seeking the maximum eigenvalue of the classical Hermitian matrix and the corresponding eigenvector in the space formed by the combination of full-polarization states. The maximum eigenvalue is related to the optimal coherence coefficient, and the eigenvector associated with the maximum eigenvalue will correspond to the optimal combination of polarization states.

The optimal coherent target decomposition theory is based on the idea of target decomposition of fully polarimetric SAR. It can be described mathematically as follows: using the eigenvalues and eigenvectors of the classical Hermitian matrix, the coherent information of the target can be decomposed into the sum of three components weighted by eigenvalues and statistically independent of each other. These three components correspond to three different scattering mechanisms or scattering centers, which are generally unidirectional, bidirectional, and volume scattering mechanisms.

In 1997, S. R. Cloude and K. P. Papathanassiou first presented the research results of optimizing the correlation of SAR interferogram by using polarimetric interferometry [59], which pioneered the research of multi-polarization SAR interferometry. According to the polarization entropy and polarization anisotropy algorithm on the basis of Cloude and Pottier's full-coherent vector wave scattering theory, they further proposed Polarimetric–Interferometric Phase Coherence Optimization (PIPCO) algorithm and gave the scattering vector formula of polarimetric Interferometry. They point out that the polarization information can be used to improve the correlation of the data and the precision of SAR interferometry. The precision of Digital Elevation Map (DEM) obtained by PolInSAR can be up to centimeters, and the scattering characteristics of terrain targets can be better understood by polarization information. They systematically studied the influence of polarization in repeated orbits interferometry and proposed the polarization interferometry coherent optimal process and the coherent optimal target decomposition theory in 1998 [73].

In the book, the polarization interferometry SAR technology is discussed in detail. The basic principle of PolInSAR is presented, and the de-correlation effect and the error of InSAR are analyzed. Scalar interferometry, vector interferometry, and optimal polarimetric interferometry theory, including optimal amplitude and phase coherence algorithms, are described. In addition, coherent scattering decomposition and coherent region theory are discussed.

1.1.4 Compact Polarization SAR [74, 75]

The advantage of the full-polarization SAR is system perfection and can fully reflect the full-polarization information of the target polarization scattering matrix. The main disadvantages are complex system and high cost, double pulse repetition frequency (PRF), and half imaging swath than single-polarization or dual-polarization SAR, four times data rate and more double average power than single-polarization radar.

In order to overcome the above shortcomings, reduce system complexity, and trade off the parameters such as imaging swath, PRF, data rate, and ambiguity, the new concept [74, 76–81] of Compact Polarization SAR (C-PolSAR) has been proposed in recent years, C-PolSAR is transmitting a single-polarization electromagnetic wave with a specific polarization state and receiving the echo with two mutually orthogonal polarization channels while reducing the complexity of the polarization system and effectively retaining the rich information of the full-polarization SAR. For compact polarization radar, the phase is extremely important. The compact polarization tries to obtain a quantitative classification of backscatter comparable to the fully polarized system while avoiding the disadvantages associated with full-polarization systems.

According to the transmitting and receiving polarization states, the main modes of Compact Polarization (CP) are the π/4 mode, the Dual Circular Polarization (DCP) mode, and the Circular Transmit Linear Receive (CTLR) mode.

In 2002, J. C. Souyris et al. introduced the π/4 CP mode, in which the transmitted polarization is the superposition of linear horizontal and vertical polarization, resulting in a linear polarization oriented 45° with respect to the horizontal polarization. The radar receives returns in horizontal and vertical polarization. The performance of π/4 mode can be comparable to that of full-polarization SAR. Polarization decomposition of scatters distribution in the scene is carried out in both horizontal and vertical directions. The data processor consists of focusing and multi-look processing. The output of focus processing is the single-look complex image, and that of multi-look processing is the 2 × 2 covariance matrix, which could be converted to a 4-unit Stokes vector [76, 77].

In 2006, N. Stacy et al. proposed the DCP mode. Rotation invariance requires that the transmitting polarization must be circular polarization. The standard DCP is a circular polarization transmit, and (Left, Right) coherent circular polarization receive. The DCP mode was shown to have good results of backscattering analysis based on Stokes parameters. Furthermore, the 2 × 2 covariance matrix can be used to reconstruct 3 × 3 quad-polarization covariance matrx [78].

In 2006, Keith Raney promoted a hybrid mode of CTLR, with a (Left or right) circular polarization transmit and two linear polarizations receive ([RH, RV] If a signal with a phase difference of 90° is fed to both the H and V antennas, the two-line polarization antenna will radiate circular polarization in space. The linear basis of the echo signal is maintained in the whole process when it arrives at the processor through the H and V receiving channels. The data processor consists of focusing and multi-look processing. The output of focus processing is the single-look complex image, and that of multi-look processing is the 2 × 2 covariance matrix, which could be converted to a 4-unit Stokes vector [74, 79].

There are two ways to deal with C-PolSAR data: first one is to reconstruct the 3 × 3 full polarization by using the 2 × 2 scattering correlation matrix obtained by the compact polarization under the assumption of the symmetric scattering and polarization state data extrapolation model. Another way of processing is to extract the polarization characteristics of the target directly from the backscattered wave.

In recent years, spaceborne PolSAR that have been launched or are under development realized multi-polarization using a compact polarization. The Mini-SAR on the Chandraayan-1 lunar probe launched in India in 2008, the Mini-RF from the US Lunar Exploration Orbiter (LRO), and the radar imaging satellite RISAT-1 launched in India in 2012, all of above PolSAR systems chose the CTLR mode of C-PolSAR. In May 2014, Japan launched the PALSAR-2 onboard ALOS-2 satellite, which also adopted the compact polarimetry as the experimental mode. Furthermore, The RadarSat Constellation Mission (RCM) follow-up of the Canadian Radarsat-2 satellite also used the CTLR mode to obtain both high resolution and wide swath.

It can be seen that the compact polarimetry is the optimal scheme for spaceborne PolSAR to obtain full polarization without increasing system complexity, degrading resolution, and decreasing swath.

The C-PolSAR system and the Compact Polarization SAR interferometry (CPolInSAR) technology will be discussed in the book, it is including the operating modes of C-PolSAR, the effect of the symmetry of terrain media, fully polarization data reconstruction, polarization classification, and the optimal coherence technology of the CPolInSAR.

1.1.5 Polarization SAR Calibration Technology [82, 83]

One of the important differences between PolSAR and conventional SAR is that PolSAR data must be calibrated. Polarization calibration is the key to realize the quantitative remote sensing of PolSAR. The objective of calibration is to measure the various uncertain factors affecting the imaging quality and remove their influence in the data processing, and establishing the accurate relationship between image intensity and the backscattering coefficient. The PolSAR calibration technology includes internal calibration, external calibration, and calibration processing [84–86].

Radiation calibration is the process of describing the overall system performance from the transmitter output to the processor output, using the SAR system to measure the amplitude and phase of a ground target backscatter signal. The purpose of radiation calibration is to establish the correct correspondence between the pixel intensity of SAR image and the target true backscattering coefficient.

The internal calibration is to measure the fluctuation of the system parameters by using the built-in sensor or calibration loop during flight and compensate for it in the imaging process, and then the relative scattering coefficient of the measured object is determined. The key parameters measured by the internal calibration are the power of transmitted pulse, receiver gain, gain/phase imbalance between different channels, crosstalk, noise, quantization error, nonlinearity, etc. The main methods of internal calibration are independent calibration of each unit of the system and power ratio calibration.

The external calibration is to measure the transfer function of radar system by using the targets with known scattering characteristics. Compared with internal calibration, the advantage of external calibration is that it can directly measure the end-to-end system performance. Therefore, external calibration can be used to measure the system parameters that are difficult to measure, such as antenna pattern, incident angle, signal propagation, and atmospheric effects.

The actual effective calibration of any SAR system requires the combination of external calibration and internal calibration. The end-to-end system performance (including absolute gain) is estimated by external calibration, and the system performance relative drift between two external calibration positions is monitored by internal calibration.

It is necessary to quantitatively analyze the relationship between different channels and different polarizations of PolSAR, that is, only need through multi-channel and multi-polarization calibration, the information contained in polarization. SAR data can be fully utilized. The objective of polarization calibration is to correct the difference of the PolSAR system on different polarizations so that the SAR images can truly reflect the dependence between the target response and the polarization state of electromagnetic waves. Polarization calibration belongs to the relative calibration since it corrects the influence of the system on the different PolSAR data in the imaging process.

The error model of PolSAR system is given by

$$M = Ke^{j\varphi}RST = Ke^{j\varphi}\begin{pmatrix} R_{hh} & R_{hv} \\ R_{vh} & R_{vv} \end{pmatrix}\begin{pmatrix} S_{hh} & S_{hv} \\ S_{vh} & S_{vv} \end{pmatrix}\begin{pmatrix} T_{hh} & T_{hv} \\ T_{vh} & T_{vv} \end{pmatrix} + \begin{pmatrix} n_{hh} & n_{hv} \\ n_{vh} & n_{vv} \end{pmatrix} \quad (1.1)$$

The calibration of PolSAR system mainly includes radiometric calibration and polarimetric calibration. Radiometric calibration is to determine the absolute amplitude gain factor of the system K, and polarization calibration is to determine the receiving distortion matrix R and transmitting distortion matrix T.

The processing of PolSAR calibration is as follows. At first, to select the specific targets with known polarization scattering matrix S, the actual scattering matrix M is measured by the system, to solve and the polarization distortion matrix R and T of the SAR system are calculated. And then, combined with the radiometric calibration of the single-polarization SAR system, the absolute gain K is obtained. Finally, the obtained R, T, and absolute gain K are applied to the imaging region unknown S, and the scattering matrix S is obtained to complete the calibration process.

In the book, the PolSAR calibration technology is introduced in detail, including the system error model, the basic calibration method, internal calibration technology, external calibration technology, calibration algorithm, appropriate conditions, and accuracy analysis of the calibration algorithm.

1.2 Polarization SAR Data Processing

With the rapid development of PolSAR technology, data processing of PolSAR has become one of the hot topics in the remote sensing field. The main subjects include an understanding of PolSAR images, statistical characteristic analysis, speckle suppression, enhancement processing, target polarization decomposition, target detection, and image classification [87, 88], etc.

1.2.1 Statistical Properties Analysis of Polarization SAR Image

The basic methods for the analysis of PolSAR data and images include span of a scattering matrix, fundamental polarization pattern, polarization synthesis, polarization signature, echo power ratio under different polarization combinations, polarization purity, power fluctuation coefficient, partial polarization coefficient, correlation analysis, and optimal polarization degree of PolSAR images, etc. All the above methods focus on analyzing and understanding of the full-polarization scattering information.

The statistical characteristics of PolSAR images are different between the single-look images and multi-look images. The statistical characteristics of multi-look images include the covariance matrix and its statistical distribution, the phase distribution, the statistical distribution of image amplitude, the joint probability distribution of the two-channel echo power, the probability distribution of the two-channel echo power ratio, and the probability distribution of the two-channel echo amplitude ratio, etc.

The analysis shows that the scattering vector of homogeneous terrain has an approximate zero-mean complex Gaussian distribution. The scattering correlation matrix approximately obeys Wishart distribution. The amplitude and phase of elements in the scattering matrix are approximately subject to Rayleigh distribution and uniform distribution, respectively. The scattering vector of target in texture region approximately obeys K distribution. All these statistical characteristics are the basis of PolSAR data processing.

1.2.2 Polarization SAR Image Speckle Suppression

Speckle noise is due to the fading of SAR coherent echoes. The phases of echoes scattered from different targets are different in the same resolution unit so that the coherent superposition of the echo phases produces signal fading. Speckle noise seriously affects understanding and feature extraction of SAR images. The objective of SAR speckle suppression is to preserve the target characteristics by minimizing the effect of speckle noise [30, 31, 46–57, 89].

Conventional method of speckle suppression is based on the multi-look averaging that can suppress image speckle at the cost of degrading the resolution.

The multiplicative noise model and the statistical distribution characteristics of SAR data are the basis of the speckle suppression technique. In 1976, Goodman put forward the concept of fully developed multiplicative speckle noise. Lee built a multiplicative noise model and proposed many spatial filtering methods such as Lee filtering, fine Lee filtering, Kuan filtering, and Frost filtering, etc. [89].

PolSAR can obtain more information by means of polarization coherent channels. Due to the correlation between polarization channels, speckle noise can be suppressed by filtering the correlation of multi-channels and also by the covariance matrix.

In 1990, Novak and Burl et al. proposed a Polarization Whitening Filter (PWF) method for speckle suppression of full-polarization SAR images, which suppresses the speckle by the optimal weighting three complex elements S_{hh}, S_{hv} and S_{vv} of the scattering matrix. The PWF method can obtain the image of "minimum deviation-average ratio" This method suppresses speckle without degrading image resolution.

The effect of PWF on speckle suppression directly depends on the accuracy of parameter estimation of scattering vector covariance matrix. Initially, some parameters of the part image are applied to speckle suppression of the whole image, so a big error is introduced obviously. Furthermore, some algorithms such as sliding rectangular window, Prewitt edge operator parameter estimation method, improved parameter estimation method based on PWF algorithm, and fine Prewitt operator parameter estimation method, were developed. Although PWF can suppress the speckle without degrading the resolution, it is a weighting algorithm, which inevitably leads to the loss of polarization information.

Another method for speckle suppression of PolSAR images is speckle vector filtering, including Lee filter algorithm based on the Minimum Mean Square Error (MMSE) error, Refined Lee polarization vector filtering algorithm, and Refined Lee filtering algorithm based on the fine Prewitt operator, etc.

Lee filter, based on the MMSE local statistical filter, is able to avoid crosstalk between different polarization channels. Moreover, Lee filter is an adaptive statistical filtering method, which can well preserve the polarization information of the filtered data well.

Based on the data selection criterion of Prewitt operator parameter estimation, Refined Lee filtering algorithm applies the local statistical minimum RMS error criterion to perform vector filtering on the PolSAR data [89]. The Span figure can be looked like the edge detection image, and the pixel at the edge of the image window is used to calculate the mean value of each element in the covariance matrix. The filter weighting coefficient calculated by the Span image can be equally independently used for filtering each element of the covariance matrix. The use of span figure preserves the scattering characteristics of the homogeneous scattering region and avoids the effect of multi-channels crosstalk on PolSAR images.

Refined Lee filtering algorithm based on the fine Prewitt operator takes the data selection criteria of the fine Prewitt operator parameter estimation method as the basis and applies Refined Lee filtering algorithm in vector filtering of PolSAR data processing [88].

1.2.3 Polarization SAR Image Enhancement

In order to better distinguish two types of targets in SAR image, the signal processing method used to enhance the contrast of the image is called contrast enhancement of SAR image. One of the typical applications of polarization synthesis technology is the Optimal Polarization Contrast Enhancement (OPCE), which is to select the optimal polarization state to make the comparative energy between the target and the clutter more prominent. The typical method is to maximize the ratio of the average received power of the two types of targets by selecting the optimal polarization state of the transmitting and receiving antennas, and then obtain the images with the maximum contrast of the two types of targets by the polarization synthesis technology [87, 88].

In 1979, Ioannidis et al. used the Kennaugh matrix and the Stokes vector to construct an optimization model for polarization contrast enhancement. The maximum signal-to-clutter ratio is obtained by the Lagrangian algorithm.

In 1989, Dubois and van Zyl proposed a simple method for optimal polarization for echo power ratio. At first, the optimal receiving polarization mode under special transmitting polarization mode is obtained by using the analytical method, and then the power ratio of echo from the optimal receiving polarization antenna can be calculated under different transmitting polarization modes by using numerical method. Finally, the optimal transmitting and receiving polarization modes are found.

The contrast enhancement based on the maximum average receiving power ratio is to improve the contrast of two types of different scattering targets by using the full-polarization information, that means to choose the optimal combination of reflecting and receiving polarization to maximize the average receiving power ratio of the two types of targets.

The polarization contrast enhancement method, based on Sequential Unconstrained Minimization Technology (SUMT), maximizes the echo energy of the other target under the special circumstance of suppressing one target (or clutter), so the maximum received energy contrast of the two targets is achieved. At first, this method converts the objective function of the optimal problem into bi-linear form, and then the numerical method is used to solve the problem by using SUMT [88]. In this mathematical model, the polarization states of transmitting and receiving antennas are independent of each other, and higher contrast can be obtained.

Professor Yang Jian from Tsinghua University has made outstanding contributions to the study of polarization contrast enhancement [90, 91]. He extended the target zero theory to two independent polarization states, improved the optimal polarization using the similarity parameter and polarization entropy, and proposed generalized relative optimal polarization concept.

Clutter suppression of PolSAR image based on the Target Null theory can suppress two types of clutter at the same time, which is more conducive to highlighting the target. Target Null (i.e., co-polarization Null or cross-polarization Null) refers to the polarization state of the transmitting and receiving antennas with null received energy. According to the Target Null theory, as long as the scattering vectors of the two types of targets are linearly independent, the transmitting polarization state a_1 and receiving polarization state a_2 of antenna that make the echo energy of the two types of targets null can be obtained. Using polarization synthesis technology, SAR image under transmitting antenna a_1 and receiving antenna a_2 can be synthesized, and the two types of clutter can be suppressed at the same time.

In this book, the contrast enhancement technology of PolSAR image is discussed, including contrast enhancement based on Stokes matrix, contrast enhancement of maximum average received power ratio, optimal contrast enhancement based on two types of Lagrangian operators and contrast-enhanced numerical solution based on SUMT. In addition, the clutter suppression methods of PolSAR image based on polarization synthesis and Target Null theory are presented.

1.2.4 Polarization SAR Target Decomposition

PolSAR target decomposition [22, 23, 92–96] is one of the important contents of polarization microwave imaging. Polarization radar emits electromagnetic waves in different polarization states, and the echoes of ground targets show different scattering characteristics. In order to understand the scattering components and mechanism, PolSAR target decomposition focuses on how to decompose the polarization information into many known and typical combinations of the scattering components such as plane, dihedral angle, helix, straight line, and sphere scattering, etc. Target decomposition simplifies the data analysis and becomes the basis of analysis, understanding the application of PolSAR data.

There are two types of PolSAR target decomposition technology. One is coherent decomposition based on target scattering matrix, and the other is incoherent decomposition based on polarization covariance matrix and coherence matrix.

Coherent decomposition is to express the polarization scattering matrix by a combination of scattering responses of some simple targets. The main methods include Pauli-basis decomposition, Krogager decomposition [92], Cameron decomposition [93], and target similarity parameter decomposition [95], etc.

Pauli-basis decomposition decomposes the target into a combination of scattering responses of a plane, a dihedral angle and a 45° dihedral angle.

Krogager decomposition decomposes the symmetrical scattering matrix into the sum of the response of a sphere, a diplane with a rotation angle of 45° and a helix three components.

Cameron decomposition is a coherent method based on the reciprocity and symmetry of radar target. Cameron decomposition extracts the maximum symmetric scattering component of the target by using the target-oriented angular rotation. Cameron decomposition divides one target into two symmetric sub-targets, one of which is with the maximum symmetric component, and the other is with the minimum symmetric component. The process of obtaining the maximum symmetric component is actually the process of maximizing the polarization energy by using the directional angle rotation.

Target similarity parameter decomposition is to obtain the description of scattering characteristics by calculating the similarity parameter between the surveyed target and the typical target. The typical targets include a plane angle, a dihedral angle, a helix and a straight line, etc.

Incoherent decomposition is to decompose the polarization covariance matrix or coherence matrix into a combination of second-order descriptors that can be easier to be described by the physical interpretation. The main methods include eigenvector–eigenvalue-based incoherent decomposition and model-based incoherent decomposition for multi-look data.

The basic method of eigenvector-eigenvalue decomposition is to decompose the target incoherent matrix into three components and represent the polarization information by using eigenvector and eigenvalue. The main incoherent methods include Huynen decomposition, Cloude–Pottier decomposition, Holm decomposition, and van Zyl decomposition, etc. Because natural features are usually represented as a distributed target, so the incoherent decomposition method is widely used.

Cloude–Pottier decomposition [22] is one of eigenvector–eigenvalue decomposition methods by the polarization covariance matrix, so it extracts the polarization information effectively. Cloude–Pottier decomposition can be widely applied in PolSAR image classification.

The basic principle of the Huynen decomposition is to decompose the echo data into two parts. That means to decompose the coherent matrix T of distributed target into the sum of deterministic target coherence matrix T0 and residual target coherence matrix TN. Huynen decomposition relates the physical properties of the target such as symmetry, irregularity, and non-uniformity, etc.

Holm decomposition interprets the eigenvalue spectrum as the sum of a scattering matrix (rank 1 coherent matrix) and two noise or residual components. Holm decomposition is a hybrid approach that combines eigenanalysis (providing unitary transformation invariance) with single target and noise model.

Van Zyl decomposition was initially used to describe the 3×3 covariance matrix of the single-base azimuth symmetry natural target. Therefore, van Zyl decomposition is based on the assumption that the coherence coefficient of co-polarization and cross-polarization is zero.

Model-based incoherent decomposition for multi-look data is to decompose one target into the incoherent superposition of several simple targets which correspond to the physical scattering model of some simple known targets. Target decomposition includes Freeman–Durden three-component decomposition, Freeman two-component decomposition, Yamaguchi four-component decomposition, and multi-component scattering model decomposition.

Freeman–Durden decomposition [23, 94] is an incoherent method based on the three-component scattering mechanism model. In this method, the target polarization covariance matrix is modeled as the sum of three scattering mechanisms of volume scattering, even scattering, and first-order Bragg surface scattering.

In 2007, Freeman proposed a new two-component decomposition theory for forest target decomposition [94]. This method model the scattering covariance matrix as the sum of two scattering mechanisms. Different from the three-component decomposition, the even scattering and direct scattering are expressed uniformly except for the volume scattering component.

Yamaguchi proposed a modified decomposition method which introduced four types of the scattering matrices as the candidates [96]. Based on Freeman–Durden model decomposition, the elements of the covariance matrix for the volume scattering components are derived by a cloud of randomly oriented dipoles, and the helical scattering components are introduced to adapt to the general non-reflective symmetric condition.

Multi-component scattering decomposition is an improved incoherent decomposition method based on Freeman–Durden decomposition. This method extends the three-component scattering model proposed by Freeman–Durden from to multi-component including surface scattering, even scattering, volume scattering, helix scattering, and line scattering, etc.

1.2.5 Polarization SAR Target Detection

PolSAR target detection [56, 73, 82, 95, 97, 98] has always been an important topic in military application. Single-polarization SAR data is usually modeled by log-normal distribution, Weibull distribution, and K distribution, etc., and the constant false alarm rate (CFAR) detector is used to discover and detect target.

PolSAR can obtain the target information by multiple receiving channels at the same time. How to detect target, suppress interference and enhance detection performance using the multi-channel information, has become a research hot topic. Chaney proposed the Power Maximum Synthesis (PMS) detector [99], and Novak proposed Optimal Polarization Detector (OPD), Polarization Whitening Filter (PWF), Polarization Matched Filter (PMF) [82, 97]. In addition, Sciotti et al. come up with Polarization Gaussian Likelihood Ratio (PGLR) detector, Larson and Novak gave two new polarization subspace target detectors for SAR data based on the Huynen dihedral model.

Target detection method based on polarization scattering characteristics has also been paid more attention. Touzi and Could et al. applied the polarization entropy and scattering angle to detect surface targets over water [56]. Professor Yang Jian also proposed the generalized similarity method to detect targets on the land and over the sea [95, 97].

This book focuses on the target detection technology of PolSAR. The classical target detectors are analyzed, including PMS detector, PWF detector, PMF detector, Span detector, and PMS detector, etc. Furthermore, the detection methods based on polarization scattering characteristics, generalized similarity, polarization target decomposition, and PolInSAR data, are discussed.

1.2.6 Polarization SAR Image Classification

Classification is a very important approach to analysis and process PolSAR image [1, 22–25, 29–37, 88, 89, 100]. With the development of PolSAR system, classification and interpretation techniques of full-polarization SAR data have become an international research hot topic, and many theories and methods based on polarization scattering characteristics have been proposed. PolSAR image classification technology are mainly divided into two categories: the supervised classification technology based on prior knowledge and unsupervised classification technology based on statistical law or polarization scattering characteristics.

The supervised classification method includes statistical classifier, neural network classifier, and Wishart Maximum Likelihood classifier, etc.

The statistical classifier is the simplest. First, speckle suppression preprocessing is applied to PolSAR data, and then Bayesian method (e.g., maximum likelihood estimation) is used to classify the image in specific polarization state.

The neural network classifier mainly includes Back Propagation (BP) neural network, self-organizing competition neural network, and dynamic learning neural network, etc.

Although the supervised classification has high accuracy, it needs enough training samples which are high quality and representative. On the contrary, the unsupervised classification method is fast and simple under certain accuracy, which helps to better understand the scattering characteristics of ground targets.

Unsupervised classification method consists of two categories. One is based on the inherent physical scattering mechanisms of PolSAR data, and the other uses the image processing technology. The main methods include H/α classifier based on Cloude–Pottier decomposition, H/γ classifier based on the enhancement of Cloude–Pottier decomposition information, Maximum Likelihood (ML) classifier using the complex Wishart distribution, iterative classification, and H/α–Wishart Classifier of dual-frequency and full-polarization SAR Images.

In 2001, van Zyl et al. proposed the unsupervised classification algorithm based on polarization scattering mechanism, and Moghaddam and Freeman et al. improved and extended it later. Van Zyl pointed out that the PolSAR data can identify the main scattering mechanism in the average Mueller matrix, and a set of top-down criteria can be used to distinguish some basic target types.

In 1997, Cloude proposed a target scattering entropy scattering angle (H/α) classifier based on Cloude–Pottier decomposition [101]. H/α classifier combines the nine zones of H–α space with the envelope of H–α curve to classify the targets into eight representative scattering categories. It is a good unsupervised classification method to understand the intrinsic physical properties of natural media by estimating the polarization scattering mechanism of the target. The results of classification are in good agreement with the real scene and can basically reflect the scattering characteristics of the real targets.

The H/α classifier based on Cloude–Pottier decomposition divides H–α space into nine zones for image classification. In 1998, Lee et al. [100] developed this method and used the supervised classification of ML estimation to improve the results of unsupervised classification. The basic idea of Lee et al. is to take the results of H/α decomposition classification as the initial training samples of ML method and reclassify pixels by using Maximum Likelihood (ML) classifier based on the complex Wishart distribution for higher classification accuracy.

The iterative classification method takes the scattering characteristics of ground targets as the basis for selecting valid data in the window. The last classification result is used as the input of the current iteration to improve the classification result incrementally.

H/α–Wishart Classifier of dual-frequency and full-polarization SAR images performs H/α classification and ML classification on the two images separately, and then combines the classification results of the two images. Finally, the category reduction polymerization technology is adopted to reduce the number of categories and improve the aggregation of species characteristics.

1.3 Applications of Polarization SAR

1.3.1 Surface Parameter Inversion of Polarization SAR Data

The rich information of the polarization scattering matrix can be used to study the physical properties of the target such as direction, shape, roughness, dielectric constant, and soil moisture, etc.

Soil moisture inversion using high-resolution microwave remote sensing data [73, 83, 88] is an important contribution of polarization microwave imaging radar in the applications of hydro geography, meteorology, and economic optimization of agricultural process. The main difficulty in quantitative estimation of soil moisture or surface roughness is how to separate these two effects on radar backscattering. Because the surface roughness and soil moisture can be estimated independently by PolSAR data, so the polarization information can be used to separate the effects of surface roughness and soil moisture on radar backscattering directly or indirectly. Therefore, polarization information plays an important role in surface parameter inversion.

Small Perturbation Model (SPM) is the most commonly used approximate method to extract parameters of random rough surface of microwave imaging radar. The SPM decomposes the random surface into many ideal sinusoidal sub-surfaces by Fourier spectral decomposition, the scattering is mainly caused by the surface that resonates with the incident wavelength and the incident angle, so SPM can obtain the backscattering coefficient.

Another promising surface parameter extraction method is the polarization surface scattering model. Based on the analysis of eigenvalues and eigenvectors of the polarization correlation matrix, the model is an extended model of the SPM.

SPM, the second-order distribution model and surface scattering model are discussed first, and the applications of PolSAR data in surface roughness and soil moisture inversion are presented in this book.

1.3.2 Information Processing of Polarization Interference SAR

Polarization Interference SAR has been widely applied in forest vegetation parameter inversion, farmland vegetation parameter inversion, surface parameter estimation, glacier parameter inversion, DEM inversion, terrain classification, forest surface biomass extraction, target detection, regional change monitoring, and urban area analysis, etc. [38, 58, 80, 102–104]. Among all the applications, vegetation parameter inversion, DEM inversion, and polarization coherence tomography have been hot topics.

(1) **Forest Parameter Inversion**

Forest height inversion is one of the important applications of Polarization Interference SAR. The multi-layer vegetation structure models of forest parameter inversion include Random Volume over Ground (RVoG) model with ground scattering component proposed by Treuhaft and the independent scattering center model.

RVoG model simulates vegetation as a random isotropic particle layer, and the backscattering of vegetation layer is independent of polarization state. The ground has a horizontal orientation and is sensitive to electromagnetic polarization.

The three-stage inversion method is based on the RVoG model containing ground scattering component and uses the linear distribution of complex coherence coefficients to invert the vertical structure parameters of vegetation in three stages.

Forest height inversion method based on ML estimation uses RVoG model to decompose the coherent matrix of samples to obtain its log-likelihood function. Vegetation height, surface phase, extinction coefficient, and other vegetation parameters can be deduced by the ML value of the matrix.

Based on the independent scattering center model, the scattering center resolution method considers the multi-layer structure as several stable independent scattering centers distributed in each layer. The interferometric phase of each scatter can be estimated according to the assumption of the independent scattering center model.

ESPRIT inversion algorithm uses the total least squares ESPRIT algorithm for forest height inversion.

CAPON method seeks the vector with minimum noise interference and constant signal power, locates the peak position of spectral function and estimates the information of independent scattering center and the phase corresponding to the local maximum value of spectral function, namely the interference phase of scattering center.

The covariance matrix of Polarization Interference SAR data can be decomposed into the signal of scattering center and the noise. Based on the assumption that signal and noise are uncorrelated, the MUSIC method can obtain the information of each independent scattering center by solving the peak of spectral function.

(2) **Farmland Vegetation Parameter Inversion**

Unlike forest vegetation, the height of crop vegetation is much smaller than that of forest, and the effect of ground scattering on Polarization Interference SAR data is greatly enhanced. The most commonly used scattering model for crop parameter inversion is the Oriented Volume over Ground (OVoG) model containing ground scattering component. According to the estimated coherence coefficients of different polarization channels, the OVoG model inversion method is to find the coherence coefficients and fit the total least square line. According to the intersection point between the straight line and the ground, the surface phase estimation value is obtained, and then the average vegetation height is estimated.

(3) **DEM Extraction**

Polarization Interference SAR combines the polarization technology with the interference technology to improve the performance of the interference measurement. After image registration, using the six polarization image data of the two channel image pairs, three optimal coherence coefficients and optimal interferograms are obtained. DEM data will be extracted through these steps such as flat earth spectral shift filtering, phase filtering, the phase unwrapping, and elevation inversion.

(4) **Polarization Interference SAR Target Detection**

Target detection technology under the surface cover: L. Sagues et al. proposed the optimal scattering mechanism that maximizes the coherence coefficient by using the Polarization Subspace Coherence Optimal Method (PSM), so the height information of the target can be obtained, and the target under the ground surface cover can be determined [102, 103].

Target detection technology under forest cover: S. r. loude et al. used interference phase to detect the targets. Based on the parameters estimated from interference coherence coefficients of different polarization channels, the detection algorithm performs coherence coefficient estimation and least square linear fitting, selecting the reference scattering coherent channels, estimating the ground scattering component and filtering intensity, and detecting the target.

(5) **Polarization Coherence Tomography (PCT)**

Polarization coherence tomography is a new type of radar imaging technology. The basic principle is to carry out Fourier–Legendre series expansion on unknown vertical profiles within a known finite interval and to determine the expansion coefficient by the linear inversion of coherence coefficient with the change of baseline and polarization mode. Using the variation of the interference coherence coefficient with the polarization state, the profile will be reconstructed and applied to forest classification and target detection under forest cover effectively.

1.3.3 *Information Processing of Compact Polarization SAR and Compact Polarization SAR Interferometry*

Information processing of CP and C-PolInSAR has been a research hotspot in recent years. The research focuses on vegetation height inversion and terrain classification, etc.

(1) **Vegetation Height Inversion**

At present, the vegetation height inversion methods based on the compact polarization interference SAR data mainly include: to reconstruct pseudo-full-polarization interference SAR data method and to ergodically solve for compact polarization interference SAR data coherent region method.

To reconstruct pseudo-full-polarization interference SAR data method uses the reconstructed data for inversion, which is based on the assumption of symmetry, there are errors to ergodically solve for compact polarization interference SAR data coherent region method can only obtain the interference coefficient of any receiving polarization state under the fixed transmitting polarization state.

The coherent optimal and coherent regions of C-PolInSAR data are used to extract the vegetation height. The ground surface phase estimation using the coherent polarization coherent optimal, the decoherence coefficient estimation using the coherent region boundary extraction and the vegetation height estimation are mainly carried out.

Comparison of tree height inversion performance of different modes of C-PolInSAR mainly uses phase difference based on optimal interference phase and phase difference based on ESPRIT method to compare.

(2) **Terrain Classification** [75, 105–107]

Similar to the PolInSAR, the C-PolInSARcan obtain abundant scattering information and contributes to target classification. The key is to extract useful parameters from the C-PolInSAR data for terrain classification. In the book, an unsupervised classification method is proposed to make the initial classification by using the substitution consistency parameters extracted from the C-PolInSAR data, and then further classification is made by introducing Shannon entropy parameters of C-PolInSAR.

The substitution consistency parameter can reflect the polarization information contained in the C-PolInSAR data. The interference Shannon entropy parameter and the Shannon entropy parameter of scattering intensity part can reflect the influence of scattering intensity and interference. Therefore, these three parameters are used to classify the C-PolInSAR data.

At first, the substitution consistency parameter, interference part Shannon entropy parameter, and scattering intensity part Shannon entropy parameter of each pixel are estimated. And then the C-PolInSAR data are divided into six parts with the same number of pixels according to the substitution consistency coefficient. Wishart ML method was used to cluster and the initial classification was obtained. Then, each class of data obtained from the initial segmentation is further divided into 36 subclasses by using interference part Shannon entropy parameter and scattering intensity part Shannon entropy parameter. Finally, in each initial class, the clustering and merging method is used to merge each initial class into the required number of classes and the final classification is obtained.

1.3.4 *Effect of Terrain Azimuth Slope on Polarization SAR*

The completely polarized electromagnetic waves can be represented by a polarization ellipse. For any polarization ellipse, it can be described by three geometric parameters: ellipse inclination θ, ellipse rate angle χ, and ellipse size A.

The terrain azimuth slope deviates the elliptic inclination θ in the polarization ellipse so that the measured PolSAR data contains terrain azimuth slop information.

The terrain slope angle can be extracted from polarization ellipse inclination information. At the same time, terrain azimuth slope will affect the relative amplitude and phase of all the terms of the polarization covariance matrix and correlation matrix except $\langle |S_{HH} + S_{VV}|^2 \rangle$. This will have a great impact on target decomposition, information extraction, and other applications based on covariance matrix and correlation matrix of PolSAR data such as H/a decomposition, feature classification, crop density, soil moisture, surface roughness estimation. In order to reduce the effect of terrain azimuth slope on the application, PolSAR data can be used for azimuth slope estimation and compensation [88, 101, 106–109].

In the book, the effect of terrain azimuth slope on PolSAR data is discussed in detail, and the method of azimuth angle offset estimation is presented. The main factors affecting the accuracy of azimuth slope estimation are analyzed, and polarimetric SAR data compensation for terrain azimuth slope variation and its application are discussed.

1.4 Book Content Arrangement

In this book, the polarization microwave imaging and the polarization information processing technologies, including PolSAR system technology, PolSAR information processing technology, and application techniques of PolSAR image are studied, a total of 17 chapters.

Chapter 1 Introduction

The development of polarization SAR (PolSAR) system is presented, and then PolSAR system technologies such as polarization hierarchy and system operating architecture, polarization SAR interferometry (PolInSAR), compact polarization SAR (C-PolSAR), and polarization SAR calibration are introduced. The PolSAR data processing technologies, including understanding and statistical analysis of PolSAR images, speckle suppression and enhancement, target decomposition, target detection and image classification, are discussed. Finally, the polarization SAR data application technologies, including surface parameter inversion, information processing of PolInSAR and C-PolSAR, effect of terrain azimuth slope on PolSAR, are presented. The content and structure of the book are given.

Chapter 2 The Principle of Polarization SAR

The basic theory of electromagnetic wave, the characteristics, and field equation are described first, and then the presentation of partially polarized electromagnetic wave, antenna transmitting and receiving polarization state, the presentation of polarization scattering characteristics, the relationships of various polarization representation methods, the transformation of polarization basis, polarization synthesis, and polarization characteristic diagram are discussed. All of these are the theoretical foundation of the book.

Chapter 3 Polarization SAR

The principle of SAR is described, and the polarization hierarchy and system parameters of polarization SAR are described. The operating architectures of polarization SAR such as Time-Division Multiplexing polarization pulses, Frequency-Division Multiplexing polarization pulses, Code-Division Multiplexing polarization pulses, Space-Division Twin-burst polarization pulses in Azimuth and Compact Polarization are discussed in detail.

Chapter 4 Polarization SAR Interferometry

The basic principle of the Polarization SAR Interference (PolInSAR) is presented, and the de-correlation effect and the error of Interference SAR are analyzed. Scalar interferometry, vector interferometry and optimal polarization interferometry theory, including optimal amplitude and phase coherence algorithms, are described. In addition, coherent scattering decomposition and coherent region theory are discussed.

Chapter 5 Compact Polarization and Compact Polarization SAR Interferometry

The modes of C-PolInSAR system are presented firstly, and then the effect of the symmetry of terrain and the method of reconstructing the C-PolInSAR data under the reflection symmetry condition are presented. In addition, the methods of polarization classification of C-PolSAR and the reconstruction of fully polarimetric SAR data are discussed. The coherence optimization of C-PolInSAR, including the coherence amplitude and the coherence phase is analyzed.

Chapter 6 Polarization SAR Calibration

SAR Calibration technologies, including calibration model, internal calibration, external calibration, and calibration algorithm, are discussed. In addition, polarization SAR calibration technologies such as error model, basic method, internal calibration, external calibration, calibration algorithm and its applicable condition and precision are studied.

Chapter 7 Polarization SAR Image Analysis

The basic methods of analyzing polarization SAR image such as the span of a scattering matrix, polarization combination figure, polarization synthesis, polarization signature, echo power ratio under different polarization combinations, echo polarization purity, power fluctuation coefficient, partial polarization coefficient are discussed. The correlation analysis and optimal polarization degree of polarization SAR image are studied.

Chapter 8 Statistical Characteristics of Polarization SAR Image

The statistical characteristics of single-look Polarization SAR image are presented, and then that of the multi-look polarization SAR image is given by the covariance matrix and its statistical distribution, the phase distribution, the statistical distribution of the amplitude product, the joint probability distribution of the two-channel echo power, the probability distribution of the two-channel echo power ratio and amplitude ratio are discussed in this chapter.

Chapter 9 Polarization SAR ImageSpeckle Suppression

The Polarization SAR echo model with speckle noise is presented, and then Polarization Whitening Filter (PWF) and vector filtering for Speckle suppression are discussed in detail. PWF includes algorithm analysis and filter parameter estimation methods such as rectangular sliding window, Prewitt operator non-rectangular window, improved Prewitt operator, and fine Prewitt operator parameter. Vector filtering method, including Lee filtering algorithm based on Minimum Mean Square Error (MMSE) error, Refined Lee polarization vector filter algorithm, Lee filtering based on the refined fine Prewitt operator algorithm, are described.

Chapter 10 Polarization SAR Image Enhancement

This chapter focuses on the polarization data preprocessing method. The contrast enhancement techniques based on Stokes matrix, maximum average received power ratio, Lagrangian multiplier, and Sequential Unconstrained Minimization Technology (SUMT), are described in detail. Clutter suppression based on polarization synthesis and Target Null Theory is studied.

Chapter 11 Polarization SAR Target Decomposition

Polarization target decomposition techniques can be divided into coherent target decomposition and incoherent target decomposition. The coherent target decomposition mainly includes Pauli decomposition, Krogager decomposition, Cameron decomposition, and target similarity parameter decomposition. The incoherent target decomposition can be based on polarization covariance matrix and polarization coherence matrix. Polarization target decomposition based on feature vector, including Cloude–Pottier decomposition, Holm decomposition, and van Zyl decomposition. The polarization target decomposition based on the model, including Freeman–Durden decomposition, Freeman two-component decomposition, Yamaguchi four-component model, and multi-component scattering model decomposition, are discussed in depth.

Chapter 12 Polarization SAR Target Detection

This chapter focuses on the target detection technology of Polarization SAR. The classical polarization target detectors such as Optimal Polarization Detector (OPD), Polarization Whitening Filter (PWF), Polarization Matched Filter (PMF), SPAN detector, and Power Maximum Synthesis (PMS) detector are studied. The polarization target detection methods based on target characteristic parameters, polarization similarity parameters, polarization target decomposition, and polarization SAR inteference data are discussed.

Chapter 13 Polarization SAR Image Classification

This chapter focuses on the classification of Polarization SAR image. The supervised classifier includes the statistical classifier, the neural network classifier, and Wishart Maximum Likelihood classifier, etc. The unsupervised classification will depend on the Polarization scattering mechanism. The main unsupervised classifier includes H/α classifier Cloude–Pottier decomposition, H/γ classifier based on the enhancement of Cloude–Pottier decomposition information, Maximum Likelihood (ML) classifier based on the Wishart distribution et al. In addition, the unsupervised classification algorithm of PolInSAR and multi-band PolInSAR images is discussed.

Chapter 14 Surface Parameter Inversion of Polarization SAR Data

Small Perturbation Model (SPM), the second-order distribution of surface scatters and polarized surface scattering model are presented in this chapter. And surface roughness inversion and soil moisture inversion are discussed.

Chapter 15 Polarization–Interferometry SAR Information Processing

Forest parameter inversion methods of Polarization–Inteferometry SAR such as multi-layer vegetation structure model, three-stage inversion, Maximum Likelihood inversion, and scattering center resolution method are discussed in detail. In addition, information processing technologies of PolInSAR such as farmland vegetation parameter inversion, DEM inversion, and target detection are studied.

Chapter 16 Compact Polarization SAR Interferometry Information Processing

The vegetation height inversion technology focuses on the three-stage inversion of vegetation parameters, and the performances of tree height inversion under different C-Polarization modes are compared. In addition, the terrain classification technology of C-PolInSAR is discussed.

Chapter 17 Influence and Compensation of Terrain Azimuth Slopes on Polarimetric Synthetic Aperture Radar Data

The effect of terrain azimuth slope on polarization SAR data is discussed. The azimuth offset introduced by terrain azimuth slope is estimated, and the main factors affecting the estimation accuracy are analyzed. In addition, the compensation and the application of azimuth offset of polarization SAR data are presented.

References

1. J.-S. Lee, M.R. Grunes, E. Pottier, Quantitative comparison of classification capability: fully polarimetric versus dual and single-polarization SAR. IEEE. Trans. GRS **39**(11), 2343–2351 (2001)
2. Cosmo-Skymed System HandBook, 30 April, 2007
3. G. Sinclair, The transmission and reception of elliptically polarized waves. Proc. IRE **38**, 148–151 (1950)
4. J.R. Huynen, Phenomenological theory of radar targets, Ph. D. Dissertation, 1970
5. A.B. Kostinski, W.M. Boerner, On foundation of radar polarimetry. IEEE Trans. Antennas Propag. **34**(12), 1395–1403
6. F.T. Ulaby, C. Elachi, in *Radar Polarimetry for Geoscience Applications* (Artech House Inc, Boston, London, 1990), pp. 281–295

7. G.R. Valenzuela, SAR imaging of ocean waves in the presence of variable currents, in *Proceedings of the Workshop on Modulation of Short Wind Waves in the Gravity-Capillary Range by Non-Uniform Currents*, The Netherlands, 24–26 May 1988 (pp. 141–153)
8. R.L. Jordan, B.L. Huneycutt, M. Werner, The SIR-C\X-SAR synthetic aperture radar system. Proc. IEEE **79**(6) (1991)
9. Y.-L. Desnos, C. Buck, J. Guijarro, G. levrini, J.-L. Suchail, R. Torres, H. Laur, J. Closa, B. Rosich, The ENVISAT advance synthetic aperture radar system (IEEE, 2000)
10. R.K. Hawkins, R. Touzi, J. Wolfe, P.J. Meadows, ASAR AP mode performance and applications potential (IEEE, 2003)
11. A. Rosenqvist, M. Shimada, M. Watanabe, ALOS PALSAR: technical outline and mission concepts, in *4th International Symposium on Retrieval of Bio- and Geophysical Parameters from SAR Data for Land Applications Innsbruck*, Austria, 16–19 Nov 2004
12. P.A. Fox, A.P. Luscombe, A.A. Thompson, Radarsat-2 SAR modes development and utilization. Can. J. Remote Sens. **30**(3), 258–264 (2004)
13. J. Mittermayer, H. Runge, Conceptual studies for exploiting the TerraSAR-X dual receive antenna (IEEE, 2003)
14. M. Stangl, R. Werninghaus, R. Zahn, The TerraSAR-X active phased array antenna,2003 IEEE
15. Cosmo-Skymed System Description & User Guide, 4 May, 2007
16. T. Misra, S.S. Rana, V.H. Bora, N.M. Sesai, C.V.N. Rao, Rajeevjyothi, SAR payload of radar imaging satellite (RISAT) of ISRO, in *Proceedings of EUSAR 2006*
17. T. Misra, S.S. Rana, N.M. Desai, D.B. Dave, Rajeevjyoti et al., Synthetic aperture radar payload on-board RISAT-1: configuration, technology and performance. Curr. Sci. **104**(4) (2013)
18. F. RidhaTouzi, Charbonneay, characterization of target symmetric scattering using polarimetric SAR. IEEE Trans. GRS **40**(11), 2507–2516 (2002)
19. Y. Dong, B.C. Forster, C. Ticehurst, A new decomposition of radar polarization signatures. IEEE Trans. GRS **36**(3), 933–939 (1998)
20. Y.-Q. Jin, S.R. Cloude, Numerical eigenanalysis of the coherency matrix for a layer of random nonspherical scatterers. IEEE Trans. GRS **32**(6), 1179–1185 (1994)
21. T.L. Ainsworth, S.R. Cloude, J.S. Lee, Eigenvector analysis of polarimetric SAR data, in *Geoscience and Remote Sensing Symposium, IGARSS'02*, vol. 1 (pp 626–628)
22. S.R. Cloude, E. Pottier, A review of target decomposition theorems in radar polarimetry. IEEE Trans. GRS **34**(2), 498–518 (1996)
23. A. Freeman, S.L. Darden, A three-component scattering model for polarimetric SAR data. IEEE Trans. GRS **36**(3), 963–973 (1998)
24. U. Bem, E. Pottier, Object based analysis of polarimetric SAR data in alpha entropy-anisotropy decomposition using fuzzy classification by cognition. IEEE Trans. GRS **30**(4), 697–704 (1992)
25. M. Qong, T. Tadono, H. Wakabayashi, M. Shimada, Mueller matrix based classification of polarimetric SAR data, Geoscience and Remote Sensing Symposium, 2000. Proceedings. IGARSS **1**, 375–377 (2000)
26. J.R. Huynen, A new extended target decomposition scheme, in *Geoscience and Remote Sensing Symposium, IGARSS'94*, vol. 2 (1994), pp. 1124–1125
27. I. Hajnsek, K.P. Papathanassiou, S.R. Cloude, Removal of additive noise in polarimetric eigenvalue processing, in *Geoscience and Remote Sensing Symposium, IGARSS'01*, vol. 6 (2001), pp. 2778–2780
28. M. Qong, A new scattering mechanism enhancement scheme of polarimetric SAR images. IEEE Trans. GRS **40**(12), 2582–2596 (2002)
29. S.R. Cloude, E. Pottier, An entropy based classification scheme for land application of polarimetric SAR. IEEE Trans. GRS **35**(1), 68–78 (1997)
30. G. Liu, S. Huang, Bayesian classification of multi-look polarimetric SAR images with a generalized multiplicative speckle model and adaptive a priori probabilities. Int. J. Remote Sens. **19**(1), 161–170 (1998)

31. J.-S. Lee, M.R. Grunes, G. Grandi, Polarimetric SAR speckle filtering and its implication for classification. IEEE Trans. GRS **37**(5), 2363–2373 (1999)
32. L.E. Pierce, F.T. Ulaby, K. Sarabandi, M.C. Dobson, Knowledge based classification of polarimetric SAR images. IEEE Trans. GRS **32**(5), 1081–1086 (1994)
33. J.S. Lee, M.K. Grunes, Classification of multi-look polarimetric SAR imagery based on complex wishart distribution. Int. J. Remote Sens. **15**(11), 2299–2311 (1994)
34. P. Lombardo, T. Macri Pellizzeri, A. Tomasuolo, Classification of polarimetric SAR images of suburban areas using joint annealed segmentation and H/A/α decomposition, in *IEEE/ISPRS Joint Workshop on Remote Sensing and Data Fusion over Urban Areas* (2001), pp 117–121
35. F. Mattia, T. Toan, J.-C. Souyris, The effect of surface roughness on multifrequency polarimetric SAR data. IEEE Trans. GRS **35**(4), 954–965 (1997)
36. Y. Dong, A.K. Milne, B.C. Forster, Segmentation and classification of vegetated areas using polarimetric SAR image data. IEEE. Trans. GRS **39**(2), 321–329 (2001)
37. D.B. Trizna, C. Bachmann, M. Sletten, Projection pursuit classification of multiband polarimetric SAR land images. IEEE. Trans. GRS **39**(11), 2380–2385 (2001)
38. I. Hajnsek, S.R. Cloude, J.-S. Lee, E. Pottier, Inversion of surface parameters from polarimetric sar data, in *Proceedings of Geoscience and Remote Sensing Symposium, IGARSS 2000*, vol. 3 (2000), pp. 1095–1097
39. M. Moghaddam, S. Saatchi, An inversion algorithm applied to SAR data to retrieve surface parameters, in *Geoscience and Remote Sensing Symposium, IGARSS'93*, vol. 2 (1993), pp. 587–589
40. E.D. Hernandez-Baquero, J.R. Schott, Atmospheric and surface parameter retrievals from multispectral thermal imagery via reduced-rank multivariate regression, in *Proceedings of Geoscience and Remote Sensing Symposium, IGARSS 2000*, vol. 4 (2000), pp. 1525–1527
41. Y. Oh, K. Sarabandi, F.T. Ulaby, An inversion algorithm for retrieving soil moisture and surface roughness from polarimetric radar observation, in *Geoscience and Remote Sensing Symposium, IGARSS'94*, vol. 3 (1994), pp. 1582–1584
42. I. Hajnsek, E. Pottier, S.R. Cloude, Inversion of surface parameters from polarimetric SAR. IEEE Trans. GRS **41**(4) 727–744 (2003)
43. D.L. Schuler, J.S. Lee, D. Kasilingam, G. Nesti, Remote sensing of surface roughness using polarimetric SAR data, in *Geoscience and Remote Sensing Symposium, IGARSS '01*, vol. 2 (2001), pp. 916–918
44. I. Hajnsek, K.P. Papathanassiou, S.R. Cloude, L- and P-band for surface parameter estimation, in *Geoscience and Remote Sensing Symposium, IGARSS '01*, vol. 6 (2001), pp. 2775–2777
45. I. Hajnsek, K.P. Papathanassiou, A. Moreira, S.R. Cloude, Surface parameter estimation using interferometric and polarimetric SAR, in *Geoscience and Remote Sensing Symposium, IGARSS '02*, vol. 1 (2002), pp. 420–422
46. E. Jakeman, R.J.A. Tough, Generalized K distribution: a statistical model for weak scattering. J. Opt. Soc. Am. A **4**(9), 1764–1772 (1987)
47. J.K. Jaa, Amplitude distribution of composite terrain radar clutter and the K-distribution. IEEE Trans. Antennas Propag. **32**(10), 1044–1061 (1984)
48. A. Baraldi, F. Panniggiani, A refined gamma MAP SAR speckle filter with improved geometrical adaptivity. IEEE Trans. GRS **33**, 1245–1257 (1995)
49. L.M. Novak, M.C. Burl, Optimal speckle suppression in polarimetric SAR imagery. IEEE Trans. On AES **26**(2), 293–305 (1990)
50. J.-S. Lee, M.R. Grunes, S.A. Mango, Speckle suppression in multipolarization multifrequency SAR imagery. IEEE Trans. GRS **39**(4), 535–544 (1991)
51. L M. Novak, M.C. Burl, W.W. Irving, Optimal speckle suppression in polarimetric SAR target detection. IEEE Trans. AES **29**(1), 234–243 (1993)
52. G. Liu, S. Huang, A. Torre, F. Rubertone, The multilook polarimetric whitening filter (MPWF) for intensity speckle suppression in polarimetric SAR images. IEEE Trans. GRS **36**(3), 1020–1061 (1998)
53. Y. Domg, A.K. Milne, B.C. Forster, Toward edge sharpening: ASAR speckle filtering algorithm. IEEE Trans. GRS **39**(4), 851–863 (2001)

54. C. Ju, C.R. Moloney, An edge-enhanced segmentation method for SAR images, in *Proceedings of Electrical and Computer Engineering*, 25–28 May 1997, vol. 2, pp. 599–602
55. J. Schou, H. Skriver, Restoration of polarimetric SAR images using simulated annealing. IEEE Trans. GRS **39**(9), 2005–2016 (2001)
56. A Lopes, E. Nezry, R. Touzi, H. Laur, Structure detection and statistical adaptive speckle filtering in SAR images. Int. J. Remote Sens. **14**(9), 1735–1758 (1993)
57. R. Touzi, A. Lopes, The principle of speckle filtering in polarimetric SAR imagery. IEEE. Trans. GRS **32**(5), 1110–1114 (1994)
58. Y. Zhen, Research on synthetic aperture radar interferometry and polarimetric interferometry, Ph.D. Dissertation, Chinese Academy of Sciences, June 2003
59. K.P. Papathanassiou, S.R. Cloude, Polarimetric effects in repeat-pass SAR interferometry, in *Proceedings of IGARSS' 97*, Singapore, Singapore (1997), pp. 1926–1928
60. M. Neumann, A. Reigber, L. Ferro-Famil, PolInSAR coherence set theory and application, in *Proceedings of EUSAR 2006*, Dresden, Germany (2006)
61. M. Moghaddam, Effect of medium symmetries on parameter estimation with polarimetric interferometry. J. Electromagn. Waves Appl. **14**(2), 173–184 (2000)
62. X. Tao, Y. Jian, P. Yingning, Approach to reversion of DEM based on polarimetric SAR interferometry. J. Tsinghua Univ. (Sci. Technol.) **47**(7), 1170–1173 (2007)
63. C. Erxue, L. Zengyuan, P. Yong et al., Polarimetric synthetic aperture radar interferometry based mean tree height extraction technique. Sci. Silvae Sin. **43**(4), 66–70 (2007)
64. B. Lu, Research on PolInSAR vertical structure parameter estimation method based on multi-layer model and its application in forest covered area, Graduate School of Chinese Academy of Sciences, Beijing
65. A.E.E. Rogers, R.P. Ingalls, Venus: mapping the surface reflectivity by radar interferometry. Science **165**, 797–799 (1969)
66. L.C. Graham, Synthetic Interferometrer Radar for Topographic Mapping. Proceedings of IEEE **62**, 763–768 (1974)
67. H.A. Zebker, R.M. Goldstein, Topographic mapping from interferometric SAR observations. J. Geophys. Res. **91**, 4993–4999 (1986)
68. F.K. Li, R.M. Goldstein, Studies of multibaseline space-borne interferometric synthetic aperture radars. IEEE Trans. GRS **28**(1), 88–97 (1990)
69. E. Rodriguez, J.M. Martin, Theory and design of interferometric synthetic aperture radars. IEE Proc F **139**(2), 147–159 (1992)
70. H.A. Zebker, J. Villasenor, Decorrelation in interferometric radar echoes. IEEE Trans. GRS **30**(5), 950–959 (1992)
71. H.A. Zebker, C.L. Werner, P.A. Rosen, S. Hensley, Accuracy of topographic maps derived from ERS-1 interferometric radar. IEEE Trans. GRS **32**(4), 823–836 (1994)
72. C.L. Werner, S. Hensley et al., Techniques and applications of SAR interferometry for ERS-1: topographic mapping, change detection, and slope measurement, in *Proceedings First ERS-1 Symposium—Space at the Service of our Environment*, 4–6 Nov 1992, pp. 205–210
73. S.R. Cloude, K.P. Papathanassiou, Polarimetric SAR interferometry. IEEE Trans. Geosci. Remote Sens. **36**(5), 1551–1565 (1998)
74. K. Raney, Dual-polarized SAR and Stokes parameters. IEEE Geosci. Remote Sens. Lett. **3**, 319–417 (2006)
75. M.-L. Truong-Loï et al., The conformity coefficient or how to explore the scattering behavior from compact polarimetry mode, in *2009 IEEE Radar Conference* (2009)
76. T. Lulu, Study on information processing of polarimetric and compact polarimetric SAR interferometry, Ph.D. Dissertation, Chinese Academy of Sciences, 2010
77. J.C. Souyris, S. Mingot, Polarimetry based on one transmitting and two receiving polarizations: the p/4 mode, in *Proceedings IGARSS'02*, Toronto, Canada, July 2002
78. N. Stacy, M. Preiss, Compact polarimetric analysis of X band SAR data, in *Proceedings of EUSAR'06*, Dresden, Germany, May 2006
79. K. Raney, Hybrid polarimetric SAR architecture, in *Proceedings of IGARSS'06*, July 2006

80. P.C. Dubois-Fernandez et al., The compact polarimetry alternative for space-borne SAR at low frequency. IEEE Trans. Geosci. Remote. Sens. **46**(10), 3208–3222 (2008)
81. J.C. Souyris, P. Imbo, R. Fjortoft, S. Mingot, J.S. Lee, Compact polarimetry based on symmetry properties of geophysical media: the p/4 mode. IEEE Trans. Geosci. Remote. Sens. **43**(3), 634–646 (2005)
82. L.M. Novak, M.C. Burl, Optimal polarimetric processing for enhanced target detection. IEEE Trans. Geosci. Remote Sens. **29**(1), 234–243 (1993)
83. S.R. Cloude, K.P. Papathanassiou, Three-stage inversion process for polarimetric SAR interferometry. IEE Proc.-Radar Sonar Navig. **150**(3), 125–134 (2003)
84. M.W. Whitt, F.T. Ulaby, P. Polatin, V.V. Liepa, A general polarimetric radar calibration technique. IEEE Trans. Antennas Propag. **39**(1), 62–67 (1991)
85. A. Guissard, Phase calibration of polarimetric radars from slightly rough surfaces. IEEE Trans. Geosci. Remote Sens. **32**(3), 712–715 (1994)
86. X. Yuxiao, Multi-polarimetric synthetic aperture radar calibration technology, Master's Thesis, Institute of Electronics, Chinese Academy of Sciences, 2002
87. D. Bowei, Research on multi-polarimetric synthetic aperture radar system and polarimetric information processing, Ph. D. Dissertation, Chinese Academy of Sciences, June 2000
88. L. Xiuqing, Research on polarimetric information processing technology of fully polarization synthetic aperture radar, Ph.D. Dissertation, Chinese Academy of Sciences, Jan 2004
89. J.-S. Lee, E. Pottier, in *Polarimetric Radar Imaging: From Basics to Applications 2008* (CRC Press, New York, 2008)
90. J. Yang, Y. Yamaguchi, H. Yamada, Development of target null theory. IEEE Trans. GRS **39**(2), 330–338 (2001)
91. J. Yang, Y. Yamaguchi, W.M. Boerner, Numerical methods for solving the optimal problem of contrast enhancement. IEEE Trans. GRS **38**(2), 965–971 (2000)
92. E. Krogager, New decomposition of the radar target scattering matrix. Electron. Lett. **26**(18), 1525–1527 (1990)
93. W.L. Cameron, L.K. Leung, Feature motivated polarization scattering matrix decomposition, in *Proceedings of IEEE International Radar Conference*, Arlington, VA (1990)
94. A. Freeman, Fitting a two-component scattering model to polarimetric SAR data from forests. IEEE Trans. Geosci. Remote Sens. **45**(8), 10 (2007)
95. An Wentao, Study on target decomposition and scattering feature extraction of polarimetric SAR, Ph.D. thesis of Tsinghua University, 2010
96. H. Yamada, H. Onoda, Y. Yamaguchi, On scattering model decomposition with PolInSAR data, in *Proceedings of EUSAR 2008*, Friedrichshafen, Germany
97. L.M. Novak, M.B. Sechtin et al., Studies of target detection algorithms that use polarimetric radar data. IEEE Trans. Aerosp. Electron. Syst. **25**(2), 150–165 (1989)
98. X. Shunping, W. Xuesong, D. Dahai et al., in *Polarimetric Radar Imaging Processing and Application* (Science Press, Beijing, 2013)
99. B. Chaney, On the performance of polarimetric target detection algorithms. IEEE Trans. Aerosp. Electron. Syst. **29**(1), 234–244 (1993)
100. L. Ferro-Famil, E. Pottier, J.S. Lee, Classification and interpretation of polarimetric interferometric SAR data, in *Proceedings of IGARSS'02*, Toronto, Canada (2002)
101. J.S. Lee, E. Krogager, D. L. Schuler, T. L, Ainsworth, On the estimation of polarization orientation angles induced form azimuth slopes using polarimetric SAR data, in *Geoscience and Remote Sensing Symposium, Proceedings, IGARSS 2000*, vol. 3 (2000), pp. 1310–1312
102. L. Sagues et al., Indoor experiments on polarimetric SAR interferometry. IEEE Trans. Geosci. Remote Sens. **38**(2) (2000)
103. L. Sagues et al., Polarimetric radar interferometry for improved mine detection and surface clutter rejection. IEEE Trans. Geosci. Remote Sens. **39**(6), 1271–1278 (2001)
104. A. Reigber et al., Multi-baseline coherence optimization in partial and compact polarimetric modes, in *Proceedings of IGARSS 2008*, Boston, USA (2008)
105. L. Tan, P. Huang, Improved investigation on unsupervised classification of compact PolInSAR data, in *Proceedings of EUSAR 2012* (2012)

106. J.J. Van Zyl, The effect of topography on radar scattering from vegetated areas. IEEE Trans. GRS **31**(1), 153–160 (1993)
107. J.-S. Lee, D.L. Schuler, T.L. Ainsworth, On the estimation of radar polarization orientation shifts induced by terrain slopes. IEEE Trans. GRS **40**(1), 30–40 (2001)
108. D.L. Schuler, J.-S. Lee, D. Kasilingam, Surface roughness and slope measurements using polarimetric SAR data. IEEE Trans. GRS **40**(3), 687–698 (2002)
109. D.L. Schuler, J.-S. Lee, Measurement of topography using polarimetric SAR images. IEEE Trans. GRS **34**(5), 1266–1276 (1996)

Chapter 2
Theoretical Basics of Polarization Synthetic Aperture Radar

2.1 Theory of Electromagnetic Wave

2.1.1 The Characteristics of Electromagnetic Wave and Maxwell Equations

People have realized and been studying electromagnetic phenomena for thousands of years. However, electricity and magnetism have been treated independently for a long time, and knowledge has never been beyond the scope of qualitative understanding. Until the middle of the nineteenth century, the British physicist J. C. Maxwell inherited the field method of Faraday's electromagnetism, and creatively proposed the concept of displacement current. Therefore, the macroscopic electromagnetic field equations—Maxwell's equations and the electromagnetic wave theory of light are established. His achievements laid the foundation for the study of electromagnetic field theory and opened up a new era of knowing electromagnetic wave and making benefit for human society [1–3].

2.1.1.1 Time-Varying Maxwell Equations

The relationship between the time-varying electromagnetic field and the media constitutes a complete set of equations. In principle, if the source distribution, the nature and geometry shape of the media are known, the electromagnetic field problem always can be solved. For any point at the range $\boldsymbol{r}$ in space, the electric field vector satisfies Maxwell's equation at any time forever.

$$\nabla \times \mathbf{E}(\mathbf{r}, t) = -\frac{\partial}{\partial t}\mathbf{B}(\mathbf{r}, t) \tag{2.1}$$

$$\nabla \times \mathbf{H}(\mathbf{r}, t) = \frac{\partial}{\partial t}\mathbf{D}(\mathbf{r}, t) + \mathbf{J}(\mathbf{r}, t) \tag{2.2}$$

R. Yang et al., *Polarimetric Microwave Imaging*,
https://doi.org/10.1007/978-981-15-8897-6_2

$$\nabla \times \mathbf{D} = \rho \tag{2.3}$$

$$\nabla \cdot \mathbf{B} = 0 \tag{2.4}$$

where the **E** and **H** are the electric field intensity **E** (V/m) and the magnetic field intensity **H** (A/m), respectively. **B** and **D** are the magnetic flux density **B** (Wb/m^2) and electric flux density **D** (C/m^2), respectively. **J** is the current density **J** (A/m^2). $\boldsymbol{\rho}$ *is* the free charge density (C/m^2). **r** expresses the location vector. t expresses the time. The law of conservation of electronic charge is given by

$$\nabla \times \mathbf{J}(\mathbf{r}, t) + \frac{\partial}{\partial t}\rho(\mathbf{r}, t) = 0 \tag{2.5}$$

Equation (2.5) is also called the current continuity equation.

The macrography properties of media for electric field propagation are characterized by permittivity (dielectric constant), permeability, and conductivity

$$\mathbf{D} = \varepsilon \mathbf{E} \tag{2.6}$$

$$\mathbf{H} = \frac{1}{\mu}\mathbf{B} \tag{2.7}$$

$$\mathbf{J} = \sigma \mathbf{E} \tag{2.8}$$

where permittivity, permeability, and conductivity are usually denoted by ε (F/m), μ (H/m), and σ (Ω/m), respectively. Maxwell equations show that the electric field vector and magnetic field vector of electromagnetic wave are not independent of each other so that the electric field vector can only be used to completely represent the electromagnetic wave propagation in free space.

2.1.1.2 The Basic Equations of Harmonic Varying Electromagnetic Field

The harmonic varying electromagnetic field is a special case of time-varying electromagnetic field. For monochromatic plane waves, the instantaneous value of the harmonic varying field is

$$\dot{\mathbf{E}}(\boldsymbol{r}, t) = \mathrm{Re}[\dot{\mathbf{E}}(\boldsymbol{r})e^{j\omega t}] \tag{2.9}$$

$$\dot{\mathbf{H}}(\boldsymbol{r}, t) = \mathrm{Re}[\dot{\mathbf{H}}(\boldsymbol{r})e^{j\omega t}] \tag{2.10}$$

$$\dot{\mathbf{D}}(\boldsymbol{r}, t) = \mathrm{Re}[\dot{\mathbf{D}}(\boldsymbol{r})e^{j\omega t}] \tag{2.11}$$

$$\dot{\mathbf{B}}(\boldsymbol{r},t) = \mathrm{Re}[\dot{\mathbf{B}}(\boldsymbol{r})e^{j\omega t}] \tag{2.12}$$

where $\dot{\mathbf{E}}(\boldsymbol{r})$, $\dot{\mathbf{H}}(\boldsymbol{r})$, $\dot{\mathbf{D}}(\boldsymbol{r})$, and $\dot{\mathbf{B}}(\boldsymbol{r})$ are the complex amplitudes which are the function of the coordinate $\boldsymbol{r}$ in space. ω is the angle frequency. Re is for the real part of the complex variable. By using the simplified complex symbol, leaving out the points on Re and vectors, harmonic varying Maxwell Equations can be written as

$$\nabla \cdot \mathbf{E} = \rho/\varepsilon \tag{2.13}$$

$$\nabla \times \mathbf{E}(\mathbf{r},t) = -j\omega\mu\mathbf{H}(\mathbf{r},t) \tag{2.14}$$

$$\nabla \cdot \mathbf{H} = 0 \tag{2.15}$$

$$\nabla \times H(\boldsymbol{r},t) = (\sigma - j\omega\varepsilon)\mathbf{E}(\boldsymbol{r},t) \tag{2.16}$$

The law of conservation of electronic charge is given by

$$\nabla \cdot \mathbf{J}(\boldsymbol{r},t) + j\omega\rho(\boldsymbol{r},t) = 0 \tag{2.17}$$

In a passive medium ($\rho = 0, \mathbf{J} = 0$), the vector equations in the frequency domain can be obtained from Maxwell's equations as follows:

$$\nabla^2\mathbf{E}(\boldsymbol{r}) + k^2\mathbf{E}(\boldsymbol{r}) = 0 \tag{2.18}$$

$$\nabla^2 H(\boldsymbol{r}) + k^2 H(\boldsymbol{r}) = 0 \tag{2.19}$$

2.1.2 The Electromagnetic Wave Propagation

In a lossless region, with zero current and zero charge densities, the Maxwell curl equations of time-invariant are

$$\nabla \times \mathbf{E} = -j\omega\mu_0\boldsymbol{H} \tag{2.20}$$

$$\nabla \times \mathbf{H} = j\omega\varepsilon_0\boldsymbol{E} \tag{2.21}$$

where **E** and **H** are the functions of $\boldsymbol{r}$, and the $\boldsymbol{r}$ is the vector range from the origin to the point in the determining field. However, in a local scope for an important class

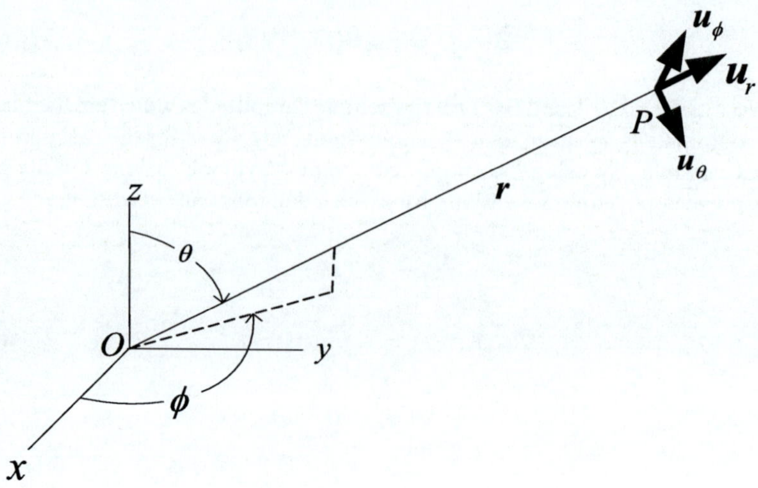

Fig. 2.1 The coordinate system for the traveling wave

of electromagnetic fields, **E** and **H** only depend on the scalar range from one point to the other point in the field. Figure 2.1 shows the coordinates.

In the vicinity of point P, we get

$$\boldsymbol{E}, \boldsymbol{H}=\boldsymbol{E}(\boldsymbol{r}), \boldsymbol{H}(\boldsymbol{r}) \tag{2.22}$$

where assuming that the variation of **E** and **H** with θ and φ is can be neglected compared to the variation with $\boldsymbol{r}$. The variation of the fields functional accurately describes configurations with sources or reflecting targets in the vicinity of the origin if point P is far away from sources of scattering. However, it is incorrect in the vicinity of sources or reflectors.

When the field is considered as only functions of $\boldsymbol{r}$, the curl Eqs. (2.20) and (2.21) are expanded as

$$\begin{aligned} \nabla \times \mathbf{E} &= -\frac{1}{r}\frac{d(rE_\phi)}{dr}\mathbf{u}_\theta + \frac{1}{r}\frac{d(rE_\theta)}{dr}\mathbf{u}_\phi = -j\omega\mu_0(H_r\mathbf{u}_r + H_\theta\mathbf{u}_\theta + H_\phi\mathbf{u}_\phi) \\ \nabla \times \mathbf{H} &= -\frac{1}{r}\frac{d(rH_\phi)}{dr}\mathbf{u}_\theta + \frac{1}{r}\frac{d(rH_\theta)}{dr}\mathbf{u}_\phi = j\omega\varepsilon_0(E_r\mathbf{u}_r + E_\theta\mathbf{u}_\theta + E_\phi\mathbf{u}_\phi) \end{aligned} \tag{2.23}$$

where $\mathbf{u}_r$, $\mathbf{u}_\theta$, $\mathbf{u}_\phi$ are the real unit vectors shown in Fig. 2.1. Let the coefficients equal for the same unit vectors in the first equation, differentiating the Eq. (2.23), and then substituting to the result equation using the coefficient of $\mathbf{u}_\theta$, $\mathbf{u}_\phi$ from the curl equation of **H**, and obtain:

$$\frac{d^2(rE_{\theta,\phi})}{dr^2} + k^2 r E_{\theta,\phi} = 0 \tag{2.24}$$

where $k^2 = \omega^2 \mu_0 \varepsilon_0$. Solutions of these equations are

$$E_{\theta,\phi} = \frac{C_{\theta,\phi}}{r} e^{\pm jkr} \tag{2.25}$$

Using the similar method, **H** can be found, and its components are

$$\begin{aligned} H_\theta &= \pm \frac{C_\phi / Z_0}{r} e^{\pm jkr} \\ H_\phi &= \mp \frac{C_\theta / Z_0}{r} e^{\pm jkr} \end{aligned} \tag{2.26}$$

where $Z_0 = \sqrt{\mu_0/\varepsilon_0}$.

If taking a component of the electric field, such as E_θ, and write its corresponding time-varying form as follows:

$$\widetilde{E}_\theta = \mathrm{Re}(E_\theta e^{j\omega t}) = \frac{|C_\theta|}{r} \cos(\omega t \pm kr + \Phi)$$

where Φ is the phase angle. This equation represents a component of the electromagnetic wave. If the electromagnetic wave propagates in the direction of increasing range, there is a negative sign in the cosine function. If the electromagnetic wave propagates in the range decreasing direction, there is a positive sign in the cosine function. The wave propagating toward the origin can be rejected for physical reasons and the negative sign in the cosine wave will be retained.

When the curl equations of Maxwell were expanded in spherical coordinates, with the assumption that the fields vary only with r, the curl of **E** had no radial component. This is from Eq. (2.21) that **H** does not have a radial component. Likewise, since the curl of **H** has no radial component, **E** does not have a radial component.

The components of these field vectors vary only with the radial distance r. This solution represents a specific electromagnetic field that is important in remote sensing. The fields are

$$\mathbf{E} = \frac{1}{r}(C_\theta \mathbf{u}_\theta + C_\phi \mathbf{u}_\phi) e^{-jkr} \tag{2.27}$$

$$\mathbf{H} = \frac{1}{r}\frac{1}{Z_0}(-C_\phi \mathbf{u}_\theta + C_\theta \mathbf{u}_\phi) e^{-jkr} \tag{2.28}$$

It is apparent from these equations that **E** and **H** are perpendicular to $\mathbf{u}_r$, the direction of wave travel. Further, if neglecting the phase variation with r, the scalar product of **E** and **H**, is zero, showing that **E** and **H** are perpendicular to each other. The phenomenon revealed by the above equations is a spherical wave traveling outward from the origin of coordinates.

The assumption in Eq. (2.22) is valid only in the vicinity of a selected point P. At a different point, the field coefficients will differ from the field coefficients values at P and the direction of wave travel will be different. At P, the radial range $\boldsymbol{r}$ is very large, compared with the region, in which it ensures that assumptions about the behavior of the fields to be valid. Then a surface with equal $\boldsymbol{r}$ is almost a plane surface. In general, the electromagnetic wave is described as a plane wave with amplitudes and phases constant over a plane, rather than a spherical wave with constant over a spherical surface and at P, a rectangular coordinate system with an axis pointing in the direction of wave travel is constructed. If the wave travels in the Z direction, the fields can be written as

$$\mathbf{E} = (E_x\mathbf{u}_x + E_y\mathbf{u}_y)e^{-jkz} \tag{2.29}$$

$$\mathbf{H} = \frac{1}{Z_0}(-E_y\mathbf{u}_x + E_x\mathbf{u}_y)e^{-jkz} \tag{2.30}$$

where $\mathbf{u}_x$ and $\mathbf{u}_y$ are the unit vectors. In the new coordinate system, the phase referenced to the origin of the formal coordinate can be discarded, and its variation with large $\boldsymbol{r}$ coefficient amplitudes varies so slowly, so as to be ignored.

Two other wave symbols are commonly used to describe electromagnetic waves. From $\cos(\omega t - kr)$, it can be noted that at a constant position r_0, the change of phase is 2π radians when time changes by $2\pi/\omega$. The time interval is

$$T = \frac{2\pi}{\omega} = \frac{1}{f} \tag{2.31}$$

where T is the wave period. At time t_0, as the radial range changes $2\pi/k$, the phase will change 2π radians, the increment in range $\boldsymbol{r}$ corresponding to a phase change of 2π radians is called the wavelength. Thus,

$$\lambda = \Delta r = \frac{2\pi}{k} \tag{2.32}$$

If imaging an electromagnetic wave in the space at a static time, the wavelength is the length of one complete cycle for a sine wave. In the vacuum or approximately air, thus

$$\lambda = \frac{2\pi}{2\pi f\sqrt{\mu_0\varepsilon_0}} = \frac{c}{f} \tag{2.33}$$

where c is the velocity of light in vacuum.

2.2 Polarization Characteristics and Representations of Electromagnetic Waves [3, 4]

2.2.1 *Polarization Ellipse*

The electric field intensity of a monochromatic electromagnetic wave traveling along the $+z$ direction is given by

$$\mathbf{E} = (E_x\mathbf{u}_x + E_y\mathbf{u}_y)e^{-jkz} = (\mathbf{u}_x|E_x|e^{j\phi_x} + \mathbf{u}_y|E_y|e^{j\phi_y})e^{-jkz} \tag{2.34}$$

where $\mathbf{u}_x$ and $\mathbf{u}_y$ are the unit vectors alone the x and y directions, and ϕ_x and ϕ_y are the E_x and E_y phases, respectively. The time-varying electric vector is

$$\mathbf{E}(z,t) = \mathrm{Re}(\mathbf{E}e^{j\omega t}) = |E_x|\cos(\omega t - kz + \phi_x)\mathbf{u}_x + |E_y|\cos(\omega t - kz + \phi_y)\mathbf{u}_y \tag{2.35}$$

where ω is the angular frequency, k is the wavenumber.

Let $\xi_x = |E_x|\cos(\omega t - kz + \phi_x)$, $\xi_y = |E_y|\cos(\omega t - kz + \phi_y)$, Eq. (2.35) can be rewritten as

$$\frac{\xi_x^2}{|E_x|^2} - 2\frac{\xi_x}{|E_x|}\frac{\xi_y}{|E_y|}\cos\phi + \frac{\xi_y^2}{|E_y|^2} = \sin^2\phi \tag{2.36}$$

where $\phi = \phi_y - \phi_x$. For a monochromatic wave, the trajectory of the electric field intensity victor changing with the time, and Eq. (2.36) corresponds to an ellipse. The ellipse represents the polarization characteristics of the electromagnetic wave, which is called the polarization ellipse, shown in Fig. 2.2.

The polarization ellipse can be characterized by three parameters known as the ellipse orientation angle τ, the ellipticity angle, ε and the ellipse size A:

$$\tan 2\tau = \frac{2|E_x||E_y|}{|E_x|^2 - |E_y|^2}\cos\phi \tag{2.37}$$

$$\sin 2\varepsilon = \frac{2|E_x||E_y|}{|E_x|^2 + |E_y|^2}\sin\phi \tag{2.38}$$

$$A = \sqrt{|E_x|^2 + |E_y|^2} \tag{2.39}$$

where the ellipse orientation angle ε represents the shape of the ellipse and the rotation direction of the electric field vector. If $\varepsilon > 0$, it can be got a left-handed polarization, and if $\varepsilon < 0$, it can be got a right-handed polarization.

There are some special cases, if $|E_x|$ or $|E_y|$ is zero, the ellipse collapses to a line, the wave is called the linear polarization wave. Another special case is encountered when the two amplitudes are the same ($|E_x| = |E_y|$) and the relative phase difference

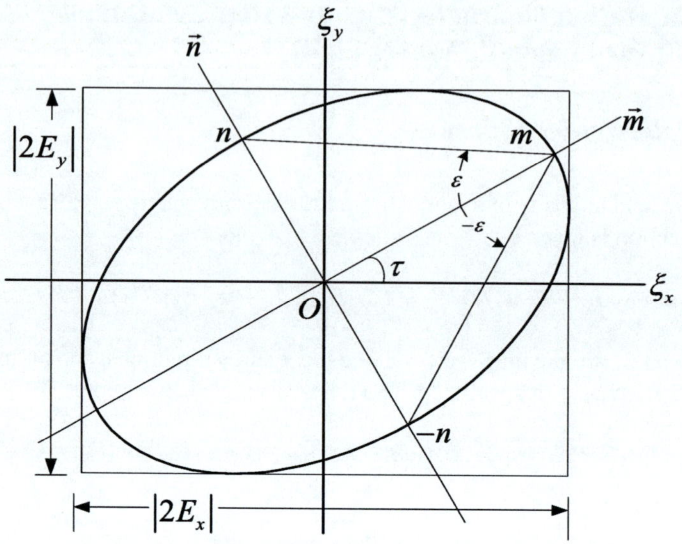

Fig. 2.2 A polarization ellipse

$\phi = \pm\pi/2$. In this case, the wave is circularly polarized. If $\phi = \pi/2$, the wave is left circularly polarized, and if $\phi = -\pi/2$, the wave is right circularly polarized.

2.2.2 *Jones Vector and Polarization Ratio*

Although the polarization ellipse can clearly represent the physical meaning of the electromagnetic wave, a more convenient mathematical description is needed. The electric field intensity of a monochromatic electromagnetic wave traveling along the $+z$ direction can be represented by a two-dimensional vector:

$$\mathbf{E} = \begin{bmatrix} |E_x| \exp(j\omega t - kz + \phi_x) \\ |E_y| \exp(j\omega t - kz + \phi_y) \end{bmatrix} \tag{2.40}$$

Equation (2.40) without regard to the absolute phase is rewritten as

$$\mathbf{E} = \begin{bmatrix} |E_x| \\ |E_y| \exp(j\phi) \end{bmatrix} \tag{2.41}$$

where $\phi = \phi_x - \phi_y$. The vector in Eq. (2.41) is called Jones vector.

A new descriptor introduced from the two orthogonal components of Jones vector is a complex polarization ratio, which describes the polarization state. The polarization ratio is defined as the ratio of two orthogonal components of Jones vector:

$$\rho = \frac{|E_y|}{|E_x|} e^{j\phi} = \tan\gamma \cdot e^{j\phi} \tag{2.42}$$

where γ and ϕ are the phase description parameters of the electromagnetic wave.

The polarization ratio can be used to represent the electric field density.

$$\mathbf{E} = |E_x| \exp(j\phi_x) \begin{bmatrix} 1 \\ \rho \exp(j\phi) \end{bmatrix} \tag{2.43}$$

2.2.3 *Stokes Vector and Poincare Sphere*

Jones vector is only applicable to fully polarized waves whose amplitudes of the two orthogonal components of the wave vector are constant and the phase difference is constant. But for partially polarized waves which do not satisfy above the condition, a new representation method has to be introduced. Stokes vector can represent both fully polarized wave and partially polarized wave. It was as early as 1852 when Stokes showed that the amplitude and polarization characteristics of quasi-monochromatic light wave may be represented by four real parameters, which are called Stokes parameters. These parameters can be organized into a vector, known as the Stokes vector. Stokes vector of the fully polarized wave is given by

$$\mathbf{G} = \begin{bmatrix} G_0 \\ G_1 \\ G_2 \\ G_3 \end{bmatrix} = \begin{bmatrix} |E_x|^2 + |E_y|^2 \\ |E_x|^2 - |E_y|^2 \\ 2|E_x||E_y|\cos\phi \\ 2|E_x||E_y|\sin\phi \end{bmatrix} \tag{2.44}$$

where $|E_x|$ and $|E_y|$ are the amplitudes of the two orthogonal components of the electric field, and ϕ is the phase difference. Equation (2.44) shows that Stokes vector can completely describe the amplitude and polarization information of electromagnetic wave. G_0 represents the amplitude of wave, $|E_x|$ and $|E_y|$ can be calculated by G_0 and G_1, and ϕ can be calculated by G_2 and G_3. For the fully polarized wave, only three of four Stokes parameters are independent since

$$G_0^2 = G_1^2 + G_2^2 + G_3^2 \tag{2.45}$$

For the fully polarized wave, the Stokes vector can be represented by the geometric parameters (ε, τ) and phase parameters (γ, ϕ)

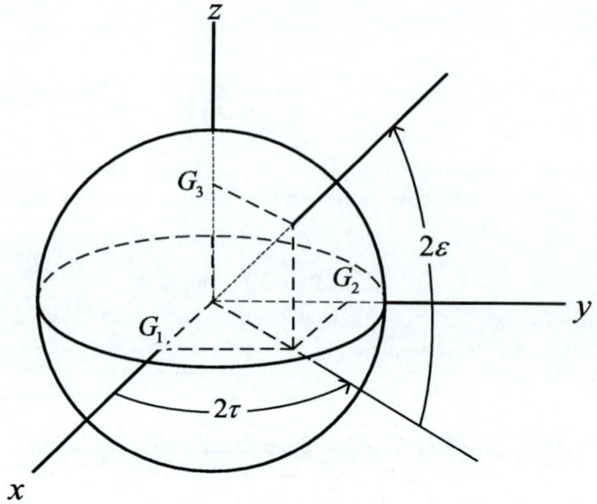

Fig. 2.3 The Poincare sphere

$$G = G_0 \begin{bmatrix} 1 \\ \cos 2\varepsilon \cos 2\tau \\ \cos 2\varepsilon \sin 2\tau \\ \sin 2\varepsilon \end{bmatrix} = G_0 \begin{bmatrix} 1 \\ \cos 2\gamma \\ \sin 2\gamma \cos \phi \\ \sin 2\gamma \sin \phi \end{bmatrix} \quad (2.46)$$

- The polarization state of the fully polarized wave can be expressed from the Poincare sphere with radius G_0. Each polarization state is corresponding to the point on the Poincare sphere, and vice versa. G_1, G_2, and G_3 are regarded as the Cartesian coordinates on the point of the Poincare sphere. 2ε and 2τ represent the latitude and the longitude of the point, respectively, as seen in Fig. 2.3. Because the sign of ε determines the handedness of the polarization state, the upper hemisphere (corresponding to $\varepsilon > 0$) displays left-handed polarization and the lower hemisphere displays right-handed polarization.

2.3 Representations of Partially Polarized Electromagnetic Waves

The polarization representation of a monochromatic wave has been discussed in previous sections. The monochromatic wave is fully polarized or is a pure polarized, so the trajectory of the electric field vector endpoint at the propagating space is a polarization ellipse with constant ellipse orientation angle and the ellipticity angle. This is a time-invariant ellipse. However, in practice, electromagnetic waves are usually non-monochromatic, and the trajectory of the electric field vector endpoint in the propagating space is no longer a time-invariant ellipse, but a curve similar to a shape and time-variant ellipse, such waves called partially polarized waves. It is apparent that a fully polarized wave is a special case of a partially polarized wave.

In the field of electromagnetic engineering, the narrow-band signal or the quasi-monochromatic wave is very common, whose bandwidth is small relative to the center frequency. In this case, the electric field vector of the quasi-monochromatic wave traveling along the $+z$ direction is given by

$$\mathbf{E}(t) = \begin{bmatrix} E_x(t) \\ E_y(t) \end{bmatrix} = \begin{bmatrix} a_x(t)\exp(\bar{\omega}t - kz + \phi_x(t)) \\ a_y(t)\exp\bar{\omega}t - kz + \phi_y(t) \end{bmatrix} \tag{2.47}$$

where $\bar{\omega}$ denotes the center frequency, $a_x(t), a_y(t)$ and $\phi_x(t), \phi_y(t)$ are the amplitudes and absolute phases of the electrical component, respectively. Relative to the term $\bar{\omega}t$, there are slowly varying procedures with time. The electric field vector is

$$\mathbf{E}(t) = \begin{bmatrix} E_x(t) \\ E_y(t) \end{bmatrix} \tag{2.48}$$

Equation (2.48) is a time-varying two-dimensional complex vector. The coherency matrix is defined as

$$\mathbf{J}_m = \left\langle \mathbf{E}(t)\mathbf{E}(t)^{\mathrm{H}} \right\rangle = \begin{bmatrix} J_{xx} & J_{xy} \\ J_{yx} & J_{yy} \end{bmatrix} = \begin{bmatrix} \langle |E_x(t)|^2 \rangle & \langle E_x(t)E_y^*(t) \rangle \\ \langle E_x^*(t)E_y(t) \rangle & \left\langle |E_y(t)|^2 \right\rangle \end{bmatrix} \tag{2.49}$$

where the superscript H represents the conjugation, $\langle \bullet \rangle$ indicates the ensemble averaging, and $\mathbf{J}_m$ is the Hermitian positive semi-definite matrix. A parameter μ can be defined by the elements in Eq. (2.49) as follows:

$$\mu = \frac{J_{xy}}{\sqrt{J_{xx}J_{yy}}} \tag{2.50}$$

μ $(0 \le \mu \le 1)$ is called a coherency factor and the module $|\mu|$ corresponds to the degree of correlation between the electric field components. If $|\mu| = 1$, the two orthogonal electric field components are completely coherent, so the wave is considered to be fully polarized. If $|\mu| < 1$, the two orthogonal electric field components are partially coherent, so the wave is considered to be partially polarized. If $|\mu| = 0$, the two orthogonal electric field components are completely incoherent, so the wave is considered to be completely unpolarized.

Arraying the elements of the coherence matrix as a vector, the coherence vector's form of the wave can be obtained as follows:

$$\mathbf{J} = \begin{bmatrix} J_{xx} \\ J_{xy} \\ J_{yx} \\ J_{yy} \end{bmatrix} \tag{2.51}$$

The relationship between the coherency vector **J** and the Stokes vector is

$$\mathbf{G} = \mathbf{QJ}, \quad \mathbf{Q} = \begin{bmatrix} 1 & 0 & 0 & 1 \\ 1 & 0 & 0 & -1 \\ 0 & 1 & 1 & 0 \\ 0 & j & -j & 0 \end{bmatrix} \tag{2.52}$$

Using Eq. (2.52), the partially polarized wave can be represented by the Stokes vector as

$$\mathbf{G} = \begin{bmatrix} G_0 \\ G_1 \\ G_2 \\ G_3 \end{bmatrix} = \begin{bmatrix} \langle |E_x(t)|^2 \rangle + \langle |E_y(t)|^2 \rangle \\ \langle |E_x(t)|^2 \rangle - \langle |E_y(t)|^2 \rangle \\ 2\mathrm{Re}(\langle E_x^*(t) E_y(t) \rangle) \\ 2\mathrm{Im}(\langle E_x^*(t) E_y(t) \rangle) \end{bmatrix} \tag{2.53}$$

The Stokes parameters of the partially polarized waves satisfy the following constraint:

$$G_0^2 > G_1^2 + G_2^2 + G_3^2 \tag{2.54}$$

It can be proved that the Stokes vector of any wave can be decomposed into the sum of the Stokes vector of a fully unpolarized wave and the Stokes vector of a fully polarized wave. The unique representation is expressed as

$$G = G^{(1)} + G^{(2)} \tag{2.55}$$

where $G = \begin{bmatrix} G_0 \\ G_1 \\ G_2 \\ G_3 \end{bmatrix}$, $G^{(1)} = \begin{bmatrix} G_0 - \sqrt{G_1^2 + G_2^2 + G_3^2} \\ 0 \\ 0 \\ 0 \end{bmatrix}$, $G^{(2)} = \begin{bmatrix} \sqrt{G_1^2 + G_2^2 + G_3^2} \\ G_1 \\ G_2 \\ G_3 \end{bmatrix}$.

The degree of polarization is defined as the ratio of the fully polarized wave intensity to the total wave intensity.

$$PD = \frac{\sqrt{G_1^2 + G_2^2 + G_3^2}}{G_0} \tag{2.56}$$

The representation of the partially polarized wave can be extended to the whole Poincare sphere and any point in the Poincare sphere has special physical meaning:

(1) $PD = 0$, the wave is fully unpolarized, corresponding to the center of the Poincare sphere;
(2) $PD = 1$, the wave is fully polarized, corresponding to the points on the surface of the Poincare sphere;
(3) $0 < PD < 1$, the wave is partially polarized, corresponding to the points inner the Poincare sphere.

2.4 Transmitting and Receiving of Electromagnetic Wave and Antenna [3, 4]

2.4.1 Transmitting of Electromagnetic Wave

In the radar remote sensing area, the electromagnetic wave with special frequency and polarization is transmitted by the antenna. The transmitting antenna can be looked as a load connected to a transmitter by means of a transmission line or waveguide, the impedance of the transmitting antenna and the power from the transmitter into the antenna can be measured. The part energy is radiated out from antenna, and the rest is converted into heat, both parts can also be measured. The transmitting antenna cannot radiate the same power in all directions, and its radiation direction characteristics can be determined.

2.4.2 Receiving of Fully Polarization Electromagnetic Wave

Due to the electric vector **E** of the incident wave is orthogonal to the propagation direction of wave, it can be represented by two components in the appropriate coordinate system. In a coordinate system with an axis opposite to the propagating direction of the incident wave, the effective length of the antenna **h** can be composed of three components, but only the component perpendicular to the direction of the incident wave will contribute to the antenna terminal voltage.

Figure 2.4 shows the receiving antenna coordinate including the incident wave. Assuming that the antenna is located at the origin of the xyz coordinate system and its effective length is known, the incidence wave is illustrated in the $\xi\eta\zeta$ coordinate system. The z axis is to be parallel to the ζ axis in the opposite direction, and the y axis is to be parallel to the η axis.

Since the z component of **h** contributes nothing to doing, so it can be ignored, writing as follows:

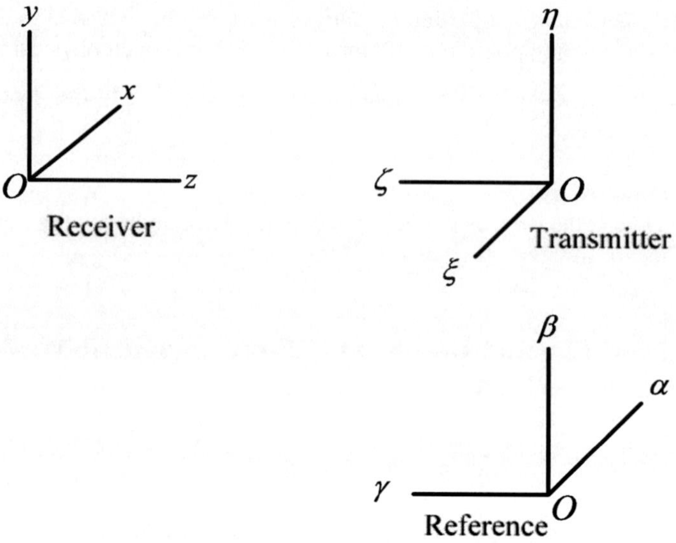

Fig. 2.4 The receiving antenna coordinate

$$\mathbf{h}(x, y) = \begin{bmatrix} h_x \\ h_y \end{bmatrix} \quad \mathbf{E}(\xi, \eta) = \begin{bmatrix} E_\xi \\ E_\eta \end{bmatrix} \tag{2.57}$$

The electric field density vector in Eq. (2.57) is called a Jones vector. Notice the signs between brackets are representing components of the antenna and electric field.

In the same coordinate system, the received voltage is satisfied as

$$V = \mathbf{h}^T \mathbf{E} \tag{2.58}$$

In the book, **h** is always shown in the right-handed coordinate system, and one axis points outward the antenna. Therefore, **E** is transformed in the xyz coordinate system as

$$\mathbf{E}(x, y) = \mathbf{diag}(-1, 1)\mathbf{E}(\xi, \eta) \tag{2.59}$$

The received voltage is rewritten as

$$V = \mathbf{h}^T(x, y)\mathbf{E}(x, y) = \mathbf{h}^T(x, y)\mathbf{diag}(-1, 1)\mathbf{E}(\xi, \eta) \tag{2.60}$$

2.4.3 Receiving of Partially Polarization Wave

When an incident wave $\mathbf{E}(t)$ falls on a receiving antenna with effective length $\mathbf{h}$, the received voltage is

$$V(t) = \mathbf{h}^T\mathbf{E}(t)$$

If the incident wave is fully polarized and if $\mathbf{E}$ and $\mathbf{h}$ are expressed in the same coordinate. With regard to a partially polarized wave, instead of the voltage, the received power should be measured. The power is the time average.

$$W = \frac{\overline{VV^*}}{8R_a} = \frac{1}{8R_a}\overline{\mathbf{h}^T\mathbf{E}(t)\mathbf{h}^\dagger\mathbf{E}^*(t)} \tag{2.61}$$

In Fig. 2.4, if it hopes the electric field intensity is represented propagating along ζ direction in nature coordinate system, then

$$\mathbf{E}(x, y, t) = \mathbf{diag}(-1, 1)\mathbf{E}(\xi, \eta, t) \tag{2.62}$$

Then, the power becomes

$$\begin{aligned} W &= \frac{1}{8R_a}\left[\mathbf{h}(x, y) \otimes \mathbf{h}^*(x, y)\right]^T \overline{\mathbf{diag}(-1, 1)\mathbf{E}(\xi, \eta, t) \otimes \mathbf{diag}(-1, 1)\mathbf{E}^*(\xi, \eta, t)} \\ &= \frac{1}{8R_a}\left[\mathbf{h}(x, y) \otimes \mathbf{h}^*(x, y)\right]^T \mathbf{diag}(1, -1, -1, 1)\overline{\mathbf{E}(\xi, \eta, t) \otimes \mathbf{E}^*(\xi, \eta, t)} \end{aligned} \tag{2.63}$$

The Kronecker product involving the electric field intensity is the coherency vector of the incident wave, and the field is represented in the standard right-handed coordinate system.

If $\mathbf{E}$ and $\mathbf{h}$ are represented in the same coordinate system, the received power is rewritten as

$$W = \frac{1}{8R_a}\mathbf{J}_A^T\overline{\mathbf{E}(t) \otimes \mathbf{E}^*(t)} \tag{2.64}$$

Here, the Kronecker product involving the electric field is not called the coherency vector of the incident wave since here the coherency vector is used to describe a wave in right-handed coordinate system. The wave is traveling along a coordinate axis.

2.4.4 Polarization Matching of Antenna

Assuming that an antenna locates at the origin of the spherical coordinate system in Fig. 2.1, and a monochromatic plane wave is traveling toward the antenna along the (θ, φ) direction. If the effective length of the antenna in this direction is **h**, and Jones vector is denoted as $\mathbf{E}_i$, it can be known from the electromagnetic network theory that the open-circuit voltage induced by the electromagnetic wave on the receiving antenna is

$$V = \mathbf{E}_i^T \mathbf{h} \tag{2.65}$$

where **h** and $\mathbf{E}_i$ are complex vectors, V is a complex voltage. **h** and $\mathbf{E}_i$ are in the same coordinate system shown in Fig. 2.1. That is defined in a coordinate system which origin point is the center of receiver antenna aperture.

According to Eq. (2.65), the open-circuit voltage and the received power can be enhanced by means of choosing the effective length of the receiving antenna reasonably. The output power of the antenna is

$$P = VV^* = \left|\mathbf{E}_i^T \mathbf{h}\right|^2 = \left|h_\theta E_{\theta i} + h_\varphi E_{\varphi i}\right|^2 \tag{2.66}$$

For a fixed incident wave $\mathbf{E}_i$, the maximum output power of the antenna depends on $|h_\theta|$ and $\left|h_\varphi\right|$, and it can be calculated by

$$P_{\max} = (\mathbf{h} \cdot \mathbf{h}^*)(\mathbf{E}_i \mathbf{E}_i^*) = |\mathbf{h}|^2 |\mathbf{E}_i|^2 \tag{2.67}$$

When the output power is maximum, the antenna is located at an optimum matched state. If changing the polarization characteristics of the antenna and keeping the impedance matching condition, the ratio of the power received actually by the antenna to the power received by the antenna under the optimal polarization matching is

$$\rho = \frac{\left|\mathbf{E}_i^T \mathbf{h}\right|^2}{|\mathbf{E}_i|^2 |\mathbf{h}|^2} \tag{2.68}$$

The parameter ρ is the polarization match coefficient, scope is from 0 to 1, the polarization match coefficient represents the degree of polarization matched between the receiving antenna with effective length **h** and the incident wave.

As the transmitting and receiving antenna system has a pair of polarization matched antenna, the polarization match coefficient $\rho = 1$. If the polarization characteristic of transmitting and receiving antenna dos not receiving any signal, $\rho = 0$ then the polarization of this pair antenna is called the orthogonal polarization. PolSAR transmits the orthogonally polarized electromagnetic wave and receives the scattered wave based on the orthogonal polarization basis. In this way, the polarization match coefficient of the two antennas in the orthogonal polarization state will be zero. However, This ideal condition is to difficult to implement in practice, and there must be some crosstalk between orthogonally polarized antennas. The polarization isolation is defined as the suppression of an antenna to the orthogonally matched electromagnetic wave to it. PolSAR requires polarization isolation as much as possible.

As we know that a fully unpolarized electromagnetic wave can be decomposed into a pair of fully polarized waves with arbitrary, independent, orthogonal, and equal power density. This means that for a fully unpolarized incident wave, the antenna can only receive the half power, it is independent of the polarization state of the antenna.

2.5 Representations of Polarization Scattering Characteristics [4, 5]

2.5.1 FSA–BSA Convention

In order to represent the scattering target, it needs to construct a coordinate system. In general, the coordinate system of the transmitting antenna satisfies the IEEE right-hand helix criterion. That means the z axis is consistent with the propagation direction of the transmitted electromagnetic wave. If the z axis of the receiving antenna is consistent with the propagation direction of the scattering electromagnetic wave, the coordinate system is called forward-scattered alignment (FSA) convention or forward-scattered alignment coordinate system, which is commonly used in forward-looking or bi-static SAR system. If the z axis of the receiving antenna is opposite to the propagation direction of the scattering electromagnetic wave, the coordinate system is called backscattered alignment (BSA) convention or BSA coordinate system, which usually is used in monostatic SAR system. Figure 2.5 shows the coordinate systems of FSA–BSA convention.

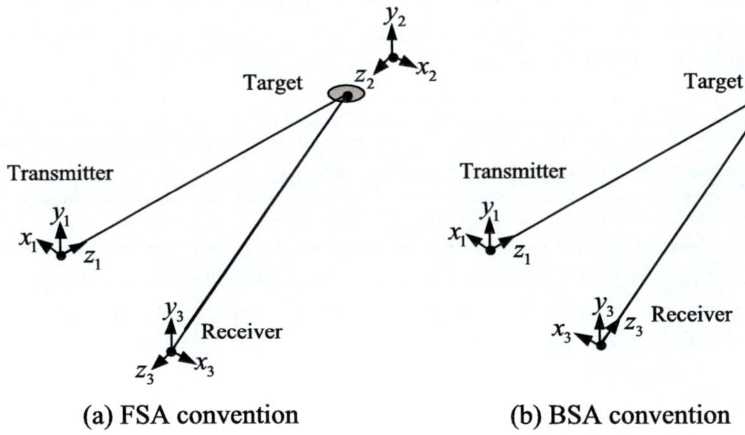

Fig. 2.5 Coordinate systems of the transmitting and receiving antenna

2.5.2 *Representation of Coherently Scattering Targets—Scattering Matrix*

When a radar wave illuminates a target, a part of the incident energy is reflected or scattered. If the incident wave is mono-frequency, the target is stationary and the radar target aspect angle is constant, the scattered wave will also be mono-frequency and fully polarized wave. Such a target is called coherently scattering target.

Scattering matrix is often called Sinclair matrix, which reflected the linear relationship between the incident wave and the Jones vector of scattering wave.

For a monostatic radar system, the BSA convention is commonly chosen, and the transmitting and receiving antennas adopt the same coordinate system (x, y, z). Let the Jones vector of incident wave under transmitting coordinate and the Jones vector of scattering wave under receiving coordinate are $\mathbf{E}_i$ and $\mathbf{E}_s$, respectively. The relationship between $\mathbf{E}_i$ and $\mathbf{E}_s$ can be defined as a scattering matrix $[\mathbf{S}]$

$$\mathbf{E}_s = [\mathbf{S}]\mathbf{E}_i = \begin{bmatrix} S_{xx} & S_{xy} \\ S_{yx} & S_{yy} \end{bmatrix} \mathbf{E}_i \tag{2.69}$$

where S_{xy} expresses a backscattering coefficient for transmitting antenna with x polarization and receiving antenna with y polarization.

Using the Sinclair matrix, we get the received voltage of the antenna.

$$V = \mathbf{h}_r^{\mathrm{T}} \mathbf{S} \mathbf{h}_t \tag{2.70}$$

where $\mathbf{h}_t$ is the effective length of the transmitting antenna and $\mathbf{h}_r$ is the effective length of the receiving antenna.

In general, the polarization scattering matrix has a complex form, which is not only related to the shape, size, structure, material, and other physical factors of the target itself but also related to the relative attitude orientation between target and transmitting receiving measuring system, geometry space location, radar frequency, etc. From the reciprocity theorem, it can be proved that the Sinclair matrix is symmetric in the backscattering, that is $\mathbf{S} = \mathbf{S}^{\mathrm{T}}$.

2.5.3 *Representation of Depolarization Target*

If the scattering characteristic of the target is time-varying, the scattered echo is a partially polarized wave, then the target is called depolarization target. Sinclair matrix can be used to describe the coherently scattering target, but not effective describe the depolarization target. The representation of the depolarization target will be discussed in the following.

2.5.3.1 Mueller Matrix

Mueller matrix reflects the linear relationship between the incident wave and the scattered wave. Based on the relationship between Stokes vector and Jones vector, the relationship between Stokes vector of the incident wave and Stokes vector of the scattered wave can be derived as follows:

$$\mathbf{J}_s = \mathbf{L}\mathbf{J}_i \tag{2.71}$$

where $\mathbf{L}$ is the Muller matrix, it is a real asymmetric matrix.

The coherency vector of the scattered wave from a target is given by

$$\mathbf{C} = \langle \mathbf{E}(t) \otimes \mathbf{E}(t)^* \rangle \tag{2.72}$$

The Muller matrix is

$$\mathbf{L} = \mathbf{Q}\mathbf{W}\mathbf{Q}^{-1} \tag{2.73}$$

$\mathbf{Q}$ is defined in Eq. (2.52), and $\mathbf{W}$ is given by

$$\mathbf{W} = \begin{bmatrix} S_{xx}S_{xx}^* & S_{xx}S_{xy}^* & S_{xy}S_{xx}^* & S_{xy}S_{xy}^* \\ S_{xx}S_{yx}^* & S_{xx}S_{yy}^* & S_{xy}S_{yx}^* & S_{xy}S_{yy}^* \\ S_{yx}S_{xx}^* & S_{yx}S_{xy}^* & S_{yy}S_{xx}^* & S_{yy}S_{xy}^* \\ S_{yx}S_{yx}^* & S_{yx}S_{yy}^* & S_{yy}S_{yx}^* & S_{yy}S_{yy}^* \end{bmatrix} \tag{2.74}$$

The Mueller matrix is usually asymmetric. There is a one-to-one relationship between Mueller matrix $\mathbf{L}$ and polarization scattering matrix $\mathbf{S}$.

2.5.3.2 Stokes Matrix

Kennaugh Matrix is derived from receiving power Eq. (2.60). Neglecting the constant terms, the received power is given by

$$W_c = (\mathbf{h}_r^{\mathrm{T}}\mathbf{S}\mathbf{h}_t) \cdot (\mathbf{h}_r^{\mathrm{T}}\mathbf{S}\mathbf{h}_t)^* = (\mathbf{h}_r^{\mathrm{T}}\mathbf{S}\mathbf{h}_t) \otimes (\mathbf{h}_r^{\mathrm{T}}\mathbf{S}\mathbf{h}_t)^* \tag{2.75}$$

where $\otimes$ symbolizes the standard Kronecker matrix product. Using the property of Kronecker matrix product, Eq. (2.75) can be rewritten as

$$W_c = (\mathbf{h}_r \otimes \mathbf{h}_r^*)^{\mathrm{T}}(\mathbf{S} \otimes \mathbf{S}^*)(\mathbf{h}_t \otimes \mathbf{h}_t^*) \tag{2.76}$$

Using Eq. (2.76), the Kronecker product target matrix $\boldsymbol{\kappa}_S$ is defined as

$$\boldsymbol{\kappa}_S = \mathbf{S} \otimes \mathbf{S}^* = \begin{bmatrix} |S_{xx}|^2 & S_{xx}S_{xy}^* & S_{xy}S_{xx}^* & |S_{xy}|^2 \\ S_{xx}S_{yx}^* & S_{xx}S_{yy}^* & S_{xy}S_{yx}^* & S_{xy}S_{yy}^* \\ S_{yx}S_{xx}^* & S_{yx}S_{xy}^* & S_{yy}S_{xx}^* & S_{yy}S_{xy}^* \\ |S_{yx}|^2 & S_{yx}S_{yy}^* & S_{yy}S_{yx}^* & |S_{yy}|^2 \end{bmatrix} \tag{2.77}$$

For the depolarization target, we only get the average received power

$$W_{\mathrm{av}} = (\mathbf{h}_r \otimes \mathbf{h}_r^*)^{\mathrm{T}}\langle \boldsymbol{\kappa}_S \rangle(\mathbf{h}_t \otimes \mathbf{h}_t^*) \tag{2.78}$$

where $\langle \boldsymbol{\kappa}_S \rangle = \frac{1}{N}\sum_{n=1}^{N} \boldsymbol{\kappa}_{Sn}$.

If using Stokes vectors express the effective lengths of transmitting antenna and receiving antenna, Eq. (2.78) can be rewritten as

$$W_{\mathrm{av}} = \mathbf{G}_{Ar}^{\mathrm{T}}\mathbf{Q}^*\langle \boldsymbol{\kappa}_S \rangle \mathbf{Q}^{-1}\mathbf{G}_{At} \tag{2.79}$$

where $\mathbf{G}_{Ar}$ and $\mathbf{G}_{At}$ are Stokes vectors of the transmitting and receiving antenna, and Q was given in Eq. (2.52). Kennaugh matrix is defined as

$$\mathbf{K} = \mathbf{Q}^*\langle \boldsymbol{\kappa}_S \rangle \mathbf{Q}^{-1} \tag{2.80}$$

Under backscattering condition, Kennaugh matrix is a symmetric matrix. Kennaugh matrix is common used to describe depolarization target. Sometimes, it is called Stokes reflection matrix. In the Kennaugh matrix, all elements are real.

2.5.3.3 Covariance Matrix and Coherency Matrix

In data analyses processing of PolSAR, in order to convenient representation, usually need to make the target polarization scattering matrix vectorizing, and obtain the scattering matrix, furthermore obtain the polarization covariance matrix and coherency matrix of target. The covariance matrix and coherency matrix can describe the second-order statistical characteristics of the scattering target. For the information capacity, this is equivalent to the Mueller and Kennaugh matrices.

Based on the linear horizontal and vertical polarization, Sinclair matrix [S] can be represented under the vectorial form as

$$\mathbf{k} = V([\mathbf{S}]) = \frac{1}{2}\mathrm{tr}([\mathbf{S}]\boldsymbol{\Psi}) = [\,k_1\ k_2\ k_3\ k_4\,]^{\mathrm{T}} \tag{2.81}$$

where tr(·) denotes the trajectory of matrix and $\boldsymbol{\Psi}$ is a group of basic matrices 2 × 2 which are orthogonal in the Hermitian scalar product space. There are two types of basic matrices, one is Lexicographic

$$\boldsymbol{\Psi}_L = \left\{2\begin{bmatrix}1 & 0\\0 & 0\end{bmatrix}, 2\begin{bmatrix}0 & 1\\0 & 0\end{bmatrix}, 2\begin{bmatrix}0 & 0\\1 & 0\end{bmatrix}, 2\begin{bmatrix}0 & 0\\0 & 1\end{bmatrix}\right\} \tag{2.82}$$

The scattering matrix [S] can be represented under the vectorial form as

$$\mathbf{k}_L = \begin{bmatrix} S_{\mathrm{hh}} & S_{\mathrm{hv}} & S_{\mathrm{vh}} & S_{\mathrm{vv}} \end{bmatrix}^{\mathrm{T}} \tag{2.83}$$

where the superscript T represents the transpose of the matrix. The superiority of the scattering vector is that it can directly provide information on each element for the scattering vector.

Using the scattering vector $\mathbf{k}_L$, the covariance matrix is defined as

$$\mathbf{C}_4 = \mathbf{k}_L \mathbf{k}_L^{\mathrm{H}} \tag{2.84}$$

where the superscript **H** represents the conjugate transpose. $\mathbf{C}_4$ is the covariance matrix of single-look image. Multi-look processing is usually used to suppress speckle of polarimetric SAR data, which averages over N adjacent pixels. The covariance matrix of multi-look image $\langle \mathbf{C}_4 \rangle$ is defined as

$$\langle \mathbf{C}_4 \rangle = \frac{1}{N} \sum_{i=1}^{N} \mathbf{k}_{Li} \mathbf{k}_{Li}^{\mathrm{H}} \tag{2.85}$$

Another basic matrix is Pauli basis:

$$\boldsymbol{\Psi}_P : \sqrt{2}\begin{bmatrix}1 & 0\\0 & 1\end{bmatrix}, \sqrt{2}\begin{bmatrix}1 & 0\\0 & -1\end{bmatrix}, \sqrt{2}\begin{bmatrix}0 & 1\\1 & 0\end{bmatrix}, \sqrt{2}\begin{bmatrix}0 & -j\\j & 0\end{bmatrix} \tag{2.86}$$

The corresponding scattering vector is given by

$$\mathbf{k}_P = \frac{1}{\sqrt{2}} \begin{bmatrix} S_{\mathrm{hh}} + S_{\mathrm{vv}} & S_{\mathrm{hh}} - S_{\mathrm{vv}} & S_{\mathrm{hv}} + S_{\mathrm{vh}} & j(S_{\mathrm{hv}} - S_{\mathrm{vh}}) \end{bmatrix}^{\mathrm{T}} \tag{2.87}$$

The coherency matrix is defined as

$$\mathbf{T}_4 = \mathbf{k}_P \mathbf{k}_P^{\mathrm{H}} \tag{2.88}$$

After multi-look processing, the coherency matrix is given by

$$\langle \mathbf{T}_4 \rangle = \frac{1}{N} \sum_{i=1}^{N} \mathbf{k}_{Pi} \mathbf{k}_{Pi}^{\mathrm{H}} \tag{2.89}$$

Under the BSA convention and satisfying reciprocity theory, the two cross-polarization measurements should be identical, $S_{hv} = S_{vh}$. The scattering matrix is a symmetric matrix. The covariance vector $\mathbf{k}_P$ and $\mathbf{k}_L$ can be simplified to a 3×3 dimensional matrix:

$$\mathbf{k}_{3L} = \begin{bmatrix} S_{hh} & \sqrt{2}S_{hv} & S_{vv} \end{bmatrix}^{T} \tag{2.90}$$

$$\mathbf{k}_{3P} = \frac{1}{\sqrt{2}} \begin{bmatrix} S_{hh} + S_{vv} & S_{hh} - S_{vv} & 2S_{hv} \end{bmatrix}^{T} \tag{2.91}$$

are given as

$$\langle \mathbf{C}_3 \rangle = \langle \mathbf{k}_{3L} \mathbf{k}_{3L}^{H} \rangle \tag{2.92}$$

$$\langle \mathbf{T}_3 \rangle = \langle \mathbf{k}_{3P} \mathbf{k}_{3P}^{H} \rangle \tag{2.93}$$

The relationship between the covariance matrix and the coherency matrix is

$$\langle \mathbf{T}_3 \rangle = \mathbf{D}_3 \mathbf{C}_3 \mathbf{D}_3^{H} \tag{2.94}$$

where

$$\mathbf{D}_3 = \frac{1}{\sqrt{2}} \begin{bmatrix} 1 & 0 & 1 \\ 1 & 0 & -1 \\ 0 & \sqrt{2} & 0 \end{bmatrix} \tag{2.95}$$

2.5.3.4 The Relationship Between Different Polarization Representations

Radar transmits a polarized electromagnetic wave to the target, and scattering happens. The polarization scattering process can be described by means of the scattering matrix and Muller matrix. The scattering matrix establishes the relationship between Jones vector of the incident waves and Jones vector of the scattering waves, however, the scattering matrix is only used to describe the single scatter. Muller matrix establishes the relationship between Stokes vector of the incident waves and Stokes vector of the scattering waves, Muller matrix is used to describe the scattering characteristics of a scatter, which is common composed from many incoherency scattering centers. Kennaugh matrix gives a method, which can seek for the receiving power of antenna according to the polarization stats of transmitting and receiving antennas. Covariance matrix transformed from the scattering vector will be widely used in polarization classification. Figure 2.6 shows the relationship between the different polarization representations.

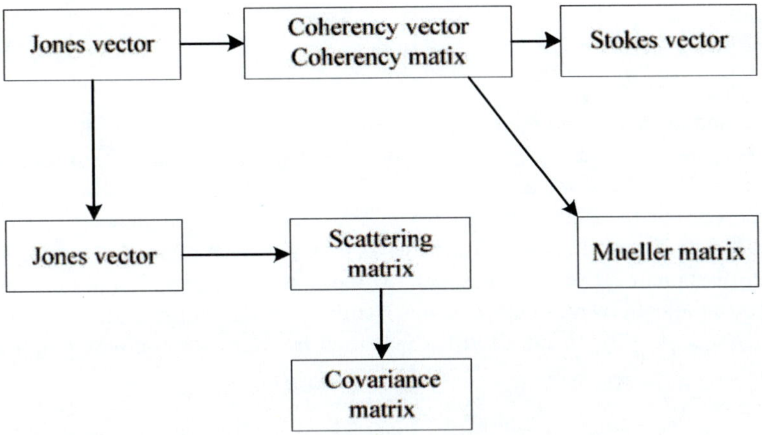

Fig. 2.6 Relationship between different polarization representations

2.6 Polarization Basis Transformation [6, 7]

Polarization basis transformation is a very important topic for the study on radar polarimetry theory and its applications. Any monochromatic plane wave can be represented as a combination of two orthogonal linear polarization states that define the reference polarization basis, and there is an infinite number of polarization bases that can be applied. The representation of electromagnetic wave depends on the selection of reference polarization basis, the representation of elements in scattering matrix is also related to the selection of reference polarization basis. The influence of the polarization basis transformation on the scattering matrix will be discussed in the following.

Make (x, y) and (i, j) two groups of orthogonal polarization basis, the Jones vectors of any monochromatic plane wave under the two bases are, respectively,

$$\mathbf{E}_{xy} = \begin{bmatrix} E_x \\ E_y \end{bmatrix}, \quad \mathbf{E}_{ij} = \begin{bmatrix} E_i \\ E_j \end{bmatrix} \tag{2.96}$$

The transformation from the (h, v) polarization basis to the (i, j) basis is a linear transformation in a two-dimensional complex space, which is controlled by the orthogonal unitary matrix $[\mathbf{U}_2]$ with the 2×2 **U** units.

$$\mathbf{E}_{ij} = [\mathbf{U}_2]\mathbf{E}_{xy} \tag{2.97}$$

And the transformation matrix $[\mathbf{U}_2]$ is given by

$$[\mathbf{U}_2] = \begin{bmatrix} e^{j\frac{1}{2}\varphi} & 0 \\ 0 & e^{-j\frac{1}{2}\varphi} \end{bmatrix} \begin{bmatrix} \cos\xi & \sin\xi\cdot e^{j\phi} \\ -\sin\xi\cdot e^{-j\phi} & \cos\xi \end{bmatrix} \tag{2.98}$$

where ξ and ϕ are the parameters describing the polarization state in Jones vectors, and φ is used to limit the relative phase between two orthogonal polarization states in a group of polarization bases. However, the absolute phase φ cannot describe the polarization state of electromagnetic wave, Thus, it is not important, usually, it can be chosen $\varphi = 0$. The orthogonal unitary matrix is $[\mathbf{U}_2]$, it guarantees that the amplitude of electromagnetic wave is independent of the polarization basis, and the phase of the electromagnetic wave keeps constant, while the polarization basis changed.

When $\varphi = 0$, $[\mathbf{U}_2]$ in Eq. (2.98) is rewritten by means of complex polarization ratio as

$$[\mathbf{U}_2] = \frac{1}{\sqrt{1+\rho\rho^*}} \begin{bmatrix} 1 & \rho \\ -\rho^* & 1 \end{bmatrix} \tag{2.99}$$

For a PolSAR system based on horizontal and vertical polarization basis, the measured scattering matrix is $[\mathbf{S}]_{\mathrm{HV}} = \begin{bmatrix} S_{\mathrm{hh}} & S_{\mathrm{hv}} \\ S_{\mathrm{vh}} & S_{\mathrm{vv}} \end{bmatrix}$.

Under (H, V) polarization basis, the relationship between Jones vectors of incident wave and scattering wave is represented by the scattering matrix as

$$\mathbf{E}^s_{\mathrm{HV}} = [\mathbf{S}]_{\mathrm{HV}} \mathbf{E}^i_{\mathrm{HV}} \tag{2.100}$$

The relationship between Jones vectors of the polarization basis (i, j) and (H, V) is

$$\begin{aligned} \mathbf{E}^i_{ij} &= [\mathbf{U}_2]\mathbf{E}^i_{\mathrm{HV}} \rightarrow \mathbf{E}^i_{\mathrm{HV}} = [\mathbf{U}_2]^{\mathrm{H}} \mathbf{E}^i_{ij} \\ \mathbf{E}^s_{ij} &= [\mathbf{U}_2]\mathbf{E}^s_{\mathrm{HV}} \rightarrow \mathbf{E}^s_{\mathrm{HV}} = [\mathbf{U}_2]^{\mathrm{H}} \mathbf{E}^s_{ij} \end{aligned} \tag{2.101}$$

Using $[\mathbf{U}_2]^{\mathrm{H}} = [\mathbf{U}_2]^{-1}$, we get

$$\mathbf{E}^s_{\mathrm{HV}} = [\mathbf{S}]_{\mathrm{HV}} \mathbf{E}^i_{\mathrm{HV}} = [\mathbf{S}]_{\mathrm{HV}} ([\mathbf{U}_2]^{\mathrm{H}} \mathbf{E}^i_{ij}) \rightarrow \mathbf{E}^s_{ij} = [\mathbf{U}_2][\mathbf{S}]_{\mathrm{HV}} [\mathbf{U}_2]^{\mathrm{H}} \mathbf{E}^i_{ij} \tag{2.102}$$

The scattering matrix under the (i, j) polarization basis is given by

$$[\mathbf{S}]_{ij} = [\mathbf{U}_2][\mathbf{S}]_{\mathrm{HV}} [\mathbf{U}_2]^{\mathrm{H}} \tag{2.103}$$

It can be seen that, if the scattering matrix of a polarization basis is known, the scattering matrix of any other orthogonal polarization basis can be obtained through the polarization basis transforming, this is the superiority of the full-polarization radar system over the conventional single-polarization radar system.

From Eq. (2.103), the polarization basis transformation formulas for three-dimensional polarization scattering vectors with different polarization basis can be obtained. The polarization basis transformation formula of the three-dimensional scattering vector $\mathbf{k}_{3L}$ in Lexicographic basis is given as follows:

$$\mathbf{k}_{3L_{ij}} = [\mathbf{U}_{3L}]\mathbf{k}_{3L_{\mathrm{HV}}},\ [\mathbf{U}_{3L}] = \frac{1}{1+\rho\rho^*}\begin{bmatrix} 1 & \sqrt{2}\rho & \rho^2 \\ -\sqrt{2}\rho^* & 1 & \sqrt{2}\rho \\ \rho^{*2} & -\sqrt{2}\rho^* & 1 \end{bmatrix} \tag{2.104}$$

The polarization basis transformation formula of the three-dimensional scattering vector $\mathbf{k}_{3P}$ in the Pauli basis is given by

$$\mathbf{k}_{3P_{ij}} = [\mathbf{U}_{3P}]\mathbf{k}_{3P_{\mathrm{HV}}},$$

$$[\mathbf{U}_{3P}] = \frac{1}{2(1+\rho\rho^*)}\begin{bmatrix} 2+\rho^2+\rho^{*2} & \rho^{*2}-\rho^2 & 2(\rho-\rho^*) \\ \rho^2-\rho^{*2} & 2-(\rho^2+\rho^{*2}) & 2(\rho+\rho^*) \\ 2(\rho-\rho^*) & -2(\rho+\rho^*) & 2(1-\rho\rho^*) \end{bmatrix} \tag{2.105}$$

2.7 Polarization Synthesis and Polarization Signatures [8–10]

2.7.1 Polarization Synthesis and Polarization Signatures

According to the scattering matrix of target the received echo power for any combination of the transmitting and receiving antennas can be calculated, this process is called polarization synthesis. If the scattering characteristic of target is known, the received power can be obtained by means of polarization synthesis.

The polarization synthesis formula of the scattering matrix [S], Stokes matrix [K], and covariance matrix [C] are given by

$$\sigma_{rt}^0(\psi_r, \chi_r, \psi_t, \chi_t) = \frac{4\pi}{A}\langle |p_r\mathbf{S}p_t|^2\rangle \tag{2.106}$$

$$\sigma_{rt}^0(\psi_r, \chi_r, \psi_t, \chi_t) = \frac{4\pi}{A}\mathbf{J}_r^T\cdot\langle\mathbf{K}\rangle\cdot\mathbf{J}_t \tag{2.107}$$

$$\sigma_{rt}^0(\psi_r, \chi_r, \psi_t, \chi_t) = \frac{4\pi}{A}\mathbf{B}^T\langle\mathbf{C}\rangle\mathbf{B} \tag{2.108}$$

where $\sigma_{rt}^0(\psi_r, \chi_r, \psi_t, \chi_t)$ is the radar backscattering coefficient, $\langle\rangle$ denotes the assemble average, A is the assemble average area, r and t represent the receiving and the transmitting subscripts, respectively. (ψ_r, χ_r) and (ψ_t, χ_t) represent the ellipse orientation angle and the ellipticity angle, respectively.

In Eq. (2.106), the polarization vectors of the receiving and transmitting field p_r and p_t are defined as

$$p_m = E_m/|E_m| \quad m = r, t \tag{2.109}$$

In Eq. (2.107), $\mathbf{J}_r$ is polarization state vector of Stokes vector format describing the receiving antenna polarization state (ψ_r, χ_r), which is independent of the target scattering. $\mathbf{J}_t$ is Stokes vector of the transmitting antenna, whose polarization state is described by (ψ_t, χ_t). The $\mathbf{J}_r$ and $\mathbf{J}_t$ are substituted into Eq. (2.107), and the coefficient is omitted, and we get

$$\sigma_n^0(\psi_r, \chi_r, \psi_t, \chi_t) = \begin{bmatrix} 1 \\ \cos 2\chi_r \cdot \cos 2\psi_r \\ \cos 2\chi_r \cdot \sin 2\psi_r \\ \sin 2\psi_r \end{bmatrix}^{\mathrm{T}} \cdot \langle \mathbf{K} \rangle \cdot \begin{bmatrix} 1 \\ \cos 2\chi_t \cdot \cos 2\psi_t \\ \cos 2\chi_t \cdot \sin 2\psi_t \\ \sin 2\psi_t \end{bmatrix} \tag{2.110}$$

In Eq. (2.108), antenna vector $\mathbf{B}$ is defined as

$$\mathbf{B} = \left[p_v^r p_v^t \; p_v^r p_h^t \; p_h^r p_v^t \; p_h^r p_h^t \right]^T \tag{2.111}$$

Although [**S**], [**K**], and [**C**] can be used for polarization synthesis, However, [**K**] and [**C**] have a special advantage related to [**S**], it is that [**K**] and [**C**] can only use one matrix to represent an assemble of the resolution element set, it does not need to establish each matrix for each element to represent. The other advantage is needless calculated amount for polarization synthesis.

When using Eqs. (2.106)–(2.108) to calculate the scattering coefficient of target (echo power), the polarization states of the transmitting–receiving antenna can be chosen arbitrarily in the valid range.

The combination of the same polarization states $(\psi_r = \psi_t, \chi_r = \chi_t)$ is called the co-polarization mode, and the combination of the orthogonal polarization states $(\psi_r = \psi_t + \pi/2, \chi_r = -\chi_t)$ is called the cross-polarization mode. In the co- and cross-polarization modes, Eqs. (2.106)–(2.108) is reduced into two variables, so the relationship between the received power and the polarization state can be represented by the 3D picture, which is called co-polarization signatures and cross-polarization signatures.

For a certain extent, the polarization signature reflects the variation of the received echo power under the specific polarization combination. Usually, different types of target echoes have different polarization signatures, so they can be used to analyze the polarization characteristics of different targets.

2.7.2 *Scattering Model and Polarization Signatures*

2.7.2.1 Scattering Models and Polarization Signatures of Point Targets

This section will introduce several typical point targets and their polarization scattering characteristics. Typical point targets include conducting sphere, dihedral corner reflector, trihedral corner reflector, and conducting helix, etc.

(a) **Conducting Sphere**

The conducting sphere is shown in Fig. 2.7. The scattering matrix is

$$\mathbf{S} = \frac{a}{2}\begin{pmatrix} 1 & 0 \\ 0 & 1 \end{pmatrix} \tag{2.112}$$

From Eq. (2.112), the covariance matrix and the Stokes scattering are

$$\mathbf{C} = \frac{a^2}{4}\begin{pmatrix} 1 & 0 & 0 & 1 \\ 0 & 0 & 0 & 0 \\ 0 & 0 & 0 & 0 \\ 1 & 0 & 0 & 1 \end{pmatrix} \tag{2.113}$$

and

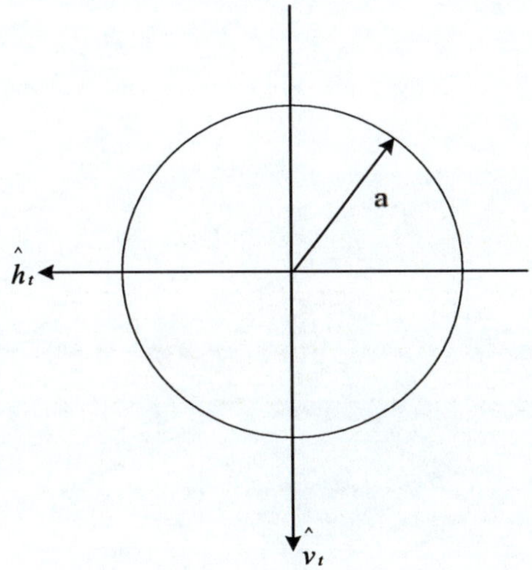

Fig. 2.7 A conducting sphere with radius a

$$\mathbf{K} = \frac{a^2}{8}\begin{pmatrix} 1 & 0 & 0 & 0 \\ 0 & 1 & 0 & 0 \\ 0 & 0 & 1 & 0 \\ 0 & 0 & 0 & -1 \end{pmatrix} \tag{2.114}$$

The polarization signatures of the conducting sphere to co-polarization and cross-polarization are shown in Fig. 2.8. For the co-polarization response, the maximum of the backscattering coefficient appears in linear polarization state, and independent of the linear polarization orientation angle. This independent is an inevitable result of that the sphere is non-directivity. For the cross-polarization response, the maximum of the backscattering coefficient appears at circular polarization and the minimum appears at linear polarization. the maximum appears at circular polarization is due to that for a circular polarization wave, scattering wave is also a circular polarization wave, only rotation direction is opposite, that is same as the polarization of receiving antenna. So, the maximum appears.

If a plate is large enough (so large that the transmitting antenna cannot irradiate its edge), the polarization response is the same as that of a conducting sphere when the electromagnetic wave is used at a normal incidence.

(b) A Dihedral Corner Reflector

A dihedral corner reflector with side length (a) and (2b) is shown in Fig. 2.9. Because the electromagnetic wave will be reflected once on each surface of the dihedral reflector, so the direction of the scattered wave is parallel to that of the incident wave. The scattering matrix is given by

$$S = \frac{k_0 ab}{\pi}\begin{pmatrix} -\cos 2\alpha & \sin 2\alpha \\ \sin 2\alpha & \cos 2\alpha \end{pmatrix} \tag{2.115}$$

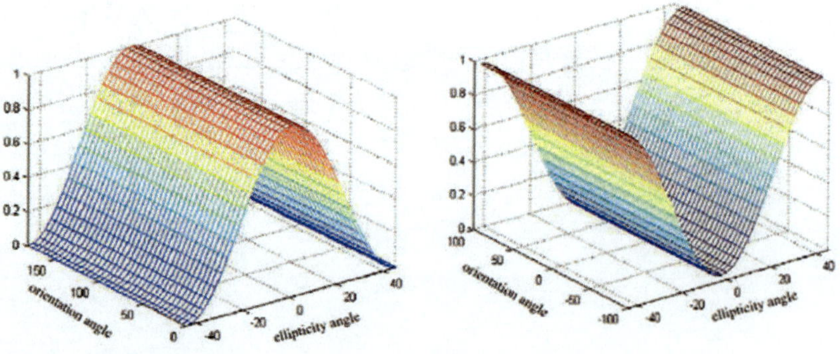

(a) co-polarization signature (b) cross-polarization signature

Fig. 2.8 Polarization signature of a conducting sphere

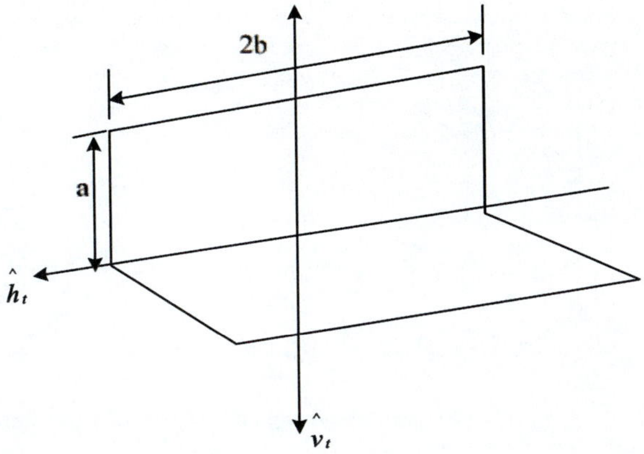

Fig. 2.9 A trihedral corner reflector

where α is deflection angle of the corner reflector relative to the incident wave.

When $\alpha = 0°$, Eq. (2.115) is rewritten as

$$S = \frac{k_0 ab}{\pi}\begin{pmatrix} -1 & 0 \\ 0 & 1 \end{pmatrix} \tag{2.116}$$

The Stokes matrix **K** and the covariance matrix **C** are

$$\mathbf{K} = \frac{k_0^2 a^2 b^2}{2\pi^2}\begin{pmatrix} 1 & 0 & 0 & 0 \\ 0 & 1 & 0 & 0 \\ 0 & 0 & -1 & 0 \\ 0 & 0 & 0 & 1 \end{pmatrix} \tag{2.117}$$

$$\mathbf{C} = \frac{k_0^2 a^2 b^2}{2\pi^2}\begin{pmatrix} 1 & 0 & 0 & -1 \\ 0 & 0 & 0 & 0 \\ 0 & 0 & 0 & 0 \\ -1 & 0 & 0 & 1 \end{pmatrix} \tag{2.118}$$

The polarization signatures for a dihedral corner reflector are shown in Fig. 2.10. Figure 2.10a is co-polarization signature and Fig. 2.10b is cross-polarization signature.

It is can be seen that the difference of the polarization responses between the dihedral reflector and the conducting sphere is very great. For a dihedral corner reflector at ±45°, linear polarizations deviated from the horizontal polarization and vertical polarization the two minimum values happen for the co-polarization response, and corresponding the two maximum values happen for the cross-polarization response. The difference of the polarization responses between the dihedral reflector and the conducting sphere, as well as the difference between the co- and cross-polarization responses of the dihedral reflector, are caused that the dihedral reflector reflects two times for the incident wave. This extra reflection causes an additional 180° phase

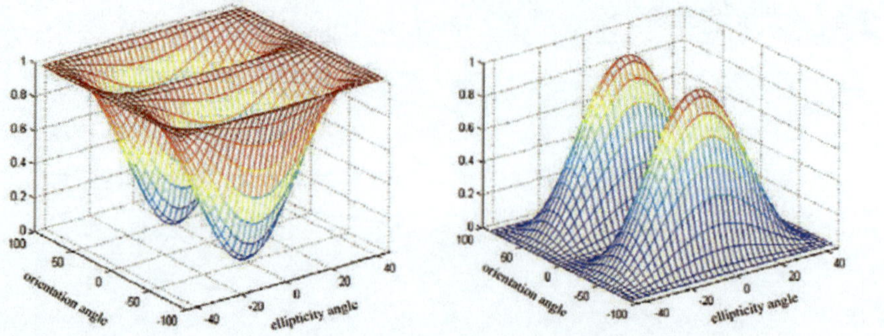

(a) co-polarization signature (b) cross-polarization signature

Fig. 2.10 Polarization signature of a dihedral corner reflector

shift between S_{hh} and S_{vv} of the scattering matrix, resulting in a difference of the polarization response.

(c) A Trihedral Corner Reflector

A trihedral corner reflector with side length of *l* is shown in Fig. 2.11. For a triangle reflector, the incident wave must be reflected three times before returning to the radar.. Each reflection of the additional two reflections will cause the additional 180° phase shift between scattering matrix S_{hh} component and S_{vv} component. So, the total additional 360° phase shift will be caused. Therefore, when the incident wave is along the alignment of the triangular reflector, its scattering matrix is similar to that of the conducting sphere, except for the different multiplicity factor. The scattering matrix is given by

Fig. 2.11 A trihedral corner reflector

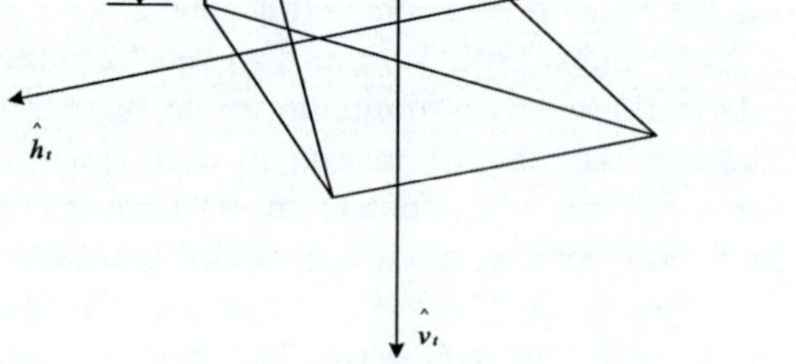

$$S = \frac{k_0 l^2}{\sqrt{12\pi}} \begin{pmatrix} 1 & 0 \\ 0 & 1 \end{pmatrix} \tag{2.119}$$

The Stokes matrix **K** and the covariance matrix **C** are

$$K = \frac{k_0^2 l^4}{24\pi^2} \begin{pmatrix} 1 & 0 & 0 & 0 \\ 0 & 1 & 0 & 0 \\ 0 & 0 & 1 & 0 \\ 0 & 0 & 0 & -1 \end{pmatrix} \tag{2.120}$$

$$C = \frac{k_0^2 l^4}{12\pi^2} \begin{pmatrix} 1 & 0 & 0 & 1 \\ 0 & 0 & 0 & 0 \\ 0 & 0 & 0 & 0 \\ 1 & 0 & 0 & 1 \end{pmatrix} \tag{2.121}$$

The co- and cross-polarization responses are shown in Fig. 2.12.

Figure 2.12 Polarization signature of a trihedral corner reflector. Figure 2.12a is co-polarization signature, Fig. 2.12b is the cross-polarization signature of a trihedral corner reflector.

It can be seen from Fig. 2.12 that, similar to a conductor sphere, the polarization response of the trihedral reflector is independent of the ellipse orientation angle. This non-directional future means the polarization response is robust related to the rotation around the line between radar antenna and the trihedral reflector,which makes the trihedral corner reflector to be a good polarization calibration target of PolSAR.

(d) **Conducting Helix**

The left-handed and right-handed helices are shown in Fig. 2.13, and their scattering matrices are given by

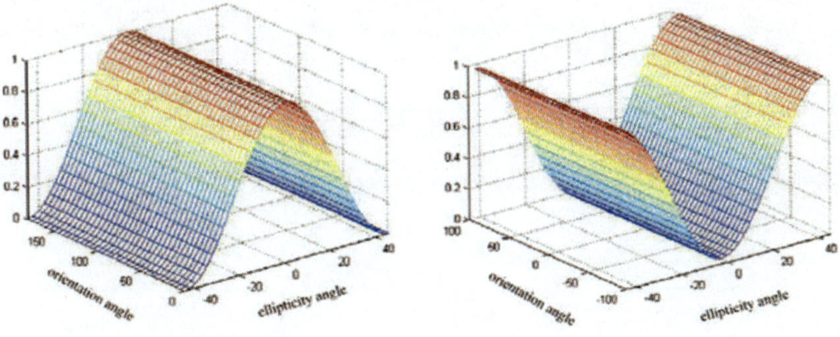

(a) co-polarization signature (b) cross-polarization signature

Fig. 2.12 Polarization signature of a trihedral corner reflector

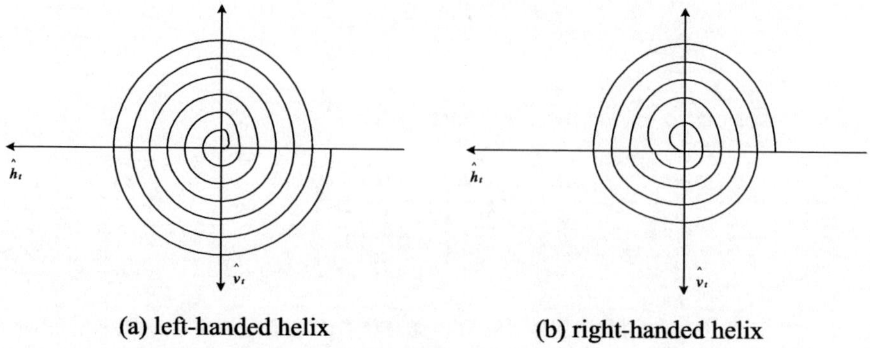

Fig. 2.13 Left-handed and right-handed helices

$$S_l = \frac{1}{2}\begin{bmatrix} -1 & i \\ i & 1 \end{bmatrix} \tag{2.122}$$

$$S_r = \frac{1}{2}\begin{bmatrix} -1 & -i \\ -i & 1 \end{bmatrix} \tag{2.123}$$

The Stokes matrix **K** is

$$K_l = \frac{1}{4}\begin{pmatrix} 1 & 0 & 0 & 1 \\ 0 & 0 & 0 & 0 \\ 0 & 0 & 0 & 0 \\ 1 & 0 & 0 & 1 \end{pmatrix} \tag{2.124}$$

$$K_r = \frac{1}{4}\begin{pmatrix} 1 & 0 & 0 & -1 \\ 0 & 0 & 0 & 0 \\ 0 & 0 & 0 & 0 \\ -1 & 0 & 0 & 1 \end{pmatrix} \tag{2.125}$$

The covariance matrix **C** is

$$C_l = \frac{1}{4}\begin{pmatrix} 1 & -i & -i & -1 \\ -i & 1 & 1 & i \\ -i & 1 & 1 & i \\ -1 & -i & -i & 1 \end{pmatrix} \tag{2.126}$$

$$C_r = \frac{1}{4}\begin{pmatrix} 1 & -i & -i & -1 \\ i & 1 & 1 & -i \\ i & 1 & 1 & -i \\ -1 & i & i & 1 \end{pmatrix} \tag{2.127}$$

The co- and cross-polarization responses of the left-handed and right-handed helices are shown in Figs. 2.14 and 2.15, where (a) shows the co-polarization response and (b) shows the cross-polarization response.

It can be seen that the co-polarization maxima of the left-handed and right-handed helices have occurred at the circular polarization with the same rotation direction of the helices. In this case, the coupling between the incident wave electric field and the helix is greatest. The scattered wave has the same spin direction as the incident wave, which means that the scattered wave has the same polarization as the receiving antenna. Therefore, since the polarization of scattering wave is same as that of the incident wave, the orthogonal polarization antenna will not receive any energy, which can be verified from the position where the minimum value of the cross-polarization response occurs. If the incident wave is in the opposite direction from the helix, there is no coupling between them and no energy scattered. That is the reason for

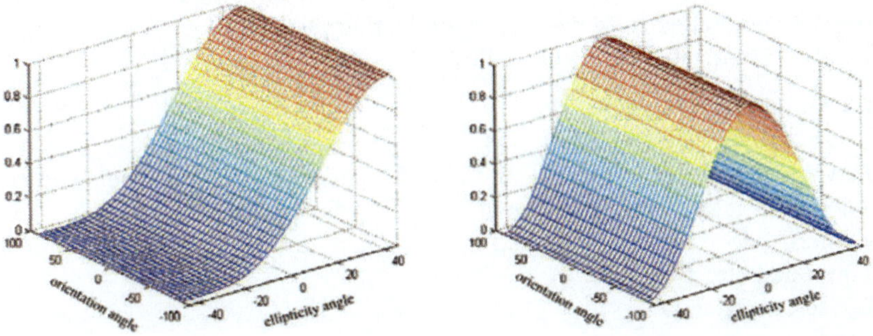

(a) co-polarization signature (b) cross-polarization signature

Fig. 2.14 Polarization signature of a left-handed helix

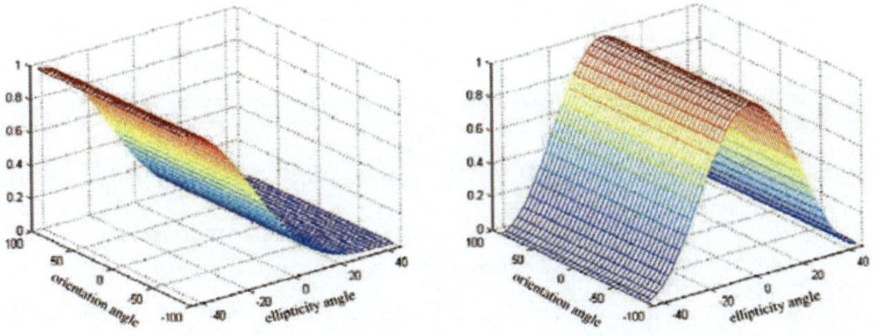

(a) co-polarization signature (b) cross-polarization signature

Fig. 2.15 Polarization signature of a right-handed helix

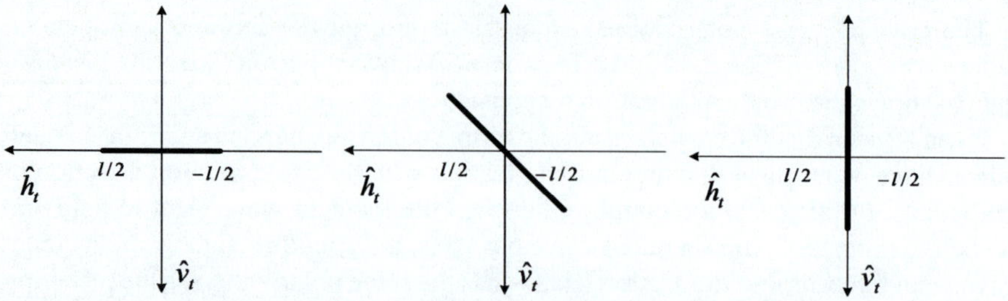

(a) 0° (b) 45°clockwise (c) 90°

Fig. 2.16 A short and thin cylinder oriented at different angles from the horizontal direction

the minima of the co-polarization and cross-polarization responses will occur at the circular polarization opposite to the rotation direction. In addition, the cross-polarization response of the two helices is the same, so the spin direction of the helix can be not determined from the cross-polarization response but can be determined from the co-polarization response easily.

(e) A Short and Thin Cylinder

A short and thin cylinder with length l and diameter a, placed perpendicular to the direction of traveling wave, is shown in Fig. 2.16, where (a) 0° oriented the horizontal direction, (b) 45° clockwise oriented the horizontal direction, and (c) 90° oriented the horizontal direction.

The scattering matrix is

$$S = \frac{k_0^2 l^3}{3[\ln(4l/a) - 1]} \begin{pmatrix} \sin^2\alpha & -\sin\alpha\cos\alpha \\ -\sin\alpha\cos\alpha & \cos^2\alpha \end{pmatrix} \tag{2.128}$$

where α is an angle of the cylinder oriented clockwise with respect to the $\hat{h}_t$ axis.

The Stokes matrix and the covariance matrix are given by

$$K = \frac{k_0^4 l^6}{36[\ln(4l/a) - 1]^2} \begin{bmatrix} 1 & -\cos 2\alpha & -\sin 2\alpha & 0 \\ -\cos 2\alpha & \cos^2 2\alpha & -\sin\alpha\cos 2\alpha & 0 \\ -\sin 2\alpha & -\sin\alpha\cos 2\alpha & \sin^2 2\alpha & 0 \\ 0 & 0 & 0 & 0 \end{bmatrix} \tag{2.129}$$

$$C = \frac{k_0^4 l^6}{9[\ln(4l/a) - 1]^2} \begin{bmatrix} \sin^4 2\alpha & -\sin^3\alpha\cos\alpha & -\sin^3\alpha\cos\alpha & \sin^2\alpha\cos^2\alpha \\ -\sin^3\alpha\cos\alpha & \sin^2\alpha\cos^2\alpha & \sin^2\alpha\cos^2\alpha & -\sin\alpha\cos^3\alpha \\ -\sin^3\alpha\cos\alpha & \sin^2\alpha\cos^2\alpha & \sin^2\alpha\cos^2\alpha & -\sin\alpha\cos^3\alpha \\ \sin^2\alpha\cos^2\alpha & -\sin\alpha\cos^3\alpha & -\sin\alpha\cos^3\alpha & \cos^4\alpha \end{bmatrix} \tag{2.130}$$

The polarization signatures of a short and thin cylinder oriented at 0°, 45° clockwise and 90° from the horizontal direction, are shown in Figs. 2.17, 2.18 and 2.19.

It can be seen that the maximum of the co-polarization response and the minimum of the cross-polarization response have occurred for the linear polarization at the same orientation angle as the cylinder. Therefore, It is explaining that the radar polarization can be used to deduce the orientation angle The minima of polarized responses of the co-polarization and cross-polarization are occurred at the linear polarization orthogonal to the cylinder because the polarized incident wave is orthogonal to the cylinder, so there is no induced current inside the cylinder and no scattered energy can be produced. The cross-polarization signature of the cylinder oriented horizontally is as same as that of the cylinder oriented perpendicularly. Although the co-polarization response can clearly show the orientation angle of the cylinder, but alone using the cross-polarization response is not allowed.

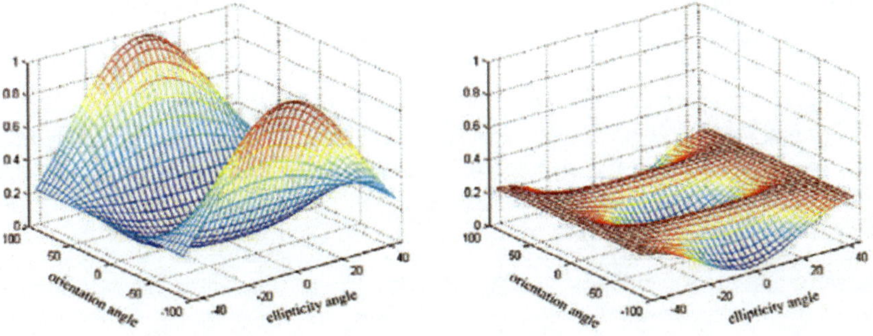

(a) co-polarization signature (b) cross-polarization signature

Fig. 2.17 Polarization signature of a short and thin cylinder oriented at 0°

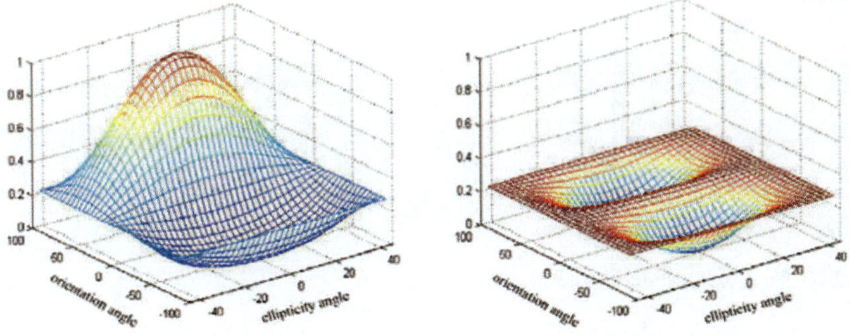

(a) co-polarization signature (b) cross-polarization signature

Fig. 2.18 Polarization signature of a short and thin cylinder oriented at 45°

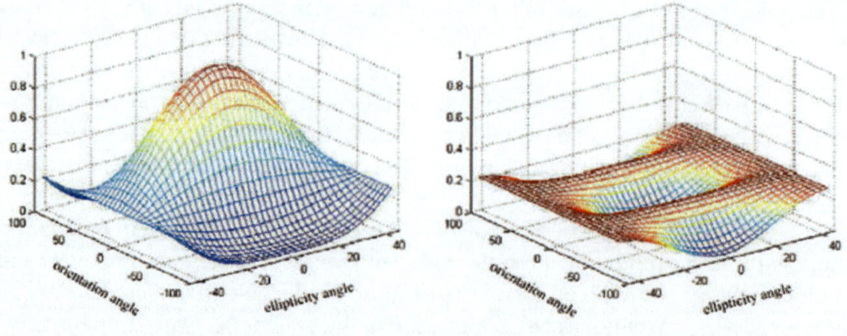

(a) co-polarization signature (b) cross-polarization signature

Fig. 2.19 Polarization signature of a short and thin cylinder oriented at 90°

2.7.2.2 Scattering Models and Polarization Signatures of Distributed Targets

There are many kinds of scattering process in nature, but they are usually composed of four basic scattering mechanisms: Surface scattering (Odd scattering), Bragg scattering (diffuse scattering), 2-Bounce scattering (even scattering), and Volume scattering. The scattering models are shown in Fig. 2.20, and the main characteristics will be introduced as follows.

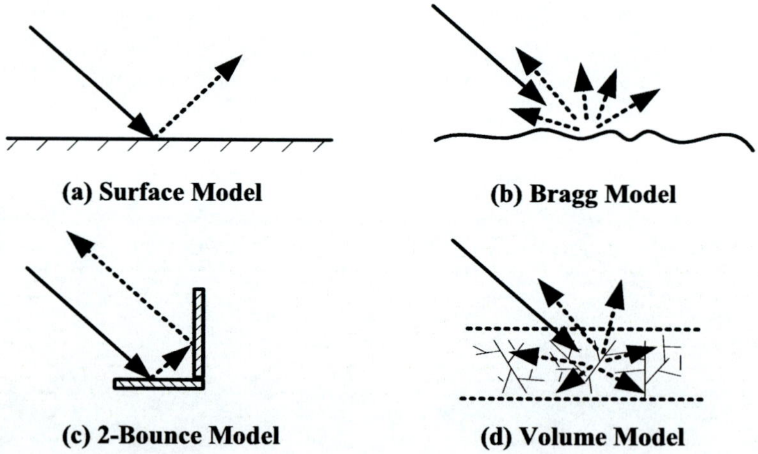

Fig. 2.20 Four basic scattering mechanisms

(1) **Surface Scattering (Odd scattering)**

The surface scattering model is shown in Fig. 2.20a. The surface scattering is similar to the mirror reflection of visible light, which is common in nature, such as calm water, dry riverbed, some flat and smooth rocks, or wasteland, etc. The echo energy is related to the angle between the incident wave and the scattering surface. The normalization scattering matrix, Stokes matrix, and covariance matrix are

$$S_{odd} = \begin{bmatrix} 1 & 0 \\ 0 & 1 \end{bmatrix} \tag{2.131}$$

$$K = \begin{pmatrix} 1 & 0 & 0 & 0 \\ 0 & 1 & 0 & 0 \\ 0 & 0 & 1 & 0 \\ 0 & 0 & 0 & -1 \end{pmatrix} \tag{2.132}$$

$$C = \begin{pmatrix} 1 & 0 & 0 & 1 \\ 0 & 0 & 0 & 0 \\ 0 & 0 & 0 & 0 \\ 1 & 0 & 0 & 1 \end{pmatrix} \tag{2.133}$$

The co-polarization and cross-polarization signatures of surface scattering are shown in Fig. 2.21.

(2) **Bragg Scattering (diffuse scattering)**

Since the surface of terrain in nature is often rough and undulating, Bragg scattering model is also called as diffuse scattering model. Bragg scattering model shown in Fig. 2.20b can be used to describe the scattering mechanism. The terrain in nature,

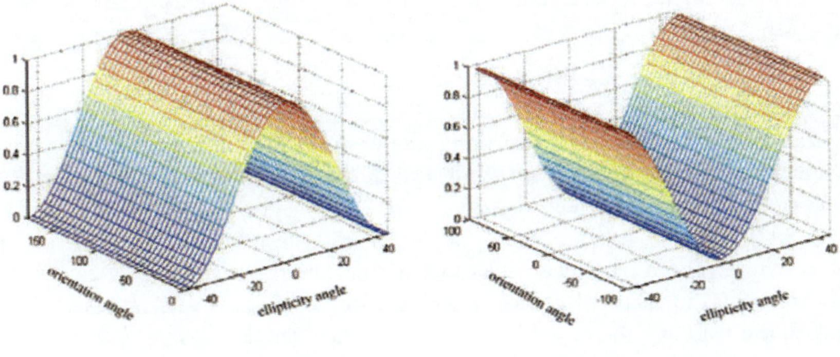

(a) co-polarization signature (b) cross-polarization signature

Fig. 2.21 Polarization signature of surface scattering

such as solidified volcanic lava, fluctuated water, crops covered surface, and some rocks, etc., can be described in Bragg scattering model, which is represented in SAR images as the targets ranging from gray to white.

When the surface roughness is restricted that the rms surface roughness is much smaller than a wavelength, the first-order Bragg surface model can be applicable to describe the echo scattering characteristics. The normalized scattering matrix, Stokes matrix, and covariance matrix are

$$S_{bragg} = \begin{bmatrix} 1 & 0 \\ 0 & \sqrt{\beta} \end{bmatrix} \tag{2.134}$$

$$K = \begin{pmatrix} 1 & (1-\beta)/(1+\beta) & 0 & 0 \\ (1-\beta)/(1+\beta) & 1 & 0 & 0 \\ 0 & 0 & 2\sqrt{\beta}/(1+\beta) & 0 \\ 0 & 0 & 0 & 2\sqrt{\beta}/(1+\beta) \end{pmatrix} \tag{2.135}$$

$$K = \begin{pmatrix} 1 & (1-\beta)/(1+\beta) & 0 & 0 \\ (1-\beta)/(1+\beta) & 1 & 0 & 0 \\ 0 & 0 & 2\sqrt{\beta}/(1+\beta) & 0 \\ 0 & 0 & 0 & 2\sqrt{\beta}/(1+\beta) \end{pmatrix} \tag{2.136}$$

$$C = \begin{pmatrix} |\beta|^2 & 0 & 0 & \beta \\ 0 & 0 & 0 & 0 \\ 0 & 0 & 0 & 0 \\ \beta^* & 0 & 0 & 1 \end{pmatrix} \tag{2.137}$$

where β is the real polarization coefficient, related to the incident angle of the electromagnetic wave and the dielectric constant of the scattering surface.

The co-polarization and cross-signatures of Bragg scattering are shown in Fig. 2.22.

(3) **2-Bounce Scattering (even scattering)**

The scatter with the 2-bounce scattering model is usually composed of two perpendicular scattering surfaces, as shown in Fig. 2.20c. Under this model, the received power of the scatter is greater, usually for HH or VV components are very strong. In the SAR image, it is represented the target with the gray level scope from light gray to white. The typical even scattering process is the scattering of electromagnetic wave on the dihedral corner reflector. Moreover, the scattering mechanisms between the wall and the ground in city and the tree trunk and the ground in forest are approximated by the even scattering model.

Assuming the scattering coefficients of H polarization and V polarization on the perpendicular surface (such as the wall and the tree trunk) are $R_{\perp h}$ and $R_{\perp v}$,

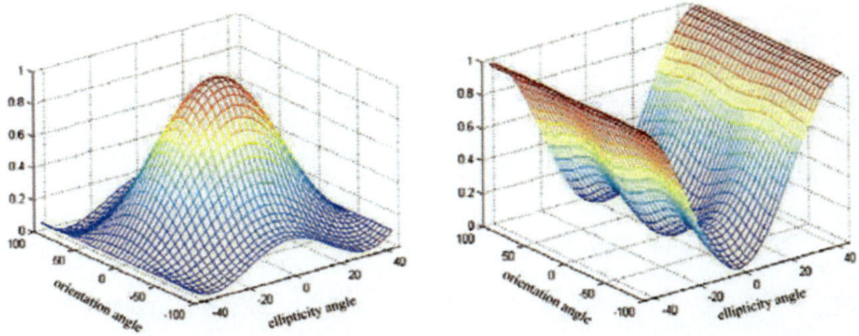

(a) co-polarization signature (b) cross-polarization signature

Fig. 2.22 Polarization signature of Bragg scattering

the scattering coefficients of H polarization and V polarization on the horizontal surface (such as the ground) are $R_{//h}$ and $R_{//v}$, respectively. Considering the phase attenuation of electromagnetic waves in the propagation and multiple scattering, the normalized scattering matrix, Stokes matrix, and the covariance matrix are

$$S_{even} = \begin{bmatrix} 1 & 0 \\ 0 & \alpha \end{bmatrix} \tag{2.138}$$

$$K = \begin{pmatrix} 1 & (1-\alpha^2)/(1+\alpha^2) & 0 & 0 \\ (1-\alpha^2)/(1+\alpha^2) & 1 & 0 & 0 \\ 0 & 0 & 2\alpha\cos\delta/\left(1+\alpha^2\right) & 2\alpha\sin\delta/\left(1+\alpha^2\right) \\ 0 & 0 & -2\alpha\sin\delta/\left(1+\alpha^2\right) & 2\alpha\cos\delta/\left(1+\alpha^2\right) \end{pmatrix} \tag{2.139}$$

$$C = \begin{pmatrix} |\beta|^2 & 0 & 0 & \beta \\ 0 & 0 & 0 & 0 \\ 0 & 0 & 0 & 0 \\ \beta^* & 0 & 0 & 1 \end{pmatrix} \tag{2.140}$$

where

$$\alpha = e^{j2(\gamma_h-\gamma_v)}\frac{R_{\perp h}R_{//h}}{R_{\perp v}R_{//v}} \tag{2.141}$$

$\delta = \gamma_h - \gamma_v$ is the phase difference between the HH and VV components, and γ_h and γ_v are the phase loss of the H polarization and V polarization electromagnetic wave, respectively. The scattering amplitudes of HH and VV components in the 2-Bounce scattering model are not always equal, and the phase difference between them is not necessarily $\pm\pi$.

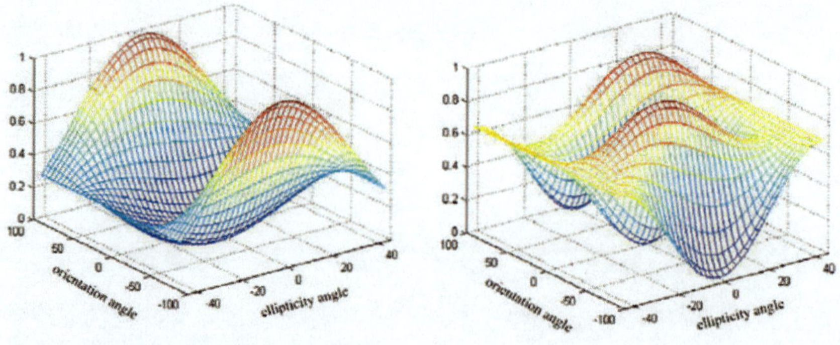

(a) co-polarization signature (b) cross-polarization signature

Fig. 2.23 Polarization signature of 2-Bounce scattering

The co- and cross-polarization signatures of 2-Bounce scattering are shown in Fig. 2.23.

(4) **Volume Scattering**

The volume scattering model indicates that the radar echo is reflected from the particle cloud composed by a very thin cylindrical scatters randomly distributed in spatial direction. The typical representation of this model is the vegetation area composed of large branches in Fig. 2.20d, through some simplified assumptions, the second-order statistics representation of the scattering mechanism can be obtained and the Stokes matrix and covariance matrix are

$$K = \begin{pmatrix} 1 & 0 & 0 & 0 \\ 0 & 1/2 & 0 & 0 \\ 0 & 0 & 1/2 & 0 \\ 0 & 0 & 0 & 0 \end{pmatrix} \tag{2.142}$$

$$\langle C \rangle = \begin{bmatrix} 1 & 0 & 0 & 1/3 \\ 0 & 1/3 & 1/3 & 0 \\ 0 & 1/3 & 1/3 & 0 \\ 1/3 & 0 & 0 & 1 \end{bmatrix} \tag{2.143}$$

The co- and cross-polarization signatures of Volume scattering are shown in Fig. 2.24.

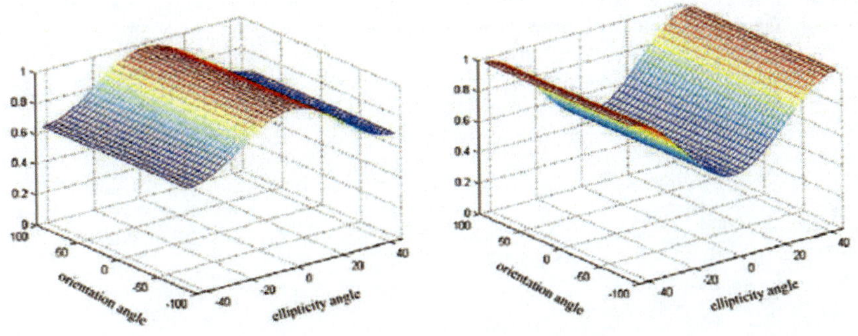

(a) co-polarization signature (b) cross-polarization signature

Fig. 2.24 Polarization signature of volume scattering

2.8 Conclusion

In this chapter, some basic concepts and fundamental theory of PolSAR are introduced, and some methods of polarization representation and scattering characteristic of electromagnetic waves are described. Moreover, some methods of polarization analysis such as polarization basis transformation, polarization synthesis, and polarization signatures are also presented. In order to understand the polarization scattering mechanism, the scattering characteristics of the typical point targets and basic distribution targets in nature are introduced. All the contents discussed in this chapter provide a theoretical basis for the content of the following chapters.

References

1. A.B. Kostinski, W.M. Boerner, On the foundations of radar polarimetry. IEEE Trans. Antennas Propog. **34**(12), 1395–1404 (1986)
2. F.T. Ulaby, C. Elachi, *Radar Polarimetry for Geoscience Applications* (Artech House Inc., Boston, London, 1990), pp. 281–295
3. H. Mott, *Remoter Sensing with Polarimetric Radar* (Wiley, Hoboken, New Jersey, 2007)
4. L.M.J. Brown, J.A. Conway, J.T, Macklin, Polarimetric synthetic aperture radar fundamental concepts and analysis tools. Geo J. Res. **9**(1), 23–35 (1991)
5. Z. Zhaowen, X. Shunping, W. Xuesong, *Radar Polarization Information Processing and Application* (National Defense Industry Press, Beijing, 1999)
6. K.P. Papathanassiou, *Polarimetric SAR Interferometry* (Technical University Graz, 1999), p. 149
7. A.B. Kostinski, W.M. Boerner, On foundation of radar polarimetry. IEEE Trans. Antennas Propag. **AP-34**(12), 1395–1403 (1986)
8. D. Bowei, *Research on Multi-polarimetric Synthetic Aperture Radar System and Polarimetric Information Processing*. Ph.D. thesis of Chinese Academy of Sciences, June 2000
9. Z. Xianda, *Matrix Analysis and Applications* (Tsinghua University Press, Beijing, 2004)

10. S.R. Cloude, E. Pottier, A review of target decomposition theorems in radar polarimetry. IEEE Trans. Geosci. Remote Sens. **34**(2), 498–518 (1996)

Chapter 3
Polarimetric Synthetic Aperture Radar

Polarimetric Synthetic Aperture Radar (PolSAR) acquires the full-polarization scattering matrix of each pixel in the scene by means of transmitting and receiving electromagnetic waves with different polarizations. Comparing with the conventional single PolSAR, the polarimetric scattering matrix contains more information which can represent the response of the terrain to different electromagnetic waves. Therefore, PolSAR has been widely applied in target detection, target recognition, texture feature extraction, terrain classification, and target parameter inversion [1], etc., and has become an important research direction in the microwave remote sensing.

In this chapter, the PolSAR system technique is focused in more detail. The principle of PolSAR is introduced, and then the polarization hierarchy and system parameters are described. The operating architectures of PolSAR, including polarization time-division multiplexing pulses, polarization frequency-division multiplexing pulses, polarization code-division multiplexing pulses, polarization space-division twin-bursts in azimuth and compact polarimetry, are discussed.

3.1 The Principle of Synthetic Aperture Radar

Synthetic Aperture Radar (SAR) obtains the two-dimensional high-resolution image by means of pulse compression in range dimension and synthetic aperture processing in azimuth dimension. The high-range resolution is achieved by transmitting Linear Frequency Modulation (LFM) signal and pulse compression technique. Due to the movement of the vehicle, radar echo in azimuth also has the same characteristic as LFM signal, so the azimuth resolution is realized by azimuth focusing. LFM signal and pulse compression technique are the basis of SAR imaging. The schematic diagram of the side-looking SAR is shown in Fig. 3.1.

It is assumed that the radar onboard the vehicle moves in a uniform straight line with the vehicle. The flight direction of the vehicle is defined as azimuth dimension and vertical to the azimuth as range dimension. In Fig. 3.1, v_s is the speed of the

R. Yang et al., *Polarimetric Microwave Imaging*,
https://doi.org/10.1007/978-981-15-8897-6_3

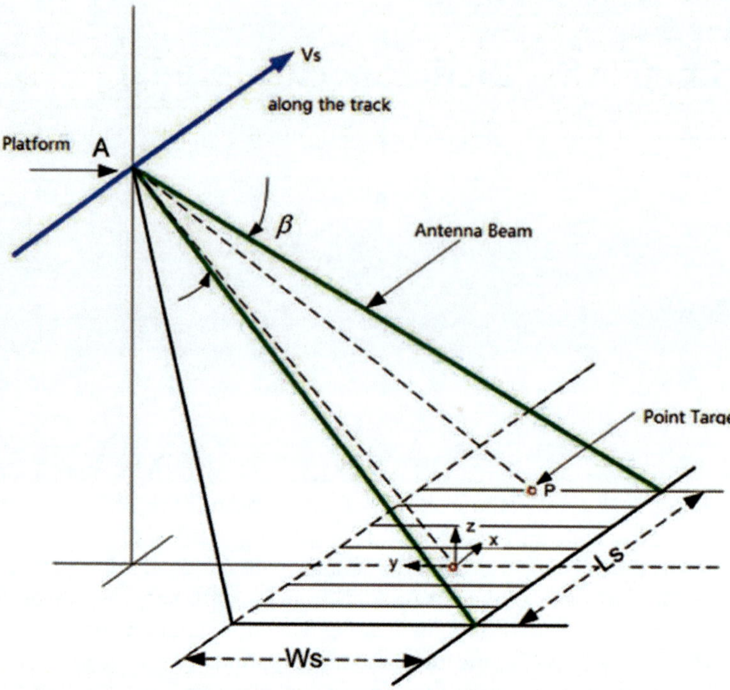

Fig. 3.1 The schematic diagram of the side-looking SAR

vehicle, W_s is the swath width. The azimuth extent of radar 3 dB beam at the slant range R is L_s, called synthetic aperture length, and T_s is the synthetic aperture time.

SAR transmits LFM pulses to the ground with a certain Pulse Repetition Frequency (PRF) along the azimuth track. After pulse compression of the echo in the range, the main-lobe width of the output signal is much narrower than that of the transmitting pulse, so the high-range resolution can be obtained. In azimuth, the spatial positions of the transmitting signals are arranged to synthesize a large line array antenna, thus the high azimuth resolution can be obtained.

Since SAR is a kind of microwave imaging radar, SAR image is required to accurately reflect the backscattering coefficients of the terrain. If the backscattering coefficient of the terrain is expressed as $\sigma_0(x, r)$ and the SAR image is expressed as $\widehat{\sigma}_0(x, r)$, SAR system can be equivalent to a two-dimensional filtering network. The two-dimensional compression diagram of SAR system is shown in Fig. 3.2.

Based on the network theory, it is assumed that the Impulse Response Function (IRF) of a linear network is $h(x, r)$, and the output of SAR system is the 2D image $\widehat{\sigma}_0(x, r)$.

$$\widehat{\sigma}_0(x, r) = \sigma_0(x, r) \otimes h(x, r) \tag{3.1}$$

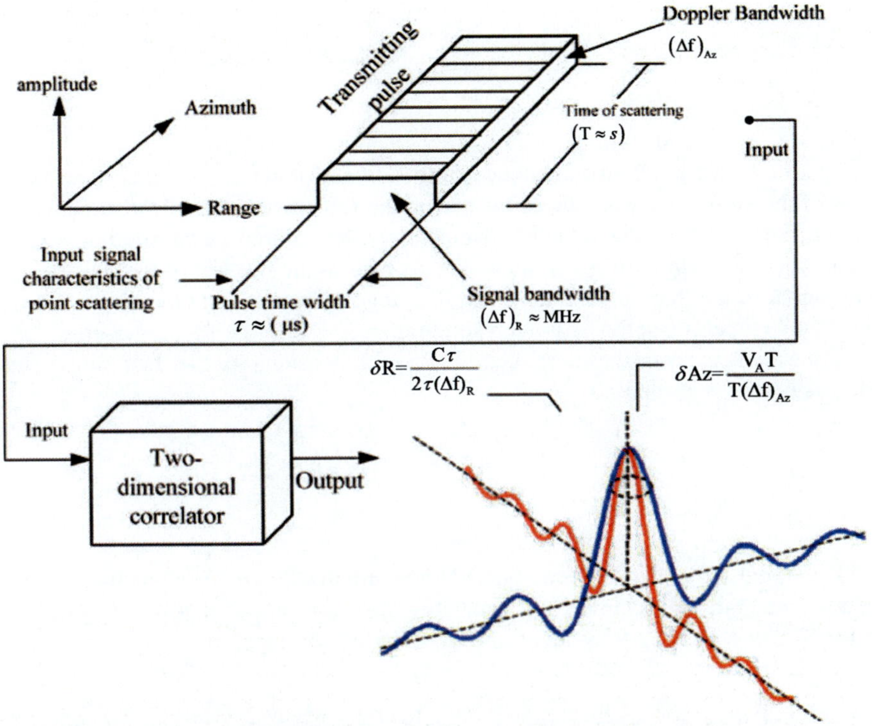

Fig. 3.2 Two-dimensional compression diagram of SAR system

Only when the IRF of the linear network is a δ function, the SAR image can accurately reproduce the backscattering coefficient of the terrain. In order to make the SAR image reflect the backscattering coefficient as accurately as possible, the IRF of SAR system is required to be as close to the δ function as possible. The IRF of SAR system will be derived as follows.

The transmitting signal is LFM waveform given as Eq. (3.2):

$$S(t) = \mathrm{Re}\left[\mu(t)\exp(j2\pi f_0 t)\right] = \mathrm{Re}\left[A\exp\left(j\cdot\frac{1}{2}k_r t^2\right)\exp(j2\pi f_0 t)\right] \tag{3.2}$$

where $u(t)$ is complex amplitude and it is a function of time. A is the amplitude of LFM signal and k_r is the chirp rate.

The echo scattered from a point target is

$$S_r(t) = \mathrm{Re}\left[k\mu(t - t_d)\exp(j2\pi f_0(t - t_d))\right] \tag{3.3}$$

where t_d is the delay time.

$$t_d = \frac{2R}{c} = \frac{2R_0}{c} + \frac{(x - x_p)^2}{cR_0} = \frac{2R_0}{c}\left[1 + \frac{(x - x_p)^2}{2R_0^2}\right] \tag{3.4}$$

where $x = v_s t$ is a function of time.

There are two kinds of time concepts in Eqs. (3.3) and (3.4). One is the 'fast' time in the complex amplitude in Eq. (3.3), and the other is the 'slow' time delay in Eq. (3.4). Because the vehicle movement can be ignored in transmitting signal duration, so the 'slow' time can be looked as zero in the interval of the 'fast' time. The signal in the 'fast' time determined the range resolution, and the movement of the vehicle determined the azimuth resolution.

The echo is expressed as the function of two dimensions for the 'fast' time t and the position of the vehicle x.

$$S_r(x,t) = \mathrm{Re}\left[k\mu\left(t - \frac{2R_0}{c}\left[1 + \frac{(x - x_p)^2}{2R_0^2}\right]\right)\exp\left(j2\pi f_0\left(t - \frac{2R_0}{c}\left[1 + \frac{(x - x_p)^2}{2R_0^2}\right]\right)\right)\right] \tag{3.5}$$

The signal in Eq. (3.5) is demodulated and normalized first, and then the autocorrelation integration is achieved by the matched filter. The output of pulse compression is the IRF $h(x, t)$ of SAR system as follows:

$$\begin{aligned} h(x,t) &= \mathrm{Re}\left\{\iint\left[e^{j\frac{1}{2}k_r(t'-t_d)^2}\cdot e^{-j\frac{2\pi}{\lambda R_0}(x'-x_p)^2}\right]\left[e^{-j\frac{1}{2}k_r(t+t'-t_d)^2}e^{-j\frac{2\pi}{\lambda R_0}(x+x'-x_p)^2}\right]dx'dt'\right\} \\ &= \mathrm{Re}\left\{\left[T_p\cdot e^{-j\frac{1}{2}k_r t^2}\cdot\frac{\sin\left[\pi\left(\frac{t}{\tau_r}\right)\right]}{\pi\left(\frac{t}{\tau_r}\right)}\right]\cdot\left[L_s e^{j\frac{2\pi}{\lambda R_0}x^2}\frac{\sin\left[\pi\left(\frac{x}{\rho_a}\right)\right]}{\pi\left(\frac{x}{\rho_a}\right)}\right]\right\} \end{aligned} \tag{3.6}$$

where T_p is transmitting pulse duration, τ_r is the main-lobe width of the IRF which determines the range resolution, L_s is synthetic aperture length, ρ_a is the azimuth resolution that is the main-lobe width processed by Doppler beam sharpening technique.

The envelope of the IRF for SAR system is given as

$$E\{h(x,t)\} = \left|\frac{\sin[\pi(t/\tau_r)]}{\pi(t/\tau_r)}\right|\cdot\left|\frac{\sin[\pi(x/\rho_a)]}{\pi(x/\rho_a)}\right| \tag{3.7}$$

Equation (3.7) shows that the shape of the unweighted envelope of the IRF is *sinc* function both in azimuth and range dimensions. Due to the independence of two variables, the IRF of SAR system can be regarded as the product of two impulse responses in range and azimuth dimensions. That is,

$$h(x,t) = h_r(x,t)\cdot h_a(x,t) \tag{3.8}$$

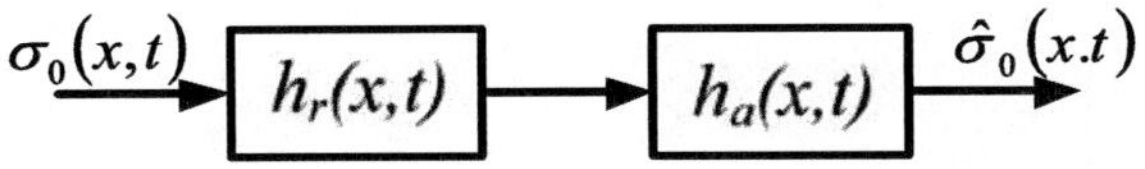

Fig. 3.3 The theoretical model of SAR

The theoretical model of SAR system is represented in Fig. 3.3. SAR signal processing can be divided into two steps: pulse compression in range and beam sharpening in azimuth.

If the position of vehicle can be ignored in the transmitting pulse duration, the echo signal will be divided into the independent two-dimensional parts with the 'fast' time in range and the 'slow' time in azimuth. The echo data from the two dimensions can be independently processed. However, the coupling between the range and the azimuth really exists, and it cannot be ignored, especially for spaceborne SAR.

3.2 Polarization Hierarchy of Polarimetric Synthetic Aperture Radar

PolSAR can obtain the polarimetric scattering matrix by transmitting and receiving the electromagnetic waves with different polarizations. Because the polarimetric scattering matrix contains the abundant information of the target, so PolSAR improves the ability of SAR to acquire the target scattering information greatly. According to the difference in polarization hierarchy between the polarization scattering matrixes, PolSAR can be divided into single polarization, dual polarization, full polarization, and compact polarimetry, etc. A comparison of different polarization hierarchies of PolSAR is shown in Table 3.1.

(1) Single-polarization SAR

Single-polarization SAR is the simplest microwave imaging radar. An HH or VV single complex image is obtained by the single-polarization antenna, single transmitting channel, and single co-polarization receiving channel. A large number of spaceborne SAR systems, such as Seasat, ERS-1, JERS-1, Radarsat-1, etc., are single polarization.

(2) Dual-polarization SAR

The dual-polarization SAR has a horizontal (H) and vertical (V) dual-polarization antenna, H or V orthogonal single-polarization transmitting channel. If the echo will be received in H or V single-polarization channel, the real image of HH or VV can be obtained. A typical example of dual polarization is AP (Alternating Polarization) Mode of Envisat-ASAR. And if the echo will be received in H and V dual-polarization channels, the images of HH/HV or VH/VV can be obtained. The H or V polarization on transmit and H and V dual polarization on receive has been used to acquire the dual-polarization strip mode in most of spaceborne SAR systems, such as SIR-C, RadarSat-2, RiSat, and ALOS-PALSAR.

Table 3.1 Polarization hierarchy of PolSAR

Polarization hierarchy		Antenna	Transmit	Receive	Processing condition	Image
Single-polarization		Single H or V linear polarization antenna	Single-polarization transmitting	Single-channel receiving	No hypothesis	HH or VV 1 amplitude image
Dual-polarization		H and V dual-linear polarization antenna	Single-polarization transmitting	Single-channel receiving	No hypothesis	HH and VV 2 amplitude images
		H and V dual-linear polarization antenna	Single-polarization transmitting	Dual-channel receiving	Reciprocity and symmetry hypothesis	HH, HVHV or VV, VH 2 × 2 matrix 2 amplitude images
Full-polarization	Quad-polarization	H and V dual-linear polarization antenna	Single-channel orthogonal transmitting	Dual-channel receiving	Reciprocity and symmetry hypothesis	HH, HV, VV 3 × 3 scattering matrix
	Full-polarization	H and V dual-linear polarization antenna	Two-channel orthogonal transmitting	Dual-channel receiving	No hypothesis	HH/HV/HH/VV 4 × 4 matrix
Compact polarimetry	π/4 model	H and V dual-linear polarization antenna	Transmitting 45° linear polarization	Dual-linear polarization coherent receiving	Reciprocity and symmetry hypothesis	2 × 2 covariance matrix, reconstruction of full-polarization SAR matrix
	Dual circular polarimetric mode (DCP)	H and V dual-linear polarization antenna	Left or right circular polarization transmitting	Left and right circular polarization coherent receiving	Reciprocity and symmetry hypothesis	2 × 2 covariance matrix, reconstruction of 3 × 3 quad-polarization covariance matrix

(continued)

Table 3.1 (continued)

Polarization hierarchy		Antenna	Transmit	Receive	Processing condition	Image
	Circular transmit linear receive (CTLR)	H and V dual-linear polarization antenna	Left or right circular polarization transmitting	H and V dual-linear polarization coherent receiving	Reciprocity and symmetry hypothesis	2 × 2 covariance matrix, reconstruction of full-polarization matrix

(3) Full-polarization SAR

The standard PolSAR is full-polarization SAR system which consists of the H and V orthogonal dual-polarization antenna, H and V orthogonal transmitting dual channels, and the H and V orthogonal receiving dual channels. Transmitting polarization and receiving polarization should be the orthogonal polarization pairs. The data produced by the full-polarization SAR is the 2 × 2 scattering matrix of HH, HV, VH, and VV for each unit of the scene. On the assumptions of reciprocity and symmetry, the 3 × 3 covariance matrix of HH, HV, and VV of each pixel in the scene will be obtained.

The full-polarization SAR requires the double PRF of dual-polarization or single-polarization SAR to satisfy the lower limit of Nyquist sampling frequency. The echo of full-polarization SAR can be described by the four-element Stokes vector. All information obtained from the observation scenario are applicable for 2 × 2 (symmetric Sinclair or Kennaugh) scattering matrix and 3 × 3 covariance matrix.

The advantage of full-polarization SAR is system perfect, which can obtain the full-polarization information to reflect the scattering matrix of target overall.

The main disadvantages of full-polarization SAR are complex system and high cost. Compared with single-polarization and dual-polarization SARs, the PRF is doubled, and the swath width is halved. The data rate is four times that single-polarization SAR and the average data rate is a similarity as dual-polarization SAR. The average power is more than twice that of single-polarization SAR. For linear polarization system, the range ambiguity of the cross-polarization echo at a large incident angle is depraved seriously. In order to restrain the range ambiguity, the incident angle must be much smaller than that of the single-polarization SAR.

In order to overcome above disadvantages, reduce system complex, treat of the indexes and parameters such as swath width, data rate, PRF, and range ambiguity ratio, etc., it is necessary to find a new configuration of PolSAR system. Compact Polarimetry SAR (C-PolSAR) is a new proposal technology for reducing the complexity of full-polarization SAR system.

(4) Compact Polarimetry SAR [2]

The C-PolSAR is transmitting a single-polarization electromagnetic wave and receiving the dual orthogonal polarization echoes and their relative phases. Compared with the conventional full-polarization SAR, the C-PolSAR is able to preserve the polarization information comprehensively and reduce requirement the system complexity effectively. In order to describe the polarization field completely, the 2 × 2 covariance matrix is needed to be measured at first, and then the four-element Stokes vector is reconstructed.

According to the transmitting and receiving polarization states, the main modes of CPolconfiguration are the π/4 mode, the Dual Circular Polarimetric (DCP) mode, and the Circular Transmit Linear Receive (CTLR) mode.

(1) The π/4 mode

In the π/4 mode, CPolSAR is composed of the H and V linear polarization antenna, and the transmitting polarization is the superposition of linear H and V polarization, which results in a linear polarization oriented 45° with respect to the H polarization. And it is coherent receiving backscattering components of H and V polarization. The performance of π/4 compact polarimetry mode can be comparable to that of full-polarization SAR. In π/4 mode, polarization decomposition is carried out to the distribution scatters in the scene on both horizontal and vertical directions. The data processing includes focus processing and multi-look processing. The output of focus processing is the single-look complex image data, and the output of multi-look processing is the 2 × 2 covariance matrix which could be converted to four-element Stokes vector.

The π/4 mode has a great superiority on the selection of parameters, such as PRF, swath width, radiation balance of dual receiving channels, power consumption, data rate of downlink, polarization decomposition, and calibration processing methods, etc. Unlike full-polarization SAR, the π/4 mode can not distinguish the orientation of view angle for a scattering point (odd reflect, even reflect, dipole) relative to the radar. However, the π/4 mode has the same ability as the full-polarization SAR to distinguish the distribution targets, such as crop and forest, etc.

(2) The Dual Circular Polarimetric (DCP) mode

The DCP mode is composed of the H and V dual-polarization antenna, the transmitting channel with left or right circular polarization, and the coherent receiving dual channels with left and right circular polarization. According to the rotation invariance requirement, transmitting polarization must be circular polarization. The standard DCP method is transmitting circular polarization and receiving coherent the left and right DCP, to obtain the 2 × 2 covariance matrix, through processing to reconstruct 3 × 3 quad-polarization.

The circular polarization direction of the scattering echo is opposite to that of the transmitting wave. For example, if transmitting the right circular polarization, usually the backscattering of left circular polarization is stronger, the right circular polarization becomes the receiving cross-polarization state.

Transmitting the circular polarization and receiving DCP will be a compact polarimetry configuration which can be adopted for spaceborne SAR. However, the additional hardware will be increased.

(3) The Circular Transmitting Linear Receiving (CTLR) mode

The CTLR mode is composed of the H and V dual-polarization antenna, a (Left or Right) circular polarization transmitting and dual orthogonal linear polarization coherent receiving. The linear dual-polarization antenna feeds a signal with a phase difference of 90° at the same time and radiates circular polarization in space. The linear basis of the echo signal is maintained in the whole process when it arrives at the processor through the H and V receiving channels. The data processing includes focus processing and multi-look processing. The output of focus processing is the single-look complex image, and the output of multi-look processing is the 2 × 2 covariance matrix which could be converted to a four-element Stokes vector.

Related to the 2 × 2 scattering matrix of the full-polarization SAR, only the 2 × 2 covariance matrix can be obtained by the CPolSAR. The results obtained by the C-PolSAR will be equivalent to that obtained by the full-polarization SAR in many applications, the results obtained by the C-PolSAR is much better than that by the single-polarization SAR and the conventional dual-polarization radar. The data of the C-PolSAR can be completely represented by 2 × 2 covariance matrix, which is well adapted to modern polarization decomposition strategies. The existing quad-polarization data files can be easily converted from a C-PolSAR data file to the data product of four-element Stokes vector. Table 3.1 shows the different polarization hierarchies of PolSAR.

3.3 The System Parameters of Polarization Synthetic Aperture Radar

The system parameters of PolSAR consist of technical parameters and polarization characteristic parameters. The technical parameters mainly include polarization mode, resolution, Peak-to Side-lobe Ratio (PSLR), Integrated Side-lobe Ratio (ISLR), system sensitivity, ambiguity ratio, polarization isolation, amplitude–phase imbalance between the polarization channels, radiation resolution, radiation accuracy, etc. The polarization characteristic parameters are extracted from fully PolSAR data, including the span of a scattering matrix, alpha angle, polarization entropy, and similarity parameters, they can represent the polarization characteristics of different targets to some extent.

3.3.1 The Technical Parameters of Polarization Synthetic Aperture Radar

3.3.1.1 Polarization Mode

The polarization is used to describe the spatial orientation of the electromagnetic field radiated by the antenna. In the polarization plane perpendicular to the direction of electromagnetic wave propagation, the trajectory at the end of the electric field vector is defined as the polarization of the electromagnetic wave. According to the different direction, polarization can be divided into linear polarization (horizontal polarization, vertical polarization) and circular polarization (left circular polarization, right circular polarization). The polarization parameters mainly include axial ratio, inclination angle, and rotation direction.

Polarization mode refers to the antenna polarization adopted by transmitting and receiving, and the polarization hierarchies such as single polarization, dual polarization, quad-polarization, full polarization, and compact polarimetry, etc.

Polarization mode is the important parameter of PolSAR. Depending on the different polarization modes, the incident angle, and the operating frequency, the backscattered energy will very different. The polarization mode determines the main polarimetric information for data processing, feature extraction, ground feature classification, target detection, target identification, and parameter inversion, etc.

3.3.1.2 Resolution

Resolution is defined as the ability of the imaging radar to distinguish the two closely spaced point targets. SAR is the two-dimensional high-resolution imaging radar, so resolution can be divided into the range resolution and the azimuth resolution.

(1) The range resolution [3, 4]

The range resolution of SAR is mainly determined by the bandwidth of LFM signal, that is

$$\rho_r = \frac{c}{2B} \tag{3.9}$$

In order to obtain the high ground range resolution, it needs to increase the transmitting signal bandwidth B, where ρ_r is the slant range resolution. However, the people always pay attention to the ground range resolution ρ_g. The relationship between the ground range resolution ρ_g and the slant range resolution ρ_r is

$$\rho_g = \frac{\rho_r}{\sin \eta} \tag{3.10}$$

where η is the incident angle.

(2) The azimuth resolution [5]

The azimuth resolution of SAR, independent to the slant range and wavelength, is the half of antenna length.

$$\rho_a = \frac{D_a}{2} \tag{3.11}$$

The synthetic aperture length illuminated by the 3 dB main lobe of the uniform aperture antenna is

$$L_S = K_A \cdot \frac{\lambda R_0}{D_a} \tag{3.12}$$

where K_A is the antenna pattern factor in azimuth, and $K_A = 0.886$. The Doppler bandwidth is

$$B_a = K_A \cdot \frac{2v_s}{D_a} \tag{3.13}$$

K_P is the ratio of signal processing bandwidth B_P to Doppler bandwidth B_a, and $K_P \leq 1$, then B_P is

$$B_P = K_P \cdot B_a = K_P K_A \frac{2v_s}{D_a} \tag{3.14}$$

The pulse broadening factor of azimuth compression output is expressed as K_a, and $K_a > 1$, the azimuth resolution is

$$\rho_a = K_a \frac{v_s}{B_P} = \frac{K_a}{K_A K_P} \cdot \frac{D_a}{2} \tag{3.15}$$

3.3.1.3 Peak-To Side-Lobe Ratio and Integrated Side-Lobe Ratio [3, 4]

(1) Peak-to Side-lobe Ratio (PSLR)

PSLR is the power ratio between the main peak power to the largest side-lobe power located in an interval of ten times peak width, which is usually expressed in dB. Figure 3.4 shows the definition of PSLR.

(2) Second Side-lobe Ratio (SSLR)

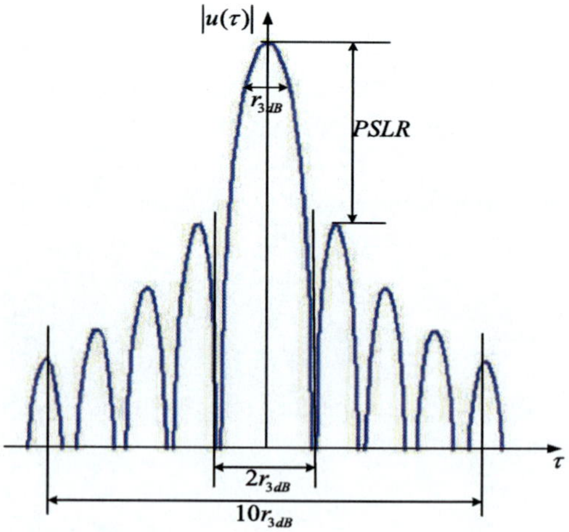

Fig. 3.4 The definition of PSLR

SSLR is the power ratio of the main peak power to the largest side-lobe power in the interval of 10–20 times peak width, which is usually expressed in dB. Figure 3.5 shows the definition of SSLR.

(3) Integrated Side-lobe Ratio (ISLR)

ISLR is a power ratio of the peak energy to the side-lobe integration energy, which is usually expressed in dB. Figure 3.6 shows the definition of ISLR.

$$ISLR = \frac{\int_{1/B}^{-1/B} |u(\tau)|^2 d\tau}{\int_{-T_p}^{-1/B} |u(\tau)|^2 d\tau + \int_{-1/B}^{T_p} |u(\tau)|^2 d\tau} \tag{3.16}$$

where $|u(\tau)|$ is the Impulse Response Function (IRF) at the output of the matched filter, T_p is the pulse width, and $2T_p$ is the length of the Impulse Response Function (IRF). In fact, the far side-lobe is too low to be measured, so it necessary to introduce

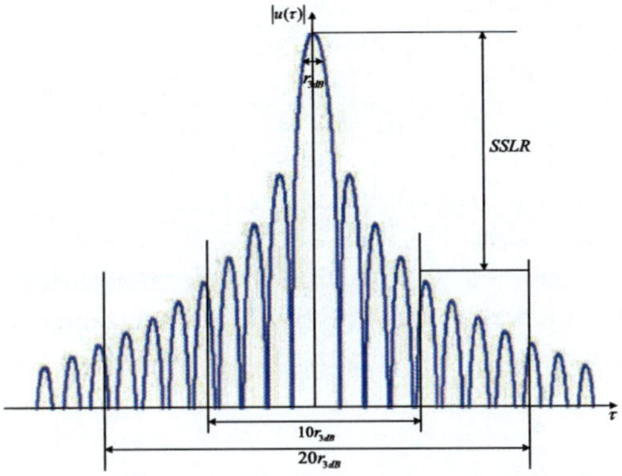

Fig. 3.5 The definition of SSLR

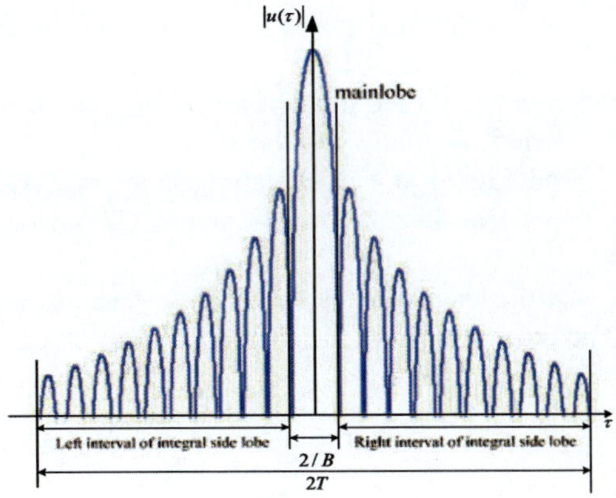

Fig. 3.6 The definition of ISLR

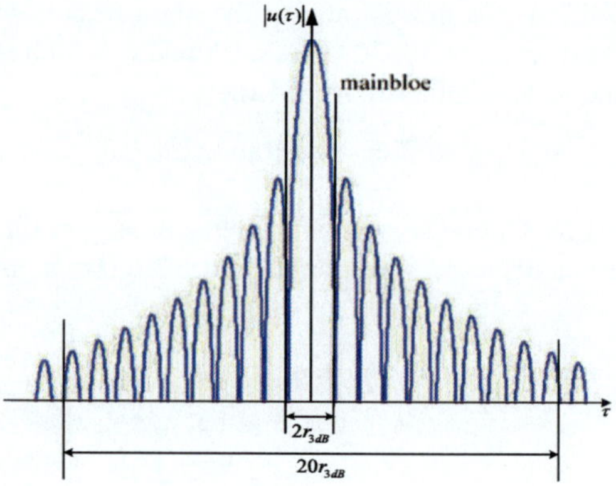

Fig. 3.7 The measurement of ISLR

the second definition of ISLR. The side-lobe integration energy is limited in the interval 20 times peak width, which is illustrated in Fig. 3.7.

3.3.1.4 System Sensitivity (Noise Equivalent Sigma Zero) [5, 6]

System sensitivity, also known as Noise Equivalent Signal Zero (NESZ), which is an index to estimate the ability of SAR imaging to the weak targets. The physical meaning is the corresponding backscattering coefficient as the output signal-to-noise ratio is zero.

Radar equation is the basic formula in radar system design, the peak power of the transmitter and the SNR of the system can be estimated by means of the radar equation. The general radar equation is expressed as

$$\frac{S}{N} = \frac{P_t G^2 \lambda^2 \sigma}{(4\pi)^3 R^4 (k_0 T_0 B_n F_n)} \tag{3.17}$$

where P_t is the peak power, G is antenna Gain, σ is the Radar Cross Section (RCS) of target, $k_0 = 1.38 \times 10^{-23}$J/K is the Boltzmann constant, T_0 is the absolute temperature, B_n is the equivalent noise bandwidth that approximately equals to system bandwidth B_r, and F_n is the noise figure.

The Signal-to Noise Ratio (SNR) of SAR image can be improved by pulse compression in range and azimuth dimensions, the signal processing gain is G_{pr} G_{pa}.

On the range dimension, SNR can be improved by means of the pulse compressing. The processing gain in range G_{pr} is expressed as

$$G_{pr} = B_r \cdot T_p \tag{3.18}$$

where T_p is the pulse width of the transmitting, B_r is the system bandwidth.

On the azimuth dimension, the azimuth focusing processing is equivalent to the coherent accumulation of pulses, and the coherent accumulation gain can be expressed as

$$G_{pa} = T_S \cdot PRF \tag{3.19}$$

where T_s is the synthetic aperture time.

In Eq. (3.17), σ is the Radar Cross Section RCS of the minimum resolution cell, it can be given by

$$\sigma = \sigma_0 \cdot \rho_g \rho_a \tag{3.20}$$

where σ_0 is the normalized backscattering coefficient, ρ_g is the ground resolution, and ρ_a is the azimuth resolution. Equations (3.18)–(3.20) are substituted into Eq. (3.17), the radar equation of SAR can be expressed as

$$\frac{S}{N} = \frac{P_{av} G^2 \lambda^3 \sigma_0 \rho_g}{2(4\pi)^3 R^3 v_s (k_0 T_0 F_n) L_{oss}} \tag{3.21}$$

where L_{oss} is the system loss, P_{av} is the average power of the transmitter. According to Eq. (3.21), the SNR of radar output can be calculated under a certain average transmitting power. Equation (3.21) can be rewritten as

$$P_{av} = \frac{2(4\pi)^3 R^3 v_s (k T_0 F_n) L_{oss}}{G^2 \lambda^3 \sigma_0 \rho_g} \cdot \frac{S}{N} \tag{3.22}$$

Equation (3.22) can be used to calculate the average transmitting power for satisfying the SNR requirement of radar system output.

Usually, the NESZ is used to express the system sensitivity, that is, the lowest limit of scattering coefficient for SNR = 1. Any backscattering coefficient lower than this value, the target characteristic will very difficult to be discovered. NESZ is defined as

$$NE\sigma^0 = \frac{\sigma_0}{S/N} = \frac{2(4\pi)^3 R^3 v_s (k_0 T_0 F_n) L_{oss}}{P_{av} G^2 \lambda^3 \rho_g} \tag{3.23}$$

The SNR of radar output means that how many times the actual backscattering coefficient σ_0 is higher than NESZ.

3.3.1.5 Ambiguity Ratio [5, 6]

Since SAR is operating in the pulse mode, ambiguity is inevitable as same as the pulse radar. That means that the radar will receive both the useful signal and the ambiguity interference at the same time. The ambiguity signal is the echo from the outside of the image area, combined with the useful signal to enter the receiver, and it will degrade the image quality at the output of SAR signal processor. Ambiguity problem is particularly prominent in the spaceborne SAR system. Ambiguity ratio is defined as the ratio of power between the ambiguity interference and the useful signal, including Range Ambiguity to Signal Ratio (RASR) and Azimuth Ambiguity to Signal Ratio (AASR). RASR and AASR are the parameters to evaluate the ambiguity interference degree of SAR system.

(1) Azimuth Ambiguity to Signal Ratio (AASR)

Since the antenna pattern exists in the side-lobes, the Doppler spectrum of azimuth pattern is non-bandlimited. When the echo signal spectrum in azimuth is sampled by limited frequency (PRF), it will lead to overlap for the Doppler spectrum of azimuth. It is called the ambiguity of azimuth. The signal generation of SAR in the azimuth is shown in Fig. 3.8.

Figure 3.9 shows the schematic diagram of azimuth ambiguity, and B_p is Doppler processing bandwidth.

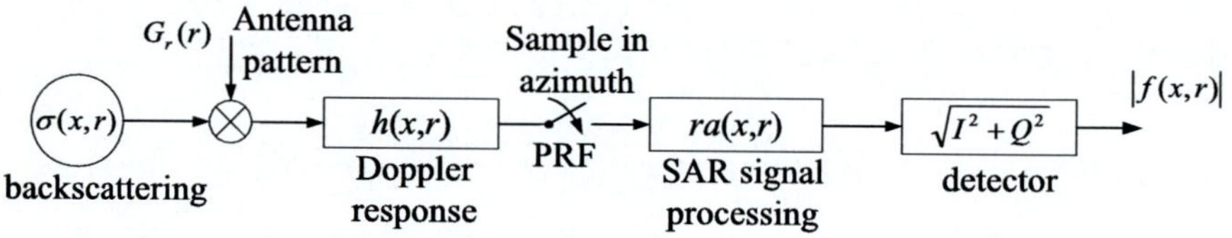

Fig. 3.8 Signal generation of SAR in the azimuth

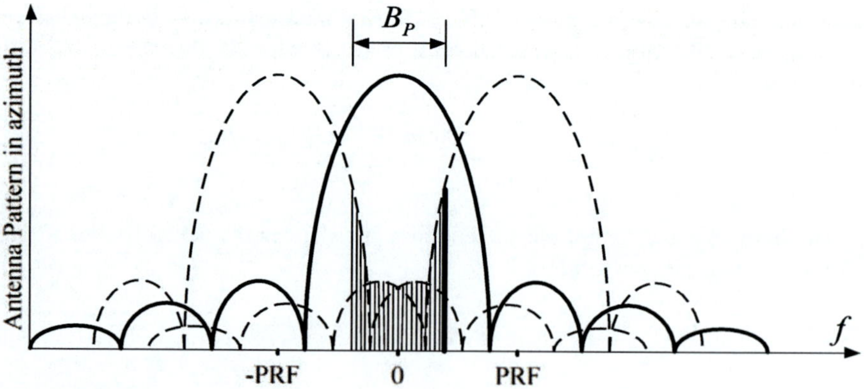

Fig. 3.9 The schematic diagram of azimuth ambiguity

In order to convenient for calculation, assuming that the reflecting characteristics of a target for different positions in azimuth at the same range are uniform and the azimuth antenna pattern is the same in each range gate. AASR can be calculated by

$$AASR = \frac{\sum\limits_{\substack{m=-\infty \\ m\neq 0}}^{+\infty} \int_{-B_p/2}^{+B_p/2} G^2(f + m \cdot PRF)df}{\int_{-B_p/2}^{+B_p/2} G^2(f)df} \tag{3.24}$$

where f is Doppler frequency and $G(f)$ is the gain of azimuth antenna pattern.

(2) Range Ambiguity to Signal Ratio (RASR)

The generation mechanism of range ambiguity is the same as that of conventional pulse radar. Since the range antenna pattern exists in the side-lobes, the echo signals from different ranges are sampled in the same window. Figure 3.10 shows a schematic diagram of range ambiguity. Duo to the far range, the fast velocity, and high PRF requirement of spaceborne SAR, the echo of this pulse only can be received after a burst of radar transmitting pulses.

In this way, as soon as the radar receives the echo of the target within the imaging swath for the certain transmitting pulse (such as the target 2 in Fig. 3.10). the echo of the target located at the near end outside the imaging swath for the next transmitting pulse (such as the target 3 in Fig. 3.10), and the echo of the target located at the far end outside the imaging swath for the previous transmitting pulse (such as the target 1 in Fig. 3.10) may be received at the same time, that is, the origin of the range ambiguity. The echo from the outside of the swath will interfere in the useful echo from the inside of the swath.

In order to descript the range ambiguity, Fig. 3.11 shows the relationship of spatial positions between the swath and the ambiguity region. The swath is width illuminated by the 3 dB beam width of the antenna pattern on range, and the first ambiguity region

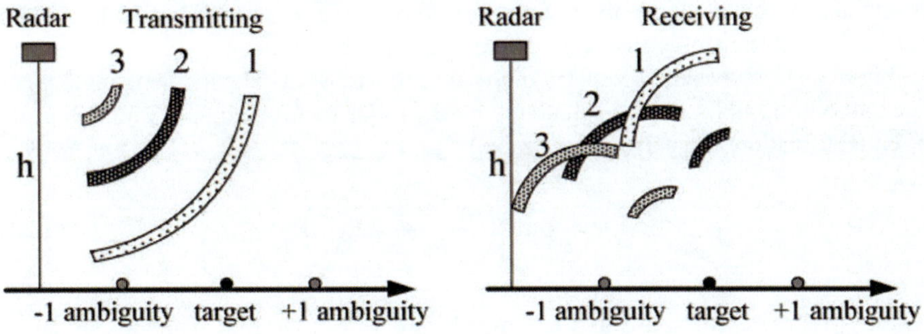

Fig. 3.10 The schematic diagram of range ambiguity

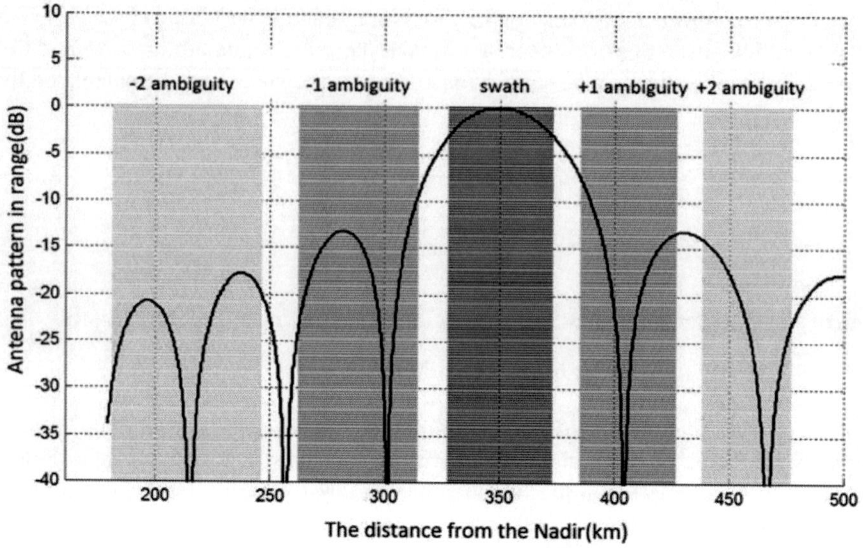

Fig. 3.11 Relationship between the swath and the ambiguity region

(±1 right side and left side) and the second ambiguity region (±2 right side and left side) are given in Fig. 3.11.

In order to evaluate the range ambiguity, the position of ambiguity signal should be determined at first. If the echo delay time of a target in the swath is t_i, the ambiguity signal can be at the following range:

$$R_{ij} = \frac{c}{2}\left(t_i + \frac{j}{PRF}\right) \qquad j = \pm 1,\ \pm 2,\ \ldots,\ \pm n_h \tag{3.25}$$

If j is positive, the ambiguity signal comes from the far range outside the imaging swath for the previous transmitting pulse. If j is negative, the ambiguity signal comes from the near range outside the imaging swath for the next transmitting pulse. When j equals to n_h, the ambiguity target is located at the horizon.

Figure 3.12 shows the geometry of the spaceborne SAR. The incident angle η_{ij} of each ambiguity point can be calculated by the radar looking angle γ_{ij} at ambiguity point (Earth adopt the part sphere model).

$$\eta_{ji} = \sin^{-1}\left(\frac{R_e + h}{R_e}\sin\gamma_{ij}\right) \tag{3.26}$$

where R_e is the earth radius and h is the altitude of the satellite. The looking angle γ_{ij} can be calculated by cosine law:

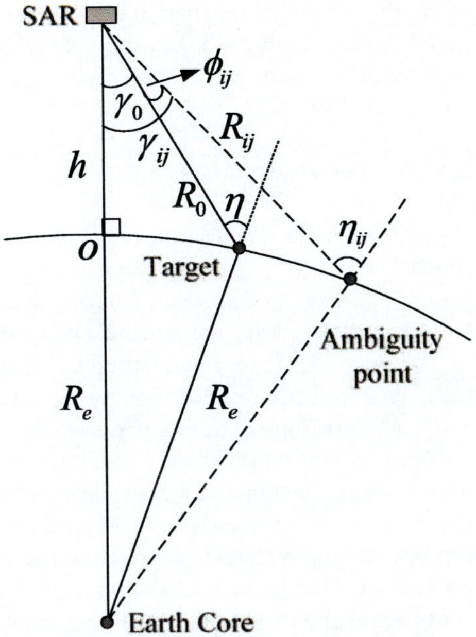

Fig. 3.12 The geometry of the spaceborne SAR

$$\gamma_{ij} = \cos^{-1}\left[\frac{R_{ij} + (R_e + h)^2 - R_e^2}{2R_{ij}(R_e + h)}\right] \tag{3.27}$$

The RASR of the swath can be calculated by

$$RASR = \frac{\sum_{i=1}^{N} S_{ai}}{\sum_{i=1}^{N} S_i} \tag{3.28}$$

where S_{ai} is the ambiguity signal power and S_i is the useful signal power at the ith sampling point in the echo window, N is the total number of the samples in the swath.

According to the above equation, as calculating the ambiguity signal and useful signal power, it only needs to consider the parameters can't be countervailed in the radar equation.

Equation (3.29) gives the expression.

$$\begin{cases} S_i = \frac{\sigma_{ij}^0 \cdot G_{ij}^2}{R_{ij}^3 \sin \eta_{ij}} & j = 0 \\ S_{ai} = \sum\limits_{\substack{j=-n_h \\ j \neq 0}}^{n_h} \frac{\sigma_{ij}^0 \cdot G_{ij}^2}{R_{ij}^3 \sin \eta_{ij}} & j \neq 0 \end{cases} \tag{3.29}$$

where G_{ij} is the antenna gain in the range direction. ${\sigma_{ij}}^0$ is the normalized backscattering at the incident angle η_{ij}, which is usually approximated by a negative exponential function $\sigma_{ij}^0 = e^{-\eta_{ij}/\eta_0}$.

3.3.1.6 Polarization Isolation [7, 8]

Polarization isolation describes the level of crosstalk between different polarization channels in PolSAR system. The level of crosstalk is loswer, the system will be better. If the level of crosstalk is so large that will affect data application, the crosstalk between polarization channels must be calibrated.

The main reasons for generating the crosstalk between polarization channels are the polarization impurity of antenna, the polarization isolation of the transmit switch and the polarization coupling between the receiving channels.

There is crosstalk between antennas with different polarizations. The antenna neither can only transmit pure electromagnetic waves containing one polarization component nor can completely separate only one polarization component from the echo containing different polarization components, the crosstalk of different polarizations will lead to the measurement error of the scattering matrix. The polarization component impurity of the receiving echo is common in PolSAR system, which cannot be completely eliminated by improving the antenna technology. The crosstalk caused by the antenna takes effect on the transmitting and receiving, which will affect the amplitude and phase measurement of each component in the scattering matrix. The polarization coupling caused by the polarization switch is only effective on transmit.

If the actual scattering matrix of the target is S, the scattering matrix measured by the system is

$$\begin{aligned} M &= Ke^{jj}\begin{bmatrix} 1 & d_2 \\ d_1 & f_1 \end{bmatrix}\begin{bmatrix} S_{HH} & S_{HV} \\ S_{VH} & S_{VV} \end{bmatrix}\begin{bmatrix} 1 & d_3 \\ d_4 & f_2 \end{bmatrix} + \begin{bmatrix} N_{HH} & N_{HV} \\ N_{VH} & N_{VV} \end{bmatrix} \\ &= Ke^{jj}RST + N \end{aligned} \tag{3.30}$$

where K is the total absolute gain factor of PolSAR system, j is the total absolute phase, N is the additive noise, and the matrices T and R represent the radar polarimetric characteristics on transmit and receive, respectively. d_1, d_2 are the HV and VH crosstalk terms, respectively, on receive, d_3, d_4 are the HV and VH crosstalk terms on transmit, respectively, which are caused by cross-polarization antenna pattern, transmitter switch and circulator isolation. $d_1 = R_{HV}/R_{HH}$ is the normalized crosstalk from H channel into V channel on receive, $d_2 = R_{VH}/R_{HH}$ is the normalized crosstalk from V channel into H channel on receive, $d_3 = T_{HV}/T_{HH}$ is the normalized crosstalk from V channel into H channel on transmit, $\delta_4 = T_{VH}/T_{HH}$ is the normalized crosstalk from H channel into V channel on transmit.

3.3.1.7 Amplitude–Phase Imbalance Between the Polarization Channels [7, 8]

In an ideal PolSAR system the gains between different polarization channesls always are same, but in practical PolSAR system they are not exactly equal. In order to quantitatively describe the gain imbalance between the polarization channels of PolSAR system, the amplitude imbalance is introduced. The amplitude imbalance represents the relative amplitude of the V polarization electromagnetic wave to that of the H polarization, and the phase imbalance represents the phase difference between the polarization dual channels.

In order to analyze the influence of the imbalance, the relationship between measurement matrix and scattering matrix can be simplified as

$$\begin{bmatrix} Z_{HH} & Z_{HV} \\ Z_{VH} & Z_{VV} \end{bmatrix} = Ke^{jj} \begin{bmatrix} 1 & 0 \\ 0 & f_1 \end{bmatrix} \begin{bmatrix} S_{HH} & S_{HV} \\ S_{VH} & S_{VV} \end{bmatrix} \begin{bmatrix} 1 & 0 \\ 0 & f_2 \end{bmatrix} \tag{3.31}$$

where $f_1 = R_{VV}/R_{HH}$ is the channel imbalance (amplitude–phase) between the H and V channels on receive, and $f_2 = T_{VV}/T_{HH}$ is the channel imbalance (amplitude–phase) between the H and V channels on transmit.

The imbalance on transmit will mainly change the amplitude and the phase of HH and VV channels, it don't create the phase error. The measurement error and the amplitude error have similar variation low, and they don't increase with the increase of the absolute value of the amplitude–phase imbalance.

The imbalance on receive mainly creates the error between VH and VV channels. Since the symmetry of the scattering matrix, the effect of the imbalance on receive is equivalent to that on transmit. Therefore, the amplitude–phase imbalance mainly influences on the HV and VV polarimetric data processing, especially for the target detection or target classification.

3.3.2 *The Polarization Characteristic Parameters*

The polarization characteristic parameters is a kind of polarization parameters which is extracted from fully polarimetric SAR data and can reflect the polarization characteristic of the target to a certain extent. It including the span of a scattering matrix, alpha angle, polarization entropy, and similarity parameters, which can represent the polarization characteristics of different targets to some extent. The polarimetric characteristic parameters have been widely used in polarization feature extraction, target classification, target detection, and parameter inversion, etc.

3.3.2.1 The Span (Total Power)

The Sinclair matrix obtained by Jones vector is

$$S = \begin{bmatrix} S_{HH} & S_{HV} \\ S_{VH} & S_{VV} \end{bmatrix} \tag{3.32}$$

The span of a scattering matrix is defined as

$$P_{span} = [S_{HH}]^2 + [S_{HV}]^2 + [S_{VH}]^2 + [S_{VV}]^2 \tag{3.33}$$

The monostatic radar has one antenna for transmitting and receiving. According to the reciprocity theorem $S_{HV} = S_{VH}$, the span of a scattering matrix can be rewritten as

$$P_{span} = [S_{HH}]^2 + 2[S_{HV}]^2 + [S_{VV}]^2 \tag{3.34}$$

Elements of the Mueller matrix are

$$m_{11} = \frac{1}{2}\left([S_{HH}]^2 + [S_{HV}]^2 + [S_{VH}]^2 + [S_{VV}]^2\right) \tag{3.35}$$

The span is

$$P_{span} = 2m_{11} \tag{3.36}$$

3.3.2.2 The Polarization Entropy (H) and Scattering Angle (α)

Entropy is a concept in information theory used to indicate the degree of disorder. The polarization entropy is introduced into polarization information processing to measure the consistency of polarization characteristics in the neighborhood area around the target. The volume of the polarization entropy is greater, the disorder and the consistency around the target are worse.

In general, the sum of the coherent matrices in the square neighborhood area centered on the target $T = \sum T_i$ is a complex conjugate matrix with three non-negative real eigenvalues. The eigenvalue decomposition is given by

$$T = U \begin{bmatrix} \lambda_1 & 0 & 0 \\ 0 & \lambda_2 & 0 \\ 0 & 0 & \lambda_3 \end{bmatrix} U^H, \lambda_1 \geq \lambda_2 \geq \lambda_3 \geq 0 \tag{3.37}$$

$$U = \begin{bmatrix} \cos\alpha_1 \cdot e^{j\varphi_1} & \cos\alpha_2 \cdot e^{j\varphi_2} & \cos\alpha_3 \cdot e^{j\varphi_3} \\ \sin\alpha_1 \cos\beta_1 \cdot e^{j\delta_1} & \sin\alpha_2 \cos\beta_2 \cdot e^{j\delta_2} & \sin\alpha_3 \cos\beta_3 \cdot e^{j\delta_3} \\ \sin\alpha_1 \sin\beta_1 \cdot e^{j\gamma_1} & \sin\alpha_2 \sin\beta_2 \cdot e^{j\gamma_2} & \sin\alpha_3 \sin\beta_3 \cdot e^{j\gamma_3} \end{bmatrix} \quad (3.38)$$

where $\lambda_1, \lambda_2, \lambda_3$ are the eigenvalues of the coherent matrix and U is the eigenvector corresponding to the eigenvalue.

The polarization entropy H is defined as follows:

$$H = \sum_{i=1}^{3} -P_i \log_3 P_i \quad 0 \le H \le 1 \quad (3.39)$$

where $p_i = \frac{\lambda_i}{\sum_{j=1}^{3} \lambda_j}$.

The polarization entropy H describes the randomness of the scattering process, and H scope is from 0 to 1, $0 \le H \le 1$. As $H = 0$, the coherency matrix has only one non-zero eigenvalue, the system stands the complete polarization state and has only one scattering matrix, which corresponding to the deterministic scattering process. With the increase of entropy, the randomness of target polarization state also increases. Under $H = 1$ limited condition, the coherency matrix has three identical eigenvalues, the scattering of target is complete degenerated a random noise. It stands a complete non-polarization state, but it can't obtain any polarization information of targets.

The scattering angle α is defined as

$$\alpha = \sum_{i=1}^{3} \alpha_i p_i \quad (3.40)$$

The scattering angle is a continuous variation parameter from 0° to 90°, it represents the types of scattering mechanism of the target to a certain extent. When $\alpha = 0°$, the isotropic surfaces, such as a calm water surface or a homogeneous conducting sphere can be obtained. As the scattering angle increases, the scattering will become anisotropic surface scattering mechanisms, it is $|s_{hh}| \neq |s_{vv}|$. If $\alpha = 45°$ corresponding to a dipole scattering model, among a co-polarization scattering coefficient will become zero. If $\alpha > 45°$, the reflected scattering mechanism of target is the scattering of the anisotropic dihedral corner reflector. Under $\alpha = 90°$ extreme condition, the most representative scattering mechanisms are scattering models of the dihedral corner reflector or the helix.

3.3.2.3 Similarity Parameters

Professor Yang Jian defined the similarity parameters between the two scattering matrices to represent the degree of approximation between the two scatters [26].

For a reciprocal case, the scattering matrix of the target is

$$S = \begin{bmatrix} S_{HH} & S_{HV} \\ S_{VH} & S_{VV} \end{bmatrix} \tag{3.41}$$

Let ψ denote the orientation angle of the target, then the following scattering matrix is defined

$$S^0 = J(-\psi)SJ(\psi) = \begin{bmatrix} S^0_{HH} & S^0_{HV} \\ S^0_{VH} & S^0_{VV} \end{bmatrix} \tag{3.42}$$

where $J(\psi) = \begin{bmatrix} \cos\psi & -\sin\psi \\ \sin\psi & \cos\psi \end{bmatrix}$.

Using the modified Pauli scattering vector $\overline{k^0}$ to represent the scattering matrix S^0 as

$$\overline{k^0} = \frac{1}{\sqrt{2}}\left[S^0_{HH} + S^0_{VV}, S^0_{HH} - S^0_{VV}, 2S^0_{HV}\right]^T \tag{3.43}$$

The similarity parameter of the two scattering matrices is defined as

$$r(S_1, S_2) = \frac{\left|\left(\overline{k^0_1}\right)^H\left(\overline{k^0_2}\right)\right|^2}{\left\|\overline{k^0_1}\right\|^2\left\|\overline{k^0_2}\right\|^2} \quad 0 \le r(S_1, S_2) \le 1 \tag{3.44}$$

where $\| \ \|$ denotes the vector 2-norm. Obviously, the larger the r, the closer the two scatters are.

The parameter of similarity to the scattering matrix of a plate (or sphere) is defined as

$$r_1 = r(S_1, diag(1, 1)) = \frac{\left|S^0_{HH} + S^0_{VV}\right|^2}{2\left(\left|S^0_{HH}\right|^2 + \left|S^0_{VV}\right|^2 + 2\left|S^0_{HV}\right|^2\right)} \tag{3.45}$$

The parameter of similarity to the dihedral corner reflector scattering matrix is defined as

$$r_2 = r(S, diag(1, -1)) = \frac{\left|S^0_{HH} - S^0_{VV}\right|^2}{2\left(\left|S^0_{HH}\right|^2 + \left|S^0_{VV}\right|^2 + 2\left|S^0_{HV}\right|^2\right)} \tag{3.46}$$

By means of calculating the similarity parameter between any target and a specific target, it can be used to extract some scattering characteristics of the specific targets, such as the plate, the dihedral corner reflector, etc., and used for target classification and target detection.

3.4 The Operating Architectures of Polarization Synthetic Aperture Radar

In order to obtain the full-polarization scattering matrix of the target, multi-polarization SAR system is required to have the ability to obtain quad-polarization (HH/VH/HV/VV) echoes. For the simultaneous quad-polarization, the dual-channel parallel transmitting and the four-channel parallel receiving should be adopted, which will greatly increase the complexity of the radar instrument. For a multi-polarization SAR system, dual-polarization antenna, single transmitting channel, single or dual receiving channels usually are adopted, by means of combining different polarizations on transmit and receive to obtain the quad-polarization echoes quasi-simultaneously. The combining model between transmitting and receiving channels includes the time-division multiplexing pulses, the frequency-division multiplexing pulses, the code-division multiplexing pulses, the space-division twin-bursts in azimuth, and compact polarimetry.

3.4.1 The Time-Division Multiplexing Pulses

The time-division PolSAR system is composed of an orthogonal polarization (H and V) antenna, single transmitting channel (H or V), dual receiving channels (H and V), and polarization switch. Figure 3.13 shows the schematic diagram of time-division PolSAR system.

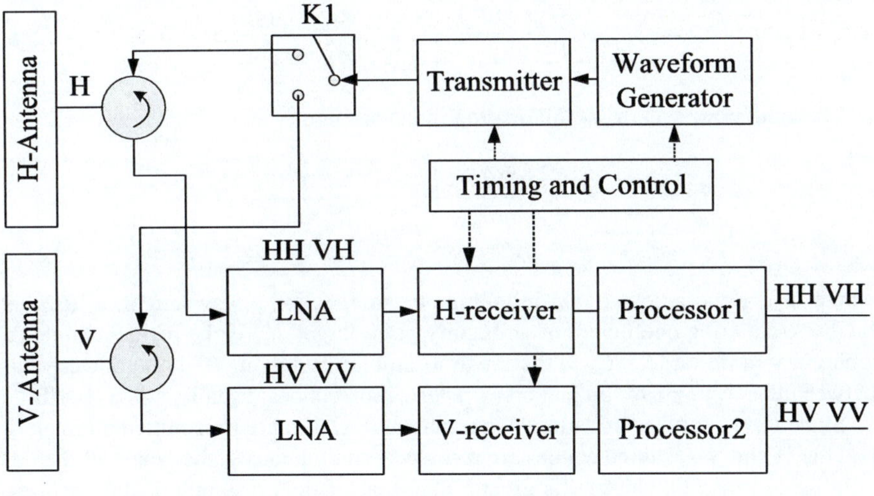

Fig. 3.13 The schematic diagram of time-division PolSAR system

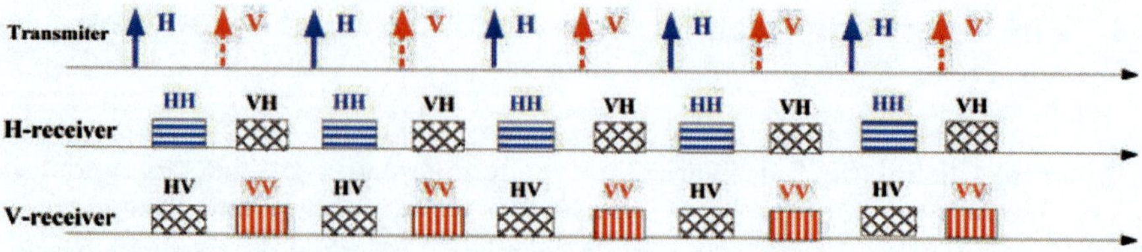

Fig. 3.14 The sequence diagram of the time-division multiplexing pulses

The sequence diagram of the time-division multiplexing pulses for multi-polarization is shown in Fig. 3.14. In the interval of Pulse Repetition Time (PRT), the radar transmits the H or V electromagnetic waves alternately and receives the H and V scattered echoes simultaneously. For example, H electromagnetic wave is transmitted in the first PRT, H channel receives the HH co-polarization echo and V channel receives the HV cross-polarization echo. And then, V polarization electromagnetic wave is transmitted in the second PRT, H channel receives the VH cross-polarization echo and V channel receives the VV co-polarization echo. Based on the time-division multiplexing pulses architecture, the quasi-quad-polarization (HH/VH/HV/VV) can be obtained to reconstruct the full-polarization scattering matrix of target.

Range ambiguity of the time-division multiplexing pulses

Since the H or V polarization electromagnetic waves are transmitted alternately and the H and V scattered echoes are received simultaneously, the echo powers of range ambiguity for co-polarization and cross-polarization are different.

The echo power of co-polarization range ambiguity is

$$S_a = \sum_{i=\text{odd}} \frac{1}{K} \frac{G_{\text{ai}}^2 \sigma^0(\theta_{\text{ai}})}{R_{\text{ai}}^3 \sin(\theta_{\text{ai}})} + \sum_{i=even} \frac{G_{\text{ai}}^2 \sigma^0(\theta_{\text{ai}})}{R_{\text{ai}}^3 \sin(\theta_{\text{ai}})}, i \neq 0$$

The echo power of cross-polarization range ambiguity is

$$S_a = \sum_{i=odd} K \frac{G_{\text{ai}}^2 \sigma^0(\theta_{\text{ai}})}{R_{\text{ai}}^3 \sin(\theta_{\text{ai}})} + \sum_{i=even} \frac{G_{\text{ai}}^2 \sigma^0(\theta_{\text{ai}})}{R_{\text{ai}}^3 \sin(\theta_{\text{ai}})}, i \neq 0$$

where G_{ai} is the antenna gain in ambiguity area, λ is the wavelength, $\sigma^0(\theta_{ai})$ is the backscattering coefficient of ambiguity point target, θ_{ai} is the incident angle of ambiguity point target, R_{ai} is the range to ambiguity target, K is the power ratio between the co-polarization and cross-polarization echoes; usually, it is 6–10 dB.

Since the H or V polarization electromagnetic waves are transmitted alternately and the H and V scattered echoes are received simultaneously, the power of the co-polarization range ambiguity is greater than that of the cross-polarization echoes. The weak cross-polarization echo will be interfered by the strong co-polarization signal scattered from the range ambiguity zone so that the quality of the cross-polarization image is degraded greatly. This is a serious disadvantage of the time-division multiplexing pulses operating architecture.

The time-division multiplexing pulses with double PRF is the common operating architecture of full-polarization SAR, such as AIRSAR of JPL, CV990 and DC-8 airborne SAR of NASA, SIR-C/X SAR of NASA and DLR, RadarSat-2 of CSA, TerraSAR of DLR, and PALSAR of JAXA. In spite of simple structure and easy implementation, double PRF makes the average power doubled, the swath width half, and the RASR degraded. The ambiguity problem of spaceborne PolSAR can't be ignored, so other operating architectures must be proposed to suppress the ambiguity for meeting the requirement of large incident angle.

3.4.2 The Frequency-Division Multiplexing Pulses

The frequency-division PolSAR system is composed of an orthogonal polarization (H and V) antenna with different center frequencies, dual transmitting channels, and dual receiving channels. In the frequency-division PolSAR system, through the orthogonal polarization (H and V) antenna, the pulse signals with different polarization and different center frequencies are transmitted with the dual channels at the same time, the echoes are received and recorded with the dual channels at the same time, and then the echoes are separated according to the frequency by the filter, so the four kinds of the full-polarization information can be obtained. Considering the transmitting power is limited by the platform, the H and V polarization signals only can be transmitted by pair. And need has a very narrow time slot between two transmitting pulses to convenient for power conversion. Figure 3.15 shows the sequence schematic diagram of the frequency-division multiplexing pulses.

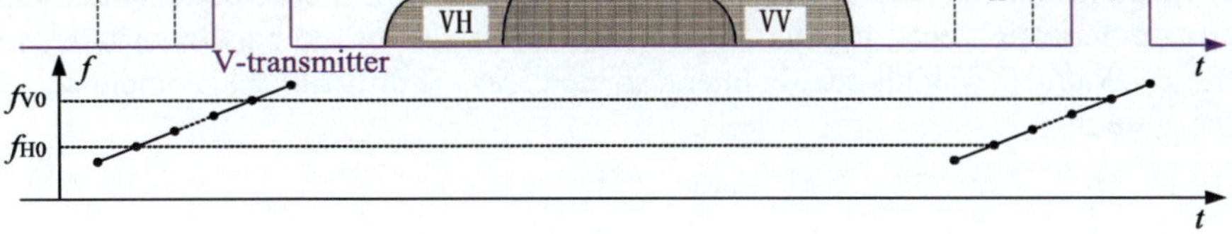

Fig. 3.15 The sequence diagram of frequency-division multiplexing pulses

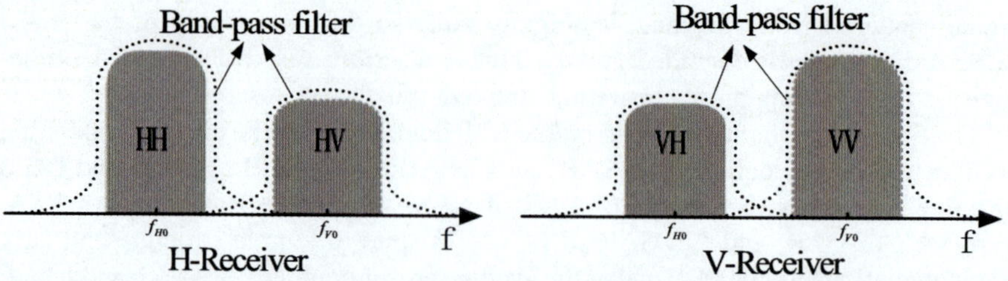

Fig. 3.16 H and V polarization frequency-division signals

where f_{H0}, f_{V0} are the central frequencies of the H and V transmitting LFM signals, respectively, with the same chirp rate, same pulse duration and same spectrum bandwidth.

$$B_{HH} = B_{HV} = B_{VH} = B_{VV} = k \cdot T_p \tag{3.47}$$

Therefore, the band-pass filters with different center frequencies can separate the co-polarization signal and the cross-polarization signal effectively as shown in Fig. 3.16.

(1) Azimuth Ambiguity to Signal Ratio (AASR)

When the antenna azimuth size and the satellite altitude are given, the AASR only depends on PRF. If the band-pass filter is an ideal filter, the AASR of the frequency-division architecture should be the same as that of the time-division architecture, but the PRF is the half of the time-division system.

(2) Range Ambiguity to Signal Ratio (RASR)

Since the PRF of frequency-division architecture is the same as that of conventional single-polarization SAR, the RASR got a great improvement than that of time-division architecture. This allows selecting the higher PRF to optimize the azimuth ambiguity, and the high quality of two-dimensional SAR image can be obtained.

Since the H polarization and V polarization pulses transmitted by the frequency-division PolSAR are continuous in the frequency domain, so the bandwidth of transmitting signal will increase more than two times, it will bring great difficulties to the design and manufacture of the antenna. In addition, the co-polarization and cross-polarization echo signals are separated by means of the band-pass filter, so the development of high-quality filter also increases the difficulty and complexity of the system.

3.4.3 *The Code-Division Multiplexing Pulses*

The code-division PolSAR system is composed of the orthogonal polarization (H and V) antenna, dual orthogonal coded pulses transmitting channels, and dual orthogonal decoded receiving channels. Figure 3.17 shows the schematic diagram of code-division PolSAR system. The code-division PolSAR system through the orthogonal polarization (H and V) antenna, transmitting H and V linear polarization pulse signals at the same time, the transmitting pulse signals are orthogonally coded. In dual receiving channels, the echoes are orthogonally decoded to separate the co-polarization and cross-polarization, so the scattering matrix can be obtained.

The selection of the transmitted pulse waveform for the code-division PolSAR is introduced as follows. Suppose $t_H(t)$ and $t_V(t)$ are transmitted through dual channels at the same time, and the echoes scattered from a point target are received simultaneously by the dual channels are

$$\begin{aligned} r_H(t) &= t_H(t-\tau)\cdot h_{HH} + t_V(t-\tau)\cdot h_{VH} \\ r_V(t) &= t_H(t-\tau)\cdot h_{HV} + t_V(t-\tau)\cdot h_{VV} \end{aligned} \tag{3.48}$$

where τ is the time delay.

If the transmitted signal $t_H(t)$ and $t_V(t)$ meets the following conditions

$$\begin{aligned} &R_{HV}(\tau) = R^*_{VH}(-\tau) = \int_{-\infty}^{\infty} t^*_H(t)\cdot t^*_V(t+\tau)dt = 0 \qquad \forall\tau \\ &R_{ii}(\tau) = \int t^*_i(t)\cdot t_i(t+\tau)dt = \begin{cases} 1 & for\ \tau = 0 \\ 0 & for\ \tau \neq 0 \end{cases} \quad where: i = H\ or\ V \end{aligned} \tag{3.49}$$

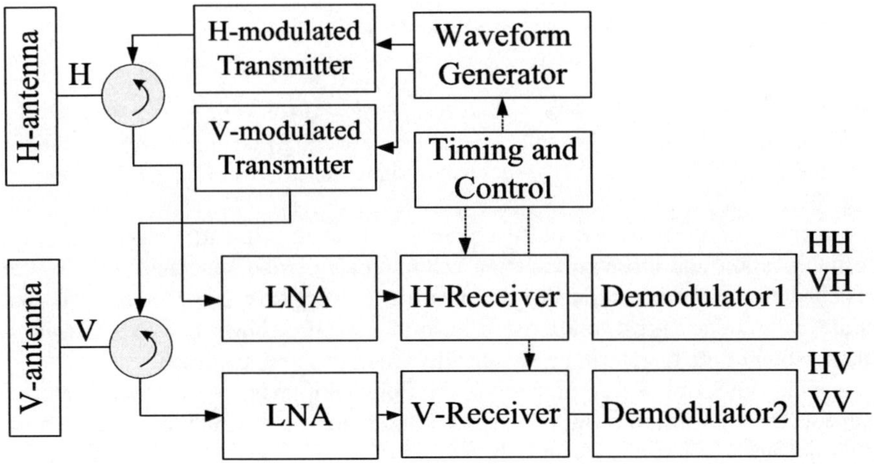

Fig. 3.17 The schematic diagram of code-division PolSAR system

If $t_H(t)$ and $t_V(t)$ is orthogonal, the polarization parameters of the target can be obtained by the matched filter. In fact, the condition of Eq. (3.49) is too ideal to be possible, and it can only be approximated to some extent.

For the practice code-division PolSAR, LFM signal can still be considered. Suppose the two LFM pulses have the same frequency bandwidth but have opposite chirp rate:

$$\begin{aligned} t_H(t) &= rect(t/T_p) \cdot \exp(jkt^2/2) \\ t_V(t) &= rect(t/T_p) \cdot \exp(-jkt^2/2) \end{aligned} \tag{3.50}$$

where $rect(t/T) = \begin{cases} 1 & |t| \le T_p \\ 0 & |t| > T_p \end{cases}$.

Equation (3.50) shows that the bandwidth of two LFM pulses are same, $B = kT_p/2\pi$, and the gain ratio of pulse compression is $R_c = TB$.

$$|R_{HH}(\tau)| = |R_{VV}(\tau)| = T_p^2 \left| \sin c[\pi R_c \frac{\tau}{T_p} \cdot f(\tau)] \cdot f(\tau) \right| \tag{3.51}$$

$$\begin{aligned} |R_{HV}(\tau)| = |R_{VH}(\tau)| = \frac{T_p}{\sqrt{R_c}} \Bigg| & \cos\left[\frac{\pi}{2} R_c (\frac{\tau}{T_p})^2\right] \cdot C_i\left[\sqrt{R_c} \cdot f(\tau)\right] \\ & + \sin\left[\frac{\pi}{2} R_c \left(\frac{\tau}{T_p}\right)^2\right] \cdot S_i\left[\sqrt{R_c} \cdot f(\tau)\right] \\ & + j \left\{ \begin{array}{l} \sin\left[\frac{\pi}{2} R_c (\frac{\tau}{T_p})^2\right] \cdot C_i\left[\sqrt{R_c} \cdot f(\tau)\right] \\ - \cos\left[\frac{\pi}{2} R_c \left(\frac{\tau}{T_p}\right)^2\right] \cdot S_i\left[\sqrt{R_c} \cdot f(\tau)\right] \end{array} \right\} \Bigg| \end{aligned} \tag{3.52}$$

where $f(\tau) = 1 - \frac{|\tau|}{T_p}$, $|\tau| \le T_p$, $C_i(x) = \int_0^{\pi} \cos(\frac{\pi}{2}t^2)dt$, $S_i(x) = \int_0^{\pi} \sin(\frac{\pi}{2}t^2)dt$.

It can be seen that the LFM pulses with opposite chirp rate approximately satisfy the condition of Eq. (3.50). The sequence of the code-division PolSAR is shown in Fig. 3.18.

Figure 3.19 presents the pulse compression results. It can be seen that the co-polarization and cross-polarization echo signals can be separated well by the matched filters corresponding to the positive and negative slope of LFM pulses while maintaining a good compression performance. This shows that the polarization information of the target can be obtained by using this architecture.

Ideally, AASR and RASR of the code-division multiplexing pulses PolSAR are the same as those of the frequency-division architecture. Meanwhile, this architecture does not need to switch, which is beneficial to improve the life and reliability of PolSAR system. This is very important for spaceborne SAR.

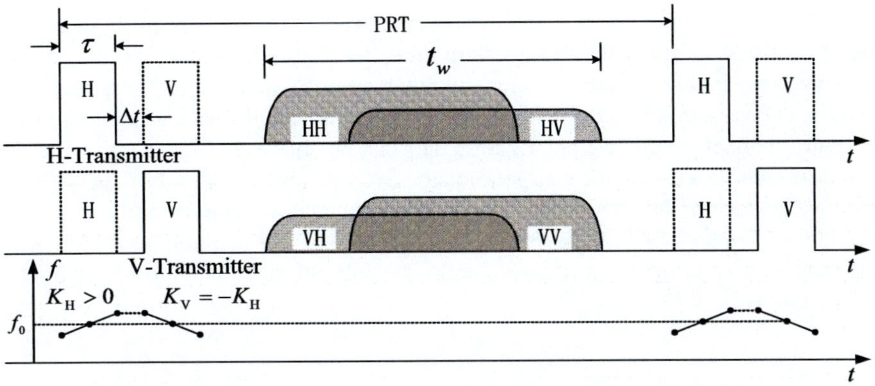

Fig. 3.18 The sequence diagram of code-division multiplexing pulses

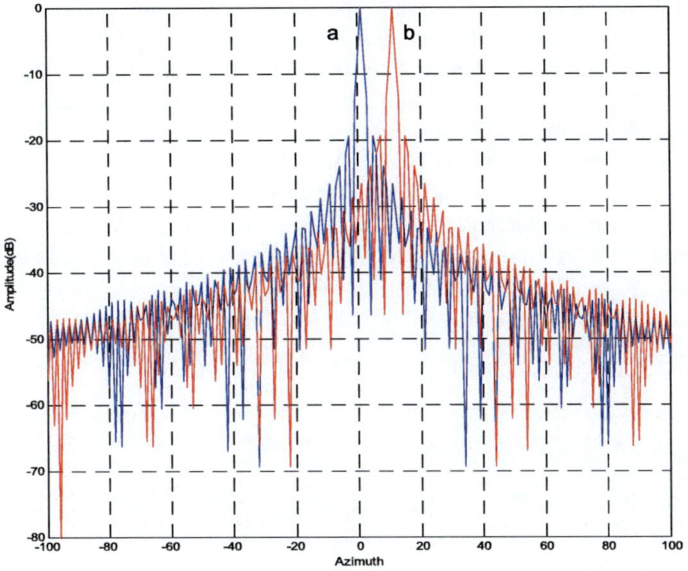

Fig. 3.19 The pulse compression output of co-polarization (**a**) and cross-polarization (**b**)

3.4.4 The Space-Division Twin-Bursts in Azimuth

The polarization architecture of space-division twin-bursts in azimuth has been widely used in many spaceborne PolSAR such as Envisat SAR of ESA, COSMO-SkyMed of ASI, and RiSAT of ISRO. The experimental DRA (Dual Receive Antenna) Mode of TerraSAR of DLR also adapted the space-division twin-bursts architecture to enable quad-polarization measure [9–14].

This polarization architecture divides the antenna aperture into some sub-apertures along the track. During each sub-aperture time, the H polarization or V polarization twin-burst pulses are radiated through the polarization switch to the H or V antenna, and the echoes are simultaneously received by the H and V dual channels or H or V single channel, so the full-polarization data can be obtained. This operating mode simplifies the structure of the full-polarization SAR system, reduces the data rate and the cost. The principle of the space-division twin-bursts in azimuth is similar to that of ScanSAR. The ScanSAR expands the swath width at cost of azimuth resolution, and the space-division architecture also obtains the full polarization at cost of azimuth resolution.

3.4.4.1 The Principle of ScanSAR [15]

ScanSAR is a common mode for wide swath. In a synthetic aperture time, the antenna aperture is divided into several sub-apertures. By means of the antenna beam, cyclically scanning the multiple sub-swathes of range dimension in the interval of sub-aperture times and the wide imaging swath can be obtained. Figure 3.20 shows the diagram of the antenna beam scanning between the three sub-swathes in the range dimension. The scanning order is 1–2–3.

At the beginning, the antenna beam points to the first sub-swath and transmits burst pulses at PRF in the interval of the dwell time T_{d1}, receives the echo of target. And then, the beam is switched to the second sub-swath and dwells in it for the time of T_{d2}. At last, the beam is switched to the third sub-swath and dwells in it for the

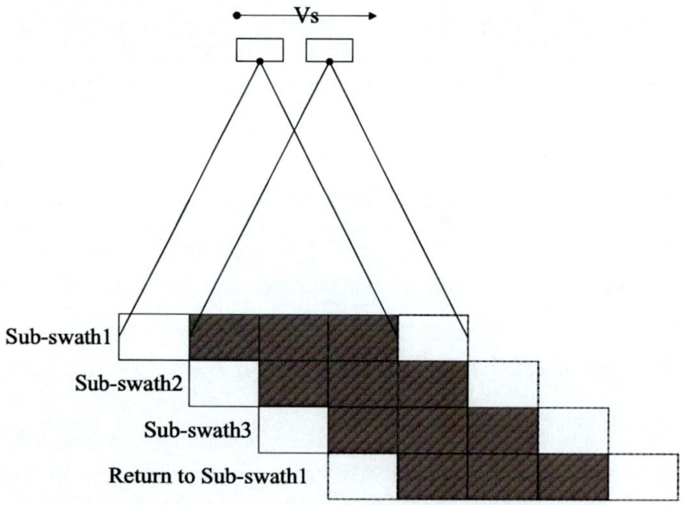

Fig. 3.20 The diagram of three beams (1–2–3) scanning in range dimension

time of T_{d3}. After scanning the three sub-swathes (1–2–3), the antenna beam returns to the first sub-swath, and a new cycle begins.

The time sequence parameters are as follows:

Imaging cycle T_{Fj}, in common is the 3 dB width of azimuth beam, for general SAR, it is synthetic aperture time.

The dwelling time T_{dj} is the continuous imaging time in one sub-swath. It is the time from the beam illuminating into one sub-swath to leaving the sub-swath.

The returning time T_{Rj} is a time interval between double continuous imaging for each imaging swath. It is a time interval which the beam from the sub-swath scanning once again returning to that sub-swath.

For all the sub-swathes, Eq. (3.53) must be satisfied

$$\sum_{j=1}^{N_s} T_{dj} \leq T_{Rj} \tag{3.53}$$

N_s is the number of sub-swathes. T_{dj} is the dwelling time in the jth sub-swath, and T_{Rj} is the returning time from the antenna beam illuminating the jth sub-swath to returning this sub-swath.

The azimuth resolution of ScanSAR is limited by

$$\rho_{opt} = \frac{D_a V_g}{1.7718 V_s}\left(\frac{\sum_{j=1}^{N_s} R_j}{R_k} + 1\right) \tag{3.54}$$

Therefore, the deterioration ratio of azimuth resolution of ScanSAR to that of conventional SAR can be expressed as

$$\frac{\rho_{scan}}{\rho_{strip}} \geq \frac{V_g}{V_s}\left(\frac{\sum_{j=1}^{N_s} R_j}{R_k} + 1\right) \tag{3.55}$$

For low orbit SAR, the ground velocity v_g of the satellite equals to the flight velocity v_s approximately. When the swath is narrow, the slant range $R_j \approx R_k$, and $\frac{\rho_{scan}}{\rho_{strip}} \geq (N+1)$. Therefore, It can be seen that under idea single look condition, the single-look azimuth resolution of ScanSAR is worse than $N+1$ times that of conventional StripSAR.

For ScanSAR with N sub-swathes, the synthetic aperture time T_s needs to be divided into $N+1$ parts in the azimuth dimension, and the average dwelling time of each sub-swath is $T_s/(N+1)$. The single-look azimuth resolution of ScanSAR is $(N+1)$ times worse than that of conventional StripSAR, but the swath width is extended to N times. It can be seen that ScanSAR expands the swath at a cost of the degradation of azimuth resolution.

3.4.4.2 The Principle of the Space-Division Twin-Bursts in Azimuth [17–21]

The principle of the space-division twin-bursts in azimuth is similar to that of ScanSAR, and the synthetic aperture time is also divided into several segments. The difference is that the antenna switches the polarization states along with the flight in azimuth instead of the sub-swathes in range dimension. During the segmented dwelling time, the radar transmits H or V polarization bursts alternately and receives the co- and cross-polarization echoes through the dual channels or single channel. According to the number of the receiving channels, the space-division architecture has two configurations on receive, dual channels and single channel.

(1) Space-division twin-bursts mode with dual channels on receive

The space-division PolSAR with dualcchannels on receive is composed of the orthogonal polarization (H and V) antenna, a transmitter and a polarization switch, dual orthogonal receivers, and dual data processors. The schematic diagram is shown in Fig. 3.21.

The twin-bursts are transmitted sequentially to the H polarization antenna or V polarization antenna by means of controlling the polarization switch $K1$. The HH and VH polarization echoes are received by the H polarization channel, and HV and VV polarization echoes are received by the V polarization channel. Finally, the HH/HV/VH/VV full-polarization data are obtained by the data processors.

Figure 3.22 shows the sequence and polarization state of space-division PolSAR system with dual-receivers. The synthetic aperture time T_s in azimuth should be divided into three segments, and the dwell time of each segment should be $T_s/3$. The radar transmits H polarization and V polarization burst alternately within the

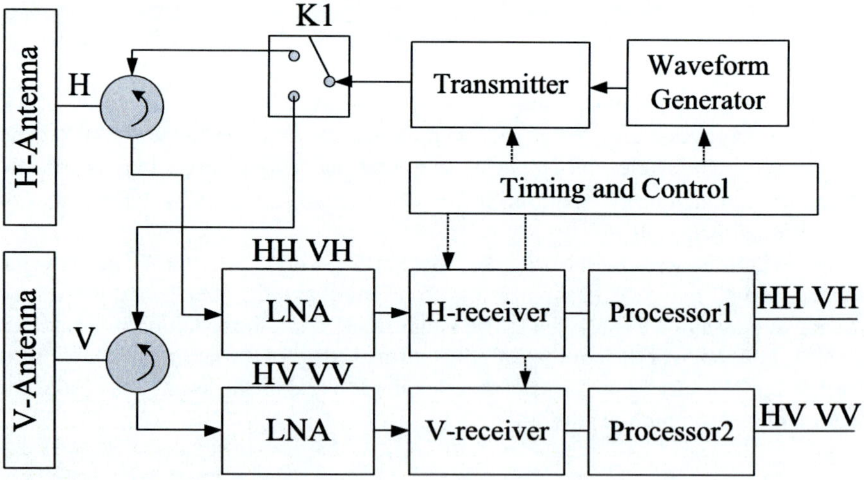

Fig. 3.21 The schematic diagram of space-division PolSAR with dual receivers

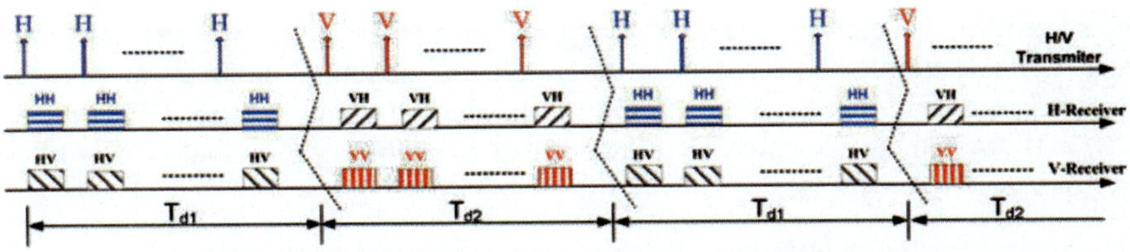

Fig. 3.22 The sequence and polarization state of space-division PolSAR with dual receivers

dwelling time T_{d1} and T_{d2} of adjacent beams and receives H polarization and V polarization echoes simultaneously.

Figure 3.23a shows the imaging segments in the dwelling time T_{d1} on the condition of the H polarization transmission and H and V dual-polarization reception. It can be seen that HH and HV polarization images can be continued seamlessly in multiple dwelling times T_{d1} in the azimuth dimension.

Figure 3.23b shows the imaging segments in the dwelling time T_{d2} on the condition of the V polarization transmission and H and V dual-polarization reception. It can be seen that HV and VV polarization images can be continued seamlessly in multiple dwelling times T_{d2} in the azimuth dimension.

The above analysis shows that the HH/HV/VH/VV fully polarimetric SAR image of the same segment can be obtained quasi-simultaneously by means of the alternate

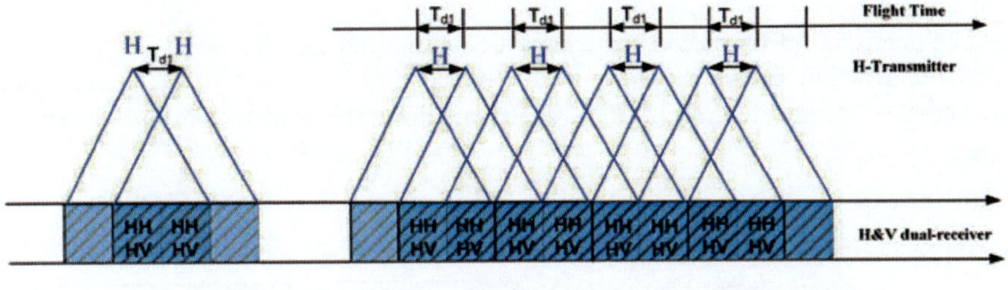

(a) H-Transmitting and H&V dual-receiving in the dwelling time T_{d1}

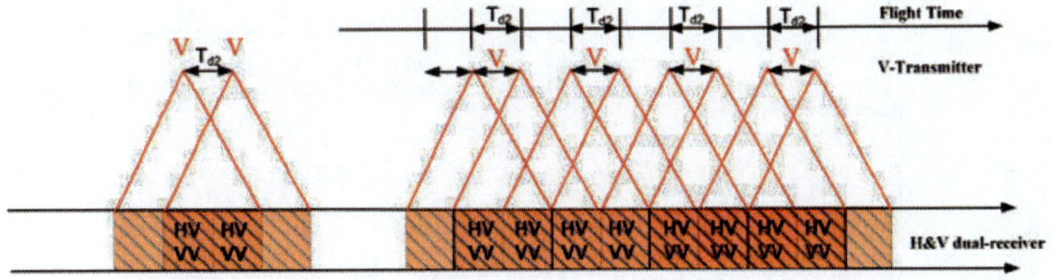

(b) V-Transmitting and H&V dual-receiving in the dwelling time T_{d2}

Fig. 3.23 The polarization states of the imaging segments in azimuth transmission of H polarization and V polarization burst in adjacent dwelling times and dual channels simultaneous reception of H polarization and V polarization echoes. However, the single-look azimuth resolution of the fully polarimetric (HH/HV and VH/VV) SAR image is three times worsened than that of the dual-polarization (HH/HV or VH/VV) SAR.

(2) Space-division twin-bursts mode with single channel on receive

The space-division PolSAR with single channel on receive is composed of the orthogonal polarization (H and V) antenna, a transmitter, a T/R ($K1/K2$) Switch, a receiver and a data processor. The schematic diagram is shown in Fig. 3.24.

The twin-bursts are sent sequentially to the H polarization or V polarization antenna through the transmit switch $K1$. The reflected echo signal is received by H polarization and V polarization antenna by means of controlling the receive switch $K2$. The HH orVH polarization echo received by the H polarization antenna and HV or VV polarization echo received by the V polarization antenna are sent to single coherence receiver. After data processing, the HH/HV/VH/VV full-polarization data can be obtained finally.

Figure 3.25 shows the sequence and polarization state of space-division PolSAR with single channel on receive. The synthetic aperture time T_s in azimuth should be divided into five segments, and the dwelling time of each segment should be $T_s/5$.

The radar transmits H and V polarization burst alternately within the adjacent dwelling time of $T_{d1} + T_{d2}$ and $T_{d3} + T_{d4}$, and receives H or V polarization echo, respectively.

The polarization states and sequences are configured as follows:

H polarization transmission in the dwelling time $T_{d1} + T_{d2}$, H polarization reception in the dwell time of T_{d1}, V polarization reception in the dwell time of T_{d2}.

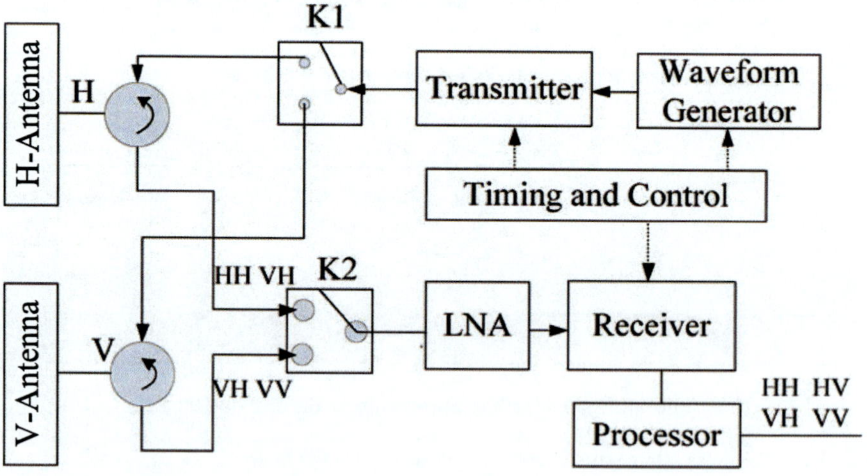

Fig. 3.24 The schematic diagram of space-division PolSAR with single channel

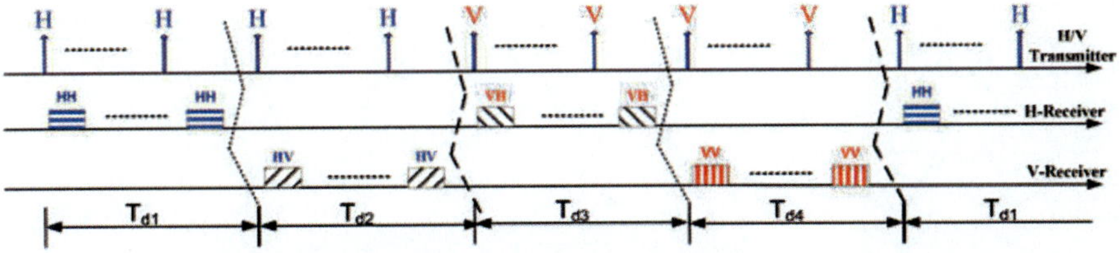

Fig. 3.25 The sequence and polarization state of space-division PolSAR with single receiver

V polarization transmission in the dwelling time $T_{d3} + T_{d4}$, H polarization reception in the dwell time of T_{d3}, V polarization reception in the dwell time of T_{d4}.

Figure 3.26a shows the imaging segment in the dwelling time of T_{d1} on the condition of the H polarization transmission and H polarization reception. HH polarization image can be continued seamlessly within the adjacent dwelling time of T_{d1} in the azimuth dimension.

Figure 3.26b shows the imaging segment in the dwelling time of T_{d2} on the condition of the H polarization transmission and V polarization reception. HV polarization image can be continued seamlessly within the adjacent dwelling time of T_{d2} in the azimuth dimension.

Figure 3.26c shows the imaging segment in the dwelling time of T_{d3} on the condition of the V polarization transmission and H polarization reception. VH polarization image can be continued seamlessly within the adjacent dwelling time of T_{d3} in the azimuth dimension.

Figure 3.26d shows the imaging segment in the dwelling time of T_{d4} on the condition of the V polarization transmission and V polarization reception. VV polarization image can be continued seamlessly within the adjacent dwelling time of T_{d4} in the azimuth dimension.

The above analysis shows that the HH/HV/VH/VV full-polarization SAR images in the same segment can be obtained quasi-simultaneously by means of the alternate transmission of H and V polarization burst in adjacent dwelling times and in turn reception of H and V polarization echo on single channel. The azimuth resolution of the full-polarization (HH/HV/VH/VV) SAR image is five times worsened than that of the single polarization (HH or HV or VH or VV) SAR.

The space-division PolSAR with single receiver is more flexible, which can flexibly configure the dwelling time and transmitting and receiving polarization states according to the polarization and resolution requirements. Figure 3.26 shows the HH/HV/VH/VV full-polarization states of the imaging segments in azimuth. If dual-polarization HH/VV or HH/HV or VH/VV radar images are required, only need to adopt Fig. 3.26a–d polarization combinations. If HH/VV image is needed, the polarization configuration of (a) and (d) can be used. If the HH/HV image is needed, the polarization configuration of (a) and (b) can be used. If the VH/VV image is needed, the polarization configuration of (c) and (d) can be used. For the configurations mentioned above, the single-look azimuth resolution is three times worsened.

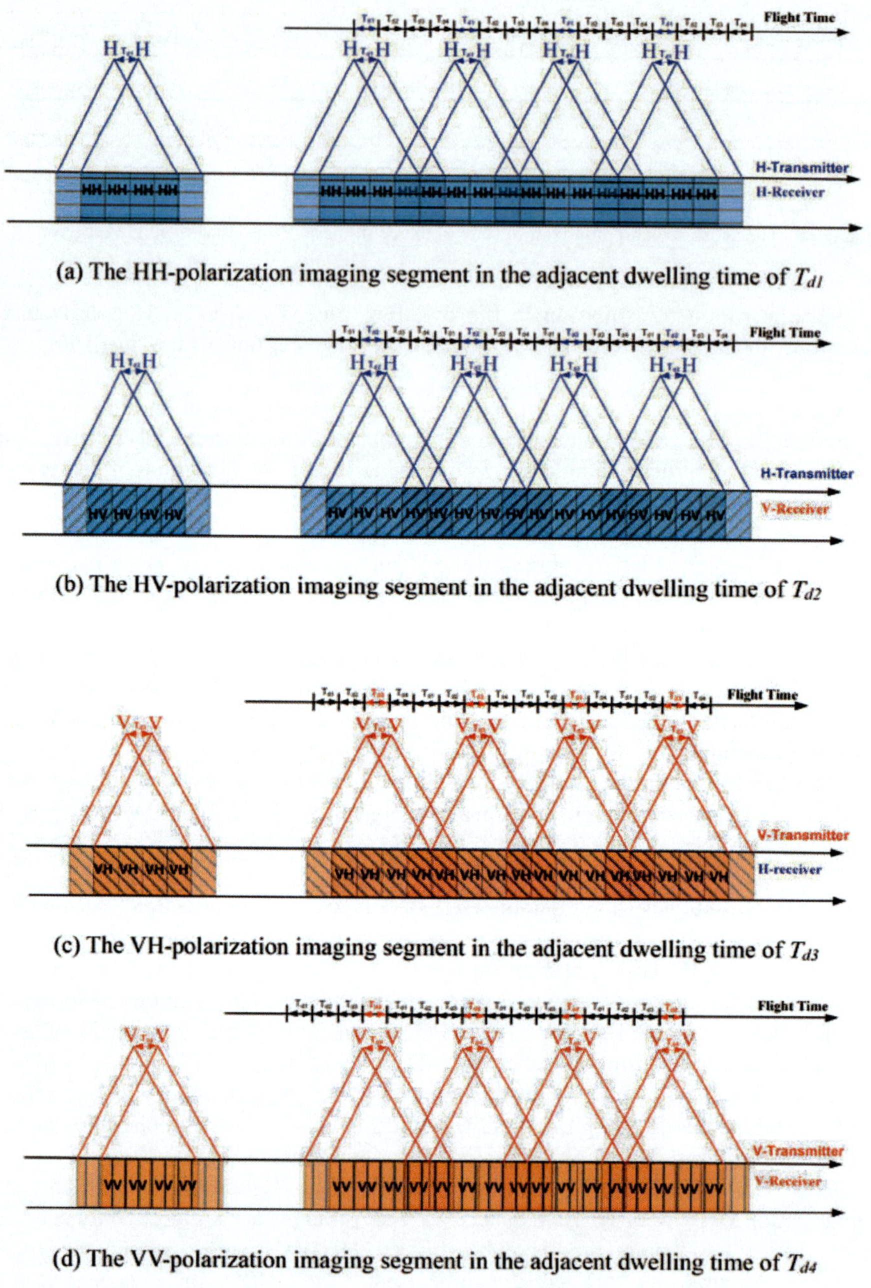

(a) The HH-polarization imaging segment in the adjacent dwelling time of T_{d1}

(b) The HV-polarization imaging segment in the adjacent dwelling time of T_{d2}

(c) The VH-polarization imaging segment in the adjacent dwelling time of T_{d3}

(d) The VV-polarization imaging segment in the adjacent dwelling time of T_{d4}

Fig. 3.26 The polarization states of the imaging segments in azimuth

3.4.5 Compact Polarimetry Synthetic Aperture Radar

In order to overcome the disadvantages of the full-polarization SAR systems mentioned above, reduce the system complexity and trade off the parameter selection such as swath width, PRF, data rate, and ambiguity ratio, C-PolSAR has been developing rapidly in recent years. C-PolSARSAR only transmits single-polarization signal and receives the orthogonal polarization signals and corresponding phases through the dual channels. The 2×2 covariance matrix was measured and used as a four-element Stokes vector to reconstruct the full-polarization data. The main modes of Compact Polarimetry (CP) are the π/4 mode, the DCP mode and the CTLR mode [22–25].

3.4.5.1 The π/4 CP Mode

The π/4 mode C-PolSAR consists of the H and V dual-polarization antenna, a transmitter, a phase shifter, H and V dual-polarization coherent receivers and dual data processors. The schematic diagram of the π/4 CP mode is shown in Fig. 3.27.

In the π/4 CP mode, the H and V polarization antenna is excited by the electromagnetic wave shifted by 0°, the superposition of linear H and V polarization results in a linear polarization oriented 45° with respect to the H polarization in space. The π/4 polarization echo signals reflected from the ground surface are passing through the H polarization and V polarization antenna, and going into the H polarization and V polarization dual orthogonal receivers, respectively. By means of data processing, the 2×2 average covariance matrix is obtained, which can be converted to four-element Stokes vector. Then full-polarization data can be reconstructed.

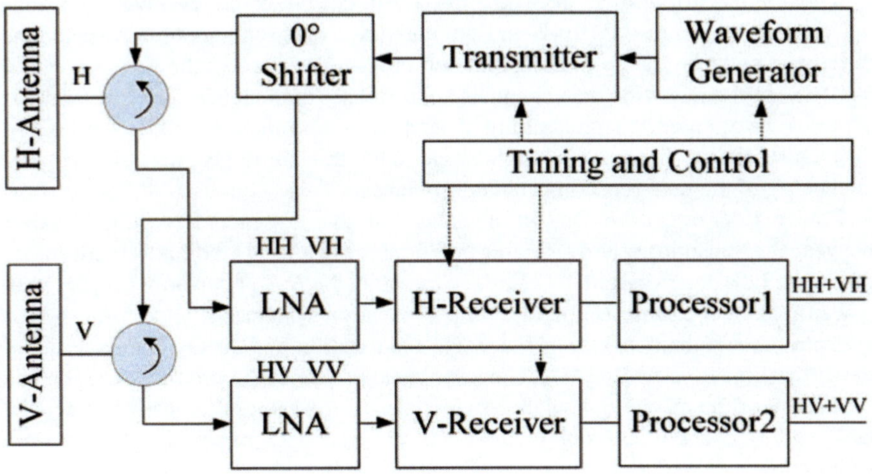

Fig. 3.27 The schematic diagram of the π/4 CP mode

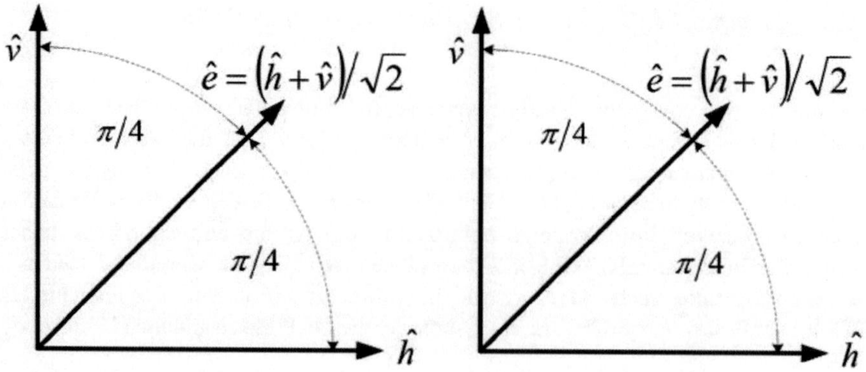

Fig. 3.28 The relationship between the transmitting and receiving signals in π/4 mode

The relationship between the transmitting and receiving signals in π/4 model is shown as Fig. 3.28. The H and V two linear polarization signals of receiving are located at π/4 on both sides of the transmitting π/4 linear polarization signal.

Compared to full-polarization SAR system, the π/4 mode increases the preserving capability of the information greatly. Based on the assumption of symmetry of terrain, the reconstruction method of full-polarization data for distributed target in π/4 mode is obtained.

The main characteristics of the π/4 mode are as follows:

PRF and Imaging swath width: the full-polarization model needs the dual-polarization transmitting which increases system complication greatly. Comparing with the standard single polarization system, the full-polarization model needs double PRF. In order to suppress the range ambiguity, the swath width has to be decreased to half of single-polarization system. However, PRF and swath width in the π/4 mode are as same as that of the single-polarization SAR.

The signal imbalance between the dual channels on receive: The full-polarization SAR transmits the H and V polarization electromagnetic waves, and the difference between the co-polarization and cross-polarization echo power is about 6–10 dB, so the receiving dual channels are seriously imbalanced. The π/4 model is adopting the oriented 45° transmitting linear polarization signals so that the echoes of the dual receiving channels are symmetrical to the transmitting signal, which makes the energy of the dual receiving channels balanced.

Power Consumption: One disadvantage of the π/4 model is the imbalance between the transmitting and receiving fields, which results in a 3 dB power consumption and a 3 dB deterioration of $NE\sigma^0$. To maintain the $NE\sigma^0$ performance, the peak power must be the same as that of a full-polarization system.

Data rate: If the data rate of the single-polarization SAR is D_R, the data rate of the full-polarization SAR is $4D_R$. However, the data rate of the π/4 mode is $2D_R$, and the energy of the receiving dual channels is almost the same. By means of a special BAQ coding data compression method, the real-time data processing on-board can be implemented.

3.4.5.2 The Dual Circular Polarimetric (DCP) Mode

The DCP mode C-PolSAR consists of the H and V dual-polarization antenna, a transmitter, a transmitting phase shifter, two receiving phase shifters 1 and 2, the H and V dual-polarization coherent receivers, and dual data processors. The schematic diagram of the DCP mode is shown in Fig. 3.29.

In the DCP mode, the H and V polarization antennas are excited by the electromagnetic wave phase shifted by +90° (or −90°), the linear superposition of the antenna patterns results in a right circle polarization (or left circle polarization). The echo is received by the H and V dual-polarization antennas and dual orthogonal receivers.

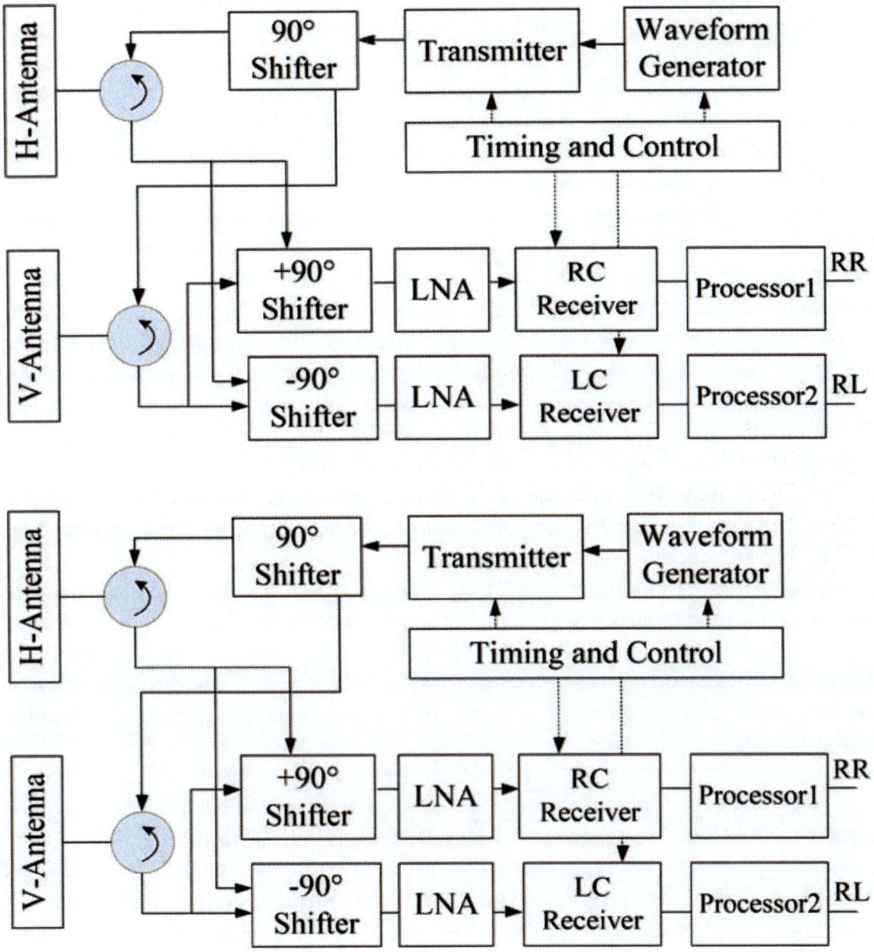

Fig. 3.29 The schematic diagram of the DCP mode

The echo received by the H polarization antenna is phase shifted by +90° through the first receiving phase shifter 1, the right circle polarization signal is obtained and sent to the right circle polarization receiver.

The echo received by the V polarization antenna is phase shifted by −90° through the receiving phase shifter 2, the left circle polarization signal is obtained and sent to the left circle polarization receiver.

The 2×2 average covariance matrix will be converted to four-element Stokes vector for the reconstruction of full-polarization data by dual data processors.

If the backscattered field is rotationally invariant related to the scene geometry, transmitting the circular polarization is required. Corresponding to the circular polarization transmitting field, the selected mode of CP is the coherent orthogonal DCP mode on receive. The DCP mode has a wide inheritance in radar astronomy. A famous example of the circular polarization transmission and orthogonal circular polarization reception is the Arecibo Radar telescope.

The DCP mode had good results in the analysis of backscatters based on Stocks matrix parameters. The main Stocks parameters are polarization purity *m*, circle polarization ratio, the relative phase of backscattering field with respect to two linear electric field vectors, etc. The formula of Stokes parameters based on the circular polarization basis usually corresponds to the normal case.

The received circular polarization echo is in inverse phase with the transmitted signal. For example, if the right circular polarization signal is transmitted, in general, the backscattering echo of the left circular polarization backscattering is more powerful. So that, the right circle polarization becomes the cross-polarization state on receive. This is mainly due to the fact that these odd scatterings from mirror scattering, Bragg scattering and trihedral angle scattering are usually dominant. The circular polarization is more sensitive to the relative phase of the even scattered echo. The phase difference between the ideal dihedral angle scattering (lossless) and the single surface scattering or odd surface scattering is 180°.

The DCP mode has superiority in Stokes parameter calculation, the power balance between the receiving dual channels, crosstalk of dual-polarization models, calibration, and decomposition strategy, etc.

The transmitting circular polarization, receiving DCP will be an optional polarization structure for spaceborne SAR. The specific contents considered in the design are including the measurement of circular polarization ratio, the maximum voltage that can distinguish different scattering types, the robustness of random orientation to the distribution state of dihedron scatters, also including the minimization of weight and power, etc.

3.4.5.3 The Circular Transmit Linear Receive (CTLR) Mode

The CTLR mode C-PolSAR consists of the H and V dual-polarization antennas, a transmitter, a 90° phase shifter on transmit, the H and V dual-polarization coherent receivers, and dual data processors. The schematic diagram of the CTLR mode is shown in Fig. 3.30.

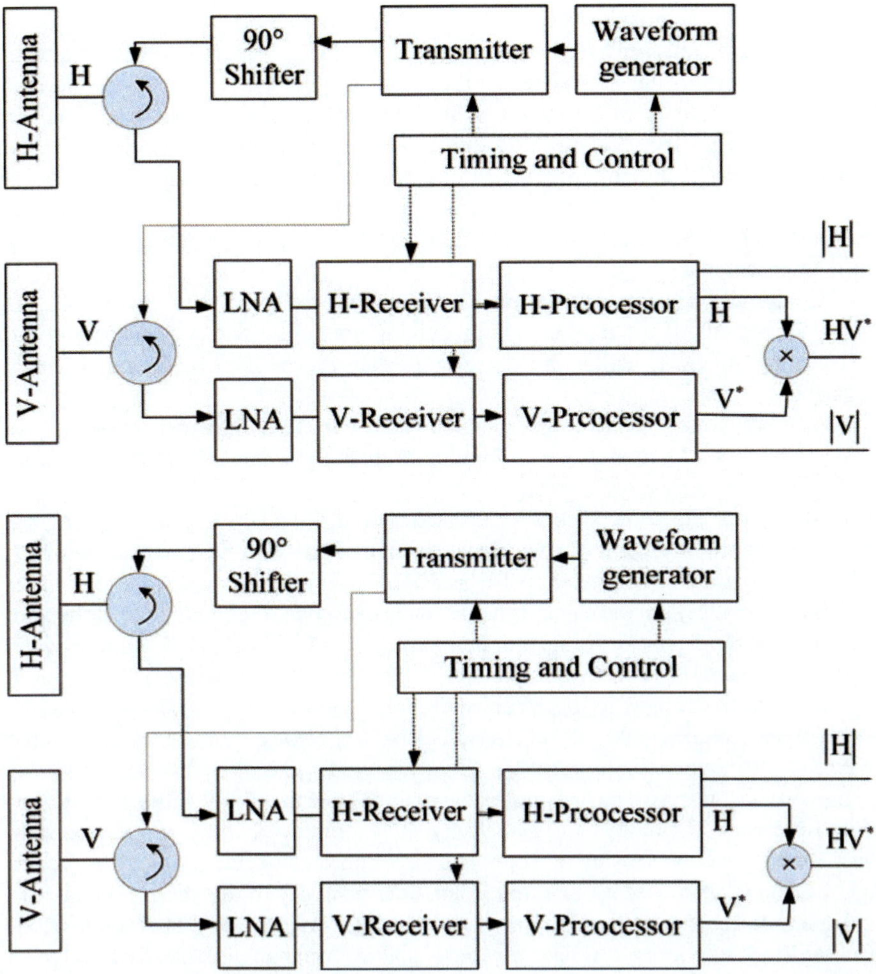

Fig. 3.30 The schematic diagram of the CTLR mode

In the CTLR mode, the H and V polarization antennas is excited by the electromagnetic wave shifted by +90° (or −90°) at the same time, the superposition of the linear H and V polarization result in a right (or left) circle polarization. The scattered echo is received by the H and V dual-polarization antennas and dual orthogonal receivers, respectively. Through data processing, the 2×2 average covariance matrix can be obtained, and the four-element Stokes vector reconstruction of full-polarization data is realized.

The main characteristics of the CTLR mode are as follows:

Stokes parameter calculation: Since the circular polarization on transmitting is rotationally invariant in relation to the scene geometry, the four parameters of Stokes can be calculated by means of the dual-polarization echoes based on the coherent linear polarization basis. The data product preserves all information of the backscattered field, and independent of the observed polarization basis. The Single-Llook Complex (SLC) data can also be synthesized to produce the Right Circular Polarization (RCP) image or the Left Circular Polarization (LCP) image directly by the data processors.

The signal imbalance between the dual-channel receiving: In the CTLR mode, the signals of the dual-channel receiving are symmetric with respect to the transmitting signal. It makes the energy collected on the dual receiving channels balanced.

Crosstalk: Since the average signal levels of the dual-channel receiving are similar, the risk of crosstalk or ambiguity from the stronger polarization leaking to the weaker polarization is small.

Calibration: Since the second-order statistics of the dual-channel receiving are the same, and independent of the geometric characteristics of the backscatter cell. When the backscatter coefficients of H and V polarizations are equal, the first-order statistics (averages) are the same. If deviation is caused by the imbalance of the gain or spectrum, the deviation value can be measured and compensated. Similarly, the relative phases of the H and V components in the transmitting (circular) polarization, and the random changes in the received H and V polarization components, can be detected and compensated, respectively. This self-checking characteristic is unique to the CTLR mode.

Sensitivity for errors: The significant superiority of the CTLR mode is the sensitivity for errors. The analysis of sensitivity show, compared with the tradition ratio between co-direction sensing and opposite direction sensing, the ratio of the circular polarization is decreased by two times than the sensitivity of the channel balance.

Instrument complexity:Compared with the DCP model, the CTLR mode needs less hardware of equipment, lighter weight, and less channel mismatched source. It is an optimum architecture of a system for the planetary or lunar mission.

Decomposition strategy: The data of the circular polarization transmitting linear polarization receiving model are represented completely by the 2×2 matrix, which is conducive to improve the decomposition strategy. Similar to the $H - \alpha$ decomposition of the full-polarization data, the $m - \delta$ decomposition method can be applied to the circular polarization transmitting linear polarization receiving model for data acquisition.

The CTLR mode can simplify the system structure. It needs less hardware, lighter weight and lower power consumption, and it has a lot of superiority in Stokes parameter calculation, channels balance, the robustness to the relative error and crosstalk, etc. The CTLR mode optimizes the relative phase calibration and amplitude calibration, and the echo data is fully represented by the 2×2 matrix, more suitable for the improved decomposition strategy. Therefore, the CTLR mode is the optimal architecture for a lunar or planetary mission and it is an optional attracted technology for the Ground Penetration Radar (GPR). Mini-SAR carried by the lunar prober Chandraayan-1 launched by India in 2008 and the mini-RF loaded by LRO (Lunar Reconnaissance Orbiter) launched by the United States in 2009, both radars also adopted the CTLR mode [16].

3.4.6 Conclusion

The earliest airborne and low-orbit spaceborne PolSAR systems in the world all adopted the time-division multiplexing pulses to obtain the multi-polarization information, Since the simple structure and easy implementation. However, as soon as the orbit is higher, for a large incident angle, to obtain the wide imaging swath this operating architecture becomes very difficult.

The frequency-division multiplexing pulses and code-division multiplexing pulses both operating architectures can solve the range ambiguity problem of spaceborne PolSAR to a certain extent. However the frequency-division PolSAR system requires the double signal bandwidth and the high-quality band-pass filter for separating the signals, its implementation is more difficult. In contrast, the code-division multiplexing pulses can also effectively suppress the range ambiguity without increasing the signal bandwidth and data rate, but the orthogonal-coded signal generator and the improved transmitter are needed, so the difficulty of implementation is increased greatly. Although both operating architectures overcome their shortcomings to some extent, the complexity of the PolSAR system also is increased correspondingly. The principle of the space-division twin-bursts in azimuth is similar to that of ScanSAR, and the full-polarization is obtained at cost of azimuth resolution.

The C-PolSAR is transmitting a single polarization and receiving two orthogonal polarization echoes and their relative phases. The 2×2 covariance matrix is measured and used as a four-element Stokes vector to reconstruct for the full-polarization data. The C-PolSAR overcomes the shortcomings of the operating architectures mentioned above, reduces the system complexity and cost, and trade off the parameter selection such as the imaging swath width, data rate of downlink, PRF and ambiguity, etc. Table 3.2 compares the operating architectures of PolSAR systems.

Table 3.2 Compares the operating architectures of PolSAR systems

Operating architectures	Antenna	Transmit	Receive	PRF	Band width	Data rate	Output
Time-division	H and V dual-polarization antenna	Single-channel polarization switch alternately transmitting H and V polarization waves, transmitting energy is double	Dual-channel coherent receiving	2PRF	B	$4D_R$	HH/HV/VH/VV full-polarization 2 × 2 scattering matrix imaging swath is reduce
Frequency-division	H and V dual-polarization antennas	Transmitting frequency dividing signals of two channels at the same time	Dual-channel coherent receiving	PRF	≥ 2B	$4D_R$	HH/HV/VH/VV full-polarization 2 × 2 scattering matrix
Code-division	H and V dual-polarization	Transmitting orthogonal coding signals of two channels	Dual-channel coherent receiving	PRF	B	$2D_R$	HH/HV/VH/VV full-polarization 2 × 2 scattering matrix
Space-division	H and V dual-polarization antennas	Single-channel polarization switch alternately transmitting H and V polarization waves, according to space division	Single-channel coherent receiving	PRF	B	D_R	HH/HV/VH/VV full-polarization 2 × 2 scattering matrix The azimuth resolution is deteriorated about five times
	H and V dual-polarization antennas	Single-channel polarization switch alternately transmitting H and V-polarization waves, according to space division	Dual-channel coherent receiving	PRF	B	$2D_R$	HH/HV/VH/VV full-polarization 2 × 2 scattering matrix The azimuth resolution is deteriorated about 3 times

(continued)

Table 3.2 (continued)

Operating architectures	Antenna	Transmit	Receive	PRF	Band width	Data rate	Output
$\pi/4$ CP	H and V dual-polarization antennas	Transmitting $\pi/4$ linear polarization	H and V dual-channel coherent receiving	PRF	B	$2D_R$	2×2 covariance matrix, reconstruct the full-polarization scattering matrix
DCP CP	H and V dual-polarization antennas	Transmitting LC polarization or RC polarization	two-channel coherent receiving LC and RC polarization	PRF	B	$2D_R$	
CTLR CP	H and V dual-polarization antenna	Transmitting LC polarization or RC polarization	H and V linear polarization two-channel coherent receiving	PRF	B	$2D_R$	

Note PRF, is the Pulse Repetition Frequency; B and is the signal bandwidth; D_R is the data rate of the single-polarization SAR

References

1. Y. Wu, The current status and implications of polarimetric SAR interferometry. J. Electron. Inf. Technol. **29**(5), 1258–1262 (2007)
2. M.E. Nord et al., Comparison of compact polarimetric synthetic aperture radar modes. IEEE Trans. Geosci. Remote Sens. **47**(1), 174–188 (2009)
3. P. Lacomme, J.-P. Hardange, J.-C. Marchais, E. Normant, *Air and Spaceborne Radar System: An Introduction* (William Andrew Pubilishing, LLC, New York, USA, 2001)
4. D.R. Wehner, *High Resolution Radar* (Artech House, Boston, London, 1995)
5. J.C. Clurlander, *Synthetic Aperture Radar System and Processing*, California Institute of Technology, JPL, Pasadena, California (Wiley, 1991)
6. C. Zhang, S.A.R. Principle, *System Analysis and Application* (Science Press, Beijing, 1989)
7. P. Dubois, D. Evans, A. Freeman, J. van Zyl, Approach to derivation of SIR-C science requirements for calibration, in *IEEE Proceedings of IGARSS'89*, pp. 243–246
8. A. Freeman, M. Alves, B. Chapman, J. Cruz, Y. Kim, S. Shaffer, J. Sun, E. Turner, K. Sarabandi, SIR-C data quality and calibration results. IEEE Trans. Geosci. Remote Sens. **4**(33), 848–857 (1995)
9. Y.-L. Desnos, C. Buck, J. Guijarro, G. Levrini, J.-L. Suchail, R. Torres, H. Laur, J. Closa, B. Rosich, The Envisat advance synthetic aperture radar system (IEEE, 2000)
10. R.K. Hawkins, R. Touzi, J. Wolfe, P.J. Meadows, ASAR AP mode performance and applications potential (IEEE, 2003)
11. *COSMO-Skymed System Description & User Guide*, 4 May 2007
12. *COSMO-Skymed System Hand Book*, 30 Apr 2007
13. T. Misra, S.S. Rana, V.H. Bora, N.M. Sesai, C.V.N. Rao, Rajeevjyothi, SAR payload of radar imaging satellite (RISAT) of ISRO, in *Proceedings of EUSAR 2006*
14. A. Currie, M.A. Brown, Wide-swath SAR. IEE Proc. F **139**(2), 122–135 (1992)
15. P. Spudis, S. Nozette, B. Bussey, K. Raney, H. Winters, C.L. Lichtenberg, W. Marinelli, J.C. Crusan, M.M. Gates, Mini-SAR: an imaging radar experiment for the Chandrayaan-1 mission to the Moon. Curr. Sci. **96**(4), 533–539 (2009)
16. B. Dai, Research on multi-polarimetric synthetic aperture radar system and polarimetric information processing, Ph.D. Thesis of Chinese Academy of Sciences, June 2000
17. R.L. Jordan, B.L. Huneycutt, M. Werner, The SIR-C\X-SAR synthetic aperture radar system. Proc. IEEE **79**(6) (1991)
18. R.L. Jordan, B.L. Huneycutt, M. Werner, The SIR-C/X-SAR synthetic aperture radar system. IEEE Trans. Geosci. Remote Sens. **33**(4), 829–839 (1995)
19. P.A. Fox, A.P. Luscombe, A.A. Thompson, Radarsat-2 SAR modes development and utilization. Can. J. Remote Sens. **30**(3), 258–264 (2004)
20. T. Fujimra, T. Kimura, Guidance and electro-optics division, compact polarimetric observation using phased array antenna and its case study for PALSAR, NEC Corporation, Japan, in *Proceedings of EUSAR 2008*, vol. 1
21. M. Stangl, R. Werninghaus, R. Zahn, TerraSAR-X active phased array antenna (IEEE, 2003)
22. R.K. Raney, A perspective on compact polarimetric, in *IEEE Geoscience and Remote Sensing Society Newsletter*, Sept 2011, pp. 12–18
23. R.K. Raney, Hybrid-polarity SAR architecture, in *CD-ROM Proceedings, IEEE International and Geoscience Remote Sensing Symposium IGARSS*, Denver, CO (2006)
24. R.K. Raney, Hybrid-quad-pol SAR, in *Proceedings, International Geoscience and Remote Sensing Symposium IGARSS*, Boston (IEEE, 2008)
25. M.E. Nord et al., Comparison of compact polarimetric synthetic aperture radar modes. IEEE Trans. Geosci. Remote Sens. **47**, 174–188 (2009)
26. J. Yang, Y. Peng, S. Lin, Similarity between two scattering matrices. Electron. Lett. **37**(3), 193–194 (2001)

Chapter 4
Polarimetric Interferometric Synthetic Aperture Radar

4.1 Overview

Polarimetric Interferometry SAR (PolInSAR) is a new earth observation technology developed on the basis of polarimetric SAR and interferometric SAR technology [1]. It uses the effective combination of polarization and interference information to obtain three-dimensional structural feature information and scattering information of the observation object. PolInSAR achieves the expansion of space observation through the effective combination of interference information and polarization information. It has both the characteristics of interferometric SAR being sensitive to the position, distribution, motion, and change information of scatterers, as well as the characteristics of polarimetric SAR being sensitive to the structure, direction, symmetry, texture, and dielectric of scatterers [2]. Compared with polarimetric SAR and interferometric SAR, PolInSAR can provide richer information, and has huge application potential in fields such as agriculture, forestry, hydrogeography, urban infrastructure, volcanoes, seismology, archeology, and military reconnaissance [3].

The concept of PolInSAR was first proposed by S. R. Cloude and K. P. Papathanassiou. In 1997, they used SIR-C/X-SAR data to study the influence of polarization on the coherence of SAR interferograms and carried out research on the application of polarization interference to optimize the coherence of SAR interferograms [4, 5]. In 1998, the traditional scalar interference was extended to vector interference. At the same time, the coherent optimal theory of polarization interference and the target decomposition theory of polarization interference were proposed [1]. Then the theoretical foundation of polarimetric interference radar remote sensing was built up and a new direction of SAR earth observation was opened up.

The technology of PolInSAR has the incomparable characteristics contrast of conventional SAR, which has broad development prospects [6]. Since its inception, it has received widespread attention and has become a hot spot in the field of SAR remote sensing.

R. Yang et al., *Polarimetric Microwave Imaging*,
https://doi.org/10.1007/978-981-15-8897-6_4

4.2 Basic Principles of Polarimetric Interferometric SAR

The interferometric measurement of polarimetric SAR combines polarization technology and interference technology and uses full polarization observation for interference processing. It combines the characteristics of interference information sensitive to spatial distribution and polarization information sensitive to the physical properties of scatterers. At the same time, the fine structural characteristics of the target and the spatial distribution characteristics are combined to improve the performance of interference applications and to distinguish the vertical distribution information of different scattering mechanisms in the resolution unit.

4.2.1 Basic Principles of Interferometric SAR Measurement

Interferometric SAR is a technology that uses phase information extracted from SAR complex data as an information source to obtain three-dimensional information and change information on the surface [7]. Obtain complex image pairs of the same area on the ground through two antennas simultaneously (single-track mode) or two nearly parallel observations (repeated orbit mode). Due to the position geometric relationship between the target and the two antennas, a phase difference is generated on the complex image to form an interferogram. From the interferogram, three-dimensional information and change information of each point on the image can be extracted.

The repetitive orbital interferometry is taken as an example to introduce the principle of interferometric SAR measurement [8]. Figure 4.1 shows a schematic diagram of radar interferometry. In the figure, the point S_1 and S_2 is the position of the two antennas. The distance between the antennas is expressed by the baseline distance

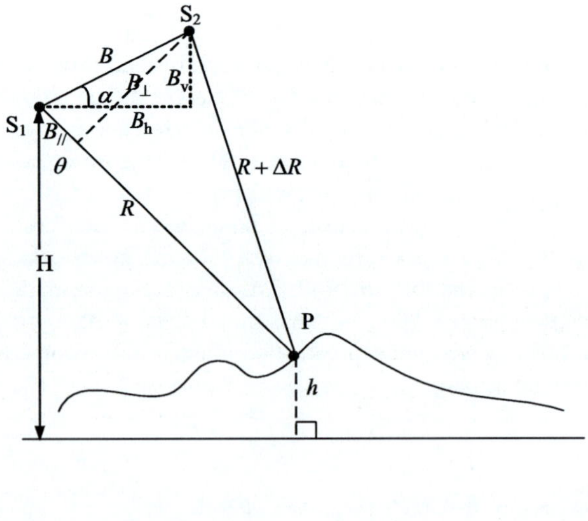

Fig. 4.1 Interferometric SAR imaging geometry

B. The angle between the baseline and the horizontal direction is α. The component of the baseline perpendicular to the Line of Sight (LOS) is $B_{\perp}$, and the component of baseline parallel the LOS is $B_{//}$. The component of baseline parallel to the horizontal direction is B_h, and the component of baseline perpendicular to the horizontal direction is B_v. H is the platform height. h represents the height of the ground target P. The distance from the antenna S_1 and S_2 to the point P is R and $R + \Delta R$.

The signals received at S_1 and S_2 are as follows [8]:

$$u_1(R) = |u_1(R)| \exp(j\phi_1(R)) \tag{4.1}$$

$$u_2(R + \Delta R) = |u_2(R + \Delta R)| \exp(j\phi_2(R + \Delta R)) \tag{4.2}$$

The phase of the received signal consists of two parts, including the phase proportional to the transmission path and the random phase caused by the scattering characteristics, that is,

$$\phi_1 = 2\frac{2\pi}{\lambda}R + \arg\{u_1\} \tag{4.3}$$

$$\phi_2 = 2\frac{2\pi}{\lambda}(R + \Delta R) + \arg\{u_2\} \tag{4.4}$$

where $\arg\{u_1\}$ and $\arg\{u_2\}$ represent the random phase caused by different scattering characteristics. The coefficient "2" indicates two-way transmission and reception. If the single flight mode is used, this coefficient should be removed. λ is the radar wavelength.

After registration of the two complex images, the interferogram can be obtained by multiplying the complex conjugates.

$$v = u_1(R)u_2^*(R + \Delta R) = |u_1||u_2| \exp(j(\phi_1 - \phi_2)) \tag{4.5}$$

In Eq. (4.5), assume that the random phase contribution in the two images is the same, that is, $\arg\{u_1\} = \arg\{u_2\}$. The phase of the interferogram is

$$\phi = -\frac{4\pi}{\lambda}\Delta R + 2\pi N, N = 0, \pm 1, \pm 2, \ldots \tag{4.6}$$

Usually, resolving the phase can only obtain the main value of the phase, so a key step in interference processing is to use the phase unwrapping to obtain the true phase value.

From Fig. 4.1, there is

$$\sin(\theta - \alpha) = \frac{(R + \Delta R)^2 - R^2 - B^2}{2RB} \tag{4.7}$$

Ignore $(\Delta R)^2$ of Eq. (4.7). There is

$$\Delta R = B\sin(\theta - \alpha) \tag{4.8}$$

The baseline B is decomposed along the LOS to obtain the components $B_{\|}$ parallel to the LOS and the components $B_{\perp}$ perpendicular to the LOS as follows:

$$B_{\|} = B\sin(\theta - \alpha) \tag{4.9}$$

$$B_{\perp} = B\cos(\theta - \alpha) \tag{4.10}$$

From Eqs. (4.9) and (4.10), there is

$$\theta = \alpha - \arcsin\left(\frac{\lambda\phi}{4\pi B}\right) \tag{4.11}$$

Therefore, if the platform height, slope distance, angles between the baseline and horizontal direction, and radar angle are known, the height of point P can be determined as follows:

$$h = \mathrm{H} - R\cos\theta \tag{4.12}$$

4.2.2 *Decoherence Effect of Interferometric SAR Measurement*

SAR interference image is the result of coherent processing of echo signals obtained by two antennas detecting the same area. The basis of interference processing is that the echo signals of two observations must have high coherence. This section will analyze the factors affecting the coherence of radar signals.

The coherence of the two complex SAR images s_1 and s_2 can be measured by the complex coherence coefficient [9].

$$\gamma = \frac{E\{s_1 \cdot s_2^*\}}{\sqrt{E\{|s_1|^2\} \cdot E\{|s_2|^2\}}} \tag{4.13}$$

The modulus $|\gamma|$ of the coherence coefficient can be used as a similarity measure of SAR signal. It can be seen from Eq. (4.13) that it is the normalized expected value of the interferogram pixels $i = s_1 s_2^*$. The phase of the coherence coefficient γ is the interference phase and $|\gamma| \in [0, 1]$. If the scattering characteristics of the scatterer are traversed in various states, the spatial average can be used instead of the overall average. Then the N point sample window can be used to estimate the coherence coefficient.

$$\hat{\gamma} = \frac{\sum_N s_1 s_2^*}{\sqrt{\sum_N |s_1|^2 \sum_N |s_2|^2}} \tag{4.14}$$

There are many factors affecting the coherence of two SAR images [8, 10], including SAR system noise (thermal noise and quantization noise), imaging geometry, volume scattering effects, changes in surface scatterers with time, data processing processes, etc. These factors affect the final coherence coefficient.

$$\gamma = \gamma_{\text{SNR}} \gamma_{\text{geom}} \gamma_{\text{temporal}} \gamma_{\text{vol}} \gamma_{\text{process}} \tag{4.15}$$

where γ_{SNR} represents the system noise decoherence, γ_{geom} represents the geometric decoherence, γ_{temporal} represents temporal decoherence, γ_{vol} represents volume scattering decoherence, γ_{process} represents data processing decoherence.

4.2.3 Interferometric SAR Measurement Error Analysis

The elevation inversion error of the interferometric SAR system mainly comes from slope measurement error, baseline estimation error, phase estimation error, platform height measurement error, etc. [11].

From Eqs. (4.7) and (4.9), we obtain

$$\phi = -\frac{4\pi}{\lambda} B_{//} = -\frac{4\pi}{\lambda} (B_{\text{h}} \sin\alpha - B_{\text{v}} \cos\alpha) \tag{4.16}$$

From Eqs. (4.12) and (4.16), the accuracy of R, B_{h}, B_{v}, H and ϕ will limit the accuracy of the elevation inversion [12].

$$\frac{\partial h}{\partial R} = -\cos\theta \Rightarrow \sigma_h^{(1)} = \sigma_R \cos\theta \tag{4.17}$$

$$\frac{\partial h}{\partial B_{\text{h}}} = \frac{\partial h}{\partial \theta} \cdot \frac{\partial \theta}{\partial B_{\text{h}}} = \frac{-R\sin^2\theta}{B_{\text{v}}\sin\theta + B_{\text{h}}\cos\theta} \Rightarrow \sigma_h^{(2)} = \frac{R\sin^2\theta}{B_{\text{v}}\sin\theta + B_{\text{h}}\cos\theta} \sigma_{B_{\text{h}}} \tag{4.18}$$

$$\frac{\partial h}{\partial B_{\text{v}}} = \frac{\partial h}{\partial \theta} \cdot \frac{\partial \theta}{\partial B_{\text{v}}} = \frac{R\sin\theta\cos\theta}{B_{\text{v}}\sin\theta + B_{\text{h}}\cos\theta} \Rightarrow \sigma_h^{(3)} = \frac{R\sin\theta\cos\theta}{B_{\text{v}}\sin\theta + B_{\text{h}}\cos\theta} \sigma_{B_{\text{v}}} \tag{4.19}$$

$$\frac{\partial h}{\partial \text{H}} = 1 \Rightarrow \sigma_h^{(4)} = \sigma_{\text{H}} \tag{4.20}$$

$$\frac{\partial h}{\partial \phi} = \frac{\partial h}{\partial \theta}\frac{\partial \theta}{\partial \phi} = \frac{-R\lambda \sin\theta}{4\pi(B_\mathrm{v}\sin\theta + B_\mathrm{h}\cos\theta)} \Rightarrow \sigma_h^{(5)} = \frac{R\lambda \sin\theta}{4\pi(B_\mathrm{v}\sin\theta + B_\mathrm{h}\cos\theta)}\sigma_\phi \tag{4.21}$$

Equation (4.17) indicates that the elevation inversion error $\sigma_h^{(1)}$ comes from the instability of the system clock and data sampling clock, or the distance range deviation σ_R of the antenna to a specific distance range unit due to the electromagnetic wave propagation delay in the atmosphere and ionosphere. Equations (4.18)–(4.21) represent the relationship between horizontal baseline error σ_{B_h}, vertical baseline error σ_{B_v}, platform height measurement error σ_H, phase error σ_ϕ, and elevation measurement error σ_h, respectively.

4.2.4 Interferometric SAR Data Processing Flow

Interference signal processing is to obtain a DEM map of an observation area from a single-look complex image (SLC) pair. Interferometric SAR signal processing mainly includes data registration, flat earth effect elimination, interference fringe filtering, phase unwrapping, and elevation inversion. Figure 4.2 shows the interferometric SAR signal processing flow.

The main steps of the interferometric SAR signal processing process are as follows [13, 14].

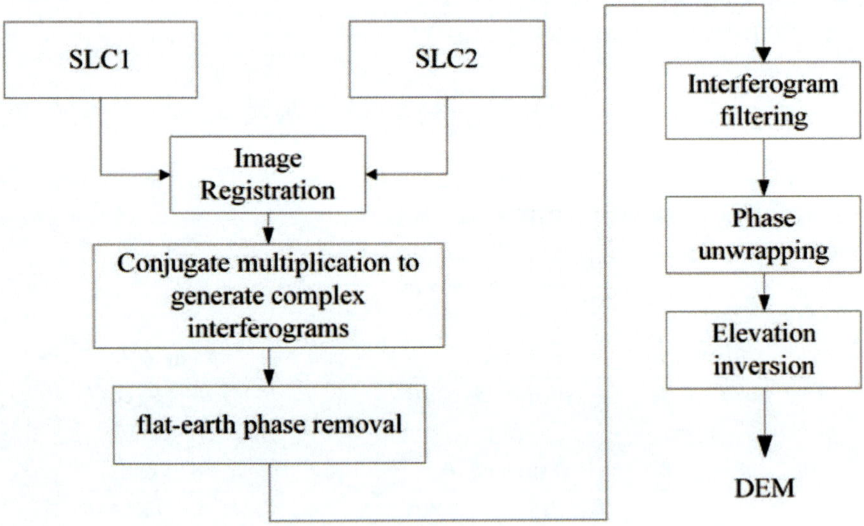

Fig. 4.2 Interferometric SAR signal processing flow

(1) Image registration: Two images of the same area obtained by the interferometric SAR system have certain offset and misalignment in the distance range and azimuth directions. The conjugate multiplication of the two images directly can only obtain uniformly distributed random noise. Only after registering the two images and achieving certain registration accuracy, an effective interferogram can be obtained.
(2) Flat earth effect elimination: The interference fringe generated after registration is not a true reflection of the natural changes in the terrain. The flat earth effect refers to the phenomenon that the interference fringe displayed by the flat earth in the interferogram with constant height changes periodically with distance range and azimuth. It must be removed before subsequent processing.
(3) Interferogram filtering: There is a lot of noise in the unfiltered interferogram, and the signal-to-noise ratio is very low. If it is not processed, it will seriously affect the subsequent phase unwrapping.
(4) Phase unwrapping: The process of recovering the phase from the main value or the phase difference value to the true value is called phase unwrapping. Phase unwrapping is a key step in the process of interferometric SAR signals and has a significant impact on the accuracy of the digital elevation map.
(5) Elevation inversion: Elevation inversion is the last step to achieve the digital elevation of surveying and mapping area from interference phase. In order to reduce inversion errors, a series of control points are needed to obtain accurate imaging geometric parameters, which can ensure digital elevation graph accuracy.

4.3 Scalar Interference and Vector Interference

Interferometry is to obtain two complex scalar signals of the same resolution unit from two antennas separated in space. For the analysis convenience, conventional scalar interference is characterized in a different way. The scalar interferogram generation process can be described as the average Hermitian product of two complex scalar signals s_1 and s_2 from the same resolution unit obtained from two antennas separated in space. The complex interference signals s_1 and s_2 are combined into a complex vector $\begin{bmatrix} s_1 & s_2 \end{bmatrix}^{\mathrm{T}}$. A 2×2 semi-positive definite Hermitian matrix **J** [1] is defined as

$$[\mathbf{J}] = \begin{bmatrix} j_{11} & j_{12} \\ j_{21} & j_{22} \end{bmatrix} = \left\langle \begin{bmatrix} s_1 \\ s_2 \end{bmatrix} \begin{bmatrix} s_1^* & s_2^* \end{bmatrix} \right\rangle = \begin{bmatrix} \langle s_1 s_1^* \rangle & \langle s_1 s_2^* \rangle \\ \langle s_2 s_1^* \rangle & \langle s_2 s_2^* \rangle \end{bmatrix} \tag{4.22}$$

Interference phase can be expressed as

$$\phi = \arg(s_1 s_2^*) \tag{4.23}$$

Phase ϕ contains information related to range and terrain. The interference coherence coefficient can be expressed as follows:

$$\xi = \frac{\left|\langle s_1 s_2^* \rangle\right|}{\sqrt{\langle s_1 s_1^* \rangle \langle s_2 s_2^* \rangle}} = \frac{|j_{12}|}{\sqrt{j_{11} j_{22}}} \tag{4.24}$$

Unlike single polarimetric interferometric system, a full-polarimetric interferometric system obtains two scattering matrices for each resolution unit in the scene from two different perspectives. Conventional interferometry processes scalar images, while full-polarimetric interferometry requires interference processing on two vector images. Assume $[\mathbf{S}_1]$ and $[\mathbf{S}_2]$ to be the scattering matrices of image pairs obtained by polarimetric interferometry, and $\mathbf{k}_1$ and $\mathbf{k}_2$ are their corresponding three-dimensional scattering matrices, respectively. $\mathbf{k}_1$ and $\mathbf{k}_2$ can be Pauli-based scattering vectors $\mathbf{k}_{1P}$ and $\mathbf{k}_{2P}$, or dictionary-based scattering vectors $\mathbf{k}_{1L}$ and $\mathbf{k}_{2L}$. Because the elements of the Pauli scattering vector are very close to the physical characteristics of wave scattering, which can better explain the scattering process, the Pauli scattering vector is selected here as an example. A 6×6 semi-definite Hermitian matrix can be defined by the outer product of $\mathbf{k}_1$ and $\mathbf{k}_2$ as follow [1, 15]:

$$[\mathbf{T}_6] = \left\langle \begin{bmatrix} \mathbf{k}_{1P} \\ \mathbf{k}_{2P} \end{bmatrix} \begin{bmatrix} \mathbf{k}_{1P}^{\mathrm{H}} & \mathbf{k}_{2P}^{\mathrm{H}} \end{bmatrix} \right\rangle = \begin{bmatrix} \mathbf{T}_{11} & \boldsymbol{\Omega}_{12} \\ \boldsymbol{\Omega}_{12}^{\mathrm{H}} & \mathbf{T}_{22} \end{bmatrix} \tag{4.25}$$

where

$$\mathbf{T}_{11} = \langle \mathbf{k}_{1P} \mathbf{k}_{1P}^{\mathrm{H}} \rangle \tag{4.26}$$

$$\boldsymbol{\Omega}_{12} = \langle \mathbf{k}_{1P} \mathbf{k}_{2P}^{\mathrm{H}} \rangle \tag{4.27}$$

$$\mathrm{T}_{22} = \langle \mathbf{k}_{2P} \mathbf{k}_{2P}^{\mathrm{H}} \rangle \tag{4.28}$$

$\mathbf{T}_{11}$ and $\mathbf{T}_{22}$ contain the full polarization information observed at different positions, respectively. $\boldsymbol{\Omega}_{12}$ is not only describing the polarization information but also the interference information.

By analyzing the interference information of different polarization states, it is found that the interference coherence coefficient and interference phase information are strongly dependent on the polarization state. In order to describe the close relationship between the interference information and the polarization state, the unit projection vectors $\boldsymbol{\omega}_1$ and $\boldsymbol{\omega}_2$ corresponding to the polarization scattering mechanism is introduced. The projection of the scattering vector $\mathbf{k}_1$ and $\mathbf{k}_2$ on $\boldsymbol{\omega}_1$ and $\boldsymbol{\omega}_2$ is the complex scattering coefficient.

$$\mu_1 = \boldsymbol{\omega}_1^{\mathrm{H}} \mathbf{k}_1 \qquad \mu_2 = \boldsymbol{\omega}_2^{\mathrm{H}} \mathbf{k}_2 \tag{4.29}$$

Similar to Eq. (4.22), there can be obtained from complex scalar μ_1 and μ_2 as follows:

$$[J] = \left\langle \begin{bmatrix} \mu_1 \\ \mu_2 \end{bmatrix} \begin{bmatrix} \mu_1^* & \mu_2^* \end{bmatrix} \right\rangle = \begin{bmatrix} \boldsymbol{\omega}_1^{\mathrm{H}} \mathbf{T}_{11} \boldsymbol{\omega}_1 & \boldsymbol{\omega}_1^{\mathrm{H}} \boldsymbol{\Omega}_{12} \boldsymbol{\omega}_2 \\ \boldsymbol{\omega}_2^{\mathrm{H}} \boldsymbol{\Omega}_{12}^{\mathrm{H}} \boldsymbol{\omega}_1 & \boldsymbol{\omega}_2^{\mathrm{H}} \mathbf{T}_{22} \boldsymbol{\omega}_2 \end{bmatrix} \tag{4.30}$$

The vector interferogram is expressed as follows:

$$\mu_1 \mu_2^* = (\boldsymbol{\omega}_1^{\mathrm{H}} \mathbf{k}_1)(\boldsymbol{\omega}_2^{\mathrm{H}} \mathbf{k}_2)^* = \boldsymbol{\omega}_1^{\mathrm{H}} [\boldsymbol{\Omega}_{12}] \boldsymbol{\omega}_2 \tag{4.31}$$

Similar to Eqs. (4.23)–(4.24), the interference coherence coefficient and interference phase in vector are derived as follows:

$$\gamma(\boldsymbol{\omega}_1, \boldsymbol{\omega}_2) = \frac{\left| \langle \mu_1 \mu_2^{\mathrm{H}} \rangle \right|}{\sqrt{\langle \mu_1 \mu_1^{\mathrm{H}} \rangle \langle \mu_2 \mu_2^{\mathrm{H}} \rangle}} = \frac{\left| \boldsymbol{\omega}_1^{\mathrm{H}} \boldsymbol{\Omega}_{12} \boldsymbol{\omega}_2 \right|}{\sqrt{\boldsymbol{\omega}_1^{\mathrm{H}} \mathbf{T}_{11} \boldsymbol{\omega}_1} \sqrt{\boldsymbol{\omega}_2^{\mathrm{H}} \mathbf{T}_{22} \boldsymbol{\omega}_2}} \tag{4.32}$$

$$\phi(\boldsymbol{\omega}_1, \boldsymbol{\omega}_2) = \arg\left(\langle \mu_1 \mu_2^{\mathrm{H}} \rangle \right) = \arg(\boldsymbol{\omega}_1^{\mathrm{H}} \boldsymbol{\Omega}_{12} \boldsymbol{\omega}_2) \tag{4.33}$$

This formula is more general than the commonly used scalar representation. It introduces a normalized projection vector representing the scattering mechanism to calculate the interference phase. If $\boldsymbol{\omega}_1 \neq \boldsymbol{\omega}_2$, the coherence coefficient contains a decoherence term γ_{Pol} related to the polarization state in addition to the mono-polarization interference coefficient γ_{Int}.

$$\gamma = \gamma_{\mathrm{Int}} \cdot \gamma_{\mathrm{Pol}} \tag{4.34}$$

In addition, an interference diagram can be formed between any polarization states using the polarization basis transformation [1, 16]. After transforming the scattering vector $\mathbf{k}_1$ and $\mathbf{k}_2$ from the basis (H, V) to the basis (i, j), the complex scattering coefficient under the newly polarization basis is obtained as follows:

$$\mu_1 = \boldsymbol{\omega}_1^{\mathrm{H}} \mathbf{k}_{1ij} = \boldsymbol{\omega}_1^{\mathrm{H}} [\mathbf{U}_3] \mathbf{k}_{1\mathrm{HV}} \tag{4.35}$$

where $[\mathbf{U}_3]$ is 3×3 unitary matrix.

The interferogram obtained by polarization basis conversion is

$$\begin{aligned} \mu_1 \mu_2^* = (\boldsymbol{\omega}_1^{\mathrm{H}} \mathbf{k}_{1ij})(\boldsymbol{\omega}_2^{\mathrm{H}} \mathbf{k}_{2ij})^{\mathrm{H}} &= (\boldsymbol{\omega}_1^{\mathrm{H}} [\mathbf{U}_3] \mathbf{k}_{1\mathrm{HV}})(\boldsymbol{\omega}_2^{\mathrm{H}} [\mathbf{U}_3] \mathbf{k}_{2\mathrm{HV}})^{\mathrm{H}} \\ &= \boldsymbol{\omega}_1^{\mathrm{H}} [\mathbf{U}_3][\boldsymbol{\Omega}_{12}][\mathbf{U}_3]^{\mathrm{H}} \boldsymbol{\omega}_2 \end{aligned} \tag{4.36}$$

By comparing Eqs. (4.36) and (4.31), it can be seen that the interferogram under any polarization basis can be expressed as the unit similarity transformation of matrix $[\boldsymbol{\Omega}_{12}]$. The interference phase under the new basis is

$$\phi = \arg(\mu_1 \mu_2^*) = \arg(\boldsymbol{\omega}_1^{\mathrm{H}} [\mathbf{U}_3][\boldsymbol{\Omega}_{12}][\mathbf{U}_3]^{\mathrm{H}} \boldsymbol{\omega}_2) \tag{4.37}$$

The corresponding interference coherence coefficient is

$$\gamma = \frac{|\langle \mu_1 \mu_2^* \rangle|}{\sqrt{\langle \mu_1 \mu_1^* \rangle \langle \mu_2 \mu_2^* \rangle}} = \frac{|\langle \boldsymbol{\omega}_1^{\mathrm{H}} [\mathbf{U}_3][\boldsymbol{\Omega}_{12}][\mathbf{U}_3]^{\mathrm{H}} \boldsymbol{\omega}_2 \rangle|}{\sqrt{\langle \boldsymbol{\omega}_1^{\mathrm{H}} [\mathbf{U}_3][\mathbf{T}_{11}][\mathbf{U}_3]^{\mathrm{H}} \boldsymbol{\omega}_1 \rangle \langle \boldsymbol{\omega}_2^{\mathrm{H}} [\mathbf{U}_3][\mathbf{T}_{22}][\mathbf{U}_3]^{\mathrm{H}} \boldsymbol{\omega}_2 \rangle}} \tag{4.38}$$

4.4 The Theory of Polarization Interference Coherence Optimization

Since the polarization mode has a great influence on the coherence of SAR interferograms, it is important to find the polarization mode making polarization interference processing optimally. The polarization interference coherence optimization technology embodies the advantages in the combination of polarization and interference technology and plays an important role in the application of polarization interference.

Polarization interference optimal algorithms are divided into two types, including optimal amplitude coherent and optimal phase coherent. Optimal amplitude method is to seek the largest coherence value using different polarization combination in the polarization state space. Optimal phase method is to seek the max phase shift of the coherence coefficients using different polarization combination in the polarization state space.

4.4.1 Amplitude Coherent Optimal Algorithm

The commonly used coherent optimal algorithms for polarization interference amplitude are coherent optimal method of double scattering mechanism proposed by SR Cloude and KP Papathanassiou [5, 17], coherent optimal method of single scattering mechanism proposed by Colin et al. [18], polarization subspace method proposed by Sagues et al. [19, 20], and so on.

4.4.1.1 Coherent Optimal Method of Double Scattering Mechanism

From Eq. (4.32), the coherent optimal mathematical of the double scattering mechanism is expressed as follows:

$$\begin{gathered} \underset{\boldsymbol{\omega}_1, \boldsymbol{\omega}_2}{\text{maximize}}\ \gamma = \frac{|\boldsymbol{\omega}_1^{\mathrm{H}} \boldsymbol{\Omega}_{12} \boldsymbol{\omega}_2|}{\sqrt{\boldsymbol{\omega}_1^{\mathrm{H}} \mathbf{T}_{11} \boldsymbol{\omega}_1} \sqrt{\boldsymbol{\omega}_2^{\mathrm{H}} \mathbf{T}_{22} \boldsymbol{\omega}_2}} \\ |\boldsymbol{\omega}_1| = |\boldsymbol{\omega}_2| = 1 \end{gathered} \tag{4.39}$$

Let the complex Lagrangian function as follows:

$$L = \boldsymbol{\omega}_1^{\mathrm{H}} \boldsymbol{\Omega}_{12} \boldsymbol{\omega}_2 + \lambda_1 (\boldsymbol{\omega}_1^{\mathrm{H}} \mathbf{T}_{11} \boldsymbol{\omega}_1 - C_1) + \lambda_2 (\boldsymbol{\omega}_2^{\mathrm{H}} \mathbf{T}_{22} \boldsymbol{\omega}_2 - C_2) \tag{4.40}$$

where C_1 an C_2 are the constants, λ_1 and λ_2 are Lagrangian multipliers. The problem to find the maximum value of L can be simplified as follows:

$$\frac{\partial L}{\partial \boldsymbol{\omega}_1^{\mathrm{H}}} = \boldsymbol{\Omega}_{12} \boldsymbol{\omega}_2 + \lambda_1 \mathbf{T}_{11} \boldsymbol{\omega}_1 = 0 \tag{4.41}$$

$$\frac{\partial L}{\partial \boldsymbol{\omega}_2^{\mathrm{H}}} = \boldsymbol{\Omega}_{12}^{\mathrm{H}} \boldsymbol{\omega}_1 + \lambda_2 \mathbf{T}_{22} \boldsymbol{\omega}_2 = 0 \tag{4.42}$$

Eliminating $\boldsymbol{\omega}_1$ and $\boldsymbol{\omega}_2$ in Eqs. (4.41) and (4.42), respectively, and let $\nu = \lambda_1 \lambda_2^*$. There are

$$\mathbf{T}_{22}^{-1} \boldsymbol{\Omega}_{12}^{\mathrm{H}} \mathbf{T}_{11}^{-1} \boldsymbol{\Omega}_{12} \boldsymbol{\omega}_2 = \nu \boldsymbol{\omega}_2 \tag{4.43}$$

$$\mathbf{T}_{11}^{-1} \boldsymbol{\Omega}_{12} \mathbf{T}_{22}^{-1} \boldsymbol{\Omega}_{12}^{\mathrm{H}} \boldsymbol{\omega}_1 = \nu \boldsymbol{\omega}_1 \tag{4.44}$$

From Eqs. (4.43) to (4.44), this extreme value problem is reduced to the problem of finding the eigenvalues and eigenvectors from two 3 × 3 matrices. The two matrices in Eqs. (4.43)–(4.44) are not Hermitian matrices, but they both have non-negative real eigenvalues ν, and the maximum correlation coefficient value is equal to the square root of the maximum eigenvalue.

$$\gamma_{\max} = \sqrt{\nu}_{\max} \tag{4.45}$$

From the maximum eigenvalues $\nu_{\max}$, two optimal eigenvectors $\boldsymbol{\omega}_{1\mathrm{opt}}$ and $\boldsymbol{\omega}_{2\mathrm{opt}}$ can be obtained, corresponding to the main and auxiliary images. These are the best scattering mechanisms. By projecting the scattering vector $\mathbf{k}_1$ and $\mathbf{k}_2$ onto $\boldsymbol{\omega}_{1\mathrm{opt}}$ and $\boldsymbol{\omega}_{2\mathrm{opt}}$, two optimized scalar complex images $\mu_{1\mathrm{opt}}$ and $\mu_{2\mathrm{opt}}$ can be obtained. And an interferogram can be formed between the two optimized scalar complex images to obtain the interferogram with the largest correlation coefficient.

$$\mu_{1\mathrm{opt}} \mu_{2\mathrm{opt}}^* = \left(\boldsymbol{\omega}_{1\mathrm{opt}}^{\mathrm{H}} \mathbf{k}_1\right) \left(\boldsymbol{\omega}_{2\mathrm{opt}}^{\mathrm{H}} \mathbf{k}_2\right)^* = \omega_{1\mathrm{opt}}^{\mathrm{H}} \boldsymbol{\Omega}_{12} \boldsymbol{\omega}_{2\mathrm{opt}} \tag{4.46}$$

It should be noted that, since the absolute phase changes of the eigenvectors all satisfy Eqs. (4.43)–(4.44), the solution $(\boldsymbol{\omega}_{1\mathrm{opt}}, \boldsymbol{\omega}_{2\mathrm{opt}})$ of the optimal projection vector is not unique. The relative phase relationship between the eigenvectors $\boldsymbol{\omega}_{1\mathrm{opt}}$ and $\boldsymbol{\omega}_{2\mathrm{opt}}$ is uncertain, which can introduce additional polarization phases into the generated interferogram and then affect the accuracy of PolInSAR processing. Therefore, additional conditions must be added to determine the phase difference between $\boldsymbol{\omega}_{1\mathrm{opt}}$ and $\boldsymbol{\omega}_{2\mathrm{opt}}$. Usually requires the following condition.

$$\arg(\boldsymbol{\omega}_{1\mathrm{opt}}^{\mathrm{H}} \boldsymbol{\omega}_{2\mathrm{opt}}) = 0 \tag{4.47}$$

4.4.1.2 Coherent Optimal Method for Single Scattering Mechanism

The coherent optimal method of the double scattering mechanism is mainly concerned with the magnitude of the interference coherence coefficient. The projection vector is not limited, which may cause the optimal scattering mechanism of neighboring resolution units to change drastically. Then, it will appear large differences in the projection vectors and interference phases of the independent scattering centers in uniform area. To overcome these shortcomings, Colin et al. proposed a coherent optimal method with a single scattering mechanism that restricts the scattering projection vectors of the primary and secondary images to be same [18].

From Eq. (4.32), assume $\boldsymbol{\omega}_1 = \boldsymbol{\omega}_2 = \boldsymbol{\omega}$, there is

$$|\gamma|^2 = \gamma\gamma^* = \frac{(\boldsymbol{\omega}^{\mathrm{H}}\boldsymbol{\Omega}_{12}\boldsymbol{\omega})(\boldsymbol{\omega}^{\mathrm{H}}\boldsymbol{\Omega}_{12}^{\mathrm{H}}\boldsymbol{\omega})}{(\boldsymbol{\omega}^{\mathrm{H}}\mathbf{T}_{11}\boldsymbol{\omega})(\boldsymbol{\omega}^{\mathrm{H}}\mathbf{T}_{22}\boldsymbol{\omega})} \tag{4.48}$$

Let $f = (\omega^{\mathrm{H}}\boldsymbol{\Omega}_{12}\omega)(\omega^{\mathrm{H}}\boldsymbol{\Omega}_{12}^{\mathrm{H}}\omega)$, $g = \boldsymbol{\omega}^{\mathrm{H}}\mathbf{T}_{11}\boldsymbol{\omega}$, and $h = \boldsymbol{\omega}^{\mathrm{H}}\mathbf{T}_{22}\boldsymbol{\omega}$. There are

$$\frac{\partial f}{\partial \boldsymbol{\omega}^{\mathrm{H}}} = 2(\boldsymbol{\Omega}_{12}\boldsymbol{\omega})(\boldsymbol{\omega}^{\mathrm{H}}\boldsymbol{\Omega}_{12}^{\mathrm{H}}\boldsymbol{\omega}) \tag{4.49}$$

$$\frac{\partial g}{\partial \boldsymbol{\omega}^{\mathrm{H}}} = 2\mathbf{T}_{11}\boldsymbol{\omega} \tag{4.50}$$

$$\frac{\partial h}{\partial \boldsymbol{\omega}^{\mathrm{H}}} = 2\mathbf{T}_{22}\boldsymbol{\omega} \tag{4.51}$$

Let $\frac{\partial|\gamma|^2}{\partial \boldsymbol{\omega}^{\mathrm{H}}} = 0$. There is

$$\frac{\partial f}{\partial \boldsymbol{\omega}^{\mathrm{H}}}hg = \frac{\partial g}{\partial \boldsymbol{\omega}^{\mathrm{H}}}hf + \frac{\partial h}{\partial \boldsymbol{\omega}^{\mathrm{H}}}gf \tag{4.52}$$

Let $f = cc^{\mathrm{H}}$, where $c = \boldsymbol{\omega}^{\mathrm{H}}\boldsymbol{\Omega}_{12}\boldsymbol{\omega}$ and $c^{\mathrm{H}} = \boldsymbol{\omega}^{\mathrm{H}}\boldsymbol{\Omega}_{12}^{\mathrm{H}}\boldsymbol{\omega}$. There is

$$(h\mathbf{T}_{11} + g\mathbf{T}_{22})^{-1}(c^{\mathrm{H}}\mathbf{T}_{12} + c\mathbf{T}_{12}^{\mathrm{H}})\boldsymbol{\omega} = |\gamma|^2\boldsymbol{\omega} \tag{4.53}$$

Using approximate conditions $g \approx h \approx c \approx c^{\mathrm{H}}$, the coherent optimization problem is transformed into an eigendecomposition problem as follows:

$$(\mathbf{T}_{11}+\mathbf{T}_{22})^{-1}(\boldsymbol{\Omega}_{12}+\boldsymbol{\Omega}_{12}^{\mathrm{H}})\omega = |\gamma|^2\boldsymbol{\omega} \tag{4.54}$$

Make eigendecomposition on the matrix of Eq. (4.54), and the eigenvector corresponding to the largest eigenvalue is the best scattering vector $\boldsymbol{\omega}_{\mathrm{opt}}$.

4.4.1.3 Polarization Subspace Method

Polarization Subspace Method (PSM) [21, 22] exploits the principle of polarization basis conversion synthesize interference coherence coefficients under any polarization state, and uses the ergodic search method to find the polarization combination that maximizes the interference coherence coefficients.

In polarization interferometry, the polarization information of two SAR images are contained in their scattering vectors $\mathbf{k}_1$ and $\mathbf{k}_2$, respectively, and 3×3 interferogram can be calculated as follows:

$$[\Phi]_{\text{int}} = \arg(\mathbf{k}_1\mathbf{k}_2^{\text{H}}) \tag{4.55}$$

Use $[\mathbf{T}_{11}] = \langle \mathbf{k}_1\mathbf{k}_1^{\text{H}} \rangle$, $[\mathbf{T}_{22}] = \langle \mathbf{k}_2\mathbf{k}_2^{\text{H}} \rangle$, and $[\mathbf{\Omega}_{12}] = \langle \mathbf{k}_1\mathbf{k}_2^{\text{H}} \rangle$, define Interference coherence matrix as

$$[\mathbf{\Gamma}] = [\mathbf{\Omega}_{12}]. \Big/ \sqrt{[\mathbf{T}_{11}]. * [\mathbf{T}_{22}]} \tag{4.56}$$

In the above equation, the symbols ./ and .* represent, respectively, division and multiplication by elements. In the linear polarization basis (*H*, *V*), the interference coherence of each polarization combination is the amplitude of the corresponding elements in $[\mathbf{\Gamma}]$.

$$\|[\Gamma]\| = \begin{bmatrix} \gamma_{HH-HH} & \gamma_{HH-HV} & \gamma_{HH-VV} \\ \gamma_{HV-HH} & \gamma_{HV-HV} & \gamma_{HV-VV} \\ \gamma_{VV-HH} & \gamma_{VV-HV} & \gamma_{VV-VV} \end{bmatrix} \tag{4.57}$$

Choose the polarization combination with largest coherence coefficient under (*H*, *V*) basis, and extract the interference phase of the corresponding element in the phase matrix $[\Phi]_{\text{int}}$.

According to the transformation theorem of polarization basis, the interference coherence matrix at an arbitrary polarization basis (*X*, *Y*) can be synthesized by the polarization interferometry observations under (*H*, *V*) basis.

$$[T]_{(\tau,\varepsilon)} = \begin{bmatrix} \gamma_{XX-XX} & \gamma_{XX-XY} & \gamma_{XX-YY} \\ \gamma_{XY-XX} & \gamma_{XY-XY} & \gamma_{XY-YY} \\ \gamma_{YY-XX} & \gamma_{YY-XY} & \gamma_{YY-YY} \end{bmatrix} = [\Omega_{12}]_{(\tau,\varepsilon)}. \Big/ \sqrt{[T_{11}]_{(\tau,\varepsilon)}. * [T_{22}]_{(\tau,\varepsilon)}} \tag{4.58}$$

where ε and τ are the ellipticity angle and pointing angle corresponding to the polarization base to be searched, respectively.

$$[\mathbf{T}_{11}]_{(\tau,\varepsilon)} = \langle [\mathbf{U}_k]\mathbf{k}_1\mathbf{k}_1^{\text{H}}[\mathbf{U}_k]^{\text{H}} \rangle = \langle [\mathbf{U}_k][\mathbf{T}_{11}][\mathbf{U}_k]^{\text{H}} \rangle \tag{4.59}$$

$$[\mathbf{T}_{22}]_{(\tau,\varepsilon)} = \left\langle [\mathbf{U}_k]\mathbf{k}_2\mathbf{k}_2^{\mathrm{H}}[\mathbf{U}_k]^{\mathrm{H}} \right\rangle = \left\langle [\mathbf{U}_k][\mathbf{T}_{22}][\mathbf{U}_k]^{\mathrm{H}} \right\rangle \tag{4.60}$$

$$[\boldsymbol{\Omega}_{12}]_{(\tau,\varepsilon)} = \left\langle [\mathbf{U}_k]\mathbf{k}_1\mathbf{k}_2^{\mathrm{H}}[\mathbf{U}_k]^{\mathrm{H}} \right\rangle = \left\langle [\mathbf{U}_k][\boldsymbol{\Omega}_{12}][\mathbf{U}_k]^{\mathrm{H}} \right\rangle \tag{4.61}$$

Here, $[\mathbf{U}_k]$ is the polarization basis transformation matrix. The optimal interference coherence is to search for the polarization state corresponding to the local highest value by traversing the interference coherence coefficient under the arbitrary polarization basis (X, Y).

It is assumed to be under the case of short baseline with no time decorrelation. If the two images have the same polarization state, the maximum coherence value can be obtained. Therefore, the diagonal elements of the matrix $\left|[\boldsymbol{\Gamma}]_{(\tau,\varepsilon)}\right|$ have larger values than other elements, and the search process of the optimal coherence coefficient can be expressed as follows:

$$\gamma_{\mathrm{opt}} = \max\{\gamma_{XX-XX}(\tau,\varepsilon), \gamma_{XY-XY}(\tau,\varepsilon), \gamma_{YY-YY}(\tau,\varepsilon)\} \tag{4.62}$$

In addition, according to the symmetry of the polarization state, there is

$$\gamma_{XX-XX}(\tau,\varepsilon) = \gamma_{YY-YY}\left(-\tau, \varepsilon + \frac{\pi}{2}\right) \tag{4.63}$$

The search process can be further simplified to the following equation.

$$\gamma_{\mathrm{opt}} = \max\{\gamma_{XX-XX}(\tau,\varepsilon), \gamma_{XY-XY}(\tau,\varepsilon)\} \tag{4.64}$$

Full-polarization space can be constructed by equally spaced sampling (τ, ε), and the coherence coefficients $\gamma_{XX-XX}(\tau,\varepsilon)$ and $\gamma_{XY-XY}(\tau,\varepsilon)$ at each sampling point are calculated. Search for local maximums from co-polarization results $\gamma_{XX-XX}(\tau,\varepsilon)$ and cross-polarization results $\gamma_{XY-XY}(\tau,\varepsilon)$ at all sampling points. The PSM method can directly provide information on the scattering mechanism, which is easy to analyze and interpret. However, this method has a large amount of calculation and takes a long time.

4.4.2 Phase Coherent Optimal Algorithm

The phase coherence optimal algorithm for polarization interference SAR [23] is based on the assumption that the scattering mechanisms of the main and auxiliary images are the same. And the algorithm seeks a set of polarization state combinations to make the largest interference phase offset of the produced complex interference image pair.

Since the phase coherent optimal algorithm only processes the phase and the scattering mechanism of the main and auxiliary images is the same, the interference coherence coefficient used here can be expressed as follows [13, 23]:

$$\gamma = \boldsymbol{\omega}^{\mathrm{H}}\boldsymbol{\Omega}_{12}\boldsymbol{\omega} \tag{4.65}$$

Here, the tangent function is used to measure the phase angle. The phase angle needs to be shifted to the main value interval of the tangent function in $\left[-\frac{\pi}{2}, \frac{\pi}{2}\right]$.

$$\hat{\gamma} = \boldsymbol{\omega}^{\mathrm{H}}\boldsymbol{\Omega}_{12}\exp(j\varphi)\boldsymbol{\omega} \tag{4.66}$$

where $\varphi = -\angle\mathrm{tr}(\boldsymbol{\Omega}_{12})$. $\angle$ is an argument, and $\mathrm{tr}(\cdot)$ is the trace of a matrix.

Let $\hat{\Omega}_{12} = \boldsymbol{\Omega}_{12}\exp(j\varphi)$. Then, there are

$$\hat{\gamma} = \boldsymbol{\omega}^{\mathrm{H}}\hat{\Omega}_{12}\boldsymbol{\omega} \tag{4.67}$$

$$\tan\angle\hat{\gamma} = \mathrm{Im}(\hat{\gamma})/\mathrm{Re}(\hat{\gamma}) \tag{4.68}$$

In Eq. (4.68), $\mathrm{Im}(\cdot)$ and $\mathrm{Re}(\cdot)$ are the real and imaginary parts of complex numbers, respectively.

$$\mathrm{Im}(\hat{\gamma}) = \frac{1}{2j}(\boldsymbol{\omega}^{\mathrm{H}}\hat{\Omega}_{12}\boldsymbol{\omega} - \boldsymbol{\omega}^{\mathrm{H}}\hat{\Omega}_{12}^{\mathrm{H}}\boldsymbol{\omega}) \tag{4.69}$$

$$\mathrm{Re}(\hat{\gamma}) = \frac{1}{2}(\boldsymbol{\omega}^{\mathrm{H}}\hat{\Omega}_{12}\boldsymbol{\omega} + \boldsymbol{\omega}^{\mathrm{H}}\hat{\Omega}_{12}^{\mathrm{H}}\boldsymbol{\omega}) \tag{4.70}$$

Therefore, the problem is actually to obtain the combination of polarization states making $\tan\angle\hat{\gamma}$ maximum. According to the Lagrange multiplier method, this problem can be transformed into the matrix eigenvalue problem.

$$-j(\hat{\Omega}_{12} + \hat{\Omega}_{12}^{\mathrm{H}})^{-1}(\hat{\Omega}_{12} - \hat{\Omega}_{12}^{\mathrm{H}})\boldsymbol{\omega} = \lambda\boldsymbol{\omega} \tag{4.71}$$

The maximum eigenvalue $\lambda_{\max}$ and minimum eigenvalue $\lambda_{\min}$ of the matrix $-j(\hat{\Omega}_{12} + \hat{\Omega}_{12}^{\mathrm{H}})^{-1}(\hat{\Omega}_{12} - \hat{\Omega}_{12}^{\mathrm{H}})$ correspond to the lowest and highest phase centers, respectively, and the corresponding eigenvectors $\boldsymbol{\omega}_1$ and $\boldsymbol{\omega}_2$ are the phase optimal scattering vectors.

4.5 Coherent Scattering Decomposition

Equations (4.43)–(4.44) contain two 3×3 complex eigenvalue equations, which result three eigenvalues, i.e., ν_1, ν_2, and ν_3, and have $\nu_1 \geq \nu_2 \geq \nu_3 \geq 0$. Each eigenvalue corresponds to a pair of eigenvectors $\boldsymbol{\omega}_{ij}$, where $i = 1, 2$, respectively, represent the main and auxiliary images, and $j = 1, 2, 3$ represent different optimal scattering vectors. The first vector pair $\{\boldsymbol{\omega}_{11}, \boldsymbol{\omega}_{21}\}$ corresponds to the maximum eigenvalue and represents the optimal scattering mechanism obtained from the completely three-dimensional complex space of the vector $\mathbf{k}_i$. The second vector pair $\{\boldsymbol{\omega}_{12}, \boldsymbol{\omega}_{22}\}$ corresponds to the next largest eigenvalue and can be interpreted as the optimal scattering mechanism in a two-dimensional subspace orthogonal to the first eigenvector. Similarly, the third eigenvector pair $\{\boldsymbol{\omega}_{13}, \boldsymbol{\omega}_{23}\}$ corresponds to the smallest eigenvalue and represents the optimal scattering mechanism in a one-dimensional subspace orthogonal to the two-dimensional space in which the previous two vectors are located. In this way, the matrices in Eqs. (4.43)–(4.44) can be expressed as the weighted sum of three matrices, and each matrix is weighted by its corresponding eigenvalue [1].

$$\mathbf{T}_{22}^{-1}\boldsymbol{\Omega}_{12}^{\mathrm{H}}\mathbf{T}_{11}^{-1}\boldsymbol{\Omega}_{12} = \nu_1(\boldsymbol{\omega}_{11}\boldsymbol{\omega}_{11}^{\mathrm{H}}) + \nu_2(\boldsymbol{\omega}_{12}\boldsymbol{\omega}_{12}^{\mathrm{H}}) + \nu_3(\boldsymbol{\omega}_{13}\boldsymbol{\omega}_{13}^{\mathrm{H}}) \tag{4.72}$$

$$\mathbf{T}_{11}^{-1}\boldsymbol{\Omega}_{12}\mathbf{T}_{22}^{-1}\boldsymbol{\Omega}_{12}^{\mathrm{H}} = \nu_1(\boldsymbol{\omega}_{21}\boldsymbol{\omega}_{21}^{\mathrm{H}}) + \nu_2(\boldsymbol{\omega}_{22}\boldsymbol{\omega}_{22}^{\mathrm{H}}) + \nu_3(\boldsymbol{\omega}_{23}\boldsymbol{\omega}_{23}^{\mathrm{H}}) \tag{4.73}$$

Equations (4.72) and (4.73) are the expressions of coherent scattering decomposition, and a total of six scattering mechanisms are decomposed, which are statistically independent. From this, three interferograms can be formed. Each corresponds to a pair of scattering mechanisms and is not related to the other two pairs.

4.6 Coherent Region Theory of Polarization Interference SAR

The coherence region of polarization interference SAR is also called coherence set, which refers to the distribution range of complex interference coherence coefficients in the complex plane among all polarization states [24, 25].

4.6.1 Coherent Region Theory

According to Eq. (4.32), the expression of complex interference coherence coefficients under arbitrary transmitting and receiving polarization states is

$$\gamma(\boldsymbol{\omega}_1, \boldsymbol{\omega}_2) = |\gamma| e^{j\phi} = \frac{\boldsymbol{\omega}_1^{\mathrm{H}}\boldsymbol{\Omega}_{12}\boldsymbol{\omega}_2}{\sqrt{\boldsymbol{\omega}_1^{\mathrm{H}}\mathbf{T}_{11}\boldsymbol{\omega}_1}\sqrt{\boldsymbol{\omega}_2^{\mathrm{H}}\mathbf{T}_{22}\boldsymbol{\omega}_2}} \tag{4.74}$$

Assume the equivalent coherence matrix is

$$\tilde{\mathbf{T}}_6 = \begin{bmatrix} \mathbf{I}_3 & \boldsymbol{\Pi} \\ \boldsymbol{\Pi}^{\mathrm{H}} & \mathbf{I}_3 \end{bmatrix} = \mathbf{R}\mathbf{T}_6\mathbf{R}^{\mathrm{H}} \tag{4.75}$$

where $\mathbf{R} = \begin{bmatrix} \mathbf{T}_{11}^{-\frac{1}{2}} & 0 \\ 0 & \mathbf{T}_{22}^{-\frac{1}{2}} \end{bmatrix}$ and $\Pi = \mathbf{T}_{11}^{-\frac{1}{2}} \mathbf{\Omega}_{12} \mathbf{T}_{22}^{-\frac{1}{2}}$. Let $\mathbf{w}_i = \frac{\sqrt{\mathbf{T}_{ii}} \boldsymbol{\omega}_i}{\boldsymbol{\omega}_i^{\mathrm{H}} \sqrt{\mathbf{T}_{ii}} \boldsymbol{\omega}_i}$. Equation (4.74) can be equal to

$$\gamma(\boldsymbol{\omega}_1, \boldsymbol{\omega}_2) = |\gamma| e^{j\phi} = \mathbf{w}_1^{\mathrm{H}} \Pi \mathbf{w}_2, \quad \mathbf{w}_i^{\mathrm{H}} \mathbf{w}_i = 1, \ i = 1, 2 \tag{4.76}$$

4.6.1.1 Generalized Coherent Set

From Eqs. (4.75) to (4.76), the normalized coherent set can be expressed as follows:

$$\begin{aligned} \Gamma &= \{\gamma(\boldsymbol{\omega}_1, \boldsymbol{\omega}_2) : \boldsymbol{\omega}_1, \boldsymbol{\omega}_2 \in \mathbf{C}^3, \boldsymbol{\omega}_1^{\mathrm{H}} \boldsymbol{\omega}_1 = \boldsymbol{\omega}_2^{\mathrm{H}} \boldsymbol{\omega}_2 = 1\} \\ &= \{\mathbf{w}_1^{\mathrm{H}} \Pi \mathbf{w}_2 : \mathbf{w}_1, \mathbf{w}_2 \in \mathbf{C}^3, \mathbf{w}_1^{\mathrm{H}} \mathbf{w}_1 = \mathbf{w}_2^{\mathrm{H}} \mathbf{w}_2 = 1\} \end{aligned} \tag{4.77}$$

For a square matrix $\mathbf{A} \in \mathbf{C}^{n \times n}$, it's numeric field (value domain) $W(\mathbf{A})$ is a set of pluralities represented by Hermitian products:

$$W(\mathbf{A}) = \{\mathbf{x}^{\mathrm{H}} \mathbf{A} \mathbf{x} : \mathbf{x} \in \mathbf{C}^n, \mathbf{x}^{\mathrm{H}} \mathbf{x} = 1\} \tag{4.78}$$

Substitute $\mathbf{y}^{\mathrm{H}} \mathbf{A} \mathbf{x}$ for Hermitian product $\mathbf{x}^{\mathrm{H}} \mathbf{A} \mathbf{x}$ in Eq. (4.78). Obtain $q-$ numeric field.

$$W_q(\mathbf{A}) = \{\mathbf{y}^{\mathrm{H}} \mathbf{A} \mathbf{x} : \mathbf{x}, \mathbf{y} \in \mathbf{C}^n, \mathbf{y}^{\mathrm{H}} \mathbf{y} = \mathbf{x}^{\mathrm{H}} \mathbf{x} = 1, \mathbf{y}^{\mathrm{H}} \mathbf{x} = q\} \tag{4.79}$$

By comparing Eqs. (4.77) and (4.79), it can be seen that the expression of normalized coherent set is the $q-$ number field of matrix $\mathbf{\Pi}$, namely:

$$\Gamma = \bigcup_{q \in \mathbf{C}, |q| \le 1} W_q(\mathbf{\Pi}) \tag{4.80}$$

For $W_q(\mathbf{\Pi}) = e^{i\theta} W_{|q|}(\mathbf{\Pi})$ and $q = |q| e^{i\theta}$, Eq. (4.80) exists phase ambiguity. In order to solve the problem, restrict $\arg(\boldsymbol{\omega}_1, \boldsymbol{\omega}_2) = 0$, which is equal to limit the q to the real plane in $q-$ number field.

To sum up, the limitation of revising the Eq. (4.76) is $\mathbf{w}_1^{\mathrm{H}} \mathbf{w}_2 = q, \ q \in [0, 1]$, and is denoted as $\gamma_{2MC}(\mathbf{w}_1, \mathbf{w}_2)$. This expression represents the two-mechanism coherence coefficient.

$$\mathbf{\Gamma}_{\mathrm{2MC}} = \bigcup_{q \in [0,1]} W_q(\mathbf{\Pi}) \tag{4.81}$$

4.6.1.2 Single Mechanism Coherent Set

When the projection vectors of the main and auxiliary images are the same, that is $\boldsymbol{\omega}_1 = \boldsymbol{\omega}_2$, the expression of the coherence coefficient $\gamma_{1\mathrm{MC}}$ is

$$\gamma_{1\mathrm{MC}} = \frac{\boldsymbol{\omega}^{\mathrm{H}}\boldsymbol{\Omega}_{12}\boldsymbol{\omega}}{\sqrt{\boldsymbol{\omega}^{\mathrm{H}}\mathbf{T}_{11}\boldsymbol{\omega}\boldsymbol{\omega}^{\mathrm{H}}\mathbf{T}_{22}\boldsymbol{\omega}}}, \quad \boldsymbol{\omega}^{\mathrm{H}}\boldsymbol{\omega} = 1 \tag{4.82}$$

Let $\mathbf{w} = \frac{\sqrt{\mathbf{T}_{22}\boldsymbol{\omega}}}{\sqrt[4]{\boldsymbol{\omega}^{\mathrm{H}}\mathbf{T}_{11}\boldsymbol{\omega}\boldsymbol{\omega}^{\mathrm{H}}\mathbf{T}_{22}\boldsymbol{\omega}}}$ and $\mathbf{M} = \mathbf{T}_{22}^{-\frac{1}{2}}\mathbf{T}_{11}^{-\frac{1}{2}}$. Equation (4.82) can be

$$\gamma_{1MC}(\mathbf{w}) = \mathbf{w}^{\mathrm{H}}\mathbf{M}\Pi\mathbf{w}, \quad \mathbf{w}^{\mathrm{H}}\mathbf{M}\mathbf{w} = 1 \tag{4.83}$$

In Eq. (4.83), **M** is neither a Hermitian matrix nor a positive definite matrix. Since $\mathbf{T}_{11}$ and $\mathbf{T}_{22}$ are both positive definite Hermitian matrices, **M** are approximately positive definite and approximately Hermitian matrices. From the analysis in the previous section, the single-mechanism coherence coefficient can be summarized as a normalized inner product number domain problem.

$$W_{\mathbf{M}}(\mathbf{A}) = \{\mathbf{x}^{\mathrm{H}}\mathbf{M}\mathbf{A}\mathbf{x};\ \mathbf{x} \in \mathbf{C}^n, \mathbf{x}^{\mathrm{H}}\mathbf{M}\mathbf{x} = 1\} \tag{4.84}$$

From the above equation, a single-mechanism coherent set $\boldsymbol{\Gamma}_{1\mathrm{MC}}$ can be defined as

$$\Gamma_{1\mathrm{MC}} = \{\mathbf{w}^{\mathrm{H}}\mathbf{M}\Pi\mathbf{w} : \mathbf{w} \in \mathbf{C}^3, \mathbf{w}^{\mathrm{H}}\mathbf{M}\mathbf{w} = 1\} = W_{\mathbf{M}}(\boldsymbol{\Pi}) \tag{4.85}$$

4.6.1.3 Approximate Single Mechanism Coherent Set

In general, the coherence coefficient can be expressed as follows:

$$\gamma_{\mathrm{app}}(\boldsymbol{\omega}) = \frac{\boldsymbol{\omega}^{\mathrm{H}}\boldsymbol{\Omega}_{12}\boldsymbol{\omega}}{\boldsymbol{\omega}^{\mathrm{H}}\mathbf{T}\omega}, \quad \boldsymbol{\omega}^{\mathrm{H}}\boldsymbol{\omega} = 1, \mathbf{T} = \frac{\mathbf{T}_{11} + \mathbf{T}_{22}}{2} \tag{4.86}$$

Let $\bar{\Pi} = T^{-\frac{1}{2}}\boldsymbol{\Omega}_{12}\mathbf{T}^{-\frac{1}{2}}$ and $\mathbf{w} = \frac{\sqrt{\mathbf{T}\boldsymbol{\omega}}}{\boldsymbol{\omega}^{\mathrm{H}}\sqrt{\mathbf{T}\boldsymbol{\omega}}}$. There is

$$\gamma_{\mathrm{app}}(\boldsymbol{\omega}) = \mathbf{w}^{\mathrm{H}}\bar{\Pi}w, \quad \mathbf{w}^{\mathrm{H}}\mathbf{w} = 1 \tag{4.87}$$

From this coherence coefficient expression, a coherence set can be defined as follows:

$$\Gamma_{\mathrm{app}} = W(\bar{\Pi}) \tag{4.88}$$

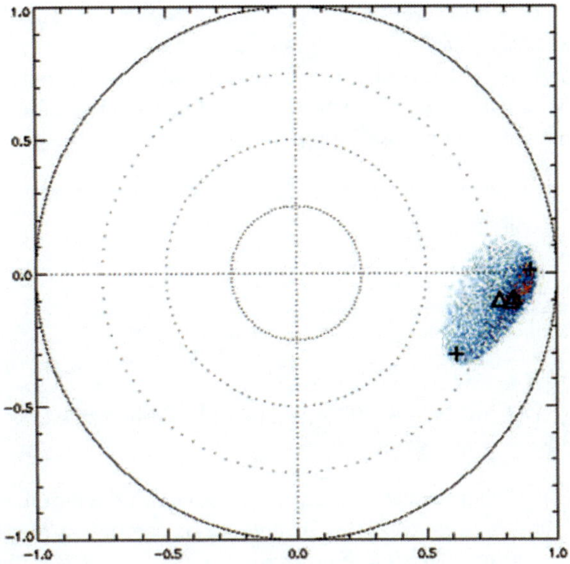

Fig. 4.3 Representation of a coherent region shape within a complex unit circle

4.6.2 Coherent Region Shape Parameter Extraction

From the above analysis, it is known that coherence coefficients under different polarization channels can be obtained by changing the complex projection vector $\boldsymbol{\omega}_i$ in the entire scattering mechanism space. These coherence coefficients form the coherence region of the polarimetric interference SAR data and can express the shape of the coherence region in a complex unit circle area shape. Figure 4.3 shows the shape of the coherent region represented within the complex unit circle. The following describes a numerical method for calculating the shape distribution of coherent regions when the scattering mechanism of the main and auxiliary images is the same, i.e., $\boldsymbol{\omega}_1 = \boldsymbol{\omega}_2$.

In general, the shape of a coherent region is approximately elliptical, which is called a coherent ellipse. Let $\boldsymbol{\Upsilon} = [\gamma_1, \gamma_2, \ldots, \gamma_N]^{\mathrm{T}}$, where $\gamma_i, (i = 1, \ldots, N)$ denotes the coherence coefficient of N polarization channels. Solve the following eigenvalue problem.

$$\mathbf{M}\mathbf{v} = \lambda\mathbf{v} \tag{4.89}$$

where $\mathbf{M} = \frac{1}{N}\left[\mathrm{Re}(\boldsymbol{\Upsilon}') \ \mathrm{Im}(\boldsymbol{\Upsilon}')\right]^{\mathrm{T}}\left[\mathrm{Re}(\boldsymbol{\Upsilon}') \ \mathrm{Im}(\boldsymbol{\Upsilon}')\right]$ and $\boldsymbol{\Upsilon}' = \Upsilon - \frac{1}{\mathrm{N}}\sum_{i}^{N}\boldsymbol{\Upsilon}'$. $\mathrm{Re}(\cdot)$ and $\mathrm{Im}(\cdot)$ represent the real and imaginary parts of a plurality, respectively. The shape and distribution of a coherent ellipse can be described by five parameters.

$$|\tilde{\gamma}|, \tilde{\phi}, \lambda_1, \lambda_2, \alpha_1 \tag{4.90}$$

$|\tilde{\gamma}|$ and $\tilde{\phi}$ are the average absolute coherence coefficient and phase, respectively. λ_1 and λ_2 are two real eigenvalues obtained by the eigendecomposition of Eq. (4.89 and $\lambda_1 \geq \lambda_2$. α_1 is the gradient of the first eigenvalue relative to $\mathbf{v}_1$. Using these parameters, the characteristic parameters representing the geometric properties of the coherent ellipse can be obtained.

(1) Pseudo-ellipse rate: It indicates the slenderness of a coherent ellipse, which is defined as

$$\chi = \frac{\lambda_1 - \lambda_2}{\lambda_1 + \lambda_2} \tag{4.91}$$

Obviously, $\chi \in [0, 1]$. And the closer the value of χ is to 1, the closer the coherent ellipse is to a straight line.

(2) Coherence coefficient/phase standard deviation: After calculating the standard deviation $\text{SDEV}_{|\gamma|}$ of coherence coefficient amplitude and the standard deviation $\text{SDEV}_{|\phi|}$ of coherence coefficient phase, the coherence coefficient/phase standard deviation can be obtained as follows:

$$\rho = \frac{\text{SDEV}_{|\gamma|}}{\text{SDEV}_{|\gamma|} + \frac{\text{SDEV}_{\phi}}{\pi}} \tag{4.92}$$

(3) External tendency: It indicates the tendency of the coherent ellipse to expand outwards, and its expression is

$$\vartheta = \chi \cdot \frac{\text{SDEV}_{|\gamma|}}{\text{SDEV}_{|\gamma|} + \frac{\text{SDEV}_{\phi}}{\pi}} \cdot |\tilde{\gamma}| \tag{4.93}$$

It can be seen that the smaller the external tendency, the more the coherent region converges.

4.7 Summary

This chapter outlines the basic principles of PolInSAR and elaborates vector interference and coherent optimality theory of polarimetric interference. These basic theories are the basis for subsequent polarimetric interference SAR information processing

References

1. S.R. Cloude, K.P. Papathanassiou, Polarimetric SAR interferometry. IEEE Trans. Geosci. Remote Sens. **36**(5), 1551–1565 (1998)
2. R.N. Treuhaft, P.R. Siqueira, Vertical structure of vegetated land surfaces from interferometric and polarimetric radar. Radio Sci. **35**(1), 141–177 (2000)
3. Z. Bin, Z. La-mei, S. De-ming, W. Wei, Information extraction using polarimetric interferometric SAR data: present and future. J. Electron. Inform. Technol. **28**(10), 1979–1984 (2006)
4. K.P. Papathanassiou, S.R. Cloude, *Polarimetric Effects in Repeat-Pass SAR Interferometry* (IGARSS97, Singapore, Singapore, 1997), pp. 1926–1928
5. S.R. Cloude, K.P. Papathanassiou, Polarimetric optimisation in radar interferometry. Electron. Lett. **33**(13), 1176–1178 (1997)
6. W. Yi-rong, H. Wen, W. Yan-ping, The current status and implications of polarimetric SAR interferometry. J. Electron. Inform. Technol. **29**(5), 1258–1262 (2007)
7. W. Chao, Z. Hong, L. Zhi, *Spaceborne Synthetic Aperture Radar Interferometry* (Science Press, Beijing, 2002), p. 235
8. R. Bamler, P. Hartl, Synthetic aperture radar interferometry. Inverse Prob. **14**, R1–R54 (1998)
9. R. Touzi et al., Coherence estimation for SAR imagery. IEEE Trans. Geosci. Remote Sens. **37**(1), 135–149 (1999)
10. P.A. Rosen et al., Synthetic aperture radar interferometry. Proc. IEEE **88**(3), 333–382 (2000)
11. F.K. Li, R.M. Goldstein, Studies of multibaseline spaceborene interferometric synthetic aperture radars. IEEE Trans. Geosci. Remote Sens. **28**(1), 88–97 (1990)
12. W. Nan, *Signal Processing Techniques for Interferometric Synthetic Aperture Radar* (Xidian University, Xi'an, 2007), p. 119
13. C. Bing, *Research on Signal Processing Technology of Interference and Polarization Interference Synthetic Aperture Radar* (Graduate School of Chinese Academy of Sciences, Beijing, 2008), p. 125
14. Zhao Zhiwei, *Spaceborne Scanning Synthetic Aperture Radar System and Signal Processing Technology* (Graduate School of Chinese Academy of Sciences, Beijing, 2007)
15. K.P. Papathanassiou, Single-baseline polarimetric SAR interferometry. IEEE Trans. Geosci. Remote Sens. **39**(11), 2352–2363 (2001)
16. K.P. Papathanassiou, *Polarimetric SAR Interferometry* (Technical University Graz, 1999), p. 149
17. S.R. Cloude, K.P. Papathanassiou, Coherence optimisation in polarimetric SAR Interferometry, in *Proceedings of the IGARSS'97* (1997), pp. 1932–1934
18. E. Colin, C. Titin-Schnaider, W. Tabbara, Investigation on different interferometric coherence optimization methods, in *POLinSAR 03 Proceedings* (Roma, 2003)
19. L. Sagues, et al., Indoor experiments on polarimetric SAR interferometry. IEEE Trans. Geosci. Remote Sens. **38**(2), 2000
20. L. Sagues et al., Polarimetric radar interferometry for improved mine detection and surface clutter rejection. IEEE Trans. Geosci. Remote Sens. **39**(6), 1271–1278 (2001)
21. M. Tabb, et al., Phase Diversity: A decomposition for vegetation parameter estimation using polarimetric SAR interferometry, in *Proceedings of EUSAR 2002* (Cologne, Germany, 2002)
22. T. Flynn, M. Tabb, R. Carande, Coherence region shape extraction for vegetation parameter estimation in polarimetric SAR interferometry, in *Proceedings of the IGARSS'02* (2002), pp. 2596–2598
23. M. Neumann, A. Reigber, L. Ferro-Famil, PolInSAR coherence set theory and application, in *Proceedings of the EUSAR 2006* (Dresden, Germany, 2006)

Chapter 5
Compact Polarimetry SAR and Compact Polarimetric SAR Interferometry

5.1 Overview

Polarimetric SAR can obtain more abundant target information compared to ordinary single-polarized synthetic aperture radar. However, the addition of polarimetric ability increases the complexity of SAR system greatly, which imposes demanding requirements for antenna technology, data download rate, and power consumption. Compact polarimetry (CP) technology is a new technique proposed to reduce the complexity of full PolSAR system [1–4]. Instead of alternately transmitting two orthogonally polarized pulses and receiving the returns in two orthogonal polarizations, compact PolSAR (C-PolSAR) system transmits single polarized pulses of specific polarization state and receives the returns in two orthogonal polarization. CP system reduces the requirement of data download rate and system complexity effectively while keeping most of the polarimetric information obtained by full polarimetric SAR system. Compact Polarimetric SAR Interferometry (C-PolInSAR), which combines CP and interferometry techniques together, contains abundant polarimetric and interferometry information.

Performances of dual polarimetric SAR, compact polarimetric SAR, and full polarimetric SAR in terrain classification are compared by Ainsworth, etc [5]. Recently, CP technology is applied to PolInSAR system, and more attentions are paid to the information processing of compact PolInSAR. The potentials of the application of compact PolInSAR in forest height retrieval are analyzed by Angelliaume and Lavalle. The system design problems of spaceborne P-band compact PolInSAR are discussed by Dubois-Fernandez, etc., and tree height retrieval performance with P-band C-PolInSAR data is analyzed.

R. Yang et al., *Polarimetric Microwave Imaging*,
https://doi.org/10.1007/978-981-15-8897-6_5

5.2 Compact Polarimetric SAR Modes

As we have mentioned, C-PolSAR system transmits single polarized pulses of specific polarization state and receives the returns in two orthogonal polarization. Different compact polarimetry modes are proposed based on the polarization state of transmitting and receiving electromagnetic wave [1, 4, 6]. There are three in commonly used CP modes.

(1) $\pi/4$ mode. The transmitting polarization is linear polarization oriented at 45° and receiving polarization is horizontal and vertical linear polarization. The scattering vector for the CP mode is given by

$$\boldsymbol{k}_{\pi/4} = \left[\, S_{\mathrm{HH}} + S_{\mathrm{HV}} \;\; S_{\mathrm{HV}} + S_{\mathrm{VV}} \,\right]^T / \sqrt{2} \tag{5.1}$$

The corresponding covariance matrix is

$$\begin{aligned} \boldsymbol{J}_{\pi/4} = \langle \boldsymbol{k}_{\pi/4} \boldsymbol{k}_{\pi/4}^{\mathrm{H}} \rangle &= \frac{1}{2}\begin{bmatrix} \langle |S_{\mathrm{HH}}|^2 \rangle & \langle S_{\mathrm{HH}} \cdot S_{\mathrm{VV}}^* \rangle \\ \langle S_{\mathrm{VV}} \cdot S_{\mathrm{HH}}^* \rangle & \langle |S_{\mathrm{VV}}|^2 \rangle \end{bmatrix} + \frac{1}{2}\begin{bmatrix} \langle |S_{\mathrm{HV}}|^2 \rangle & \langle |S_{\mathrm{HV}}|^2 \rangle \\ \langle |S_{\mathrm{HV}}|^2 \rangle & \langle |S_{\mathrm{HV}}|^2 \rangle \end{bmatrix} \\ &+ \frac{1}{2}\begin{bmatrix} 2\mathrm{Re}(\langle S_{\mathrm{HH}} \cdot S_{\mathrm{VV}}^* \rangle) & \langle S_{\mathrm{HH}} \cdot S_{\mathrm{HV}}^* + S_{\mathrm{VV}}^* \cdot S_{\mathrm{HV}} \rangle \\ \langle S_{\mathrm{HV}}^* \cdot S_{\mathrm{HV}} + S_{\mathrm{VV}} \cdot S_{\mathrm{HV}}^* \rangle & 2\mathrm{Re}(\langle S_{\mathrm{VV}} \cdot S_{\mathrm{HV}}^* \rangle) \end{bmatrix} \end{aligned} \tag{5.2}$$

(2) Dual circular polarimetric (DCP) mode. The compact polarimetric mode is based on right circular polarization (RCP) transmit, and right and left circular polarization receives. The scattering vector for DCP mode is

$$\boldsymbol{k}_{\mathrm{DCP}} = \left[\, S_{\mathrm{RR}} \;\; S_{\mathrm{RL}} \,\right]^T = \left[\, (S_{\mathrm{HH}} - S_{\mathrm{VV}} + i2S_{\mathrm{HV}}) \;\; i(S_{\mathrm{HH}} + S_{\mathrm{VV}}) \,\right]^{\mathrm{T}} / 2 \tag{5.3}$$

In Eq. (5.3), R and L represent right circular and left circular polarization, respectively.

The corresponding covariance matrix is

$$\begin{aligned} \boldsymbol{J}_{\mathrm{DCP}} = \langle \boldsymbol{k}_{\mathrm{DCP}} \boldsymbol{k}_{\mathrm{DCP}}^{\mathrm{H}} \rangle &= \frac{1}{4}\begin{bmatrix} \langle |(S_{\mathrm{HH}} - S_{\mathrm{VV}}|^2 \rangle & -i\langle (S_{\mathrm{HH}} - S_{\mathrm{VV}}) \cdot (S_{\mathrm{HH}} + S_{\mathrm{VV}})^* \rangle \\ i\langle (S_{\mathrm{HH}} + S_{\mathrm{VV}}) \cdot (S_{\mathrm{HH}} - S_{\mathrm{VV}})^* \rangle & \langle |S_{\mathrm{HH}} + S_{\mathrm{VV}}|^2 \rangle \end{bmatrix} \\ &+ \frac{1}{4}\begin{bmatrix} 4\langle |S_{\mathrm{HV}}|^2 \rangle & 0 \\ 0 & 0 \end{bmatrix} + \frac{1}{4}\begin{bmatrix} 4Im(\langle (S_{\mathrm{HH}} - S_{\mathrm{VV}}) \cdot S_{\mathrm{HV}}^* \rangle) & 2\langle (S_{\mathrm{HH}} + S_{\mathrm{VV}})^* \cdot S_{\mathrm{HV}} \rangle \\ 2\langle (S_{\mathrm{HH}} + S_{\mathrm{VV}}) \cdot S_{\mathrm{HV}}^* \rangle & 0 \end{bmatrix} \end{aligned} \tag{5.4}$$

(3) Circular transmit and linear receive (CTLR) mode. The compact polarization mode is based on right circular polarization transmit, and linear horizontal and vertical polarization receive. The scattering vector is given by (taking right circular polarization for example)

$$\boldsymbol{k}_{\mathrm{CTLR}} = \begin{bmatrix} S_{\mathrm{HH}} - iS_{\mathrm{HV}} & -iS_{\mathrm{VV}} + S_{\mathrm{HV}} \end{bmatrix}^{\mathrm{T}} / \sqrt{2} \tag{5.5}$$

The corresponding covariance matrix is

$$J_{\mathrm{CTLR}} = \langle \boldsymbol{k}_{\mathrm{CTLR}} \boldsymbol{k}_{CTLR}^{\mathrm{H}} \rangle = \frac{1}{2}\begin{bmatrix} \langle |S_{\mathrm{HH}}|^2 \rangle & i\langle S_{HH} \cdot S_{\mathrm{VV}}^* \rangle \\ -i\langle S_{VV} \cdot S_{\mathrm{HH}}^* \rangle & \langle |S_{\mathrm{VV}}|^2 \rangle \end{bmatrix} + \frac{1}{2}\begin{bmatrix} \langle |S_{\mathrm{HV}}|^2 \rangle & -i\langle |S_{\mathrm{HV}}|^2 \rangle \\ i\langle |S_{\mathrm{HV}}|^2 \rangle & \langle |S_{\mathrm{HV}}|^2 \rangle \end{bmatrix}$$
$$+ \frac{1}{2}\begin{bmatrix} -2Im(\langle S_{\mathrm{HH}} \cdot S_{\mathrm{HV}}^* \rangle) & \langle S_{\mathrm{HH}} \cdot S_{\mathrm{VV}}^* \rangle + \langle S_{\mathrm{VV}}^* \cdot S_{\mathrm{HV}} \rangle \\ \langle S_{\mathrm{HH}}^* \cdot S_{\mathrm{HV}} \rangle + \langle S_{\mathrm{VV}} \cdot S_{\mathrm{HV}}^* \rangle & 2Im(\langle S_{\mathrm{VV}} \cdot S_{\mathrm{HV}}^* \rangle) \end{bmatrix} \tag{5.6}$$

In Eqs. (5.2), (5.4), and (5.6), $\langle \cdot \rangle$ denotes spatial averaging. From the covariance matrices expressed in Eqs. (5.2), (5.4), and (5.6) and some symmetric assumption, the 3×3 pseudo-quad-pol covariance matrix can be reconstructed.

5.3 Reconstruction of Full Polarimetric Data from Compact Polarimetric Data

Reconstruction of full polarimetric (FP) information from CP data, that is, reconstruction of 3×3 FP covariance matrix from 2×2 CP covariance matrix is very important for the processing of CP information.

5.3.1 Effect of Symmetry Properties of Geophysical Media

First, reflection and rotation symmetries are considered. Figure 5.1 shows the layout of symmetry plane involved in reflection, rotation symmetries, with respect to the incident wave plane.

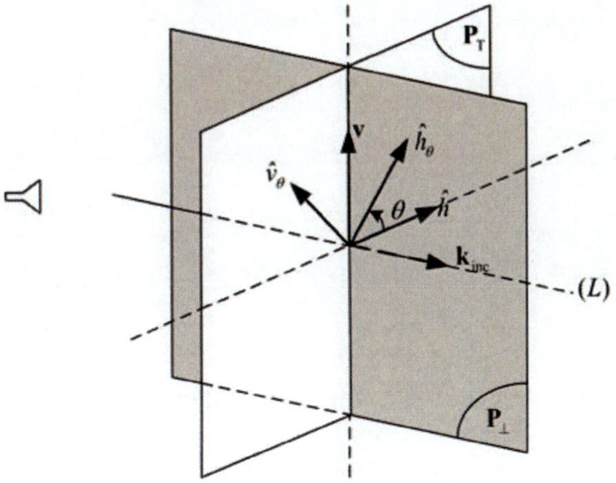

Fig. 5.1 Layout of symmetry plane involved in reflection, rotation symmetries, with respect to the incident wave

In Fig. 5.1, $\hat{h}$ and $\hat{v}$ are the horizontal and vertical directions belonging to the transverse plane $\boldsymbol{P}_{\mathrm{T}}$ of an incidence wave, respectively. The incidence wave plane contains $\mathbf{v}$ and $\mathbf{k}_{\mathrm{inc}}$. Reflection symmetry occurs when the observed media is vertical with respect to $\boldsymbol{P}_{\perp}$. Under this assumption, there is a complete de-correlation of the co-polarized and the cross-polarized backscattering coefficients

$$\langle S_{\mathrm{HH}} \cdot S_{\mathrm{HV}}^{*} \rangle = \langle S_{\mathrm{HV}} \cdot S_{\mathrm{VV}}^{*} \rangle \tag{5.7}$$

Generally, if the following conditions are satisfied, then the symmetry assumption is valid.

$$\begin{aligned} &2 \cdot \mathrm{Re}\langle S_{\mathrm{HH}} \cdot S_{\mathrm{HV}}^{*} \rangle \ll \langle |S_{\mathrm{HH}}|^2 \rangle + \langle |S_{\mathrm{HV}}|^2 \rangle \\ &2 \cdot \mathrm{Re}\langle S_{\mathrm{VV}} \cdot S_{\mathrm{HV}}^{*} \rangle \ll \langle |S_{\mathrm{VV}}|^2 \rangle + \langle |S_{\mathrm{HV}}|^2 \rangle \\ &|\langle S_{\mathrm{HH}} \cdot S_{\mathrm{HV}}^{*} \rangle + \langle S_{\mathrm{VV}} \cdot S_{\mathrm{HV}}^{*} \rangle| \ll \langle S_{\mathrm{HH}} \cdot S_{\mathrm{HV}}^{*} \rangle + \langle |S_{\mathrm{HV}}|^2 \rangle \end{aligned} \tag{5.8}$$

Experiments with real SAR data show that most L-band and C-band data around 10 m resolution satisfies the conditions in Eq. (5.8) even in the case of small samples. Hence, as reconstruct full polarimetric data, this symmetric condition can be used.

When a medium is rotation symmetry about an axis L, the covariance matrix remains unchanged after a rotation by an angle θ in the transverse plane $\boldsymbol{P}_{\mathrm{T}}$, the covariance matrix terms obey

$$\begin{aligned} &\langle |S_{\mathrm{HH}}|^2 \rangle = \langle |S_{\mathrm{VV}}|^2 \rangle \\ &\frac{\langle |S_{\mathrm{HV}}|^2 \rangle}{\langle |S_{\mathrm{HH}}|^2 \rangle} = \frac{\langle |S_{\mathrm{HV}}|^2 \rangle}{\langle |S_{\mathrm{VV}}|^2 \rangle} = \frac{\left(1 - \frac{\langle S_{\mathrm{HH}} \cdot S_{\mathrm{VV}}^{*} \rangle}{\sqrt{\langle |S_{\mathrm{HH}}|^2 \rangle \cdot \langle |S_{\mathrm{VV}}|^2 \rangle}}\right)}{2} = \frac{1-\rho}{2} \\ &\mathrm{Re}\langle S_{\mathrm{HH}} \cdot S_{\mathrm{HV}}^{*} \rangle = \mathrm{Re}\langle S_{\mathrm{HV}} \cdot S_{\mathrm{VV}}^{*} \rangle = 0 \\ &\mathrm{Im}\langle S_{\mathrm{HH}} \cdot S_{\mathrm{HV}}^{*} \rangle = \mathrm{Im}\langle S_{\mathrm{HV}} \cdot S_{\mathrm{VV}}^{*} \rangle \end{aligned} \tag{5.9}$$

In Eq. (5.8), ρ represents the degree of coherence between S_{HH} and S_{VV}. It is generally a complex value, which becomes real under rotation symmetry condition.

$$\rho = \frac{\langle S_{\mathrm{HH}} \cdot S_{\mathrm{VV}}^{*} \rangle}{\sqrt{\langle |S_{\mathrm{HH}}|^2 \rangle \cdot \langle |S_{\mathrm{VV}}|^2 \rangle}} \tag{5.10}$$

5.3.2 *Reconstruction of FP Data Using Reflection Symmetry Properties*

From the reflection symmetry assumption, it can be induced that $\langle S_{\mathrm{HH}} \cdot S_{\mathrm{HV}}^{*} \rangle = \langle S_{\mathrm{HV}} \cdot S_{\mathrm{VV}}^{*} \rangle$. As a result, the last term in Eqs. (5.2), (5.4), and (5.6) is 0, and the covariance matrix of FP can be simplified as

$$\langle[C]\rangle = \begin{bmatrix} \langle|S_{HH}|^2\rangle & 0 & \langle S_{HH}\cdot S_{VV}^*\rangle \\ 0 & 2\langle|S_{HV}|^2\rangle & 0 \\ \langle S_{VV}\cdot S_{HH}^*\rangle & 0 & \langle|S_{VV}|^2\rangle \end{bmatrix} \tag{5.11}$$

From the above analysis, it can be seen that the reconstruction process can be simplified into solving five unknowns from four measurements equation set, no matter what CP mode is used. It is an underdetermined problem and some extra constraints have to be added to perform the reconstruction. Souyris et al. proposes using the relation between linear coherence coefficient $|\rho|$ and cross-polarization ratio $\langle|S_{HV}|^2\rangle/(\langle|S_{HH}|^2\rangle + \langle|S_{VV}|^2\rangle)$ to solve the equations iteratively.

$$\frac{\langle|S_{HV}|^2\rangle}{\langle|S_{HH}|^2\rangle + \langle|S_{VV}|^2\rangle} = \frac{1-\rho}{4},\ \rho = \frac{\langle S_{HH}\cdot S_{VV}^*\rangle}{\sqrt{\langle|S_{HH}|^2\rangle\cdot\langle|S_{VV}|^2\rangle}} \tag{5.12}$$

5.3.3 Reconstruction of FP Data from $\pi/4$ Mode CP Data

From Eq. (5.2), and ignoring constant item 1/2, we can get

$$\boldsymbol{J}_{\pi/4} = \begin{bmatrix} J_{11} & J_{12} \\ J_{21} & J_{22} \end{bmatrix} = \begin{bmatrix} \langle|S_{HH}|^2\rangle & \langle S_{HH}\cdot S_{VV}^*\rangle \\ \langle S_{VV}\cdot S_{HH}^*\rangle & \langle|S_{VV}|^2\rangle \end{bmatrix} + \begin{bmatrix} \langle|S_{HV}|^2\rangle & \langle|S_{HV}|^2\rangle \\ \langle|S_{HV}|^2\rangle & \langle|S_{HV}|^2\rangle \end{bmatrix} \tag{5.13}$$

That is

$$J_{11} = \langle|S_{HH}|^2\rangle + \langle|S_{HV}|^2\rangle \tag{5.14}$$

$$J_{12} = \langle S_{HH}\cdot S_{VV}^*\rangle + \langle|S_{HV}|^2\rangle \tag{5.15}$$

$$J_{22} = \langle|S_{VV}|^2\rangle + \langle|S_{HV}|^2\rangle \tag{5.16}$$

$$\rho = \frac{\langle S_{HH}\cdot S_{VV}^*\rangle}{\sqrt{\langle|S_{HH}|^2\rangle\cdot\langle|S_{VV}|^2\rangle}} = \frac{J_{12} - \langle|S_{HV}|^2\rangle}{\sqrt{J_{11} - \langle|S_{HV}|^2\rangle}\sqrt{J_{22} - \langle|S_{HV}|^2\rangle}} \tag{5.17}$$

Let $\langle|S_{HV}|^2\rangle_{(0)} = 0$, and

$$\rho_{(0)} = \frac{J_{12}}{\sqrt{J_{11}\cdot J_{22}}} \tag{5.18}$$

From Eq. (5.12), we can get

$$\frac{(1-\rho_{(0)})}{4}=\frac{\langle|S_{\mathrm{HV}}|^2\rangle}{\langle|S_{\mathrm{HH}}|^2\rangle+\langle|S_{\mathrm{VV}}|^2\rangle} \tag{5.19}$$

Let $\langle|S_{\mathrm{HV}}|^2\rangle = X$, so we can get $\langle|S_{\mathrm{HH}}|^2\rangle = J_{11}-X$, $\langle|S_{\mathrm{VV}}|^2\rangle = J_{22}-X$, and Eq. (5.19) is

$$\frac{(1-\rho_{(0)})}{4}=\frac{X}{J_{11}-X+J_{22}-X} \tag{5.20}$$

And so forth

$$X_{(1)}=\langle|S_{\mathrm{HV}}|^2\rangle_{(1)}=\frac{(J_{11}+J_{22})}{2}\left(\frac{1-|\rho_{(0)}|}{3-|\rho_{(0)}|}\right) \tag{5.21}$$

Replacing Eq. (5.21) into Eq. (5.17), we can get

$$\rho_{(1)}=\frac{J_{12}-\left(\langle|S_{\mathrm{HV}}|^2\rangle\right)_0}{\sqrt{(J_{11}-\left(\langle|S_{\mathrm{HV}}|^2\rangle\right)_{(0)}\cdot(J_{22}-\left(\langle|S_{\mathrm{HV}}|^2\rangle\right)_{(0)}}} \tag{5.22}$$

So,

$$\langle|S_{\mathrm{HV}}|^2\rangle_{(2)}=\frac{(C_{11}+C_{22})}{2}\cdot\left(\frac{1-|\rho_{(1)}|}{3-|\rho_{(1)}|}\right). \tag{5.23}$$

From Eq. (5.14)–(5.23), an iterative process is set up to solve the equations,

(1) Initialization: let $\langle|S_{\mathrm{HV}}|^2\rangle = 0$, $\rho=\frac{J_{12}}{\sqrt{J_{11}\cdot J_{22}}}$.
(2) Perform the following equations iteratively until they are convergent.

$$|S_{\mathrm{HV}}|^2{}_{(i+1)}=\frac{(J_{11}+J_{22})}{2}\cdot\left(\frac{1-|\rho_{(i)}|}{3-|\rho_{(i)}|}\right), \tag{5.24}$$

$$\rho_{(i+1)}=\frac{J_{12}-\left(\langle|S_{\mathrm{HV}}|^2\rangle\right)_{(i+1)}}{\sqrt{(J_{11}-\left(\langle|S_{\mathrm{HV}}|^2\rangle\right)_{(i+1)}\cdot(J_{22}-\left(\langle|S_{\mathrm{HV}}|^2\rangle\right)_{(i+1)}}}. \tag{5.25}$$

We can get the estimation of $\langle|S_{\mathrm{HV}}|^2\rangle$.

(3) Reconstructed pseudo-quad-pol covariance matrix is

$$[C_{\pi/4}]_{\text{pseudoquad}} = \begin{bmatrix} J_{11} - |S_{\text{HV}}|^2 & 0 & J_{12} - |S_{\text{HV}}|^2 \\ 0 & 2|S_{\text{HV}}|^2 & 0 \\ (J_{12} - |S_{\text{HV}}|^2)^* & 0 & J_{22} - |S_{\text{HV}}|^2 \end{bmatrix}. \tag{5.26}$$

5.3.4 Reconstruction of FP Data from DCP Mode CP Data

From Eq. (5.4), and ignoring constant item, we can get

$$J_{\text{DCP}} = \begin{bmatrix} J_{11} & J_{12} \\ J_{21} & J_{22} \end{bmatrix} = \begin{bmatrix} \langle |(S_{\text{HH}} - S_{\text{VV}})|^2 \rangle & -i\langle (S_{\text{HH}} - S_{\text{VV}}) \cdot (S_{\text{HH}} + S_{\text{VV}})^* \rangle \\ i\langle (S_{\text{HH}} + S_{\text{VV}}) \cdot (S_{\text{HH}} - S_{\text{VV}})^* \rangle & \langle |S_{\text{HH}} + S_{\text{VV}}|^2 \rangle \end{bmatrix} + \begin{bmatrix} 4\langle |S_{\text{HV}}|^2 \rangle & 0 \\ 0 & 0 \end{bmatrix}. \tag{5.27}$$

Let H $= \langle |S_{\text{HH}}|^2 \rangle$, V $= \langle |S_{\text{VV}}|^2 \rangle$, X $= \langle |S_{\text{HV}}|^2 \rangle$, $I = \text{Re}(\langle S_{\text{HH}} \cdot S_{\text{VV}}^* \rangle)$, $Q = \text{Im}(\langle S_{\text{HH}} \cdot S_{\text{VV}}^* \rangle)$, so

$$J_{11} = \langle |(S_{\text{HH}} - S_{\text{VV}})|^2 \rangle + 4\langle |S_{\text{HV}}|^2 \rangle = H + V - 2I + 4X \tag{5.28}$$

$$J_{12} = -i\langle (S_{\text{HH}} - S_{\text{VV}}) \cdot (S_{\text{HH}} + S_{\text{VV}})^* \rangle = -iH + iV + 2Q \tag{5.29}$$

$$J_{22} = \langle |S_{\text{HH}} + S_{\text{VV}}|^2 \rangle = H + V + 2X. \tag{5.30}$$

We can get

$$H = \frac{J_{11} + J_{22} - 4X}{4} - \frac{1}{2}\text{Im}(J_{12}) \tag{5.31}$$

$$V = \frac{J_{11} + J_{22} - 4X}{4} + \frac{1}{2}\text{Im}(J_{12}) \tag{5.32}$$

$$I = \frac{J_{22} - J_{11}}{2} \tag{5.33}$$

$$Q = \frac{\text{Re}(J_{12})}{2}. \tag{5.34}$$

Then

$$\rho = \frac{\langle S_{\text{HH}} \cdot S_{\text{VV}}^* \rangle}{\sqrt{\langle |S_{\text{HH}}|^2 \rangle \cdot \langle |S_{\text{VV}}|^2 \rangle}} = \frac{I + iQ}{\sqrt{H \cdot V}} \tag{5.35}$$

Let $\langle |S_{\text{HV}}|^2 \rangle_{(0)} = 0$, we can get

$$\rho_{(0)} = \frac{\langle S_{HH} \cdot S_{VV}^* \rangle}{\sqrt{\langle |S_{HH}|^2 \rangle \cdot \langle |S_{VV}|^2 \rangle}} = \frac{\frac{J_{22}-J_{11}}{2} + i\frac{Re(J_{12})}{2}}{\sqrt{\frac{J_{11}+J_{22}}{4} - \frac{1}{2}Im(J_{12})} \cdot \sqrt{\frac{J_{11}+J_{22}}{4} + \frac{1}{2}Im(J_{12})}}. \tag{5.36}$$

From Eq. (5.12),

$$\frac{(1-\rho_{(0)})}{4} = \frac{\langle |S_{HV}|^2 \rangle}{\langle |S_{HH}|^2 \rangle + \langle |S_{VV}|^2 \rangle} = \frac{X}{H+V}, \tag{5.37}$$

and so forth

$$X_{(1)} = \frac{J_{11}+J_{22}}{4}\frac{1-|\rho_{(0)}|}{3-|\rho_{(0)}|} \tag{5.38}$$

Replacing Eq. (5.38) into Eq. (5.35), we can get

$$\rho_{(1)} = \frac{\frac{J_{22}-J_{11}}{2} + i\frac{\mathrm{Re}(J_{12})}{2}}{\sqrt{\frac{J_{11}+J_{22}-4X_{(1)}}{4} - \frac{1}{2}\mathrm{Im}(J_{12})} \cdot \sqrt{\frac{J_{11}+J_{22}-4X_{(1)}}{4} + \frac{1}{2}\mathrm{Im}(J_{12})}}. \tag{5.39}$$

So,

$$X_{(2)} = \frac{J_{11}+J_{22}}{4}\frac{1-|\rho_{(1)}|}{3-|\rho_{(1)}|} \tag{5.40}$$

From Eqs. (5.36)–(5.40), an iterative process is set up to solve the equations,

(1) Intialization: let $\langle |S_{HV}|^2 \rangle_{(0)} = 0$, $\rho_{(0)} = \frac{\frac{J_{22}-J_{11}}{2} + i\frac{\mathrm{Re}(J_{12})}{2}}{\sqrt{\frac{J_{11}+J_{22}-4X_{(1)}}{4} - \frac{1}{2}\mathrm{Im}(J_{12})} \cdot \sqrt{\frac{J_{11}+J_{22}-4X_{(1)}}{4} + \frac{1}{2}\mathrm{Im}(J_{12})}}$.

(2) Perform the following equations iteratively until they are convergent.

$$|S_{HV}|^2{}_{(i+1)} = \frac{J_{11}+J_{22}}{4}\frac{1-|\rho_{(i)}|}{3-|\rho_{(i)}|} \tag{5.41}$$

$$\rho_{(i+1)} = \frac{\frac{J_{22}-J_{11}}{2} + i\frac{\mathrm{Re}(J_{12})}{2}}{\sqrt{\frac{J_{11}+J_{22}-4X_{(i)}}{4} - \frac{1}{2}\mathrm{Im}(J_{12})} \cdot \sqrt{\frac{J_{11}+J_{22}-4X_{(i)}}{4} + \frac{1}{2}\mathrm{Im}(J_{12})}} \tag{5.42}$$

We can get the estimation of $\langle |S_{HV}|^2 \rangle$.

(3) Reconstructed pseudo-quad-pol covariance matrix is

$$[C_{\mathrm{DCP}}]_{\mathrm{pseudoquad}} = \begin{bmatrix} \frac{J_{11}+J_{22}-4\langle|S_{\mathrm{HV}}|^2\rangle}{4} - \frac{1}{2}\mathrm{Im}(J_{12}) & 0 & \frac{J_{22}-J_{11}+4\langle|S_{\mathrm{HV}}|^2\rangle}{4} + i\frac{\mathrm{Re}(J_{12})}{2} \\ 0 & 2\langle|S_{\mathrm{HV}}|^2\rangle & 0 \\ \frac{J_{22}-J_{11}}{2} - i\frac{Re(J_{12})}{2} & 0 & \frac{J_{22}-J_{11}-4\langle|S_{\mathrm{HV}}|^2\rangle}{4} + \frac{1}{2}\mathrm{Im}(J_{12}) \end{bmatrix} \tag{5.43}$$

5.3.5 Reconstruction of FP Data from CTLR Mode CP Data

From Eq. (5.6), and ignoring constant item, we can get

$$\boldsymbol{J}_{\mathrm{CTLR}} = \begin{bmatrix} J_{11} & J_{12} \\ J_{21} & J_{22} \end{bmatrix} = \begin{bmatrix} \langle|S_{\mathrm{HH}}|^2\rangle & i\langle S_{HH}\cdot S_{\mathrm{VV}}^*\rangle \\ -i\langle S_{VV}\cdot S_{\mathrm{HH}}^*\rangle & \langle|S_{\mathrm{VV}}|^2\rangle \end{bmatrix} + \begin{bmatrix} \langle|S_{\mathrm{HV}}|^2\rangle & -i\langle|S_{\mathrm{HV}}|^2\rangle \\ i\langle|S_{\mathrm{HV}}|^2\rangle & \langle|S_{\mathrm{HV}}|^2\rangle \end{bmatrix}. \tag{5.44}$$

That is

$$J_{11} = \langle|S_{\mathrm{HH}}|^2\rangle + \langle|S_{\mathrm{HV}}|^2\rangle \tag{5.45}$$

$$J_{12} = i\langle S_{HH}\cdot S_{\mathrm{VV}}^*\rangle - i\langle|S_{\mathrm{HV}}|^2\rangle \tag{5.46}$$

$$J_{22} = \langle|S_{\mathrm{VV}}|^2\rangle + \langle|S_{\mathrm{HV}}|^2\rangle. \tag{5.47}$$

$$\rho = \frac{\langle S_{\mathrm{HH}}\cdot S_{\mathrm{VV}}^*\rangle}{\sqrt{\langle|S_{\mathrm{HH}}|^2\rangle\cdot\langle|S_{\mathrm{VV}}|^2\rangle}} = \frac{-iJ_{12} + \langle|S_{\mathrm{HV}}|^2\rangle}{\sqrt{J_{11} - \langle|S_{\mathrm{HV}}|^2\rangle}\sqrt{J_{22} - \langle|S_{\mathrm{HV}}|^2\rangle}} \tag{5.48}$$

Let $\langle|S_{\mathrm{HV}}|^2\rangle_{(0)} = 0$, we can get

$$\rho_{(0)} = \frac{\langle S_{\mathrm{HH}}\cdot S_{\mathrm{VV}}^*\rangle}{\sqrt{\langle|S_{\mathrm{HH}}|^2\rangle\cdot\langle|S_{\mathrm{VV}}|^2\rangle}} = \frac{-iJ_{12}}{\sqrt{J_{11}\cdot J_{22}}} \tag{5.49}$$

From Eq. (5.12), we can get

$$\frac{(1-\rho_{(0)})}{4} = \frac{\langle|S_{\mathrm{HV}}|^2\rangle}{\langle|S_{\mathrm{HH}}|^2\rangle + \langle|S_{\mathrm{VV}}|^2\rangle} \tag{5.50}$$

Let $\langle|S_{\mathrm{HV}}|^2\rangle = X$, and we can get $\langle|S_{\mathrm{HH}}|^2\rangle = J_{11} - X$, $\langle|S_{\mathrm{VV}}|^2\rangle = J_{22} - X$, so

$$X_{(1)} = \langle|S_{\mathrm{HV}}|^2\rangle_{(1)} = \frac{J_{11}+J_{22}}{2}\frac{1-|\rho_{(0)}|}{3-|\rho_{(0)}|} \tag{5.51}$$

Replacing Eq. (5.51) into Eq. (5.48), we can get

$$\rho_{(1)} = \frac{\langle S_{HH} \cdot S_{VV}^{*} \rangle}{\sqrt{\langle |S_{HH}|^2 \rangle \cdot \langle |S_{VV}|^2 \rangle}} = \frac{-iJ_{12} + \left(\langle |S_{HV}|^2 \rangle\right)_{(1)}}{\sqrt{(J_{11} - \left(\langle |S_{HV}|^2 \rangle\right)_{(1)} \cdot (J_{22} - \left(\langle |S_{HV}|^2 \rangle\right)_{(1)}}} \tag{5.52}$$

So,

$$X_{(2)} = \frac{J_{11} + J_{22}}{2} \frac{1 - |\rho_{(1)}|}{3 - |\rho_{(1)}|} \tag{5.53}$$

From Eqs. (5.49)–(5.53), an iterative process is set up to solve the equations,

(1) Intialization: let $\langle |S_{HV}|^2 \rangle = 0$, $\rho = \frac{-iJ_{12}}{\sqrt{J_{11} \cdot J_{22}}}$.
(2) Perform the following equations iteratively until they are convergent.

$$|S_{HV}|^2{}_{(i+1)} = \frac{(J_{11} + J_{22})}{2} \cdot \left(\frac{1 - |\rho_{(i)}|}{3 - |\rho_{(i)}|} \right), \tag{5.54}$$

$$\rho_{(i+1)} = \frac{-iJ_{12} + \left(\langle |S_{HV}|^2 \rangle\right)_{(i)}}{\sqrt{(J_{11} - \left(\langle |S_{HV}|^2 \rangle\right)_{(i)} \cdot (J_{22} - \left(\langle |S_{HV}|^2 \rangle\right)_{(i)}}}. \tag{5.55}$$

We can get the estimation of $\langle |S_{HV}|^2 \rangle$.

(3) Reconstructed pseudo-quad-pol covariance matrix is

$$[C_{CTLR}]_{pseudoquad} = \begin{bmatrix} J_{11} - \langle |S_{HV}|^2 \rangle & 0 & \langle |S_{HV}|^2 \rangle - iJ_{12} \\ 0 & 2\langle |S_{HV}|^2 \rangle & 0 \\ \left(|S_{HV}|^2 - iJ_{12}\right)^* & 0 & J_{22} - \langle |S_{HV}|^2 \rangle \end{bmatrix}. \tag{5.56}$$

It is noted that the reconstruction is based on the hypotheses of Eq. (5.12). It may occur that $|\rho| > 1$ or $\langle |S_{HV}|^2 \rangle < 0$ for points can't satisfy Eq. (5.12). In these cases, we regularize by setting $|\rho| = 1$, $\langle |S_{HV}|^2 \rangle = 0$ and halt the iteration.

5.3.6 Experiments and Result Analysis

Full PolSAR data of Los Angeles Bay acquired by AIRSAR is used to perform the experiment. This scene contains various types of media such as ocean, urban area, grassland, streets, and Golden Gate Bridge with metal structure. The size of the image is 900 × 800 pixels. $\pi/4$, DCP, and CTLR mode CP data is generated from the PolSAR data based on Eqs. (5.1)–(5.6). Comparison between reconstructed pseudo-quad-pol data and full PolSAR data of different CP modes are illustrated in Figs. 5.2, 5.3 and 5.4, where the X axis represents full PolSAR data and the Y axis represents reconstructed pseudo-quad-pol data. In Figs. 5.2, 5.3 and 5.4, the diagonal line represents the reconstructed data is identical to the full PolSAR data. Obviously, the closer the distribution of the points to the diagonal line, the better the reconstruction effect. Different colors show the frequency of the occurrence of the points, in which red color shows high frequency, while blue color shows low frequency.

As we can see from Figs. 5.2, 5.3 and 5.4, reconstructed $\langle |S_{\mathrm{HH}}|^2\rangle$ and $\langle |S_{\mathrm{VV}}|^2\rangle$ are quite close to full PolSAR data, while $\langle |S_{\mathrm{HV}}|^2\rangle$ and $\rho_{\mathrm{H-V}}$ have a larger deviation.

Figure 5.5 shows the comparison of scattering entropy H and average scattering angle α by Cloude–Pottier decomposition with reconstructed pseudo-quad-pol data and full PolSAR data, respectively.

As we can see from Fig. 5.5, H/α decomposition from reconstructed CP data is of quite great agreement with FP data. Among them, decomposition result of CTLR mode data is most similar to FP data (Fig. 5.6).

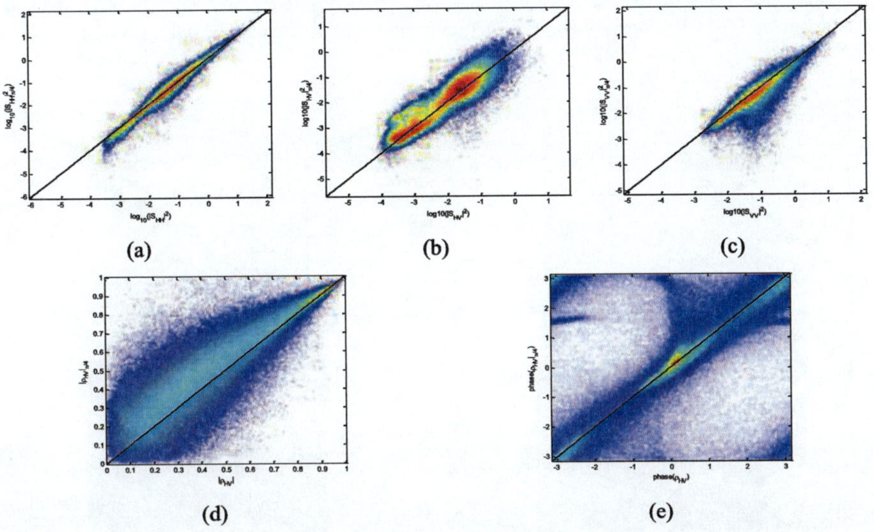

Fig. 5.2 Reconstructed pseudo-quad-pol data compared with full PolSAR data of $\pi/4$ CP mode. **a** Reconstructed $\log_{10}(\langle |S_{\mathrm{HH}}|^2\rangle)$ compared with PolSAR data, **b** Reconstructed $\log_{10}(\langle |S_{\mathrm{HV}}|^2\rangle)$ compared with PolSAR data, **c** Reconstructed $\log_{10}(\langle |S_{\mathrm{VV}}|^2\rangle)$ compared with PolSAR data, **d** Reconstructed $|\rho_{\mathrm{H-V}}|$ compared with PolSAR data, and **e** Reconstructed $\arg(\rho_{\mathrm{H-V}})$ compared with PolSAR data

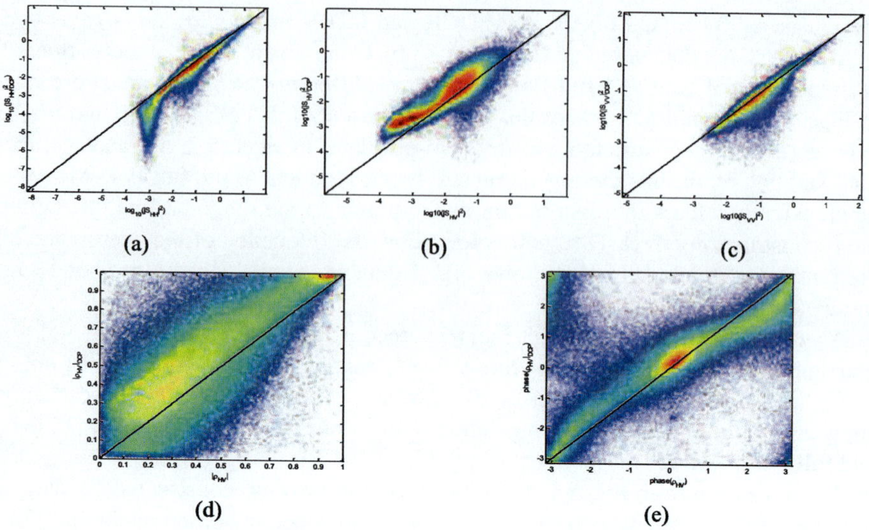

Fig. 5.3 Reconstructed pseudo-quad-pol data compared with full PolSAR data of DCP mode. **a** Reconstructed $\log_{10}(\langle|S_{\mathrm{HH}}|^2\rangle)$ compared with PolSAR data, **b** Reconstructed $\log_{10}(\langle|S_{\mathrm{HV}}|^2\rangle)$ compared with PolSAR data, **c** Reconstructed $\log_{10}(\langle|S_{\mathrm{VV}}|^2\rangle)$ compared with PolSAR data, **d** Reconstructed $|\rho_{\mathrm{H-V}}|$ compared with PolSAR data and **e** Reconstructed $\arg(\rho_{\mathrm{H-V}})$ compared with PolSAR data

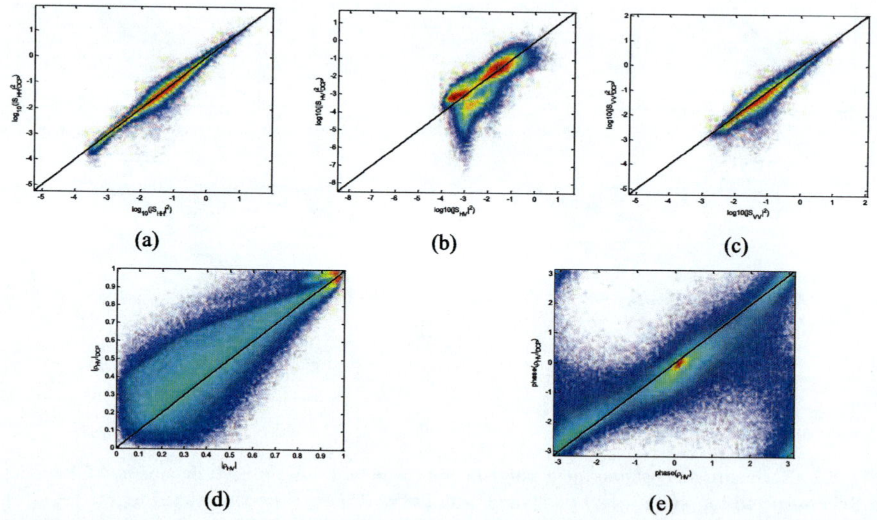

Fig. 5.4 Reconstructed pseudo-quad-pol data compared with full PolSAR data of CTLR mode. **a** Reconstructed $\log_{10}(\langle|S_{\mathrm{HH}}|^2\rangle)$ compared with PolSAR data, **b** Reconstructed $\log_{10}(\langle|S_{\mathrm{HV}}|^2\rangle)$ compared with PolSAR data, **c** Reconstructed $\log_{10}(\langle|S_{\mathrm{VV}}|^2\rangle)$ compared with PolSAR data, **d** Reconstructed $|\rho_{\mathrm{H-V}}|$ compared with PolSAR data and **e** Reconstructed $\arg(\rho_{\mathrm{H-V}})$ compared with PolSAR data

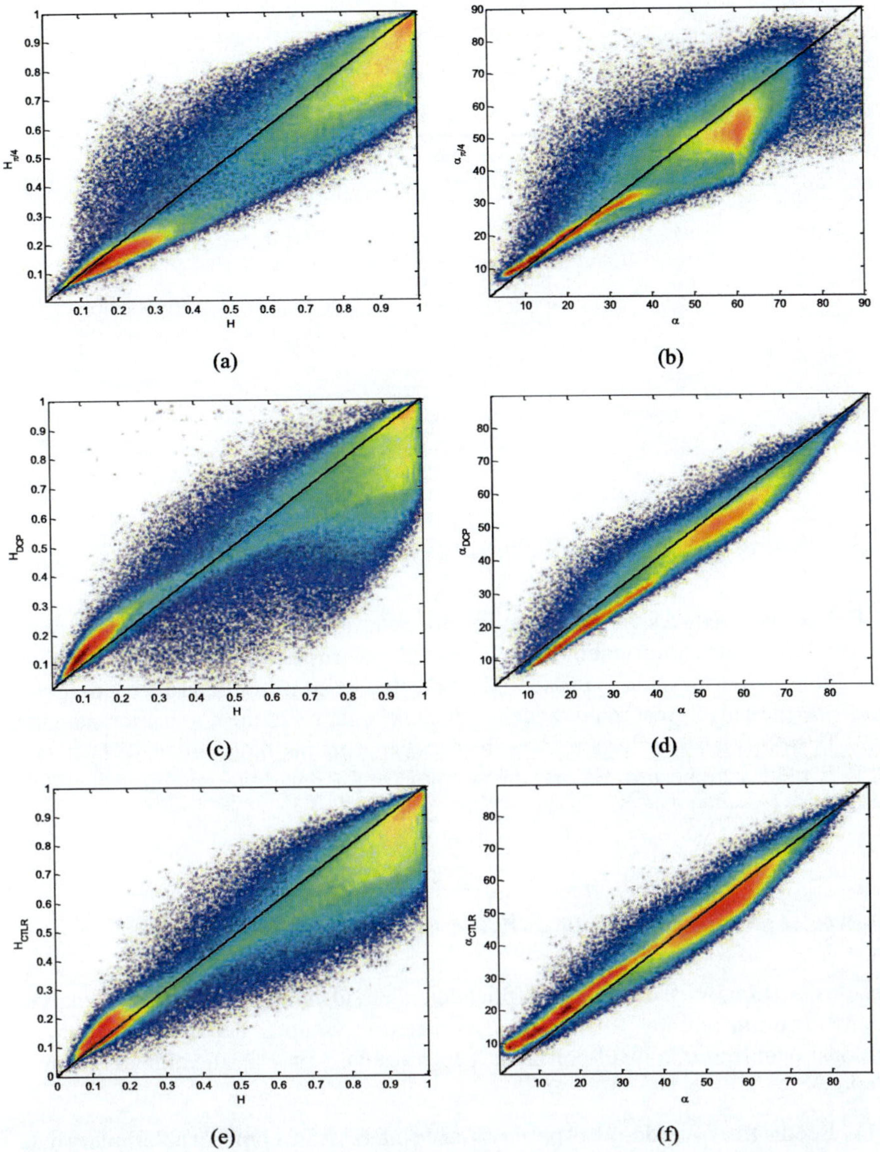

Fig. 5.5 H/α decomposition results comparison of different pseudo-quad-pol data reconstructed from different CP modes. **a** H decomposed of reconstructed $\pi/4$ mode CP data versus H decomposed of FP data. **b** α decomposed of reconstructed $\pi/4$ mode CP data versus α decomposed of FP data. **c** H decomposed of reconstructed DCP mode CP data versus H decomposed of FP data. **d** α decomposed of reconstructed DCP mode CP data versus α decomposed of FP data. **e** H decomposed of reconstructed CTLR mode CP data versus H decomposed of FP data. **f** α decomposed of reconstructed CTLR mode CP data versus α decomposed of FP data

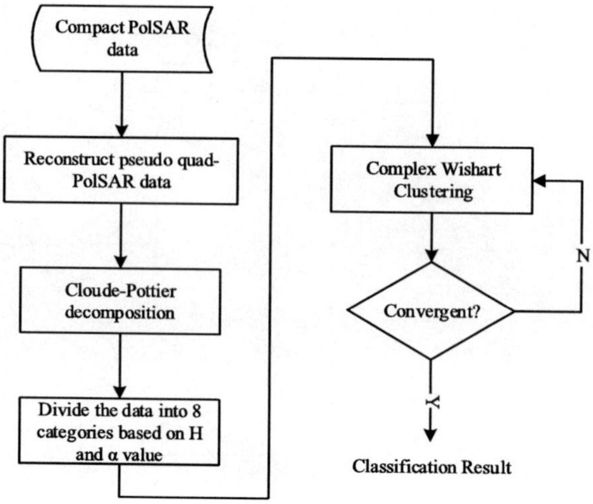

Fig. 5.6 Classification scheme of compact PolSAR data

5.4 Analysis of the Classification of Compact PolSAR

CP effectively reduces the complexity of full polarimetry system while keeping most of the polarimetric information obtained by full polarimetric system. However, some polarimetric information is missed. The potential of compact polarimetry in terrain classification is of great importance for the investigation of the application potential of CP. In this section, terrain classification experiments based on $\pi/4$, DCP, and CTLR modes are performed, and the potential of CP data in terrain classification is discussed.

5.4.1 The Classification Scheme of Compact PolSAR

In this section, Wishart H/α classification method is applied with the polarization scattering entropy H and average scattering angle α decomposed from the reconstructed pseudo-quad-pol data. The procedure of classification is given as follows:

(1) Reconstruct pseudo-quad-pol covariance matrix from compact polarimetry data.
(2) Cloude–Pottier decomposition is performed on the reconstructed pseudo-quad-pol covariance matrix. The polarization scattering entropy H and average scattering angle α is obtained.
(3) Based on H/α plane, all the pixels in the image are divided into and categories.
(4) Maximum likelihood classification based on Wishart clustering is performed until the iteration termination condition is reached.

It is noted that for the maximum likelihood classification of compact polarimetry SAR data, information obtained by CP system can be described by the 2 × 2 covariance matrix $\langle C_2 \rangle$, which follows complex Wishart distribution. The probability density function for the covariance matrix is

$$p(\langle C_2 \rangle) = \frac{|\langle C_2 \rangle|^{n-q} \exp(-\mathrm{tr}(V^{-1}\langle C_2 \rangle))}{K(n, q)|V|^n} \tag{5.57}$$

In Eq. (5.57), $K(n, q) = \pi^{\frac{q(q-1)}{2}} \Gamma(n) \cdots \Gamma(n - q + 1)$, $\Gamma(n) = (n - 1)!$, $\mathrm{tr}(\cdot)$ is the trace of a matrix, $|\cdot|$ is the determinant of a matrix, and q is the dimension of the scattering matrix, where q = 2. An equivalent distance measure can be expressed by

$$d(\langle C_2 \rangle, \mathbf{V}_\mathrm{m}) = \ln|\mathbf{V}_\mathrm{m}| + tr(\mathbf{V}_m^{-1}\langle C_2 \rangle) \tag{5.58}$$

In Eq. (5.58), $\mathbf{V}_\mathrm{m} = E[\langle C_2 \rangle | \omega_m]$, is defined as the mean of $\langle C_2 \rangle$ for all pixels belong to the mth class.

Mathematically, a pixel is assigned to class ω_m, if

$$d(\langle C_2 \rangle, \mathbf{V}_\mathrm{m}) \le d\big(\langle C_2 \rangle, \mathbf{V}_\mathrm{j}\big), \forall j \ne m. \tag{5.59}$$

5.4.2 Experiment Results and Analysis

NASA/JPL AIRSAR L-band PolSAR data of San Fransisco bay is used for illustration. $\pi/4$, DCP, and CTLR mode CP data are generated from the PolSAR data. Then the classification experiment is performed.

Figure 5.7 shows the (H, α) value distributed in the $\mathrm{H} - \alpha$ plane. Figure 5.8 shows classification results of different data based on the (H, α) value, in which (a) is the classification result of full PolSAR data, while (b), (c), and (d) shows the classification result of $\pi/4$, DCP, and CTLR mode CP data.

As we can see from Fig. 5.7, the (H, α) distribution of reconstructed quad-pol PolSAR data is similar to the original PolSAR data. Among them, decomposition result of CTLR mode data is the most similar to the original PolSAR data. The classification result shown in Fig. 5.8 also proves it.

Figure 5.9 shows the Wishart H/α classification results of this area. As we can see from Fig. 5.9, ideal classification results can be provided by the reconstructed pseudo-quad-pol data from three different CP modes.

In order to analyze the classification performance of different CP modes quantitatively, the classification result of full PolSAR data is taken as a reference to compare the similarity of classification between CP data and FP data. The results are shown in Tables 5.1, 5.2 and 5.3. Take the second column in Tables 5.1, 5.2 and 5.3, for example, it means the distribution of the pixels in the classification result of CP data,

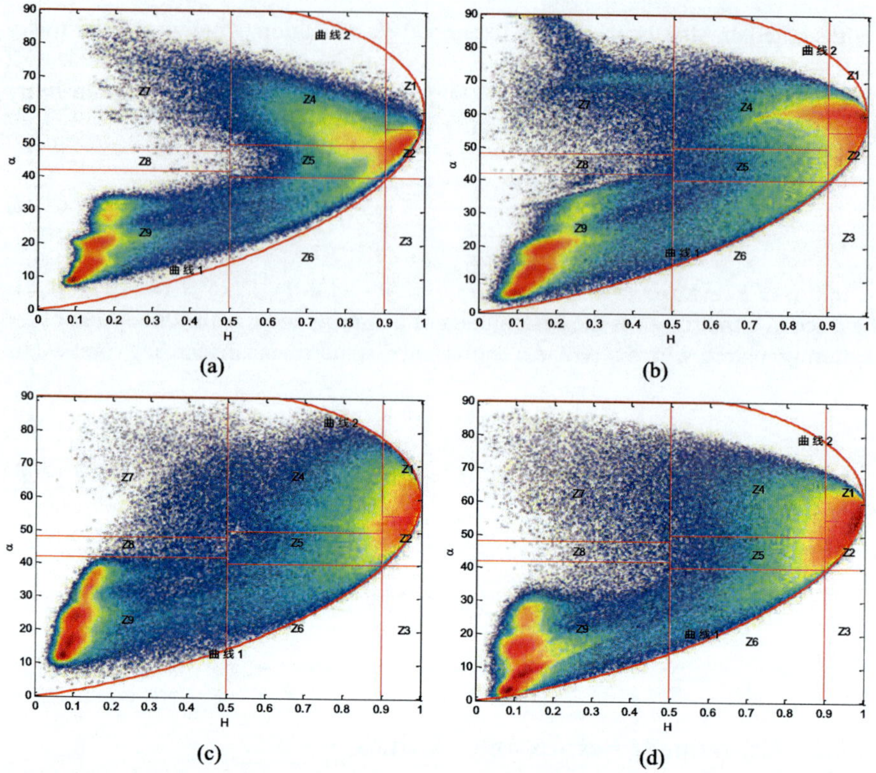

Fig. 5.7 H/α distribution of the experiment data. **a** H/α distribution of full PolSAR data, **b** H/α distribution of pseudo-quad-PolSAR data reconstructed from $\pi/4$ mode CP data, **c** H/α distribution of pseudo-quad-PolSAR data reconstructed from DCP mode CP data, and **d** H/α distribution of pseudo-quad-PolSAR data reconstructed from CTLR mode CP data

while in the classification of FP data, they are classified as Class 1 (C1). As we can see form Tables 5.1, 5.2 and 5.3, there is some inconsistency between CP data and FP data since CP introduces some information loss.

As we can see from Fig. 5.9 and Tables 5.1, 5.2 and 5.3, classification based on three different CP modes can provide ideal classification results. Among them, CTLR mode data can provide better results than other modes.

5.5 Compact Polarimetric SAR Interferometry

C-PolInSAR is a new technique that combines C-PolSAR and interferometry SAR techniques. C-PolInSAR system measures the compact polarimetric scattering

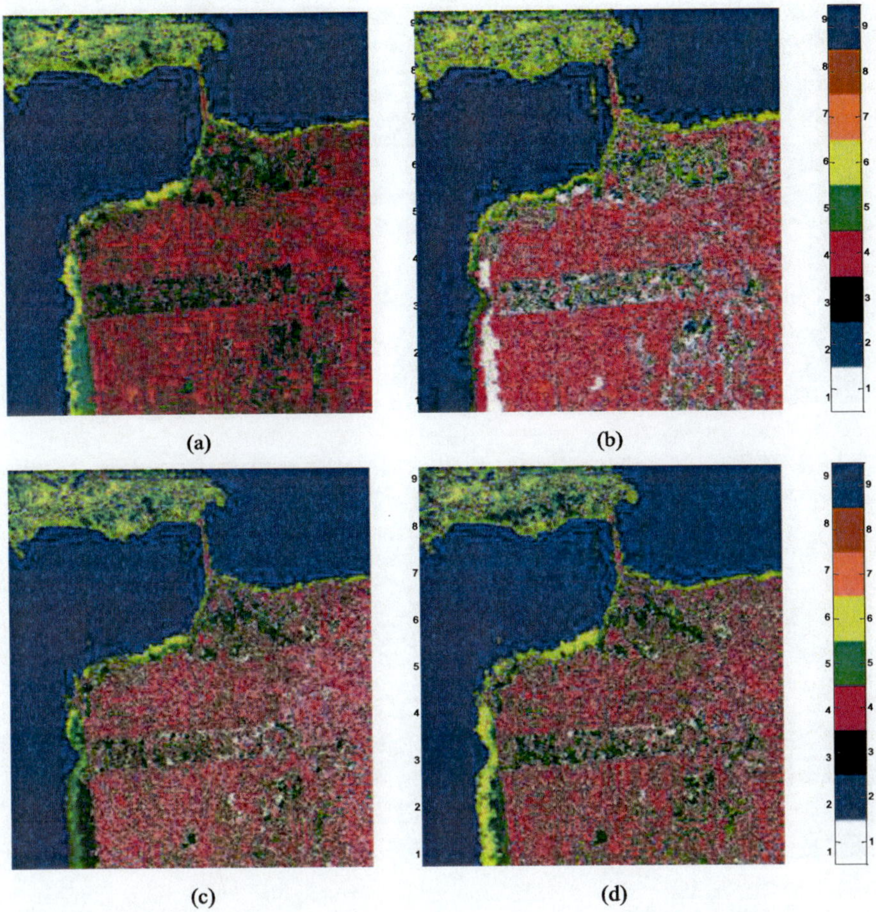

Fig. 5.8 H/α classification of the experiment data. **a** H/α classification of full PolSAR data, **b** H/α classification of $\pi/4$ mode CP data, **c** H/α classification of DCP mode CP data, and **d** H/α classification of CTLR mode CP data

matrices of the observed scene from two slightly different look angles. Like C-PolSAR, C-PolInSAR has different modes. There are also three in common applied C-PolInSAR modes based on the CP modes.

(1) $\pi/4$ mode. The scattering vector of $\pi/4$ mode is

$$\boldsymbol{k}_{\pi/4} = \left[\boldsymbol{k}_{\pi/4,1}^{\mathrm{T}} \ \boldsymbol{k}_{\pi/4,2}^{\mathrm{T}} \right]^{\mathrm{T}} \tag{5.60}$$

where $\boldsymbol{k}_{\pi/4,\mathrm{m}} = \left[S_{\mathrm{HH},m} + S_{\mathrm{HV},m} \right]^{T} / \sqrt{2}, m = 1, 2.$

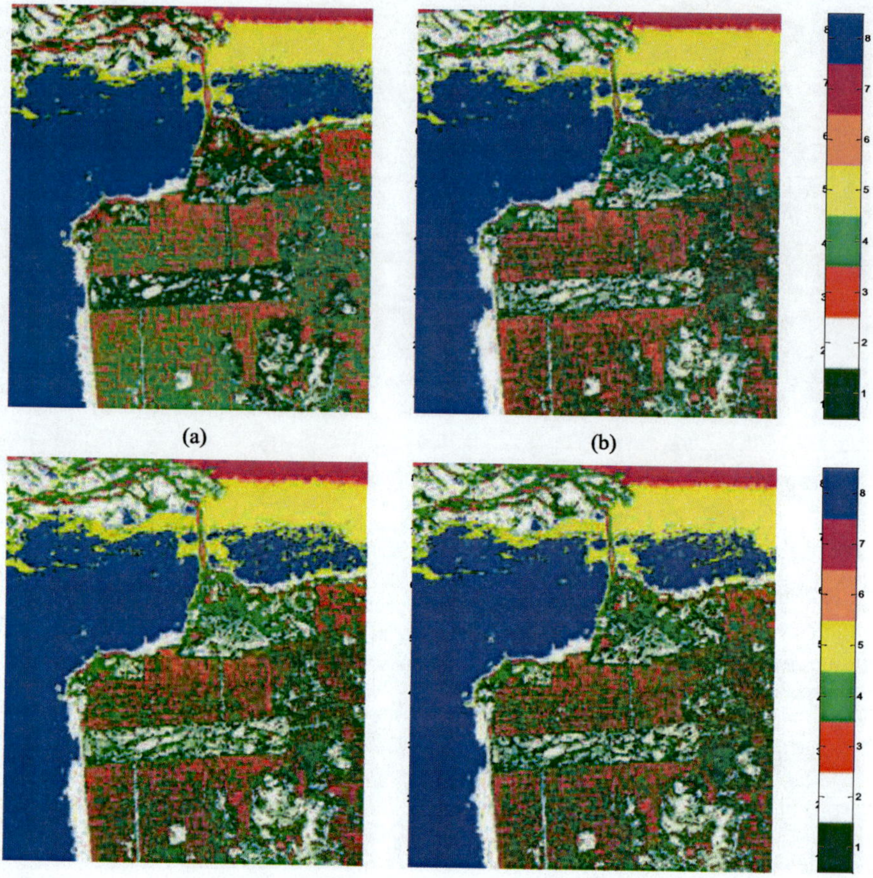

Fig. 5.9 Wishart H/α classification result of the experiment data. **a** Classification result of full PolSAR data, **b** Classification result of $\pi/4$ mode CP data, **c** Classification result of DCP mode CP data, and **d** Classification result of CTLR mode CP data

In Eq. (5.59), $m = 1, 2$ denotes the measurements of master and slave antenna, respectively. The information collected by the $\pi/4$ mode C-PolInSAR system can be expressed by the 4×4 covariance matrix $\boldsymbol{J}_{4,\pi/4}$

$$\boldsymbol{J}_{4,\pi/4} = \langle \boldsymbol{k}_{\pi/4} \boldsymbol{k}_{\pi/4}^{\mathrm{H}} \rangle = \begin{bmatrix} \boldsymbol{J}_{11} & \boldsymbol{J}_{12} \\ \boldsymbol{J}_{12}^{H} & \boldsymbol{J}_{22} \end{bmatrix} \tag{5.61}$$

(2) DCP mode. The scattering vector of DCP mode is

Table 5.1 Comparison between $\pi/4$ mode CP data and FP data

FP								
$\pi/4$ mode	C1	C2	C3	C4	C5	C6	C7	C8
C1	0.6481	0.0435	0.0232	0.5645	0.0015	2.64e−4	0	0
C2	0.2218	0.9360	2.04e−5	9.35e−6	0.0649	0	0	0.0312
C3	0	0	0.6472	0.1572	0	0.0621	0.0058	0
C4	0.1195	0	0.2836	0.2759	0.0051	0.0185	0.1377	0
C5	0.0106	0.0061	1.12e−4	0.0023	0.8924	4.39e−5	0.0051	0.0126
C6	0	0	0.0351	0	0	0.8272	0.0176	0
C7	0	0	0.0108	0	0.0179	0.0919	0.8338	0
C8	0	0.0144	0	0	0.0182	0	0	0.9561
合计	1.000	1.000	1.000	1.000	1.000	1.000	1.000	1.000

Table 5.2 Comparison between DCP mode CP data and FP data

FP								
DCP mode	C1	C2	C3	C4	C5	C6	C7	C8
C1	0.5137	0.0042	0.0719	0.5991	0	6.59e−4	0	0
C2	0.2969	0.9690	5.10e−5	0.0061	0.0195	0	0	0.0110
C3	4.27e−4	0	0.7077	0.0457	0	0.2445	0.0718	0
C4	0.1875	0.0012	0.2179	0.3486	0.0114	0.0030	0.0430	0
C5	0.0015	0.0079	0	4.49e−4	0.8867	0	3.11e−4	0.0765
C6	0	0	0.0014	0	0	0.7276	0.0643	0
C7	0	0	0.0010	0	0.0768	0.0242	0.8205	0
C8	0	0.0177	0	0	0.0056	0	0	0.9125
合计	1.000	1.000	1.000	1.000	1.000	1.000	1.000	1.000

Table 5.3 Comparison between CTLR mode CP data and FP data

FP								
CTLR mode	C1	C2	C3	C4	C5	C6	C7	C8
C1	0.6565	0.0149	0.0596	0.6370	0.0027	1.76e−004	0	0
C2	0.2279	0.9750	0	0.0096	0.0303	0	0	0.0103
C3	0.0023	0	0.6361	0.0421	0	0.1984	0.0265	0
C4	0.1128	0.0012	0.2962	0.3110	0.0127	0.0043	0.0993	0
C5	5.25e−4	0.0025	1.2e−5	2.71e−4	0.8979	0	1.56e−4	0.0589
C6	0	0	0.0060	0	0	0.7665	0.0443	0
C7	0	0	0.0021	0	0.0531	0.0306	0.8297	0
C8	0	0.0075	0	0	0.0033	0	0	0.9307
合计	1.000	1.000	1.000	1.000	1.000	1.000	1.000	1.000

$$\boldsymbol{k}_{\mathrm{DCP}} = \left[\boldsymbol{k}_{\mathrm{DCP},1}^{\mathrm{T}}\ \boldsymbol{k}_{\mathrm{DCP},2}^{\mathrm{T}}\right]^{\mathrm{T}} \tag{5.62}$$

where $\boldsymbol{k}_{\mathrm{DCP,m}} = \left[(S_{\mathrm{HH},m} - S_{\mathrm{VV},m} + i2S_{\mathrm{HV},m}\ i\left(S_{\mathrm{HH},m} + S_{\mathrm{VV},m}\right)\right]^{T}/\sqrt{2}, m = 1, 2.$

The information collected by the DCP mode C-PolInSAR system can be expressed by the 4×4 covariance matrix $\boldsymbol{J}_{4,\mathrm{DCP}}$

$$\boldsymbol{J}_{4,\mathrm{DCP}} = \langle \boldsymbol{k}_{\mathrm{DCP}} \boldsymbol{k}_{\mathrm{DCP}}^{\mathrm{H}} \rangle = \begin{bmatrix} \boldsymbol{J}_{11} & \boldsymbol{J}_{12} \\ \boldsymbol{J}_{12}^{H} & \boldsymbol{J}_{22} \end{bmatrix} \tag{5.63}$$

(3) CTLR mode. The scattering vector of CTLR mode is

$$\boldsymbol{k}_{\mathrm{CTLR}} = \left[\boldsymbol{k}_{\mathrm{CRLR},1}^{\mathrm{T}}\ \boldsymbol{k}_{\mathrm{CTLR},2}^{\mathrm{T}}\right]^{\mathrm{T}} \tag{5.64}$$

where $\boldsymbol{k}_{\mathrm{CTLR,m}} = \left[(S_{\mathrm{HH},m} - i S_{\mathrm{HV},m}\ -i S_{\mathrm{VV},m} + S_{\mathrm{HV},m}\right]^{T}/\sqrt{2}, m = 1, 2.$

The information collected by the CTLR mode C-PolInSAR system can be expressed by the 4×4 covariance matrix $\boldsymbol{J}_{4,\mathrm{CTLR}}$

$$\boldsymbol{J}_{4,\mathrm{CTLR}} = \langle \boldsymbol{k}_{\mathrm{CTLR}} \boldsymbol{k}_{\mathrm{CTLR}}^{\mathrm{H}} \rangle = \begin{bmatrix} \boldsymbol{J}_{11} & \boldsymbol{J}_{12} \\ \boldsymbol{J}_{12}^{H} & \boldsymbol{J}_{22} \end{bmatrix} \tag{5.65}$$

In Eqs. (5.61), (5.63), and (5.65), J_{11} and J_{22} are the compact polarimetric covariance matrix of master and slave antenna, respectively, and they contain compact polarimetric information collected from different positions. $\boldsymbol{J}_{12}$ is the covariance matrix which contains both compact polarimetric and interferometric information.

5.6 Reconstruction of Quad PolInSAR Covariance Matrix from Compact PolInSAR Data

The reconstruction of quad-PolInSAR data from C-PolInSAR data [6–11] is the reconstruction of 6×6 PolInSAR convariance matrix from 4×4 compact PolInSAR covariance matrix.

From Eqs. (5.61), (5.63), and (5.65), we can conclude that $\boldsymbol{J}_{11}$ and $\boldsymbol{J}_{22}$ are the compact polarimetric covariance matrix of master and slave antenna, respectively, and their reconstruction is described in Sect. 5.3. As a result, we have to focus on the reconstruction of 3×3 cross-covariance matrix $\boldsymbol{C}_{12}$ from $\boldsymbol{J}_{12}$ [12, 13].

The reconstruction of $\boldsymbol{C}_{12}$ is based on the reflection symmetry and the rotation symmetry condition. In PolInSAR, the media of reflection symmetry satisfy the following equations:

$$\langle S_{\mathrm{HH},m} S_{\mathrm{HV,n}}^{*} \rangle = \langle S_{\mathrm{VV},m} S_{\mathrm{HV,n}}^{*} \rangle = 0, m, n = 1, 2. \tag{5.66}$$

The rotation invariant media satisfy following equations [8–11],

$$4\langle S_{\mathrm{HV},1} S_{\mathrm{HV},2}^{*} \rangle = \langle S_{\mathrm{HH},1} S_{\mathrm{HH},2}^{*} \rangle + \langle S_{\mathrm{VV},1} S_{\mathrm{VV},2}^{*} \rangle - \langle S_{\mathrm{HH},1} S_{\mathrm{VV},2}^{*} \rangle - \langle S_{\mathrm{VV},1} S_{\mathrm{HH},2}^{*} \rangle \tag{5.67}$$

where m, n denote the measurements of master and slave antenna, respectively.

In the following, the reconstruction of PolInSAR covariance matrix $\boldsymbol{C}_{12}$ from different mode C-PolInSAR covariance matrix is introduced. The expression of $\boldsymbol{C}_{12}$ is.

$$\begin{bmatrix} S_{\mathrm{HH},1} S_{\mathrm{HH},2}^{*} & \sqrt{2} S_{\mathrm{HH},1} S_{\mathrm{HH},2}^{*} & S_{\mathrm{HH},1} S_{\mathrm{VV},2}^{*} \\ \sqrt{2} S_{\mathrm{HV},1} S_{\mathrm{HH},2}^{*} & 2 S_{\mathrm{HV},1} S_{\mathrm{HV},2}^{*} & \sqrt{2} S_{\mathrm{HV},1} S_{\mathrm{VV},2}^{*} \\ S_{\mathrm{VV},1} S_{\mathrm{HH},2}^{*} & \sqrt{2} S_{\mathrm{VV},1} S_{\mathrm{HV},2}^{*} & S_{\mathrm{VV},1} S_{\mathrm{VV},2}^{*} \end{bmatrix} \tag{5.68}$$

5.6.1 *Reconstruction from π/4 Mode Compact POLInSAR Covariance Matrix*

From Eqs. (5.60) and (5.61), and ignoring constant item, we can get the expression of π/4 mode C-PolInSAR cross-covariance matrix

$$\boldsymbol{J}_{12} = \begin{bmatrix} j_{11} & j_{12} \\ j_{21} & j_{22} \end{bmatrix} \tag{5.69}$$

$$j_{11} = \langle S_{\mathrm{HH},1} S_{\mathrm{HH},2}^{*} \rangle + \langle S_{\mathrm{HH},1} S_{\mathrm{HV},2}^{*} \rangle + \langle S_{\mathrm{HV},1} S_{\mathrm{HH},2}^{*} \rangle + \langle S_{\mathrm{HV},1} S_{\mathrm{HV},2}^{*} \rangle \tag{5.70}$$

$$j_{12} = \langle S_{\mathrm{HH},1} S_{\mathrm{VV},2}^{*} \rangle + \langle S_{\mathrm{HH},1} S_{\mathrm{HV},2}^{*} \rangle + \langle S_{\mathrm{HV},1} S_{\mathrm{VV},2}^{*} \rangle + \langle S_{\mathrm{HV},1} S_{\mathrm{HV},2}^{*} \rangle \tag{5.71}$$

$$j_{21} = \langle S_{\mathrm{VV},1} S_{\mathrm{HH},2}^{*} \rangle + \langle S_{\mathrm{VV},1} S_{\mathrm{HV},2}^{*} \rangle + \langle S_{\mathrm{HV},1} S_{\mathrm{HH},2}^{*} \rangle + \langle S_{\mathrm{HV},1} S_{\mathrm{HV},2}^{*} \rangle \tag{5.72}$$

$$j_{22} = \langle S_{\mathrm{VV},1} S_{\mathrm{VV},2}^{*} \rangle + \langle S_{\mathrm{VV},1} S_{\mathrm{HV},2}^{*} \rangle + \langle S_{\mathrm{HV},1} S_{\mathrm{VV},2}^{*} \rangle + \langle S_{\mathrm{HV},1} S_{\mathrm{HV},2}^{*} \rangle \tag{5.73}$$

There are four measurements and nine unknowns in Eqs. (5.70)–(5.73). Based on reflection symmetry assumption, Eqs. (5.70)–(5.73) can be expressed as

$$j_{11} \approx \langle S_{\mathrm{HH},1} S_{\mathrm{HH},2}^{*} \rangle + \langle S_{\mathrm{HV},1} S_{\mathrm{HV},2}^{*} \rangle \tag{5.74}$$

$$j_{12} \approx \langle S_{\mathrm{HH},1} S_{\mathrm{VV},2}^{*} \rangle + \langle S_{\mathrm{HV},1} S_{\mathrm{HV},2}^{*} \rangle \tag{5.75}$$

$$j_{21} \approx \langle S_{\mathrm{VV},1} S^*_{\mathrm{HH},2} \rangle + \langle S_{\mathrm{HV},1} S^*_{\mathrm{HV},2} \rangle \tag{5.76}$$

$$j_{22} = \langle S_{\mathrm{VV},1} S^*_{\mathrm{VV},2} \rangle + \langle S_{\mathrm{HV},1} S^*_{\mathrm{HV},2} \rangle \tag{5.77}$$

Let $\langle S_{\mathrm{HH},1} S^*_{\mathrm{HH},2} \rangle = x$, $\langle S_{\mathrm{HV},1} S^*_{\mathrm{HV},2} \rangle = m$, $\langle S_{\mathrm{HH},1} S^*_{\mathrm{VV},2} \rangle = y$, $\langle S_{\mathrm{VV},1} S^*_{\mathrm{HH},2} \rangle = z$, and $\langle S_{\mathrm{VV},1} S^*_{\mathrm{VV},2} \rangle = s$, we can get the following expressions through the rotation invariance assumption of the cross-polarization item,

$$\begin{aligned} x + m &= j_{11} \\ y + m &= j_{12} \\ z + m &= j_{21} \\ s + m &= j_{22} \\ 4m &= x + s - y - z. \end{aligned} \tag{5.78}$$

Solving Eq. (5.78), we can get

$$\langle S_{\mathrm{HH},1} S^*_{\mathrm{HH},2} \rangle = \frac{3j_{11} - j_{22} + j_{12} + j_{21}}{4} \tag{5.79}$$

$$\langle S_{\mathrm{HH},1} S^*_{\mathrm{VV},2} \rangle = \frac{5j_{12} - j_{11} - j_{22} + j_{21}}{4} \tag{5.80}$$

$$\langle S_{\mathrm{VV},1} S^*_{\mathrm{HH},2} \rangle = \frac{5j_{21} - j_{11} - j_{22} + j_{12}}{4} \tag{5.81}$$

$$\langle S_{\mathrm{HV},1} S^*_{\mathrm{HV},2} \rangle = \frac{j_{11} + j_{22} - j_{12} - j_{21}}{4} \tag{5.82}$$

$$\langle S_{\mathrm{VV},1} S^*_{\mathrm{VV},2} \rangle = \frac{3j_{22} - j_{11} + j_{12} + j_{21}}{4} \tag{5.83}$$

The expression of reconstructed $\boldsymbol{C}_{12}$ can be easily obtained by Eq. (5.79)–(5.83), which is

$$\boldsymbol{C}_{12} = \begin{bmatrix} \langle S_{\mathrm{HH},1} S^*_{\mathrm{HH},2} \rangle & 0 & \langle S_{\mathrm{HH},1} S^*_{\mathrm{VV},2} \rangle \\ 0 & 2\langle S_{\mathrm{HV},1} S^*_{\mathrm{HV},2} \rangle & 0 \\ \langle S_{\mathrm{VV},1} S^*_{\mathrm{HH},2} \rangle & 0 & \langle S_{\mathrm{VV},1} S^*_{\mathrm{VV},2} \rangle \end{bmatrix} \tag{5.84}$$

5.6.2 Reconstruction from DCP Mode Compact POLInSAR Covariance Matrix

The expression of DCP mode C-PolInSAR cross-covariance matrix is

$$\boldsymbol{J}_{12} = \begin{bmatrix} j_{11} & j_{12} \\ j_{21} & j_{22} \end{bmatrix} \tag{5.85}$$

$$\begin{aligned} j_{11} = & \langle S_{\mathrm{HH},1} S_{\mathrm{HH},2}^{*} \rangle - \langle S_{\mathrm{HH},1} S_{\mathrm{VV},2}^{*} \rangle - 2i \langle S_{\mathrm{HH},1} S_{\mathrm{HV},2}^{*} \rangle \\ & - \langle S_{\mathrm{VV},1} S_{\mathrm{HH},2}^{*} \rangle + \langle S_{\mathrm{VV},1} S_{\mathrm{VV},2}^{*} \rangle + 2i \langle S_{\mathrm{VV},1} S_{\mathrm{HV},2}^{*} \rangle \\ & + 2i \langle S_{\mathrm{HV},1} S_{\mathrm{HH},2}^{*} \rangle - 2i \langle S_{\mathrm{HV},1} S_{\mathrm{VV},2}^{*} \rangle + 4 \langle S_{\mathrm{HV},1} S_{\mathrm{HV},2}^{*} \rangle \end{aligned} \tag{5.86}$$

$$\begin{aligned} j_{12} = & 2 \langle S_{\mathrm{HV},1} S_{\mathrm{HH},2}^{*} \rangle - \mathrm{i} \langle S_{\mathrm{HH},1} S_{\mathrm{HH},2}^{*} \rangle - i \langle S_{\mathrm{HH},1} S_{\mathrm{VV},2}^{*} \rangle \\ & + i \langle S_{\mathrm{VV},1} S_{\mathrm{HH},2}^{*} \rangle + i \langle S_{\mathrm{VV},1} S_{\mathrm{VV},2}^{*} \rangle + 2 \langle S_{\mathrm{HV},1} S_{\mathrm{VV},2}^{*} \rangle \end{aligned} \tag{5.87}$$

$$\begin{aligned} j_{21} = & i \langle S_{\mathrm{HH},1} S_{\mathrm{HH},2}^{*} \rangle - i \langle S_{\mathrm{HH},1} S_{\mathrm{VV},2}^{*} \rangle + 2 \langle S_{\mathrm{HH},1} S_{\mathrm{HV},2}^{*} \rangle \\ & + i \langle S_{\mathrm{VV},1} S_{\mathrm{HH},2}^{*} \rangle - i \langle S_{\mathrm{VV},1} S_{\mathrm{VV},2}^{*} \rangle + \langle S_{\mathrm{VV},1} S_{\mathrm{HV},2}^{*} \rangle \end{aligned} \tag{5.88}$$

$$j_{22} = \langle S_{\mathrm{HH},1} S_{\mathrm{HH},2}^{*} \rangle + \langle S_{\mathrm{HH},1} S_{\mathrm{VV},2}^{*} \rangle + \langle S_{\mathrm{VV},1} S_{\mathrm{HH},2}^{*} \rangle + \langle S_{\mathrm{VV},1} S_{\mathrm{VV},2}^{*} \rangle \tag{5.89}$$

Based on the reflection symmetry assumption, Eq. (5.86)–Eq. (5.89) can be expressed as

$$j_{11} = \langle S_{\mathrm{HH},1} S_{\mathrm{HH},2}^{*} \rangle - \langle S_{\mathrm{HH},1} S_{\mathrm{VV},2}^{*} \rangle - \langle S_{\mathrm{VV},1} S_{\mathrm{HH},2}^{*} \rangle + \langle S_{\mathrm{VV},1} S_{\mathrm{VV},2}^{*} \rangle + 4 \langle S_{\mathrm{HV},1} S_{\mathrm{HV},2}^{*} \rangle \tag{5.90}$$

$$j_{12} = -\mathrm{i} \langle S_{\mathrm{HH},1} S_{\mathrm{HH},2}^{*} \rangle - i \langle S_{\mathrm{HH},1} S_{\mathrm{VV},2}^{*} \rangle + i \langle S_{\mathrm{VV},1} S_{\mathrm{HH},2}^{*} \rangle + i \langle S_{\mathrm{VV},1} S_{\mathrm{VV},2}^{*} \rangle \tag{5.91}$$

$$j_{21} = i \langle S_{\mathrm{HH},1} S_{\mathrm{HH},2}^{*} \rangle - i \langle S_{\mathrm{HH},1} S_{\mathrm{VV},2}^{*} \rangle + i \langle S_{\mathrm{VV},1} S_{\mathrm{HH},2}^{*} \rangle - i \langle S_{\mathrm{VV},1} S_{\mathrm{VV},2}^{*} \rangle \tag{5.92}$$

$$j_{22} = \langle S_{\mathrm{HH},1} S_{\mathrm{HH},2}^{*} \rangle + \langle S_{\mathrm{HH},1} S_{\mathrm{VV},2}^{*} \rangle + \langle S_{\mathrm{VV},1} S_{\mathrm{HH},2}^{*} \rangle + \langle S_{\mathrm{VV},1} S_{\mathrm{VV},2}^{*} \rangle \tag{5.93}$$

Let $\langle S_{\mathrm{HH},1} S_{\mathrm{HH},2}^{*} \rangle = x$, $\langle S_{\mathrm{HV},1} S_{\mathrm{HV},2}^{*} \rangle = m$, $\langle S_{\mathrm{HH},1} S_{\mathrm{VV},2}^{*} \rangle = y$, $\langle S_{\mathrm{VV},1} S_{\mathrm{HH},2}^{*} \rangle = z$, and $\langle S_{\mathrm{VV},1} S_{\mathrm{VV},2}^{*} \rangle = s$, we can get the following expressions through the rotation invariance assumption in Eq. (5.67):

$$\begin{aligned} &x - y - z + s + 4m = j_{11} \\ &-ix - iy + iz + is = j_{12} \\ &ix - iy + iz - is = j_{21} \\ &x + y + z + s = j_{22} \\ &4m = x + s - y - z. \end{aligned} \tag{5.94}$$

Solving Eq. (5.94), we can get

$$\langle S_{\mathrm{HH},1} S^*_{\mathrm{HH},2} \rangle = \frac{2j_{22} + 2ij_{12} - 2ij_{21} + j_{11}}{8} \tag{5.95}$$

$$\langle S_{\mathrm{HH},1} S^*_{\mathrm{VV},2} \rangle = \frac{2ij_{12} + 2ij_{21} + 2j_{22} - j_{11}}{8} \tag{5.96}$$

$$\langle S_{\mathrm{VV},1} S^*_{\mathrm{HH},2} \rangle = \frac{2j_{22} - j_{11} - 2ij_{12} - 2ij_{21}}{8} \tag{5.97}$$

$$\langle S_{\mathrm{HV},1} S^*_{\mathrm{HV},2} \rangle = \frac{j_{11}}{8} \tag{5.98}$$

$$\langle S_{\mathrm{VV},1} S^*_{\mathrm{VV},2} \rangle = \frac{j_{11} + 2j_{22} - 2ij_{12} + 2ij_{21}}{8} \tag{5.99}$$

The expression of reconstructed $\boldsymbol{C}_{12}$ can be easily obtained by Eq. (5.95)–(5.99), which is shown in Eq. (5.84)

5.6.3 Reconstruction from CTLR Mode Compact POLInSAR Covariance Matrix

From Eqs. (5.64) and (5.65), and ignoring constant item, we can get the expression of CTLR mode C-PolInSAR cross-covariance matrix

$$\boldsymbol{J}_{12} = \begin{bmatrix} j_{11} & j_{12} \\ j_{21} & j_{22} \end{bmatrix} \tag{5.100}$$

$$j_{11} = \langle S_{\mathrm{HH},1} S^*_{\mathrm{HH},2} \rangle + i\langle S_{\mathrm{HH},1} S^*_{\mathrm{HV},2} \rangle - i\langle S_{\mathrm{HV},1} S^*_{\mathrm{HH},2} \rangle + \langle S_{\mathrm{HV},1} S^*_{\mathrm{HV},2} \rangle \tag{5.101}$$

$$j_{12} = i\langle S_{\mathrm{HH},1} S^*_{\mathrm{VV},2} \rangle + \langle S_{\mathrm{HH},1} S^*_{\mathrm{HV},2} \rangle + \langle S_{\mathrm{HV},1} S^*_{\mathrm{VV},2} \rangle - i\langle S_{\mathrm{HV},1} S^*_{\mathrm{HV},2} \rangle \tag{5.102}$$

$$j_{21} = -i\langle S_{\mathrm{VV},1} S^*_{\mathrm{HH},2} \rangle + \langle S_{\mathrm{VV},1} S^*_{\mathrm{HV},2} \rangle + \langle S_{\mathrm{HV},1} S^*_{\mathrm{HH},2} \rangle + i\langle S_{\mathrm{HV},1} S^*_{\mathrm{HV},2} \rangle \tag{5.103}$$

$$j_{22} = \langle S_{\mathrm{VV},1} S^*_{\mathrm{VV},2} \rangle - \mathrm{i}\langle S_{\mathrm{VV},1} S^*_{\mathrm{HV},2} \rangle + i\langle S_{\mathrm{HV},1} S^*_{\mathrm{VV},2} \rangle + \langle S_{\mathrm{HV},1} S^*_{\mathrm{HV},2} \rangle \tag{5.104}$$

Based on the reflection symmetry assumption, we can get

$$j_{11} = \langle S_{\mathrm{HH},1} S^*_{\mathrm{HH},2} \rangle + \langle S_{\mathrm{HV},1} S^*_{\mathrm{HV},2} \rangle \tag{5.105}$$

$$j_{12} = i \langle S_{\mathrm{HH},1} S^*_{\mathrm{VV},2} \rangle - i \langle S_{\mathrm{HV},1} S^*_{\mathrm{HV},2} \rangle \tag{5.106}$$

$$j_{21} = -i \langle S_{\mathrm{VV},1} S^*_{\mathrm{HH},2} \rangle + i \langle S_{\mathrm{HV},1} S^*_{\mathrm{HV},2} \rangle \tag{5.107}$$

$$j_{22} = \langle S_{\mathrm{VV},1} S^*_{\mathrm{VV},2} \rangle + \langle S_{\mathrm{HV},1} S^*_{\mathrm{HV},2} \rangle \tag{5.108}$$

Let $\langle S_{\mathrm{HH},1} S^*_{\mathrm{HH},2} \rangle = x$, $\langle S_{\mathrm{HV},1} S^*_{\mathrm{HV},2} \rangle = m$, $\langle S_{\mathrm{HH},1} S^*_{\mathrm{VV},2} \rangle = y$, $\langle S_{\mathrm{VV},1} S^*_{\mathrm{HH},2} \rangle = z$, and $\langle S_{\mathrm{VV},1} S^*_{\mathrm{VV},2} \rangle = s$, we can get the following expressions through the rotation invariance assumption in Eq. (5.67):

$$\begin{aligned} x + m &= j_{11} \\ iy - ix &= j_{12} \\ -iz + im &= j_{21} \\ s + m &= j_{22} \\ 4m &= x + s - y - z \end{aligned} \tag{5.109}$$

Solving Eq. (5.109), we can get

$$\langle S_{\mathrm{HH},1} S^*_{\mathrm{HH},2} \rangle = \frac{7j_{11} - j_{22} - j_{12} + ij_{21}}{8} \tag{5.110}$$

$$\langle S_{\mathrm{HH},1} S^*_{\mathrm{VV},2} \rangle = \frac{j_{11} + j_{22} - 7j_{12} - ij_{21}}{8} \tag{5.111}$$

$$\langle S_{\mathrm{VV},1} S^*_{\mathrm{HH},2} \rangle = \frac{j_{11} + j_{22} + j_{12} + 7ij_{21}}{8} \tag{5.112}$$

$$\langle S_{\mathrm{HV},1} S^*_{\mathrm{HV},2} \rangle = \frac{j_{11} + j_{22} + \mathrm{j}_{12} - ij_{21}}{8} \tag{5.113}$$

$$\langle S_{\mathrm{VV},1} S^*_{\mathrm{VV},2} \rangle = \frac{7j_{22} - j_{11} - j_{12} + ij_{21}}{8} \tag{5.114}$$

The expression of reconstructed $\boldsymbol{C}_{12}$ can be obtained by Eqs. (5.110)–(5.114).

5.6.4 Experiment Results and Analysis

L-band compact PolInSAR data generated by the PolInSAR data simulated by the simulator of the PolSAR processing software is used for illustration. Since the most important variables in the PolInSAR covariance matrix $\boldsymbol{C}_{12}$ are the interferometric phase of different channels, thus interferometric phases of different channels obtained from reconstructed data and PolInSAR data are compared mainly.

Figures 5.10, 5.11 and 5.12 show the comparison of interferometric phases of different polarimetric channels between reconstructed data and full PolInSAR data. Among them, (a), (b), (c) indicate the interferometric phases of HH, HV, and VV channels, respectively.

As we can see from Figs. 5.10, 5.11 and 5.12, the interferometric phases of HH and VV channels obtained from reconstructed data are very similar to the phases of full PolInSAR data, while the interferometric phases of HV channels are a little biased.

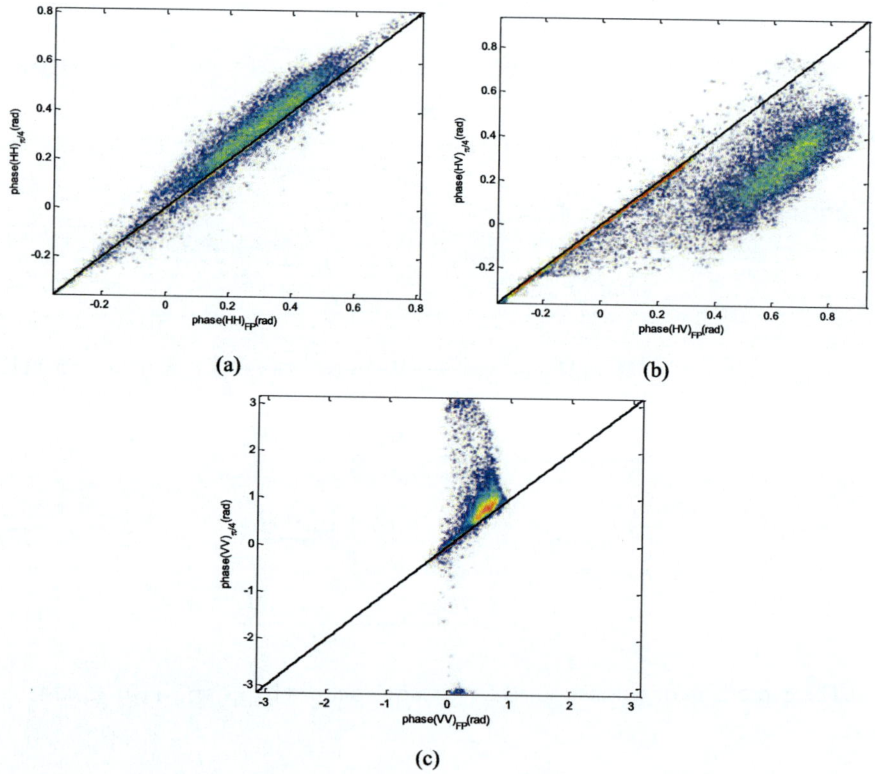

Fig. 5.10 Interfereometric phases of reconstructed data compared with full PolInSAR data of $\pi/4$ mode; **a** Comparison of the phase of HH channel, **b** Comparison of the phase of HV channel, **c** Comparison of the phase of VV channel

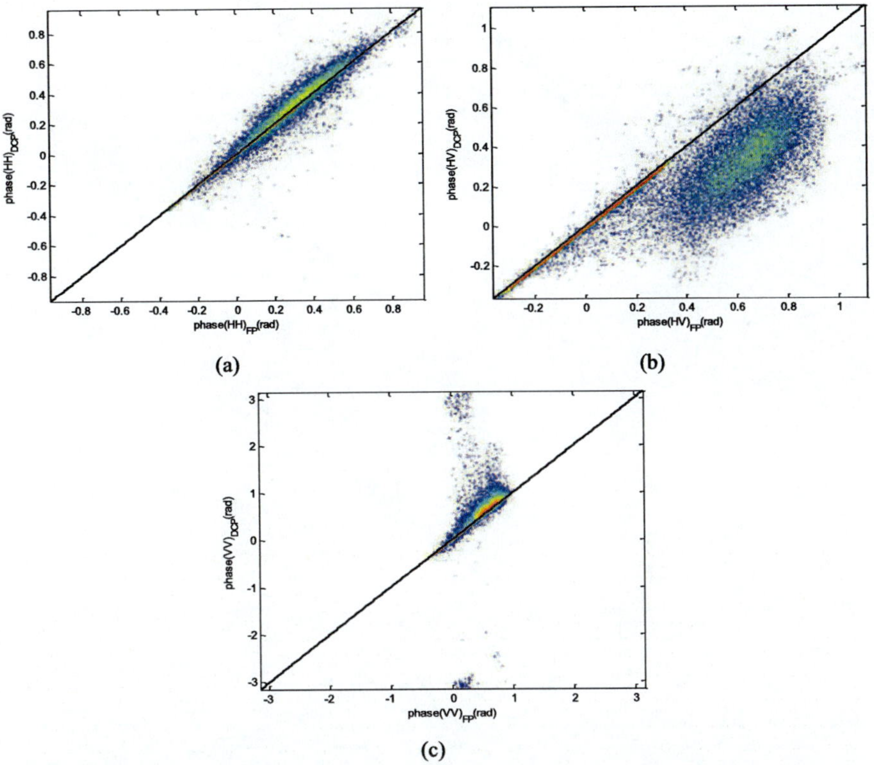

Fig. 5.11 Interfereometric phases of reconstructed data compared with full PolInSAR data of DCP mode; **a** Comparison of the phase of HH channel, **b** Comparison of the phase of HV channel, **c** Comparison of the phase of VV channel

5.7 Coherence Optimization of Compact PolInSAR Data

As in the C-PolInSAR architecture, the polarization state of the transmitting signal is fixed, we can only get the interferometric coefficient of arbitrary receiving polarization of specific transmitting polarization. Like PolInSAR system, the coherence optimization of C-PolInSAR can be categorized as amplitude coherence optimization and phase coherence optimization. Amplitude coherence optimization is to find the receiving polarization state which gets the maximum value of the amplitude of coherence. Phase coherence optimization is to find the receiving polarization state which makes the phases of coherence most distinct.

During the coherence optimization processing of C-PolInSAR, reconstruction of the 6×6PolInSAR covariance matrix from 4×4 C-PolInSAR covariance matrix is meaningless, since the reconstruction only expand the observation space artificially while the information is not increased. As a result, the coherence optimization algorithm investigated in this section is performed to 4×4 C-PolInSAR covariance matrix directly.

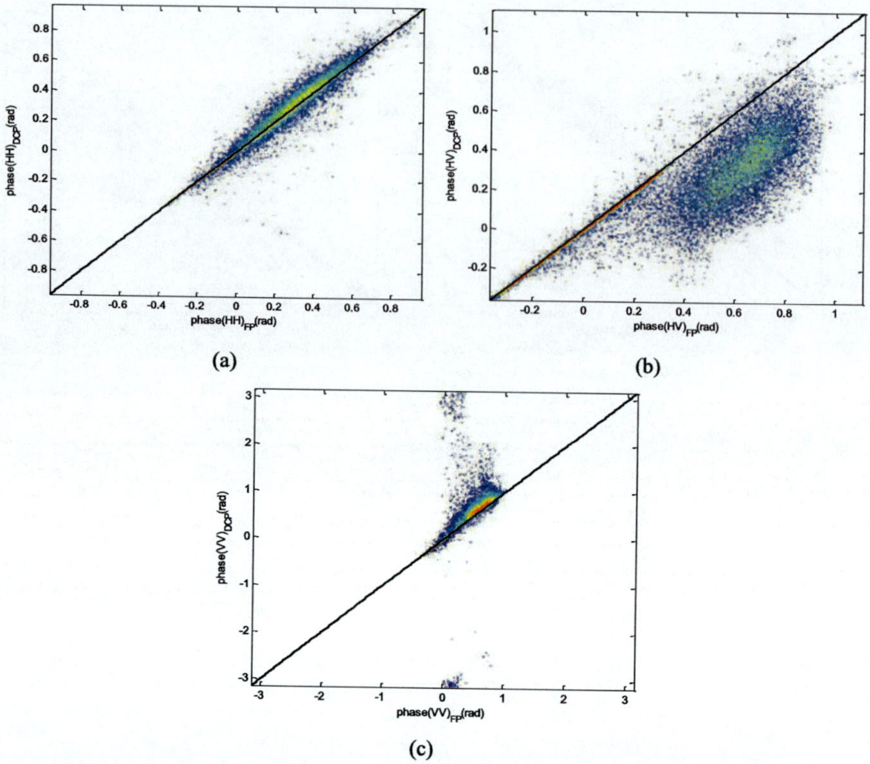

Fig. 5.12 Interfereometric phases of reconstructed data compared with full PolInSAR data of CTLR mode; **a** Comparison of the phase of HH channel, **b** Comparison of the phase of HV channel, **c** Comparison of the phase of VV channel

5.7.1 Amplitude Coherence Optimization of Compact PolInSAR

The coherence coefficient of C-PolInSAR data can be written as

$$\gamma(\boldsymbol{\omega}_1, \boldsymbol{\omega}_2) = \frac{\boldsymbol{\omega}_1^{\mathrm{H}} \boldsymbol{J}_{12} \boldsymbol{\omega}_2}{\sqrt{\boldsymbol{\omega}_1^{\mathrm{H}} \boldsymbol{J}_{11} \boldsymbol{\omega}_1} \sqrt{\boldsymbol{\omega}_2^{\mathrm{H}} \boldsymbol{J}_{22} \boldsymbol{\omega}_2}}, \tag{5.115}$$

where $\boldsymbol{\omega}_1$ and $\boldsymbol{\omega}_2$ are the 2×1 unitary projection vector representing the main and slave images.

Similar to the coherence optimization processing of PolInSAR, the Lagrangian function can be defined as

$$L = \boldsymbol{\omega}_1^{\mathrm{H}} \boldsymbol{J}_{12} \boldsymbol{\omega}_2 + \lambda_1 \left(\boldsymbol{\omega}_1^{\mathrm{H}} \boldsymbol{J}_{11} \boldsymbol{\omega}_1 - C_1 \right) + \lambda_2 \left(\boldsymbol{\omega}_2^{\mathrm{H}} \boldsymbol{J}_{22} \boldsymbol{\omega}_2 - C_2 \right), \tag{5.116}$$

where C_1 and C_2 are the constants, and λ_1 and λ_1 are the Lagrange multipliers introduced.

We can simplify the maximization problem by setting the partial derivatives to zero

$$\frac{\partial L}{\partial \boldsymbol{\omega}_1^{\mathrm{H}}} = \boldsymbol{J}_{12}\boldsymbol{\omega}_2 + \lambda_1 \boldsymbol{J}_{11}\boldsymbol{\omega}_1 = 0 \tag{5.117}$$

$$\frac{\partial L}{\partial \boldsymbol{\omega}_2^{\mathrm{H}}} = \boldsymbol{J}_{12}^{\mathrm{H}}\boldsymbol{\omega}_1 + \lambda_2 \boldsymbol{J}_{22}\boldsymbol{\omega}_2 = 0 \tag{5.118}$$

We can obtain the optimized projection vectors $\boldsymbol{\omega}_{1\text{opt}}$ and $\boldsymbol{\omega}_{2\text{opt}}$ corresponding to maximum $|\gamma(\boldsymbol{\omega}_1, \boldsymbol{\omega}_2)|$ by the eigenvalue decomposition of the matrix

$$\boldsymbol{J}_{11}^{-1}\boldsymbol{\Omega}_{12}\boldsymbol{J}_{22}^{-1}\boldsymbol{\Omega}_{12}^{H}\boldsymbol{\omega}_{1\text{opt}} = \boldsymbol{\nu}\boldsymbol{\omega}_{1\text{opt}} \tag{5.119}$$

$$\boldsymbol{J}_{22}^{-1}\boldsymbol{\Omega}_{12}^{H}\boldsymbol{J}_{11}^{-1}\boldsymbol{\Omega}_{12}\boldsymbol{\omega}_{2\text{opt}} = \boldsymbol{\nu}\boldsymbol{\omega}_{2\text{opt}} \tag{5.120}$$

The amplitude optimized coherence coefficient can be obtained by the decomposed eigenvectors.

5.7.2 Phase Coherence Optimization of Compact PolInSAR

Similar to the phase coherence optimization processing of PolInSAR, the phase coherence optimization of C-PolInSAR is assuming the scattering mechanism of the master and slave images are identical, and to find a pair of polarization combinations to make the generated interferogram have the largest phase shift.

The processing of phase coherence optimization is performed on the phase only, and the scattering mechanism of the master and slave antennas are the same, so the coherence coefficient can be written as

$$\gamma = \boldsymbol{\omega}^{\mathrm{H}}\boldsymbol{J}_{12}\boldsymbol{\omega} \tag{5.121}$$

As the tangent function is used to measure the phase angle, the phase angle have to be shifted to the principal value interval of the tangent function, which is $[-\pi/2, \pi/2]$.

$$\gamma = \boldsymbol{\omega}^{\mathrm{H}}\boldsymbol{J}_{12}\exp(j\varphi)\boldsymbol{\omega}, \tag{5.122}$$

where $\varphi = -\angle\text{trace}(\boldsymbol{J}_{12})$, and $\angle$ denotes the argument, trace$(\cdot)$ denotes the trace of a matrix.

Let $\widehat{\boldsymbol{J}}_{12} = \boldsymbol{J}_{12}\boldsymbol{J}_{12}\exp(j\varphi)$, and

$$\widehat{\gamma} = \boldsymbol{\omega}^{\mathrm{H}}\widehat{\boldsymbol{J}}_{12}\boldsymbol{\omega}, \tag{5.123}$$

so $\tan\angle\widehat{\gamma} = \mathrm{Im}(\widehat{\gamma})/\mathrm{Re}(\widehat{\gamma})$, where $\mathrm{Im}(\cdot)$ and $\mathrm{Re}(\cdot)$ denote the real part and imaginary part of a complex, respectively.

$$\mathrm{Im}(\widehat{\gamma}) = \frac{1}{2j}(\boldsymbol{\omega}^{\mathrm{H}}\widehat{\boldsymbol{J}}_{12}\boldsymbol{\omega} - \boldsymbol{\omega}^{\mathrm{H}}\widehat{\boldsymbol{J}}_{12}^{\mathrm{H}}\boldsymbol{\omega}) \tag{5.124}$$

$$\mathrm{Re}(\widehat{\gamma}) = \frac{1}{2}(\boldsymbol{\omega}^{\mathrm{H}}\widehat{\boldsymbol{J}}_{12}\boldsymbol{\omega} + \boldsymbol{\omega}^{\mathrm{H}}\widehat{\boldsymbol{J}}_{12}^{\mathrm{H}}\boldsymbol{\omega}) \tag{5.125}$$

Consequently, the problem transforms to solving the polarization combination which makes the maximum $\tan\angle\widehat{\gamma}$. According to Lagrange multipliers method, the problem is solving the eigenvalue of the matrix,

$$-j\left(\widehat{\boldsymbol{J}}_{12} + \widehat{\boldsymbol{J}}_{12}^{\mathrm{H}}\right)^{-1}\left(\widehat{\boldsymbol{J}}_{12} - \widehat{\boldsymbol{J}}_{12}^{\mathrm{H}}\right)\boldsymbol{\omega} = \lambda\boldsymbol{\omega}. \tag{5.126}$$

The corresponding phase optimization coherence γ_{div1} and γ_{div2} can be obtained by the decomposition of the eigenvalues.

5.7.3 *Experiment Results and Analysis*

Three modes of C-PolInSAR data generated from full PolInSAR data of Oberpfaffenhofen area acquired by ESAR is used. Figure 5.13 shows the amplitude coherence optimization of full PolInSAR data and C-PolInSAR data of different modes. Subfigures (a)–(d) denote the optimum amplitude coherence coefficient of full PolInSAR, $\pi/4$ mode, DCP mode, and CTLR mode compact PolInSAR data, respectively.

Figure 5.14 shows the histogram of the optimum amplitude coherence coefficient of different C-PolInSAR modes and full PolInSAR data.

As we can see from Figs. 5.13 and 5.14, the optimum amplitude coherence coefficients of compact PolInSAR data is quite similar but smaller to full PolInSAR data. It is because, in C-PolInSAR, the polarization state of the transmitting signal is fixed, while in full PolInSAR, the polarization state of the transmitting signal can be adjusted. The observation space of C-PolInSAR is smaller than full PolInSAR.

Figure 5.15 shows the histogram of the phase difference between the phase optimized coherence coefficients of full PolInSAR data and C-PolInSAR data, which is the histogram of maximum phase difference can be distinguished by full PolInSAR data and C-PolInSAR data.

It is shown in Fig. 5.15 that the phase difference distinguished by full PolInSAR data is larger than C-PolInSAR data.

In order to further analyze the optimum coherence coefficients of different C-PolInSAR modes, the optimum coherence coefficients with different channels of

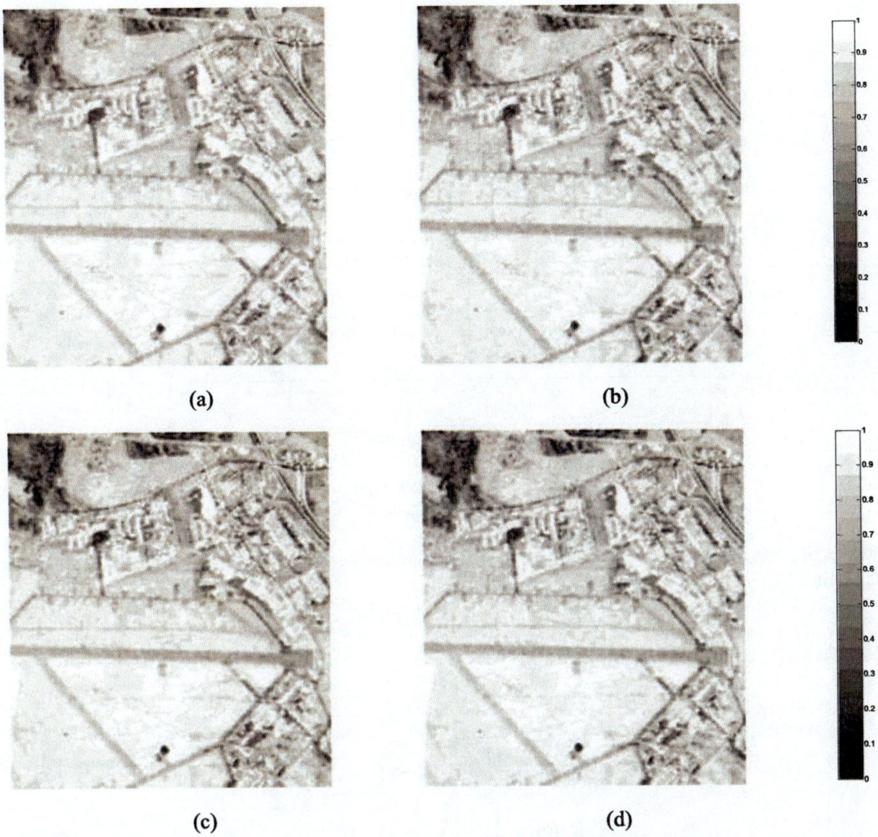

Fig. 5.13 The amplitude optimum coherence coefficients of full PolInSAR data and C-PolInSAR data. **a** Full PolInSARdata, **b** $\pi/4$ mode C-PolInSAR data, **c** DCP mode C-PolInSAR data, and **d** CTLR mode C-PolInSAR data

different classes of terrains are used for statistics. Figure 5.16 shows the selected forest, runway, and urban areas.

The amplitude optimum coherence coefficients and phase optimum coherence coefficients of the selected areas are listed in Tables 5.4 and 5.5.

As we can see from Figs. 5.4 and 5.5, in the forest and urban areas, the performance of $\pi/4$ mode is the best for it gives the maximum amplitude and the most phase difference, while in the runway area, CTLR mode gives the best performance.

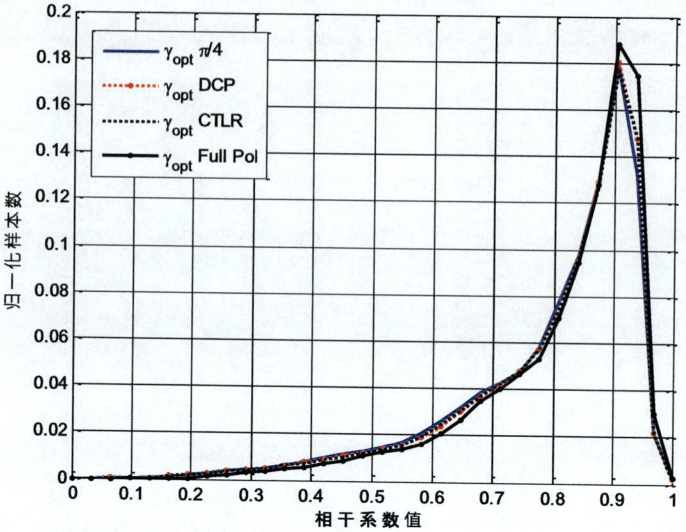

Fig. 5.14 Histogram of the amplitude optimum coherence coefficients of full PolInSAR data and C-PolInSAR data

Fig. 5.15 Histogram of the phase optimum coherence coefficients of full PolInSAR data and C-PolInSAR data

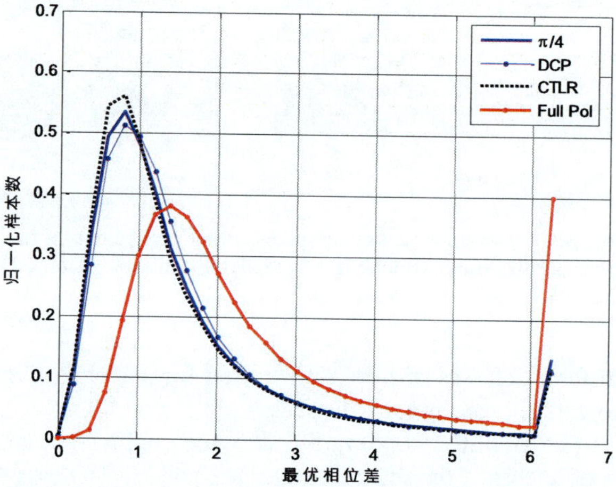

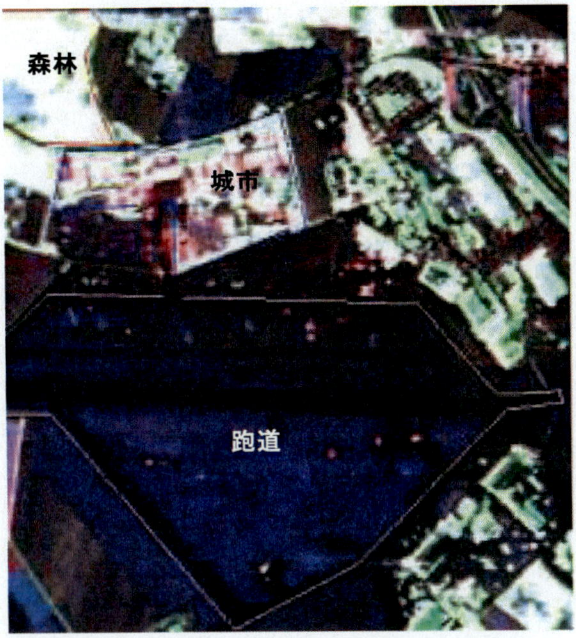

Fig. 5.16 Selected different classes of terrains

Table 5.4 The optimum amplitude coherence coefficients of different areas

Areas	π/4	DCP	CTLR	Full PolInSAR
Forest	0.4087	0.3990	0.3998	0.4501
Runway	0.8640	0.8714	0.8710	0.8779
Urban area	0.7887	0.7881	0.7879	0.8110

Table 5.5 The optimum phase coherence coefficients of different areas

Areas	π/4	DCP	CTLR	Full PolInSAR
Forest	0.8420	0.7403	0.7115	1.2234
Runway	0.2598	0.2472	0.2744	0.4786
Urban area	0.3064	0.3000	0.3081	0.4841

References

1. J.C. Souyris, S. Mingot, Polarimetry based on one transmitting and two receiving polarizations: the p/4 mode, in *Proceedings IGARSS '02* (Toronto, Canada, July 2002)
2. J.C. Souyris, P. Imbo, R. Fjortoft, S. Mingot, J.S. Lee, Compact polarimetry based on symmetry properties of geophysical media: the p/4 mode. IEEE Trans. Geosci. Remote Sens. **43** (3), 634–646 (2005)
3. K. Raney, Hybrid polarimetric SAR architecture, in *Proceedings IGARSS '06* (July 2006)

4. P.C. Dubois-Fernandez et al., The compact polarimetry alternative for spaceborne SAR at low frequency. IEEE Trans. Geosci. Remote Sens. **46**(10), 3208–3222 (2008)
5. T.L. Ainsworth, J.P. Kelly, J.-S. Lee, Classification comparison between dual-pol, compact polarimetric and quad-pol SAR imagery. ISPRS J. Photogramm. Remote Sens. (2009)
6. T. Lulu, Study on information processing of polarimetric and compact polarimetric SAR interferometry (Institute of Electronics, Chinese Academy of Sciences, 2010)
7. S. Angelliaume, P. Dubois-Fernandez, J.-C. Souyris, Compact PolInSAR for vegetation characterisation, in *Proceedings of IGARSS '07* (Barcelona, Spain, 2007)
8. M. Lavalle, E. Pottier, Y.-L. Desnos, Investigation on the performance of compact polarimetry in SAR interferometry, in *Proceedings of EUSAR 2008* (Friedrichshafen, Germany, 2008)
9. M. Moghaddam, Effect of medium symmetries on parameter estimation with polarimetric interferometry. J. Electromag. Waves Appl. **14**(2), 173–184
10. A. Reigber et al., Multi-baseline coherence optimisation in partial and compact polarimetric modes, in *Proceedings of IGARSS 2008* (Boston, USA, 2008)
11. P. Dubois-Fernandez et al., The specificity of P band PolInSAR data over vegetation, in *Proceedings of PolInSAR 2007* (2007)
12. N. Stacy, M. Preiss, Compact polarimetric analysis of X band SAR data, in *Proceedings of EUSAR '06* (Dresden, Germany, May 2006)
13. K. Raney, Dual-polarized SAR and stokes parameters. IEEE Geosci. Remote Sens. Lett. **3**, 319–417 (2006)

Chapter 6
Polarization Calibration Technology for Polarimetric Synthetic Aperture Radar

SAR calibration is used to establish the exact relationship between image intensity and ground backscattering coefficient, which is the key to realize quantitative remote sensing of SAR. In order to fully utilize the information contained in multi-channel multi-polarization data, quantitatively analyzing the correlation between different channels and different polarizations is needed, so that different polarimetric SAR images truly reflect the dependence of the target response on the polarization state of the electromagnetic wave, that is, the multi-channel multi-polarization calibration is required.

A very important difference between multi-polarization SAR and conventional SAR is that the data acquired by the former must be calibrated before it can be used for surface scattering characteristics analysis. Multi-polarization radar systems require that the data of different polarization channels must be combined with appropriate amplitude and phase weights to complete the polarization process. It is impossible to complete the operation without calibrating the system.

Multi-polarization SAR calibration techniques include internal calibration, external calibration, and calibration processing. Early polarization calibrations were focused on the development of point target devices for external calibration. Although the calibration algorithm based on the target device of the calibration point is relatively simple, it relies entirely on these devices for calibrating, and it is necessary to place the large number of devices inside the swath. This is not only time-consuming and labor-intensive, but also costly. Therefore, there is a calibration algorithm that derives the parameters describing the system performance from the data itself and correcting the data defects.

The main significance of SAR calibration (including multi-polarization SAR calibration) is

(1) Quantitative information on radar remote sensing targets can be obtained to improve the efficiency and quality of information extracted from remote sensing data, so that the target and its changes can be understood more deeply and comprehensively.

R. Yang et al., *Polarimetric Microwave Imaging*,
https://doi.org/10.1007/978-981-15-8897-6_6

(2) Solve major problems of quantitative remote sensing in radar ground observation applications, such as large area soil moisture measurement, sea surface measurement, terrestrial target statistical characteristics measurement, accurate topographic geometric parameter measurement, high-precision vegetation crop classification and biomass estimation, classification of tactical objectives, etc.
(3) Compare and match target information of different time, different radars, different remote sensing parameters, and different regions to achieve observation and dynamic monitoring of long-term changes in global targets, and greatly improve the utilization of remote sensing data.
(4) Promote the development of new radar Earth observation technologies.

6.1 Synthetic Aperture Radar Calibration Technology

In the early stage of the development of SAR, the application of remote sensing was mainly to obtain qualitative information on the position, shape, and relative change of echo intensity from SAR image. With the continuous development of SAR technology, there are more and more airborne and spaceborne SAR systems with different frequency bands, polarizations, and operating modes, and all have stable data acquisition capabilities. A large number of all-weather, day and night, high-precision remote sensing data acquisitions have enabled the application of SAR to rapidly expand to include marine, forest, sea ice monitoring, planetary exploration, geology, and crop and terrain imaging. In these applications, it is required to establish a quantitative relationship between radar image intensity and target scattering characteristics.

SAR calibration technology is the key technology to realize quantitative remote sensing of SAR. The purpose of the calibration is to measure various uncertain factors affecting the imaging quality of SAR, remove its influence during data processing, and establish an accurate relationship between image intensity and backscatter coefficient. The calibrated SAR image product should be repeatable and stable, and its accuracy is known and understandable.

In this section, the error sources of SAR systems and the related concepts of calibration are discussed, including internal calibration and external calibration, absolute calibration and relative calibration, calibration accuracy, and so on. The calibration models of SAR system are analyzed, and the internal calibration and external calibration techniques of SAR system are discussed in detail.

6.1.1 Error Sources for Synthetic Aperture Radar Systems [1]

Radiation calibration accuracy of SAR systems is not only related to the stability of the sensor subsystem, but also to the downlink and ground processor. This section will review the errors that may be introduced in each process from the generation of

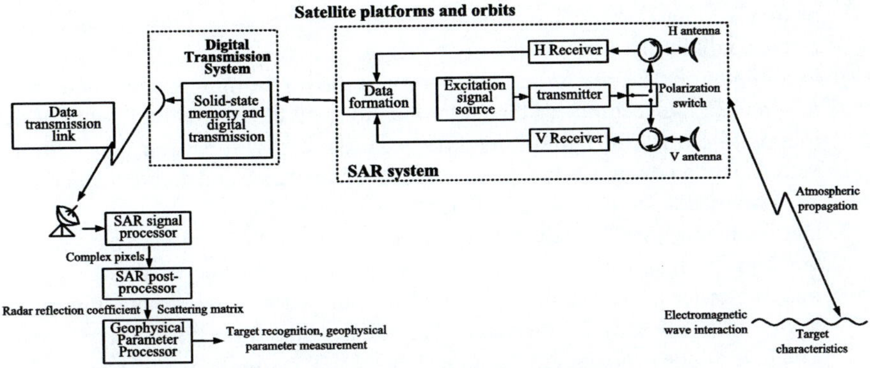

Fig. 6.1 Schematic diagram of imaging and data processing of SAR system

the SAR signal to the acquisition of the target geophysical parameters, and analyze its impact on the calibration of the entire system.

Figure 6.1 is a schematic diagram of imaging and data processing of SAR system. The figure shows the whole process of SAR system imaging of ground object, SAR image processing, and finally obtaining geophysical parameters. As can be seen from Fig. 6.1, SAR radar observation is a very complicated process. There are many uncertain factors in the whole process from signal transmission to image acquisition, which will cause signal distortion. The performance of geophysical parameter measurements is not only related to sensors, but also to data transmission links and ground processors. The error sources in SAR systems are mainly as follows:

(1) Target characteristics and atmospheric propagation: After the radar emission wave and echo propagate through the atmosphere, its electromagnetic parameters will change, such as signal attenuation, propagation delay, and polarization state changes, and mainly affect the scattering characteristics, geometric uncertainty, and propagation loss of the object to be measured.
(2) Antenna: The antenna is a key part of the SAR system, and its performance seriously affects the imaging quality of SAR. The main effects include antenna pattern uncertainty, antenna gain, phase, polarization isolation, pointing error, crosstalk, etc.
(3) Central electronic equipments: The performance of the devices in the transmitting and receiving channels can vary greatly as soon as the aging of components and ambient temperature change, resulting in changes in transmit power, receive gain, system noise, and so on. For a polarization SAR system, insufficient isolation between polarization channels can lead to crosstalk between different polarization signals. Unbalanced amplitude and phase characteristics of each polarization channel will result in intensity differences between different polarization images. The data forming unit generates quantization errors. The transmit power variation, gain imbalance of polarization channel or different frequency

channels, polarization channel phase difference, crosstalk, noise, quantization error, nonlinearity, etc. will mainly concern.

(4) Satellite or aircraft platform: Uncertainty of the platform in position and speed will reduce the accuracy of the target location and the geometric fidelity of the image. Platform roll motion, feed degradation, etc. will cause changes in the actual antenna pattern, while platform attitude variables and their ephemeris are key parameters for determining Doppler parameters of echo data.

(5) Data transmission: A severe bit rate error during data transmission can degrade the impulse response function. If an entire echo data line is lost during the imaging process, the internal fidelity of the data set will deteriorate. For multi-channel systems such as interference or polarization, this effect is very serious. The loss of one echo data line in one channel will cause the relative phase error of the channel to the channel.

(6) Data recording, imaging, and signal processing: During imaging processing, errors in the Doppler center frequency and frequency modulation slope estimation of the echo signal, as well as quantization errors in digital processing, can cause errors in the imaging processor gain. The gain uncertainty, noise, digitization, nonlinearity, interpolation error, etc. are mainly considered.

The error sources of SAR system are analyzed above. The purpose of calibration of SAR system is to eliminate the influence of these error sources to the maximum extent, accurately describe the SAR system, and establish the correct correspondence between the SAR image intensity and the target true backscatter coefficient (σ^0). The specific application requirements for σ^0 measurement accuracy determine the accuracy of the calibration of the required data. In general, the more means to obtain data (i.e., multiple incidence angles, multiple polarizations, and multiple frequencies), the more strong robust the analysis process is required, and the higher the requirements for system calibration performance.

6.1.2 *Related Concepts of SAR Calibration Technology [1]*

In this section, the basic calibration terminology and calibration performance parameters of SAR will be given, and the system calibration process will be further discussed. After the system error is partially corrected in the processor, the overall characteristic from the transmitting terminal to the receiving terminal can be used to evaluate the system performance.

6.1.2.1 Radiation Calibration

Radiation calibration is the process of measuring the amplitude and phase of a ground target backscatter signal using SAR system, describing the overall system performance from the transmitter output to the processor output. The purpose of radiometric calibration is to establish the correct correspondence between the pixel intensity of the SAR image and the true backscattering coefficient of the target. It should be noted that calibration and system testing are different concepts: calibration is a step in the regular operation of the system, while system testing is performed before or after regular operation.

6.1.2.2 Internal Calibration and External Calibration

According to the positional relationship between the calibration equipment (or target) and the radar system, the radiation calibration is divided into internal calibration and external calibration.

(1) Internal calibration

The internal calibration is a process of using the internal equipment of the system to inject a calibration signal (such as a calibration point frequency, a frequency modulation signal, etc.) into the radar data stream to monitor the performance of the radar system. Through the built-in sensor or calibration loop of the radar system, the parameters of each part of the system are measured during flight and compensated during the imaging process to determine the relative scattering coefficient of the measured object. The internal calibration can be used to estimate the relative change in transmit power and receiver gain due to temperature changes or component aging, and to measure noise levels.

Internal calibration requires no external calibration equipment, and can measure critical equipment parameter change throughout the whole life cycle of the system (e.g., receiver gain, transmitter output power, gain imbalance for channel to channel, phase difference and crosstalk, noise, quantization error, and nonlinearity). However, the monitoring range of the internal calibration is limited, and it is impossible to calibrate the parts that cannot be reached by the internal calibration circuit (such as the antenna radiation unit), and it is not possible to calibrate the external characteristics of the antenna, atmospheric propagation, etc.

(2) External calibration

The external calibration is a process of calibrating the performance of SAR system using a calibration signal generated by the ground active target itself or by passive target scattering. These ground targets can be artificial point targets with known radar cross sections (such as various corner reflectors and active transponders), or distributed targets with known and uniform scattering characteristics (such as tropical rain forests, large area of uniform grass, etc.)

The external calibration is used to determine the absolute scattering coefficient of the target being measured. The external calibration process is the same as the normal imaging process, except that there is a reference target with known scattering characteristics in the imaging region. Therefore, the external calibration can be used for SAR radar signal transmission, atmospheric propagation, echo reception, data downlink, and processing end-to-end calibration of the entire process. For example, antenna pattern, axial gain and pointing, linear dynamic range of the system, and atmospheric propagation effects can only be determined by external calibration technology.

(3) Internal calibration combined with external calibration

External calibration technology requires simple radar equipment. However, to complete the external calibration, it is necessary to place mass calibration reference targets and have higher requirements for placement (position, pointing accuracy, etc.); the RCS requiring the target is much larger than the background area, and in many areas (such as mountain, sea, etc.) the calibrators cannot be placed. Since the performance of the radar system is related to the temperature and time state, and it is a relatively fast change process, while the external calibration is limited by the calibration field, the time and number of external calibrations can be limited. Therefore, it is difficult to meet the actual calibration requirements only by external calibration.

Although the internal calibration technology can standardize the system at any time, it can only achieve relative calibration. Therefore, usually the complete calibration should be combined with the internal calibration technology and the external calibration technology and every time the system is externally calibrated using the observation area with the reference target, and the internal calibration technology is used to monitor the system performance change.

6.1.2.3 Absolute Calibration and Relative Calibration

Radar system performance parameters can be divided into absolute terms (correctness) and relative terms (accuracy), which can be described by a series of calibration parameters, namely, absolute calibration and relative calibration. Absolute calibration requires determining the system gain; however, relative calibration reflects the proportional relationship between the data acquired by the system, which don't need to determine the system gain.

(1) Absolute calibration

Absolute calibration is the process of comparing the backscattering coefficient measured by SAR system with its actual value, obtaining the corresponding systematic error parameters, and using these error parameters to correct the backscattering coefficient value of the arbitrary ground target acquired by the system. The purpose of the absolute calibration of SAR systems is to standardize the measured data so that the data obtained by the same SAR system in different frequency bands, different imaging modes of the same SAR system, different SAR systems, and even different types of sensors are comparable.

(2) Relative calibration

The relative calibration is a process of measuring the relative change of the performance parameters of SAR system in a certain time period, and is a process of performing corresponding correction in the data processing process. By eliminating the effects of system performance parameter changes by means of relative calibration, the ratio between the SAR image data truly reflects the ratio of the ground object target scattering coefficient of the imaged area.

Generally, according to the two sets of data compared with each other, the length of the time interval can be obtained, and the relative calibration can be divided into long-term relative calibration and short-term relative calibration.

(1) **Long-term relative calibration**: The time interval for the system to acquire two (or two groups) corresponding pixel data in two imaging regions is long enough, so that the main error sources of the system (such as temperature changes, attitude changes, etc.) are irrelevant in time. The process of measuring the ratio of the scattering cross-sectional area between such corresponding pixels is referred to as long-term relative calibration. The two imaging areas here may or may not be the same area.
(2) **Short-term relative calibration**: The time interval for the system to obtain the two (or two groups) corresponding pixel data in the same imaging area is relatively short, so that the error introduced by the main error source of the system can be neglected. The short-term relative calibration is a measure of the ratio of the cross-sectional area of the corresponding pixels.

The difference between short-term and long-term relative calibration is different in nature and is based on different scientific applications of the data. In data analysis like target classification, the ratio of the average power of two similar target regions within an image is a key parameter. In other applications, it may be necessary to compare and analyze two images flying over the same target area at different times for change detection. In short-term relative calibration (such as within an image), many sources of error are negligible. If the time interval between measurements is short enough, many relative errors, such as temperature changes and errors introduced by platform instability, can be ignored.

6.1.2.4 Calibration Accuracy

SAR can directly obtain a radar image reflecting the backscattering coefficient of the ground object after radiometric calibration, but it cannot accurately represent the backscattering coefficient of the object. The calibration accuracy is used to indicate the uncertainty of the backscatter coefficient after calibration. With the development and improvement of the calibration operation, the accuracy of calibration has also been continuously improved from the 3 dB absolute calibration accuracy of the US SIR-C to better than 0.9 dB absolute calibration accuracy of the German TerraSAR-X. The high calibration accuracy promotes the development of SAR data processing from qualitative research to quantitative research. Different SAR image applications require different radiometric calibration accuracies.

Some geophysical parameters and corresponding calibration requirements are given in Tables 6.1 and 6.2 [2–4].

Table 6.1 Calibration requirements for single-channel SAR calibration

Geophysical parameters	Measurement requirements	Calibration accuracy requirements		
		Absolute calibration/dB	Relative calibration (short term)/dB	Relative calibration (long term)/dB
Sea ice classification (three categories, winter)	Correct classification probability >80%	2.0	0.5	2.0
Sea ice movement	Correct match probability >95%	–	0.5	–
Snow thickness	<20 cm, Assumed snow depth >20 cm	1.0	1.0	1.0
Ocean wind speed	<20%	1.0	1.0	1.0
Ocean wave height	System transfer function	1.0	0.5	0.5
Ocean wave energy	<5%			
Soil moisture (5 humidity levels)	20% variance for each humidity level	1.0	<0.5	1.0
Ground surface roughness	–	1.0	–	–
Vegetation mapping/monitoring	Biomass density is within 25%	1.0	<0.5	0.5
Lava age classification	Accurate classification of three eras	3.0	1.0	1.0

Table 6.2 Calibration accuracy requirements for multi-frequency multi-polarization SAR SIR-C

Long-term and short-term relative calibration	±1 dB
Absolute calibration of any channel	±3 dB
Frequency calibration	±1.5 dB
Polarization channel amplitude imbalance	±0.4 dB (two way)
Polarization channel relative phase calibration	±10° (two way)
Polarization crosstalk error (polarization isolation)	−30 dB

6.1.3 Calibration Model of Synthetic Aperture Radar System

Assuming that the SAR system is linear, the receiver output power can be written as follows:

$$P_r = P_s + P_n \tag{6.1}$$

where P_r is the received power, P_s is the echo signal power, and P_n is the additive (thermal) noise power.

Ignoring the influence of the ambiguity, the relationship between the echo signal power and the target backscatter coefficient is

$$P_s = K(t,\theta)\sigma^0 \tag{6.2}$$

where $K(t,\theta)$ is the scalar factor related to imaging time, antenna pointing θ, etc.

This gives the following formula:

$$\sigma^0 = \frac{P_r - P_n}{K(t,\theta)} \tag{6.3}$$

The calibration process of SAR system is to estimate the system noise power and scalar factor to obtain the true backscatter coefficient of the target σ^0.

According to the distributed target radar equation, if the antenna satisfies reciprocity, the expression of the scalar factor $K(t,\theta)$ is [1]

$$K(t,\theta) = \frac{P_t(t)G_r(t)G^2(t,\theta)\lambda^3 cR\tau_p}{2(4\pi R)^3 L_a \sin\eta} \tag{6.4}$$

where t is the imaging moment, θ is the pitch angle, $P_t(t)$ is the transmitted signal power, $G_r(t)$ is the receiving channel gain, $G(t,\theta)$ is the antenna pattern, λ is the wavelength of the emitted electromagnetic wave, c is the speed of light, R is the slant range, τ_p is pulse width, L_a is antenna length, and η is the incidence angle of the electromagnetic wave to the target.

In Eq. (6.4), the signal wavelength λ, the slant range R, the pulse width τ_p, and the antenna length L_a can all be considered to be constant. Due to temperature variations and component aging, the transmit power, receive channel gain, and antenna pattern change over time, and the antenna pattern is also related to the pitch angle. Therefore, Eq. (6.4) can be rewritten as

$$K(t,\theta) = K_S \frac{P_t(t)G_r(t)G^2(t,\theta)}{\sin\eta} \tag{6.5}$$

where K_S consists of the determined items in $K(t,\theta)$. In this way, the calibration problem is reduced to the estimate of transmit power $P_t(t)$, receive channel gain $G_r(t)$, antenna pattern $G(t,\theta)$, and angle of incidence η. The estimate of the incidence angle depends on the platform roll angle. In addition, the noise power term in Eq. (6.3) must be estimated.

These parameters can be estimated by internal and external calibration measurements, respectively. The internal calibration uses the internal calibration device to track the drift of the transmitter output power and the receive channel gain over time.

Setting the transmitter to the standby state and recording only the thermal noise can measure the drift of the system noise power. The external calibration technique generally uses the point target in which the known scattering characteristics are placed in the calibration field to obtain the image data of the calibration field or to obtain the image data of the distribution target whose backscatter coefficient is known, thereby estimating the absolute gain of the system and antenna pattern.

6.1.4 Internal Calibration of Synthetic Aperture Radar Systems

The internal calibration is to use the internal equipment of the radar system and the calibration path to measure the changes of various parameters of the system during the imaging process, and compensate in the imaging process to determine the relative scattering coefficient of the measured object. The accuracy depends on the internal calibration technique used and the average number of independent samples, and also depends on the measurement accuracy that can be achieved by the internal calibration equipment.

The key parameters measured by the internal calibration technology are the power of the transmitted pulse, the receiver gain, the channel gain imbalance, the phase difference of the polarization channel, the crosstalk between channels, the noise, the quantization error, the system nonlinearity, etc.

There are two main methods for internal calibration:

(1) Each part of the system is independently calibrated. This method requires frequent interruption of the measurement sequence. Calibration of each part of the system will result in an increase in the measurement error term, and the errors will overlap each other, which will reduce the overall calibration accuracy.
(2) Ratio calibration method. The sampling of the transmitted signal is introduced into the receiving and data processing system for internal calibration. This method has less chance of generating errors and is allowed to be processed within frequently fluctuation intervals.

Figure 6.2 is a schematic diagram of a full-phase array antenna SAR system and internal calibration path.

The thick solid arrow in the figure shows the path of the radar signal transmission and reception during the imaging process: the chirp signal generated by the FM signal source is amplified by the pre-amplifier, passed through the circulator, the signal feeding network, and then is amplified by the *T* channel of the *T*/*R* component and transmitted through the antenna array. The echo signal of the ground object received by the antenna array is amplified by the *R* channel of the *T*/*R* component, and then reaches the radar receiver via the signal feeding network and the circulator, and then sent to the data former for sampling after being frequency converted, and then packaged with the auxiliary data, and pass data link to the ground receiving station.

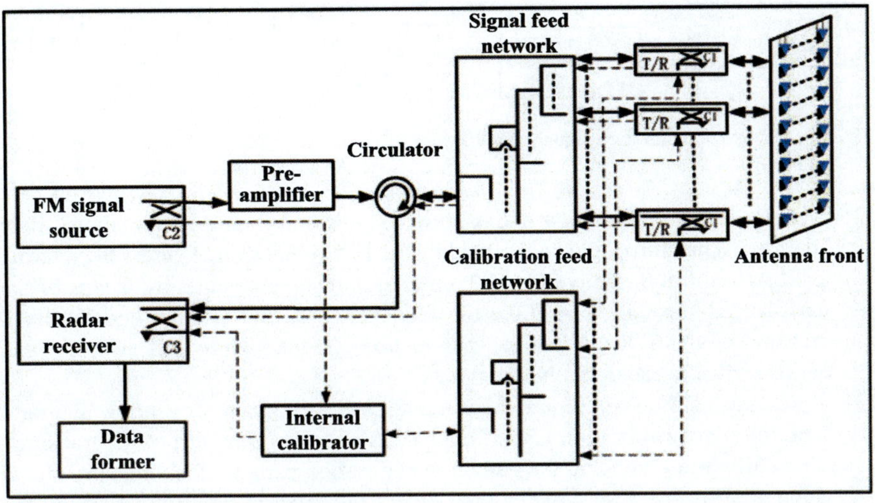

Fig. 6.2 Schematic diagram of the full-phase array antenna spaceborne SAR system and internal calibration path

The thin dotted arrows in the figure depict the signal path during calibration in a SAR system. For a clearer description, the signal flow diagrams for transmit calibration, receive calibration, and reference calibration are shown in Figs. 6.3, 6.4 and 6.5, respectively. The three calibrations are introduced separately as follows:

(1) Transmit calibration: Figure 6.3 shows the flowchart of the transmit calibration signal. The chirp signal generated by the FM signal source is amplified by the pre-amplifier, passed through the circulator, the signal feeding network, and then amplified by the *T* channel of the *T*/*R* component, and a small part of the transmitting signal is coupled to the calibration feed through the coupler C1. The network, through the internal calibrator, sends the internal calibration signal to the radar receiver and data former through the coupler C3.

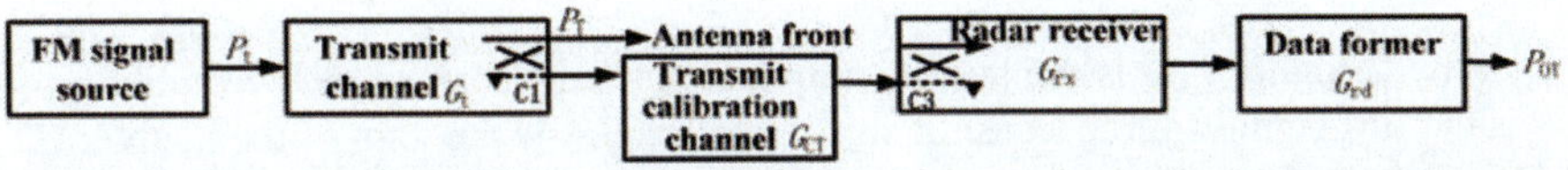

Fig. 6.3 Transmit calibration signal flow diagram

Fig. 6.4 Receive calibration signal flow diagram

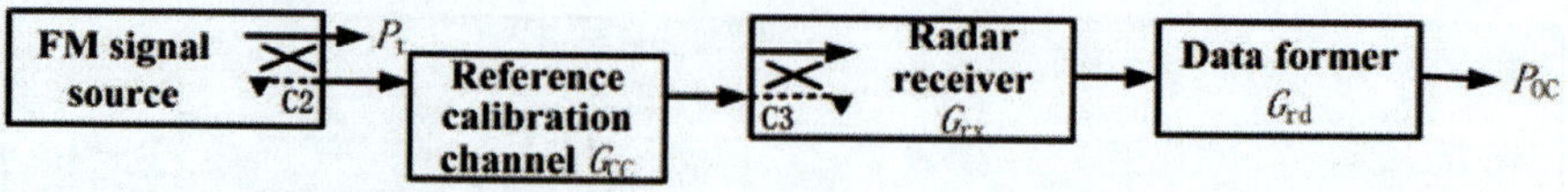

Fig. 6.5 Reference calibration signal flow diagram

(2) Receive calibration: Figure 6.4 shows the receive calibration signal flow diagram. The chirp signal generated by the FM signal source enters the internal calibrator through the coupler C2, enters the receive channel by means of the coupler C1 through the calibration feed network, and is amplified by the *R* channel of the *T*/*R* component, passes through the signal feed network, and loops arrival to radar receiver and data former.
(3) Reference calibration: Figure 6.5 is a reference calibration signal flow diagram. The reference calibration path is composed of a common part of the transmission calibration path and the reception calibration path, that is, the chirp signal generated by the FM signal source enters the internal calibrator through the coupler C2, enters the radar receiver through the coupler C3, and then data former.

From Fig. 6.3, the system output signal power can be obtained when transmit calibration.

$$P_{OT} = P_t \bullet G_t \bullet G_{C1} \bullet G_{CT} \bullet G_{C3} \bullet G_{rx} \bullet G_{rd} \tag{6.6}$$

In the formula, P_{OT} is the system output signal power when transmit calibration; P_t is the FM signal source transmit signal power; G_t is the system transmit channel gain; G_{C1} is the gain of the coupler C1; G_{CT} is the internal calibration transmit calibration path gain (including the gain of the internal calibrator path, the calibration feed network, and the interconnecting cable); G_{C3} is the gain of the coupler C3; G_{rx} is the gain of the radar receiver; and G_{rd} is the processing gain of the data former.

From Fig. 6.4, the output signal power of the system when receive calibration is

$$P_{OR} = P_t \bullet G_{C2} \bullet G_{CR} \bullet G_{C1} \bullet G_r \bullet G_{rx} \bullet G_{rd} \tag{6.7}$$

In the formula, P_{OR} is the system output signal power when receive calibration; P_t is the FM signal source transmit signal power; G_{C2} is the gain of the coupler C2; G_{CR} is the internal calibration receive calibration path gain (including the gain of the internal calibrator path, the calibration feed network, and the interconnecting cable); G_{C1} is the gain of the coupler C1; G_r is the gain of the receive channel except the radar receiver; G_{rx} is the gain of the radar receiver; and Grd is the processing gain of the data former.

From Fig. 6.5, the output signal power of the system when reference calibration is

$$P_{OC} = P_t \bullet G_{C2} \bullet G_{CC} \bullet G_{C3} \bullet G_{rx} \bullet G_{rd} \tag{6.8}$$

In the formula, P_{OC} is the system output signal power when reference calibration; P_t is the FM signal source transmit signal power; G_{C2} is the gain of the coupler C2; G_{CC} is the internal calibration reference calibration path gain (including the gain of the internal calibrator path and the interconnecting cable); G_{C3} is the gain of the coupler C3; G_{rx} is the gain of the radar receiver; and G_{rd} is the processing gain of the data former.

From Eqs. (6.6)–(6.8), we can get

$$\frac{P_{OT}}{P_{OC}} = G_t \cdot \frac{G_{C1}G_{CT}}{G_{C2}G_{CC}} = G_t \cdot G_{CL\text{-}T} \tag{6.9}$$

$$\frac{P_{OR}}{P_{OC}} = G_r \cdot \frac{G_{C1}G_{CR}}{G_{C3}G_{CC}} = G_r \cdot G_{CL\text{-}R} \tag{6.10}$$

$$\frac{P_{OT} \cdot P_{OR}}{P_{OC}} = (P_t G_t \cdot G_r G_{rx} G_{rd}) \cdot \frac{(G_{C1})^2 G_{CT} G_{CR}}{G_{CC}} = (P_T \cdot G_R) \cdot G_{CL} \tag{6.11}$$

In the formula, $P_T = P_t G_t$ is the transmit signal power of SAR system, $G_R = G_r G_{rx} G_{rd}$ is the total receive gain of SAR system, $G_{CL\text{-}T} = \frac{G_{C1}G_{CT}}{G_{C2}G_{CC}}$ is the internal calibration gain when transmit calibration of SAR system, $G_{CL\text{-}R} = \frac{G_{C1}G_{CR}}{G_{C3}G_{CC}}$ is the internal calibration gain when receive calibration of SAR system, and $G_{CL} = \frac{(G_{C1})^2 G_{CT} G_{CR}}{G_{CC}}$ is the total internal calibration gain of SAR system.

Expressing the above formula in dB, we can get

$$P_{OT}(\text{dB}) - P_{OC}(\text{dB}) = G_t(\text{dB}) + G_{CL\text{-}T}(\text{dB}) \tag{6.12}$$

$$P_{OR}(\text{dB}) - P_{OC}(\text{dB}) = G_r(\text{dB}) + G_{CL\text{-}R}(\text{dB}) \tag{6.13}$$

$$P_{OT}(\text{dB}) + P_{OR}(\text{dB}) - P_{OC}(\text{dB}) = P_T(\text{dB}) + G_R(\text{dB}) - G_{CL}(\text{dB}) \tag{6.14}$$

Through the internal calibration test at two different times during the imaging process, we can get

$$\Delta P_{OT}(\text{dB}) - \Delta P_{OC}(\text{dB}) = \Delta G_t(\text{dB}) + \Delta G_{CL\text{-}T}(\text{dB}) \tag{6.15}$$

$$\Delta P_{OR}(\text{dB}) - \Delta P_{OC}(\text{dB}) = \Delta G_r(\text{dB}) + \Delta G_{CL\text{-}R}(\text{dB}) \tag{6.16}$$

$$\Delta P_{OT}(\text{dB}) + \Delta P_{OR}(\text{dB}) - \Delta P_{OC}(\text{dB}) = \Delta P_T(\text{dB}) + \Delta G_R(\text{dB}) - \Delta G_{CL}(\text{dB}) \tag{6.17}$$

In the formula, ΔP_{OT} (dB), ΔP_{OR} (dB), and ΔP_{OC} (dB) can be obtained by processing the calibration test data in the SAR system, and $\Delta G_{CL\text{-}T}$ (dB), $\Delta G_{CL\text{-}R}$

(dB), and ΔG_{CL} (dB) can be obtained by pre-launch test. In this way, the relative changes ΔG_t(dB) in the transmit channel gain, the relative changes ΔG_r(dB) in the receive channel gain, and the relative change (ΔP_T(dB) + ΔG_R(dB)) of total system power (including transmit power and receive channel gain) of the SAR system can be calculated separately. Then the corresponding parameters of the system are compensated and corrected to complete the internal calibration of the system.

To perform accurate internal calibration, the performance of the system's transceiver channel must be measured in real time. However, during the actual system operation, it is difficult to perform normal imaging and internal calibration at the same time. System performance can be considered to be linear in a short period of imaging, so an internal calibration is usually performed before and after imaging, i.e., a complete internal calibration process including reference calibration, receive calibration, and transmit calibration. The relative change in system performance at the beginning and end of imaging process is obtained. Using linear interpolation, the relative change of system transmit channel gain, receive channel gain, and total system power at any two moments can be got.

6.1.5 *External Calibration of Synthetic Aperture Radar Systems [1]*

The internal calibration described above can complete the relative calibration of SAR image data. The SAR system must also use an external calibration to achieve a complete calibration process. The external calibration obtains the radar system transfer function using ground targets with known scattering characteristics. Compared with the internal calibration, the advantage of the external calibration is that the end-to-end system performance can be directly measured. Therefore, external calibration technology can be used to measure system parameters that are difficult to measure, such as antenna patterns, look angles, and signal propagation atmospheric effects. The shortcoming of the external calibration is that the measuring of the system transfer function is not sufficient due to the inability to frequently image the external calibration field, and it is impossible to measure the effects of short-term system instability or platform motion. The actual effective calibration of any spaceborne SAR system requires the combination of external calibration and internal calibration. The end-to-end system performance (including absolute gain) is estimated by external calibration. The relative drift in system performance between two external calibration measurements is monitored by internal calibration.

External calibration typically uses two types of targets: (1) Point targets or specular scatterers with known radar cross-sectional area (RCS). (2) Relatively stable large area and uniformly distributed target with known scattering characteristics.

6.1.5.1 Point Target Calibration

Point targets are generally man-made devices such as corner reflectors, transponders, audio signal generators, and receivers. These devices occupy a much smaller geometric area than a resolution unit of SAR image, but their radar cross-sectional area is much larger than the total backscatter power of the surrounding background area within the resolution unit.

In order to minimize the calibration error in the backscattering region, the RCS of the point target should be at least 20 dB greater than the total scattered power (i.e., $\sigma^0 \delta x \delta R_g$ SAR) of the SAR image resolution unit. When placing the calibration target, in addition to considering the background power, the main considerations are as follows:

(1) Pointing angle: The pointing angle of the calibration device relative to the radar must be accurately measured (for example, uncertainty <1.0°). Generally, the radar cross-sectional area is very large in relation to its actual orientation.
(2) Multipath effect: When the transmitted or reflected signal is scattered from the local topography or nearby structures, the scattered signal and the reflected signal of the calibrator are received by SAR antenna almost simultaneously, thus multipath effect occurs.
(3) Scattering characteristics of the calibration equipment: The scattering characteristics of the calibration equipment need to be measured in a controlled environment (such as a microwave darkroom) over a range of temperatures and look angles to obtain the RCS characteristics of the equipment. It should be noted that for passive devices such as corner reflectors, the RCS is very sensitive to the deformation of the reflector side plane. Manufacturing errors or warpage deformation caused by thermal cycling may cause the actual RCS value of the device to deviate significantly from the theoretical design value.

Point target calibration equipment includes passive calibration equipment and active calibration equipment, which are introduced separately below.

A. Passive calibration equipment

The most commonly used equipment for SAR external calibration is the corner reflector, the most used of which is the triangular trihedral reflector. The RCS of the triangular trihedral reflector is

$$\sigma_{\mathrm{TT}} = \frac{4\pi a^4}{3\lambda^2} \tag{6.18}$$

where a is the side length. It has a relatively large and stable RCS and a wide 3 dB beamwidth (about 40°), unrelated to wavelength and reflector size.

The reflector must be constructed such that its error tolerance is less than the radar wavelength. If the RCS loss is required to not exceed 0.1 dB, the root mean square error of the surface irregularity is required to be less than 0.1 λ, the root mean square error of the plane curvature is less than 0.2 λ, and the orthogonality requires that the planar calibration is better than 0.2°. If the device direction introduces an additional uncertainty of 0.2 dB, the accuracy of the device can typically be on the order of 0.5 dB. Since the uncertainty in background backscatter estimation or the effect of multipath effects will also introduce additional calibration errors, it is necessary to find a suitable location to set the calibrators so that the effect of these errors is small relative to the RCS value of the corner reflector (e.g., less than 20 dB).

If a suitable dark background (low σ^0) is not found, a larger reflector can be used to increase the RCS. However, this usually increases the error in production and layout. Other types of reflectors with a certain aperture but with a larger RCS can be used. The RCS values and beamwidth parameters of various reflectors are shown in Table 6.3. The square trihedral RCS with the same beam width as the triangular trihedral angle increases, but is not as stable as the triangular trihedral angle, especially plane bending is prone to occur especially in high-wind environments. One solution to these problems with passive reflectors is to utilize an active system whose RCS value can be controlled by adjustment of the device gain.

B. Active calibration equipment

Such devices include transponders, receivers, and audio generators. The transponder is similar to the reflector in that it transmits the delayed signal back to the radar. The advantage of the transponder is that it can increase the signal strength through the amplifier. The ground receiver is actually a half transponder with a certain recording capability. It can be used to directly estimate the transmit power and antenna pattern. These two parameters are key parameters of the system that cannot be measured by the internal calibration device. The audio generators are used in pairs to produce two consecutive audio signals that are respectively located at two shifted frequencies within the system bandwidth and that their polarizations are mutually orthogonal. The audio generator is mainly used to measure the cross-polarization isolation of the radar. A comprehensive ground calibration field should generally include all three types of active calibration equipment.

(1) Active Transponders

The functional block diagram of the transponder is shown in Fig. 6.6, and its peak RCS is

$$\sigma_{TP} = \lambda^2 G^{\mathrm{t}} G^{\mathrm{r}} G_{\mathrm{e}} / 4\pi \tag{6.19}$$

Table 6.3 Scattering characteristics of several common reflectors

Reflector	Max RCS	3 dB beam width	Shape
Spherical	πr^2	2π	Radius r
Square plate	$4\pi a^4/\lambda$	$0.44\lambda/a$	Side length a
Robert lens	$4\pi^3 r^4/\lambda^2$	~40°	Radius r
Triangle trihedral	$4\pi a^4/3\lambda^2$	~40°	Side length a
Square trihedral	$12\pi a^4/\lambda^2$	~40°	Side length a

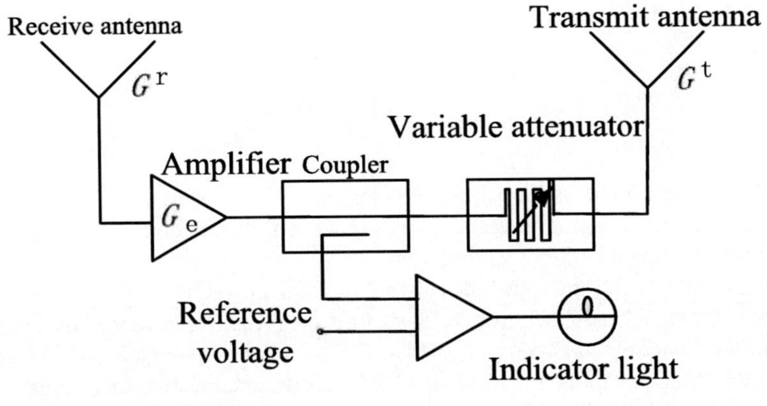

Fig. 6.6 An active transponder

where G^t, G^r are the transmit and receive antenna gains and G_e is the net gain of the transponder electronics. This design has greater flexibility to achieve the desired RCS by controlling the amplifier gain. The choice of antenna mainly considers the requirements of cross-polarization isolation and beamwidth, and secondly considers the gain. For the dual antenna design, since the signal is amplified by the transponder, it is important to consider the cross-coupling between the antennas. By spatially separating the two antennas, the required cross-coupling performance (less than −80 dB) can be achieved. A standard gain horn antenna or a microstrip antenna is typically used. A corrugated horn antenna can be used if greater cross-polarization isolation and lower side lobes are required.

(2) Ground receiver

The block diagram of the ground receiver is shown in Fig. 6.7. The system includes a receiving antenna, an envelope detection circuit capable of locking the radar PRF, and an AD sampling and storage device. The system can be integrated to a transponder for dual functionality. The ground receiver can be used to directly measure the azimuth antenna pattern. Deploying a certain number of ground receivers in range dimension can also measure the range antenna pattern indirectly. If the beam center of SAR

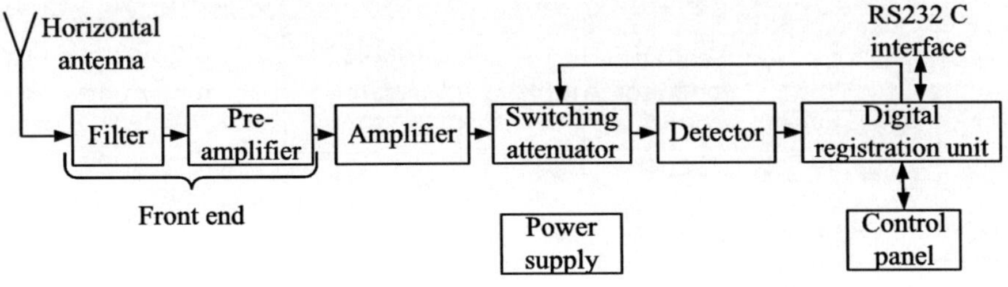

Fig. 6.7 A ground calibration receiver

is aligned with the beam center of the receiver antenna, the peak transmit power of SAR is

$$\text{EIRP} = \frac{(4\pi R)^2 P_r}{G^r G_e \lambda^2} \tag{6.20}$$

In the formula, EIRP is the effective isotropic transmission power, R is the slope distance, P_r is the received power after the signal is digitized, and G^r and G_e are the gain of the antenna and the receiving unit.

The ground receiver measures the forward transmit power, which is much stronger than the reflected RCS or background. Therefore, the ground receiver can measure the SAR antenna pattern very accurately. However, if the SAR antenna is not reciprocal, the ground receiver cannot determine the SAR receive antenna pattern because it can only measure the characteristics of the entire synthetic aperture radar transmit link.

(3) Continuous wave audio signal generator

A continuous wave audio generator typically includes a linear polarization antenna and a signal generator, as shown in Fig. 6.8.

Continuous wave audio generators are typically used in pairs. The frequency of the signals transmitted by the two audio generators is staggered within the system bandwidth, and one unit transmits a signal of one polarization and the other unit transmits a signal of its orthogonal polarization. The cross-polarization isolation of SAR receive antenna can be determined from the original signal data by the following formula:

$$I_{xp} = \frac{G^r(f_x)}{G^r(f_l)} \tag{6.21}$$

where $G^r(f_l)$ and $G^r(f_x)$ are the gains of the co-polarization and cross-polarization of the SAR receive antenna, respectively.

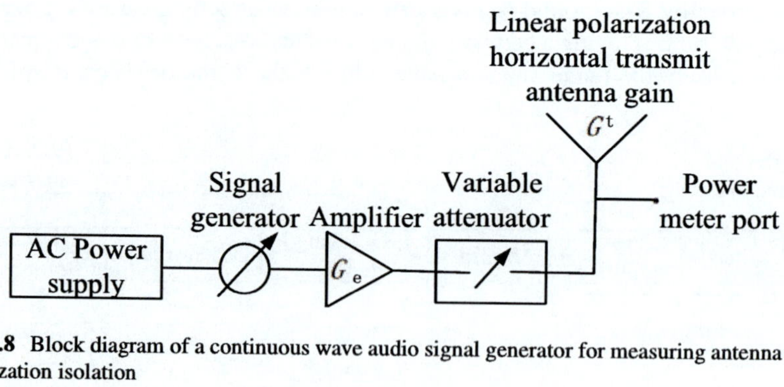

Fig. 6.8 Block diagram of a continuous wave audio signal generator for measuring antenna cross-polarization isolation

A signal with a frequency difference $f_x - f_1$ will produce a one-way Doppler shift of the sensor relative to the target position on range line. By performing a Fourier transform on each range line in the ground processor, the values of the variables in Eq. (6.21) can be obtained. Due to the Doppler shift, the received audio signal may have several FFT window migrations throughout the entire azimuth aperture of SAR. If it is necessary to azimuthally sum the signals of adjacent distance lines to reduce the signal estimation error, it must be noted that for each distance line used, the audio signal falls in discrete different FFT windows.

6.1.5.2 Distribution Target Calibration

Distribution target calibration refers to external calibration using large area natural targets with uniform backscatter characteristics. A basic assumption of the distribution target calibration is that the scattering characteristics of these regions are stable, or the changes are known. This then can separate the image features associated with the target scattering characteristics from the sensor performance.

An important benefit of using distributed target calibration is that radar performance can be measured at various operating points throughout the system's dynamic range. In the aforementioned point target calibration, it is always required that the RCS of the calibration device is much higher than σ^0 of the surrounding background to reduce the background estimation error. Therefore, the point target device can only measure high-end system performance within the linear dynamic range of the system. The distribution target calibration point has a wide range of σ^0 values, which can estimate system performance at many points within the linear dynamic range of the system.

The spaceborne SAR system usually uses the Amazon Rainforest as the calibration reference target, mainly because the scattering characteristics are stable all the year, the coverage area is large, the scattering characteristics of the large area are uniform, and the relationship between the echo intensity and the incident angle has been massive measurement.

The main steps for external calibration using the Amazon Rainforest are [5]

(1) Select a large area with uniform scattering characteristics.
(2) Remove uneven areas such as rivers.
(3) Average the image of the selected area along the azimuth.
(4) Smoothing processing of the obtained range antenna pattern.

6.2 Multi-polarization Synthetic Aperture Radar Calibration Technology

Compared with the single-polarization SAR system, the multi-polarization SAR system has the ability to transmit and receive different polarization electromagnetic waves, which is more complicated in the system structure, so the calibration method is correspondingly complicated.

In order to fully utilize the information contained in the multi-channel multi-polarization SAR data, it is necessary to quantitatively analyze the correlation between different channels and different polarizations, that is, multi-channel multi-polarization calibration is required. The purpose of polarization calibration is to correct the difference of the effects of SAR system on different polarization data, so that the SAR images with different polarizations truly reflect the dependence of the target response on the polarization state of electromagnetic waves. In this sense, polarization calibration is a category of relative calibration because its correction object is the influence of the system on different polarization data during imaging.

In this section, the calibration model for polarimetric SAR and the basic method for calibration of a polarimetric SAR system are discussed.

6.2.1 Polarization Synthetic Aperture Radar System Calibration Model

Multi-polarization imaging radars are used to measure the target scattering matrix, Stokes matrix, or Mueller matrix. By taking measurements of any these matrices, the target RCS at any transmit and receive polarizations can be synthesized. For multi-polarization SAR, the system model can be expressed by [1, 2]

$$\begin{aligned}\mathbf{M} &= Ke^{j\phi}\begin{pmatrix} R_{hh} & R_{hv} \\ R_{vh} & R_{vv} \end{pmatrix}\begin{pmatrix} S_{hh} & S_{hv} \\ S_{vh} & S_{vv} \end{pmatrix}\begin{pmatrix} T_{hh} & T_{hv} \\ T_{vh} & T_{vv} \end{pmatrix} + \begin{pmatrix} N_{hh} & N_{hv} \\ N_{vh} & N_{vv} \end{pmatrix} \\ &= Ke^{j\phi}\mathbf{RST} + \mathbf{N}\end{aligned} \tag{6.22}$$

where $\mathbf{M}$ is the scattering matrix measured by the multi-polarization radar system; K is the total absolute amplitude gain factor of the multi-polarization radar system; ϕ is the total phase shift (absolute phase factor) of the system, including all possible sources (e.g., the transmit channel transmission delay phase, receive channel transmission delay phase, and effects of SAR radar correlators); S is the target scattering matrix; R is the radar receiving distortion matrix describing the multi-polarization radar receive channel; T is the radar transmitting distortion matrix describing the multi-polarization radar transmit channel; and N is the additive noise matrix.

Ideally, R and T are unit matrices. In the actual measurement of the polarimetric SAR system, there are many sources of error, including (1) error in the transmit channel and the receive channel, causing amplitude and phase distortion of the signal and (2) the crosstalk between system H/V channels. There is a mutual coupling between the H transmitting channel and the V transmitting channel, so that when the H (or V) polarization wave is transmitted, in addition to the main component of the H (or V) polarization, the V (or H) polarization component is also included. Similarly, there is also a mutual coupling between the H receiving channel and the V receiving channel, so that when the H channel (or V channel) receives the H (or V) polarization wave, in addition to the H (or V) polarization main component, contains V (or H) polarization major components. Therefore, the polarization data **M** measured by the polarimetric SAR is not equal to the polarization scattering matrix S of the target. As soon as the distortion matrices R and T are not unit matrices, the relative relationship between the elements of the measurement matrix **M** is distorted compared to the relationship between the elements of the target scattering matrix S. Therefore, it is necessary to calibrate the polarization measurement data before using it.

The polarization radar data calibration problem can be described in the following. To solve the scattering matrix S from the measured matrix **M**, it is necessary to estimate the matrix R and T representing the radar polarization characteristics of the receive end and the transmit end. Once the estimates of R and T are obtained, it is easy to invert them, resulting in an estimate of S. Since the absolute phase factor of the system is lost in the polarization SAR, the item can be uncalibrated; and the additive noise term can be measured when the radar system is in the simple receive mode. Therefore, the contents of the calibration of the polarization system mainly include (1) radiation calibration, determining the system absolute amplitude gain factor K and (2) polarization calibration, determining the reception distortion matrix R and the transmit distortion matrix T. The polarization system error model can be described as channel imbalance and cross-channel crosstalk, i.e.,

$$\mathbf{R} = A_r e^{j\phi_r}\begin{pmatrix} 1 & \delta_2 \\ \delta_1 & f_1 \end{pmatrix}, \quad \mathbf{T} = A_t e^{j\phi_t}\begin{pmatrix} 1 & \delta_3 \\ \delta_4 & f_2 \end{pmatrix} \tag{6.23}$$

where A_r, A_t represent the gains introduced by the receive channel and the transmit channel, respectively, and ϕ_r and ϕ_t indicate the absolute phase introduced by the receive channel and the transmit channel, respectively.

The crosstalk at the receive end δ_1, δ_2 and the crosstalk at the transmit end δ_3, δ_4 indicate the crosstalk due to the cross-polarization antenna pattern and the poor isolation of the transmit switch and the circulator. $\delta_1 = R_{vh}/R_{hh}$ is the crosstalk of the normalized H receive channel into the V receive channel, $\delta_2 = R_{hv}/R_{hh}$ is the crosstalk of the normalized V receive channel into the H receive channel, and $\delta_3 = T_{hv}/T_{hh}$ is the crosstalk of the normalized V transmit channel into the H transmit channel, $\delta_4 = T_{vh}/T_{hh}$ is the normalized H transmit channel which enters the crosstalk of the V transmit channel. $f_1 = R_{vv}/R_{hh}$ is the imbalance of the V channel of the receive end with respect to the H channel, and $f_2 = T_{vv}/T_{hh}$ is the imbalance of the V channel of the transmit end with respect to the H channel. Channel imbalances are usually complex, and for many polarization applications, the amplitude and phase characteristics must be known accurately. Thus, the calibration model of the polarimetric SAR system can be simplified as

$$\mathbf{M} = K e^{j\phi}\begin{pmatrix} 1 & \delta_2 \\ \delta_1 & f_1 \end{pmatrix}\begin{pmatrix} S_{hh} & S_{hv} \\ S_{vh} & S_{vv} \end{pmatrix}\begin{pmatrix} 1 & \delta_3 \\ \delta_4 & f_2 \end{pmatrix} + \begin{pmatrix} N_{hh} & N_{hv} \\ N_{vh} & N_{vv} \end{pmatrix} \tag{6.24}$$

According to the system model given by Eq. (6.24), some targets with known polarization scattering matrices S are selected. After the actual scattering matrix M is measured by the system, the polarization distortion matrices R and T can be solved, and then the obtained R and T are applied to the unknown imaging area of S to complete the system calibration.

6.2.2 *Basic Method for Calibration of Polarimetric Synthetic Aperture Radar System*

In order to calibrate the scattering matrix data, the system distortion matrices R and T need to be estimated. Expanding (6.24), the four equations related to the actual measured value of the scattering matrix can be got, see (6.25a–6.25d):

$$M_{\mathrm{hh}} = S_{\mathrm{hh}} + \delta_2 S_{\mathrm{vh}} + \delta_4 S_{\mathrm{hv}} + \delta_2\delta_4 S_{\mathrm{vv}} + N_{\mathrm{hh}} \tag{6.25a}$$

$$M_{\mathrm{hv}} = \delta_3 S_{\mathrm{hh}} + \delta_2\delta_3 S_{\mathrm{vh}} + f_2 S_{\mathrm{hv}} + f_2\delta_2 S_{\mathrm{vv}} + N_{\mathrm{hv}} \tag{6.25b}$$

$$M_{\mathrm{vh}} = \delta_1 S_{\mathrm{hh}} + f_1 S_{\mathrm{vh}} + \delta_1\delta_4 S_{\mathrm{hv}} + f_1\delta_4 S_{\mathrm{vv}} + N_{\mathrm{vh}} \tag{6.25c}$$

$$M_{\mathrm{vv}} = \delta_1\delta_3 S_{\mathrm{hh}} + f_1\delta_3 S_{\mathrm{vh}} + f_2\delta_1 S_{\mathrm{hv}} + f_1 f_2 S_{\mathrm{vv}} + N_{\mathrm{vv}} \tag{6.25d}$$

These four equations contain ten unknowns (ignoring noise), namely, four (complex) scattering matrix elements and six (complex) radar system unknowns (δ and f). To solve these ten unknowns, at least six additional equations containing these variables are needed. There are three ways to get the required six equations: (1) using a man-made point target with a known scattering matrix, (2) making certain assumptions about the scattering characteristics of the measured object and the polarimetric SAR system, and (3) using a scattering matrix (or equivalent Mueller matrix) known distribution target.

6.2.2.1 Calibration Algorithm Using Artificial Point Targets

The earliest polarimetric SAR calibration method was the method using artificial point targets. The calibration targets used in these methods are very limited, and commonly used include conductive balls, dihedral corner reflectors, triangular reflectors, and polarimetric active calibrators. These calibration targets can be used alone or in combination. In order to solve the calibration problem, generally three point targets are necessary because it is difficult to obtain more than two equations from each calibration target.

Although the point target calibration method is relatively simple and does not require any prior knowledge of the imaging area, each point target can only calibrate a small imaging area near it. For areas far from it, the calibration error will be very high. In order to calibrate the image of the entire swath, a large number of point targets are required to form an array covering the entire swath. Moreover, the larger size of the point target (not the ideal point) and its interaction with the background (including specular and secondary scattering of background ground-to-point targets and then to radar, or vice versa) can lead to measurement errors in the RCS. In order to improve the signal-to-noise ratio, the target size needs to be increased, but this will cause difficulties in the manufacture of the calibrator and the field layout. Improper background selection can also lead to unacceptable calibration errors. Then the point target is very sensitive to the error of the radar pointing, and a small pointing error may result in a large calibration error. Although the structural design of the calibration target is constantly improving, the directional sensitivity is still a problem that needs serious consideration.

6.2.2.2 Reasonable Assumptions on the Measured Objects and Radar Systems

Reasonable assumptions for the measured objects characteristics and the polarimetric SAR system, and then a small number of artificial point targets can also be used for polarization calibration.

Van Zyl assumes that the radar system receive distortion matrix and the transmit distortion matrix are the same, that is, $R = T$, and three equations are obtained, namely, $\delta_1 = \delta_3$, $\delta_2 = \delta_4, f_1 = f_2$. And assumes that the object to be measured has reciprocal characteristics, and most of the objects have azimuthal symmetry. There is no correlation between the co-polarized component and the cross-polarized component of the scattering matrix, that is, $\langle S_{\mathrm{hh}} S_{\mathrm{hv}}^{*} \rangle = 0$ and $\langle S_{\mathrm{hv}} S_{\mathrm{vv}}^{*} \rangle = 0$ are obtained, and the other two equations are obtained. The ratio of HH to VV measured using a trihedral reflector is then used to estimate f_1 and f_2. This will solve 10 equations and complete the entire calibration process.

Klein pointed out that the reciprocity assumption of the object can completely replace van Zyl's $R = T$ hypothesis, and pointed out that if the backscattering property of the object is assumed, only two man-made point targets are needed to solve the polarization calibration problem. Klein et al. simplified the distributed target calibration algorithm, the Sarabandi algorithm, assuming that the four crosstalk terms of the radar system are all zero, and then assumed that the ground object has reciprocity characteristics, and only one conductive ball or triangular reflector can solve the polarization calibration problem.

This type of calibration algorithm is very simple to use, the number of point targets used is less than that of a simple point target algorithm, and the performance requirements for point targets are reduced, and it has been successfully applied to the calibration of SIR-C multi-polarization SAR systems. But the problem with such algorithms is that their accuracy cannot be determined. Since the assumptions about the objects are really suitable for the measured imaging area, it is not known before the test, and its validity is verified by the point target algorithm. Like the calibration algorithm that uses the distributed target purely, this kind of algorithm also needs people to obtain a large amount of ground object polarization scattering knowledge to support, which can be regarded as the transition type from point target calibration algorithm to distributed target calibration algorithm.

6.2.2.3 Calibration Algorithm Using Distributed Targets

If a standard distribution calibration target exists, the problem of the point target calibration method can be avoided, and there is no need to make any assumptions about the radar system parameters and the object to be tested. Kamal pointed out that for the distributed target calibration algorithm, Sarabandi algorithm, the polarization scatterometer can be used to accurately measure the Mueller matrix of the distribution target, while the Mueller matrix fully characterizes the polarization scattering characteristics of the object. Simply compare the calibrated data with the data obtained from the corresponding polarization scatterometer, and the accuracy of such algorithms can be detected. With the increasing polarization data of the ground objects obtained by people, the understanding of the polarization characteristics of the ground objects is becoming more and more profound. This calibration algorithm using distributed targets will gradually become the mainstream of the calibration algorithms. It is unrealistic to place artificial point targets for some areas with poor conditions. At present, the biggest problem faced by this algorithm is that people do not know deeply about the polarization scattering characteristics of various ground objects, and it is not easy to choose the ground object that can be used as the target of polarization calibration.

6.3 Internal Calibration of Polarimetric Synthetic Aperture Radar Systems

The internal calibration is a process of using the internal equipment of the system to inject a calibration signal into the radar data stream to describe the performance of the radar system. The multi-polarization SAR system has two polarization channels, and the system structure is more complicated than the single-polarization SAR system. And because of the increase of the channel, in addition to the relative change of the system transmit and receive channel gains through internal calibration, it is also desirable to analyze the channel consistency and channel crosstalk through the internal calibration. In this section, the internal calibration path of the polarimetric SAR system is given, and the various crosstalks that may exist between the channels are analyzed. The internal calibration method for channel consistency and crosstalk in the polarimetric SAR system is discussed in detail.

6.3.1 *Analysis of Internal Calibration Path in Polarimetric Synthetic Aperture Radar System*

The composition and internal calibration path of the full phased array antenna polarimetric SAR system is shown in Fig. 6.9. This is a typical time-division operating mode polarimetric SAR system.

The thick solid arrows in the figure show the path of radar signal transmit and receive during imaging. Two transmitting and two receiving are required in a complete full polarization measurement cycle to complete the four groups of HH, HV, VH, and VV measurement of polarimetric echo data. The signal flow for each imaging is as follows: The chirp signal generated by the FM signal source is amplified after the polarization selection switch (H or V), passed through the circulator, signal feed network (H or V), and then amplified by the *T* channel of *T*/*R* component, and it is transmitted through the corresponding antenna array radiating element (H or V). The antenna array H and V radiating elements receive the ground echo signals (HH, HV and VH, VV) simultaneously, and are amplified by the *R* channel of the *T*/*R* component, and then passed through the signal feeding network (H and V) and the circulator, arrive at the radar receiver (H and V), send the data to the generator to sample, and package the auxiliary data and then by means of the downlink transmit it to the ground receiving station.

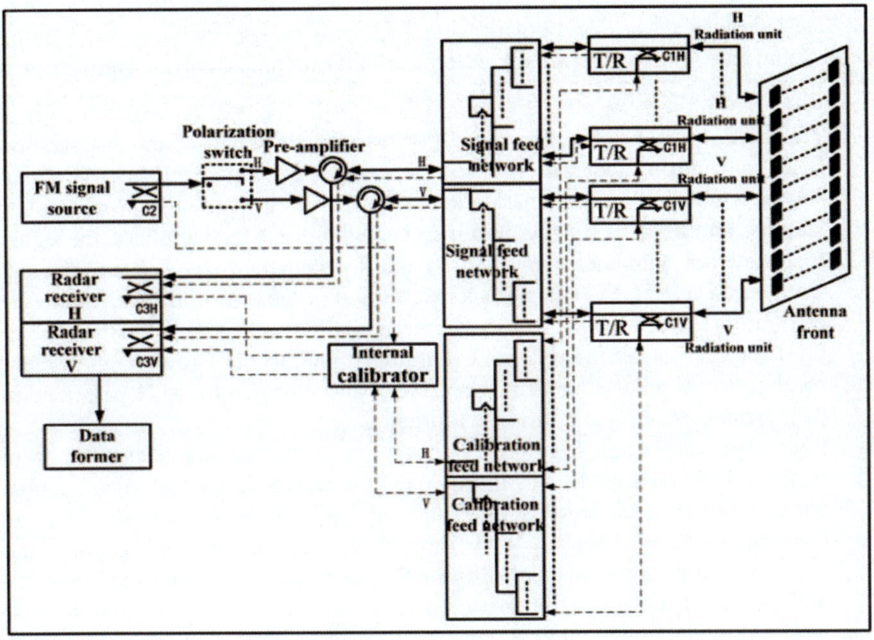

Fig. 6.9 Full phased array antenna polarization SAR system and internal calibration path diagram

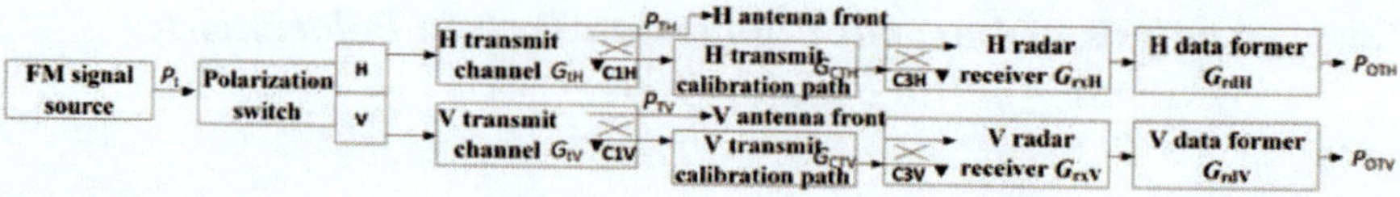

Fig. 6.10 Polarization SAR transmitting calibration signal flow diagram

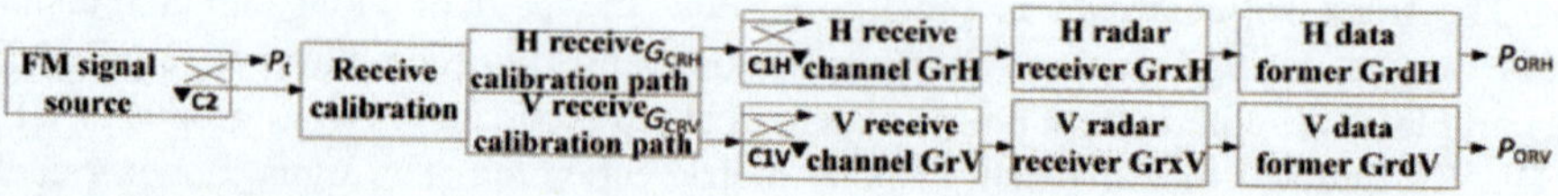

Fig. 6.11 Polarimetric SAR receiving calibration signal flow diagram

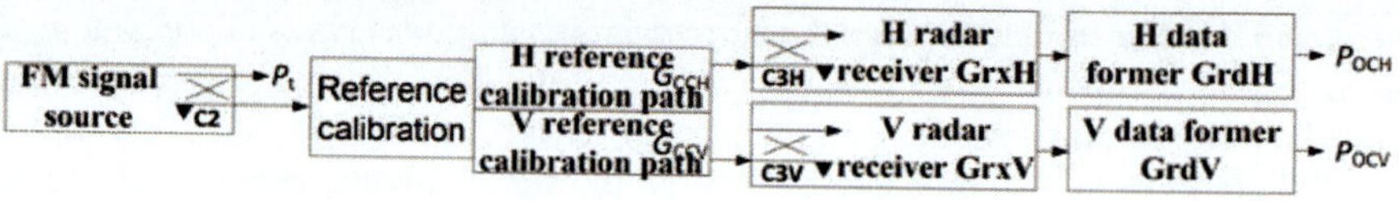

Fig. 6.12 Polarimetric SAR reference calibration signal flow diagram

The thin dashed arrows in the figure depict the signal paths in the calibration of the polarization SAR system. Similar to the single-polarization SAR system, the internal calibration of the polarization SAR system also includes transmitting calibration, receiving calibration, and reference calibration. The corresponding signal flow diagram is shown in Figs. 6.10, 6.11 and 6.12.

(1) Transmitting calibration: Figure 6.10 shows the flowchart of the polarization SAR transmitting calibration. The chirp signal generated by the FM signal source is selected by the polarization switch to be H polarization (or V polarization), amplified by the pre-amplifier, passed through the circulator, the signal feed network, and then amplified by the T channel of the *T*/*R* component. Coupler C1H (or C1V) couples a small portion of the transmitting signal to the calibration feed network, through the internal calibrator, and through the coupler C3H (or C3V) send the internal calibration signal to the H polarization radar receiver (or V polarization radar receiver), and then send it to H polarization data former (or V polarization data former).

(2) Receiving calibration: Figure 6.11 shows the calibration signal flow diagram of the polarimetric SAR. The chirp signal generated by the FM signal source enters the internal calibrator through the coupler C2, enters the receiving channel through the coupler C1H or C1V through the calibration feed network, and is amplified by the *R* channel of the *T*/*R* component via signal feed network and circulator, reaches H polarization radar receiver (or V polarization radar receiver), and then is sent to H polarization data former (or V polarization data former).

(3) Reference calibration: Figure 6.12 shows the reference calibration signal flow diagram of the polarization SAR. The chirp signal generated by the FM signal source enters the internal calibrator through coupler C2, and enters the H polarization radar receiver and data former (or V polarization radar receiver and data former) through coupler C3H (or C3V).

6.3.2 *Crosstalk Analysis of a Polarimetric Synthetic Aperture Radar System*

The influence of the polarimetric SAR system on the acquired data is mainly reflected in the polarization channel imbalance and cross-channel crosstalk. Various crosstalks present in a polarimetric SAR system are analyzed as follows.

(1) **Transmitting channel crosstalk**

The crosstalk of the polarimetric SAR transmitting channel mainly includes the following:

(1) Crosstalk caused by insufficient polarization switch isolation: When the switch is switched at H (or V)-polarization, the V (or H)-polarization path will generate a certain signal leakage, that is, crosstalk signal, due to insufficient polarization switch isolation. Generally, the isolation of the polarization switch is greater than 40 dB, so the crosstalk into the V (or H)-channel is small. Even if a part of the crosstalk signal enters the V (or H)-polarization channel, since the SAR system uses the time-division mode of operation, when operating on the H (or V)-channel, the high power amplifier of the V (or H)-channel is off, and crosstalk from the polarization switch into the V (or H)-channel is suppressed. That is to say, the crosstalk caused by the polarization switch can be neglected unless a polarization switch failure or a V (or H)-channel amplifier failure cannot be turned off.

(2) Crosstalk between paths after two transmitting channel power amplifiers: If the H polarization channel and the V polarization channel share a single calibration feed network, i.e., the signals of the two channels H and V pass through the C1H, C1V coupler to enter a common calibration feed network. They are mixed and sent to the internal calibrator, and this part of the crosstalk cannot be obtained by the internal calibration test. In this book, this situation is not discussed. However, the system shown will be mainly analyzed in this book for Fig. 6.9, that is, the calibration feed network is also divided into two networks corresponding to H polarization and V polarization. Crosstalk measurement between different polarization transmitting channels can be performed by switching in the internal calibrator.

(3) Crosstalk between the antenna array H polarization radiating element and the V polarization radiating element: This part of the crosstalk is the most important. However, as can be seen from Fig. 6.9, it is outside the couplers C1H and C1V

and is the part that the internal calibration loop cannot reach. Therefore, the measurement cannot be performed using the internal calibration and can only be measured by means of the external calibration.

(2) **Receiving channel crosstalk**

Similarly, the crosstalk of the receiving channel of polarimetric SAR system includes the following:

(1) Crosstalk between the antenna array's H polarization radiating element and the V polarization radiating element: The crosstalk is not measured by the internal calibration outside the internal calibration loop. And the multi-polarization SAR system crosstalk mainly occurs in this part. This crosstalk can only be measured by external calibration.
(2) Crosstalk between receiving channel paths after antenna reception and before radar receiver: Similarly, if the H polarization channel and the V polarization channel share a calibration feed network, this part of the crosstalk cannot be measured by internal calibration. This section will discuss in detail the system shown in Fig. 6.9, how to measure crosstalk between the transmit and receive channels by internal calibration.
(3) Crosstalk between H polarization radar receivers and V polarization radar receivers: These two receivers are usually two independent shielding boxes, and the influence between them is usually small, so the crosstalk is negligible.

6.3.3 Calibration of Channel Crosstalk and Unbalance in a Polarimetric Synthetic Aperture Radar System

The above analysis of various channel crosstalks that may exist in a polarimetric SAR system is described below. For the system shown in Fig. 6.9, how to measure crosstalk and imbalance between channels by internal calibration is analyzed.

1. **Transmitter crosstalk, δ_3 and δ_4 unbalance f_2**

From Fig. 6.10, when the H polarization transmit calibration is obtained, the H polarization signal output by the system is

$$P_{\mathrm{OT\_HH}} = P_{\mathrm{t}} \bullet G_{\mathrm{t\_HH}} \bullet G_{\mathrm{C1H}} \bullet G_{\mathrm{CTH}} \bullet G_{\mathrm{C3H}} \bullet G_{\mathrm{rxH}} \bullet G_{\mathrm{rdH}} \tag{6.26}$$

The V polarization signal output by the system at the same time is

$$P_{\mathrm{OT\_HV}} = P_{\mathrm{t}} \bullet G_{\mathrm{t\_HV}} \bullet G_{\mathrm{C1V}} \bullet G_{\mathrm{CTV}} \bullet G_{\mathrm{C3V}} \bullet G_{\mathrm{rxV}} \bullet G_{\mathrm{rdV}} \tag{6.27}$$

where $P_{\mathrm{OT\_HH}}$ and $P_{\mathrm{OT\_HV}}$ are the H polarization and V polarization signal outputs by the system when H polarization transmitting is calibrated; P_{t} is the FM signal source transmitted signal; G_{t_HH} and G_{t_HV} are the H channel gain and V channel gains of the system during H-polarized transmission, respectively; G_{C1H} and G_{C1V} are the gains of the H channel and V channel coupler C1, respectively; G_{CTH} and G_{CTV} are the internal calibration H transmit calibration path and the V transmit calibration path gain, respectively, (including the path gain of the internal calibrator, calibration feed network, and interconnect cable gain); G_{C3H} and G_{C3V} are the gains of the H channel and V channel coupler C3, respectively; G_{rxH} and G_{rxV} are the H channel and V channel radar receiver gains, respectively; and G_{rdH} and G_{rdV} are the H channel and V channel data former processing gain, respectively.

Similarly, when the V polarization transmit calibration is obtained, the V polarization signal and the H polarization signal output by the system are, respectively,

$$P_{OT\_VV} = P_t \bullet G_{t\_VV} \bullet G_{C1V} \bullet G_{CTV} \bullet G_{C3V} \bullet G_{rxV} \bullet G_{rdV} \tag{6.28}$$

$$P_{OT\_VH} = P_t \bullet G_{t\_VH} \bullet G_{C1H} \bullet G_{CTH} \bullet G_{C3H} \bullet G_{rxH} \bullet G_{rdH} \tag{6.29}$$

where P_{OT_VV} and P_{OT_VH} are the V polarization and H polarization signal outputs by the system when V polarization transmitting is calibrated, respectively, and G_{t_VV} and G_{t_VH} are the V channel gain and H channel gain of the system when V polarization transmitting, respectively.

According to the definition of 6.2.1, the crosstalk δ_3 of the normalized H transmit channel into the V transmit channel can be obtained from Eqs. (6.26) to (6.29), the crosstalk δ_4 of the normalized V transmit channel into the H transmit channel, and the transmit end unbalance f_2 of V channel relative to H channel

$$\delta_3 = \frac{|P_{OT\_HV}|}{|P_{OT\_HH}|} \tag{6.30}$$

$$\delta_4 = \frac{|P_{OT\_VH}|}{|P_{OT\_VV}|} \tag{6.31}$$

$$f_2 = \frac{P_{OT\_VV}}{P_{OT\_HH}} \tag{6.32}$$

2. **Receiver crosstalk δ_1, δ_2 and unbalance f_1**

From Fig. 6.11, when the H polarization receive calibration is obtained, the H polarization signal output by the system is

$$P_{OR\_HH} = P_t \bullet G_{C2} \bullet G_{CRH} \bullet G_{C1H} \bullet G_{r\_HH} \bullet G_{rxH} \bullet G_{rdH} \tag{6.33}$$

The V polarization signal output by the system at the same time is

$$P_{OR\_HV} = P_t \bullet G_{C2} \bullet G_{CRV} \bullet G_{C1V} \bullet G_{r\_HV} \bullet G_{rxV} \bullet G_{rdV} \tag{6.34}$$

where P_{OR_HH} and P_{OR_HV} are the H polarization and V polarization signals output by the system during the H polarization receive calibration; G_{C2} is the gain of the coupler C2; G_{CRH} and G_{CRV} are the internal calibration H receive calibration channel and V receive calibration channel gain (including gain of the internal calibrator, gain of the calibration feed network, and interconnecting cable); and G_{r_HH} and G_{r_HV} are the gain of the H receive channel and V receive channel of the system except the radar receiver, respectively.

Similarly, when the V polarization receive calibration is performed, the V polarization signal and the H polarization signal outputs by the system are, respectively,

$$P_{OR\_VV} = P_t \bullet G_{C2} \bullet G_{CRV} \bullet G_{C1V} \bullet G_{r\_VV} \bullet G_{rxV} \bullet G_{rdV} \tag{6.35}$$

$$P_{OR\_VH} = P_t \bullet G_{C2} \bullet G_{CRH} \bullet G_{C1H} \bullet G_{r\_VH} \bullet G_{rxH} \bullet G_{rdH} \tag{6.36}$$

where P_{OR_VV} and P_{OR_VH} are the V polarization and H polarization signal outputs by the system during V polarization receive calibration, respectively, and G_{r_VV} and G_{r_VH} are the V channel gain and H channel gain of the system when V polarization receives, respectively.

Also according to the definition of 6.2.1, the crosstalk δ_1 of the normalized H receive channel entering the V receive channel can be obtained by Eqs. (6.33)–(6.36), the crosstalk δ_2 of the normalized V receive channel entering the H receive channel, and the receiving imbalance f_1 of the terminal V channel relative to the H channel.

$$\delta_1 = \frac{|P_{OR\_HV}|}{|P_{OR\_HH}|} \tag{6.37}$$

$$\delta_2 = \frac{|P_{OR\_VH}|}{|P_{OR\_VV}|} \tag{6.38}$$

$$f_1 = \frac{P_{OR\_VV}}{P_{OR\_HH}} \tag{6.39}$$

(a) **Reference calibration**

From Fig. 6.12, when the H polarization reference calibration is obtained, the H polarization signal output by the system is

$$P_{OC\_HH} = P_t \bullet G_{C2} \bullet G_{CCH} \bullet G_{C3H} \bullet G_{rxH} \bullet G_{rdH} \tag{6.40}$$

The V polarization signal output by the system at the same time is

$$P_{OC\_HV} = P_t \bullet G_{C2} \bullet G_{CCV} \bullet G_{C3V} \bullet G_{rxV} \bullet G_{rdV} \tag{6.41}$$

where P_{OC_HH} and P_{OC_HV} are the H polarization and V polarization signal outputs by the system during the H polarization reference calibration, and G_{CCH} and G_{CCV} are the internal calibration H reference calibration path and the V reference calibration path gains, respectively (including the path gain of the internal calibrator and the gain of the interconnect cable).

Similarly, when the V polarization reference calibration is obtained, the V polarization signal and the H polarization signal outputs by the system are, respectively,

$$P_{OC\_VV} = P_t \bullet G_{C2} \bullet G_{CCV} \bullet G_{C3V} \bullet G_{rxV} \bullet G_{rdV} \tag{6.42}$$

$$P_{OC\_VH} = P_t \bullet G_{C2} \bullet G_{CCH} \bullet G_{C3H} \bullet G_{rxH} \bullet G_{rdH} \tag{6.43}$$

where P_{OC_VV} and P_{OC_VH} are the V polarization and H polarization signal outputs by the system during the V polarization reference calibration, respectively.

The time-division mode of operation shown in Fig. 6.9 is performed by switching the H calibration path and the V calibration path in each calibration mode to measure crosstalk and unbalance in various channels. However, since each calibration channel signal has a short duration, it can be considered that the amplitude and phase characteristics of the internal calibration channel are unchanged, and the relative calibration of crosstalk and unbalance between different polarization channels at the same time is not affected. But in fact, each parameter in the system is a time-varying function. For accurate internal calibration, the performance changes of each channel of the system must be measured in real time. As with the internal calibration of the single-polarization SAR system, the reference calibration is used, and the ratio calibration method is used to obtain the relative change of the gain of each transmit channel and receive channel of the system and the relative change of the total power of the system. The corresponding parameters are compensated and corrected to complete the internal calibration of the system.

6.4 External Calibration Algorithm for Polarimetric Synthetic Aperture Radar System

At present, there are many calibration algorithms for various types of polarimetric SAR, which have been successfully applied to the calibration of specific airborne or spaceborne SAR systems. This section mainly introduces van Zyl algorithm [6–9], Quegan calibration method [10], Sarabandi algorithm [11, 12], posterior calibration method [13], Whitt algorithm [14], JPL polarization external calibration algorithm, etc. [15].

6.4.1 *Calibration Algorithm for Mixed Point Targets and Distributed Targets—van Zyl Algorithm*

The van Zyl algorithm assumes the system model and the feature of the ground object, so that the phase difference and crosstalk between the channels no longer require the man-made point target to be calibrated, but only a triangular reflector is needed to calibrate the amplitude imbalance between the channels and the system absolute gain. The entire process of the van Zyl algorithm is as described above, and the calibration flow diagram is shown in Fig. 6.13. The whole algorithm is divided into four parts, including phase calibration, crosstalk calibration, relative amplitude and relative phase calibration of the channel imbalance, radiation calibration, etc.

6.4.1.1 Phase Calibration

The coordinate system uses horizontal (h) and vertical (v) basis vectors, and for a resolution unit, the scattering matrix is

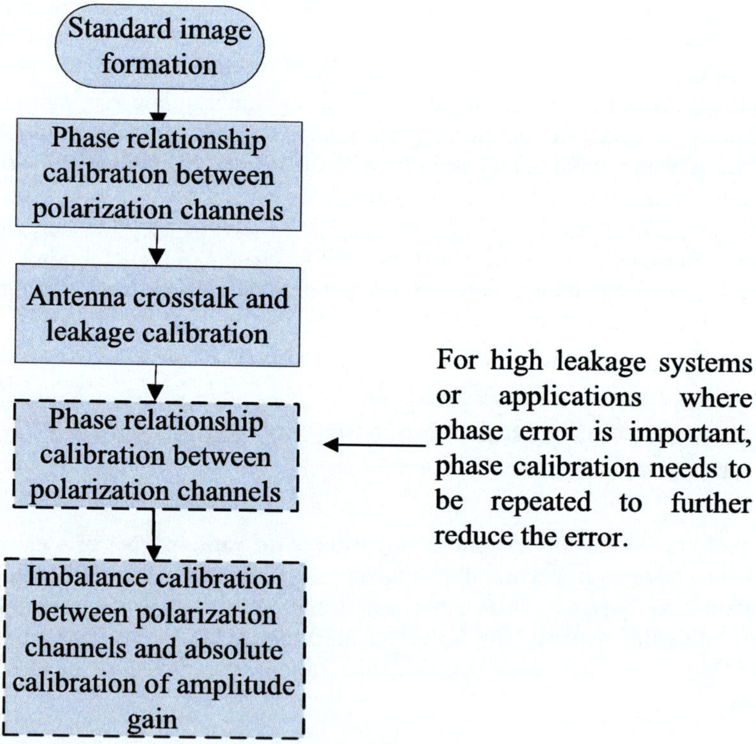

Fig. 6.13 van Zyl algorithm polarization calibration flowchart

$$\mathbf{S}_{\mathrm{JPL}} = \begin{pmatrix} S_{\mathrm{hh}} & S_{\mathrm{hv}} \\ S_{\mathrm{vh}} & S_{\mathrm{vv}} \end{pmatrix} \tag{6.44}$$

Here, S_{ij} represents a complex scattered signal of j-transmit polarization and i-receive polarization. Due to the path difference between the radar channels, the receive matrix $\mathbf{R}_{\mathrm{JPL}}$ obtained by the actual system is

$$\mathbf{R}_{\mathrm{JPL}} = \begin{pmatrix} S_{\mathrm{hh}} \exp \mathrm{j}(\phi_{\mathrm{t,h}} + \phi_{\mathrm{r,h}}) & S_{\mathrm{hv}} \exp \mathrm{j}(\phi_{\mathrm{t,v}} + \phi_{\mathrm{r,h}}) \\ S_{\mathrm{vh}} \exp \mathrm{j}(\phi_{\mathrm{t,h}} + \phi_{\mathrm{r.v}}) & S_{\mathrm{vv}} \exp \mathrm{j}(\phi_{\mathrm{t,v}} + \phi_{\mathrm{r,v}}) \end{pmatrix} \tag{6.45}$$

For simplification, extract the factor relative to the VV term

$$\mathbf{R}_{\mathrm{JPL}} = \exp \mathrm{j}(\phi_{\mathrm{t,v}} + \phi_{\mathrm{r,v}}) \begin{pmatrix} S_{\mathrm{hh}} \exp \mathrm{j}(\phi_{\mathrm{t}} + \phi_{\mathrm{r}}) & S_{\mathrm{hv}} \exp \mathrm{j}\phi_{\mathrm{r}} \\ S_{\mathrm{vh}} \exp \mathrm{j}\phi_{\mathrm{t}} & S_{\mathrm{vv}} \end{pmatrix} \tag{6.46}$$

Here $\phi_{\mathrm{t}} = \phi_{\mathrm{t,h}} - \phi_{\mathrm{t,v}}$, $\phi_{\mathrm{r}} = \phi_{\mathrm{r,h}} - \phi_{\mathrm{r,v}}$.

At least two equations are needed to solve ϕ_{t} and ϕ_{r}. From the reciprocity $S_{\mathrm{hv}} = S_{\mathrm{vh}}$, it is possible to check the averaged product on the entire image by means of checking

$$R^*_{\mathrm{JPL\text{-}hv}} R_{\mathrm{JPL\text{-}vh}} = S^*_{\mathrm{hv}} S_{\mathrm{vh}} \exp \mathrm{j}(\phi_{\mathrm{t}} - \phi_{\mathrm{r}}) \tag{6.47}$$

To obtain $\phi_{\mathrm{t}} - \phi_{\mathrm{r}}$. Since the system noise between the radar channels is uncorrelated, this complex product is slightly different for each pixel. However, by means of averaging a large number of pixels, an average value can be got, which is mainly related to the phase path difference of the channel that is interested by us.

The JPL multi-polarization image data is stored in the form of a Stokes matrix, so that the factor $\exp \mathrm{j}(\phi_{\mathrm{t,v}} + \phi_{\mathrm{r,v}})$ in (6.46) can be omitted, since the calculation of the Stokes matrix does not take this one into account—the Stokes matrix element only uses the relative phase between the polarization channels. Using Stokes matrix to store data can also reduce the amount of data, which makes the application of multi-polarization radar data more convenient.

The first step in reducing the amount of data storage is to symmetry the scattering matrix S_{JPL}. The average value of $R^*_{\mathrm{JPL\text{-}hv}} R_{\mathrm{JPL\text{-}vh}}$ in the entire image is estimated first, and the resulting phase difference $\phi_{\mathrm{t}} - \phi_{\mathrm{r}}$ is stored in the header of the data file for use in subsequent calibration procedures. Applying this phase difference to the VH term in Eq. (6.46) for each pixel causes the diagonal elements of the scattering matrix to be in same phase. In this way, the new matrix $\mathbf{Z}_{\mathrm{JPL}}$ corrected from $\mathrm{R}_{\mathrm{JPL}}$ is

$$\mathbf{Z}_{\mathrm{JPL}} = \begin{pmatrix} S_{\mathrm{hh}} \exp \mathrm{j}(\phi_{\mathrm{t}} + \phi_{\mathrm{r}}) & S_{\mathrm{hv}} \exp \mathrm{j}\phi_{\mathrm{r}} \\ S_{\mathrm{vh}} \exp \mathrm{j}\phi_{\mathrm{r}} & S_{\mathrm{vv}} \end{pmatrix} \tag{6.48}$$

Obtain symmetric matrix $\mathbf{W}_{\mathrm{JPL}}$ from symmetric matrix $\mathbf{Z}_{\mathrm{JPL}}$

$$\mathbf{W}_{\mathrm{JPL}} = \frac{1}{2}\left(\mathbf{Z} + \mathbf{Z}^{\mathrm{T}}\right) \tag{6.49}$$

Using $\mathbf{W}_{\mathrm{JPL}}$, the Stokes matrix [16] $\mathbf{M}_{\mathrm{JPL}}$ can be calculated according to the following formula:

$$\mathbf{M}_{\mathrm{JPL}} = \left(\mathbf{V}^{\mathrm{T}}\right)^{-1}\mathbf{U}\mathbf{V}^{-1} \tag{6.50}$$

where

$$\mathbf{U} = \begin{pmatrix} W_{\text{JPL-hh}}W^*_{\text{JPL-hh}} & W_{\text{JPL-hv}}W^*_{\text{JPL-hv}} & W_{\text{JPL-hh}}W^*_{\text{JPL-hv}} & W_{\text{JPL-hv}}W^*_{\text{JPL-hh}} \\ W_{\text{JPL-vh}}W^*_{\text{JPL-vh}} & W_{\text{JPL-vv}}W^*_{\text{JPL-vv}} & W_{\text{JPL-vh}}W^*_{\text{JPL-vv}} & W_{\text{JPL-vv}}W^*_{\text{JPL-vh}} \\ W_{\text{JPL-hh}}W^*_{\text{JPL-vh}} & W_{\text{JPL-hv}}W^*_{\text{JPL-vv}} & W_{\text{JPL-hh}}W^*_{\text{JPL-vv}} & W_{\text{JPL-hv}}W^*_{\text{JPL-vh}} \\ W_{\text{JPL-vh}}W^*_{\text{JPL-hh}} & W_{\text{JPL-vv}}W^*_{\text{JPL-hv}} & W_{\text{JPL-vh}}W^*_{\text{JPL-hv}} & W_{\text{JPL-vv}}W^*_{\text{JPL-hh}} \end{pmatrix} \tag{6.51}$$

$$\mathbf{V} = \begin{pmatrix} 1 & 1 & 0 & 0 \\ 1 & -1 & 0 & 0 \\ 0 & 0 & 1 & \mathrm{i} \\ 0 & 0 & 1 & -\mathrm{i} \end{pmatrix} \tag{6.52}$$

The elements that make up the matrix U and their relationship to the true scattering matrix values are listed below. The element values of W_{JPL} and S_{JPL} are not consistent because the measured data has not been phase-calibrated.

$$W_{\text{JPL-hh}}W^*_{\text{JPL-hh}} = S_{\mathrm{hh}}S^*_{\mathrm{hh}} \tag{6.53a}$$

$$W_{\text{JPL-vv}}W^*_{\text{JPL-vv}} = S_{\mathrm{vv}}S^*_{\mathrm{vv}} \tag{6.53b}$$

$$W_{\text{JPL-hv}}W^*_{\text{JPL-hv}} = S_{\mathrm{hv}}S^*_{\mathrm{hv}} \tag{6.53c}$$

$$W_{\text{JPL-hh}}W^*_{\text{JPL-vv}} = S_{\mathrm{hh}}S^*_{\mathrm{vv}}\exp \mathrm{j}(\phi_{\mathrm{t}} + \phi_{\mathrm{r}}) \tag{6.53d}$$

$$W_{\text{JPL-hh}}W^*_{\text{JPL-hv}} = S_{\mathrm{hh}}S^*_{\mathrm{hv}}\exp \mathrm{j}\phi_{\mathrm{t}} \tag{6.53e}$$

$$W_{\text{JPL-vh}}W^*_{\text{JPL-vv}} = S_{\mathrm{vh}}S^*_{\mathrm{vv}}\exp \mathrm{j}\phi_{\mathrm{r}} \tag{6.53f}$$

$$W_{\text{JPL-hv}}W^*_{\text{JPL-hh}} = S_{\mathrm{hv}}S^*_{\mathrm{hh}}\exp \mathrm{j}(-\phi_{\mathrm{t}}) \tag{6.53g}$$

$$W_{\text{JPL-vv}}W^*_{\text{JPL-vh}} = S_{\mathrm{vv}}S^*_{\mathrm{vh}}\exp \mathrm{j}(-\phi_{\mathrm{r}}) \tag{6.53h}$$

$$W_{\text{JPL-vv}}W^*_{\text{JPL-hh}} = S_{\mathrm{vv}}S^*_{\mathrm{hh}}\exp \mathrm{j}(-\phi_{\mathrm{t}} - \phi_{\mathrm{r}}) \tag{6.53i}$$

If the primary scattering mechanism of at least one point in the image is known, the other factors needed for phase calibration can be determined. A region is selected in the image whose phase difference between the HH and VV polarization scattering is known. Usually those areas where the theoretical phase difference is close to zero are selected. The value of $\phi_t + \phi_r$ by averaging $W_{hh} W_{vv}^* = S_{hh} S_{vv}^* \exp j(\phi_t + \phi_r)$ over this area can be obtained.

Knowing $\phi_t + \phi_r$, $\phi_t - \phi_r$, the ϕ_t, ϕ_r can be solved.

For single-base radar imaging, according to the reciprocity assumption, the Stokes matrix $\mathbf{M}_{JPL}$ will be a real symmetric matrix with nine independent elements. Make the $\mathbf{M}_{JPL}$ form as follows:

$$\mathbf{M}_{JPL} = \begin{pmatrix} M_{11} & M_{12} & M_{13} & M_{14} \\ M_{21} & M_{22} & M_{23} & M_{24} \\ M_{31} & M_{32} & M_{33} & M_{34} \\ M_{41} & M_{42} & M_{43} & M_{44} \end{pmatrix} \tag{6.54}$$

Then the relationship between its elements and the cross product of the elements of the scattering matrix W_{JPL} are

$$M_{11} = \frac{1}{4}[W_{JPL\text{-}hh} W_{JPL\text{-}hh}^* + W_{JPL\text{-}vv} W_{JPL\text{-}vv}^* + 2W_{JPL\text{-}hv} W_{JPL\text{-}hv}^*] \tag{6.55a}$$

$$M_{12} = \frac{1}{4}[W_{JPL\text{-}hh} W_{JPL\text{-}hh}^* - W_{JPL\text{-}vv} W_{JPL\text{-}vv}^*] \tag{6.55b}$$

$$M_{13} = \frac{1}{2}\mathrm{Re}[W_{JPL\text{-}hh}^* W_{JPL\text{-}hv}] + \frac{1}{2}Re[W_{JPL\text{-}hv}^* W_{JPL\text{-}vv}] \tag{6.55c}$$

$$M_{14} = \frac{1}{2}\mathrm{Im}[W_{JPL\text{-}hh}^* W_{JPL\text{-}hv}] - \frac{1}{2}\mathrm{Im}[W_{JPL\text{-}hv}^* W_{JPL\text{-}vv}] \tag{6.55d}$$

$$M_{22} = \frac{1}{4}[W_{JPL\text{-}hh} W_{JPL\text{-}hh}^* + W_{JPL\text{-}vv} W_{JPL\text{-}vv}^* - 2W_{JPL\text{-}hv} W_{JPL\text{-}hv}^*] \tag{6.55e}$$

$$M_{23} = \frac{1}{2}\mathrm{Re}[W_{JPL\text{-}hh}^* W_{JPL\text{-}hv}] - \frac{1}{2}\mathrm{Re}[W_{JPL\text{-}hv}^* W_{JPL\text{-}vv}] \tag{6.55f}$$

$$M_{24} = \frac{1}{2}\mathrm{Im}[W_{JPL\text{-}hh}^* W_{JPL\text{-}hv}] - \frac{1}{2}\mathrm{Im}[W_{JPL\text{-}hv}^* W_{JPL\text{-}vv}] \tag{6.55g}$$

$$M_{33} = \frac{1}{2} W_{JPL\text{-}hv} W_{JPL\text{-}hv}^* + \frac{1}{2}\mathrm{Re}[W_{JPL\text{-}hh}^* W_{JPL\text{-}vv}] \tag{6.55h}$$

$$M_{34} = \frac{1}{2}\mathrm{Im}[W_{JPL\text{-}hh}^* W_{JPL\text{-}vv}] \tag{6.55i}$$

$$M_{44} = \frac{1}{2} W_{\text{JPL-hv}} W^*_{\text{JPL-hv}} - \frac{1}{2}\text{Re}[W^*_{\text{JPL-hh}} W_{\text{JPL-vv}}] \tag{6.55j}$$

where Re[•] represents the real part of • and *Im*[•] represents the imaginary part of •.

The following relationship is satisfied between M_{11}, M_{22}, M_{33}, and M_{44}:

$$M_{44} = M_{11} - M_{22} - M_{33} \tag{6.56}$$

The uncalibrated data of the stored multi-polarization SAR Stokes matrix form is as follows:

$$\mathbf{U} = \mathbf{V}^{\text{T}} \mathbf{M}_{\text{JPL}} \mathbf{V} \tag{6.57}$$

It is possible to solve the corresponding cross product of the scattering matrix. The phase products are used for phase calibration according to the method described above to obtain ϕ_{t} and ϕ_{r}, and then applied to the matrix U to obtain a matrix X of phase calibration (consisting of a cross product of the scattering matrix L, the structure is the same as the matrix U, which is obtained by phase calibration by U), and the relationship between the elements and the elements of the matrix U is

$$L_{\text{hh}} L^*_{\text{hh}} = W_{\text{JPL-hh}} W^*_{\text{JPL-hh}} \tag{6.58a}$$

$$L_{\text{vv}} L^*_{\text{vv}} = W_{\text{JPL-vv}} W^*_{\text{JPL-vv}} \tag{6.58b}$$

$$L_{hv} L^*_{hv} = W_{JPL-hv} W^*_{JPL-hv} \tag{6.58c}$$

$$L_{\text{hh}} L^*_{\text{vv}} = W_{\text{JPL-hh}} W^*_{\text{JPL-vv}} \exp -\text{j}(\phi_{\text{t}} + \phi_{\text{r}}) \tag{6.58d}$$

$$L_{\text{hh}} L^*_{\text{hv}} = W_{\text{JPL-hh}} W^*_{\text{JPL-hv}} \exp -\text{j}\phi_{\text{t}} \tag{6.58e}$$

$$L_{\text{vh}} L^*_{\text{vv}} = W_{\text{JPL-vh}} W^*_{\text{JPL-vv}} \exp -\text{j}\phi_{\text{r}} \tag{6.58f}$$

$$L_{\text{hv}} L^*_{\text{hh}} = W_{\text{JPL-hv}} W^*_{\text{JPL-hh}} \exp \text{j}\phi_{\text{t}} \tag{6.58g}$$

$$L_{\text{vv}} L^*_{\text{vh}} = W_{\text{JPL-vv}} W^*_{\text{JPL-vh}} \exp \text{j}\phi_{\text{r}} \tag{6.58h}$$

$$L_{\text{vv}} L^*_{\text{hh}} = W_{\text{JPL-vv}} W^*_{\text{JPL-hh}} \exp \text{j}(\phi_{\text{t}} + \phi_{\text{r}}) \tag{6.58i}$$

By applying the formula $\mathbf{M}_{\mathrm{JPL}} = \left(\mathbf{V}^{\mathrm{T}}\right)^{-1}\mathbf{X}\mathbf{V}^{-1}$ to the matrix X, the Stokes matrix M_{JPL} with accurate phase calibration can be obtained to accurately describe the phase of the real scattering matrix.

In general, there are ambiguities in the determination of ϕ_{t} and ϕ_{r} by two equations, because both equations are calculated using the modulus 2π. Thus, both ϕ_{t} and ϕ_{r} may have an error of π, i.e., the value solved is the opposite of the true value. But this ambiguity has little effect in practice because it only causes the 45° linear polarization component to be deciphered to 135° linear polarization, or vice versa. For azimuthally symmetric scattering targets, no difference is perceived, in fact, most natural surfaces are like this. However, for applications where 45° linear polarization and 135° linear polarization must be distinguished, this possible ambiguity must be noticed.

6.4.1.2 Crosstalk Calibration

Consider the effects of transmit and receive on the scattering matrix. Let R be the receive system distortion matrix and T be the transmit system distortion matrix. Ignoring noise, for a symmetric scattering matrix, the W_{JPL} system model is

$$\mathbf{W}_{\mathrm{JPL}} = A\mathrm{e}^{\mathrm{j}\phi}\mathbf{R}\mathbf{S}_{\mathrm{JPL}}\mathbf{T} \tag{6.59}$$

where S_{JPL} stands for the true ground object scattering matrix and A is the total absolute amplitude factor, which is the residual gain factor after radiometric calibration. Ideally (after accurate radiometric calibration) A should be 1 and φ is the total absolute phase; since φ will be lost during polarization synthesis, it is not necessary to calibrate φ. Reciprocity requirements in the case of backscattering

$$\mathbf{S}_{\mathrm{JPL}} = \mathbf{S}_{\mathrm{JPL}}{}^{\mathrm{T}} \tag{6.60}$$

This requires

$$\mathbf{R} = \mathbf{T}^{\mathrm{T}} \tag{6.61}$$

Physically, this means that the system must be considered reciprocal. Usually this is not the case, but the symmetry of the data in the compression process forces us to accept this system model. Note that after the scattering matrix obtained by the system is symmetric, there is no simple relationship between the estimated system parameters and the actual physical system parameters. But the analysis below will show that using the system parameters obtained from these estimates, the data can still be calibrated.

Ignoring noise, the symmetric scattering matrix W_{JPL} obtained by the system can be written as follows:

$$\begin{pmatrix} W_{\text{JPL-hh}} & W_{\text{JPL-hv}} \\ W_{\text{JPL-vh}} & W_{\text{JPL-vv}} \end{pmatrix} = Ae^{j\phi}\begin{pmatrix} 1 & \delta_2 \\ \delta_1 & f \end{pmatrix}\begin{pmatrix} S_{hh} & S_{hv} \\ S_{vh} & S_{vv} \end{pmatrix}\begin{pmatrix} 1 & \delta_1 \\ \delta_2 & f \end{pmatrix} \tag{6.62}$$

The elements of these matrices are complex values, δ_1 represents crosstalk when transmit or receive a vertically polarization electric field, δ_2 represents crosstalk when transmit or receive a horizontally polarization electric field, and f represents the same polarization channel (hh, vv) imbalance in amplitude and phase on one channel (i.e., transmitter to target or target to receiver).

The crosstalk calibration can be separated from the channel imbalance calibration and the radiation calibration (i.e., calibration A).

T can be written as follows (6.63a) or (6.63b):

$$\begin{pmatrix} 1 & \delta_1 \\ \delta_2 & f \end{pmatrix} = \begin{pmatrix} 1 & 0 \\ 0 & f \end{pmatrix}\begin{pmatrix} 1 & \delta_1 \\ \frac{\delta_2}{f} & 1 \end{pmatrix} \tag{6.63a}$$

$$\mathbf{T} = \mathbf{T}_c\mathbf{T}_x \tag{6.63b}$$

Similar can be obtained

$$\begin{pmatrix} 1 & \delta_2 \\ \delta_1 & f \end{pmatrix} = \begin{pmatrix} 1 & \frac{\delta_2}{f} \\ \delta_1 & 1 \end{pmatrix}\begin{pmatrix} 1 & 0 \\ 0 & f \end{pmatrix} \tag{6.64a}$$

$$\mathbf{R} = \mathbf{R}_x\mathbf{R}_c \tag{6.64b}$$

Note that $\mathbf{T}_c$ and $\mathbf{R}_c$ are independent of system crosstalk. Substituting Eqs. (6.63a, 6.63b) and (6.64a, 6.64b) into Eq. (6.62), we get

$$\mathbf{W}_{\text{JPL}} = \mathbf{R}_x\mathbf{R}_c\mathbf{S}_{\text{JPL}}\mathbf{T}_c\mathbf{T}_x = \mathbf{R}_x\mathbf{Y}_{\text{JPL}}\mathbf{T}_x \tag{6.65}$$

Introducing the transition matrix $\mathbf{Y}_{\text{JPL}}$

$$\mathbf{Y}_{\text{JPL}} = \begin{pmatrix} Y_{hh} & Y_{hv} \\ Y_{vh} & Y_{vv} \end{pmatrix} = Ae^{j\phi}\begin{pmatrix} S_{hh} & fS_{hv} \\ fS_{vh} & f^2S_{vv} \end{pmatrix} = Ae^{j\phi}\mathbf{R}_c\mathbf{S}_{\text{JPL}}\mathbf{T}_c \tag{6.66}$$

For this matrix, only channel imbalance and radiometric calibration need to be completed.

Note that both $\mathbf{T}_x$ and $\mathbf{R}_x$ contain the element $\frac{\delta_2}{f}$. Crosstalk calibrations only require this ratio to be estimated without having to know δ_2 and f separately. Crosstalk calibration can be completed without relying on external scaling devices. Even if the manual calibration target is not placed in the imaging area, the system crosstalk effect can still be removed from the data.

For crosstalk calibration, the two assumptions for linearization are as follows:

(1) The magnitude of δ_1, δ_2 is smaller enough than 1. This means good isolation between system channels.

(2) For azimuthally symmetric random distribution targets, the co-polarization and cross-polarization components of the scattering matrix S_{JPL} are not correlated. This assumption is reasonable for a uniform, slightly rough surface in the second-order approximation. Mathematical representations are given as follows:

$$\langle S_{\mathrm{hh}}^{*} S_{\mathrm{hv}} \rangle = 0 \tag{6.67a}$$

$$\langle S_{\mathrm{vv}}^{*} S_{\mathrm{hv}} \rangle = 0 \tag{6.67b}$$

Note that for the element of the matrix $\mathrm{Y}_{\mathrm{JPL}}$, Eq. (6.67a, 6.67b) is also established, where $\langle \bullet \rangle$ represents the set averaging of $\bullet$, and on the image it represents the average processing over an image area.

Using the above assumptions, the following relationship can be obtained:

$$\langle W_{\text{JPL-hh}} W_{\text{JPL-hh}}^{*} \rangle \approx \langle Y_{\mathrm{hh}} Y_{\mathrm{hh}}^{*} \rangle \tag{6.68a}$$

$$\langle W_{\text{JPL-vv}} W_{\text{JPL-vv}}^{*} \rangle \approx \langle Y_{\mathrm{vv}} Y_{\mathrm{vv}}^{*} \rangle \tag{6.68b}$$

$$\langle W_{\text{JPL-vv}} W_{\text{JPL-hh}}^{*} \rangle \approx \langle Y_{\mathrm{vv}} Y_{\mathrm{hh}}^{*} \rangle \tag{6.68c}$$

$$\langle W_{\text{JPL-hv}} W_{\text{JPL-hh}}^{*} \rangle - \langle Y_{\mathrm{hv}} Y_{\mathrm{hh}}^{*} \rangle \approx \delta_1 \langle Y_{\mathrm{hh}} Y_{\mathrm{hh}}^{*} \rangle + \frac{\delta_2}{f} \langle Y_{\mathrm{vv}} Y_{\mathrm{hh}}^{*} \rangle + 2\left(\frac{\delta_2}{f}\right)^{*} \langle Y_{\mathrm{hv}} Y_{\mathrm{hv}}^{*} \rangle \tag{6.68d}$$

$$\langle W_{\text{JPL-hv}} W_{\text{JPL-vv}}^{*} \rangle - \langle Y_{\mathrm{hv}} Y_{\mathrm{vv}}^{*} \rangle \approx \delta_1 \langle Y_{\mathrm{hh}} Y_{\mathrm{vv}}^{*} \rangle + \frac{\delta_2}{f} \langle Y_{\mathrm{vv}} Y_{\mathrm{vv}}^{*} \rangle + 2\delta_1^{*} \langle Y_{\mathrm{hv}} Y_{\mathrm{hv}}^{*} \rangle \tag{6.68e}$$

Substituting (6.68a), (6.68b), and (6.68c) into (6.68d), (6.68e), and for $\mathrm{Y}_{\mathrm{JPL}}$ (6.67a, 6.67b) holds, we can get [17]

$$\delta_1 = \frac{\Delta P + 4P^{*} \langle Y_{\mathrm{hv}} Y_{\mathrm{hv}}^{*} \rangle \langle W_{\text{JPL-vv}} W_{\text{JPL-hh}}^{*} \rangle}{\Delta^2 - 16 \langle Y_{\mathrm{hv}} Y_{\mathrm{hv}}^{*} \rangle^2 \langle W_{\text{JPL-vv}} W_{\text{JPL-hh}}^{*} \rangle \langle W_{\text{JPL-hh}} W_{\text{JPL-vv}}^{*} \rangle} \tag{6.69a}$$

$$\frac{\delta_2}{f} = \frac{\Delta Q + 4Q^{*} \langle Y_{\mathrm{hv}} Y_{\mathrm{hv}}^{*} \rangle \langle W_{\text{JPL-hh}} W_{\text{JPL-vv}}^{*} \rangle}{\Delta^2 - 16 \langle Y_{\mathrm{hv}} Y_{\mathrm{hv}}^{*} \rangle^2 \langle W_{\text{JPL-vv}} W_{\text{JPL-hh}}^{*} \rangle \langle W_{\text{JPL-hh}} W_{\text{JPL-vv}}^{*} \rangle} \tag{6.69b}$$

where

$$\begin{aligned} \Delta = {} & \langle W_{\text{JPL-hh}} W_{\text{JPL-hh}}^{*} \rangle \langle W_{\text{JPL-vv}} W_{\text{JPL-vv}}^{*} \rangle - \langle W_{\text{JPL-vv}} W_{\text{JPL-hh}}^{*} \rangle \langle W_{\text{JPL-hh}} W_{\text{JPL-vv}}^{*} \rangle \\ & - 4 \langle Y_{\mathrm{hv}} Y_{\mathrm{hv}}^{*} \rangle \langle Y_{\mathrm{hv}} Y_{\mathrm{hv}}^{*} \rangle \end{aligned} \tag{6.69c}$$

$$P = \langle W_{\text{JPL-hv}} W_{\text{JPL-hh}}^{*} \rangle \langle W_{\text{JPL-vv}} W_{\text{JPL-vv}}^{*} \rangle - \langle W_{\text{JPL-vv}} W_{\text{JPL-hh}}^{*} \rangle \langle W_{\text{JPL-hv}} W_{\text{JPL-vv}}^{*} \rangle$$

$$-2\langle Y_{hv}Y_{hv}^{*}\rangle\langle W_{JPL\text{-}vv}W_{JPL\text{-}hv}^{*}\rangle \tag{6.69d}$$

$$\begin{aligned}Q = {} & \langle W_{JPL\text{-}hh}W_{JPL\text{-}hh}^{*}\rangle\langle W_{JPL\text{-}hv}W_{JPL\text{-}vv}^{*}\rangle - \langle W_{JPL\text{-}hv}W_{JPL\text{-}hh}^{*}\rangle\langle W_{JPL\text{-}hh}W_{JPL\text{-}vv}^{*}\rangle \\ & - 2\langle Y_{hv}Y_{hv}^{*}\rangle\langle W_{JPL\text{-}hh}W_{JPL\text{-}hv}^{*}\rangle\end{aligned} \tag{6.69e}$$

Now calculate the cross product containing the $\mathbf{Y}_{\mathrm{JPL}}$ matrix elements from $\mathbf{W}_{\mathrm{JPL}}$ and estimate the values of δ_1 and $\frac{\delta_2}{f}$.

From (6.65), we can get

$$\mathbf{Y}_{\mathrm{JPL}} = \mathbf{R}_{x}^{-1}\mathbf{W}_{\mathrm{JPL}}\mathbf{T}_{x}^{-1} \tag{6.70}$$

Therefore, assuming that the values of δ_1 and $\frac{\delta_2}{f}$ are already known and that the formula (6.67a, 6.67b) holds, we get

$$\langle Y_{hh}Y_{hh}^{*}\rangle \approx \langle W_{JPL\text{-}hh}W_{JPL\text{-}hh}^{*}\rangle \tag{6.71a}$$

$$\langle Y_{vv}Y_{vv}^{*}\rangle \approx \langle W_{JPL\text{-}vv}W_{JPL\text{-}vv}^{*}\rangle \tag{6.71b}$$

$$\langle Y_{vv}Y_{hh}^{*}\rangle \approx \langle W_{JPL\text{-}vv}W_{JPL\text{-}hh}^{*}\rangle \tag{6.71c}$$

$$\begin{aligned}\langle Y_{hv}Y_{hh}^{*}\rangle \approx {} & \langle W_{JPL\text{-}hv}W_{JPL\text{-}hh}^{*}\rangle - \delta_1\langle W_{JPL\text{-}hh}W_{JPL\text{-}hh}^{*}\rangle \\ & - \frac{\delta_2}{f}\langle W_{JPL\text{-}vv}W_{JPL\text{-}hh}^{*}\rangle - 2\delta_1^{*}\langle W_{JPL\text{-}hv}W_{JPL\text{-}hv}^{*}\rangle\end{aligned} \tag{6.71d}$$

$$\begin{aligned}\langle Y_{vv}Y_{hv}^{*}\rangle \approx {} & \langle W_{JPL\text{-}vv}W_{JPL\text{-}hv}^{*}\rangle - \delta_1^{*}\langle W_{JPL\text{-}vv}W_{JPL\text{-}hh}^{*}\rangle \\ & - \left(\frac{\delta_2}{f}\right)^{*}\langle W_{JPL\text{-}vv}W_{JPL\text{-}vv}^{*}\rangle - 2\frac{\delta_2}{f}\langle W_{JPL\text{-}hv}W_{JPL\text{-}hv}^{*}\rangle\end{aligned} \tag{6.71e}$$

$$\begin{aligned}\langle Y_{hv}Y_{hv}^{*}\rangle \approx {} & \langle W_{JPL\text{-}hv}W_{JPL\text{-}hv}^{*}\rangle + |\delta_1|^2\langle W_{JPL\text{-}hh}W_{JPL\text{-}hh}^{*}\rangle \\ & + \left|\frac{\delta_2}{f}\right|^2\langle W_{JPL\text{-}vv}W_{JPL\text{-}vv}^{*}\rangle - \delta_1\langle W_{JPL\text{-}hh}W_{JPL\text{-}hv}^{*}\rangle \\ & - \delta_1^{*}\langle W_{JPL\text{-}hv}W_{JPL\text{-}hh}^{*}\rangle - \frac{\delta_2}{f}\langle W_{JPL\text{-}vv}W_{JPL\text{-}hv}^{*}\rangle \\ & - \left(\frac{\delta_2}{f}\right)^{*}\langle W_{JPL\text{-}hv}W_{JPL\text{-}vv}^{*}\rangle + \delta_1\left(\frac{\delta_2}{f}\right)^{*}\langle W_{JPL\text{-}hh}W_{JPL\text{-}vv}^{*}\rangle \\ & + \delta_1^{*}\frac{\delta_2}{f}\langle W_{JPL\text{-}vv}W_{JPL\text{-}hh}^{*}\rangle\end{aligned} \tag{6.71f}$$

From Formula (6.69a–6.69e) to (6.71a–6.71f), an iterative method for estimating δ_1 and $\frac{\delta_2}{f}$ can be obtained. First set the initial values $\delta_1 = 0$ and $\frac{\delta_2}{f} = 0$, substitute it into Eq. (6.71f) to get the estimated value of $\langle Y_{hv}Y_{hv}^*\rangle$, and substitute this estimate into Eq. (6.69a–6.69e) to get the estimated values of δ_1 and $\frac{\delta_2}{f}$. Then substitute these two estimates of δ_1 and $\frac{\delta_2}{f}$ into Eq. (6.71f) to get a further estimate of $\langle Y_{hv}Y_{hv}^*\rangle$. This process is repeated until the values of $\langle Y_{hv}Y_{hv}^*\rangle$, δ_1, and $\frac{\delta_2}{f}$ are stable.

If noise is considered and the noise is significant, the crosstalk parameter estimate will be close to zero. This is foreseeable because the noise of the different channels is irrelevant, and the technique relies on the correlation of the crosstalk parameters between the co-polarized and cross-polarized channels. Too strong noise will cover this correlation.

6.4.1.3 Relative Amplitude and Relative Phase Calibration of the Imbalance Between Channels

In order to estimate the channel imbalance f, the following can be obtained from Eq. (6.66):

$$Y_{hh}^*Y_{vv} = f^2 S_{hh}^* S_{vv} \tag{6.72a}$$

$$\frac{Y_{vv}^*Y_{vv}}{Y_{hh}^*Y_{hh}} = |f|^4 \frac{S_{vv}^*S_{vv}}{S_{hh}^*S_{hh}} \tag{6.72b}$$

If the imaging region contains some targets, the relationship between the co-polarized components of the scattering matrix N is known, i.e., the following values are known:

$$\langle N_{hh}N_{hh}^*\rangle = \sigma_n \tag{6.73a}$$

$$\langle N_{vv}N_{vv}^*\rangle = \sigma_n \tag{6.73b}$$

$$\langle N_{hv}N_{hv}^*\rangle = \sigma_n/2 \tag{6.73c}$$

Then the amplitude and phase of f can be estimated.

For triangular reflectors, where

$$\frac{S_{vv}^*S_{vv}}{S_{hh}^*S_{hh}} = 1 \tag{6.74a}$$

$$\arg(S_{hh}^*S_{vv}) = 0 \tag{6.74b}$$

If some triangular reflectors have been placed in the imaging area before the imaging test, the amplitude and phase of f can be estimated at these placement points, as shown in the following equation:

$$|f| = \left[\frac{Y_{\mathrm{vv}}^{*} Y_{\mathrm{vv}}}{Y_{\mathrm{hh}}^{*} Y_{\mathrm{hh}}}\right]_{\mathrm{cr}}^{\frac{1}{4}} \tag{6.75a}$$

$$\arg(f) = \frac{1}{2}\arg(Y_{\mathrm{hh}}^{*} Y_{\mathrm{vv}})_{\mathrm{cr}} \tag{6.75b}$$

where subscript cr represents the scatter measurement of the triangular reflector.

Note that (6.50) only contains a single measurement of the cross product. Therefore, the appearance of background noise means that the signal and noise parts will be correlated superposition. The worst case of noise is that the noise is opposite phase with the HH(VV) channel signal and in the same phase with the other channel VV(HH) signal. In this case, assuming that the system has a high signal-to-noise ratio (SNR), then the amplitude ratio of the noisy estimate $|f|^{n}$ to f is

$$|f|^{n}/f \approx \left[1 \pm \frac{4}{\sqrt{SNR}}\right]^{1/4} \tag{6.76}$$

If SNR = 20 dB, it will result in an amplitude estimation error of +9% or −12%. The worst-case f phase error is

$$\arg(f^{n}) - \arg(f) \approx \pm\tan^{-1}\left(\frac{2}{\sqrt{SNR}}\right) \tag{6.77}$$

If SNR = 20 dB, the resulting error will be ±5.7°.

6.4.1.4 Radiation Calibration

Finally, in order to estimate A, the absolute radiation calibration factor, the system obtained a triangular reflector RCS value which is used and compared to the corresponding theoretical RCS value of the reflector to obtain an estimate of A. Assume that the theoretical value of the RCS of the triangular reflector is σ_{cr}. And assuming that θ is the local incidence angle, the estimate of A can be obtained as

$$A = \sqrt{\frac{(Y_{\mathrm{vv}}^{*} Y_{\mathrm{vv}})_{\mathrm{cr}}}{\sin(\theta_{\mathrm{cr}})\sigma_{\mathrm{cr}}}} \tag{6.78}$$

Note that after channel imbalance f correction, the values of Y_{vv} and Y_{hh} are considered the same. In the case of noise, the ratio of the worst noise estimate A^{n} to A is

$$A^{n}/A = \sqrt{1 \pm \frac{2}{\sqrt{SNR}}} \tag{6.79}$$

When SNR = 20 dB, it can cause the estimation error of A to reach +0.8 dB or −1 dB.

The placement of the triangular reflectors used in the two steps of 6.4.1.3 and 6.4.1.4 shall cover the entire swath. The aligning direction of the reflectors should be carefully measured before and after the imaging test to determine the calibration error influence.

The above algorithm has been successfully applied to the calibration of SIR-C multi-polarization SAR. The polarimetric data of the SIR-C system obtained adopts this algorithm.

6.4.2 Uniform Algorithm for Polarization Data Phase and Crosstalk Calibration—Quegan Calibration Method

The Quegan calibration method is a uniform algorithm for polarimetric SAR image distribution target parameters or scattering matrix data crosstalk calibration. It can be used for scatter matrix data calibration or for extracting characteristic factors of distributed targets [10].

The Quegan calibration method is largely based on the core idea of the van Zyl algorithm, but the algorithm is significantly different. The Quegan calibration method requires no symmetry processing of the data. When the data is not symmetrized, the covariance matrix can be constructed from single-look or multi-look Stokes matrix data. The Quegan algorithm does not require reciprocity of the system and is non-iterative. The Quegan algorithm gives a linear distortion model and introduces external parameters to eliminate the instability of the system equations. The calibration of the Quegan algorithm scattering matrix data is based on the least squares theorem. There is no separate data symmetry or phase calibration step. The Quegan calibration method takes the correction of phase distortion introduced by channel imbalance as an additional optional step.

The Quegan calibration method is straightforward, has no iteration, and is inherently stable. It can be used for distribution target calibration or complex data calibration. In this section, model assumptions, distribution target calibration, and complex data calibration methods for the Quegan calibration are discussed.

6.4.2.1 Model Assumptions

Model assumptions include two types: assumption of the target and assumption of the system.

Assumption of the Target

(1) Each pixel of the distribution target has a joint 2 × 2 complex scattering matrix, where S_{ij} indicates that the polarization state of the transmission is j and the polarization state of reception is i.
(2) According to the reciprocity theorem, $S_{12} = S_{21}$. For the vegetation area, the scattering matrix **S** can be written as a three-element vector $\mathbf{S} = (S_{11}, S_{21}, S_{22})^T$.
(3) The scene consists mainly of distributed targets that the co-polarization and cross-polarization echoes are uncorrelated, i.e., $\left\langle S_{ii}S_{ij}^{*}\right\rangle = 0$ If$j \neq i$ where $\langle\rangle$ indicates the expected value.

Assumption of the System

(4) The influence of the system can be described by a second-order linear equation, and the measured data matrix **O** can be written as

$$\mathbf{O} = \mathbf{RST} + \mathbf{N} \tag{6.80}$$

where R and T are 2 × 2 complex scattering matrices, representing phase and amplitude distortion introduced by receiving and transmitting, respectively, N is noise, t_{ij} represents the signal leaked to the i channel when the j channel transmits a signal, and r_{ij} represents the signal leaked to the i channel when the j channel receives the signal.
(5) The matrices R and T have non-diagonal terms (crosstalk) r_{ij} and t_{ij}, where $i \neq j$. The non-diagonal terms r_{ij} and t_{ij} are smaller than the diagonal terms.

Distribution Target Calibration

Equation (6.80) can be written as

$$\mathbf{O} = \mathbf{MS} + \mathbf{N} \tag{6.81}$$

where O is 4 × 1 vector $O = \left(O_{11}\ O_{21}\ O_{12}\ O_{22}\right)^T$; N is noise; and M is 4 × 3 matrix, indicating the influence of the system

$$\mathbf{M} = \begin{pmatrix} r_{11}t_{11} & r_{11}t_{21} + r_{12}t_{11} & r_{12}t_{21} \\ r_{21}t_{11} & r_{22}t_{11} & r_{22}t_{11} \\ r_{11}t_{12} & r_{11}t_{22} & r_{12}t_{22} \\ r_{22}t_{12} & r_{21}t_{22} + r_{22}t_{12} & r_{22}t_{22} \end{pmatrix}$$

What needs to be calibrated is the matrix M, not separately matrices R and T. Consider the absence of noise and then deal with the effects of noise and estimation errors.

Assumption (3) means that in the calibrated region, the covariance matrix C_s of the S vector of the main target is

$$r_{12}C_S = \begin{pmatrix} \sigma_{11} & 0 & \rho \\ 0 & \sigma_{21} & 0 \\ \rho^* & 0 & \sigma_{22} \end{pmatrix} \tag{6.82}$$

where $\rho = \langle S_{11}S_{22}^* \rangle$, $\sigma_{ij} = \left\langle |S_{ij}|^2 \right\rangle$. And the measured covariance matrix of O is

$$C = MC_SM^\dagger \tag{6.83}$$

where $M^\dagger$ is the conjugate transpose of M, where noise is ignored. When the imaging region contains more than one target type, and each target has its own covariance matrix, C_S can be understood as the spatial average of the formal matrix of Eq. (6.82). The average has the same form.

The items in the measured covariance matrix are given in the form of r_{ij} and t_{ij}. If to define a new set of parameters instead of r_{ij} and t_{ij}, these expressions can be transformed into easy-to-understand parameter forms

(i) Crosstalk ratio

$$u = r_{21}/r_{11} \quad v = t_{21}/t_{22} \quad w = r_{12}/r_{22} \quad z = t_{12}/t_{11} \tag{6.84}$$

(ii) Total system gain of channel 2

$$Y = r_{22}/t_{22} \tag{6.85}$$

(iii) Receive channel imbalance

$$k = r_{11}/r_{22} \tag{6.86}$$

(iv) Ratio of receive and transmit channel imbalances

$$\alpha = \frac{r_{22}}{r_{11}}\frac{t_{11}}{t_{22}} \tag{6.87}$$

For simplicity, introduce a parameter proportional to the true cross-polarization power

$$\beta = |r_{11}t_{22}|^2\sigma_{21} = |kY|^2\sigma_{21} \tag{6.88}$$

Note that k and α are first order. From the assumption (5), u, v, w, and z are less than 1. According to these definitions, the system distortion matrix M can be expressed as follows:

$$M = Y\begin{pmatrix} \alpha & v+\alpha w & vw \\ \alpha u & \alpha & v \\ \alpha z & 1 & w \\ \alpha uz & u+\alpha z & 1 \end{pmatrix}\begin{pmatrix} k^2 & 0 & 0 \\ 0 & k & 0 \\ 0 & 0 & 1 \end{pmatrix} \tag{6.89}$$

Thus, the items in the measured covariance matrix can be divided into two types of expressions. The first type defines the parameters required to describe the azimuthal symmetrically distributed target, i.e.,

$$\sigma_{11} = \frac{1}{|Y|^2}\frac{1}{|k|^4}\frac{C_{11}}{|\alpha|^2} \tag{6.90}$$

$$\sigma_{22} = \frac{1}{|Y|^2}C_{44} \tag{6.91}$$

$$\rho = \frac{1}{|Y|^2}\frac{1}{k^2}\frac{C_{14}}{\alpha} \tag{6.92}$$

$$\sigma_{21} = \frac{1}{|Y|^2}\frac{1}{|k|^2}\beta \tag{6.93}$$

Note that in Eqs. (6.90)–(6.93), α is a global quantity that is independent of the target type and β is a local quantity that is proportional to the cross-polarization power σ_{21} of a specific target scatter. For distribution targets that are not negligible in the correlation of co-polarization and cross-polarization echoes, some additional equations are needed.

The second type of equation is

$$C_{21} = uC_{11} + vC_{41} + \alpha\beta\left(v^* + \alpha^* w^*\right) \tag{6.94}$$

$$C_{24} = vC_{44} + uC_{14} + \alpha\beta\left(u^* + \alpha^* z^*\right) \tag{6.95}$$

$$C_{31} = zC_{11} + wC_{41} + \beta\left(v^* + \alpha^* w^*\right) \tag{6.96}$$

$$C_{34} = wC_{44} + zC_{14} + \beta\left(u^* + \alpha^* z^*\right) \tag{6.97}$$

$$C_{22} = uC_{12} + vC_{42} + |\alpha|^2\beta \tag{6.98}$$

$$C_{33} = z^*C_{31} + w^*C_{34} + \beta \tag{6.99}$$

$$\begin{aligned} C_{32} &= u^*C_{31} + v^*C_{34} + \alpha^*\beta^*\left(v^* + \alpha^* w^*\right) \\ &= zC_{12} + wC_{42} + \alpha^*\beta \end{aligned} \tag{6.100}$$

The crosstalk calibration is mainly to solve this set of equations and obtain six unknowns u, v, w,z, α, and β.

According to the definition of u, v, w, and z, the two expressions of C_{32} are completely equivalent. Therefore, this equation set has seven equations, but only six unknowns, which means that this system of equations is overdetermined. Poor independence of equations is the basis of the crosstalk calibration algorithm described in [9], as this results in a covariance matrix with a zero eigenvalue. The unique solutions of the five parameters u, v, w, z, and $\alpha^*\beta$ with just Eqs. (6.94)–(6.98) and (6.100) can be obtained. The solution of u, v, w, and z is

$$u = (C_{44}C_{21} - C_{41}C_{24})/\Delta \tag{6.101}$$

$$v = (C_{11}C_{24} - C_{21}C_{14})/\Delta \tag{6.102}$$

$$z = (C_{44}C_{31} - C_{41}C_{34})/\Delta \tag{6.103}$$

$$w = (C_{11}C_{34} - C_{31}C_{14})/\Delta \tag{6.104}$$

where $\Delta = C_{11}C_{44} - |C_{14}|^2$. As long as the co-polarization channel correlation is very good, $\Delta = 0$, these solutions are usually deterministic. From these equations, the assumption (5) can be verified. If the crosstalk is large, the calculation result will be discarded. These relationships can be combined with (6.100) to get $\alpha^*\beta$.

The values of u, v, w, and z can be uniquely determined, but α has two solutions, each of which corresponds to a solution of β. The first solution can be written as

$$\alpha_1 = \frac{C_{22} - uC_{12} - vC_{42}}{X} \tag{6.105}$$

where

$$X = C_{32} - zC_{12} - wC_{42} \tag{6.106}$$

u, v, w, and z can be given by (6.101)–(6.104). The β value associated with it is

$$\beta_1 = \frac{|X|^2}{C_{22} - uC_{12} - vC_{42}} \tag{6.107}$$

The second solution is

$$\alpha_2 = \frac{X^*}{C_{33} - z^*C_{31} - w^*C_{34}} \tag{6.108}$$

The β value associated with it is

$$\beta_2 = C_{33} - z^* C_{31} - w^* C_{34} \tag{6.109}$$

Note that for these two pairs of solutions, the value of $\alpha^*\beta$ is the same and equal to X. Since β is a real number, the phases of α two solutions are the same. The two solutions of α (and β) have different amplitudes.

Assume that the noise uncorrelates with the signal, and the noise between the different channels is also uncorrelated. The noise average is 0, and the i channel noise is recorded as N_i.

$$\langle C_{ij} \rangle = \begin{cases} \langle O_i O_j^* \rangle & \text{If } i \neq j \\ \langle |O_i|^2 \rangle + N_i & \text{If } i = j \end{cases}$$

This means that the estimate of the channel variance may be biased, while the other terms of the measured covariance matrix are unbiased estimates. The measurement stability of $Arg(\alpha)$ and the value of $Arg(\alpha)$ are determined by Eq. (6.100), which is the average value and is not affected by noise.

The effect of strictly dealing with C_{ii} deviations is quite complicated, and only simplified processing is given here. The main assumption of the simplified processing is that the observed co-polarization channel signal power is significantly higher than the cross-polarization channel and is not susceptible to noise. This means that the calculated values of u, v, w, z, and $\alpha^*\beta$ are constant, but the expressions of C_{22} and C_{33} become

$$C_{22} = uC_{12} + vC_{42} + |\alpha|^2\beta + N_2 \tag{6.110}$$

$$C_{33} = z^* C_{31} + w^* C_{34} + \beta + N_3 \tag{6.111}$$

From (6.110), the correction value of α is

$$\alpha = \frac{C_{22} - N_2 - uC_{12} - C_{42}}{\alpha^*\beta} \tag{6.112}$$

or

$$\alpha_1 = \alpha + \frac{N_2}{\alpha^*\beta} \tag{6.113}$$

Similarly, there is

$$\alpha_2 = \alpha\left(1 + \frac{N_3}{\beta}\right)^{-1} \tag{6.114}$$

Because of $N_2 > 0$, $N_3 > 0$, there is $\alpha_1 > \alpha > \alpha_2$. The difference between α_1 and α_2 increases with β decreasing.

The above argument relies on the fact that the noise effects are positive in the acquired covariance matrices C_{22} and C_{33}. If the correlation value between the co-polarization and cross-polarization backscatter coefficients of a large number of distinct regions in the image is non-zero, then another interference will occur in C_{22} and C_{33}. The effect of this interference is to reduce C_{22} and C_{33}. If $S_{11}S_{12}^* \neq 0$, the expression of C_{22} needs to add another item

$$2|k\alpha Y|^2\Re\left(ku\langle S_{11}S_{21}^*\rangle\right)$$

where $\Re$ is the real part. This correction to C_{22} depends on the values of k, u, and $S_{11}S_{12}^*$. They may be positive or negative.

The study of noise-like disturbances shows that if $N_2 \approx N_3$, a good estimate can be obtained, and the noise term between (6.113) and (6.114) can be ignored, and the following equation can be obtained:

$$f(\alpha) = |\alpha|^2 + \frac{|\alpha|}{|\alpha_2|}(1 - |\alpha_1\alpha_2|) - 1 = 0 \tag{6.115}$$

Its solution is

$$|\alpha| = \frac{|\alpha_1\alpha_2| - 1 + \sqrt{(|\alpha_1\alpha_2| - 1)^2 + 4|\alpha_2|^2}}{2|\alpha_2|} \tag{6.116}$$

Since $f(\alpha_1) = |\alpha_1|/|\alpha_2| - 1 > 0$ and $f(\alpha_2) = |\alpha_2|(|\alpha_2| - |\alpha_1|) < 0$, the solution of $|\alpha|$ is between $|\alpha_1|$ and $|\alpha_2|$. From this $|\alpha|$ value, and using known noise-independent $\alpha^*\beta$ values, $Arg(\alpha)$ and β values can be obtained, and β is a real number. The stability of the correction value α is greatly increased. If $C_{33} - z^*C_{31} - w^*C_{34} \approx 0$, the solution of α is uncertain. With (6.111), if N_3 can be ignored, and $\beta \approx 0$, i.e., $\sigma_{21} \approx 0$, this will happen. From (6.94) to (6.100) and (6.110) to (6.111), it can be seen that if $\beta = 0$, α may have no solution.

If the same processing is performed on the two cross-polarization channels, the noise power of the two cross-polarization channels can be estimated. From (6.113), we can get

$$N_2 = \alpha^*\beta(\alpha_1 - \alpha) \tag{6.117}$$

Now return to the calibration equations for the azimuth symmetrically distributed targets given by Eqs. (6.90)–(6.93). It can be seen that crosstalk can be coupled only by parameters α and β. Other calibration parameters, channel imbalance, and total system gain Y appear only in this set of four equations. Since there are six unknowns, the best ones that can be achieved are the $\sigma_{11}, \sigma_{21}, \sigma_{22}$, and ρ solutions in the form of k and Y. To get further information on the next two parameters, additional information support is required. Absolute calibration item $|Y|^2$ does not affect relative calibration. More seriously is that, between σ_{11}^2 and σ_{22}^2, the channel imbalance k increases the uncertainty to k^4 times. An error whose amplitude is Arg(k) will be introduced into the phase of ρ. As long as the $Arg(\rho)$ does not change within the scene, this does not affect the measurement of the relevant phase.

Note that σ_{21} is expressed in the form of β, unlike in Eqs. (6.90)–(6.92), which is a simple multiplication correction of the covariance matrix term. Checking the relative magnitudes of the items in (6.100), it can be seen that (6.93) can be approximated as

$$\sigma_{21} = \frac{1}{|Y|^2}\frac{1}{|k|^2}\frac{C_{23}}{\alpha} \tag{6.118}$$

Therefore, the cross-polarization power can be obtained from a certain ratio of the observed covariance of the HV and VH channels. From Eqs. (6.90) to (6.93), the modulus of the real-correlation coefficient of the co-polarization channel can be derived as

$$\frac{|\rho|}{\sqrt{\sigma_{11}\sigma_{22}}} = \frac{|C_{14}|}{\sqrt{C_{11}C_{44}}} \tag{6.119}$$

where this amount can be accurately measured from uncalibrated data and is independent of k and Y. Equation (6.119) is important because the correlation coefficient completely determines the useful phase information under the distribution target multivariate normal distribution or K distribution model. Essentially, the modulus of the correlation coefficient determines how clearly the distribution of the phase difference of the HH-VV is discernible and provides a natural measure of this distribution width.

6.4.2.2 Complex Data Calibration

The above processing relies on the acquired covariance matrix of the original data. If the Stokes matrix data is not symmetrized, it can also be used for Stokes matrix data. If the scattering matrix data is obtained, this method can be extended to the calibration of the scattering matrix data using the parameters that have been calculated from the covariance matrix. According to (6.89), the noiseless form of (6.81) can be written as

$$O = Y\begin{pmatrix} \alpha & v+\alpha w & vw \\ \alpha u & \alpha & v \\ \alpha z & 1 & w \\ \alpha uz & u+\alpha z & 1 \end{pmatrix}\begin{pmatrix} k^2 S_{11} \\ kS_{21} \\ S_{22} \end{pmatrix} \tag{6.120}$$

For vector $\overline{S} = Y\left(k^2 S_{11}, kS_{21}, S_{22}\right)^t$, this is a set of overdetermined equations. Its least squares solution is

$$\overline{S} = \left(A^{\dagger}A\right)^{-1}A^{\dagger}O \tag{6.121}$$

where A is a 4×3 matrix in (6.120), which can be simplified to

$$\overline{S} = \begin{pmatrix} 1/\alpha & P_{12} & P_{13} & 0 \\ -(\alpha^* u + z)/D & \alpha^*/D & 1/D & -(\alpha^* v + w)/D \\ 0 & P_{32} & P_{33} & 1 \end{pmatrix} O \tag{6.122}$$

where $D = 1 + |\alpha|^2$, $P_{12} = \alpha^*(u^* - \alpha^* z^* - \alpha^*[v + \alpha w])$, $P_{13} = \alpha^*(-[v + \alpha w] - \alpha[u^* - \alpha^* z^*])$, $P_{32} = |\alpha|^2(v^* - \alpha^* w^* - \alpha^*[u + \alpha z])$, $P_{33} = |\alpha|^2(-[u + \alpha z] - \alpha[v^* - \alpha^* w^*])$.

Equation (6.122) is a non-iterative calibration algorithm that recovers a vector from the acquired complex data by a unit transform. The first and third rows correspond to the correction of the co-polarization channel. P_{12} and P_{13} have the same order as the crosstalk ratio and are multiplied by the cross-polarization term in O. $1/\alpha$ is first order and is multiplied by the co-polarization term in O, so it plays a major role. The third line is similar. The important item of crosstalk calibration is on the second line. It can be seen that the estimation S_{21} contains the formation weighted sum

$$(\alpha^* O_{21} + O_{12})/D \tag{6.123}$$

And subtract the following item:

$$([\alpha^* u + z]O_{11} + [\alpha^* v + w]O_{22})/D \tag{6.124}$$

The first step approximates the step of symmetrization in [7, 18] phase calibration, because it has a multi-rotation $-Arg(\alpha)$ of O_{21}. Equation (6.123) shows that the amplitude weighting between $|\alpha|$ and 1 is also required. In comparison, the symmetry in [7, 18] simply rotates the phase of O_{21} by $Arg(C_{32})$. So to make this step operation, $Arg(C_{32}) \approx -Arg(\alpha)$ and $|\alpha| \approx 1$ are needed.

The second step can be seen as eliminating crosstalk. The weighting factors $(\alpha^* u + z)/D$ and $(\alpha^* u + z)/D$ are small, but are multiplied by the co-polarization power before being added to the cross-polarization term. This achieves an order of magnitude.

Go back to (6.124) crosstalk cancellation. The experimental data shows that $|\alpha^* u + z|$ increases with distance increasing and has large relative fluctuations. The phase also increases from $-50°$ to $+50°$ with distance increasing. The apparent undulations in these parameters can be improved by smoothing. The value of $|\alpha^* v + w|$ has a large value at the observation of the nadir angle, its maximum value appears at the intermediate distance, and the phase decreases from $40°$ to $-150°$ with the distance increasing. This opposite trend in phase has a significant impact. Tests have shown that the relative undulations of $|\alpha^* u + z|$ and $|\alpha^* v + w|$ are much larger than $|\alpha|$. The modulus of the deviation coefficient (standard deviation/average) of the crosstalk ratio $|\alpha|$ can be approximated as

$$\frac{1}{|u|}\sqrt{\frac{N_2C_{44}}{M\Delta}} \tag{6.125}$$

where M is the number of independent pixels for average. Other symbols are as defined above. The expressions of v, w, and z are the same. The modulus of the deviation coefficient of α can be approximated as

$$\sqrt{\frac{1}{M}\left(\frac{C_{22}}{C_{33}}|C_{23}|^2 - 1\right)} \tag{6.126}$$

It must be stated again that (6.120) and (6.122) directly derive the relationship of the distribution target, so it can be directly used to represent the relationship between the true value of the Stokes matrix parameter and the measured value.

6.4.3 Distribution Target Calibration Algorithm—Sarabandi Algorithm

Although the Mueller matrix W can be used to measure the distribution target very accurately, however since the matrix W is not Hermitian conjugate symmetric, the calculation process is complicated. Here the covariance matrix of the ground object can be used.

The polarization scattering vector X can be obtained from the polarization scattering matrix S, which is defined as

$$\mathbf{X} = (S_{\mathrm{vv}}\ S_{\mathrm{vh}}\ S_{\mathrm{hv}}\ S_{\mathrm{hh}})^{\mathrm{T}} \tag{6.127}$$

Then the polarization covariance matrix C is defined as

$$\mathbf{C} = \mathbf{X}(\mathbf{X}^{\mathrm{T}})^* = \begin{pmatrix} S_{\mathrm{vv}}S^*_{\mathrm{vv}} & S_{\mathrm{vv}}S^*_{\mathrm{vh}} & S_{\mathrm{vv}}S^*_{\mathrm{hv}} & S_{\mathrm{vv}}S^*_{\mathrm{hh}} \\ S_{\mathrm{vh}}S^*_{\mathrm{vv}} & S_{\mathrm{vh}}S^*_{\mathrm{vh}} & S_{\mathrm{vh}}S^*_{\mathrm{hv}} & S_{\mathrm{vh}}S^*_{\mathrm{hh}} \\ S_{\mathrm{hv}}S^*_{\mathrm{vv}} & S_{\mathrm{hv}}S^*_{\mathrm{vh}} & S_{\mathrm{hv}}S^*_{\mathrm{hv}} & S_{\mathrm{hv}}S^*_{\mathrm{hh}} \\ S_{\mathrm{hh}}S^*_{\mathrm{vv}} & S_{\mathrm{hh}}S^*_{\mathrm{vh}} & S_{\mathrm{hh}}S^*_{\mathrm{hv}} & S_{\mathrm{hh}}S^*_{\mathrm{hh}} \end{pmatrix} \tag{6.128}$$

When the polarization scattering matrix S is a symmetric matrix, C is a Hermitian matrix, and the calculation process can be simplified. In the multi-polarization SAR model, the expression of the measurement scattering matrix M can be obtained by ignoring the noise.

$$\mathbf{M} = \mathbf{RST}^{\mathrm{T}} = \begin{pmatrix} R_{\mathrm{vv}} & R_{\mathrm{vh}} \\ R_{\mathrm{hv}} & R_{\mathrm{hh}} \end{pmatrix}\begin{pmatrix} S_{\mathrm{vv}} & S_{\mathrm{vh}} \\ S_{\mathrm{hv}} & S_{\mathrm{hh}} \end{pmatrix}\begin{pmatrix} T_{\mathrm{vv}} & T_{\mathrm{hv}} \\ T_{\mathrm{vh}} & T_{\mathrm{hh}} \end{pmatrix} \tag{6.129}$$

Note that the antenna gain, transmission loss factor, and other system constants are all included in the receiving and transmitting distortion matrices R and T.

Define column vector

$$\mathbf{M} = \begin{pmatrix} M_{\mathrm{vv}} \\ M_{\mathrm{vh}} \\ M_{\mathrm{hv}} \\ M_{\mathrm{hh}} \end{pmatrix} \tag{6.130a}$$

$$\mathbf{S} = \begin{pmatrix} S_{\mathrm{vv}} \\ S_{\mathrm{vh}} \\ S_{\mathrm{hv}} \\ S_{\mathrm{hh}} \end{pmatrix} \tag{6.130b}$$

Then, by (6.129), the relationship between M and S can be obtained

$$\mathbf{M} = \mathbf{DS} \tag{6.131}$$

where

$$\mathbf{D} = \begin{pmatrix} R_{\mathrm{vv}}T_{\mathrm{vv}} & R_{\mathrm{vv}}T_{\mathrm{vh}} & R_{\mathrm{vh}}T_{\mathrm{vv}} & R_{\mathrm{vh}}T_{\mathrm{vh}} \\ R_{\mathrm{vv}}T_{\mathrm{hv}} & R_{\mathrm{vv}}T_{\mathrm{hh}} & R_{\mathrm{vh}}T_{\mathrm{hv}} & R_{\mathrm{vh}}T_{\mathrm{hh}} \\ R_{\mathrm{hv}}T_{\mathrm{vv}} & R_{\mathrm{hv}}T_{\mathrm{vh}} & R_{\mathrm{hh}}T_{\mathrm{vv}} & R_{\mathrm{hh}}T_{\mathrm{vh}} \\ R_{\mathrm{hv}}T_{\mathrm{hv}} & R_{\mathrm{hv}}T_{\mathrm{hh}} & R_{\mathrm{hh}}T_{\mathrm{hv}} & R_{\mathrm{hh}}T_{\mathrm{hh}} \end{pmatrix} \tag{6.132}$$

It is called a generalized distortion matrix.

The Mueller matrix corresponding to the ground object scattering matrix S is of the formula (2.73), where the matrix W is rewritten as

$$\mathbf{W} = \begin{pmatrix} S_{\mathrm{vv}}S_{\mathrm{vv}}^* & S_{\mathrm{vh}}S_{\mathrm{vh}}^* & S_{\mathrm{vv}}S_{\mathrm{vh}}^* & S_{\mathrm{vh}}S_{\mathrm{vv}}^* \\ S_{\mathrm{hv}}S_{\mathrm{hv}}^* & S_{\mathrm{hh}}S_{\mathrm{hh}}^* & S_{\mathrm{hv}}S_{\mathrm{hh}}^* & S_{\mathrm{hh}}S_{\mathrm{hv}}^* \\ S_{\mathrm{vv}}S_{\mathrm{hv}}^* & S_{\mathrm{vh}}S_{\mathrm{hh}}^* & S_{\mathrm{vv}}S_{\mathrm{hh}}^* & S_{\mathrm{vh}}S_{\mathrm{hv}}^* \\ S_{\mathrm{hv}}S_{\mathrm{vv}}^* & S_{\mathrm{hh}}S_{\mathrm{vh}}^* & S_{\mathrm{hv}}S_{\mathrm{vh}}^* & S_{\mathrm{hh}}S_{\mathrm{vv}}^* \end{pmatrix} \tag{6.133}$$

The accurate measured distribution target Mueller matrix used in the Sarabandi algorithm is actually a modified Mueller matrix [19], which is defined as follows:

$$\mathbf{L}_{\mathrm{m}} = \mathbf{VWV}^{-1} \tag{6.134}$$

where W is as shown in Eq. (6.133). The matrix V is defined as follows:

$$\mathbf{V}=\begin{pmatrix}1&0&0&0\\0&1&0&0\\0&0&1&1\\0&0&-\mathrm{i}&\mathrm{i}\end{pmatrix}\tag{6.135}$$

L_m and L have the following relationship:

$$\mathbf{L}_{\mathrm{m}}=\mathbf{B}\mathbf{L}\mathbf{B}^{-1}\tag{6.136}$$

where

$$\mathbf{B}=\begin{pmatrix}1&1&0&0\\1&-1&0&0\\0&0&1&0\\0&0&0&1\end{pmatrix}\tag{6.137}$$

Equation (6.128) is the polarization covariance matrix of the ground feature, rewritten as follows:

$$\mathbf{C}=\mathbf{S}(\mathbf{S}^{\mathrm{T}})^{*}=\begin{pmatrix}C_{11}&C_{12}&C_{13}&C_{14}\\C_{21}&C_{22}&C_{23}&C_{24}\\C_{31}&C_{32}&C_{33}&C_{34}\\C_{41}&C_{42}&C_{43}&C_{44}\end{pmatrix}=\begin{pmatrix}S_{vv}S_{vv}^{*}&S_{vv}S_{vh}^{*}&S_{vv}S_{hv}^{*}&S_{vv}S_{hh}^{*}\\S_{vh}S_{vv}^{*}&S_{vh}S_{vh}^{*}&S_{vh}S_{hv}^{*}&S_{vh}S_{hh}^{*}\\S_{hv}S_{vv}^{*}&S_{hv}S_{vh}^{*}&S_{hv}S_{hv}^{*}&S_{hv}S_{hh}^{*}\\S_{hh}S_{vv}^{*}&S_{hh}S_{vh}^{*}&S_{hh}S_{hv}^{*}&S_{hh}S_{hh}^{*}\end{pmatrix}\tag{6.138}$$

If the data used for calibration is stored in the form of the Mueller matrix L_m, according to Eq. (6.134), we can get

$$\mathbf{W}=\mathbf{V}^{-1}\mathbf{L}_{\mathrm{m}}\mathbf{V}\tag{6.139}$$

According to this, the corresponding matrix W can be obtained by L_m. After the element moves, the polarization covariance matrix C can be obtained.

Find the correlation matrix of M

$$\mathbf{U}=\mathbf{M}\left(\mathbf{M}^{\mathrm{T}}\right)^{*}=\mathbf{D}\mathbf{C}\widetilde{\mathbf{D}}\tag{6.140}$$

where ~ means conjugate transpose, there is

$$\mathbf{U}=\begin{pmatrix}M_{vv}M_{vv}^{*}&M_{vv}M_{vh}^{*}&M_{vv}M_{hv}^{*}&M_{vv}M_{hh}^{*}\\M_{vh}M_{vv}^{*}&M_{vh}M_{vh}^{*}&M_{vh}M_{hv}^{*}&M_{vh}M_{hh}^{*}\\M_{hv}M_{vv}^{*}&M_{hv}M_{vh}^{*}&M_{hv}M_{hv}^{*}&M_{hv}M_{hh}^{*}\\M_{hh}M_{vv}^{*}&M_{hh}M_{vh}^{*}&M_{hh}M_{hv}^{*}&M_{hh}M_{hh}^{*}\end{pmatrix}\tag{6.141}$$

In Eq. (6.140), there are a total of eight unknowns. To simplify the problem, the factor $R_{vv}T_{vv}$ is extracted from the generalized distortion matrix D, so that the unknown parameters are reduced to six complex quantities and one real quantity.

$$\mathbf{U}=\omega\begin{pmatrix}1 & x_4 & x_1 & x_1x_4\\ x_5 & x_6 & x_1x_5 & x_1x_6\\ x_2 & x_2x_4 & x_3 & x_3x_4\\ x_2x_5 & x_2x_6 & x_3x_5 & x_3x_6\end{pmatrix}\mathbf{C}\begin{pmatrix}1 & x_5^* & x_2^* & x_2^*x_5^*\\ x_4^* & x_6^* & x_2^*x_4^* & x_2^*x_6^*\\ x_1^* & x_1^*x_5^* & x_3^* & x_3^*x_5^*\\ x_1^*x_4^* & x_1^*x_6^* & x_3^*x_4^* & x_3^*x_6^*\end{pmatrix} \tag{6.142}$$

where ω is the radiation calibration constant factor (real number), and x_1, x_2, x_3, x_4, x_5, and x_6 are defined as follows:

$$x_1=\frac{R_{\mathrm{vh}}}{R_{\mathrm{vv}}}\quad x_2=\frac{R_{\mathrm{hv}}}{R_{\mathrm{vv}}}\quad x_3=\frac{R_{\mathrm{hh}}}{R_{\mathrm{vv}}}$$

$$x_4=\frac{T_{\mathrm{vh}}}{T_{\mathrm{vv}}}\quad x_5=\frac{T_{\mathrm{hv}}}{T_{\mathrm{vv}}}\quad x_6=\frac{T_{\mathrm{hh}}}{T_{\mathrm{vv}}}$$

In order to understand the unknown parameters in Eq. (6.142), the separation parameter method is used here. Equation (6.142) can be rewritten in the form of a block matrix.

$$\begin{pmatrix}\mathbf{U}_1 & \mathbf{U}_2\\ \tilde{\mathbf{U}}_2 & \mathbf{U}_3\end{pmatrix}=\omega\begin{pmatrix}\mathbf{G} & x_1\mathbf{G}\\ x_2\mathbf{G} & x_3\mathbf{G}\end{pmatrix}\begin{pmatrix}\mathbf{C}_1 & \mathbf{C}_2\\ \tilde{\mathbf{C}}_2 & \mathbf{C}_3\end{pmatrix}\begin{pmatrix}\tilde{\mathbf{G}} & x_2^*\tilde{\mathbf{G}}\\ x_1^*\tilde{\mathbf{G}} & x_3^*\tilde{\mathbf{G}}\end{pmatrix} \tag{6.143}$$

where U_1, U_3, C_1, and C_3 are 2×2 Hermitian symmetry arrays. And the expression of matrix G is

$$\mathbf{G}=\begin{pmatrix}1 & x_4\\ x_5 & x_6\end{pmatrix} \tag{6.144}$$

Three independent matrix equations can be obtained from the matrix Eq. (6.143), i.e.,

$$\mathbf{U}_i=\omega\mathbf{G}\mathbf{Z}_i\tilde{\mathbf{G}},\quad i=1,\ 2,\ 3 \tag{6.145a}$$

Here

$$\mathbf{Z}_1=\mathbf{C}_1+x_1\tilde{\mathbf{C}}_2+x_1^*(\mathbf{C}_2+x_1\mathbf{C}_3) \tag{6.145b}$$

$$\mathrm{Z}_2=x_2^*(\mathrm{C}_1+x_1\tilde{\mathrm{C}}_2)+x_3^*(\mathrm{C}_2+x_1\mathrm{C}_3) \tag{6.145c}$$

$$\mathbf{Z}_3=x_2^*(x_2\mathbf{C}_1+x_3\tilde{\mathbf{C}}_2)+x_3^*(x_2\mathbf{C}_2+x_3\mathbf{C}_3) \tag{6.145d}$$

In order to separate the two set variables (x_1, x_2, x_3) and (x_4, x_5, x_6, ω), note the following relationship:

$$\tilde{\mathbf{G}} = \frac{1}{\omega}\mathbf{Z}_2^{-1}\mathbf{G}^{-1}\mathbf{U}_2 \tag{6.146}$$

Substituting formula (6.146) into formula (6.145), we can get

$$\mathbf{U}_1\mathbf{U}_2^{-1} = \mathbf{G}\mathbf{Z}_1\mathbf{Z}_2^{-1}\mathbf{G}^{-1} \tag{6.147a}$$

$$\mathbf{U}_3\mathbf{U}_2^{-1} = \mathbf{G}\mathbf{Z}_3\mathbf{Z}_2^{-1}\mathbf{G}^{-1} \tag{6.147b}$$

It can be seen from the above two equations that $\mathbf{U}_1\mathbf{U}_2^{-1}$ and $\mathbf{Z}_1\mathbf{Z}_2^{-1}$, $\mathbf{U}_3\mathbf{U}_2^{-1}$ and $\mathbf{Z}_3\mathbf{Z}_2^{-1}$ are two pairs of similar matrices, and the similar matrices have the same eigenvalue. From Eqs. (6.145b), (6.145c), and (6.145d), it can be seen that the eigenvalues of $\mathbf{Z}_1\mathbf{Z}_2^{-1}$, $\mathbf{Z}_3\mathbf{Z}_2^{-1}$ are only the functions of x_1, x_2, and x_3, and are independent of other variables, so that these three variables have actually been separated with other variables.

Assume that the eigenvalues of $\mathbf{Z}_1\mathbf{Z}_2^{-1}$ are λ_1 and λ_2, respectively, and the eigenvalues of $\mathbf{Z}_3\mathbf{Z}_2^{-1}$ are λ_3 and λ_4, respectively. According to matrix properties, we can get

$$\mathrm{tr}\left[\mathbf{U}_1\mathbf{U}_2^{-1}\right] = \lambda_1 + \lambda_2 \tag{6.148a}$$

$$\frac{\det[\mathbf{U}_1]}{\det[\mathbf{U}_2]} = \lambda_1\lambda_2 \tag{6.148b}$$

$$\mathrm{tr}\left[\mathbf{U}_3\mathbf{U}_2^{-1}\right] = \lambda_3 + \lambda_4 \tag{6.148c}$$

$$\frac{\det[\mathbf{U}_3]}{\det[\mathbf{U}_2]} = \lambda_3\lambda_4 \tag{6.148d}$$

Here tr[] and det[] represent the trace and determinant of the matrix, respectively. For the convenience of calculation, define a new variable h as

$$h = x_3/x_2 \tag{6.149}$$

After performing algebraic operations on (6.148a–6.148d) four equations, the following relation is obtained:

$$x_1 = \frac{1}{F_1(h)}\left[\frac{\lambda_3 + \lambda_4}{\lambda_3\lambda_4}F_3(h)x_2 - F_2(h)\right] \tag{6.150}$$

$$x_2 = \frac{-K_2(h) \pm \sqrt{K_2^2(h) - 4K_1(h)K_3(h)}}{2K_1(h)} \tag{6.151}$$

Note that det[Z$_1$] is only function of x_1, and det[Z$_2$] is the function of x_1, x_2, and h. And x_1 and x_2 are both functions of h, so that two complex equations for h can be obtained as follows:

$$\lambda_1\lambda_2 x_2^*(F_4(x_1) + h^* F_5(x_1)) - (\lambda_1 + \lambda_2)\det[\mathbf{Z}_1(x_1)] = 0 \tag{6.152}$$

$$\det[\mathbf{Z}_1(x_1)] - \lambda_1\lambda_2 \det[\mathbf{Z}_2(x_1, x_2, h)] = 0 \tag{6.153}$$

In (6.150), (6.151), (6.152), and (6.153)

$$\begin{aligned} F_1(h) = {} & 2(C_{11}C_{22} - |C_{12}|^2) \\ & + 2h^*(C_{11}C_{24} + C_{13}C_{22} - C_{21}C_{14} - C_{23}C_{12}) \\ & + h(C_{31}C_{22} + C_{42}C_{11} - C_{21}C_{32} - C_{12}C_{41}) \\ & + |h|^2(2\mathrm{Re}[C_{24}C_{31} - C_{34}C_{21}] + C_{33}C_{22} + C_{44}C_{11} - |C_{23}|^2 - |C_{14}|^2) \\ & + 2(h^*)^2(C_{13}C_{24} - C_{23}C_{14}) \\ & + h^*|h|^2(C_{33}C_{24} + C_{13}C_{44} - C_{23}C_{34} - C_{14}C_{43}) \end{aligned} \tag{6.154a}$$

$$\begin{aligned} F_2(h) = {} & (C_{11}C_{42} + C_{31}C_{22} - C_{41}C_{12} - C_{32}C_{21}) \\ & + h^*(C_{11}C_{44} + C_{22}C_{33} - |C_{14}|^2 - |C_{23}|^2 + 2\mathrm{Re}[C_{13}C_{42} - C_{34}C_{21}]) \\ & + 2h(C_{31}C_{42} - C_{41}C_{32}) \\ & + 2|h|^2(C_{33}C_{42} + C_{44}C_{31} - C_{41}C_{34} - C_{43}C_{32}) \\ & + \left(h^*\right)^2(C_{44}C_{13} + C_{33}C_{24} - C_{43}C_{14} - C_{34}C_{23}) \\ & + 2h^*|h|^2(C_{33}C_{44} - |C_{34}|^2) \end{aligned} \tag{6.154b}$$

$$\begin{aligned} F_3(h) = {} & (C_{11} + 2\mathrm{Re}[C_{31}h] + C_{33}|h|^2)\cdot(C_{22} + 2\mathrm{Re}[C_{42}h] + C_{44}|h|^2) \\ & - (C_{12} + C_{32}h + C_{14}h^* + C_{34}|h|^2)\cdot(C_{21} + C_{41}h + C_{23}h^* + C_{43}|h|^2) \end{aligned} \tag{6.154c}$$

$$\begin{aligned} F_4(x_1) = {} & 2(C_{11}C_{22} - |C_{12}|^2) \\ & + 2x_1(C_{11}C_{42} + C_{31}C_{22} - C_{21}C_{32} - C_{12}C_{41}) \\ & + x_1^*(C_{13}C_{22} + C_{24}C_{11} - C_{12}C_{23} - C_{21}C_{14}) \\ & + |x_1|^2(C_{22}C_{33} + 2\mathrm{Re}[C_{13}C_{42} - C_{34}C_{21}] + C_{11}C_{44} - |C_{23}|^2 - |C_{14}|^2) \\ & + 2x_1^2(C_{31}C_{42} - C_{32}C_{41}) \\ & + x_1|x_1|^2(C_{33}C_{42} + C_{31}C_{44} - C_{41}C_{34} - C_{32}C_{43}) \end{aligned} \tag{6.155a}$$

$$\begin{aligned} F_5(x_1) = {} & (C_{11}C_{24} + C_{13}C_{22} - C_{14}C_{21} - C_{23}C_{12}) \\ & + x_1(C_{11}C_{44} + C_{22}C_{33} - |C_{14}|^2 - |C_{23}|^2 + 2\mathrm{Re}[C_{31}C_{24} - C_{34}C_{21}]) \end{aligned}$$

$$
\begin{aligned}
&+2x_1^*(C_{13}C_{24}-C_{23}C_{14})\\
&+2|x_1|^2(C_{13}C_{44}+C_{33}C_{24}-C_{23}C_{34}-C_{43}C_{14})\\
&+x_1^2(C_{44}C_{31}+C_{33}C_{42}-C_{32}C_{43}-C_{34}C_{41})\\
&+2x_1|x_1|^2(C_{33}C_{44}-|C_{34}|^2)
\end{aligned}
\tag{6.155b}
$$

$$
K_1(h)=A(h)\left(\frac{(\lambda_3+\lambda_4)F_3(h)}{\lambda_3\lambda_4F_1(h)}\right)^2-\frac{F_3(h)}{\lambda_3\lambda_4} \tag{6.156a}
$$

$$
K_2(h)=\frac{(\lambda_3+\lambda_4)F_3(h)}{\lambda_3\lambda_4F_1(h)}\left(B(h)-2A(h)\frac{F_2(h)}{F_1(h)}\right) \tag{6.156b}
$$

$$
K_3(h)=C(h)+A(h)\left(\frac{F_2(h)}{F_1(h)}\right)^2-B(h)\frac{F_2(h)}{F_1(h)} \tag{6.156c}
$$

In (6.156a), (6.156b), and (6.156c)

$$
\begin{aligned}
A(h)=&(C_{31}+h^*C_{33})(C_{42}+h^*C_{44})\\
&-(C_{32}+h^*C_{34})(C_{41}+h^*C_{43})
\end{aligned}
\tag{6.157a}
$$

$$
\begin{aligned}
B(h)=&(C_{31}+h^*C_{33})(C_{22}+h^*C_{24})\\
&+(C_{11}+h^*C_{13})(C_{42}+h^*C_{44})\\
&-(C_{12}+h^*C_{14})(C_{41}+h^*C_{43})\\
&-(C_{32}+h^*C_{34})(C_{21}+h^*C_{23})
\end{aligned}
\tag{6.157b}
$$

$$
\begin{aligned}
C(h)=&(C_{11}+h^*C_{13})(C_{22}+h^*C_{24})\\
&-(C_{12}+h^*C_{14})(C_{21}+h^*C_{23})
\end{aligned}
\tag{6.157c}
$$

Then a function according to Eqs. (6.152) and (6.153) can be constructed.

$$
\begin{aligned}
\varphi(h)=&\left(\lambda_1\lambda_2x_2^*(F_4(x_1)+h^*F_5(x_1))-(\lambda_1+\lambda_2)\det[\mathbf{Z}_1(x_1)]\right)^2\\
&+(\det[\mathbf{Z}_1(x_1)]-\lambda_1\lambda_2\det[\mathbf{Z}_2(x_1,x_2,h)])^2
\end{aligned}
\tag{6.158}
$$

The conjugate gradient method or the Newton–Gaussian method [20, 21] is used to calculate the minimum value of $\phi(h)$, so that the value of h can be solved. After h is solved, x_1, x_2, and x_3 can be easily obtained.

The elements of matrix G can be derived from the eigenvectors of $\mathbf{U}_1\mathbf{U}_2^{-1}$ and $\mathbf{U}_3\mathbf{U}_2^{-1}$. In order to simplify the expression, the following definition is used:

$$
\mathbf{N}=\begin{pmatrix}N_{11}\ N_{12}\\ N_{21}\ N_{22}\end{pmatrix}=\mathbf{Z}_1\mathbf{Z}_2^{-1}\quad \mathbf{P}=\begin{pmatrix}P_{11}\ P_{12}\\ P_{21}\ P_{22}\end{pmatrix}=\mathbf{U}_1\mathbf{U}_2^{-1} \tag{6.159a}
$$

$$\mathbf{N}' = \begin{pmatrix} N'_{11} & N'_{12} \\ N'_{21} & N'_{22} \end{pmatrix} = \mathbf{Z}_3\mathbf{Z}_2^{-1} \quad \mathbf{P}' = \begin{pmatrix} P'_{11} & P'_{12} \\ P'_{21} & P'_{22} \end{pmatrix} = \mathbf{U}_3\mathbf{U}_2^{-1} \tag{6.159b}$$

Since N and P, $\mathbf{N}'$ and $\mathbf{P}'$ are, respectively, similar matrices, the matrices $\mathbf{Q}_\mathrm{n}$ and $\mathbf{Q}_\mathrm{p}$, $\mathbf{Q}'_\mathrm{n}$ and $\mathbf{Q}'_\mathrm{p}$, respectively, composed of their eigenvectors have the following relationship:

$$\mathbf{G}\mathbf{Q}_\mathrm{n} = \mathbf{Q}_\mathrm{p}\mathbf{E} \tag{6.160a}$$

$$\mathbf{G}\mathbf{Q}'_\mathrm{n} = \mathbf{Q}'_\mathrm{p}\mathbf{E}' \tag{6.160b}$$

where $\mathbf{E}$ and $\mathbf{E}'$ are diagonal matrices.

There are a total of eight equations and seven unknowns in (6.160a) and (6.160b), which can be solved by methods similar to the above solutions of x_1, x_2, x_3, and h. After algebraic operations, we can get

$$\begin{cases} x_5 = \dfrac{N_{21}}{P_{12}}x_4 + \dfrac{N_{11} - P_{11}}{P_{12}} \\ x_6 = \dfrac{P_{22} - N_{11}}{P_{12}}x_4 + \dfrac{N_{12}}{P_{12}} \end{cases} \tag{6.161a}$$

$$\begin{cases} x_5 = \dfrac{N'_{21}}{P'_{12}}x_4 + \dfrac{N'_{11} - P'_{11}}{P'_{12}} \\ x_6 = \dfrac{P'_{22} - N'_{11}}{P'_{12}}x_4 + \dfrac{N'_{12}}{P'_{12}} \end{cases} \tag{6.161b}$$

For the solution of x_4, a method similar to solving h is used. Since the constructor for x_4 is a second-order polynomial of x_4, an analytical solution of x_4 can be obtained. The result is as follows:

$$\begin{aligned} x_4 = -&\left[\left(\frac{N'_{21}}{P'_{12}} - \frac{N_{21}}{P_{12}}\right)^*\left(\frac{N'_{11} - P'_{11}}{P'_{12}} - \frac{N_{11} - P_{11}}{P_{12}}\right)\right. \\ &\left.+\left(\frac{P'_{22} - N'_{11}}{P'_{12}} - \frac{P_{22} - N_{11}}{P_{12}}\right)^*\left(\frac{N'_{12}}{P'_{12}} - \frac{N_{12}}{P_{12}}\right)\right] \\ &/\left[\left|\frac{N'_{21}}{P'_{12}} - \frac{N_{21}}{P_{12}}\right|^2 + \left|\frac{P'_{22} - N'_{11}}{P'_{12}} - \frac{P_{22} - N_{11}}{P_{12}}\right|^2\right] \end{aligned} \tag{6.162}$$

Now, ω is only the unknown. By substituting the solved x_1, x_2, x_3, x_4, x_5, and x_6 into Eq. (6.142), ω can be solved. The calibration matrix C is obtained according to the following formula:

$$\mathbf{C} = \frac{1}{\omega|x_3 - x_1x_2|^2}\begin{pmatrix} x_3\mathbf{G}^{-1} & -x_1\mathbf{G}^{-1} \\ -x_2\mathbf{G}^{-1} & \mathbf{G}^{-1} \end{pmatrix}\mathbf{U}\begin{pmatrix} x_3^*\tilde{\mathbf{G}}^{-1} & -x_2^*\tilde{\mathbf{G}}^{-1} \\ -x_1^*\tilde{\mathbf{G}}^{-1} & \tilde{\mathbf{G}}^{-1} \end{pmatrix} \tag{6.163}$$

6.4.4 Polarimetric Synthetic Aperture Radar Posteriori Calibration Method

The posteriori calibration method for polarimetric SAR is described in this section [13]. It only has a minimum constraint on the data that satisfies reciprocity, and other additional parameters are estimated from the data itself. This is a method of performing crosstalk calibration between channels and gain imbalance calibration between channels using only observed polarimetric SAR data. Known ground targets or active radar calibrations are not available. The channel imbalance and crosstalk estimates obtained by this method do not deviate significantly from the actual polarization information.

At present, the polarization channel isolation of polarimetric SAR system can be very high, and the on-orbit active calibration technology is used to further improve the accuracy of the polarization data. Advances of polarization decomposition techniques and new methods for extracting geophysical details from polarization analysis require more accurate polarimetric SAR data. Therefore, it is expected that the relative amplitude and relative phase of the polarimetric SAR data between the polarization channels will be well calibrated.

The signal model of the posterior calibration method: The first is the relationship model between the observed polarization signal and the real calibration signal. In principle, this model combines the characteristic parameters such as crosstalk, channel imbalance, system noise, and system characteristics. Adopt a standard model that relates the transmitted horizontal polarization (H) and vertical polarization (V) with the received signal

$$\begin{bmatrix} O_{11} & O_{12} \\ O_{21} & O_{22} \end{bmatrix} = \begin{bmatrix} r_{11} & r_{12} \\ r_{21} & r_{22} \end{bmatrix} \begin{bmatrix} S_{11} & S_{12} \\ S_{21} & S_{22} \end{bmatrix} \begin{bmatrix} t_{11} & t_{12} \\ t_{21} & t_{22} \end{bmatrix} \tag{6.164}$$

Equation (6.164) shows how the actual scattering matrix [S] is changed by the crosstalk and channel imbalance of the transmitting and receiving (represented by [t] and [r], respectively). The subscripts 1, 2 are two polarization states, typically H and V. The scattering matrix can be converted into a vector form, which is rewritten as

$$\begin{bmatrix} O_{HH} \\ O_{HV} \\ O_{VH} \\ O_{VV} \end{bmatrix} = \begin{bmatrix} k^2\alpha & vk & wk\alpha & vw \\ zk^2\alpha & k & wzk\alpha & w \\ uk^2\alpha & uvk & k\alpha & v \\ uzk^2\alpha & uk & zk\alpha & 1 \end{bmatrix} \begin{bmatrix} S_{HH} \\ S_{HV} \\ S_{VH} \\ S_{VV} \end{bmatrix} \tag{6.165}$$

The k and α values are related to the channel imbalance, and u, v, w and z are the channel crosstalk parameters. It is defined as $k = r_{11}/r_{22}$, $\alpha = r_{22}t_{11}/r_{11}t_{22}$, $u = r_{21}/r_{11}$, $v = t_{21}/t_{22}$, $w = r_{12}/r_{22}$, and $z = t_{12}/t_{11}$. The total gain factor $r_{11}r_{22}$ is eliminated, so absolute radiation calibration does not have to be considered, all parameters are complex.

In order for the estimated model parameters to be available, the observed data is required to be redundant. Under the simplest reciprocity assumption, $S_{HV} = S_{VH}$. Thus, the echo differences O_{HV} and O_{VH} of the HV and VH are used to derive the calibration parameter values. Additional scattering symmetry can be used for the calibrated data, which can be determined by theoretical formula or experimental considerations. However, if the data does not satisfy the symmetry assumption, the calibration process will attempt to force symmetry and without considering polarization information. For SAR systems that produce quad-polarization data with good channel isolation, i.e., HH, HV, VH, and VV polarizations, it is assumed that only reciprocity will produce sufficient data redundancy. If the redundant data is valid, the posterior calibration can be performed.

In addition, the posterior calibration method has two other features. It is invalid for repeated calibration of the previously calibrated data. The calibration method will not destroy or deviate from the polarization information of the data.

6.4.4.1 Crosstalk Calibration

In general, the method of determining the crosstalk parameters includes solving a multi-pixel averaged polarization echo. The crosstalk parameter is determined using an average covariance matrix. Rewrite the covariance matrix items, and the calibration formula (6.165) becomes

$$[C] = [M][\Sigma][M]^T \tag{6.166}$$

where $[M]$ is the 4×4 calibration matrix of Eq. (6.165) and $[C]$ is the measurement covariance matrix. The superscript ^{T} indicates the Hermitian conjugate transformation. In the linear basis $\left[HH\ HV\ VH\ VV\right]$, assuming that the scattering matrix is reciprocal, the conventional form of the calibration covariance matrix is

$$\begin{bmatrix} \sigma_{HHHH} & A^* & A^* & \sigma_{HHVV} \\ A & \beta & \beta' & B \\ A & \beta' & \beta & B \\ \sigma_{VVHH} & B^* & B^* & \sigma_{VVVV} \end{bmatrix} \tag{6.167}$$

where the superscript * indicates conjugate, and β and β' values are real. Reciprocity means that the two cross-polarization channels HV and VH are identical, and the requirement for the post-calibration data is that they and all of their relationships are virtually identical. This form of covariance is very common, allowing non-zero rotation, direction angles, etc. Ideally, $\beta = \beta'$. However, considering system noise, let $\beta \geq \beta'$. The more stringent scattering symmetry assumption has additional requirements for the A and B values. Reflection symmetry means that there is no direction angle (rotation angle of the radar boresight), so that $A = B = 0$. However, it is not appropriate to directly make $A = B = 0$. Because this forces the reflection symmetry direction angle to be zero. Similarly, forcing reciprocity and backscattering to zero rotation at the same time requires $\widehat{m(A)} = \widehat{m(B)} = 0$. The problem becomes to estimate the A and B values with some appropriate and practical method under the assumption of scattering symmetry.

6.4.4.2 Channel Imbalance

The k and α values determine the relative gain and phase delay between the polarization channels, and α most directly affects Σ_{HVHV}, Σ_{VHVH}, and Σ_{HVVH} of the covariance elements. Assuming crosstalk is small, let $\Sigma_{HVHV} = \Sigma_{VHVH} = \beta$ and $\arg(\Sigma_{HVVH}) = 0 = \arg(\beta')$ be used to estimate the value of α. Estimate A by finding the average of Σ_{HHVH} and Σ_{HHHV}, and estimate B by finding the average of Σ_{VVHV} and Σ_{VVVH}.

Solving parameter k is more troublesome. The number of free parameters and the number of equations in (6.167) mean that equation set (6.167) is an underdetermined equation. k directly affects the co-polarization echo and its correlation. There is no additional information to determine k. Phase and amplitude of k are traditionally determined using trihedral reflectors. The k phase is well estimated by choosing any smooth surface area. The $HH - VV$ correlation coefficient ρ_{HHVV} of the odd-order scattering should be zero. Solving the amplitude of k is more troublesome. Or a model with a hypothetical scattering mechanism can be used to set the $|HH|/|VV|$ ratio, or $|HH|/|VV|$ ratio can be specifically assumed. It is also conceivable that other more or less special schemes are used to set the k value. In any case, additional information or assumptions are needed to determine the k value.

6.4.4.3 Calibration Method

Typically, calibration is required to be range dependent. The antenna gain pattern changes with the pitch angle. The relative displacement of the H and V antennas causes the phase to vary with range. This range correlation means that each range line should be independently calibrated. The calibration parameters are determined by azimuth average of the covariance matrix, and then each pixel along a given range line is calibrated by these parameters.

There is a slow phase shift in azimuth, and the phase of ρ_{HHVV} sometimes shows the amount of phase drift, which can be identified and corrected by the covariance average along the range line. Then, each pixel on that range line is calibrated. This repeated calibration of azimuth drift compensation is mainly for k, which should have less influence on α and crosstalk parameters.

A general recommendation is made to the covariance matrix averaging; nearly, 7% of the brightest pixels should be removed from the averaged data. Use the span (total power) graph to determine which pixels should be removed from the averaged data. Bright pixels tend to saturate the receiver, which causes the polarization average to deviate unrealistically. Although the 7% solution is somewhat arbitrary, the results are not sensitive to the value of the cut-off point. The polarization information of pixels with low return values in the total power graph is also poor, but as long as the pixel power near the average is sufficient and there is good polarization information, they tend not to affect the average result.

6.4.4.4 Numerical Methods

Nonlinear solutions of calibration parameters can be obtained by iterative calculations. First, the Eq. (6.166) is linearized with the current values of k, α, u, v, w, and z. The solution of this set of complex linear equations defines the updated u, v, w, and z. New k and α values are estimated, so iterative. Let $\Sigma_{HVHV} = \Sigma_{VHVH} = \beta$, $\arg(\Sigma_{HVVH}) = 0 = \arg(\beta')$, for a given crosstalk parameter, re-estimate the value of α with the new value of $[\Sigma]$. Usually, the estimates for A, B, β, and β' are not relevant with k. Only the estimates for A and B are directly related. That is, $\widehat{m(A)} = \widehat{m(B)}$, the value of k affects this part of the calibration.

These solutions of Eq. (6.166) mean that all calibration coefficients and calibration covariance matrices are self-consistent. Due to this self-consistency, there is no impact on the recalibration of the already calibrated data. Recalibrated all crosstalk parameters are zero, k and α values are real and equal to 1.

The calibrated scattering matrix is obtained directly from (6.164) and the calibration matrix $[M]^{-1}$. The calibrated HV and VH echoes are statistically equal. The HV and VH echoes may be different for any given pixel, but, assuming the target average covariance matrix is in the form (6.167), they will be equal on average. The calibrated HV and VH echoes will be directly averaged to produce a standard (calibrated) cross-polarization echo.

6.4.5 Point Target Calibration Algorithm—Whitt Algorithm

The point target calibration algorithm is the most studied among the three types of algorithms, and a large number of unique algorithms have emerged. But most algorithms either assume that the target to be tested has a certain characteristic, or that the distortion matrix of the radar system has a special form, which is where they are not satisfactory. The Whitt algorithm does not require any assumptions about the radar system distortion matrix. The target used for calibration only needs to meet a few constraints, and there is great flexibility in the selection of calibration targets. It can be used in both laboratory and field environments. So, in this section, this algorithm is introduced.

The calibration model is shown in Eq. (6.24), which is given again for convenience. The noise is ignored in the equation.

$$\mathbf{M} = Ae^{j\varphi}\begin{pmatrix} 1 & \delta_2 \\ \delta_1 & f_1 \end{pmatrix}\begin{pmatrix} S_{vv} & S_{vh} \\ S_{hv} & S_{hh} \end{pmatrix}\begin{pmatrix} 1 & \delta_3 \\ \delta_4 & f_2 \end{pmatrix} = Ae^{j\varphi}\mathbf{R}^T\mathbf{S}\mathbf{T} \tag{6.168}$$

Using the model in Eq. (6.168), consider the polarization measurements of three different point targets known as the scattering matrix. Using subscripts to distinguish different point targets and corresponding scattering matrices (measured and theoretical), three matrix equations can be obtained.

$$M_i = e^{j\varphi_i}A\mathbf{R}^{\mathrm{T}}\mathbf{S}_i\mathbf{T}, \quad i = 1, 2, 3 \tag{6.169}$$

Since the range of the phase centers of the respective targets to the radar is usually not equal, the subscript i is also used here to indicate that the absolute phases ϕ of the respective targets are different. But since most applications only need to know the relative phase of the scattering matrix S, these absolute phase calibrations are not made here.

Since the scattering matrix of at least one target is required to be reversible, without loss of generality, it is assumed that the scattering matrix of the target 1 is a reversible matrix, that is, S_1 and M_1 are reversible. Both matrices M_2 and M_3 are left multiplied by M_1^{-1}, resulting in a similar transformation as follows:

$$\mathbf{M}_T = \mathbf{M}_1^{-1}\mathbf{M}_2 = e^{j(\varphi2-\varphi1)}\mathbf{T}^{-1}\mathbf{S}_T\mathbf{T} = e^{j(\varphi2-\varphi1)}\mathbf{T}^{-1}\mathbf{S}_1^{-1}\mathbf{S}_2\mathbf{T} \tag{6.170a}$$

$$\overline{\mathbf{M}}_T = \mathbf{M}_1^{-1}\mathbf{M}_3 = e^{j(\varphi3-\varphi1)}\mathbf{T}^{-1}\overline{\mathbf{S}}_T\mathbf{T} = e^{j(\varphi3-\varphi1)}\mathbf{T}^{-1}\mathbf{S}_1^{-1}\mathbf{S}_3\mathbf{T} \tag{6.170b}$$

Note that the above two equations only contain the matrix T, but are independent with the matrix R.

$\mathbf{M}_T$ and $\mathbf{S}_T$, $\overline{\mathbf{M}}_T$ and $\overline{\mathbf{S}}_T$ are two pairs of similar matrices. The eigenvalues and eigenvectors of M_T and S_T satisfy the following relationship:

$$\mathbf{S}_T\mathbf{X}_T = \mathbf{X}_T\mathbf{\Lambda}'_T \tag{6.171a}$$

$$\mathbf{M}_T\mathbf{Y}_T = \mathbf{Y}_T\mathbf{\Lambda}_T \tag{6.171b}$$

In the above two equations, $\mathbf{\Lambda}'_T$ and $\mathbf{\Lambda}_T$ are diagonal arrays composed of eigenvalues of $\mathbf{S}_T$ and $\mathbf{M}_T$, respectively. The column vectors of $\mathbf{X}_T$ and $\mathbf{Y}_T$ are, respectively, composed of eigenvectors of S_T and M_T. Furthermore, the relationship between these eigenvalues and the eigenvector matrix is as follows:

$$\mathbf{\Lambda}'_T = \mathbf{\Lambda}_T e^{j(\varphi1-\varphi2)} \tag{6.172a}$$

$$\mathbf{Y}_T = \mathbf{T}^{-1}\mathbf{X}_T \tag{6.172b}$$

It should be noted that the order of the eigenvalues in $\mathbf{\Lambda}'_T$ and $\mathbf{\Lambda}_T$ is important. The order of the eigenvalues in $\mathbf{\Lambda}'_T$ and $\mathbf{\Lambda}_T$ must guarantee that Eq. (6.172a) establishes. In practice, the pointing error of the calibration target has a much greater influence on the amplitude of the scatter matrix eigenvalue than on its phase, so the eigenvalues should be arranged according to the phase in Eq. (6.172a). If $\mathbf{\Lambda}'_T = \mathrm{diag}(\lambda'_1, \lambda'_2)$ and the two eigenvalues of $\mathbf{M}_T$ are λ_1 and λ_2, then $\mathbf{\Lambda}_T$ should be selected as follows:

$$\mathbf{\Lambda}_T = \begin{cases} \mathrm{diag}(\lambda_1, \lambda_2), & \left|\tan^{-1}\left(\frac{\lambda'_1\lambda_2}{\lambda_1\lambda'_2}\right)\right| < \left|\tan^{-1}\left(\frac{\lambda'_1\lambda_1}{\lambda_2\lambda'_2}\right)\right| \\ \mathrm{diag}(\lambda_2, \lambda_1), & \text{Other situations} \end{cases} \tag{6.173}$$

Since the eigenvectors constituting $\mathbf{Y}_T$ and $\mathbf{X}_T$ have arbitrary scale factors, T cannot be uniquely determined by Eq. (6.172b). When the scale factor is selected, the matrix T can be uniquely determined

$$\mathbf{T} = \mathbf{X}_T\mathbf{C}\mathbf{Y}_T^{-1} \tag{6.174}$$

Here C = diag(c1, c2) is a diagonal matrix containing unknown elements c1 $\neq$ 0, c2 $\neq$ 0. In the derivation of Eq. (6.174), it is assumed that both $\mathbf{Y}_T$ and $\mathbf{X}_T$ are reversible. If the two eigenvalues of $\mathbf{S}_T$ are different, then the corresponding eigenvectors are linear independent, and the matrix $\mathbf{X}_T$ is reversible, so $\mathbf{Y}_T$ is also reversible (the necessary and sufficient condition for $\mathbf{Y}_T$ reversibility is $\mathbf{X}_T$ reversible). When the following formula is satisfied:

$$4\det(\mathbf{S}_T) \neq [\mathrm{tr}(\mathbf{S}_T)]^2 \tag{6.175}$$

then the two eigenvalues of $\mathbf{S}_T$ will be different. Therefore, when selecting a calibration target, it should select those known targets whose scattering matrix satisfies condition (6.175).

Another matrix equation for T can also be obtained from Eq. (6.170b)

$$\mathbf{T} = \overline{\mathbf{X}}_T\overline{\mathbf{C}}\overline{\mathbf{Y}}_T^{-1} \tag{6.176}$$

Here $\overline{\mathbf{C}} = \mathrm{diag}(\bar{c}_1, \bar{c}_2)$, and the eigenvalues (eigenvectors) of $\overline{\mathbf{M}}_T$ and $\overline{\mathbf{S}}_T$ are represented by $\overline{\mathbf{\Lambda}}'_T$ and $\overline{\mathbf{\Lambda}}_T$ ($\overline{\mathbf{X}}_T$ and $\overline{\mathbf{Y}}_T$), respectively. For these quantities, there are similarities between Eqs. (6.171a) and (6.172b).

Equations (6.174) and (6.176) are rearranged. It can be got

$$\mathbf{C}\mathbf{Y}_T^{-1}\overline{\mathbf{Y}}_T = \mathbf{X}_T^{-1}\overline{\mathbf{X}}_T\overline{\mathbf{C}} \tag{6.177}$$

Solve this matrix equation and get the values of the two ratios c_2/c_1 and $\bar{c}_2/\bar{c}_1$

$$\frac{c_2}{c_1} = \frac{(x_{11}\bar{x}_{21} - x_{21}\bar{x}_{11})(y_{22}\bar{y}_{11} - y_{12}\bar{y}_{21})}{(x_{22}\bar{x}_{11} - x_{12}\bar{x}_{21})(y_{11}\bar{y}_{21} - y_{21}\bar{y}_{11})} \tag{6.178a}$$

$$\frac{c_2}{c_1} = \frac{(x_{11}\bar{x}_{22} - x_{21}\bar{x}_{12})(y_{22}\bar{y}_{12} - y_{12}\bar{y}_{22})}{(x_{22}\bar{x}_{12} - x_{12}\bar{x}_{22})(y_{11}\bar{y}_{22} - y_{21}\bar{y}_{12})} \tag{6.178b}$$

$$\frac{\bar{c}_2}{\bar{c}_1} = \frac{(x_{22}\bar{x}_{11} - x_{12}\bar{x}_{21})(y_{22}\bar{y}_{12} - y_{12}\bar{y}_{22})}{(x_{22}\bar{x}_{12} - x_{12}\bar{x}_{22})(y_{22}\bar{y}_{11} - y_{12}\bar{y}_{21})} \tag{6.178c}$$

$$\frac{\bar{c}_2}{\bar{c}_1} = \frac{(x_{11}\bar{x}_{21} - x_{21}\bar{x}_{11})(y_{11}\bar{y}_{22} - y_{21}\bar{y}_{12})}{(x_{11}\bar{x}_{22} - x_{21}\bar{x}_{12})(y_{11}\bar{y}_{21} - y_{21}\bar{y}_{11})} \tag{6.178d}$$

In the above four equations, x_{mn}, y_{mn}, $\bar{x}_{mn}$, and $\bar{y}_{mn}$ are the elements of the matrix $\mathbf{X}_T$, $\mathbf{Y}_T$, $\overline{\mathbf{X}}_T$, and $\overline{\mathbf{Y}}_T$, respectively. In order to ensure that the above four equations have a confirm solution, there are some restrictions on the scattering matrix of the calibration target. Based on the constraint Eq. (6.175) and the similarity constraint equation for $\overline{\mathbf{S}}_T$, it is known that $\mathbf{X}_T^{-1}\overline{\mathbf{X}}_T$ is reversible and has a non-zero determinant. Therefore, the values of c_2/c_1 and $\bar{c}_2/\bar{c}_1$ can be confirmed if and only if the following conditions are true:

$$\overline{\mathbf{X}}_T \neq \mathbf{X}_T\alpha \quad \text{and} \quad \overline{\mathbf{X}}_T \neq \mathbf{X}_T\beta \tag{6.179}$$

Here

$$\alpha = \begin{pmatrix} \alpha_1 & 0 \\ 0 & \alpha_2 \end{pmatrix} \quad \beta = \begin{pmatrix} 0 & \beta_1 \\ \beta_2 & 0 \end{pmatrix} \tag{6.180}$$

This condition means that $\overline{\mathbf{S}}_T$ and $\mathbf{S}_T$ have at the most one common eigenvector.

At this point, the transmitting distortion matrix T can be determined by Eq. (6.174) or Eq. (6.176). The process of determining T using Eq. (6.174) is given here. Since T is a normalized matrix, its first element is represented by

$$\frac{1}{\det(\mathbf{Y}_T)}(c_1x_{11}y_{22} - c_2x_{12}y_{21}) = 1 \tag{6.181}$$

From this equation, it can be solved for c_1 and c_2 by means of which c_2/c_1 has been obtained.

$$c_1 = \det(\mathbf{Y}_T)\left(x_{11}y_{22} - \frac{c_2}{c_1}x_{12}y_{21}\right)^{-1} \tag{6.182a}$$

$$c_2 = \det(\mathbf{Y}_T)\left(\frac{c_1}{c_2}x_{11}y_{22} - x_{12}y_{21}\right)^{-1} \tag{6.182b}$$

By substituting Eq. (6.182a) and Eq. (6.182b) into Eq. (6.174), T can be solved.

The matrices M_2 and M_3 are both right multiplied by M_1^{-1}, the following similar transformation can be obtained which only contains the matrix R and does not contain the matrix T:

$$\mathbf{M}_R = \mathbf{M}_2\mathbf{M}_1^{-1} = e^{j(\varphi 2-\varphi 1)}\mathbf{R}^T\mathbf{S}_R(\mathbf{R}^T)^{-1} = e^{j(\varphi 2-\varphi 1)}\mathbf{R}^T\mathbf{S}_2\mathbf{S}_1^{-1}(\mathbf{R}^T)^{-1} \tag{6.183a}$$

$$\overline{\mathbf{M}}_R = \overline{\mathbf{M}}_3\overline{\mathbf{M}_1}^{-1} = e^{j(\varphi 3-\varphi 1)}\mathbf{R}^T\overline{\mathbf{S}}_R(\mathbf{R}^T)^{-1} = e^{j(\varphi 3-\varphi 1)}\mathbf{R}^T\mathbf{S}_3\mathbf{S}_1^{-1}(\mathbf{R}^T)^{-1} \tag{6.183b}$$

Similar to the above process of solving T, let X_R and Y_R be the matrix of the eigenvectors of S_R and M_R, respectively, then the following formula holds:

$$\mathbf{Y}_R = \mathbf{R}^T\mathbf{X}_R \tag{6.184}$$

Thus, the expression of R^T can be obtained.

$$\mathbf{R}^T = \mathbf{Y}_R\mathbf{D}\mathbf{X}_R^{-1} \tag{6.185}$$

Here D = diag(d1, d2) is a diagonal matrix containing unknown elements d1 $\neq$ 0, d2 $\neq$ 0. Another expression about R^T can also be obtained by (6.183b).

$$\overline{\mathbf{R}}^T = \overline{\mathbf{Y}}_R\overline{\mathbf{D}}\overline{\mathbf{X}_R}^{-1} \tag{6.186}$$

Here $\overline{\mathbf{D}} = \text{diag}(\overline{d}_1, \overline{d}_2)$, and the eigenvector matrix of $\overline{\mathbf{M}}_R$ and $\overline{\mathbf{S}}_R$ is represented by $\overline{\mathbf{X}}_R$ and $\overline{\mathbf{Y}}_R$, respectively. Equations (6.185) and (6.186) are considered to obtain the following equation:

$$\mathbf{D}\mathbf{X}_R^{-1}\overline{\mathbf{X}}_R = \mathbf{Y}_R^{-1}\overline{\mathbf{Y}}_R\overline{\mathbf{D}} \tag{6.187}$$

After solving, we can get

$$\frac{d_2}{d_1} = \frac{(y_{11}\bar{y}_{21} - y_{21}\bar{y}_{11})(x_{22}\bar{x}_{11} - x_{12}\bar{x}_{21})}{(y_{22}\bar{y}_{11} - y_{12}\bar{y}_{21})(x_{11}\bar{x}_{21} - x_{21}\bar{x}_{11})} \tag{6.188a}$$

$$\frac{d_2}{d_1} = \frac{(y_{11}\bar{y}_{22} - y_{21}\bar{y}_{12})(x_{22}\bar{x}_{12} - x_{12}\bar{x}_{22})}{(y_{22}\bar{y}_{12} - y_{12}\bar{y}_{22})(x_{11}\bar{x}_{22} - x_{21}\bar{x}_{12})} \tag{6.188b}$$

$$\frac{\bar{d}_2}{\bar{d}_1} = \frac{(y_{22}\bar{y}_{11} - y_{12}\bar{y}_{21})(x_{22}\bar{x}_{12} - x_{12}\bar{x}_{22})}{(y_{22}\bar{y}_{12} - y_{12}\bar{y}_{22})(x_{22}\bar{x}_{11} - x_{12}\bar{x}_{21})} \tag{6.188c}$$

$$\frac{\bar{d}_2}{\bar{d}_1} = \frac{(y_{11}\bar{y}_{21} - y_{21}\bar{y}_{11})(x_{11}\bar{x}_{22} - x_{21}\bar{x}_{12})}{(y_{11}\bar{y}_{22} - y_{21}\bar{y}_{12})(x_{11}\bar{x}_{21} - x_{21}\bar{x}_{11})} \tag{6.188d}$$

From formula (6.185), we can get

$$\frac{1}{\det(\mathbf{X}_R)}(d_1y_{11}x_{22} - d_2y_{12}x_{21}) = 1 \tag{6.189}$$

From this equation, the obtained d_2/d_1 can be used to solve d_1 and d_2.

$$d_1 = \det(\mathbf{X}_R)\left(y_{11}x_{22} - \frac{d_2}{d_1}y_{12}x_{21}\right)^{-1} \tag{6.190a}$$

$$d_2 = \det(\mathbf{X}_R)\left(\frac{d_1}{d_2}y_{11}x_{22} - y_{12}x_{21}\right)^{-1} \tag{6.190b}$$

By substituting Eqs. (6.190a) and (6.190b) into Eq. (6.185), the $\mathbf{R}^{\mathrm{T}}$ can be solved. After the system distortion matrix T and $\mathbf{R}^{\mathrm{T}}$ are both obtained, the absolute amplitude factor A can be obtained by substituting any one of Eqs. (6.169). From the *mn*th element in the scattering matrix, we get

$$A = \frac{|(\mathbf{M}_i)_{mn}|}{\left|(\mathbf{R}^{\mathrm{T}}\mathbf{S}_i\mathbf{T})_{mn}\right|} \quad i = 1, 2, 3 \tag{6.191}$$

Choosing the most accurate calibration target for the theoretical scattering matrix, the best estimate of A can be obtained.

For some single antenna systems, the transmitting and receiving distortion matrices satisfy the reciprocal relationship, i.e., $\mathbf{R}^{\mathrm{T}} = \mathbf{T}$. For the calibration of such systems, the method described above will be modified to that only two calibration targets known scattering matrices are used. The transformation relationships similar to Eqs. (6.170a) and (6.170b) can be obtained by forming matrix products $\mathbf{M} = \mathbf{M}_2^{-1}\mathbf{M}_1$ and $\overline{\mathbf{M}} = \left(\mathbf{M}_2\mathbf{M}_1^{-1}\right)^{\mathrm{T}}$. Then the above method can be used to solve the problem.

The advantages of this method are: (1) Almost any three targets can be used. (2) Do not make any assumptions about the system distortion matrix. There are only three restrictions on the known scattering matrix target: (1) At least one scattering matrix is reversible (assumed to be S_1). (2) $\mathbf{S}_1^{-1}$ S_2 and $\mathbf{S}_1^{-1}$ S_3 must have two different eigenvalues. (3) $\mathbf{S}_1^{-1}\mathbf{S}_2$ and $\mathbf{S}_1^{-1}\mathbf{S}_3$ have at the most one common eigenvector.

6.4.6 JPL Polarization Calibration Method

The polarimetric SAR system model and polarization calibration method are successfully used by NASA JPL Lab. It have been tested and verified in the calibration of airborne and spaceborne SAR systems. In this section, the system model of polarimetric SAR calibration and the JPL polarization calibration method [15] will discussed.

6.4.6.1 Polarization SAR System Calibration Model

Polarimetric radar alternately transmits signals through horizontal and vertical polarization antennas while receiving two polarization signals, using polarization time division.

The actual radar system is lossy. For example, a signal transmitted from a fully polarization antenna may have a leakage of its cross-polarization channel. Considering this leak, the actual transmitting field is written as

$$\mathbf{E}^t_{actual} = K_t(\gamma)\begin{bmatrix} 1 & \delta_1^t(\gamma) \\ \delta_1^t(\gamma) & f_t(\gamma) \end{bmatrix}\mathbf{E}^t_{ideal} \tag{6.192}$$

where γ is the radar look angle. For airborne case, the radar look angle γ of a given pixel is the same as the incidence angle η under the assumption of the Earth's plane. For spaceborne conditions, the radar look angle is different from the incidence angle, and their relationship is as follows:

$$R_t \sin\eta = R_s \sin\gamma \tag{6.193}$$

where R_s and R_t are the range from the satellite platform and the target location to the center of the Earth, respectively. Measure the slant range R between the radar and the scattering surface by the delay between the transmitted pulse and the received scatter signal.

The radar look angle is not directly measured, but is derived indirectly.

Under the assumption of the airborne earth plane, the expression is

$$\gamma = \cos^{-1}\pi\left(h/R\right) \tag{6.194}$$

Under the assumption of the spaceborne earth sphere, the expression is

$$\gamma = \cos^{-1}\pi\left(\frac{R^2 + R_s^2 - R_t^2}{2RR_s}\right) \tag{6.195}$$

As can be seen from formula (6.192), the crosstalk coefficient represented by δ is not only a function of the actual antenna. That is, if the radar platform motion pitch angle is not 0, the antenna coordinate system will have the rotation of the pitch angle. This will cause crosstalk in the transmitted signal.

In formula (6.192), f_t is the result of the difference between the two polarization antenna patterns, including the difference in gain between the two transmitting channels with polarization switches, and the difference in path length.

Ignore the Faraday rotation, the radar measurement scattering matrix can be written as

$$\begin{bmatrix} S_{hh} & S_{hv} \\ S_{vh} & S_{vv} \end{bmatrix}_{measuerd} = K(\gamma)[\mathbf{R}]\begin{bmatrix} S_{hh} & S_{hv} \\ S_{vh} & S_{vv} \end{bmatrix}_{actural}[\mathbf{T}] \tag{6.196}$$

where

$$[\mathbf{R}] = \begin{bmatrix} f_r(\gamma) & \delta_1^r(\gamma) \\ \delta_2^r(\gamma) & 1 \end{bmatrix}$$
$$[\mathbf{T}] = \begin{bmatrix} f_t(\gamma) & \delta_1^t(\gamma) \\ \delta_2^t(\gamma) & 1 \end{bmatrix} \quad (6.197)$$

The f_r in the formula (6.197) includes not only the difference between the two polarization antenna patterns, but also the difference of the gain and path length of the two receiving links.

The transmitting and receiving system distortion matrix can be written as the product of two matrices

$$[\mathbf{R}] = \begin{bmatrix} f_r(\gamma) & \delta_1^r(\gamma) \\ \delta_2^r(\gamma) & 1 \end{bmatrix} = \begin{bmatrix} 1 & \delta_1^r(\gamma) \\ \delta_2^r(\gamma)/f_r(\gamma) & 1 \end{bmatrix} \begin{bmatrix} f_r(\gamma) & 0 \\ 0 & 1 \end{bmatrix} = [\mathbf{R}_x][\mathbf{R}_c]$$
$$[\mathbf{T}] = \begin{bmatrix} f_t(\gamma) & \delta_1^t(\gamma) \\ \delta_2^t(\gamma) & 1 \end{bmatrix} = \begin{bmatrix} f_r(\gamma) & 0 \\ 0 & 1 \end{bmatrix} \begin{bmatrix} \alpha & \alpha[\delta_1^t(\gamma)/f_t(\gamma)] \\ \delta_2^t(\gamma) & 1 \end{bmatrix} = [\mathbf{R}_c][\mathbf{T}_x]$$
$$\alpha = f_t(\gamma)/f_r(\gamma) \quad (6.198)$$

Using the naming of Quegan (1994) [10], introduce the following symbols:

$$[\mathbf{R}_x][\mathbf{R}_c] = \begin{bmatrix} 1 & w \\ u & 1 \end{bmatrix} \begin{bmatrix} k & 0 \\ 0 & 1 \end{bmatrix}$$
$$[\mathbf{R}_c][\mathbf{T}_x] = \begin{bmatrix} k & 0 \\ 0 & 1 \end{bmatrix} \begin{bmatrix} \alpha & \alpha z \\ v & 1 \end{bmatrix}$$
$$\alpha = f_t(\gamma)/f_r(\gamma) \quad (6.199)$$

Therefore, by substituting formula (6.198) into formula (6.196), we can get

$$\begin{bmatrix} S_{hh} & S_{hv} \\ S_{vh} & S_{vv} \end{bmatrix}_{measuerd} = K(\gamma)[\mathbf{R}_x][\mathbf{R}_c] \begin{bmatrix} S_{hh} & S_{hv} \\ S_{hv} & S_{vv} \end{bmatrix}_{actural} [\mathbf{R}_c][\mathbf{T}_x] \quad (6.200)$$

Noise error will be ignored here.

The scattering matrix of the linear reciprocal scattering medium is symmetrical. The signal-to-noise ratio (SNR) of the cross-polarization channel is increased by 3 decibels (dB) by means of the related average of the cross-polarization measurements in the scattering matrix. Since the noise measured by the cross-polarization channel is obtained at different times in different channels, the increased 3 dB is obtained because the cross-polarization signals are correlated and the noises of the two channels are not correlated superimposed. The symmetric matrix is written as

$$\begin{bmatrix} S_{hh} & S_{hv} \\ S_{hv} & S_{vv} \end{bmatrix}_{sym} = \frac{1}{2}\left\{[\mathbf{S}]_{mensured} + \left[\mathbf{S}^T\right]_{mensured}\right\}$$

$$= K(\gamma)\frac{1}{2}\left\{[\mathbf{R}] + \left[\mathbf{T}^T\right]\right\}\begin{bmatrix} S_{hh} & S_{hv} \\ S_{hv} & S_{vv} \end{bmatrix}_{actural}\left\{[\mathbf{T}] + \left[\mathbf{R}^T\right]\right\} \quad (6.201)$$

In the formula, the superscript T indicates matrix transposition. The formula can be written as

$$\begin{bmatrix} S_{hh} & S_{hv} \\ S_{hv} & S_{vv} \end{bmatrix}_{sym} = K_C(\gamma)\left[\mathbf{D}^T\right]\begin{bmatrix} S_{hh} & S_{hv} \\ S_{hv} & S_{vv} \end{bmatrix}_{actural}[\mathbf{D}] \quad (6.202)$$

The distortion matrix $[\mathbf{D}]$ can be written as

$$[\mathbf{D}] = \begin{bmatrix} 1 & \delta_1(\gamma) \\ \delta_2(\gamma) & f(\gamma) \end{bmatrix} \quad (6.203)$$

If the matrix is symmetrized to improve the SNR, there is no need to know each individual component in the distortion matrix of the transmitting and receiving systems. Estimating the distortion matrix element of (6.203) is sufficient for the calibration of scattering matrix element.

The system model of Eq. (6.202) can also be written in a form similar to Eq. (6.200).

$$\begin{bmatrix} S_{hh} & S_{hv} \\ S_{hv} & S_{vv} \end{bmatrix}_{sym} = K_c(\gamma)\left[\mathbf{D}_x^T\right]\left[\mathbf{D}_c^T\right]\begin{bmatrix} S_{hh} & S_{hv} \\ S_{hv} & S_{vv} \end{bmatrix}_{actural}[\mathbf{D}_c][\mathbf{D}_x] \quad (6.204)$$

where

$$[\mathbf{D}] = \begin{bmatrix} 1 & \delta_1(\gamma) \\ \delta_2(\gamma) & f(\gamma) \end{bmatrix} = \begin{bmatrix} 1 & 0 \\ 0 & f(\gamma) \end{bmatrix}\begin{bmatrix} 1 & \delta_1(\gamma) \\ \delta_2(\gamma)/f_r(\gamma) & 1 \end{bmatrix} = [\mathbf{D}_c][\mathbf{D}_x] \quad (6.205)$$

The usual polarization calibration process includes

- Estimating crosstalk parameters and correcting crosstalk.
- Estimating and correcting relative phase and amplitude of co-polarization.
- Estimating the absolute radiometric calibration factor to convert the measured power to a normalized radar cross-sectional area.

6.4.6.2 Estimating Crosstalk Parameters and Correcting Crosstalk

The first step in calibration is to estimate the crosstalk parameters. Formula (6.200) of the polarimetric SAR system is written in vector form as follows:

$$\begin{bmatrix} S_{hh} \\ S_{vh} \\ S_{hv} \\ S_{vv} \end{bmatrix}_{measuerd} = K(\gamma)\begin{bmatrix} \alpha & v+\alpha w & vw \\ \alpha u & \alpha & v \\ \alpha z & 1 & w \\ \alpha uz & \alpha & 1 \end{bmatrix}\begin{bmatrix} k^2 & 0 & 0 \\ 0 & k & 0 \\ 0 & 0 & 1 \end{bmatrix}\begin{bmatrix} S_{hh} \\ S_{hv} \\ S_{vv} \end{bmatrix}_{actural} \quad (6.206)$$

where $\alpha = f_t(\gamma)/f_r(\gamma)$. Its equivalent form is as follows:

$$\begin{bmatrix} S_{hh} \\ S_{vh} \\ S_{hv} \\ S_{vv} \end{bmatrix}_{measuerd} = K(\gamma)[\mathbf{X}][\mathbf{Q}]\begin{bmatrix} S_{hh} \\ S_{hv} \\ S_{vv} \end{bmatrix}_{actural} \quad (6.207)$$

The covariance matrix is formed by the measurement vector of Eq. (6.207)

$$[\mathbf{C}]_{measuerd} = |K(\gamma)|^2[\mathbf{X}][\mathbf{Q}][\mathbf{C}]_{actural}\left[\mathbf{Q}^H\right]\left[\mathbf{X}^H\right] \quad (6.208)$$

The superscript H represents the complex conjugate transpose of the matrix. In order to estimate the distortion matrix, especially the matrix [**X**], assuming that the imaged surface has scattering symmetry, then

$$[\mathbf{C}]_{actural} = \begin{bmatrix} \langle S_{hh}S_{hh}^*\rangle & 0 & \langle S_{hh}S_{vv}^*\rangle \\ 0 & \langle S_{hv}S_{hv}^*\rangle & 0 \\ \langle S_{hh}^*S_{vv}\rangle & 0 & \langle S_{vv}S_{vv}^*\rangle \end{bmatrix} \quad (6.209)$$

where the angle brackets ⟨⟩ represent the spatial average of the image data. The inner multiplication in formula (6.208) can be written as

$$[\mathbf{O}] = [\mathbf{Q}][\mathbf{C}]_{actural}\left[\mathbf{Q}^H\right] = \begin{bmatrix} |k^2|^2\langle S_{hh}S_{hh}^*\rangle & 0 & k^2\langle S_{hh}S_{vv}^*\rangle \\ 0 & |k|^2\langle S_{hv}S_{hv}^*\rangle & 0 \\ k^{2*}\langle S_{hh}^*S_{vv}\rangle & 0 & \langle S_{vv}S_{vv}^*\rangle \end{bmatrix} \quad (6.210)$$

The matrix still has the same basic form as the original measured covariance matrix, in the sense several values in the matrix are zero. Matrix multiplication can be used to estimate crosstalk parameters

$$[\mathbf{T}] = [\mathbf{X}][\mathbf{O}]\left[\mathbf{X}^H\right] \quad (6.211)$$

Ignore second-order terms compared to other items of the same element, and the result is as follows:

$$\begin{aligned} T_{11} &\approx |\alpha|^2 O_{11} \\ T_{21} &\approx |\alpha|^2 uO_{11} + \alpha^* vO_{13}^* + \alpha\left(v^* + \alpha^* w^*\right)O_{22} \\ T_{31} &\approx |\alpha|^2 zO_{11} + \alpha^* wO_{13}^* + \left(v^* + \alpha^* w^*\right)O_{22} \end{aligned}$$

$$
\begin{aligned}
T_{41} &\approx \alpha^* O_{13}^* \\
T_{22} &\approx |\alpha|^2|u|^2 O_{11} + \alpha^* u^* v O_{13}^* + \alpha u v^* O_{13} + |\alpha|^2 O_{22} + |v|^2 O_{33} \\
T_{32} &\approx |\alpha|^2 u^* z O_{11} + \alpha^* u^* w O_{13}^* + \alpha z v^* O_{13} + \alpha O_{22} + w v^* O_{33} \\
T_{42} &\approx \alpha^* u^* O_{13}^* + v^* O_{33} + \alpha^*(u + \alpha z) O_{22} \\
T_{33} &\approx |\alpha|^2 |z|^2 O_{11} + \alpha^* z^* w O_{13}^* + \alpha z w^* O_{13} + O_{22} + |w|^2 O_{33} \\
T_{43} &\approx \alpha^* z^* O_{13}^* + w^* O_{33} + (u + \alpha z) O_{22} \\
T_{44} &\approx O_{33}
\end{aligned}
\tag{6.212}
$$

We can get the following four simplified expressions:

$$
\begin{aligned}
T_{21} &\approx u T_{11} + v T_{41} + \alpha\left(v^* + \alpha^* w^*\right) O_{22} \\
T_{31} &\approx z T_{11} + w T_{41} + \left(v^* + \alpha^* w^*\right) O_{22} \\
T_{24} &\approx u T_{14} + v O_{33} + \alpha\left(u^* + \alpha^* z^*\right) O_{22} \\
T_{34} &\approx z T_{14} + w O_{33} + \left(u^* + \alpha^* z^*\right) O_{22}
\end{aligned}
\tag{6.213}
$$

The expression is based on the actual correlation matrix **[T]** as the Hermitian matrix. From the above four expressions u, v, z, and w can be solved, and the process is as follows:

$$
\begin{bmatrix} T_{21} - \alpha T_{31} \\ T_{24} - \alpha T_{34} \end{bmatrix} = \begin{bmatrix} T_{11} & T_{41} \\ T_{14} & T_{44} \end{bmatrix} \begin{bmatrix} u - \alpha z \\ v - \alpha w \end{bmatrix}
\tag{6.214}
$$

From the above, we can obtain

$$
\begin{aligned}
u - \alpha z &= (T_{44} T_{21} - T_{41} T_{24}) / (T_{11} T_{44} - T_{14} T_{41}) \\
&\quad - \alpha (T_{11} T_{34} - T_{31} T_{14}) / (T_{11} T_{44} - T_{14} T_{41}) \\
v - \alpha w &= (T_{11} T_{24} - T_{21} T_{14}) / (T_{11} T_{44} - T_{14} T_{41}) \\
&\quad - \alpha (T_{44} T_{31} - T_{41} T_{34}) / (T_{11} T_{44} - T_{14} T_{41})
\end{aligned}
\tag{6.215}
$$

After simplified derivation, we obtain

$$
\begin{aligned}
u &= (T_{44} T_{21} - T_{41} T_{24}) / (T_{11} T_{44} - T_{14} T_{41}) \\
v &= (T_{11} T_{24} - T_{21} T_{14}) / (T_{11} T_{44} - T_{14} T_{41}) \\
w &= (T_{44} T_{31} - T_{41} T_{34}) / (T_{11} T_{44} - T_{14} T_{41}) \\
z &= (T_{11} T_{34} - T_{31} T_{14}) / (T_{11} T_{44} - T_{14} T_{41})
\end{aligned}
\tag{6.216}
$$

Knowing these four parameters can't get matrix **[O]** from matrix **[T]**, and it is also necessary to know the α value. α can be solved by the following equation:

$$
T_{32} = u^* z T_{11} + u^* w T_{41} + z v^* T_{14} + w v^* T_{44} + \alpha O_{22}
$$

$$T_{33} = |z|^2 T_{11} + z^* w T_{41} + z w^* T_{14} + |w|^2 T_{44} + O_{22} \tag{6.217}$$

Combine the two formulas to get

$$\alpha = \frac{T_{32} - u^*(zT_{11} + wT_{41}) - v^*(zT_{14} + wT_{44})}{T_{33} - z^*(zT_{11} + wT_{41}) - w^*(zT_{14} + wT_{44})} \tag{6.218}$$

The [**X**] in the matrix (6.211) can be reconstructed by Eqs. (6.216) and (6.218). It can be used to estimate the matrix [**O**] as follows:

$$\left[\hat{\mathbf{O}}\right] = \left[\mathbf{X}^H \mathbf{X}\right]^{-1} \left[\mathbf{X}^H\right] [\mathbf{T}] [\mathbf{X}] \left[\mathbf{X}^H \mathbf{X}\right]^{-1} \tag{6.219}$$

Using a scattering matrix, the formula is

$$\begin{bmatrix} k^2 S_{hh} \\ k S_{hv} \\ S_{vv} \end{bmatrix}_{actural} = \left[\mathbf{X}^H \mathbf{X}\right]^{-1} \left[\mathbf{X}^H\right] \begin{bmatrix} S_{hh} \\ S_{vh} \\ S_{hv} \\ S_{vv} \end{bmatrix}_{measuerd} \tag{6.220}$$

In order to obtain the actual scattering matrix, the channel imbalance parameters still need to be estimated. Before that, at first, the 3 dB SNR increased for the symmetry of the measured scattering matrix of the cross-polarization channel can be seen.

Once the scattering matrix is symmetrized, the system model can be rewritten as

$$\begin{bmatrix} S_{hh} & S_{hv} \\ S_{hv} & S_{vv} \end{bmatrix}_{sym} = K_C(\gamma) \begin{bmatrix} 1 & b \\ a & 1 \end{bmatrix} \begin{bmatrix} 1 & 0 \\ 0 & f \end{bmatrix} \begin{bmatrix} S_{hh} & S_{hv} \\ S_{hv} & S_{vv} \end{bmatrix}_{actual} \begin{bmatrix} 1 & 0 \\ 0 & f \end{bmatrix} \begin{bmatrix} 1 & a \\ b & 1 \end{bmatrix} \tag{6.221}$$

After multiplication operation, it can be written as

$$\begin{gathered} \begin{bmatrix} S_{hh} \\ S_{hv} \\ S_{vv} \end{bmatrix}_{sym} = \begin{bmatrix} 1 & 2b & b^2 \\ a & 1+ab & b \\ a^2 & 2a & 1 \end{bmatrix} \begin{bmatrix} W_{hh} \\ W_{hv} \\ W_{vv} \end{bmatrix} \\ \begin{bmatrix} W_{hh} \\ W_{hv} \\ W_{vv} \end{bmatrix} = \begin{bmatrix} 1 & 0 & 0 \\ 0 & f & 0 \\ 0 & 0 & f^2 \end{bmatrix} K_C(\gamma) \begin{bmatrix} S_{hh} \\ S_{hv} \\ S_{vv} \end{bmatrix}_{actual} \end{gathered} \tag{6.222}$$

This separates crosstalk correction from co-polarization channel imbalance and absolute radiometric calibration. The resulting covariance matrix is

$$[\mathbf{T}] = \begin{bmatrix} 1 & 2b & b^2 \\ a & 1+ab & b \\ a^2 & 2a & 1 \end{bmatrix} \left\langle \begin{bmatrix} W_{hh} \\ W_{hv} \\ W_{vv} \end{bmatrix} \begin{bmatrix} W_{hh} & W_{hv} & W_{vv} \end{bmatrix}^* \right\rangle \begin{bmatrix} 1 & a & a^2 \\ 2b & 1+ab & 2a \\ b^2 & b & 1 \end{bmatrix}^* \quad (6.223)$$

Superscript * indicates conjugate.

Next, assume that the covariance matrix of the surface satisfies (6.209). Under this assumption, we can get

$$\begin{aligned} T_{11} &\approx \langle W_{hh} W_{hh}^* \rangle \\ T_{21} &\approx a\langle W_{hh} W_{hh}^* \rangle + b\langle W_{vv} W_{hh}^* \rangle + 2b\langle W_{hv} W_{hv}^* \rangle \\ T_{31} &\approx \langle W_{vv} W_{hh}^* \rangle \\ T_{22} &\approx \langle W_{hv} W_{hv}^* \rangle + |a|^2\langle W_{hh} W_{hh}^* \rangle + |b|^2\langle W_{vv} W_{vv}^* \rangle + ab^*\langle W_{hh} W_{vv}^* \rangle + a^*b\langle W_{vv} W_{hh}^* \rangle \\ T_{32} &\approx a^*\langle W_{vv} W_{hh}^* \rangle + b^*\langle W_{hh} W_{vv}^* \rangle + 2a\langle W_{hv} W_{hv}^* \rangle \\ T_{33} &\approx \langle W_{vv} W_{vv}^* \rangle \end{aligned} \quad (6.224)$$

Therefore, we can get

$$\begin{aligned} T_{21} &\approx aT_{11} + bT_{31} + 2b^*\langle W_{hv} W_{hv}^* \rangle \\ T_{23} &\approx aT_{13} + bT_{33} + 2a^*\langle W_{hv} W_{hv}^* \rangle \end{aligned} \quad (6.225)$$

This can solve two complex crosstalk parameters. First, from the second equation of (6.225), it can be found out

$$bT_{33} = T_{23} - aT_{13} - 2a^*\langle W_{hv} W_{hv}^* \rangle \quad (6.226)$$

Multiplying the first equation of the formula (6.225) by T_{33}, we can get

$$\begin{aligned} T_{21}T_{33} &= aT_{11}T_{33} + bT_{31}T_{33} + 2b^*T_{33}\langle W_{hv} W_{hv}^* \rangle \\ &= aT_{11}T_{33} + T_{31}\left(T_{23} - aT_{13} - 2a^*\langle W_{hv} W_{hv}^* \rangle\right) \\ &\quad + 2\langle W_{hv} W_{hv}^* \rangle\left(T_{23} - aT_{13} - 2a^*\langle W_{hv} W_{hv}^* \rangle\right)^* \end{aligned} \quad (6.227)$$

Here, formula (6.226) is substituted into the second line of the formula, and the related items are merged to obtain the following formula:

$$\begin{aligned} &a\left(T_{11}T_{33} - T_{31}T_{13} - 4\langle W_{hv} W_{hv}^* \rangle^2\right) - 4a^*T_{31}\langle W_{hv} W_{hv}^* \rangle \\ &= T_{21}T_{33} - T_{31}T_{23} - T_{23}^*\langle W_{hv} W_{hv}^* \rangle \end{aligned} \quad (6.228)$$

From the real and imaginary parts of the above formula, the real and imaginary parts of a can be solved, and the obtained a is as follows:

$$
\begin{aligned}
a &= \frac{\Delta P + 4P^{*}T_{31}\langle W_{hv}W_{hv}^{*}\rangle}{\Delta^{2} - 16T_{13}T_{31}\langle W_{hv}W_{hv}^{*}\rangle^{2}} \\
P &= T_{21}T_{33} - T_{31}T_{23} - 2T_{23}^{*}\langle W_{hv}W_{hv}^{*}\rangle \\
\Delta &= T_{11}T_{33} - T_{31}T_{13} - 4\langle W_{hv}W_{hv}^{*}\rangle^{2}
\end{aligned} \tag{6.229}
$$

Use the same method to find b as follows:

$$
\begin{aligned}
b &= \frac{\Delta Q + 4Q^{*}T_{13}\langle W_{hv}W_{hv}^{*}\rangle}{\Delta^{2} - 16T_{13}T_{31}\langle W_{hv}W_{hv}^{*}\rangle^{2}} \\
Q &= T_{11}T_{23} - T_{13}T_{21} - 2T_{21}^{*}\langle W_{hv}W_{hv}^{*}\rangle
\end{aligned} \tag{6.230}
$$

As soon as the crosstalk parameter is known, its effect can be reversed and eliminated from Eq. (6.223) as follows:

$$
\left\langle \begin{bmatrix} W_{hh} \\ W_{hv} \\ W_{vv} \end{bmatrix} \begin{bmatrix} W_{hh} & W_{hv} & W_{vv} \end{bmatrix}^{*} \right\rangle = \begin{bmatrix} 1 & 2b & b^{2} \\ a & 1+ab & b \\ a^{2} & 2a & 1 \end{bmatrix}^{-1} [\mathbf{T}] \begin{bmatrix} 1 & a & a^{2} \\ 2b & 1+ab & 2a \\ b^{2} & b & 1 \end{bmatrix}^{*-1} \tag{6.231}
$$

The expressions of a and b cannot be solved in a simple way. To understand the crosstalk parameters, the cross-polarization echo energy $\langle W_{hv}W_{hv}^{*}\rangle$ need to be known. The cross-polarization echo energy here refers to the value after the crosstalk effect is eliminated. At first using formula (6.231), let $a = b = 0$ to estimate the initial value of $\langle W_{hv}W_{hv}^{*}\rangle$. Then Eqs. (6.229) and (6.230) are used to estimate the initial value of the crosstalk parameter. Use this initial estimate and formula (6.231) to improve the estimate of $\langle W_{hv}W_{hv}^{*}\rangle$. The crosstalk parameter is more accurately estimated with the newly improved $\langle W_{hv}W_{hv}^{*}\rangle$ value. So that iteration continues until a, b, and $\langle W_{hv}W_{hv}^{*}\rangle$ reach a stable solution.

These estimates are all dependent on the calculation of the mean value of the measured covariance matrix. The question is: How many pixel averaging are needed to estimate these parameters? The actual answer depends on the basic assumptions of the true covariance matrix as described in Eq. (6.209). The expression of the covariance matrix as described in Eq. (6.209) exhibits reflection symmetry. One aspect that does not satisfy the reflection symmetry is that there is a local tilt in azimuth. This can be modeled with local rotation of the polarization base.

$$
\begin{aligned}
\begin{bmatrix} S_{hh} & S_{hv} \\ S_{hv} & S_{vv} \end{bmatrix}_{tilt} &= \begin{bmatrix} \cos\beta & \sin\beta \\ -\sin\beta & \cos\beta \end{bmatrix} \begin{bmatrix} S_{hh} & S_{hv} \\ S_{hv} & S_{vv} \end{bmatrix}_{actual} \begin{bmatrix} \cos\beta & -\sin\beta \\ \sin\beta & \cos\beta \end{bmatrix} \\
&= \cos^{2}\beta \begin{bmatrix} 1 & \tan\beta \\ -\tan\beta & 1 \end{bmatrix} \begin{bmatrix} S_{hh} & S_{hv} \\ S_{hv} & S_{vv} \end{bmatrix}_{actual} \begin{bmatrix} 1 & -\tan\beta \\ \tan\beta & 1 \end{bmatrix}
\end{aligned} \tag{6.232}
$$

This equation is compared with formula (6.221), and the effect of tilt appears with $a = -b = -\tan\beta$ as crosstalk. Schuler et al. (1996) proposed a method for estimating azimuth tilt using correlations between co-polarization and cross-polarization echoes observed in polarimetric SAR images. The crosstalk parameter is a function of the antenna angle, and the values of the antenna angle to different slant range are different. Therefore, the crosstalk parameter is estimated on average by means of the entire line covariance matrix of the fixed slant range. The azimuth local tilt is often averaged and eliminated, resulting in better crosstalk parameter estimation.

6.4.6.3 Co-polarization Channel Unbalanced Calibration

The co-polarization channel imbalance includes amplitude imbalance and phase imbalance. If the radar hardware is stable for a long time, it can rely more on the internal calibration signal calibration. In this case, the external calibration device will only be used to periodically correct the system calibration. Conversely, if there is no internal calibration signal, or if the hardware is only stable in the short term, it must rely on external calibrator to calibrate.

The triangular reflector is the ideal calibrator for this task. The scattering matrix of the triangular reflector is

$$[S] = \frac{k_0 l^2}{\sqrt{12}\pi}\begin{bmatrix} 1 & 0 \\ 0 & 1 \end{bmatrix} \tag{6.233}$$

where the parameter l is the length of the reflector.

Since the co-polarization channel imbalance may be a function of the radar look angle, several calibrators are typically placed at different locations throughout the imaging zone perpendicular to the radar track in the calibration image to correct relative uniformity over different range. Since it is necessary to estimate the radar cross-sectional area of the calibrator from the radar image data, it is usually chosen to place the reflector on the smooth bare surface so that the radar echo signal of the background adjacent to the calibrator is as low as possible.

Co-polarization Phase Calibration

Figure 6.14 shows the system model for co-polarization phase calibration. It contains an internal calibration signal to help measure the path delay in the system.

At first, the internal calibration signal which is used to perform phase difference calibration is discussed. Then the phase difference calibration when there is no calibration signal will be discussed later. After eliminating crosstalk, the scattering matrix can be written as

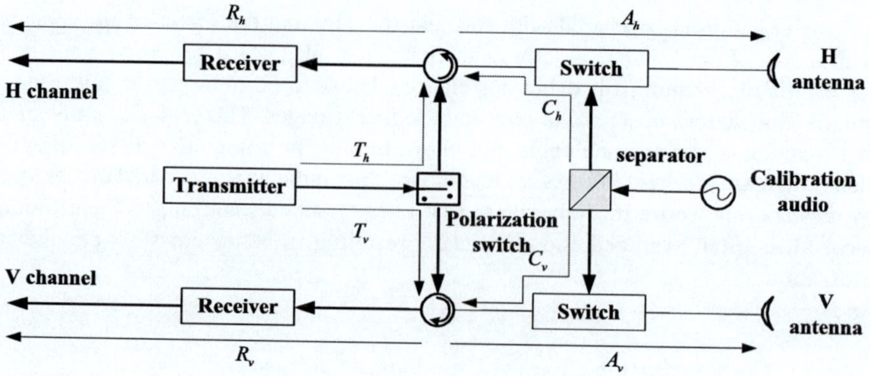

Fig. 6.14 System model for co-polarization phase calibration

$$[R] = \begin{bmatrix} S_{hh}\exp\{i(\phi_{Th} + 2\phi_{Ah} + \phi_{Rh})\} & |f|S_{hv}\exp\{i(\phi_{Tv} + \phi_{Ah} + \phi_{Av} + \phi_{Rh})\} \\ |f|S_{hv}\exp\{i(\phi_{Th} + \phi_{Ah} + \phi_{Av} + \phi_{Rv})\} & |f|^2 S_{vv}\exp\{i(\phi_{Tv} + 2\phi_{Av} + \phi_{Rv})\} \end{bmatrix} \quad (6.234)$$

As shown in formula (6.203), $f = f(\gamma)$. In each case, taking the ϕ_{Th} term as an example, refer to the signal phase generated along the point line path T_h in Fig. 6.14. For example, for R_{hh}, the signal is transmitted from the transmitter to the circulator along the T_h path and then along the A_h at the circulator to the antenna (before being radiated), which is transmission along the A_h path again upon reception. From the antenna to the circulator, and then the circulator and the remaining horizontal receiving links are transmission along the R_h path. Extract all absolute phases, formula (6.234) can be rewritten as follows:

$$[R] = \exp\{i(\phi_{Tv} + 2\phi_{Av} + \phi_{Rv})\} \begin{bmatrix} S_{hh}\exp\{i(\phi_T + 2\phi_A + \phi_R)\} & |f|S_{hv}\exp\{i(\phi_A + \phi_R)\} \\ |f|S_{hv}\exp\{i(\phi_T + \phi_A)\} & |f|^2 S_{vv} \end{bmatrix} \quad (6.235)$$

where $\phi_T = \phi_{Th} - \phi_{Tv}$, $\phi_A = \phi_{Ah} - \phi_{Av}$, $\phi_R = \phi_{Rh} - \phi_{Rv}$. The absolute phase outside the matrix of the formula (6.235) is not important in the polarization. The item should disappear after various cross-multiplications of the covariance matrix. The components of the covariance matrix are

$$\begin{aligned} R_{hh}R_{hh}^* &= S_{hh}S_{hh}^* \\ R_{hv}R_{hv}^* &= R_{vh}R_{vh}^* = |f|^2 S_{hv}S_{hv}^* \\ R_{vv}R_{vv}^* &= |f|^4 R_{vv}R_{vv}^* \\ R_{hh}R_{hv}^* &= |f|^2 S_{hh}S_{hv}^* \exp\{i(\phi_T + \phi_A)\} \\ R_{hh}R_{vh}^* &= |f|^2 S_{hh}S_{vh}^* \exp\{i(\phi_R + \phi_A)\} \\ R_{hh}R_{vv}^* &= |f|^2 S_{hh}S_{vv}^* \exp\{i(\phi_T + 2\phi_A + \phi_R)\} \end{aligned}$$

$$
\begin{aligned}
R_{hv}R_{vh}^* &= |f|^2 S_{hv}S_{vh}^* \exp\{i(\phi_R - \phi_T)\} \\
R_{hv}R_{vv}^* &= |f|^3 S_{hv}S_{vv}^* \exp\{i(\phi_R + \phi_A)\} \\
R_{vh}R_{vv}^* &= |f|^3 S_{vh}S_{vv}^* \exp\{i(\phi_T + \phi_A)\}
\end{aligned} \tag{6.236}
$$

The three unknown phases ϕ_T, ϕ_A, and ϕ_R need to be calibrated. The antenna phase difference ϕ_A can be measured after the radar system is mounted on a satellite or aircraft platform. This path consists only of fiber optic cables (or perhaps waveguides) and antennas, which will remain stable for long periods of time. The measurement of the receiving phase difference allows the calibration signal to be injected into both channels via a switching network. Since the signal is co-sourced with the radar, any phase difference (e.g., ϕ_{cal}) measured at the output of the receiving link is related to the difference in the length of the transmission path of the homologous signal in the system. According to Fig. 6.14, we can get

$$\phi_{cal} = \phi_R + \phi_C \Rightarrow \phi_R = \phi_{cal} - \phi_C \tag{6.237}$$

The ϕ_C term can be minimized by means of careful design, and since it contains the path length difference through the cable, the splitter, and the coupler, ϕ_C is stable for a long time. Using formula (6.237), the phase difference caused by the length of the receiving path can be obtained by measuring the phase difference between the two receiving link calibration signals.

Thus, only one path length phase difference ϕ_T is unknown. This phase difference can be obtained from the fact that $S_{hv} = S_{vh}$ is reciprocity. From formula (6.236), it can be found

$$R_{hv}R_{vh}^* = |fS_{hv}|^2 \exp\{i(\phi_R - \phi_T)\} \tag{6.238}$$

Due to system noise, the phase of the complex multiplication $R_{hv}R_{vh}^*$ of each pixel in the image changes slightly. $\phi_R - \phi_T$ can be estimated well by means of averaging the product of the entire scene. The phase difference of the transmission path is obtained by the following formula:

$$\phi_T = \phi_R - \arg\{\langle R_{hv}R_{vh}^* \rangle\} \tag{6.239}$$

Once the relative phase is known, all elements of the scattering matrix can be calibrated. This method only makes a hypothesis about the scattering situation that is satisfying the reciprocity condition.

If there is no calibration signal in the radar system, it is only necessary to make a simple assumption about the scattering situation, and the co-polarization phase calibration can still be completed. The basic assumption is that there are extended regions of the known co-polarization phase difference in the image. This phase difference is expressed as $\widehat{\phi}_{co}$. Extract the measured co-polarization phase difference from the image data of a hypothetical known co-polarization phase difference area.

$$\arg\{\langle R_{hh}R_{vv}^*\rangle\} = \phi_T + 2\phi_A + \phi_R = \widehat{\phi}_{co} \Rightarrow \phi_T + \phi_R = \widehat{\phi}_{co} - 2\phi_A \tag{6.240}$$

ϕ_T and ϕ_R can be solved by formulas (6.239) and (6.240). The average value of the co-polarization phase of the slightly rough surface scattering is about 0, which can be used for phase calibration. Of course, if there are triangular reflectors in the image, they can also be used to measure the co-polarization phase difference.

Correction of the Co-polarization Amplitude Imbalance

After the phase calibration is completed, the co-polarization amplitude imbalance correction is performed. This can be completed by measuring the power of the triangular reflector. Ideally, the co-polarization response of the calibrator is consistent. After the phase calibration, the covariance matrix is expressed as

$$\begin{aligned}
W_{hh}W_{hh}^* &= R_{hh}R_{hh}^* = S_{hh}S_{hh}^* \\
W_{hv}W_{hv}^* &= R_{hv}R_{hv}^* = R_{vh}R_{vh}^* = |f|^2 S_{hv}S_{hv}^* \\
W_{vv}W_{vv}^* &= R_{vv}R_{vv}^* = |f|^4 R_{vv}R_{vv}^* \\
W_{hh}W_{hv}^* &= R_{hh}R_{hv}^* \exp\{-i(\phi_T + \phi_A)\} = |f|^2 S_{hh}S_{hv}^* \\
W_{hh}W_{vh}^* &= R_{hh}R_{vh}^* \exp\{-i(\phi_R + \phi_A)\} = |f|^2 S_{hh}S_{vh}^* \\
W_{hh}W_{vv}^* &= R_{hh}R_{vv}^* \exp\{-i(\phi_T + 2\phi_A + \phi_R)\} = |f|^2 S_{hh}S_{vv}^* \\
W_{hv}W_{vh}^* &= R_{hv}R_{vh}^* \exp\{-i(\phi_R - \phi_T)\} = |f|^2 S_{hv}S_{vh}^* \\
W_{hv}W_{vv}^* &= R_{hv}R_{vv}^* \exp\{-i(\phi_R + \phi_A)\} = |f|^3 S_{hv}S_{vv}^* \\
W_{vh}W_{vv}^* &= R_{vh}R_{vv}^* \exp\{-i(\phi_T + \phi_A)\} = |f|^3 S_{vh}S_{vv}^*
\end{aligned} \tag{6.241}$$

The amplitude imbalance of co-polarization channel can be measured by means of extracting the HH polarization and the VV polarization image intensity of the small rectangular area in the center of the reflector. Typically, the small rectangular area energy is superimposed to obtain the total power of the corner reflector point response. This method is insensitive to processor focus performance compared to using the point target response peak method. For the co-polarization channel imbalance, the absolute value of the total power is not important. What is important is the relative value of the HH and VV images.

The mean square root of the HH and VV image power measurement ratio is a function of the incidence angle and is used to derive a calibration curve for the image. This curve characterizes the value $f(\gamma)$ in Eq. (6.215) or $f_r(\gamma)$ in Eq. (6.198). Use this curve to calculate the co-polarization amplitude correction value for each range line. Applying the correction value to the image, the relative correction of all polarization channels can be completed.

6.4.6.4 Absolute Radiometric Calibration

Absolute radiometric calibration is a comparison of the measured image radar cross-sectional area RCS with a theoretical model. According to the radar equation, the received power of the scattering area A is

$$P_r = \frac{P_t G_t(\gamma)}{4\pi R^2} A\sigma_0 \frac{\lambda^2 G_r(\gamma)}{(4\pi R)^2} \tag{6.242}$$

where

$$\begin{aligned} P_t &= \text{transmitting power}, \\ \lambda &= \text{radar wavelength}, \\ R &= \text{scattering area range}, \\ G_t(\gamma) &= \text{antenna transmitting gain at angle } \gamma, \\ G_r(\gamma) &= \text{antenna receiving gain at angle } \gamma, \\ \gamma &= \text{radar look angle } \gamma, \text{ and} \\ \sigma_0 &= \text{normalized radar cross - sectional area RCS for area } A, \end{aligned} \tag{6.243}$$

where A is the floor area of the scattering region.

Since the radar signal is sampled in the time domain, it is equivalent to the sight line (slant range) direction. When calculating the scattering area in SAR calibration, it is usually approximated as follows:

$$A = \frac{\delta_r \delta_a}{\sin \eta} \tag{6.244}$$

where δ_r and δ_a are the pixel distances of the slant range (vertical track) and the azimuth (along track). The scattering area is determined by the pixel distance, is the distance sampled by the radar signal, and is not the resolution of the radar.

Rewriting (6.242), we can get

$$\sigma_0 = K(R, \gamma) / A \tag{6.245}$$

where

$$K(R, \gamma) = \frac{P_r}{P_t} \frac{(4\pi)^3 R^4}{\lambda^2 G_t(\gamma) G_r(\gamma)} \tag{6.246}$$

In addition to the transmitting power and receiving power, it is necessary to estimate the parameters R and γ to determine $K(R, \gamma)$. At first, the slant range and look angle of each line on a fixed range line in the data can be estimated. Then the antenna pattern gain of this line can be determined from the look angle, and the $K(R, \gamma)$ of each line can be determined by combining other parameters of formula (6.246).

Finally, the scattering area A needs to be removed. In most postprocessing of SAR, the plane or the earth spherical assumption is still used to approximate the area A, and as is formula (6.244). If the actual digital topographic map is available, the actual scattering area can be estimated more accurately, resulting in more accurate calibration.

Even if the antenna gain pattern and system parameter information are well understood, the best absolute calibration result is obtained by placing an external calibrator in the imaging scene to correct the entire image. The radar cross-sectional area of the triangular reflector is

$$\sigma(\theta, \phi) = \frac{4\pi l^4}{\lambda^2}\left[\cos\theta + \sin\theta(\sin\phi + \cos\phi) - \frac{2}{\cos\theta + \sin\theta(\sin\phi + \cos\phi)}\right]^2 \tag{6.247}$$

where θ is the angle between the boresight axis of the triangular reflector and the vertical direction, and ϕ is the angle between the projection of boresight axis in horizontal plane and the straight edge of the bottom surface. The maximum value of the triangular reflector response is at $\theta = 54.74°$, $\phi = 45°$. The half power response is approximately in a 40° cone centered on the boresight. Therefore, when the reflector is placed at an incident angle of 54.74, the reflected energy of the radar signal is the maximum.

Comparing the theoretical value of the reflector with the measured image reflector signal, the absolute calibration constant of the radar image is obtained. Note that the radar image is given by the normalized radar cross-sectional area (i.e., power divided by the scattering area). In order to obtain higher measurement accuracy, usually, as many reflectors as possible are placed in the calibration field. The measured values of each reflector are averaged to obtain the final calibration constant. The normalized radar cross-sectional area for each pixel in the image is corrected with this value, thus ensuring that the average value of the reflector is the corrected theoretical value.

For most SAR processing, the calibration is now completed.

In the actual calibration, the influence of the terrain on the scattering area, the influence of the terrain on the antenna pattern correction, and the Faraday rotation effect should also be considered. (1) Influence of terrain on scattering area: The errors caused by the inclination angle of the surface, the incidence angle, and the scattering area are mainly considered. (2) Influence of terrain on antenna pattern correction: In order to eliminate the influence of antenna pattern in radiation calibration, it is necessary to estimate the position of the pixel in the antenna beam, and the terrain influence errors of different polarizations may be different. Sometimes the effects of terrain can cause large calibration errors. (3) Faraday polarization rotation effect: When linearly polarimetric electromagnetic waves pass through the ionosphere, measuring the scattering matrix produces a large error. For low-frequency radar systems, the estimation and elimination of Faraday rotation is an important issue. These are the problems that need to be solved one by one after the polarization calibration.

6.5 Applicable Conditions and Accuracy Analysis of Polarimetric Synthetic Aperture Radar External Calibration Algorithm

6.5.1 Applicable Conditions and Accuracy Analysis of Mixed Point Target and Distribution Target Calibration Algorithm

The main research here is the applicable conditions and accuracy problem of the van Zyl algorithm in Sect. 6.4.1.

6.5.1.1 Applicable Conditions and Accuracy Analysis of Phase Calibration Algorithm

The phase calibration in the van Zyl algorithm is to obtain phase calibration parameters using slightly rough surface signal echoes. It is generally considered that the VV and HH echoes of such regions are in the same phase. This method is much simpler than laying many manual calibrators in the calibration field.

The small perturbation model [19] is used to analyze the scattering of slightly rough surfaces [22]. It can be considered that this surface is horizontal, which is expressed as

$$z = \zeta(x, y) \tag{6.248}$$

That is, it is assumed that such a surface can be described as a two-dimensional random process. The small perturbation model has the following assumptions: (1) in the sense of root mean square, the surface displacement h is small relative to the electromagnetic wavelength λ and (2) the average slope of the surface is small, i.e.,

$$kh < 0.3 \tag{6.249}$$

$$\sqrt{2}\frac{h}{l} < 0.3 \tag{6.250}$$

where $k = 2\pi/\lambda$ is the electromagnetic wave number and l is the surface correlation length.

The ratio of the HH to the VV component of the electric field in the first-order approximation obtained by the small perturbation model is

$$\frac{E_{\mathrm{HH}}}{E_{\mathrm{VV}}}=\frac{\alpha_{\mathrm{HH}}}{\alpha_{\mathrm{VV}}} \tag{6.251}$$

$\alpha_{\mathrm{HH}}, \alpha_{\mathrm{VV}}$ are called the scattering function, in the case of backscattering

$$\alpha_{\mathrm{HH}}=\frac{\varepsilon-1}{(\cos\theta_i+b)^2},\qquad \alpha_{\mathrm{VV}}=\frac{(\varepsilon-1)(\varepsilon\sin^2\theta_i+b^2)}{(\varepsilon\cos\theta_i+b)^2} \tag{6.252}$$

where $b=\left(\varepsilon-\sin^2\theta_i\right)^{1/2}$, θ_i is the incident angle, ε is the complex relative dielectric constant, which can be expressed as $\varepsilon=\varepsilon'-\mathrm{j}\varepsilon''=\varepsilon'-\mathrm{j}60\lambda\sigma$, where σ is the dielectric conductivity.

It can be seen from Eqs. (6.251) and (6.252) that for complex dielectric constants, the co-polarization electric fields E_{HH} and E_{VV} are different in phase unless $\theta_i=0$.

Equation (6.251) can be written as

$$\frac{E_{\mathrm{HH}}}{E_{\mathrm{VV}}}=\left|\frac{E_{\mathrm{HH}}}{E_{\mathrm{VV}}}\right|\exp(\mathrm{j}\phi) \tag{6.253}$$

where φ is the phase difference between the electric fields E_{HH} and E_{VV}.

There have been many experimental studies abroad to measure the values of various ground objects at various incident angles at various frequencies from (6.251) to (6.253). Table 6.4 shows the dielectric constant and conductivity of the five types of surface at four common frequencies [22, 23], and Fig. 6.15 shows the corresponding HH-VV polarization phase difference curve.

It can be seen from Fig. 6.15 that the phase difference of HH-VV increases with the increase of the incident angle in the four graphs. In general, the phase difference of HH-VV increases with the increase of frequency. The drier the soil in all frequency ranges, the smaller the HH-VV phase difference. The phase difference of seawater HH-VV at all frequencies is greater than the phase difference of freshwater HH-VV, which indicates that the phase difference of HH-VV will change with the change of

Table 6.4 Dielectric constants and conductivity of five types of surface at four frequencies

Frequency/GHz	0.44			1.25			5.3			15		
	ε'	σ	ε''	ε'	σ	ε''	ε'	σ	ε''	ε'	σ	ε''
A: seawater	73.4	2.93	120	73.1	3.24	46.6	67.6	7.32	29	44.1	31.2	37.4
B: freshwater	80	0.046	1.9	79.7	0.38	5.45	73.5	6.24	21.2	47.5	31.1	37.3
C: wet soil	30	0.055	2.2	30	0.49	7	18	1.5	5.1	7	6	7.2
D: dry soil	15	0.01	0.4	15	0.88	3	14	0.8	2.7	7	3	3.6
E: very dry soil	3	10^{-4}	0.004	3	2×10^{-4}	0.003	3	0.015	0.05	3	0.1	0.12

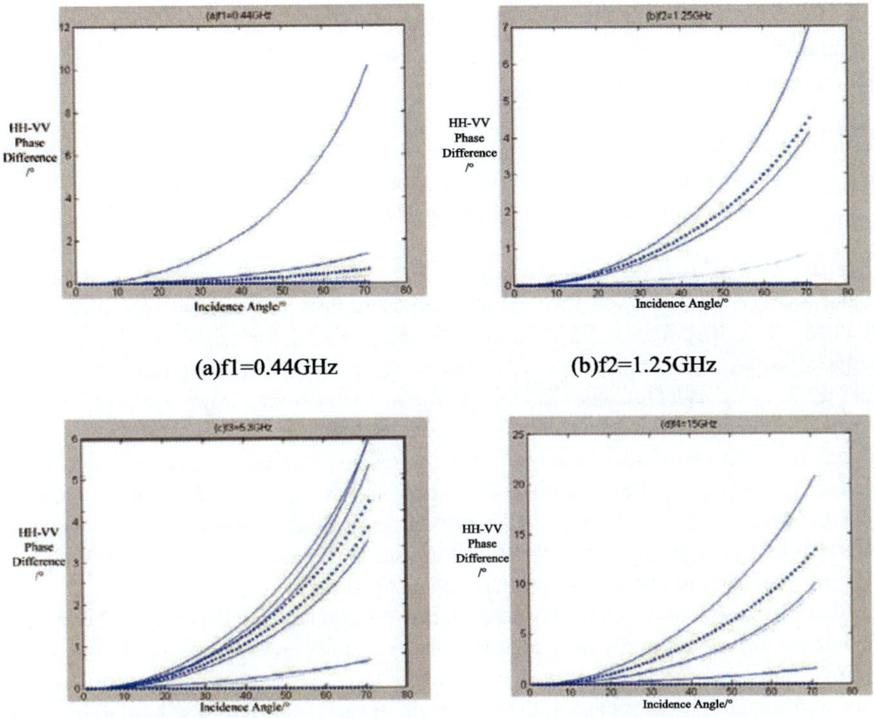

(a)f1=0.44GHz (b)f2=1.25GHz

(c)f3=5.3GHz (d)f4=15GHz

Fig. 6.15 HH-VV phase difference of five types of surface at four frequencies. *Note* The order of the curves in Fig. 6.15 (from the top to the bottom) are: **a** A, C, D, B, E. **b** A, D, C, B, E. **c** C, A, D, B, E. **d** C, D, A, B, E. The A, B, C, D, and E are listed in Table 6.5

water salinity. When the incident angle is less than 40°, the phase difference of the five surface HH-VVs is not more than 2.5° except for the wet and dry soil at 15 GHz. For typical wet and dry soils without vegetation cover, the incident angle is less than 50°, at three frequencies (0.44, 1.25, and 5.3 GHz), the HH-VV phase difference is less than 3°, the incident angle is less than 70°, and the HH-VV phase difference is less than 6°.

Based on these studies and experiments, for small incident angles (near normal incidence), a zero phase difference assumption is acceptable. However, the phase difference increases with factors such as frequency, incidence angle, salinity, and humidity. Therefore, dry land without vegetation cover should be used as the calibration target, especially for larger incidence angles. Marine surface calibration should not be used unless it is close to normal incidence.

Reference [18] points out that this algorithm requires that the channel coupling of the system must be negligible. Therefore, the ratio of the cross-polarization component to the co-polarization component of the feature backscatter must be much larger than the cross-polarization isolation of the system transmitting channel or receiving channel, for example, $\frac{|S_{vh}|^2}{|S_{vv}|^2} >> \frac{|T_{vh}|^2}{|T_{vv}|^2}$, etc. For most ground objects, channel isolation more than 20 dB can meet the requirements.

6.5.1.2 Applicable Conditions and Accuracy Analysis of Crosstalk Calibration and Removal Algorithms

The crosstalk calibration method of van Zyl algorithm does not need artificial targets, and can be widely applied to polarization image crosstalk calibration, increasing the reliability of polarization data and reducing the performance requirements of radar systems. The core is that, it is assumed that the co-polarization component and the cross-polarization component of the ground object target scattering matrix are not related to each other. The argument is that the cross-polarization scatter signal is mirror-symmetrical with respect to the projection axis of the first-order dipole with respect to the same co-polarization scatter signal caused by the scattering (second-order term) of the first-order perturbation field. Therefore, if the target is isotropically distributed in the azimuthal direction, there is no correlation between its cross-polarization and co-polarized scattering signal.

The correlation between the co-polarization component and the cross-polarization component of the ground object scattering matrix may cause the crosstalk parameter estimation to be incorrect, and it is worthwhile to study the distortion degree of the cross-polarization power caused thereby. Reference [17] provides a good method of research and has obtained valuable conclusions. If the co-polarization component of the ground object is not correlated with the cross-polarization component, i.e., $\langle S_{hh}S_{hv}^*\rangle = 0$ and $\langle S_{vv}S_{hv}^*\rangle = 0$, this calibration method is very accurate. When the crosstalk is less than $|\delta| = 0.28$ (-11 dB), the cross-polarization power is less than 0.2 dB after calibration. This accuracy is greatly improved over uncalibrated data.

When $\langle S_{hh}S_{hv}^*\rangle$ and $\langle S_{vv}S_{hv}^*\rangle$ are not equal to zero, that is, there is a correlation between the co-polarization and the cross-polarization component. When the crosstalk is lower than -14 dB, the cross-polarization power accuracy of all detected ground objects is approximately independent of the crosstalk size.

If the cross-polarization signal is required to be accurate (less than 0.1 dB) after calibration using the van Zyl algorithm, the scattering characteristics of the ground object only need to satisfy that the normalized cross-polarization correlation coefficient between the co-polarization and cross- polarization scattered signals is less than -10 dB. The normalized cross-polarization correlation coefficient can be expressed as

$$\left|\left\langle S_{co}S_{cross}^*\right\rangle\right| / \sqrt{\left\langle S_{hv}S_{hv}^*\right\rangle\left\langle S_{co}S_{co}^*\right\rangle} \tag{6.254}$$

where $\langle S_{co}S_{cross}^*\rangle = \langle S_{vv}S_{hv}^*\rangle$ or $\langle S_{hh}S_{hv}^*\rangle$.

Reference [17] gives the specific conditions that such calibration algorithms should meet to achieve a certain accuracy, and the conclusions may not be suitable for all practical situations. However, it gives an effective method to study the effectiveness and accuracy of this algorithm.

6.5.1.3 Applicable Conditions and Accuracy Analysis of Relative Amplitude and Absolute Amplitude Calibration Algorithms Between Channels

The calibration algorithm for the relative amplitude and absolute amplitude between channels uses the point target calibration method. For the applicable conditions and accuracy analysis, refer to Sect. 6.5.3.

6.5.2 Accuracy Analysis of the Distribution Target Calibration Algorithm

The use of distributed target calibration is the development trend of calibration algorithms; however, the biggest difficulty it faces is the lack of a distribution target that accurately knows its scattering characteristics.

The scattering characteristics of ground objects have always been an important research topic in the field of microwave remote sensing. With the increasing number of microwave remote sensing data, the scattering properties of ground objects have been understated in deeply. In reference [24], the Mueller matrix of the distribution target is used to derive the probability density function of the co-polarization phase difference ($\phi_{\text{copol}} = \phi_{\text{hh}} - \phi_{\text{vv}}$) and cross-polarization phase difference ($\phi_{\text{cross}} = \phi_{\text{vh}} - \phi_{\text{vv}}$) when the antenna illumination surface size is much larger than the ground object correlation length. It only be the function of the ground object correlation degree (α) and polarization phase difference (ζ).

$$f(\phi) = \frac{1-\alpha^2}{2\pi[1-\alpha^2\cos^2(\phi-\zeta)]} \bullet \left\{1 + \frac{\alpha\cos(\phi-\zeta)}{\sqrt{1-\alpha^2\cos^2(\phi-\zeta)}}\left[\frac{\pi}{2} + \arctan\frac{\alpha\cos(\phi-\zeta)}{\sqrt{1-\alpha^2\cos^2(\phi-\zeta)}}\right]\right\} \tag{6.255}$$

where $\phi = \phi_{\text{cross}}$ or ϕ_{copol}

$$\alpha = \sqrt{\frac{\lambda_{13}^2 + \lambda_{14}^2}{\lambda_{11}\lambda_{33}}},\ \zeta = \arctan\frac{\lambda_{14}}{\lambda_{13}}$$

$$\lambda_{ij} = \lambda_{ji} = \langle X_i X_j \rangle \quad i, j \in [1, 2, 3, 4]$$

When seeking ϕ_{copol}, define: $S_{\text{vv}} = X_1 + \text{i}X_2\,,\ S_{\text{hh}} = X_3 + \text{i}X_4$.
When seeking ϕ_{cross}, define: $S_{\text{vv}} = X_1 + \text{i}X_2\,,\ S_{\text{vh}} = X_3 + \text{i}X_4$.
For Mueller matrix:

$$\mathbf{L} = \begin{pmatrix} L_{11} & L_{12} & L_{13} & L_{14} \\ L_{21} & L_{22} & L_{23} & L_{24} \\ L_{31} & L_{32} & L_{33} & L_{34} \\ L_{41} & L_{42} & L_{43} & L_{44} \end{pmatrix}$$
$$= \begin{pmatrix} |S_{\text{vv}}|^2 & |S_{\text{vh}}|^2 & \text{Re}[S_{\text{vh}}^* S_{\text{vv}}] & -\text{Im}[S_{\text{vh}}^* S_{\text{vv}}] \\ |S_{\text{hv}}|^2 & |S_{\text{hh}}|^2 & \text{Re}[S_{\text{hh}}^* S_{\text{hv}}] & -\text{Im}[S_{\text{hh}}^* S_{\text{hv}}] \\ 2\text{Re}[S_{\text{vv}} S_{\text{hv}}^*] & 2\text{Re}[S_{\text{hv}} S_{\text{hh}}^*] & \text{Re}[S_{\text{vv}} S_{\text{hh}}^* + S_{\text{hv}} S_{\text{vh}}^*] & -\text{Im}[S_{\text{vv}} S_{\text{hh}}^* - S_{\text{hv}} S_{\text{vh}}^*] \\ 2\text{Im}[S_{\text{vv}} S_{\text{hv}}^*] & 2\text{Im}[S_{\text{hv}} S_{\text{hh}}^*] & \text{Im}[S_{\text{vv}} S_{\text{hh}}^* + S_{\text{hv}} S_{\text{vh}}^*] & \text{Re}[S_{\text{vv}} S_{\text{hh}}^* - S_{\text{hv}} S_{\text{vh}}^*] \end{pmatrix}$$

When seeking ϕ_{copol}: $\lambda_{11} = L_{11}/2$, $\lambda_{33} = L_{22}/2$, $\lambda_{13} = (L_{33} + L_{44})/4$, $\lambda_{14} = (L_{34} - L_{43})/4$.

When seeking ϕ_{cross}: $\lambda_{11} = L_{11}/2$, $\lambda_{33} = L_{12}/2$, $\lambda_{13} = L_{13}/2$, $\lambda_{14} = L_{14}/2$.

The range of the ground object correlation degree α is [0, 1]. The range of polarization phase ζ is $[-\pi, \pi]$, which is where the maximum probability density function appears.

Reference [12] gives a method that microwave scatterometer measures the scattering properties of ground objects, which can measure the polarization scattering characteristics of ground objects with high precision.

The polarization scattering data with high-precision distributed target can be applied to the calibration of multi-polarization SAR as an ideal calibration target. Due to the known nature of the calibration target, it is easy to determine the calibration accuracy. And the calibration algorithm is more adaptable, and can avoid the accuracy uncertainty of the van Zyl method.

6.5.3 *Accuracy Analysis of Point Target Calibration Algorithm*

In the calibration problem, the methods for estimating the target backscattering coefficient σ^0 mainly include the integral method and the peak estimation method [25]. The applicable conditions and accuracy analysis of the two methods are given below.

6.5.3.1 Integration Method

Considering the calibration problem of a uniform region A_{nu} with an unknown backscattering coefficient σ^0 in SAR image, this region contains N_{nu} pixels. The point target used for calibration is located in area A_{pnu} and contains N_{pnu} pixels. This region contains all the energy of the point target reflection cross section σ, the receiver noise, and uniform surface reflection signal with a backscatter coefficient of σ^0 (unknown). When the transmitter is turned off, the received data corresponds to the area A_n and contains N_n pixels. The incident angles corresponding to the regions A_{nu}, A_{pnu}, and A_n are the same. The integrated power of the three regions is defined as E_{nu}, E_{pnu}, and E_n, respectively.

If the backscattered signal, phase noise, and receiver noise of the uniform surface are not correlated with each other, for the uniform unknown σ^0 calibration region, the following equation is established:

$$E_u = E_{nu} - E_n(A_{nu}/A_n) \tag{6.256}$$

$$E_p = E_{pnu} - E_{nu}(A_{pnu}/A_{nu}) \tag{6.257}$$

In Eq. (6.256), E_u represents the uniform ground surface backscattered energy. In Eq. (6.257), E_p represents the total energy of the point target. From the above two formulas, we can get

$$\sigma^0 = \frac{\sigma(E_u/A_u)}{E_p} = \frac{(\sigma/A_{nu})[E_{nu} - E_n(A_{nu}/A_n)]}{E_{pnu} - E_{nu}(A_{pnu}/A_{nu})} \tag{6.258}$$

The ground surface area A in the above formulas can be calculated according to the number of pixels in the region of the image, image sampling interval in range δ_r, image sampling interval in azimuth δ_a, and the incident angle θ.

For digital image data, energy integration can be estimated by simple sum.

$$E_p = \left(\sum_{N_{pnu}}^{\text{Region}A_{pnu}} a_{ij}^2 - \frac{N_{pnu}}{N_{nu}} \sum_{N_{nu}}^{\text{Region}A_{nu}} a_{ij}^2 \right) A_{pnu} \tag{6.259}$$

Here a_{ij}^2 is the image power corresponding to the pixel ij.

Region A_{nu} normalized uniform surface backscatter power is

$$\overline{E_u} = \frac{E_u}{N_{nu}} = \left(\sum_{N_{nu}}^{\text{Region}A_{nu}} a_{ij}^2 - \frac{N_{nu}}{N_n} \sum_{N_n}^{\text{Region}A_n} a_{ij}^2 \right) \frac{\delta_r \delta_a}{\sin\theta} \tag{6.260}$$

Define

$$C_{\rm i} = E_{\rm p}/\bar{E}_{\rm u} \tag{6.261}$$

Get

$$\sigma^0 = \frac{\sigma \sin\theta}{\delta_{\rm r}\delta_{\rm a}C_{\rm i}} \tag{6.262}$$

All parameters in the formula can be known accurately or can be obtained from high-precision data.

6.5.3.2 Peak Estimation Method

In contrast to the integration method described above, the image range and azimuth resolution are taken into account when using the peak estimate of the calibration point target reflector. The point target impulse response shape and peak value are related to the matching degree of the chirp signal and the weighted reference function. When the focus is good, an expression of σ^0 similar to Eq. (6.262) can be got

$$\sigma^0 = \frac{\sigma \sin\theta}{\rho_{\rm r}\rho_{\rm a}C_{\rm p}} \tag{6.263}$$

where $C_{\rm p}$ is the ratio of the point target peak estimate (the value after noise and clutter removal) to the adjacent average clutter (the estimated clutter after noise removal), and $\rho_{\rm a}$ and $\rho_{\rm r}$ are the azimuth and range resolution, respectively.

The following discusses the accuracy analysis of the integral method and peak estimation method for point target calibration without considering noise.

For the integration method, the relative mean square error is

$$\varepsilon_{\rm I}^2 = \frac{\mathrm{var}\{\bar{E}_{\rm I}\}}{\mathrm{E}\{\bar{E}_{\rm I}\}^2} = \frac{1}{r_{\rm s}^2}\frac{A}{\Delta A_{\rm s}} + 2\frac{1}{r_{\rm s}} \tag{6.264}$$

where A is the integral area, A is required to be large enough, and $\Delta A_{\rm s}$ is the equivalent area of the clutter related region.

$$\Delta A_{\rm s} = \int \frac{\phi_{\rm s}(r)}{\phi_{\rm s}(0)}\mathrm{d}A \tag{6.265}$$

where $\varphi_{\rm s}(r)$ is the clutter autocorrelation function

$$\phi_{\rm s}(r) = \sum_{i,j} \left|R_{CiCj}\right|^2 \tag{6.266}$$

$r_{\rm s}$ is the ratio of the point target energy to the product of the average clutter power and its area.

$$r_s = \frac{E}{P_{\sigma^0} \Delta A_s} = \frac{\sigma \sin \theta}{\sigma^0 \Delta A_s} \tag{6.267}$$

θ is the incident angle corresponding to the integral region, assuming that the entire integration region is unchanged.

For the peak estimation method, the relative mean square error is

$$\varepsilon_P^2 = \frac{\mathrm{var}\{\bar{E}_P\}}{\mathrm{E}\{\bar{E}_P\}^2} = \frac{1}{(\beta_s r_s)^2} + 2\frac{1}{\beta_s r_s} \tag{6.268}$$

The area of the system resolution unit is

$$\Delta A_r = \int \frac{P(r)}{P(0)} \mathrm{d}A \tag{6.269}$$

where $P(r)$ is the system impulse response function.

The definition of β_s in Eq. (6.268) is

$$\beta_s = \frac{\Delta A_s}{\Delta A_r} \tag{6.270}$$

where ΔA_s is as defined in Eq. (6.265).

System defocus, noise, and non-uniform random distribution of surface clutter have an impact on both methods. In point target calibration, the peak estimation method and the integral method are both unbiased estimates of the point target energy, the difference is that the former needs to know the equivalent rectangular resolution of the system. The peak estimation method is sensitive to small errors in system focus, especially for systems with large time-band products. For systems with good focus, the root mean square error of the peak estimation method is always smaller than the integral method. But when the system is gradually out of focus, the root mean square error of the integral method has the characteristics that are independent of the system focus. When the signal clutter ratio is large, the uniform or non-uniform random distribution characteristics of the clutter have no effect on the root mean square error of the two methods. The effect of noise on the two methods is the same as the clutter.

When the SAR system is well focused and the impulse response is known accurately, a peak estimation method is recommended, especially when the system signal-to-noise ratio is small, because the peak estimation method has a smaller root mean square error. However, when the SAR system is not well focused, or the impulse response is not known accurately, or the system has a nonlinear phase error, it is recommended to use the integration method because it does not require a detailed known system impulse response form, and its root mean square error is independent of system focus situation, integral region size, and processor characteristics.

6.6 Summary

In the near future, the problems faced by SAR calibration (including multi-polarization SAR calibration) include the following:

(1) Develop low-cost, stable, high-gain radar transponders for external calibration, because the visibility of passive corner reflectors is low for some low-resolution spaceborne SAR.
(2) Research on phase calibration of multi-polarization SAR using internal calibration method.
(3) Because after spaceborne and airborne SARs are calibrated separately, and after SAR and scatterometer are calibrated separately, the satisfactory consistent measurements still cannot be obtained, or the data obtained by various measuring instruments still is impossible to compare, so that the mutual calibration technology of spaceborne and airborne SARs and the mutual calibration technology of SAR and scatterometer should be studied continuously.
(4) The calibration technique of SAR using phased array antennas needs further research, because the radiation pattern of this antenna may change with the scene and must be corrected.

With the development of SAR systems, more and more SAR data are obtained, and the backscattering mechanism of ground objects is also being understood more and more. The distribution targets with known backscattering behavior will also be more and more applied to the calibration algorithm of SAR. With the advancement of calibration devices and calibration technology, the calibration requirements for SAR are also increasing, which further expands the application of SAR data. Finally, due to the emergence of a large number of calibrated SAR data, the microwave remote sensing community will learn how to use these data and how to apply calibration accuracy to the accuracy analysis of the physical parameters acquired from SAR data.

References

1. J.C. Curlander, R.N. Mcdonough, *Synthetic Aperture Radar Systems and Signal Processing* (Wiley, 1991)
2. A. Freeman, SAR calibration: an overview. IEEE Trans. Geosci. Remote Sens. **30**, 1107–1121 (1992)
3. A. Freeman, The need for SAR calibration, in *IEEE Proceedings of IGARSS'89* (1989), pp. 230–233
4. A. Freeman, Y. Shen, C. Werner, Polarimetric SAR calibration experiment using active radar calibrators. IEEE Trans. Geosci. Remote Sens. **28**(2), 224–240 (1990)
5. T.I. Lukowski, R.K. Hawkins, C. Cloutier et al., RADARSAT elevation antenna pattern determination. IEEE Trans. Geosci. Remote Sens. **3**(3), 1382–1384 (1997)
6. J.J. van Zyl, Calibration of polarimetric radar image using only image parameters and trihedral corner reflector responses. IEEE Trans. Geosci. Remote Sens. **28**(3), 337–348 (1990)

7. H.A. Zebker, Y. Lou, Phase calibration of imaging radar polarimeter stokes matrices. IEEE Trans. Geosci. Remote Sens. **28**(2), 246–252 (1990)
8. H.A. Zebker, J.J. van Zyl, S.L. Durden et al., Calibrated imaging radar polarimetry: technique, examples, and applications. IEEE Trans. Geosci. Remote Sens. **29**(6), 942–961 (1991)
9. A. Freeman, J.J. van Zyl, J.D. Klein, H.A. Zebker, Y. Shen, Calibration of stokes and scattering matrix format polarimetric SAR data. IEEE Trans. Geosci. Remote Sens. **29**, 531–539 (1992)
10. S. Quegan, A unified algorithm for phase and cross-talk calibration of polarimetric data-theory and observation. IEEE Trans. GRS **32**(1), 89–99 (1994)
11. K. Sarabandi, Calibration of a polarimetric synthetic aperture radar using a known distributed target. IEEE Trans. Geosci. Remote Sens. **32**(3), 575–582 (1994)
12. K. Sarabandi, O. Yisok, F.T. Ulaby, Measurement and calibration of differential Mueller matrix of distributed targets. IEEE Trans. Antenna Propag. **40**(12), 1524–1532 (1992)
13. T.L. Ainsworth, J.S. Lee, *A New Method for a Posteriori Polarimetric SAR Calibration*
14. M.W. Whitt, F.T. Ulaby, P. Polatin et al., A general polarimetric radar calibration technique. IEEE Trans. Antennas Propag. **39**(1), 62–67 (1991)
15. J. Van Zyl, Y. Kim, *Synthetic Aperture Radar Polarimetry* (Jet Propulsion Laboratory, California Institute of Technology, 2010)
16. H.A. Zebker, J.J. van Zyl, Imaging radar polarimetry: a review. Proc. IEEE **79**(11), 1583–1606 (1991)
17. R.A. Cordey, On the accuracy of crosstalk calibration of polarimetric SAR using natural clutter statistics. IEEE Trans. Geosci. Remote Sens. **31**(2), 447–454 (1993)
18. D.R. Sheen, A. Freeman, E.S. Kasischke, Phase calibration of polarimetric radar images. IEEE Trans. Geosci. Remote Sens. **27**(6), 719–731 (1989)
19. F.T. Ulaby, C. Elachi, *Radar Polarimetry for Geoscience Applications*, Chapters 3, 4, 6 (Artech House Inc., Boston, London, 1990)
20. Q. Li, Z. Guan, F. Bai, *Principles of Numerical Calculation*, 1st ed. (Tsinghua University Press, Beijing, 2000)
21. J.M. Orteja, W.C. Reinborn, *Iterative Solution for Multivariate Nonlinear Equations*, 1st ed. (Zhu Jina, trans.) (Science Press, Beijing, 1983)
22. A. Guissard, Phase calibration of polarimetric radars from slightly rough surfaces. IEEE Trans Geosci. Remote Sens. **32**(3), 712–715 (1994)
23. M. Urabi, J. Feng, *Microwave Remote Sensing (Vol. 2)—Radar Remote Sensing and Surface Target Scattering and Radiation Theory* (Huang Peikang, Wang Yifei, trans.) (Science Press, Beijing, 1987)
24. K. Sarabandi, Derivation of phase statistics from the Mueller matrix. Radio Sci. **27**(5), 553–560 (1992)
25. A.L. Gray, P.W. Vachon, C.E. Livingstone et al., Synthetic aperture radar calibration using reference reflectors. IEEE Trans. Geosci. Remote Sens. **28**(3), 374–383 (1990)

Chapter 7
Analysis of Polarimetric Synthetic Aperture Radar Image

7.1 Basic Analysis Methods for Polarimetric Synthetic Aperture Radar Image

This chapter illustrates the basic analysis methods of fully polarimetric SAR data using the example data from the San Francisco Region, which is collected by the JPL Lab CV-990 Airborne SAR system. The data mainly covers three types of ground object, including ocean, city, and vegetation, in addition to the metal scatters of Golden Gate Bridge. Figure 7.1 is a schematic diagram of the terrain distribution covered of the data. In the figure, the wavy line area on the left is the Pacific Ocean, the wavy line area on the top is the Golden Gate Bay, and the large area on the right bottom is the San Francisco City Area. The rectangular vegetation area in the middle of the city is the Golden Gate Park. The thick black line in the picture is the famous Golden Gate Bridge. The total imaging area of this data is about 100 km^2. The flight direction of the aircraft is from left to right, and the radar is right side-looking mode.

The three main ground objects included in Fig. 7.1 have different polarization scattering characteristics, which are highly representative. The basic analysis methods of polarimetric SAR data and images are described below from different aspects.

7.1.1 *Figures of Total Polarization Power (Span) and Basic Polarization Combination*

The total power of polarization scattering represents the total scattering intensity of the polarization measurement, which is calculated by the span norm value of the target scattering matrix. Therefore, the polarization total power figure is also commonly called the span figure. The calculation method is

$$Span = |S_{\text{hh}}|^2 + |S_{\text{vv}}|^2 + 2|S_{\text{hv}}|^2 \tag{7.1}$$

R. Yang et al., *Polarimetric Microwave Imaging*,
https://doi.org/10.1007/978-981-15-8897-6_7

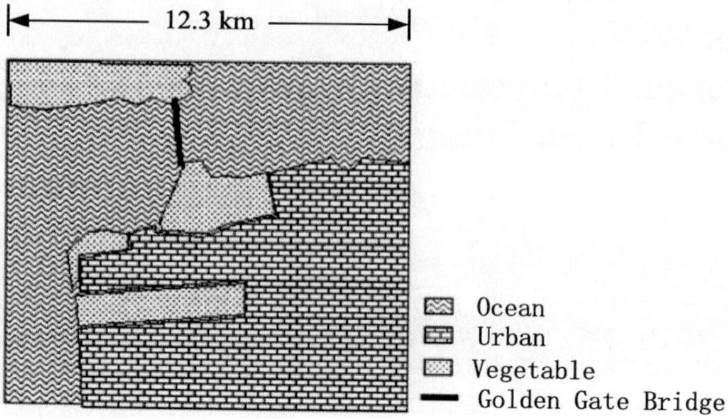

Fig. 7.1 Schematic distribution of terrain in the San Francisco area

It is assumed here that the target is reciprocal, i.e., $S_{hv} = S_{vh}$. Overall, the total power of polarization scattering reflects the electromagnetic scattering characteristics of the target. Figure 7.2 shows the total power of polarization in the San Francisco area. The gray level of the pixel is proportional to the intensity of the received power of the echo. The image resolution after four-look processing is about 40 m × 40 m.

After obtaining the target polarization scattering matrix, four basic polarization combinations can be obtained directly, namely, images of HH, HV, VH, and VV. For reciprocal scatterers, the HV echo is considered to be the same as the VH echo [1]. Figure 7.3 shows a basic polarization composite image of the San Francisco Bay Area. Each image is separately normalized, from which the scattering characteristics of different type targets can be analyzed under four basic polarization combinations.

Fig. 7.2 Polarized total power map for San Francisco Bay Area

(a) HH 极化 (b) HV(VH) 极化 (c) VV 极化

Fig. 7.3 Basic polarization combined image of the San Francisco Bay Area

7.1.2 Analysis Method Based on Polarization Synthesis

Polarization synthesis is a very efficient analytical tool for multi-polarized SAR data. From the scattering matrix of the target, only the target images under the four basic polarization combinations can be obtained. According to the Stokes matrix of the target, the radar image under arbitrary polarization combination can be obtained by polarization synthesis, which is necessary for comprehensive analysis of the target scattering characteristics. According to the calculation method of polarization synthesis, some images of specific polarization angles (polarized elliptical azimuth ψ and ellipticity angle χ) can be obtained, as shown in Fig. 7.4. The image in the upper row of the figure is the SAR image under different (ψ, χ) combinations of the co-polarization measurement. The image in the lower row of the figure is the SAR

(e)ψ=45° χ=0° (f)ψ=0° χ=45° (g)ψ=175° χ=0°

Fig. 7.4 Echo images of different polarization combinations in the San Francisco Bay Area image under different (ψ, χ) combinations in the cross-polarization measurement. The polarization angle value is shown in the text of the figure.

It can be seen from Fig. 7.4 that as the polarization state changes, the echo power of the urban area varies widely, while the brightness of vegetation and ocean regions changes less. The SAR image under arbitrary polarization combination can be obtained by polarization synthesis, and the echo power change trend of a certain object can be obtained intuitively [2]. However, it is usually only necessary to understand the trend of the ground echo power under different polarization combinations without requiring a large number of polarimetric SAR images. At this point, it is a bit cumbersome to analyze through the polarization synthesis image. So a more simple and intuitive method, the polarization map, can be used to analyze the trend of the polarization echo power of the ground target.

7.1.3 Analysis Method Based on Polarization Feature Map

According to the average Stokes matrix obtained from three typical feature areas, such as ocean, park vegetation, and city, their polarization characteristics can be obtained as shown in Fig. 7.5. The first column in the figure is the co-polarization characteristic map of the three targets, and the second column is the cross-polarization characteristic map. In addition, the third column is a normalized cross-polarization characteristic map obtained by normalizing the cross-polarization pattern to the co-polarization pattern. It can be seen that the cross-polarized echo power is much smaller than the echo power of the co-polarization [3].

The polarization map visually reflects the variation of the polarization power from different features, and provides a powerful tool for analyzing the fully polarized data. According to the polarization characteristic map, it can be determined that some type of targets have larger echo power under a certain polarization combination, and have smaller echo power under other certain polarization combination. Then the variation trend of echo power can be judged. The trend of power changes. In this way, the polarization angle that maximizes the difference between the echo powers of the two ground objects can be determined from the polarization map. By obtaining the image under the polarization angle combination by the polarization synthesis method, the discrimination ability of the two targets can be enhanced. This is a common method of polarimetric SAR image enhancement. It will be discussed in detail in Chap. 9.

In addition, since the polarization map represents the polarization scattering characteristics of the object object, it can also be used to determine the type of target. By comparing the measured polarization characteristic map of a certain target with the theoretical analysis polarization characteristic map of the scattering model, it can be determined the type of scattering mechanism of the measured target. Thus the ground targets can be roughly classified (Fig. 7.6).

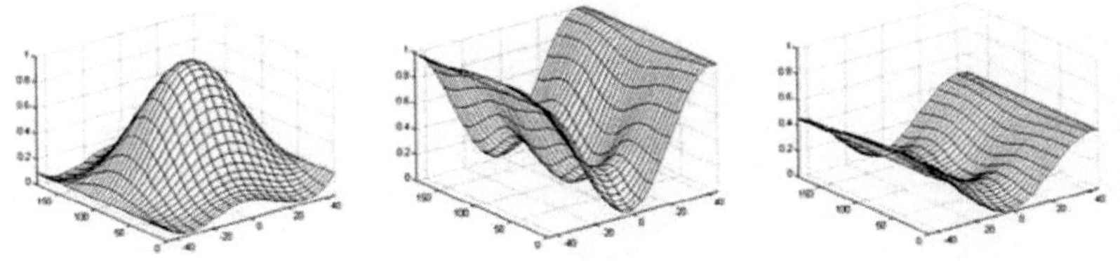

Ocean area

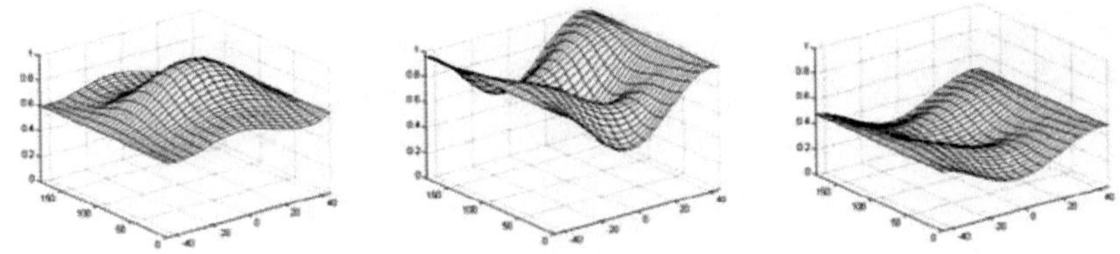

Park vegetation area

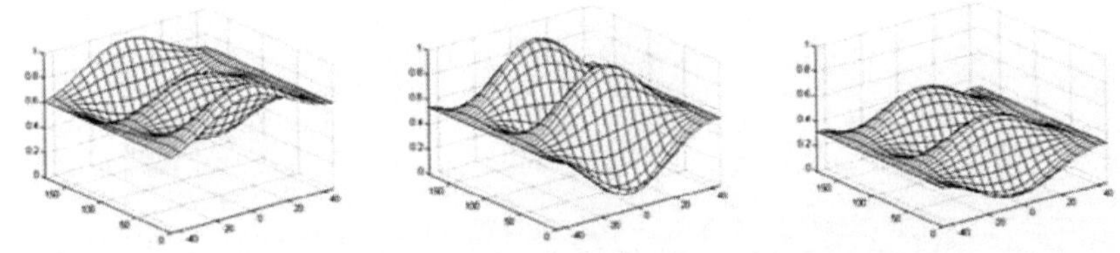

Urban area

(a) Co-polarization map (b) Cross-polarization map (c) Normalized cross-polarization map

Fig. 7.5 Polarization characteristics of three typical features in the San Francisco Bay Area

7.1.4 Echo Power Ratio Under Different Polarization Combinations

By using the ratio of the echo power of the target in an arbitrary polarization state, the relative echo power varation of different ground targets under different polarization combinations can be analyzed. As a representative, Fig. 7.7 shows an image of the ratio of echo intensities in three basic linear polarization combinations. Figure 7.7(a) shows the ratio of the two polarization components, the VV to the HH polarization; Fig. 7.7(b) and (c) show the echo intensity ratio of the cross-polarization HV and the co-polarization HH or VV. The gray value of the pixel in the figure is proportional to the ratio of the echo power form the two polarizations combination in the area.

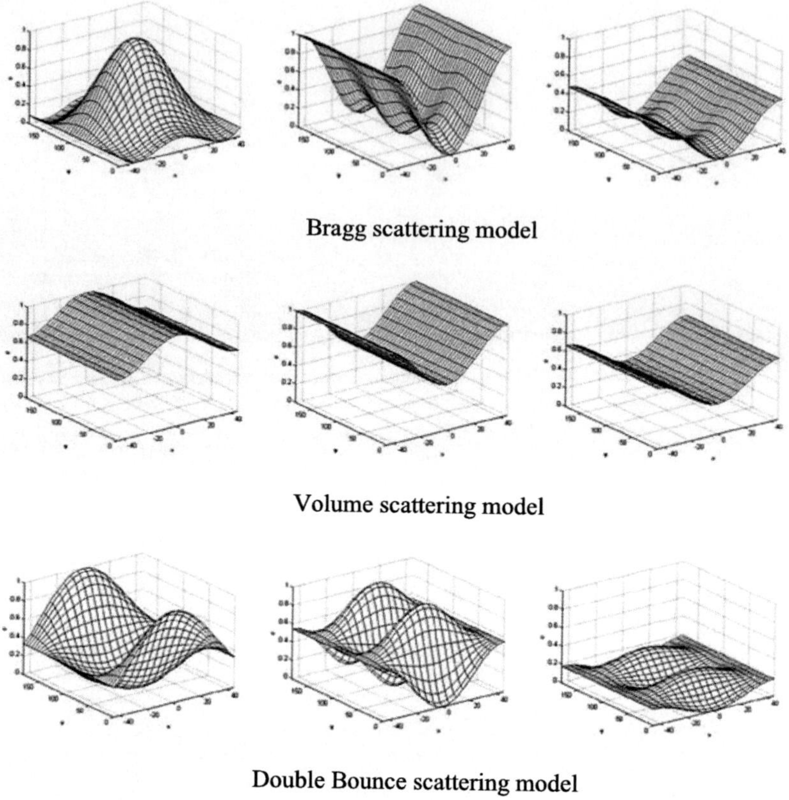

(a) Co-polarization map (b) Cross-polarization map (c) Normalized cross-polarization map

Fig. 7.6 Theoretical polarization characteristics of three typical ground scattering models

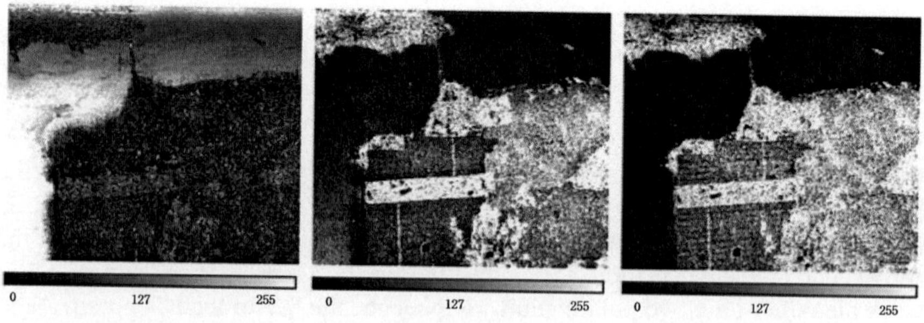

(a) VV/HH (b) HV/HH (c) HV/VV

Fig. 7.7 Ratio of echo intensity under three basic polarization combinations

From Fig. 7.7, the echo power form different parts of the ocean area are different under various polarization combinations. In the deeper Pacific region, the echo power in the VV polarization mode is much larger than that in the HH polarization mode. While in the shallower San Francisco Bay, the echo power has a larger echo in the HH polarization mode. The ocean region has the lowest echo power value under the HV polarization combination. The region containing vegetation has the highest intensity in the HV polarization, and the echo power under the VV polarization combination is slightly larger than that of HH polarization combination.

By constructing echo power ratio images under other different polarization combinations, more abundant information about the scattering characteristics from ground targets can be obtained [2–4], which will not be repeated here.

7.1.5 Polarization Purity Analysis of Echo

Any kind of polarized electromagnetic wave can be decomposed into the sum of fully polarized electromagnetic waves and completely unpolarized electromagnetic waves. Usually, the target echo received by the radar is not an ideal fully polarized wave, but a partially polarized electromagnetic wave containing a completely unpolarized component. Therefore, the fully polarized component and the completely unpolarized component can be decomposed from the measured target polarization data to analyze the polarization purity of targets' echo [5].

Figure 7.8 shows the polarization characteristics of the fully polarized component and the completely unpolarized component from three typical feature targets in the San Francisco Bay Area under the same polarization. The left column of the figure shows the polarization characteristics of the total echo from the three objects. The middle column is the polarization characteristic of the fully polarized component, and the right column is the polarization characteristic of the three completely unpolarized components. It can be seen that the completely unpolarized component of the echo from the ocean region accounts for a very small proportion of the total echo power, so it can be considered that the ocean echo is a fully polarized wave. The fully unpolarized component of the echo from the urban region is about half power of the fully polarized component. The fully polarized component of the echo from the park vegetation area is roughly equal to the completely unpolarized component.

The polarization scattering type of the ocean surface is a relatively single Bragg scattering model, so the measured polarization purity is the highest. The urban area is mainly composed of dihedral scatterers located in different directions, but it also contains other scatterers such as vegetation. So the polarization purity is low. In comparison, the scattering mechanism of the vegetation area is the most complicated. It includes both the Bragg scattering mechanism from the top of the tree and the volume scattering mechanism from the branches and leaves of the tree. It also includes the double-bounce scattering mechanism between the trunk and the ground. Therefore, the measured polarization data has the lowest purity.

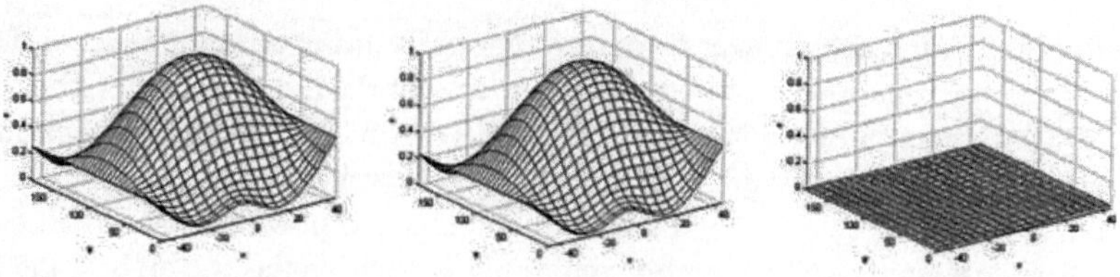

Ocean Region

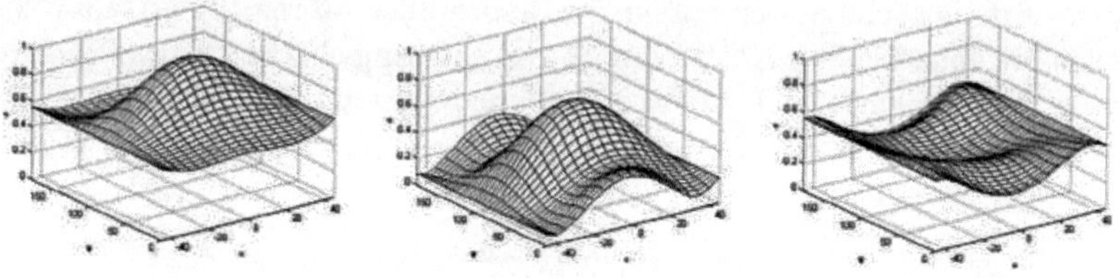

Vegetaion Region

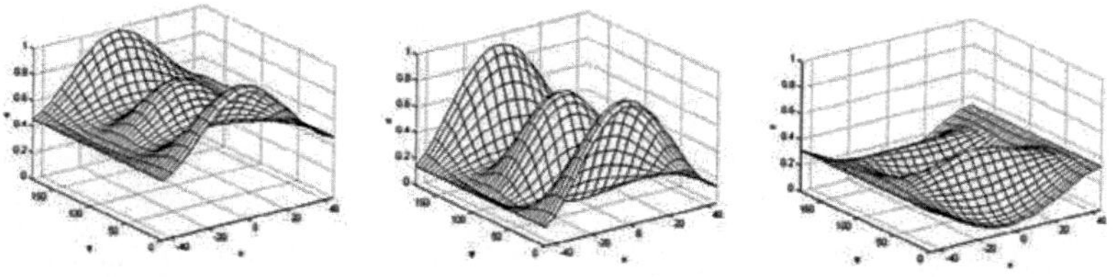

Urban Region

(a) the total polarized characteristic map (b) the fully polarized characteristic map (c) the completely unpolarized characteristic map

Fig. 7.8 Characteristic maps of different polarization components from three typical features in the San Francisco Bay Area

It is important to note that completely unpolarized echoes do not contain any polarization information and are undesirable components. Therefore, according to the polarization characteristic map of the different polarization components form the target, it is also possible to determine which polarization combination is the largest ratio of the fully polarized component to the completely unpolarized component from the target echo. Thereby we can obtain the relative higer purity image to facilitate the extraction and analysis of polarization information.

7.1.6 *Power Fluctuation Coefficient and Partial Polarization Coefficient*

The power fluctuation coefficient was first introduced by van Zyl et al. in 1987 [3] to measure the magnitude of echo non-uniformity caused by scatterers with different characteristics distributed in the same resolution unit. The power fluctuation coefficient is defined as

$$\nu_p = P_{min}/P_{max} \tag{7.2}$$

where P_{max} and P_{min} represent the maximum and minimum values of radar received power, respectively.

The smaller the value of the power fluctuation coefficient, the larger the variation of the backscattered power received by the radar under different trans-receiving polarization states. Therefore, it can be judged according to the power fluctuation coefficient whether the magnitude of the echo power from a certain type target can be effectively increased or decreased. If $\nu_p = 1$, the echo power from the target will not change under any polarization combination, which means the echo is a completely unpolarized wave. If $\nu_p = 0$, the echo power may be zero, which means the fully polarized component of echo power is high.

The partial polarization coefficient F_p was introduced by Zebker et al. [4]. This parameter is used to measure the polarization purity of the radar echo. Partial polarization coefficient is defined as

$$F_p = \frac{P_{max} - P_{min}}{P_{max} + P_{min}} \tag{7.3}$$

where P_{max} and P_{min} represent the maximum and minimum values of radar received power, respectively.

Similar to the power fluctuation coefficient, the partial polarization coefficient also indicates the degree by enhancing a certain type target echo to distinguish other targets in the image. The relationship between the partial polarization coefficient and the power fluctuation coefficient is

$$F_p = \frac{1 - \nu_p}{1 + \nu_p} \tag{7.4}$$

Figure 7.9 shows the power fluctuation coefficient and partial polarization coefficient of each resolution unit in the San Francisco Bay Area. For convenience of observation, the resulting coefficients are represented as bitmaps of 256 levels of gray.

In Fig. 7.9(a), the ocean area has the smallest average power fluctuation coefficient (0.052). The urban area average power fluctuation coefficient (0.133) is second, and the vegetation area average power fluctuation coefficient (0.729) is the largest. This

(a) Power fluctuation coefficient v_p (b) Partial polarization coefficient F_p

Fig. 7.9 Power fluctuation coefficient map and partial polarization coefficient map of the San Francisco Bay Area

also shows that the ocean region is composed of a relatively single polarization scatter model, followed by urban regions. The scattering mechanism of vegetation regions is the most complex. The large power fluctuation coefficient in the vegetation area means that the variation of polarization state angle by polarization synthesis does not significantly change the echo power of the region. This is consistent with the results of the previous analysis.

In the partial polarization coefficient map of Fig. 7.9b, the average partial polarization coefficient in the ocean area is 0.975. The average partial polarization coefficient in the city is second, which is 0.794. And the average partial polarization coefficient of the vegetation area is 0.418. This shows that the non-polar component in the vegetation area accounts for a much larger proportion than the other two types of ground targets, so its polarization purity is the lowest.

It should be particularly pointed out that it can be clearly seen from Fig. 7.9a and b that both the power fluctuation coefficient and the partial polarization coefficient are related to the angle between the urban street and the radar illumination beam. According to the city map of San Francisco, the city street in the left half of Fig. 7.9 is perpendicular to the radar beam, while the street on the right is at an acute angle to the beam. It can be seen that when the street is perpendicular to the illumination beam, the power fluctuation coefficient is small and the partial polarization coefficient is large. This indicates that the irregularity of the angle between the dihedral scatterer formed by the building and the illumination beam reduces the amount of polarization information measured.

7.2 Representation Method of Multi-frequency Multi-polarized Synthetic Aperture Radar Image

Modern Synthetic Aperture Radar is developing in the direction of multi-band multi-polarization measurement. Multi-frequency multi-polarized synthetic aperture radar can obtain more abundant ground target information, which is beneficial to analyze and understand ground targets. In general, SAR images can only provide intensity information for echo signals, but not phase information. Therefore, it must be represented by multiple images of different polarization channels to more accurately describe the richer information measured by multi-frequency multi-polarized SAR.

For multi-frequency multi-polarized SAR data, it is also possible to obtain a certain combination image by color synthesis, which can more intuitively analyze the influence of various parameters on the echo power. Since the human eye is much more sensitive to color images than grayscale images, grayscale images are often converted into color images to enhance image recognition. This image processing method is called pseudo color synthesis processing. For multi-frequency multi-polarized SAR images, their pseudo-color composite images can be obtained from different polarization channel data.

7.2.1 Color Synthesis of Multi-polarized Combined Images

The images of the three basic polarization combinations are color-synthesized. Figure 7.10(a)–(c) shows the images of the three basic polarization combinations in the Pennsylvania region by the L band SIR-C/X SAR system. It can be seen that each image shows a different polarization characteristic. At the same time, it can be seen that the differences in the different images are very small and difficult to distinguish clearly. However, if the HH polarization image is set to blue, the HV polarization image is set to green, and the VV polarization image is set to red, a polarization pseudo color image of this region can be obtained. It is possible to more clearly distinguish small differences in the image, as shown in Fig. 7.10g. Comparing the actual topographic maps, it can be found that the forest area is mainly composed of green color. This is because the forest area is mainly composed of randomly distributed scatterers with branches, trunks and leaves. The HV polarization scattering power is much stronger than that of other echo combinations under the co-polarization combination. So the forest area is mainly green. By analyzing the color of the pixels in the image, it is easier to analyze the difference in polarization scattering characteristics between each resolution unit features.

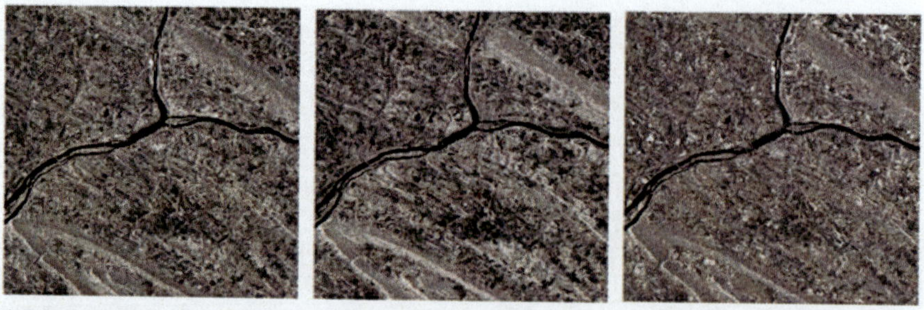

(a) L-band HH grayscale image (b) L-band HV grayscale image (c) L-band VV grayscale image

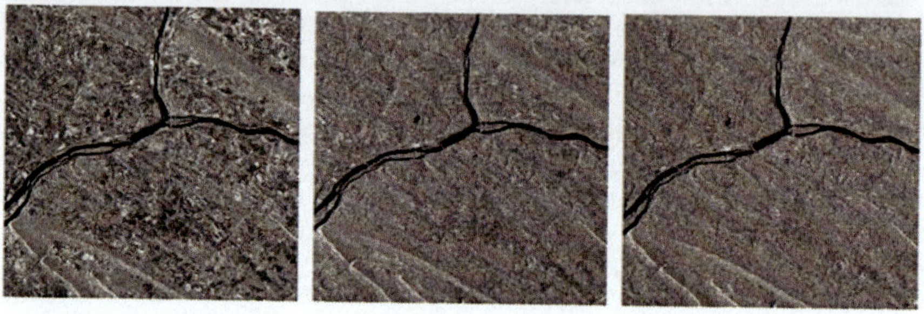

(d) L-band HH grayscale image (e) C-band HH grayscale image (f) X-band HH grayscale image

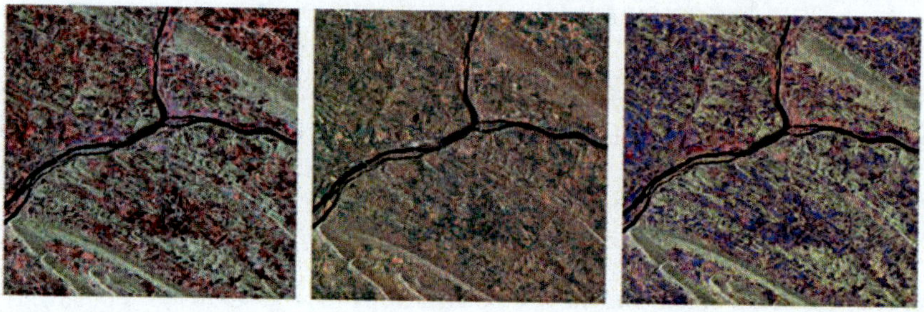

(g) Multi-polarization combined color synthesis map (h) Multi-band image color synthesis map (i)

Fig. 7.10 Color synthesis processing of multi-frequency multi-polarization SAR images

7.2.2 Color Synthesis of Multi-band Multi-polarized Synthetic Aperture Radar Images

The three grayscale images shown in Fig. 7.10(d)–(f) are the HH polarization images of the Pennsylvania region obtained in the L-band, C-band and X-band, respectively. It can be seen that only the L-band image has a large variation in gray level, which is because the wavelength of L-band is the longest. So it is most sensitive to the details of the ground target change among the three bands. A lot of targets are relatively rough relative to the C-band and X-band electromagnetic wavelengths, so SAR image details in these two bands are not as rich as in L-band images. Only the flat river region has low echo power in the three bands, and shows deep gray in the SAR image.

If the L-band image is set to red, the C-band image is set to green, and the X-band image is set to blue. The color composite map of the HH-polarized SAR image in different frequency bands can be obtained, as shown in Fig. 7.10(h). The yellower areas in the figure mean that echo powers from both C-band (green) and X-band (blue) are stronger. From the color composite image, we can distinguish the richer details than the single grayscale image.

In addition, it is also possible to color synthesize several SAR images in any frequency band and arbitrary polarization to facilitate image analysis and discrimination. Figure 7.10 (i) shows a color composite map of the L-band HH polarization (red), VV polarization (green), and the C-band VV polarization (blue) image. According to the principle of RGB, color synthesis usually requires three components to get a colorful image, which is easy to analyze and understand.

7.3 Correlation Analysis of Multi-polarized Synthetic Aperture Radar Images

Polarimetric Synthetic Aperture Radar (SAR) is a multi-channel system. It obtains the polarization scattering matrix of ground targets and other relevant information, and provides a powerful tool for understanding and analyzing the scattering properties of ground targets. For reciprocal media, the polarization scattering matrix has only three independent parameters, so there are three independent channels in polarization SAR, namely HH, HV and VV. That is, the basic parameters of polarimetric SAR measurements are the scattering amplitude and phase of HH, HV, and VV channel targets. In order to better understand the fully polarimetric SAR image, it is necessary to know the relationship between the measured data by the three independent channels, which requires analysis of the correlation among the images of polarimetric channels.

The correlation of polarimetric SAR images can be fully represented by the polarization complex covariance matrix $\mathbf{C}$ as follows:

$$\mathbf{C} = \mathbf{k}^{\mathrm{H}}\mathbf{k} = \begin{bmatrix} \sigma_1 & \rho_{12}\sqrt{\sigma_1\sigma_2} & \rho_{13}\sqrt{\sigma_1\sigma_3} \\ \rho_{12}\sqrt{\sigma_1\sigma_2} & \sigma_2 & \rho_{23}\sqrt{\sigma_2\sigma_3} \\ \rho_{13}\sqrt{\sigma_1\sigma_3} & \rho_{23}\sqrt{\sigma_2\sigma_3} & \sigma_3 \end{bmatrix} \tag{7.5}$$

where $\mathbf{k} = \begin{bmatrix} S_{\mathrm{hh}} & S_{\mathrm{hv}} & S_{\mathrm{vv}} \end{bmatrix}^{\mathrm{T}}$ and $\sigma_i = |\mathbf{k}(i)|^2, i = 1, 2, 3$.

$$\rho_{ij} = \frac{\langle \mathbf{k}(i)\mathbf{k}^*(j) \rangle}{\sqrt{\sigma_i\sigma_j}}, \ i, j = 1, 2, 3 \tag{7.6}$$

ρ_{ij} is the complex correlation coefficient between i th and j th channels in the system. The polarization covariance matrix contains all the information from radar polarization measurements. There are 9 independent parameters: three real backscatter coefficients, and the magnitude with phase of the three complex correlation coefficients. If the 9 parameters can be accurately estimated, any information on the target polarization measurement can be obtained.

Figures 7.11 and 7.12 show the results of the polarization covariance matrix parameter estimation of the L and C-band fully polarimetric SAR data in the Flevoland of Netherlands. For easy observation, the obtained values are quantized into grayscale images. This data was collected by the AIRSAR system of the JPL laboratory in August 1989 with four multi-look processing.

Subgraphs (a), (b), and (c) in Figs. 7.11 and 7.12 are HH, HV, and VV polarization images of L-band and C-band, respectively. Subgraphs (c), (d), and (e) are L are the

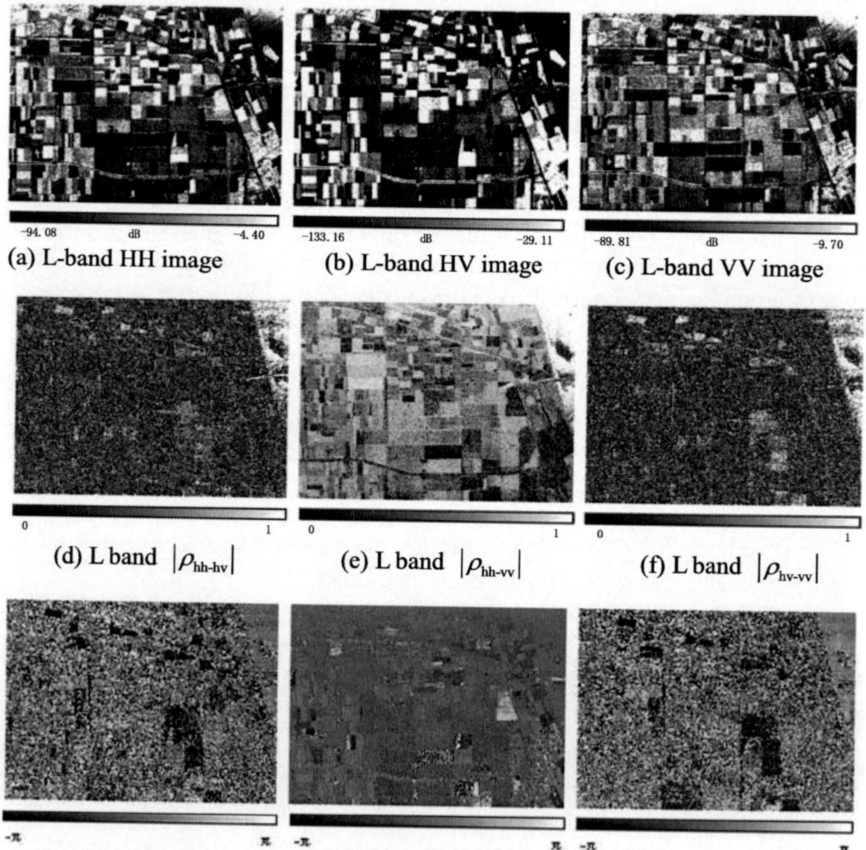

(a) L-band HH image (b) L-band HV image (c) L-band VV image

(d) L band $|\rho_{\text{hh-hv}}|$ (e) L band $|\rho_{\text{hh-vv}}|$ (f) L band $|\rho_{\text{hv-vv}}|$

(g) L band $\phi_{\text{hh-hv}}$ (h) L band $\phi_{\text{hh-vv}}$ (i) L band $\phi_{\text{hv-vv}}$

Fig. 7.11 L-band polarization covariance matrix parameters in Flevoland

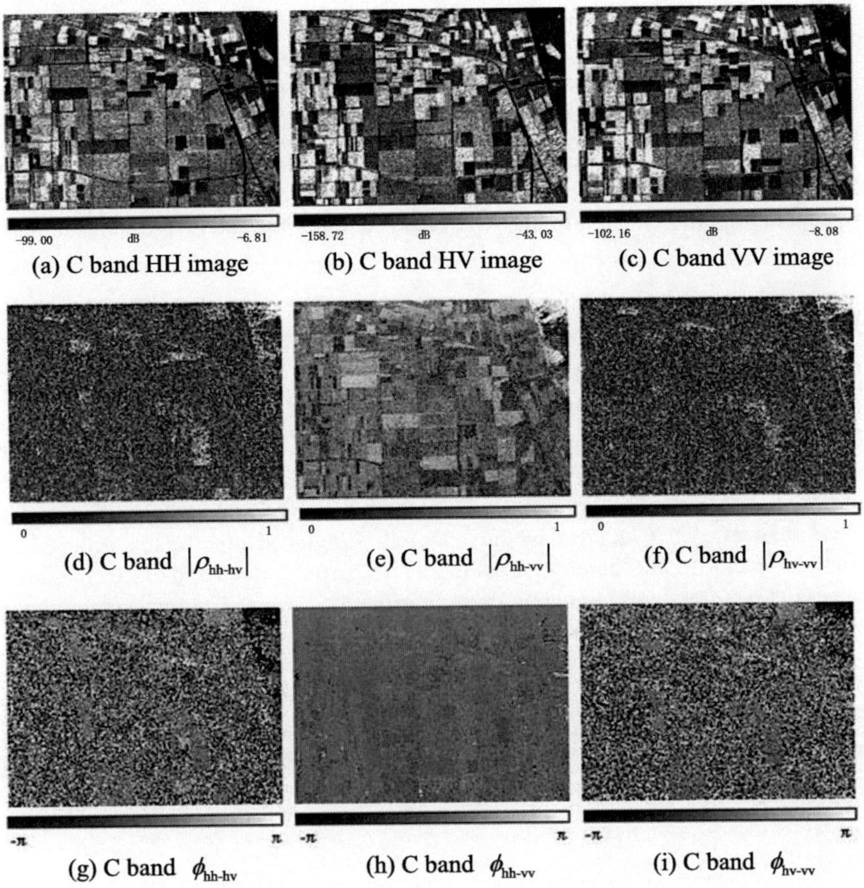

Fig. 7.12 C-band polarization covariance matrix parameters in Flevoland

amplitudes of correlation coefficients between the three independent channels of L-band and C-band, respectively. Subgraphs (f), (g), and (h) are the phase angles of correlation coefficients between the three independent channels of L-band and C-band, respectively. Analysis shows as follow.

(1) The dynamic range of cross-polarization (HV) images in two bands is larger than that of the co-polarization (HH or VV).
(2) The correlation coefficients (amplitude and phase) of co-polarizations contain more information, while correlation coefficients between the cross-polarization and the co-polarizations contain little information.
(3) When the amplitude of the correlation coefficient is large, the phase is relatively determined. While the amplitude of the correlation coefficient is small, the corresponding phase exhibits a noise-like property.

(4) The average value of the correlation coefficient between the cross-polarization channel and the co-polarization channel in the two frequency bands is very low. So it can be approximated that the cross-polarization channel and the co-polarization channel are independent.

In many aspects of polarization information processing, it is necessary to analyze the correlation characteristics of each channel in polarimetric SAR image, which is especially important in polarimetric SAR data compression and image enhancement.

7.4 Optimal Polarization Analysis of Multi-polarized Synthetic Aperture Radar Data

The fully polarized electromagnetic waves emitted by the radar are usually converted into partially polarized electromagnetic waves after being scattered by natural ground targets. In other words, the amount of polarization information contained in the electromagnetic wave received by the radar will be reduced after the scattering of the ground object [6–12]. Therefore, how to receive partially polarized electromagnetic waves in an optimal state to obtain more information becomes an important issue in multi-polarized SAR information processing.

This section focuses on the analysis of the optimal polarization of partially polarized electromagnetic waves. The polarimetric SAR image obtained by the optimization process contains the largest polarization information, and the specific target has the highest polarization. This helps to better understand the scattering mechanism of the target and to more accurately distinguish different types of features.

7.4.1 Solution of Target Optimal Polarization

The relationship between the Stokes vector $\mathbf{J}_\mathrm{t}$ of the radar-transmitted electromagnetic wave and the Stokes vector $\mathbf{J}_\mathrm{r}$ of the scattered electromagnetic wave received by the radar is established by the Mueller matrix.

$$\mathbf{J}_\mathrm{r} = \mathbf{M}\mathbf{J}_\mathrm{t} \tag{7.7}$$

where $\mathbf{M}$ is 4×4-dimensional Mueller matrix.

For convenience of presentation, let $\mathbf{J}_\mathrm{t} = \begin{bmatrix} T_0 & T_1 & T_2 & T_3 \end{bmatrix}$, and correspondingly let $\mathbf{J}_\mathrm{r} = \begin{bmatrix} R_0 & R_1 & R_2 & R_3 \end{bmatrix}$. The elements in the matrix $\mathbf{M}$ are denoted as $m_{ij}(i, j = 0, 1, 2, 3)$. Usually the emitted electromagnetic waves are fully polarized, and the elements of their Stokes vector $\mathbf{J}_\mathrm{t}$ satisfy the following relationship:

$$T_0^2 = T_1^2 + T_2^2 + T_3^2 \tag{7.8}$$

The received partially polarized electromagnetic wave $\mathbf{J}_r$ can be expressed as the sum of a fully polarized electromagnetic wave $\mathbf{J}_{pol}$ and a completely unpolarized electromagnetic wave $\mathbf{J}_{un}$.

$$\mathbf{J}_r = \mathbf{J}_{pol} + \mathbf{J}_{un} \tag{7.9}$$

where

$$\mathbf{J}_{pol} = \left[\sqrt{R_1^2 + R_2^2 + R_3^2}\ R_1\ R_2\ R_3 \right]^T \tag{7.10}$$

$$\mathbf{J}_{un} = \left[R_0 - \sqrt{R_1^2 + R_2^2 + R_3^2}\ 0\ 0\ 0 \right]^T \tag{7.11}$$

The degree of polarization in a partially polarized electromagnetic wave is defined as

$$p = \frac{\sqrt{R_1^2 + R_2^2 + R_3^2}}{R_0} \tag{7.12}$$

For convenience, the parameter symbols used here are different from Chap. 2, but their physical meanings are the same. The following process can calculate the extreme values of ground targets polarization degree and the corresponding polarization ellipse angles. Obviously, the maximum and minimum values can be obtained by calculating the degree of polarization under all polarization combinations. Then the corresponding polarization ellipse angles can be obtained. This calculation method is called the system overall calculation method. The specific calculation steps are as follows.

(1) Change the polarization ellipse angle (ψ_t, χ_t) of the emitted electromagnetic waves step by step within the valid value range. The azimuthal angle ψ_t of the polarization ellipse ranges from 0° to 180°, and the ellipticity angle χ_t ranges from −45° to 45°.
(2) Determine a scatter electromagnetic wave corresponding to each of the polarization states of the transmitted wave according to the Mueller matrix of the target. Calculate the corresponding polarization degree according to the formula (7.12).
(3) Determine the extreme value of the polarization degree and its corresponding polarization ellipse.

Although this system calculation method is simple, the calculation amount of the entire radar image is very surprising, even if the sampling interval of the polarization angle is selected as 1°. If the sampling interval is selected very large, the calculation accuracy is difficult to ensure. In order to obtain the extreme value of target polarization degree more efficiently and accurately, a mathematical analysis method is used to solve this problem.

The matrix of Eq. (7.7) is partitioned and rewritten as follows:

$$\begin{bmatrix} R_0 \\ \tilde{\mathbf{R}} \end{bmatrix} = \begin{bmatrix} m_{00} & \tilde{\mathbf{n}}^{\mathrm{T}} \\ \tilde{\mathbf{m}} & \tilde{\mathbf{M}} \end{bmatrix} \cdot \begin{bmatrix} T_0 \\ \tilde{\mathbf{T}} \end{bmatrix} \tag{7.13}$$

where $\tilde{\mathbf{R}}$, $\tilde{\mathbf{n}}$, $\tilde{\mathbf{m}}$, and $\tilde{\mathbf{T}}$ are all three-dimensional column vectors.

$$\begin{aligned} \tilde{\mathbf{R}} &= \begin{bmatrix} R_1 & R_2 & R_3 \end{bmatrix}^{\mathrm{T}} & \tilde{\mathbf{n}} &= \begin{bmatrix} m_{01} & m_{02} & m_{03} \end{bmatrix}^{\mathrm{T}} \\ \tilde{\mathbf{T}} &= \begin{bmatrix} T_1 & T_2 & T_3 \end{bmatrix}^{\mathrm{T}} & \tilde{\mathbf{m}} &= \begin{bmatrix} m_{10} & m_{20} & m_{30} \end{bmatrix}^{\mathrm{T}} \end{aligned} \tag{7.14}$$

$\tilde{\mathbf{M}}$ is a 3 × 3-dimensional Mueller block submatrix.

$$\tilde{\mathbf{M}} = \begin{bmatrix} m_{11} & m_{12} & m_{13} \\ m_{21} & m_{22} & m_{23} \\ m_{31} & m_{32} & m_{33} \end{bmatrix} \tag{7.15}$$

Thus, the degree of polarization can be expressed as

$$p = \frac{\sqrt{R_1^2 + R_2^2 + R_3^2}}{R_0} = \frac{\left\| \tilde{\mathbf{R}} \right\|}{R_0} \tag{7.16}$$

The goal of the analysis below is to get the extreme value of (7.16). Usually the emitted electromagnetic waves are fully polarized. So the following constraints apply.

$$T_0^2 = T_1^2 + T_2^2 + T_3^2 \tag{7.17}$$

where T_0 is the total power of the transmitted wave, which is a fixed value.

The total power of the scattered wave received by the radar is R_0, which is usually a function of the transmitted wave polarization state. In the expression of polarization degree p, $\left\| \tilde{\mathbf{R}} \right\|$ and R_0 is the variable quantity, which greatly increases the complexity of solving the extreme problem. After knowing Mueller matrix of the target, the relationship between transmitting and receiving electromagnetic waves is determined. Therefore, for the convenience of calculation, let R_0 unchanged to change T_0, as well as remain R_0/T_0 equal to the value in the system. Thus, the problem is reduced to calculate the extreme value of $\left\| \tilde{\mathbf{R}} \right\|$ under the constraint condition (7.17), which greatly reduces the complexity of mathematical analysis. From formula (7.13), we can get as follow:

$$F = \left\| \tilde{\mathbf{R}} \right\|^2 = T_0^2 \|\tilde{\mathbf{m}}\|^2 + 2T_0 \cdot \tilde{\mathbf{m}}^{\mathrm{T}} \tilde{\mathbf{M}} \tilde{\mathbf{T}} + \tilde{\mathbf{T}}^{\mathrm{T}} \tilde{\mathbf{G}} \tilde{\mathbf{T}} \tag{7.18}$$

In (7.18), $\tilde{\mathbf{G}} = \tilde{\mathbf{M}}^{\mathrm{T}} \cdot \tilde{\mathbf{M}}$. Constraint condition can be as follow:

$$R_0 = m_{00}T_0 + \tilde{\mathbf{n}}^{\mathrm{T}}\tilde{\mathbf{T}},\ R_0 \text{is a fixed value} \tag{7.19}$$

$$T_0^2 = T_1^2 + T_2^2 + T_3^2,\ T_0 \text{is a variate} \tag{7.20}$$

Substituting the first constraint variable (7.19) into Eq. (7.18). It is known that F is only related to three variables T_1,T_2, and T_3. Let $c = ||\tilde{\mathbf{m}}||^2$, $\tilde{\mathbf{b}} = [b_1\ b_2\ b_3] = \tilde{\mathbf{m}}^{\mathrm{T}} \cdot \tilde{\mathbf{M}}$, and $\tilde{\mathbf{B}} = [B_1\ B_2\ B_3]$, where $B_i = m_{00}b_i - cm_{0i}, i = 1, 2, 3$. F can be expressed in a more concise form.

$$F = \left(cR_0^2 + 2R_0\tilde{\mathbf{B}} \cdot \tilde{\mathbf{T}} + \tilde{\mathbf{T}}^{\mathrm{T}} \cdot \tilde{\mathbf{A}} \cdot \tilde{\mathbf{T}}\right)/m_{00}^2 \tag{7.21}$$

where $\tilde{\mathbf{A}}$ is a 3 × 3-dimensional symmetrical matrix, in which elements a_{ij} can be obtained by $\tilde{\mathbf{G}}$.

$$a_{ii} = m_{00}^2 g_{ii} - 2m_{00}m_{0i}b_i + cm_{0i}^2 \tag{7.22}$$

$$a_{ij} = -m_{00}m_{0i}b_j - m_{00}m_{0j}b_i + cm_{0i}m_{0j} + g_{ij}m_{00}^2 \quad i < j \tag{7.23}$$

$$a_{ij} = a_{ji} \quad j > i \tag{7.24}$$

Because $\tilde{\mathbf{A}}$ is a symmetric matrix, it can be diagonalized. Let $\lambda_i, i = 1, 2, 3$ to be the eigenvalue of $\tilde{\mathbf{A}}$. F can be expressed as follows:

$$F = \left(cR_0^2 + 2R_0 \sum\nolimits_i B_i\mathbf{T}_i + \sum\nolimits_i \lambda_i\mathbf{T}_i^2\right) \Big/ m_{00}^2 \tag{7.25}$$

The problem is then equivalent to the optimization of F under the following constraints:

$$\left(T_1^2 + T_2^2 + T_3^2\right) \cdot m_{00}^2 = \left(R_0 - \tilde{\mathbf{n}}^{\mathrm{T}}\tilde{\mathbf{T}}\right)^2 \tag{7.26}$$

The extreme values of F can be determined by the Lagrangian multiplier method.

$$L = F(\tilde{\mathbf{T}}) - \alpha \cdot \left[\left(T_1^2 + T_2^2 + T_3^2\right) \cdot m_{00}^2 - \left(R_0 - \tilde{\mathbf{n}}^{\mathrm{T}}\tilde{\mathbf{T}}\right)\right] \tag{7.27}$$

The above formula takes the partial derivative of T_1, T_2, and T_3 separately. And the following linear simultaneous equations are obtained:

$$\left(\lambda_1 + \alpha\left(m_{01}^2 - m_{00}^2\right)\right)T_1 + m_{01}m_{02}\alpha T_2 + \alpha m_{01}m_{03}T_3 = -R_0B_1 + \alpha R_0 m_{01} \tag{7.28}$$

$$m_{02}m_{01}\alpha T_1 + \left(\lambda_2 + \alpha\left(m_{02}^2 - m_{00}^2\right)\right)T_2 + \alpha m_{02}m_{03}T_3 = -R_0B_2 + \alpha R_0 m_{02} \tag{7.29}$$

$$m_{03}m_{01}\alpha T_1 + \alpha m_{03}m_{02}T_2 + \left(\lambda_3 + \alpha\left(m_{03}^2 - m_{00}^2\right)\right)T_3 = -R_0B_3 + \alpha R_0 m_{03} \tag{7.30}$$

Solving this simultaneous equation, you can represent T_1, T_2, and T_3 as a function of the parameters α, substituting them into the constraint (7.26) to get an equation for α. Solving this equation can obtain a set of α, and then find the corresponding F. Take the maximum and minimum values of F, and normalizing the extreme value to get the polarization degree. The optimal sub-vectors $\tilde{\mathbf{T}}_{opt}$ can be obtained from these two corresponding values α, which are used to calculate the corresponding transmit wave polarization angle and radar received power after normalizing.

Although the above mathematical analysis method is more complicated, it can greatly reduce the amount of calculation. Compared with the method of gradually searching in the two-dimensional polarization angle plane, the calculation time is saved about half. The advantages of this method are obvious when the polarimetric SAR image is large.

7.4.2 Extreme Value Analysis of Target Polarization

Take the full-polarization SAR data of San Francisco Bay Area as an example to illustrate the analysis method of the target optimal polarization degree. Figure 7.13 shows the total polarization power of the fully polarimetric SAR data for the region, and Fig. 7.14 shows the statistical histogram for Fig. 7.13. Figures 7.15 and 7.16 show the images generated by the maximum ($P_{\max}$) and minimum ($P_{\min}$) polarization degree

Fig. 7.13 Total power map of polarization in the San Francisco Bay Area

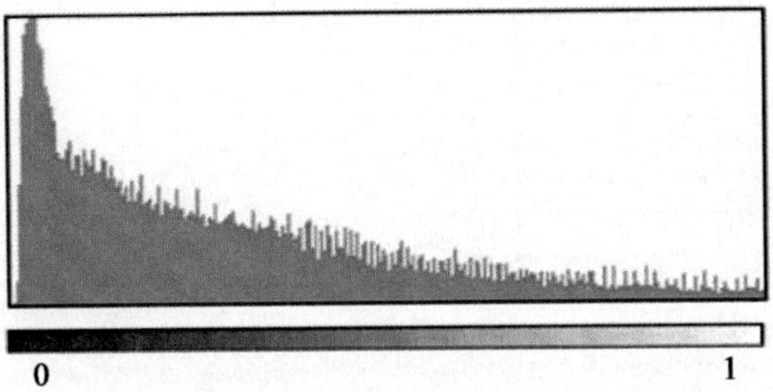

Fig. 7.14 Statistical histogram of the total power map of polarization

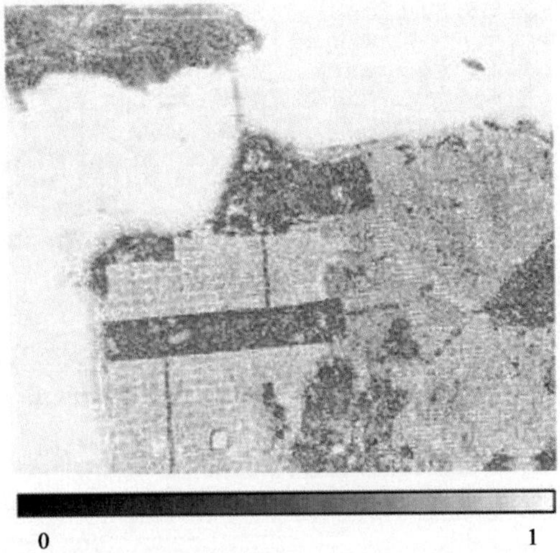

Fig. 7.15 San Francisco Bay Area maximum polarization degree map

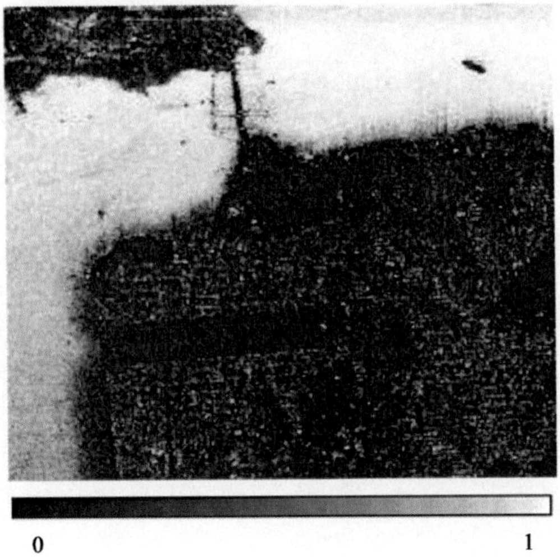

Fig. 7.16 San Francisco Bay Area minimum polarization degree map

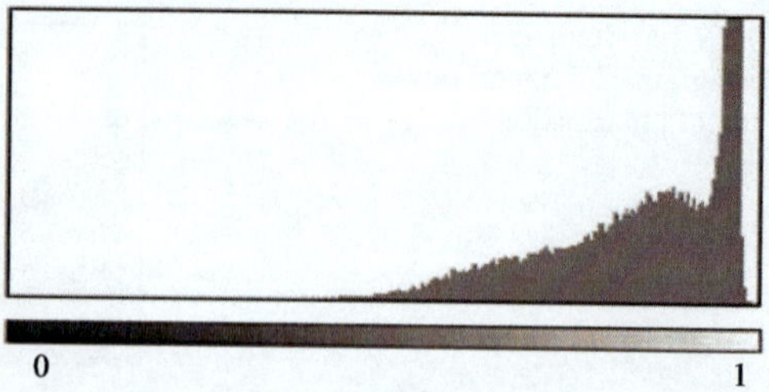

Fig. 7.17 Statistical histogram of the maximum polarization degree map

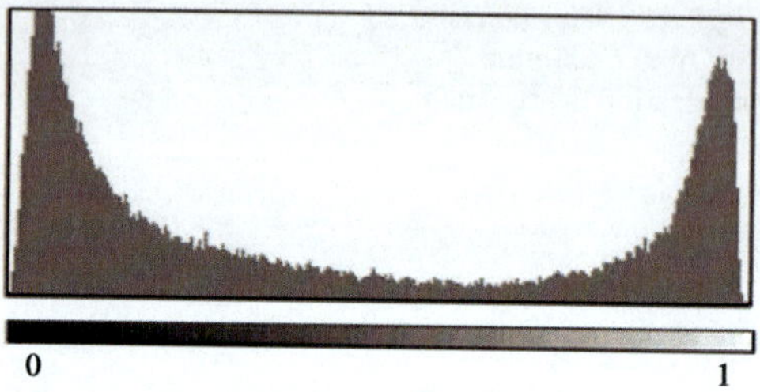

Fig. 7.18 Statistical histogram of the minimum polarization degree map

of the San Francisco Bay Area, respectively. The statistical histograms are shown in Figs. 7.17 and 7.18, respectively. Comparing the images and their histograms, you can notice the following content.

(1) The magnitude of the target polarization degree is independent of the target echo power received by the radar. Some targets with low received power, such as the ocean, have higher polarization degree values.
(2) The ocean region has a relatively high polarization degree value, and the ocean region exhibits a brighter gray value both in the maximum polarization image and in the minimum polarization image. The electromagnetic wave scattered on the ocean surface has the smallest proportion of completely non-polarized components and the highest polarization purity.
(3) The urban area has a relatively high value $P_{\max}$, which is numerically close to the ocean area. However, the value $P_{\min}$ of urban areas is very low. Therefore, it can be inferred that the polarization purity of the scattered electromagnetic waves from the urban area target can be significantly changed by using different polarization combination to send and receive electromagnetic waves.
(4) The polarization degree of parks and vegetation areas are very low, indicating that the proportion of completely unpolarized components in the scattered waves is the largest and the polarization purity is the lowest.

7.4.3 Dynamic Range Analysis of Target Polarization Degree

In order to further analyze the change of the polarization purity of the target scatter electromagnetic wave, the dynamic range parameter of the polarization degree is introduced, expressed as follow.

$$\Delta P = P_{\max} - P_{\min} \tag{7.31}$$

According to the dynamic range of polarization degree, the variation sensitivity of target scattering wave polarization purity with the change of radar transmission and reception polarization can be determined. The pixel dynamic range of polarization degree in the San Francisco Bay Area is shown in Figs. 7.19 and 7.20 is its statistical histogram.

As can be seen from the figure, the urban area has a high Δp value, while the Δp value of the ocean area is very small. This shows that the scattering echoes with higher relative polarization purity can be obtained by changing the polarization state of the electromagnetic waves transmitted and received by the radar. Thereby the information contained in the polarization information of the urban region can be increased. While the polarization purity of the ocean region is higher and the dynamic range is small, the ocean area has a high polarization information under any combination of polarization states.

Fig. 7.19 Dynamic range of polarization degree in the San Francisco Bay Area

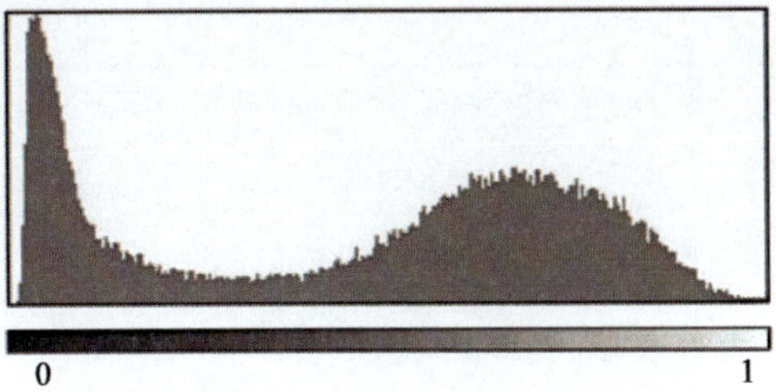

Fig. 7.20 Statistical histogram from the dynamic range of polarization degree

7.5 Summary

Polarimetric SAR images contain a wealth of target information, and effective understanding of polarimetric SAR images is an important topic in polarization information processing. This chapter analyzes the several aspects of polarization SAR image understanding, and gives the basic analysis methods of polarization SAR image with the optimal polarization degree analysis method, which lays a foundation for the subsequent polarization information processing content.

References

1. Chris Oliver, Shaun Quegan, *Understanding Synthetic Aperture Radar Images* (Artech House, Boston London, 1998)
2. Howard A. Zebker, J. van Zyl Jakob, Saniel N. Held, Imaging radar polarimetry from wave synthesis. J. Geophys. Res. **92**(B1), 683–701 (1987)
3. J.J. Van Zyl, H.A. Zebker, C. Elachi, Imaging radar polarization signatures: theory and observation. Radio Sci. **22**(4), 529–543 (1987)
4. Diane L. Evans, Tom G. Farr, J. van Zyl Jakob et al., Radar polarimetry: analysis tools and applications. IEEE Trans. GRS **26**(6), 774–789 (1988)
5. Haruto Hirosawa, Degree of polarization of radar backscatters from a mixed target. IEEE Trans. GRS **35**(2), 466–470 (1997)
6. D.L. Schuler, J.S. Lee, K.W. Hoppel, Polarmetric SAR image signatures of the Ocean and Gulf stream features. IEEE Trans. GRS **31**(6), 1210–1221 (1993)
7. Stephen L. Durden, J. van Zyl Jakob, Howard A. Zerker, Modeling and observation of the radar polarization signature of forested areas. IEEE Trans. GRS **27**(3), 290–301 (1989)
8. S.R. Cloude, J. Fortuny, J.M. Lopez-Sanchez et al., Wide-band polarimetric radar inversion studies for vegetation layers. IEEE Trans. GRS **37**(5) 430–2441 (1999)
9. Jakob J. van Zyl, The effect of topography on radar scattering from Vegetated areas. IEEE Trans. GRS **31**(1), 153–160 (1993)
10. Lee Jong-Sen, Hoppel Karl, Principal components transformation of multifrequency polarimetric SAR imagery. IEEE Trans. GRS **30**(4), 686–696 (1992)
11. P. Ferrazzoli, S. Paloscia, P. Pampaloni et al., The potential of multifrequency polarimetric SAR in assessing agricultural and arboreous biomass. IEEE Trans. GRS **35**(1), 5–16 (1997)
12. Ridha Touzi, Stephane Goze, Thuy Le Toan et al., Polarimetric discriminators for SAR images. IEEE Trans. GRS **30**(5), 973–980 (1992)

Chapter 8
Statistical Characteristics of Polarimetric SAR

8.1 Statistical Analysis of Single-Look SAR Images

Chapter 7 discusses the analysis methods of polarimetric SAR images, which focuses on the analysis and understanding of the target's fully polarimetric scatter information. This chapter analyzes the statistical properties of polarization measurement data, which can provide relevant prior knowledge to improve the effectiveness of practical applications. Before performing statistical analysis on multi-polarimetric SAR images, the statistical model of the single-look single-polarimetric SAR image is first analyzed.

For the actual distribution target, each resolution unit is composed of a lot of independent scatterers, as shown in Fig. 8.1.

In the process of interaction between electromagnetic waves and independent scattering elements, each scatter element has an effect on the amplitude and phase of the backscattered wave. For SAR, the signal received by the receiver at a certain moment is the vector sum of the backscattered signals from many distributed targets in the coverage area under the radar transmitting antenna. It can be expressed as follows:

$$R = A \cdot \mathrm{e}^{j\phi} = \sum_{k=1}^{N} A_k \cdot \mathrm{e}^{j\phi_k} \tag{8.1}$$

where A is the amplitude of the signal R received by the radar and its phase is ϕ. A_k is the signal amplitude of the kth scatter, and its instantaneous phase is ϕ_k. N is the total number of independent scatters in the resolution unit.

This summation process takes into account the weighting of the SAR impulse response function and any small effects of the scattering and propagation processes on the amplitude and phase of the signal. Equation (8.1) is only a theoretical analysis. In practice, the amplitude A_k and phase ϕ_k of each independent scatterer's echo signal cannot be measured. Because they are too small compared to the SAR resolution unit,

R. Yang et al., *Polarimetric Microwave Imaging*,
https://doi.org/10.1007/978-981-15-8897-6_8

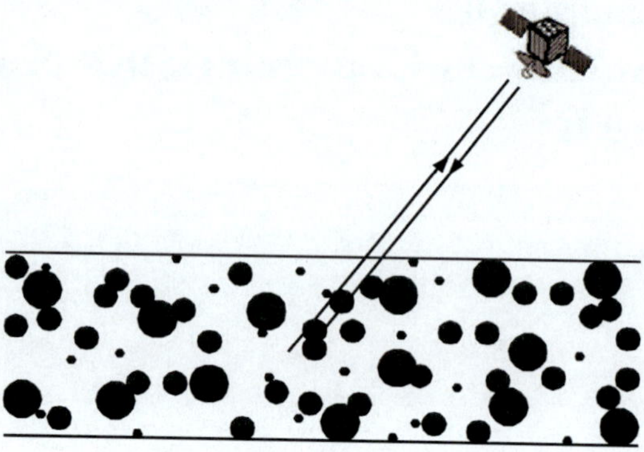

Fig. 8.1 Distributed target scattering diagram

and there are many such independent scatterers in each resolution unit. Therefore, the characteristics of SAR echo signals can only be studied by statistical analysis [1].

For scatterers with random phases, it is generally assumed that the phase ϕ_k of each independent scatterer echo signal obeys a uniform distribution between $[-\pi, \pi)$ and is independent of the amplitude A_k. When the number N of independent scattering elements is sufficiently large, according to the central limit theorem, the amplitudes A and phases ϕ of Eq. (8.1) are the random variables of statistically independent, and consistent with the Rayleigh distribution as well as uniformly distributed between $[-\pi, \pi)$. The probability density function of each feature quantity in the SAR image is given below:

1. Joint probability density function of I, Q signals

In order to save processor bandwidth, the echo signals received by SAR are usually divided into two signals as I and Q, which are $x_I = A\cos\phi$ and $x_Q = A\sin\phi$. According to the central limit theorem, they are independent and identically distributed Gaussian random variables with a mean of zero and a variance of $\sigma/2$. Its joint probability density function can be expressed as follows:

$$p_{x_I, x_Q}(x_I, x_Q) = \frac{1}{\pi\sigma}\exp\left(-\frac{x_I^2 + x_Q^2}{\sigma}\right) \tag{8.2}$$

2. Probability density function of the received signal phase

The phase ϕ of the received signal obeys the uniform distribution in the interval $[-\pi, \pi)$, and its probability density function is

$$p_\varphi(\varphi) = \frac{1}{2\Pi} \tag{8.3}$$

3. Probability density function of the received signal amplitude

The amplitude $A = \sqrt{x_1^2 + x_2^2}$ of the received signal is a random variable obeying the Rayleigh distribution, and its probability density function is

$$p_A(A) = \frac{2A}{\sigma}\exp\left(-\frac{A^2}{\sigma}\right),\ A \geq 0 \tag{8.4}$$

The Rayleigh distribution has a non-zero average $\frac{\sqrt{\pi\sigma}}{2}$, which can be regarded as the DC component of the signal envelope detected by the radar. The variance of this distribution is $\left(1 - \frac{\pi}{4}\right)\sigma$. In the statistical analysis of data, the coefficient of variation (CV) is also a commonly used statistic, which is defined as the ratio of the standard deviation to the average.

$$CV = \frac{s}{m} \tag{8.5}$$

where s is the standard deviation of the random variable and m is the average. Therefore, the variance coefficient of the received signal amplitude is

$$CV_A = \sqrt{4/\pi - 1} \tag{8.6}$$

4. Probability density function of received signal power

The power $I = A^2$ of the radar received signal obeys a negative exponential distribution, and its probability density function is

$$p_I(I) = \frac{1}{\sigma}\exp\left(-\frac{I}{\sigma}\right), I \geq 0 \tag{8.7}$$

The mean and variance of the received signal power are both σ and the variance coefficient $CV_I = 1$.

It is worth noting that all of the parameters in the probability density function can be fully characterized by a parameter σ except the phase with uniformly distributed. The parameter σ contains all the information of the target in the probability density function. As can be seen from Eq. (8.7), σ represents the average intensity of the image.

Figure 8.2 shows a single-look image of L-band airborne SAR. The image quality is poor with strong speckle noise for no multi-look processing. The statistical characteristics of this data are analyzed to verify the correctness of the probabilistic model of the SAR image feature quantity obtained above. A statistical histogram of each feature quantity from the data is given in Fig. 8.3, where figure (a) is a statistical histogram of the original data in-phase component (I-way data) x_{I}, and figure (b) is a statistical histogram of the original data quadrature phase component (Q-way Data) x_{Q}. Figure (c) is the statistical histogram of the original data phase ϕ. Figure (d) is the statistical histogram of the original data amplitude A. Figure (e) is the statistical histogram of the radar received signal power I.

Fig. 8.2 L-band airborne SAR single-look image

The theoretical statistical distribution of each feature quantity obtained according to Eqs. (8.2) to (8.7) is given in the figure, and is represented by a thick solid line. It can be seen that the probability distribution model obtained by theoretical analysis is in agreement with the statistical distribution of the actual SAR data.

8.2 Statistical Analysis of Single-Look Polarimetric SAR Data

The statistical characteristics of the traditional single-channel single-polarized SAR data are obtained in Sect. 8.1. The statistical properties of the radar measurement data are analyzed to help understand the prior knowledge and improve the practical application of radar data. The most basic data obtained by polarimetric SAR is four basic polarization combined image data represented by scattering matrix elements. For each polarization combined image, its statistical properties are not different from the statistical properties of the previously analyzed single-channel single-polarized SAR image. That is, orthogonal dual-channel sampling is a Gaussian random variable with zero mean and equal variance. The amplitude and echo power of a single-look polarimetric SAR image can be expressed by Rayleigh distribution and exponential distribution, respectively. The polarization measurement vector obeys complex Gaussian distribution. Next, the relationship between the images of different polarimetric SAR channels and their statistical properties is analyzed. This is necessary for a better understanding of multi-polarized SAR image information.

The relationship between the radar receiving electric field and the transmitting electric field can be described by a scattering matrix.

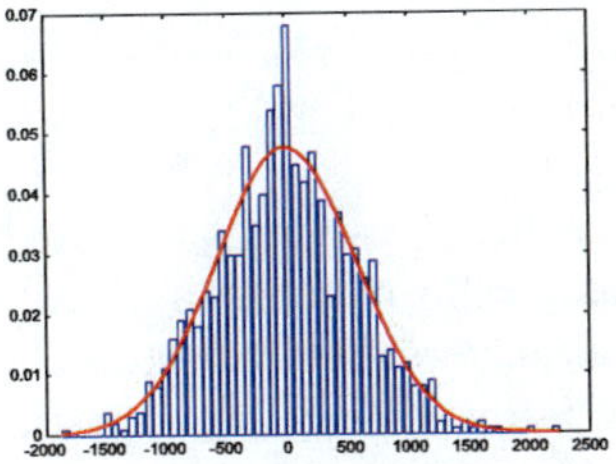

(a) statistical histogram of the original data in-phase component (I-way data) 260

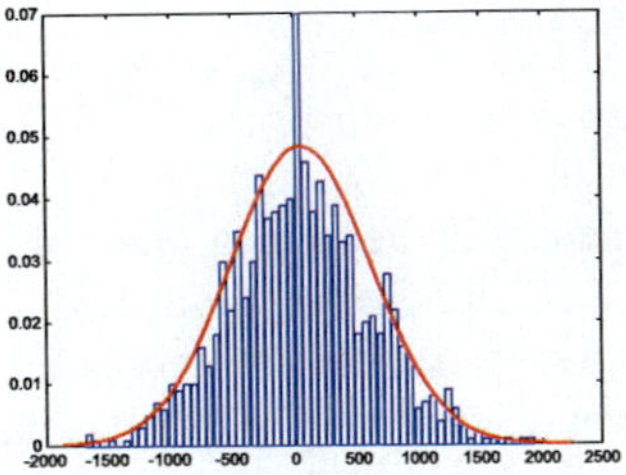

(b) statistical histogram of the original data quadrature phase component

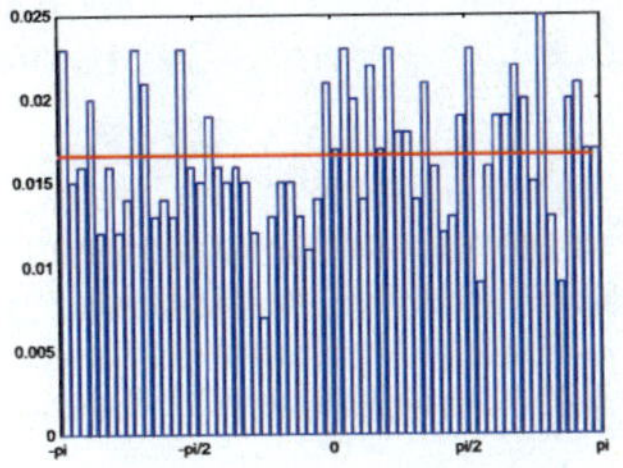

(c) statistical histogram of the original data phase

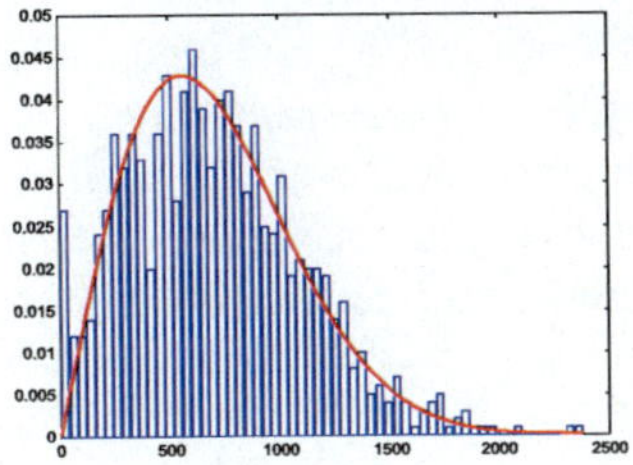

(d) statistical histogram of the original data amplitude

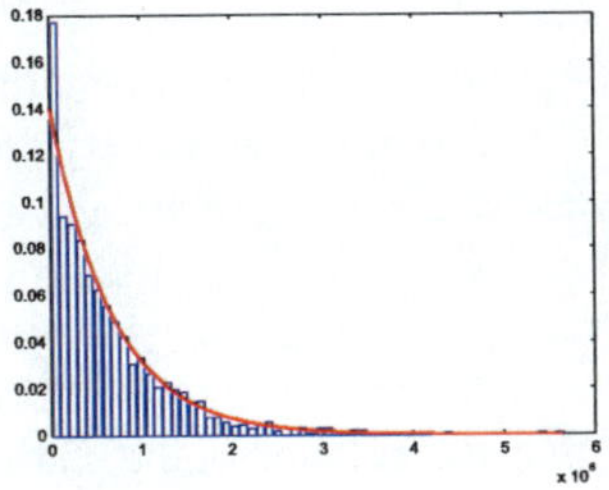

(e) statistical histogram of the radar received signal power

Fig. 8.3 Comparison between statistical distribution and theoretical distribution of each feature quantity from L-band airborne SAR data

$$\begin{bmatrix} E_{\mathrm{v}}^{\mathrm{R}} \\ E_{\mathrm{h}}^{\mathrm{R}} \end{bmatrix} = \frac{\mathrm{e}^{\mathrm{j}k_0 r}}{r} \begin{bmatrix} s_{\mathrm{vv}} & s_{\mathrm{vh}} \\ s_{\mathrm{hv}} & s_{\mathrm{hh}} \end{bmatrix} \cdot \begin{bmatrix} E_{\mathrm{v}}^{\mathrm{T}} \\ E_{\mathrm{h}}^{\mathrm{T}} \end{bmatrix} \tag{8.8}$$

where r is the distance between the scattering target and the receiving antenna, and k_0 is the wave number of the electromagnetic wave.

For reciprocal media, the scattering matrix is symmetrical, i.e., $s_{\mathrm{vh}} = s_{\mathrm{hv}}$. Each element in the scattering matrix can be fully described by its amplitude and phase. When the radar illuminates a large area of random phase medium or a randomly distributed medium surface, many point independent scatter elements in the resolution unit contribute to the total energy received by the radar [2, 3]. Therefore, each element in the scattering matrix can be expressed as

$$s_{pq} = \left|s_{pq}\right| \cdot e^{j\varphi_{pq}} = \sum_{n=1}^{N} \left|s_{pq}^{n}\right| \cdot e^{j\varphi_{pq}^{n}} \quad p, q = v, h \tag{8.9}$$

where N is the total number of independent scatterers in the resolution unit, and $|s_{pq}^{n}|$ and ϕ_{pq}^{n} are the amplitude and phase of the nth scatterer echo, respectively. The phase of the scatter element given in Eq. (8.9) includes all factors that can cause a phase change, such as phase delay due to the difference in distance between the scatter element and the center of the distribution target. If N is large enough, according to the central limit theorem, the real and imaginary parts of the scattering matrix element s_{pq} are independent and identically distributed zero-mean Gaussian random variables. Accordingly, it can be concluded that the amplitudes $|s_{pq}^{n}|$ obey Rayleigh distributions and ϕ_{pq}^{n} are independent random variables uniformly distributed in the interval $[-\pi, \pi)$, respectively.

For convenience of description, the complex polarization scattering matrix is decomposed into the sum of the real part and the imaginary parts, so the polarization measurement vector of the reciprocal medium can be expressed as follows:

$$\mathbf{X} = \begin{bmatrix} s_{\mathrm{vv}} \\ s_{\mathrm{hh}} \\ s_{\mathrm{vh}} \end{bmatrix} = \begin{bmatrix} x_1 + jx_2 \\ x_3 + jx_4 \\ x_5 + jx_6 \end{bmatrix} \tag{8.10}$$

Except the scattering amplitude, the greatest benefit of full polarization measurements is the preservation of phase information between echo signals from different polarization channels. However, it is very difficult and unnecessary to obtain the absolute value of the each element phase in the polarization scattering matrix. Usually, a matrix element is to select as the reference, such as the phase of s_{vv}, and the phase of other elements can be determined by the phase difference between them. Therefore, determining the statistical distribution of the phase difference between different polarization channel data is an important process to analyze statistical characteristic of the polarimetric SAR data.

8.2.1 *Statistical Characteristics Analysis of Cross-Polarization Terms*

From the analysis of the polarization measurement data of a large number of distributed targets, it can be known that the cross-polarization term (s_{vh}) in the scattering matrix and the co-polarization term (s_{vv} and s_{hh}) are statistically independent of each other. Therefore, the statistical properties of s_{hv} can be represented by independent parameters, which can be uniquely represented by the variance σ_{c}^2 of the $s_{\mathrm{vh}} = x_5 + jx_6$ real and imaginary parts. It is easy to get the joint probability density function of x_5 and x_6 as follows:

$$f_{x_5,x_6}(x_5,x_6)=\frac{1}{2\pi\sigma_c^2}\exp\left[-\frac{x_5^2+x_6^2}{2\sigma_c^2}\right] \tag{8.11}$$

The assumption is used here, which is zero-mean Gaussian random variables of the real and imaginary parts in the scattering matrix element. Correspondingly, the amplitude $|s_{\mathrm{vh}}|$ obeys the Rayleigh distribution, and the phase ϕ_{vh} obeys the uniform distribution in $[-\pi, \pi)$. Their joint probability density function is

$$f_{|s_{\mathrm{vh}}|,\varphi_{\mathrm{vh}}}(|s_{\mathrm{vh}}|,\varphi_{\mathrm{vh}})=\frac{1}{2\pi\sigma_{\mathrm{c}}^2}|s_{\mathrm{hv}}|\exp\left[-\frac{|s_{\mathrm{hv}}|^2}{2\sigma_{\mathrm{c}}^2}\right] \tag{8.12}$$

Since s_{vh} and s_{vv} are statistically independent of each other, and both ϕ_{vh} and ϕ_{vv} are uniformly distributed random variables, the phase difference $\phi_{\mathrm{c}}=\phi_{\mathrm{vh}}-\phi_{\mathrm{vv}}$ is also a random variable uniformly distributed between $[-\pi, \pi)$.

8.2.2 *Statistical Characteristics Analysis of the Co-polarization Term*

Since the co-polarization terms (s_{hh} and s_{vv}) of the scattering matrix are not independent of each other, the statistical property analysis of the co-polarization term is more complicated. There is

$$s_{vv}=x_1+jx_2,\ s_{hh}=x_3+jx_4 \tag{8.13}$$

Thus, each statistical component of the co-polarization term can be represented by a four-dimensional random vector.

$$\tilde{\mathbf{X}}=\begin{bmatrix}x_1 & x_2 & x_3 & x_4\end{bmatrix}^{\mathrm{T}} \tag{8.14}$$

Each element x_1, x_2, x_3, and x_4 in the random vector is a Gaussian random variable of zero mean, so their joint probability density function can be completely determined by a 4×4 symmetric positive definite matrix, i.e., covariance matrix C.

$$\mathbf{C}=\left\langle\tilde{X}\tilde{X}^{\mathrm{T}}\right\rangle \tag{8.15a}$$

Or it can be expressed as follows:

$$c_{ij}=c_{ji}=\left\langle x_i x_j\right\rangle,\ i,j=1,2,3,4 \tag{8.15b}$$

The joint probability density function based on the covariance matrix can be expressed as follows:

$$f_{\tilde{\mathbf{X}}}(x_1, x_2, x_3, x_4) = \frac{1}{4\pi^2|\mathbf{C}|^{1/2}}\exp\left[-\frac{1}{2}\tilde{\mathbf{X}}^{\mathrm{T}}\mathbf{C}^{-1}\tilde{\mathbf{X}}\right] \tag{8.16}$$

In the above equations, the superscript T represents the transposition of the column vector, and <*x*> represents the mean value of *x*.

For an element in the scattering matrix, the real part and the imaginary part are independent and identically distributed zero-mean random variables. So the elements in the covariance matrix can be expressed as follows:

$$c_{11} = c_{22} = \left\langle x_1^2 \right\rangle = \left\langle x_2^2 \right\rangle \tag{8.17a}$$

$$c_{12} = \langle x_1 x_2 \rangle = 0 \tag{8.17b}$$

$$c_{33} = c_{44} = \left\langle x_3^2 \right\rangle = \left\langle x_4^2 \right\rangle \tag{8.17c}$$

$$c_{34} = \langle x_3 x_4 \rangle = 0 \tag{8.17d}$$

Furthermore, it has been pointed out above that the absolute phase ϕ_{pp} of the scattering matrix co-polarization term is uniformly distributed and independent of the amplitude $|s_{pp}|$. Therefore, the sum of phase $\phi_{\mathrm{vv}} + \phi_{\mathrm{hh}}$ is also uniformly distributed, and is statistically independent from $|s_{\mathrm{vv}}||s_{\mathrm{hh}}|$, i.e.,

$$\langle |s_{\mathrm{vv}}||s_{\mathrm{hh}}|\cos(\varphi_{\mathrm{vv}} + \varphi_{\mathrm{hh}}) \rangle = 0 \tag{8.18a}$$

$$\langle |s_{\mathrm{vv}}||s_{\mathrm{hh}}|\sin(\varphi_{\mathrm{vv}} + \varphi_{\mathrm{hh}}) \rangle = 0 \tag{8.18b}$$

While

$$x_1 x_3 - x_2 x_4 = |s_{\mathrm{vv}}||s_{\mathrm{hh}}|\cos(\varphi_{\mathrm{vv}} + \varphi_{\mathrm{hh}}) \tag{8.19a}$$

$$x_1 x_4 - x_2 x_3 = |s_{\mathrm{vv}}||s_{\mathrm{hh}}|\sin(\varphi_{\mathrm{vv}} + \varphi_{\mathrm{hh}}) \tag{8.19b}$$

According to Eqs. (8.17) and (8.18), we can obtain

$$c_{13} = c_{24} \tag{8.20a}$$

$$c_{14} = -c_{23} \tag{8.20b}$$

The function of Mueller matrix is the connection between Stokes of scattering wave and Stokes of incidence wave, that is, from Eqs. (8.16) and (8.19), only four elements in the covariance matrix are unknown, namely, c_{11}, c_{13}, c_{14}, and c_{33}, which can be obtained directly from the Mueller matrix of the target. The role of the Mueller matrix is to relate the Stokes vector of the scattered wave with the Stokes vector of the incident wave, i.e.,

$$\mathbf{J}_s = \mathbf{M}\mathbf{J}_i \tag{8.21}$$

The Mueller matrix is represented by a polarization scattering matrix element as follows:

$$\mathbf{M} = \begin{bmatrix} |s_{\mathrm{vv}}|^2 & |s_{\mathrm{vh}}|^2 & \mathrm{Re}[s_{\mathrm{vh}}^* s_{\mathrm{vv}}] & -\mathrm{Im}[s_{\mathrm{vh}}^* s_{\mathrm{vv}}] \\ |s_{\mathrm{vh}}|^2 & |s_{\mathrm{hh}}|^2 & \mathrm{Re}[s_{\mathrm{hh}}^* s_{\mathrm{hv}}] & -\mathrm{Im}[s_{\mathrm{hh}}^* s_{\mathrm{hv}}] \\ 2\mathrm{Re}[s_{\mathrm{vv}}^* s_{\mathrm{hv}}] & 2\mathrm{Re}[s_{\mathrm{hv}}^* s_{\mathrm{hh}}] & \mathrm{Re}[s_{\mathrm{vv}}^* s_{\mathrm{hh}} + s_{\mathrm{hv}} s_{\mathrm{vh}}^*] & -\mathrm{Im}[s_{\mathrm{vv}} s_{\mathrm{hh}}^* - s_{\mathrm{hv}} s_{\mathrm{vh}}^*] \\ 2\mathrm{Im}[s_{\mathrm{vv}} s_{\mathrm{hv}}^*] & 2\mathrm{Im}[s_{\mathrm{hv}} s_{\mathrm{hh}}^*] & \mathrm{Im}[s_{\mathrm{vv}} s_{\mathrm{hh}}^* + s_{\mathrm{hv}} s_{\mathrm{vh}}^*] & \mathrm{Re}[s_{\mathrm{vv}} s_{\mathrm{hh}}^* - s_{\mathrm{hv}} s_{\mathrm{vh}}^*] \end{bmatrix} \tag{8.22}$$

where Re(x) and Im(x) represent the real and imaginary parts of the complex number x, respectively.

In practice, the measured values are usually partially polarized electromagnetic waves, so the target Mueller matrix needs to be averaged. According to the statistically independent relationship between the co-polarization term and the cross-polarization term, the average Mueller matrix can be represented by the elements of the covariance matrix obtained by Eqs. (8.15), (8.17), and (8.20).

$$\langle \mathbf{M} \rangle = \begin{bmatrix} 2c_{11} & 2\sigma_c^2 & 0 & 0 \\ 2\sigma_c^2 & 2c_{33} & 0 & 0 \\ 0 & 0 & 2c_{13} + 2\sigma_c^2 & 2c_{14} \\ 0 & 0 & -2c_{14} & 2c_{13} - 2\sigma_c^2 \end{bmatrix} \tag{8.23}$$

Then, the unknown elements in the covariance matrix and the variance of the cross-polarization terms in the target scattering matrix can be obtained from the average Mueller matrix elements.

$$\sigma_c^2 = M_{12}/2 \tag{8.24a}$$

$$c_{11} = M_{11}/2;\ c_{33} = M_{22}/2 \tag{8.24b}$$

$$c_{13} = (M_{33} + M_{44})/4;\ c_{14} = (M_{34} - M_{43})/4 \tag{8.24c}$$

Based on the resulting covariance matrix, the joint probability density function of the parameters x_1, x_2, x_3, and x_4 can be obtained by Eq. (8.15). These real and imaginary parameters are represented by the complex modulo A and the phase angle ϕ.

$$x_1 = A_1 \cos \varphi_{\mathrm{vv}},\ x_2 = A_1 \sin \varphi_{\mathrm{vv}} \tag{8.25a}$$

$$x_3 = A_2 \cos \varphi_{hh}, x_4 = A_2 \sin \varphi_{hh} \tag{8.25b}$$

Then, the joint probability density function of the amplitude and phase of each element in the co-polarization term is

$$f_{A_1,A_2,\varphi_{vv},\varphi_{hh}}(A_1, A_2, \varphi_{vv}, \varphi_{hh}) = \frac{1}{4\pi^2\sqrt{\Delta}} A_1 A_2 \exp\left\{-\frac{1}{2}[\alpha_1 A_1^2 + \alpha_2 A_{21}^2 - 2\alpha_3 A_1 A_2]\right\} \tag{8.26}$$

where

$$\Delta = |\mathbf{A}| = (c_{11}c_{33} - c_{13}^2 - c_{14}^2)^2$$

$$\alpha_1 = c_{33}/\sqrt{\Delta}, \alpha_2 = c_{11}/\sqrt{\Delta}$$

$$\alpha_3 = [c_{13}\cos(\varphi_{hh} - \varphi_{vv}) + c_{14}\sin(\varphi_{hh} - \varphi_{vv})]/\sqrt{\Delta}$$

In order to derive the statistical properties of the phase difference in the co-polarization elements, it is first necessary to know the joint probability density function of ϕ_{vv} and ϕ_{hh}. This can be obtained by taking the edge distribution of Eq. (8.26).

$$f_{\varphi_{vv},\varphi_{hh}}(\varphi_{vv}, \varphi_{hh}) = \int_0^\infty \int_0^\infty f_{A_1,A_2,\varphi_{vv},\varphi_{hh}}(A_1, A_2, \varphi_{vv}, \varphi_{hh}) \mathrm{d}A_1 \mathrm{d}A_2 \tag{8.27}$$

The parameter α_1 in Eq. (8.24) is a positive real number, and the following results are obtained after integrating A_1:

$$f_{\varphi_{vv},\varphi_{hh}}(\varphi_{vv}, \varphi_{hh}) = \frac{1}{4\pi^2\sqrt{\Delta}}\left\{\frac{1}{\alpha_1}\int_0^\infty A_2 e^{-\frac{\alpha_2}{2}\rho_2^2} dA_2 + \sqrt{\frac{\pi}{8\alpha_1^3}\alpha_3 \int_0^\infty A_2^2\left[1 \pm erf\left(\frac{|\alpha_3|}{\sqrt{8\alpha_1}}A_2\right)\right] e^{-\frac{1}{2\alpha_1}(\alpha_1\alpha_2 - \alpha_3^2)A_2^2} dA_2}\right\} \tag{8.28}$$

Here erf() is the error function, and its definition is

$$\mathrm{erf}(x) = \frac{2}{\sqrt{\pi}}\int_0^x \mathrm{e}^{-t^2}\mathrm{d}t \tag{8.29}$$

The $\pm$ sign in the equation is determined according to the $\pm$ sign of α_3. Calculating the integral in (8.28), we can get

$$f_{\varphi_{vv},\varphi_{hh}}(\varphi_{vv},\varphi_{hh}) = \frac{1}{4\pi^2\sqrt{\Delta}}\left\{\frac{1}{\alpha_1\alpha_2} + \frac{\alpha_3^2}{\alpha_1\alpha_2(\alpha_1\alpha_2-\alpha_3^2)}\right.$$
$$\left. + \frac{\sqrt{\pi}|\alpha_3|}{\sqrt{8\alpha_1}(\alpha_1\alpha_2-\alpha_3^2)}\int_0^\infty erf\left(\frac{|\alpha_3|}{\sqrt{8\alpha_1}}A_2\right)\cdot e^{-\frac{1}{2\alpha_1}(\alpha_1\alpha_2-\alpha_3^2)A_2^2}dA_2\right\} \tag{8.30}$$

where

$$\int_0^\infty \text{erf}\left(\frac{|\alpha_3|}{\sqrt{8\alpha_1}}A_2\right)\text{e}^{-\frac{1}{2\alpha_1}(\alpha_1\alpha_2-\alpha_3^2)A_2^2}\text{d}A_2 = \sqrt{\frac{2\alpha_1}{\pi(\alpha_1\alpha_2-\alpha_3^2)}}\arctan\left(\frac{|\alpha_3|}{2\sqrt{\alpha_1\alpha_2-\alpha_3^2}}\right) \tag{8.31}$$

Probability density function of the phase difference $\phi = \phi_{hh} - \phi_{vv}$ in the interval $[-\pi, \pi)$ can be further calculated as follows:

$$f_\varphi(\varphi) = \frac{c_{11}c_{33}-c_{13}^2-c_{14}^2}{2\pi(c_{11}c_{33}-D^2)}\left\{1+\frac{D}{\sqrt{c_{11}c_{33}-D^2}}\left[\frac{\pi}{2}+\arctan\frac{D}{\sqrt{c_{11}c_{33}-D^2}}\right]\right\} \tag{8.32}$$

where $D = c_{13}\cos\varphi + c_{14}\sin\varphi$. And the covariance matrix element can be represented by the Mueller matrix element as

$$c_{11} = \frac{M_{11}}{2};\ c_{33} = \frac{M_{22}}{2};\ c_{13} = \frac{M_{33}+M_{44}}{4};\ c_{14} = \frac{M_{34}-M_{13}}{4}$$

To further understand the probability density function of the phase difference in the co-polarization element, two special cases are examined below. When s_{vv} and s_{hh} are mutually independent random variables, the c_{13} and c_{14} values are 0, $f_\varphi(\varphi) = 1/(2\pi)$ is a uniform distribution between $[-\pi, \pi)$. When the measurement results is a fully polarized electromagnetic wave, the random variables s_{vv} and s_{hh} are completely correlated. In this case, the determinant value of the covariance matrix **C** is 0. $f_\varphi(\varphi)$ can be obtained as a δ function.

It is worth noting that the probability density function of the phase difference can be represented by only two parameters as follows:

$$\eta = \sqrt{\frac{c_{13}^2+c_{14}^2}{c_{11}c_{33}}},\ \zeta = \arctan\frac{c_{14}}{c_{13}} \tag{8.33}$$

Here, η and ζ vary within the range of [0–1) and $[-\pi \sim \pi$, respectively. What are the physical meanings of the parameters η and ζ? Let's examine the complex correlation coefficients between polarized channels.

$$\rho_c = \frac{E[s_i s_j^*]}{\sqrt{E[|s_i|^2]E[|s_j|^2]}} = |\rho_c| e^{j\theta} \tag{8.34}$$

Let $s_i = s_{vv}$, $s_j = s_{hh}$. According to the definition of Eq. (8.10), ρ_c can be expressed as follows:

$$\rho_c = \frac{\langle (x_1 + jx_2)(x_3 - jx_4) \rangle}{\sqrt{\langle (x_1 + jx_2)^2 \rangle \langle (x_3 + jx_4)^2 \rangle}} = \frac{\langle x_1 x_3 \rangle - j\langle x_1 x_4 \rangle}{\sqrt{\langle x_1^2 \rangle \langle x_3^2 \rangle}} = \frac{c_{13} - jc_{14}}{\sqrt{c_{11}c_{33}}} \tag{8.35}$$

It can be seen that the parameter η is numerically equal to the modulus of the complex correlation coefficient ρ_c.

$$\eta = |\rho_c| \tag{8.36}$$

In addition, the parameter ζ is actually the phase difference between the co-polarization terms when the echo is completely polarized.

Based on the parameter η and ζ, Eq. (8.32) can be expressed as

$$f_\varphi(\varphi) = \frac{1 - \eta^2}{2\pi[1 - \eta^2 \cos^2(\varphi - \zeta)]} \left\{ 1 + \frac{\eta \cos(\varphi - \zeta)}{\sqrt{1 - \eta^2 \cos^2(\varphi - \zeta)}} \left[\frac{\pi}{2} + \arctan \frac{\eta \cos(\varphi - \zeta)}{\sqrt{1 - \eta^2 \cos^2(\varphi - \zeta)}} \right] \right\} \tag{8.37}$$

Figure 8.4 shows the probability density function for the phase difference of the

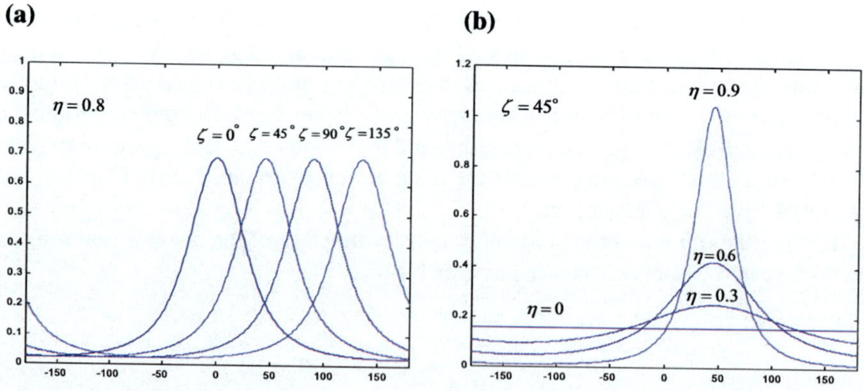

Fig. 8.4 Probability density function of phase difference in the co-polarization term under different parameters

co-polarization term under different parameters. Figure (a) shows the probability density function under different ζ values under the condition of keeping η constant. Figure (b) is the probability density function graph obtained by changing η while keeping ζ unchanged. It can be seen that the maximum value of the probability density function appears at $\varphi = \zeta$. And the peak value of the probability density function increases as the correlation coefficient increases. When the two channels are completely correlated, i.e., $\eta = |\rho_c| = 1$, the probability density function of the phase difference of the co-polarization term will become a δ function, and the phase difference between the two channels is constant. When the two channels are completely uncorrelated, i.e., $\eta = |\rho_c| = 0$, the phase difference of the co-polarization term will appear as a uniform distribution between $[-\pi, \pi)$. This is consistent with the results obtained in the previous analysis.

Finally, it should be pointed out that the statistical characteristics of the cross-polarization phase difference under the non-reciprocal condition of the scattering medium can also be obtained by a similar method, and the probability density function for backscattering can also be obtained by Eq. (8.32). The cross-polarization covariance matrix is as follows:

$$c_{11} = \frac{M_{11}}{2}, c_{33} = \frac{M_{12}}{2}, c_{13} = \frac{M_{13}}{2}, c_{14} = \frac{M_{14}}{2} \tag{8.38}$$

8.3 Statistical Characteristics of Multi-look Polarimetric SAR Images

Section 8.2 analyzes the statistical properties of single-look polarimetric SAR images and focuses on the probability density function of the phase difference in the co-polarization term. In practical applications, it is usually necessary to perform multi-look processing on SAR data. Although multi-look processing inevitably brings about negative effects such as resolution reduction, it is possible to effectively suppress speckle noise and reduce the amount of image data. Therefore, it is widely used in SAR data processing, and the processing of multi-polarimetric SAR data is no exception.

The previous analysis is known. For a uniform scene, if the orthogonal channel sampling of a single-polarimetric SAR is regarded as a Gaussian random variable with zero mean and equal variance, the signal amplitude and image intensity of the single-look SAR can be, respectively, expressed as Rayleigh distribution and the negative exponential distribution. The polarization measurement vector obeys a Gaussian distribution. For the n-look case, the signal amplitude and image intensity of each polarization channel will obey the χ distribution and χ^2 distribution of $2n$ degrees of freedom, respectively, and the multi-look covariance matrix obeys the complex Wishart distribution [3–9]. The statistical characteristics of multi-look polarimetric SAR images in uniform scenes are analyzed in detail below. Compared with the single-look case, the analysis of the statistical characteristics of the multi-look polarimetric SAR will be more general, because the former is only a special case when the visual number n is 1.

8.3.1 Covariance Matrix of Multi-look Polarimetric SAR Data and Its Statistical Distribution

The scattering matrix for reciprocal media has only three independent complex elements that can be vectorized into the target polarization measurement vector **X**:

$$\mathbf{X} = \left[s_{\mathrm{hh}}\ s_{\mathrm{hv}}\ s_{\mathrm{vv}} \right]^{\mathrm{T}} = \left[s_1\ s_2\ s_3 \right]^{\mathrm{T}} \tag{8.39}$$

For non-reciprocal media, the dimension of the polarization measurement vector is 4. The covariance matrix of the polarization measurement vector is

$$\mathbf{C} = \left\langle \mathbf{X} \cdot \mathbf{X}^{\mathrm{H}} \right\rangle \tag{8.40}$$

The superscript H represents the transposed conjugate. It is can be see that the complex covariance matrix **C** is an Emmett matrix, i.e., $\mathbf{C} = \mathbf{C}^{\mathrm{T}}$.

Ideally, the co-polarization terms (HH and VV) of the scattering matrix are correlated, while the cross-polarization term is statistically independent from the co-polarization term (such as HV and HH, HV, and VV), in which correlation coefficient is zero. However, from the actual polarimetric SAR data measurement analysis, it is found that there is a certain correlation between the cross-polarization term and the co-polarization term, but the correlation coefficient value is very low. This may be due to crosstalk between channels and system thermal noise.

As previously analyzed, when the resolution unit of the radar illumination area contains many independent scatterers, **X** can be described by a multi-dimensional complex Gaussian distribution.

$$p_{\mathbf{X}}(\mathbf{X}) = \frac{1}{\pi^3 |\mathbf{C}|} \exp(-\mathbf{X}^{\mathrm{T}} \mathbf{C}^{-1} \mathbf{X}) \tag{8.41}$$

Here $|\mathbf{C}|$ is the value of the covariance matrix determinant. The real and imaginary parts of any two complex elements $s_m = x_m + jy_m$ and $s_n = x_n + jy_n$ in vector **X** satisfy the following statistical relationship:

$$E[x_m] = E[y_m] = 0,\ E\left[x_m^2\right] = E\left[y_m^2\right],\ E[x_m y_m] = 0 \tag{8.42a}$$

$$E[x_m x_n] = E[y_m y_n],\ E[y_m x_n] = -E[x_m y_n] \tag{8.42b}$$

Non-coherent averaging of n single-look covariance matrices yields n-look covariance matrix $\mathbf{Z}$ as

$$\mathbf{Z} = \frac{1}{n}\sum_{k=1}^{n}\mathbf{X}(k)\mathbf{X}(k)^{\mathrm{H}} = \frac{1}{n}\sum_{k=1}^{n}\begin{bmatrix} |s_{\mathrm{hh}}|^2 & s_{\mathrm{hh}}s_{\mathrm{hv}}^* & s_{\mathrm{hh}}s_{\mathrm{vv}}^* \\ s_{\mathrm{hv}}s_{\mathrm{hh}}^* & |s_{\mathrm{hv}}|^2 & s_{\mathrm{hv}}s_{\mathrm{vv}}^* \\ s_{\mathrm{vv}}s_{\mathrm{hh}}^* & s_{\mathrm{vv}}s_{\mathrm{hv}}^* & |s_{\mathrm{vv}}|^2 \end{bmatrix}_k \tag{8.43}$$

where n is a multi-look number and $\mathbf{X}(k)$ is the kth single-look sample vector. Let $\mathbf{A} = n\mathbf{Z}$, then matrix $\mathbf{A}$ has the form of a complex Wishart distribution.

$$p_{\mathbf{A}}^{(n)}(\mathbf{A}) = \frac{|\mathbf{A}|^{n-q}\exp[-\operatorname{tr}(\mathbf{C}^{-1}\mathbf{A})]}{\mathbf{K}(n,q)|\mathbf{C}|^n} \tag{8.44}$$

where tr(•) expresses trace of the matrix.

$$\mathbf{K}(n,q) = \pi^{q(q-1)/2}\Gamma(n)\ldots\Gamma(n-q+1) \tag{8.45}$$

Γ(•) is the Gamma function. The parameter q is the dimension of the vector $\mathbf{X}$, and for the reciprocal medium, $q = 3$.

Thus, the distribution of the multi-look covariance matrix $\mathbf{Z}$ can be easily derived from Eq. (8.44):

$$p_{\mathbf{Z}}^{(n)}(\mathbf{Z}) = \frac{n^{qn}|\mathbf{Z}|^{n-q}\exp[-n\cdot\operatorname{tr}(\mathbf{C}^{-1}\mathbf{Z})]}{\mathbf{K}(n,q)|\mathbf{C}|^n} \tag{8.46}$$

When $q = 1$, the probability density function of the n-look image power is obtained.

$$p_I^{(n)}(I) = \frac{n^n I^{n-1}\exp[-nI/C_{11}]}{\Gamma(n)\cdot(C_{11})^n} \tag{8.47}$$

where $I = Z_{11}$ is the intensity of the n-look image, and $C_{11} = E[s_1 s_1^*]$.

8.3.2 *Statistical Distribution of Phase Difference in Multi-look Polarimetric SAR Images*

The statistical distribution law of phase difference in different channels of multi-look polarimetric SAR is analyzed below. Let s_i and s_j be two different elements in the single-look polarization scattering matrix, then the phase difference between the single-look polarimetric SAR channels can be obtained by the following equation:

$$\varphi_{(1)} = \arg(s_i s_j^*) \tag{8.48}$$

Correspondingly, the multi-look phase difference with the look number n is

$$\varphi_{(n)} = \arg\left[\frac{1}{n}\sum_{k=1}^{n} s_i(k)s_j^*(k)\right] \tag{8.49}$$

$\varphi_{(n)}$ is the angle of the non-diagonal element in the multi-look covariance matrix **Z**. For convenience of presentation, the subscript n of $\varphi_{(n)}$ is omitted. In the process of analyzing the phase of multi-polarimetric SAR image, the complex correlation coefficient is an indispensable parameter. Here is the definition of the complex correlation coefficient ρ_c.

$$\rho_c = \frac{E[s_i s_j^*]}{\sqrt{E[|s_i|^2]E[|s_j|^2]}} = |\rho_c|e^{j\theta} \tag{8.34}$$

It can be known from Eq. (8.49) that the data to be analyzed below only involves two elements in the polarization scattering matrix, i.e., $q = 2$. So matrix **A** and single-look covariance matrix **C** can be simplified as follows:

$$\mathbf{A} = \begin{bmatrix} A_{11} & \alpha e^{j\varphi} \\ \alpha e^{-j\varphi} & A_{22} \end{bmatrix} \tag{8.50}$$

$$\mathbf{C} = \begin{bmatrix} C_{11} & \sqrt{C_{11}C_{22}}|\rho_c|e^{j\theta} \\ \sqrt{C_{11}C_{22}}|\rho_c|e^{-j\theta} & C_{22} \end{bmatrix} \tag{8.51}$$

Here, α and φ are the modes and arguments of the matrix element A_{12}, respectively, φ is the phase difference of the multi-look polarimetric SAR image, and $C_{ii} = E[|s_i|^2]$. For convenience, A_{11}, A_{22}, and α are separately defined as

$$B_1 = \frac{A_{11}}{C_{11}},\ B_2 = \frac{A_{22}}{C_{22}},\ \eta = \frac{\alpha}{\sqrt{C_{11}C_{22}}} \tag{8.52}$$

Substitute the normalized new matrix **A** into Eq. (8.42).

$$p(B_1, B_2, \eta, \varphi) = \frac{(B_1B_2 - \eta^2)^{n-2}\eta}{\pi(1 - |\rho_c|^2)^n \Gamma(n)\Gamma(n-1)} \exp\left(-\frac{B_1 + B_2 - 2\eta|\rho_c|\cos(\psi - \theta)}{1 - |\rho_c|^2}\right) \tag{8.53}$$

Thus, the probability density function of the multi-look phase difference can be obtained by integrating B_1, B_2, and η in Eq. (8.53).

$$p_\psi^{(n)}(\varphi) = \frac{\Gamma(n + 1/2)(1 - |\rho_c|^2)^n \beta}{2\sqrt{\pi}\Gamma(n)(1 - \beta^2)^{n+1/2}} + \frac{(1 - |\rho_c|^2)^n}{2\pi} F(n, 1; 1/2; \beta^2),\ \pi \le \Psi \le \pi \tag{8.54}$$

$$\beta = |\rho_c|\cos(\varphi - \theta) \tag{8.55}$$

$F(n,1;1/2;\beta^2)$ is a Gaussian hypergeometric function. When the value of n is small, the hypergeometric function can be expressed as elementary algebra. For example, when $n = 1$, then

$$F(n,1;1/2;z) = (1-z)^{-1}\left[1 + \frac{\sqrt{z}\sin^{-1}\sqrt{z}}{\sqrt{1-z}}\right] \tag{8.56}$$

Substituting into Eq. (8.54), the probability density function of phase difference in single-look case can be obtained.

$$p_{\varphi}^{(1)}(\varphi) = \frac{1}{2\pi}\left(\frac{1-|\rho_c|^2}{1-\beta^2}\right)\cdot\left[1 + \frac{\beta}{\sqrt{1-\beta^2}}\cos^{-1}(-\beta)\right] \tag{8.57}$$

After simple algebraic conversion, we can find that Eq. (8.57) is exactly the same as Eq. (8.37) obtained by the Mueller matrix. This also proves the correctness of the two methods from one side. Similarly, the probability density function of the phase difference in the case of 2, 3, and 4 looks can be further obtained.

The probability density function of the phase difference in two-look image is

$$p_{\varphi}^{(2)}(\varphi) = \frac{3}{8}\frac{(1-|\rho_c|^2)^2\beta}{(1-\beta)^{5/2}} + \frac{(1-|\rho_c|^2)^2}{4\pi(1-\beta^2)^2}\cdot\left[2+\beta^2+\frac{3\beta}{(1-\beta^2)^{1/2}}\sin^{-1}(\beta)\right] \tag{8.58}$$

The probability density function of the phase difference in three-look image is

$$p_{\varphi}^{(3)}(\varphi) = \frac{15}{32}\frac{(1-|\rho_c|^2)^3\beta}{(1-\beta^2)^{7/2}} + \frac{(1-|\rho_c|^2)^3(1-\beta^2)^{-3}}{16\pi}\left[8+9\beta^2-2\beta^4+\frac{15\beta}{(1-\beta^2)^{1/2}}\sin^{-1}(\beta)\right] \tag{8.59}$$

The probability density function of the phase difference in four-look image is

$$p_{\varphi}^{(4)}(\varphi) = \frac{35}{64}\frac{(1-|\rho_c|^2)^4\beta}{(1-\beta^2)^{9/2}} + \frac{(1-|\rho_c|^2)^4}{96\pi(1-\beta^2)^4}\left[48+87\beta^2-38\beta^4+8\beta^6+\frac{105\beta}{(1-\beta^2)^{1/2}}\sin^{-1}(\beta)\right] \tag{8.60}$$

Multi-polarimetric image phase difference information has been widely used in many research fields of polarimetric SAR data, such as polarimetric SAR original data compression, polarimetric SAR image target detection and recognition. So many scholars have made in-depth research on the statistical distribution of phase difference. Except Eq. (8.54), there are other forms of mathematical expressions. The following results are given in an article published in 1994 by Ian R. Joughin et al. [4]:

$$p_{\varphi}^{(n)}(\varphi)=\frac{(1-|\rho_c|^2)^n\Gamma(2n)}{2^{n+1/2}\sqrt{\pi}\Gamma(n)}\left(\frac{1}{1-\beta^2}\right)^{(n+1/2)/2}\cdot \mathrm{P}_{n-3/2}^{-n-1/2}(-\beta) \tag{8.61}$$

$\mathrm{P}_{n-3/2}^{-n-1/2}$ is the associated Legendre function. R. J. A. Tough et al. derived another mathematical expression for the phase difference probability density function of a polarimetric SAR image as follows [10]:

$$\begin{aligned}p_{\varphi}^{(n)}=\frac{(1-|\rho_c|^2)^n}{2\pi}&\left\{\frac{(2n-2)!}{[(n-1)!]^2 2^{2(n-1)}}\times\left[\frac{(2n-1)\beta}{(1+\beta^2)^{n+1/2}}\cos^{-1}(-\beta)+\frac{1}{(1-\beta^2)^n}\right]\right.\\&\left.+\frac{1}{2(n-1)}\times\sum_{k=0}^{n-2}\left(\frac{\Gamma(n-1/2)}{\Gamma(n-k-1/2)}\frac{\Gamma(n-k-1)}{\Gamma(n-1)}\frac{1+(2k+1)\beta^2}{(1-\beta^2)^{k+2}}\right)\right\}\end{aligned} \tag{8.62}$$

The above three mathematical expressions are all based on the complex Wishart distribution of the complex covariance matrix, and the calculation methods are the same except for the different special functions. Therefore, in essence, they are different mathematical expressions of the same probability density function [11], and one of the most convenient expressions can be selected in the application.

From the expression of the phase difference probability density function, we can see that the value is only related to the look number n and the complex correlation coefficient $|\rho_c|$. The peak of the function is located at $\varphi=\theta$. The relationship between the phase difference probability density function and the look number n as well as the correlation coefficient $|\rho_c|$ is given in Fig. 8.5. Figure (a) shows the distribution at the look numbers $n=1,2,4,16$, and $|\rho_c|=0.8$. Obviously, the larger the look number, the more the probability density function tends to the δ function. So multi-look processing increases the accuracy of phase statistics. Figure (b) shows the distribution under different correlation coefficients when the look number $n=4$. It can be seen that when $|\rho_c|=0$, it is a uniform distribution, and when $|\rho_c|=1$, the probability density function will tend to function. This also shows that the higher

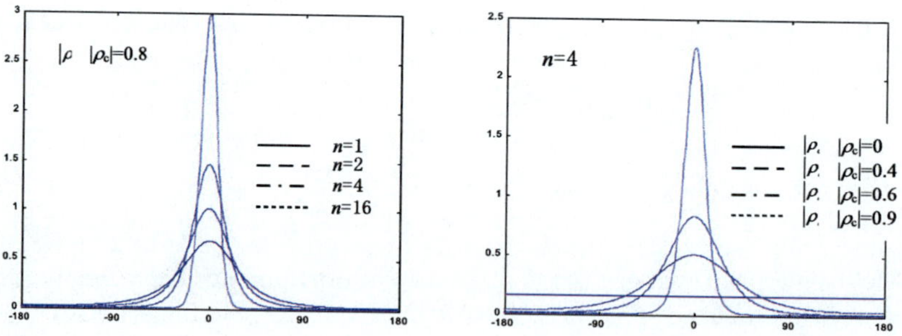

(a) (b)

Fig. 8.5 The relationship between the phase difference probability density function and the look number n as well as the correlation coefficient $|\rho_c|$

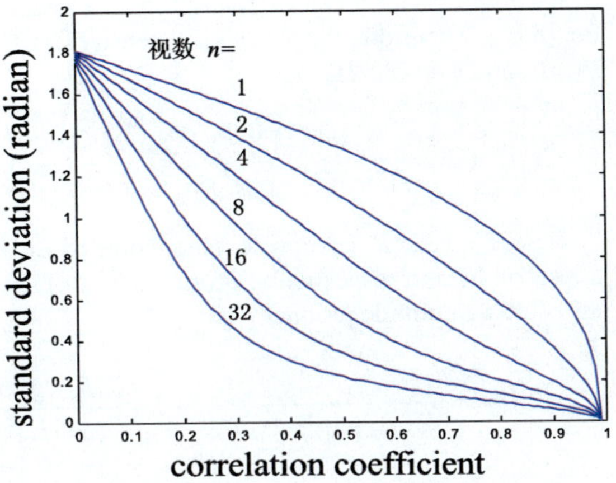

Fig. 8.6 The relationship between the standard deviation of the phase difference and the correlation coefficient

the correlation coefficient between the two images, the narrower the distribution of the phase difference. Figure 8.6 is a plot of the standard deviation of the phase difference and the correlation coefficient $|\rho_c|$. It can be seen that the multi-look processing effectively improves the accuracy of the phase statistics, especially for large multi-look numbers, such as $n = 16$ and 32. The standard deviation of the obtained phase differences is small. At the same time, it can be seen that when the correlation coefficient is zero, the standard deviation reaches the maximum value, and the phase difference distribution has the strongest randomness. When the correlation coefficient is 1, the standard variance value is zero, and the phase difference has no randomness.

8.3.3 Statistical Distribution of Multi-look Polarimetric SAR Image Amplitude Product

The product absolute value of the two different terms s_i and s_j in the polarization scattering matrix is an important physical quantity in the polarization measurement of SAR. Especially when studying the interference problem, it represents the amplitude of the phase interferogram. In the case of multi-look, the amplitude normalized value of the phase interferogram is as follows:

$$\xi = \frac{\left|\frac{1}{n}\sum_{k=1}^{n} s_i(k)s_j^*(k)\right|}{\sqrt{E[|s_i|^2]E[|s_j|^2]}} = \frac{g}{h} \tag{8.63}$$

where g is the amplitude product absolute value $|s_i s_j^*|$ of multi-look polarimetric SAR data.

By comparison, it can be found that ξ is consistent with the expression of η in Eq. (8.52). Therefore, the probability density function of ξ is obtained by integrating B_1, B_2, and φ in (8.53).

$$p_{\xi}^{(n)}(\xi) = \frac{4n^{n+1}\xi^n}{\Gamma(n)(1-|\rho_c|^2)} \cdot I_0\left(\frac{2|\rho_c|n\xi}{1-|\rho_c|^2}\right) \cdot K_{n-1}\left(\frac{2n\xi}{1-|\rho_c|^2}\right) \tag{8.64}$$

Here, $I_0(\cdot)$ and $K_n(\cdot)$ represent the modified Bessel functions of the first class and the second class, respectively. From Eq. (8.64), the probability density function of the n-look amplitude product g is

$$p_g^{(n)}(g) = \frac{4n^{n+1}g^n}{\Gamma(n)(1-|\rho_c|^2)h^{n+1}} \cdot I_0\left(\frac{2|\rho_c|ng/h}{1-|\rho_c|^2}\right) \cdot K_{n-1}\left(\frac{2ng/h}{1-|\rho_c|^2}\right) \tag{8.65}$$

8.3.4 Joint Probability Distribution of Two-Channel Echo Power in Multi-look Polarimetric SAR

The expression of the joint probability density function of the two-channel echo power in the multi-look polarimetric SAR is given below, which is necessary for polarization information processing or polarimetric SAR image interference. In addition, this joint probability density function is also the basis for the probability density function that yields the echo power ratio and amplitude ratio. According to Eq. (8.52), the two-channel echo power of the multi-look polarimetric SAR is

$$R_1 = \frac{1}{n}\sum_{k=1}^{n}|s_i(k)|^2 = \frac{B_1C_{11}}{n} \tag{8.66a}$$

$$R_2 = \frac{1}{n}\sum_{k=1}^{n}|s_j(k)|^2 = \frac{B_2C_{22}}{n} \tag{8.66b}$$

$C_{ii} = E\,[|s_i|^2]$. First, the joint probability density function of B_1 and B_2 is given, which can be obtained from the integral of η and φ in Eq. (8.53). Its probability density function is

$$p_{B_1,B_2}^{n}(B_1, B_2) = \frac{(B_1, B_2)^{(n-1)/2}\exp\left(-\frac{B_1+B_2}{1-|\rho_c|^2}\right)}{\Gamma(n)(1-|\rho_c|^2)|\rho_c|^{n-1}} \cdot I_{n-1}\left(2\sqrt{B_1B_2}\frac{|\rho_c|}{1-|\rho_c|^2}\right) \tag{8.67}$$

Further, according to Eq. (8.66), the joint probability density function of R_1 and R_2 can be obtained as

$$p(R_1, R_2) = \frac{n^{n+1}(R_1R_2)^{(n-1)/2}\exp\left[-\frac{n(R_1/C_{11}+R_2/C_{22})}{1-|\rho_c|^2}\right]}{(C_{11}C_{22})^{(n+1)/2}\Gamma(n)(1-|\rho_c|^2)|\rho_c|^{n-1}} \cdot I_{n-1}\left(2n\sqrt{\frac{R_1R_2}{C_{11}C_{22}}}\frac{|\rho_c|}{1-|\rho_c|^2}\right) \tag{8.68}$$

$I_0(\cdot)$ also represents the modified Bessel function of the first class.

8.3.5 *Probability Distribution of Two-Channel Echo Power Ratio for Multi-look Polarimetric SAR*

In polarimetric SAR research, the power ratio of echoes received by different polarization channels is a very important parameter. This parameter has a wide range of applications in polarization data quantization, compression, and polarization analysis. First, the probability density function of the normalized echo power ratio is obtained. From Eq. (8.52), the normalized echo power ratio can be obtained in Eq. (8.69).

$$\mu = \frac{B_1}{B_2} = \frac{\sum_{k=1}^{n}|s_i(k)|^2/C_{11}}{\sum|s_j(k)|^2/C_{22}} = \frac{\sum_{k=1}^{n}|s_i(k)|^2}{\tau\sum_{k=1}^{n}|s_j(k)|^2} \tag{8.69}$$

where $\tau = C_{11}/C_{22}$, which is the spectral covariance matrix diagonal element ratio of single-look polarimetric SAR.

Then the probability density function of the normalized echo power ratio can be normalized by Eq. (8.51).

$$p_\mu^{(n)}(\mu) = \frac{\Gamma(2n)(1-|\rho_c|^2)^n(1+\mu)\mu^{n-1}}{\Gamma(n)\Gamma(n)[(1+\mu)^2-4|\rho_c|^2\mu]^{(2n+1)/2}} \tag{8.70}$$

Figure 8.7(a) shows the probability density function of the normalized power ratio for different looks, when the correlation coefficient $|\rho_c| = 0.8$. It can be seen that when the look number $n = 1$, the function shape is relatively flat and forms a long tail. As the look number increases, the function shape gradually becomes sharp. Figure 8.7(b) is a plot of the standard deviation of the normalized power ratio and the correlation coefficient for of the multi-look polarimetric SAR. It can be seen from the figure that the standard deviation of the power ratio does not change much when the look number n is small. And the multi-look processing can significantly reduce the standard deviation of SAR echo power ratio when the look number is large.

Based on Eq. (8.70), it is easy to find the probability density function of the echo power ratio for any two polarization channels. Let the ratio of the echo power of the VV channel to the HH channel be as given in Eq. (8.71).

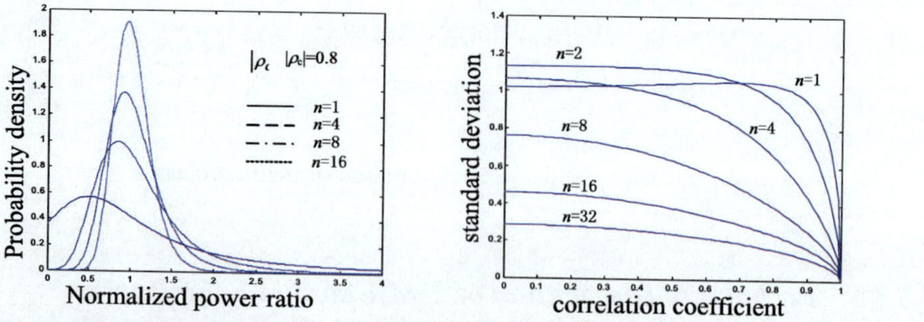

(a) Probability density function (b) Relationship between standard deviation and correlation coefficient

Fig. 8.7 Multi-look polarimetric SAR normalized power ratio

$$r_{\mathrm{vv/hh}} = \frac{\sum_{k=1}^{n} |s_{\mathrm{vv}}(k)|^2}{\sum_{k=1}^{n} |s_{\mathrm{hh}}(k)|^2} = \tau \mu_{\mathrm{vv/hh}} \tag{8.71}$$

where $\tau = C_{\mathrm{vv}}/C_{\mathrm{hh}}$.

Substitute Eqs. (8.71) to (8.70).

$$p^{(n)}_{r_{\mathrm{vv/hh}}}(r_{\mathrm{vv/hh}}) = \frac{\tau^n \Gamma(2n)(1-|\rho_{\mathrm{c}}|^2)^n(\tau+r)r^{n-1}}{\Gamma(n)\Gamma(n)[(\tau+r)^2 - 4\tau|\rho_{\mathrm{c}}|^2 r]^{(2n+1)/2}} \tag{8.72}$$

Figure 8.8(a)–(c) is probability density function plots of multi-look polarimetric SAR echo power ratio under different look n, different correlation coefficients $|\rho_{\mathrm{c}}|$, and different parameter τ, respectively. From this figure, we can have a more intuitive understanding of the probability density function (8.72) and the meaning of its parameters.

The correlation coefficient between the cross-polarized channel data and the co-polarization channel data in the polarimetric SAR is very small, which can be considered $|\rho_{\mathrm{c}}| = 0$. Substitute this condition into Eq. (8.72), and the probability density function of the echo power ratio between the HV channel and the HH channel is given in Eq. (8.73).

$$p^{(n)}_{r_{\mathrm{hv/hh}}}(r_{\mathrm{hv/hh}}) = \frac{\tau^n \Gamma(2n)(\tau+r)r^{n-1}}{\Gamma(n)\Gamma(n)(\tau+r)^{2n+1}} \tag{8.73}$$

In the case of different look n, the probability density function of the echo power ratio between the cross-polarized channel and the co-polarized channel is shown in Fig. 8.9.

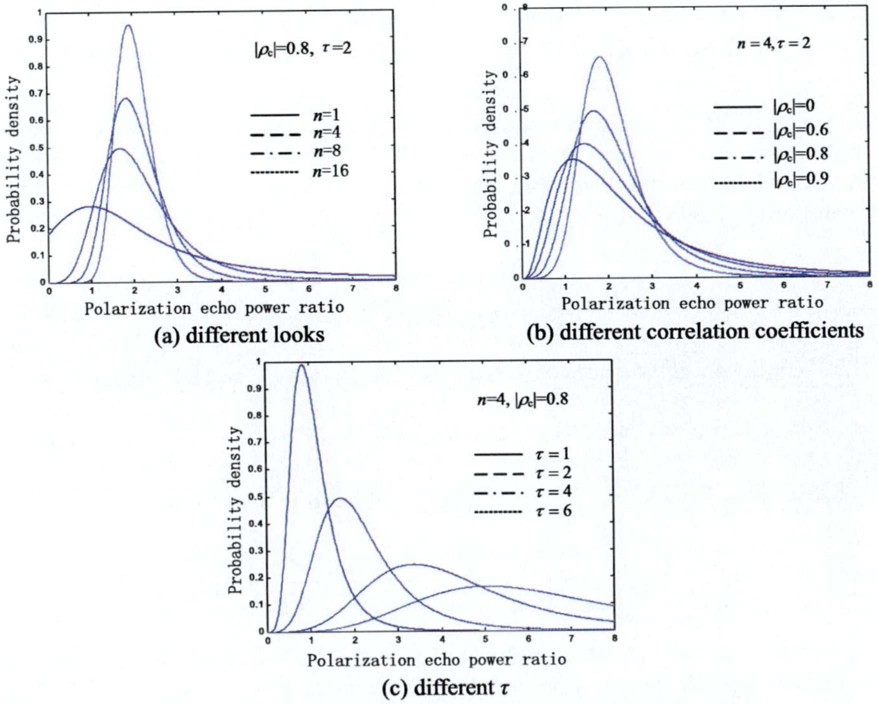

Fig. 8.8 Probability density function of echo power ratio for the co-polarization channel under different conditions

Fig. 8.9 Probability density function of echo power ratio for cross-polarization and co-polarization channel under different looks

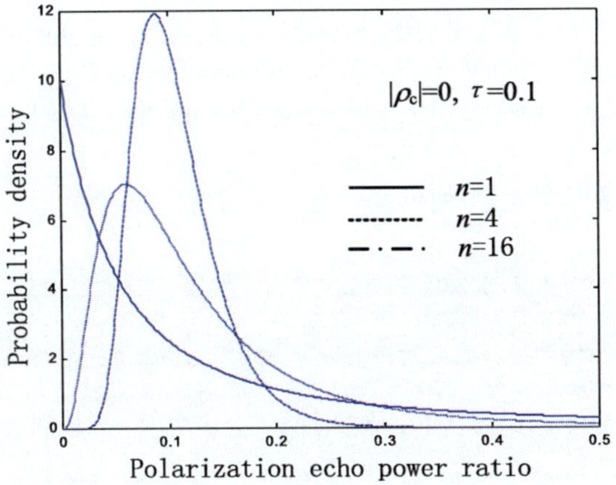

8.3.6 Probability Distribution of Amplitude Ratio for Two-Channel Echo in Multi-look Polarimetric SAR

From Eq. (8.69), calculate the square root of the normalized echo power ratio μ to get the normalized echo amplitude ratio $v = \sqrt{\mu}$. According to Eq. (8.70), the probability density function of the normalized echo amplitude ratio in multi-look polarimetric SAR is

$$p_v^{(n)}(v) = \frac{2\Gamma(2n)(1-|\rho_c|^2)^n(1+v^2)v^{2n-1}}{\Gamma(n)\Gamma(n)[(1+v^2))^2 - 4|\rho_c|^2v^2]^{n+1/2}} \tag{8.74}$$

In fact, the amplitude ratio of the echo signals in the VV and HH channels is

$$z_{\mathrm{vv/hh}} = \sqrt{r_{\mathrm{vv/hh}}} \tag{8.75}$$

Substituting Eq. (8.75) into Eq. (8.72), the probability density function is

$$p_{z_{\mathrm{vv/hh}}}^{(n)}(z_{\mathrm{vv/hh}}) = \frac{2\tau^n\Gamma(2n)(1-|\rho_c|^2)^n(\tau+z^2)z^{2n-1}}{\Gamma(n)\Gamma(n)[(\tau+z^2)^2 - 4\tau|\rho_c|^2z^2]^{(2n+1)/2}} \tag{8.76}$$

Similarly, let the correlation coefficient $|\rho_c| = 0$. The probability density function of the HV and HH channel signal amplitude ratio is

$$p_{z_{\mathrm{hv/hh}}}^{(n)}(z_{\mathrm{hv/hh}}) = \frac{2\tau^n\Gamma(2n)(\tau+z^2)z^{2n-1}}{\Gamma(n)\Gamma(n)(\tau+z^2)^{2n+1}} \tag{8.77}$$

When $n = 1$, the probability density function of the amplitude ratio in single-look polarimetric SAR can be obtained from Eqs. (8.76) and (8.77).

Figure 8.10 shows the probability density function of the normalized amplitude

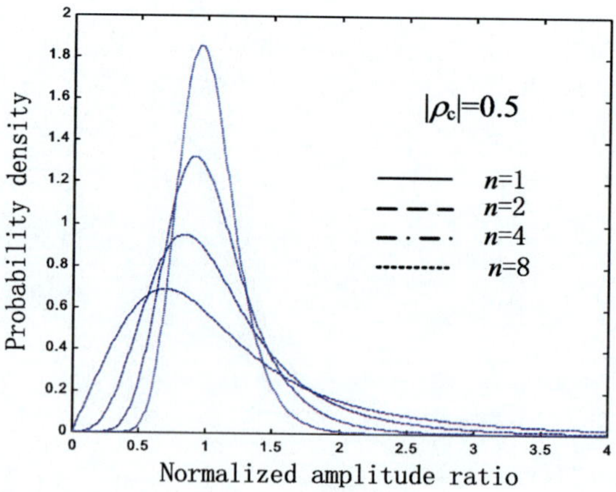

Fig. 8.10 Probability density function of normalized amplitude ratio in multi-look polarimetric SAR under different look numbers

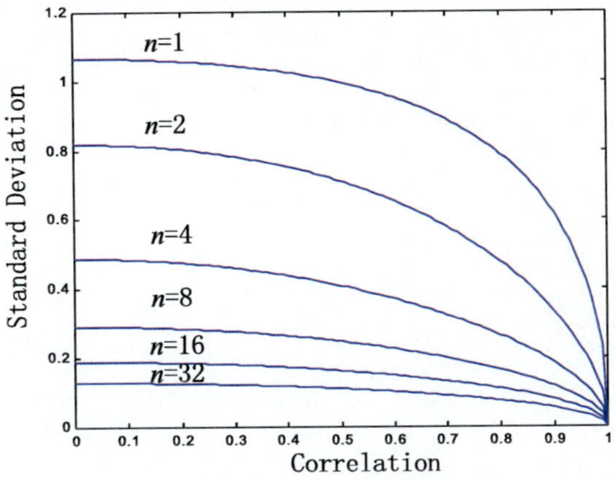

Fig. 8.11 The relationship between the standard deviation and the correlation coefficient of the amplitude ratio in multi-look polarimetric SAR under different look fields

ratio of multi-look polarimetric SAR images in different looks with the correlation coefficient $|\rho_c| = 0.5$. Consistent with the distribution of the normalized power ratio, the shape of the probability density function becomes narrower as the number of looks increases. Figure 8.11 shows the relationship between the standard deviation of the amplitude ratio and the correlation coefficient in multi-look polarimetric SAR image. It is obvious that the standard deviation of the amplitude ratio decreases significantly with the increase of the look number, which indicates that the multi-look processing has a certain smoothing effect on the polarimetric SAR data.

References

1. C. Oliver, S. Quegan, *Understanding synthetic aperture radar images* (Artech House, Boston London, 1998)
2. R. Touzi, A. Lopes, Statistics of the stokes parameters and of the complex coherence parameters in one-look and multilook speckle fields. IEEE Trans. GRS **34**(2), 519–531 (1996)
3. K. Sarabandi, Derivation of phase statistics from the mueller matrix. Radio Sci. **27**(5), 553–560 (1992)
4. I.R. Joughin, D.P. Winebrenner, D.B. Percival, Probability density functions for multilook polarimetric signatures. IEEE Trans. GRS **32**(3), 562–574 (1994)
5. D.R. Sheen, L.P. Johnston, Statistical and spatial properties of forest clutter measured with polarimetric synthetic aperture radar (SAR). IEEE Trans. GRS **30**(3), 578–588 (1992)
6. L. Jong-Sen, K.W. Hoppel, S.A. Mango et al., Intensity and phase statistics of multilook polarimetric and interferometric SAR imagery. IEEE Trans. GRS **32**(5), 1017–1027 (1994)
7. A. Lopes, F. Sery, Optimal speckle reduction for the product model in multilook polarimetric SAR imagery and the wishart distribution. IEEE Trans. GRS **35**(3), 632–647 (1997)
8. E. Rignot, R. Kwok, Characterization of spatial statistics of distributed targets in SAR data. Int. J. Remote Sensing **14**(2), 345–363 (1993)
9. G.Q. Liu, S.J. Huang, H. Xiong et al., Statistical analysis of multi-look polarimetric SAR imagery. Electron. Sci. J. **20**(1), 62–67 (1998)
10. R.J.A. Tough, D. Blacknell, S. Quegan, Estimators and distributions in single and multi-look polarimetric and interferometric data, in *Geoscience and Remote Sensing Symposium.*

IGARSS'94. Surface and Atmospheric Remote Sensing: Technologies, Data Analysis and Interpretation, vol. 4. International, pp. 2176–2178, (1994)

11. L.M.J. Brown, J.A. Conway, J.T. Macklin, Polarimetric synthetic aperture radar: fundamental concepts and analysis tools. Gec J. Res. **9**(1), 23–35 (1991)

Chapter 9
Speckle Filtering of Polarimetric SAR Image

Synthetic Aperture Radar is a coherent system. The echo phases of different scatters are different in the same resolution unit. The coherent superposition of the echo phases results in speckle noise of SAR images. Coherent speckle noise seriously affects SAR image understanding and feature extraction. The purpose of speckle filtering in SAR images is to preserve the texture features and scattering characteristics of the edges of the objects and to minimize the influence of speckle noise.

The traditional speckle filtering method is multi-looks averaging. This method can suppress the image speckle, but it loses image resolution at the same time. Multi-polarized Synthetic Aperture Radar (MPSAR) provides more ground object information than Single-polarized Synthetic Aperture Radar (SPSAR). Between the polarization channels it has the correlation, which can be used for speckle noise filtering of scattering matrix or covariance matrix.

In this chapter, speckle suppression techniques of Polarimetric Synthetic Aperture Radar (PSAR) images are studied, and the echo model of PSAR with speckle noise is analyzed. The methods of polarization whitening and vector filtering for speckle in PSAR images are discussed in detail. Polarization whitening filter parameter estimation methods include rectangular sliding window parameter estimation, non-rectangular window parameter estimation of Prewitt operator, improved Prewitt operator parameter estimation method, and fine Prewitt operator parameter estimation method. Vector filtering methods for polarimetric SAR data include Lee filtering algorithm based on least mean square error, refined Lee polarimetric vector filtering algorithm, and refined Lee filtering algorithm based on fine Prewitt operator.

R. Yang et al., *Polarimetric Microwave Imaging*,
https://doi.org/10.1007/978-981-15-8897-6_9

9.1 Polarimetric SAR Echo Signal Model with Coherent Speckle Noise

The statistical characteristics of polarimetric SAR data and the multiplicative noise model of speckle noise are the basis of speckle suppression. In 1976, Goodman proposed the concept of fully developed multiplicative speckle noise. Lee constructed a multiplicative noise model and proposed a series of spatial filtering methods. Typical methods include Lee filtering, fine Lee filtering, Kuan filtering, and Frost filtering [1–12].

The measurement vector X expression of polarimetric SAR is expressed as follows:

$$\mathbf{X} = \begin{bmatrix} S_{hh} \\ S_{hv} \\ S_{vv} \end{bmatrix}. \tag{9.1}$$

The complex elements in vector X are the joint complex Gauss distribution, and their probability density function is as follows:

$$f(\mathbf{X}) = \frac{1}{\pi^3 |C|} \exp\left(-\mathbf{X}^H C^{-1} \mathbf{X}\right) \tag{9.2}$$

where C is a polarized covariance matrix and its expression is

$$\mathrm{C} = \sigma_{\mathrm{HH}} \begin{bmatrix} 1 & 0 & \rho\sqrt{\gamma} \\ 0 & \varepsilon & 0 \\ \rho^*\sqrt{\gamma} & 0 & \gamma \end{bmatrix} \tag{9.3}$$

Here,

$$\sigma_{HH} = E\left\{|S_{hh}|^2\right\},\ \varepsilon = \frac{E\left\{|S_{hv}|^2\right\}}{E\left\{|S_{hh}|^2\right\}}, \tag{9.4}$$

$$\gamma = \frac{E\left\{|S_{vv}|^2\right\}}{E\left\{|S_{hh}|^2\right\}},\ \rho = \frac{E\left\{S_{hh} \cdot S_{vv}^*\right\}}{\left[E\left\{|S_{hh}|^2\right\} \cdot E\left\{|S_{vv}|^2\right\}\right]^{1/2}} \tag{9.5}$$

These are the polarization parameters commonly used.

Coherent speckle noise is multiplicative noise and its model is

$$\mathbf{Y} = \sqrt{g}\mathbf{X} \tag{9.6}$$

where X is a complex Gauss polarization vector and g is a spatial texture variable. It characterizes the radiation features of the surface and is related to the natural fluctuation and uniformity of the scene. Generally, it is assumed that g is Gamma distribution and its probability density function is as follows:

$$f_g(g) = \frac{1}{\overline{g}}\left(\frac{g}{\overline{g}}\right)^{(v-1)}\frac{1}{\Gamma(v)}\exp\left(\frac{-g}{\overline{g}}\right) \tag{9.7}$$

Here, $\overline{g}$ and v are parameters related to the mean and variance of random variable g.

$$E(g) = \overline{g}v \tag{9.8}$$

$$E\left(g^2\right) = \overline{g}^2 v(v+1) \tag{9.9}$$

From the above relations, it can be concluded that the measurement vector Y of the target obeys the generalized K distribution, and its probability density function is as follows:

$$f(\mathbf{Y}) = \frac{2}{\pi^3 \overline{g}^v \Gamma(v)|\Sigma|}\cdot\frac{K_{3-v}\left(2\sqrt{\frac{Y^H C^{-1} Y}{\overline{g}}}\right)}{\left(\overline{g}YC^{-1}Y\right)^{(3-v)/2}} \tag{9.10}$$

Suppose that $\overline{g} = 1/v$. That is, the mean value of texture variables is 1. When v $\rightarrow \infty$, the measured vector Y will degenerate to Gaussian model.

9.2 Polarization Whitening Filtering of Speckle in Full Polarization SAR Images

In 1990, Novak and Burl et al. proposed a Polarization Whitening Filter (PWF) method for speckle filtering in fully polarized SAR images. It reduces the SAR image speckle by optimally weighting the three complex elements S_{hh}, S_{vh}, S_{vv} in the polarization scattering vector. The aim of polarization whitening filtering is to obtain an image which has the minimum deviation-mean ratio. The speckle suppression effect of this method is the same as three-look average without reducing image resolution [1–3].

The effect of polarization whitening filter of SAR image is directly related to the accuracy of parameter estimation for target scattering vector covariance matrix. Initially, some scholars applied these parameters of an area in the image to the speckle suppression of the whole image. This way has a large error in processing the whole image. Then, some scholars used parameter estimation methods such as rectangular sliding window and Prewitt edge operator. In many cases, the estimation errors of these two methods are still very large, and the speckle suppression effect is not satisfied, sometimes even leading to the local image quality deterioration.

The improved parameter estimation method and the fine Prewitt operator parameter estimation polarization whitening filter algorithm have better filtering effect than the traditional method.

9.2.1 Polarization Whitening Filtering Algorithms

Due to the reciprocity theorem, and in backscattering coordinate system, the scattering complex vector of an isotropic target can be expressed as

$$X = \left[S_{hh} \; S_{vh} \; S_{vv} \right]^T \tag{9.11}$$

The covariance matrix is

$$C = E\left[X \cdot X^H\right] = \sigma_{HH} \begin{bmatrix} 1 & 0 & \rho\sqrt{\gamma} \\ 0 & \varepsilon & 0 \\ \rho^*\sqrt{\gamma} & 0 & \gamma \end{bmatrix} \tag{9.12}$$

$E[]$ represents the average of the data, the superscript T denotes transpose, and H denotes complex conjugate transpose. The polarization parameters are as follows:

$$\sigma_{HH} = E\left\{|S_{hh}|^2\right\},\; \varepsilon = \frac{E\left\{|S_{hv}|^2\right\}}{E\left\{|S_{hh}|^2\right\}}$$
$$\gamma = \frac{E\left\{|S_{vv}|^2\right\}}{E\left\{|S_{hh}|^2\right\}},\; \rho = \frac{E\left\{S_{hh} \cdot S_{vv}^*\right\}}{\left[E\left\{|S_{hh}|^2\right\} \cdot E\left\{S_{vv}^2\right\}\right]^{1/2}}, \tag{9.13}$$

Assuming that speckle noise is multiplicative, its model as (9.6), rewritten as follows:

$$Y = \sqrt{g}X \tag{9.14}$$

where g is a spatial texture variable, which represents the scattering characteristics of the surface, and is related to the natural fluctuation and uniformity of the scene. g obeys Gamma distribution.

The speckle measurement factor (s/m) is usually used to measure the speckle level of SAR image. It is defined as the ratio of standard deviation of pixel intensity to mean value. The calculation formula is (9.15). Equivalent look number (N_{ENL}) is defined as (9.16).

$$s/m = \frac{\sqrt{\mathrm{var}\{y\}}}{\mathrm{mean}\{y\}} \tag{9.15}$$

$$N_{ENL} = 1/(s/m)^2 \tag{9.16}$$

It can be seen the reduction of s/m is the equivalent as improving the look number N_{ENL} of the image. The purpose of polarization whitening filtering is to find a group of polarization combination of antennas, so as to minimize the s/m. In other words, the optimal weighting vector is found to minimize the (s/m) of the image.

$$y = \boldsymbol{Y}^H \boldsymbol{W}_{opt} \boldsymbol{Y} = g\boldsymbol{X}^H \boldsymbol{W}_{opt} \boldsymbol{X} \tag{9.17}$$

It is not difficult to prove that the following relationship exists with Formula (9.17):

$$E\{\boldsymbol{X}^H \boldsymbol{W}_{opt} \boldsymbol{X}\} = tr(\boldsymbol{C}\boldsymbol{W}_{opt}) = \sum_{i=1}^{3} \lambda_i \tag{9.18}$$

$$\mathrm{var}\{\boldsymbol{X}^H \boldsymbol{W}_{opt} \boldsymbol{X}\} = tr(\boldsymbol{C}\boldsymbol{W}_{opt})^2 = \sum_{i=1}^{3} \lambda_i^2 \tag{9.19}$$

Here, E denotes the mean, var denotes the variance, and tr denotes the trace of the matrix, and $\lambda_1, \lambda_2, \lambda_3$ are the eigenvalues of $\boldsymbol{C}\boldsymbol{W}_{opt}$ matrix.$\boldsymbol{C}\boldsymbol{W}_{opt}$ is Hermitian matrix.

$$\begin{aligned} \left(\frac{s}{m}\right)^2 &= \left[\frac{\mathrm{var}(y)}{E^2(y)}\right] = \left[\frac{E(g^2)}{E^2(g)} \cdot \frac{\mathrm{var}(\boldsymbol{X}^H \boldsymbol{W}_{opt} \boldsymbol{X})}{E^2(\boldsymbol{X}^H \boldsymbol{W}_{opt} \boldsymbol{X})} + \frac{\mathrm{var}(g)}{E^2(g)}\right] \\ &= \left[\frac{v+1}{v} \cdot \frac{\sum_{i=1}^{3} \lambda_i^2}{\left(\sum_{i=1}^{3} \lambda_i\right)^2} + \frac{1}{v}\right] \end{aligned} \tag{9.20}$$

where v is the variance of g and is a constant. Finding the minimum value is equivalent to finding the minimum value of the following formula:

$$\sum_{i=1}^{3} \lambda_i^2 \Big/ \left(\sum_{i=1}^{3} \lambda_i\right)^2 \tag{9.21}$$

Since $\lambda_1, \lambda_2, \lambda_3$ are all real, the following relationship is established:

$$\lambda_1^2 + \lambda_2^2 \geq 2\lambda_1\lambda_2,\ \lambda_1^2 + \lambda_3^2 \geq 2\lambda_1\lambda_3,\ \lambda_2^2 + \lambda_3^2 \geq 2\lambda_2\lambda_3 \tag{9.22}$$

Substituting Eq. (9.22) into (9.21)

$$\frac{\sum_{i=1}^{3}\lambda_i^2}{\left(\sum_{i=1}^{3}\lambda_i\right)^2}=\frac{\sum_{i=1}^{3}\lambda_i^2}{\sum_{i=1}^{3}\lambda_i^2+2\lambda_1\lambda_2+2\lambda_1\lambda_3+2\lambda_2\lambda_3}\geq\frac{\sum_{i=1}^{3}\lambda_i^2}{3\sum_{i=1}^{3}\lambda_i^2}=\frac{1}{3} \quad (9.23)$$

The medium number is established, that is, the condition that (s/m) takes the minimum value is $\lambda_1=\lambda_2=\lambda_3$. That is, the optimal weight vector $\boldsymbol{W}_{opt}$ is a matrix such that $\boldsymbol{C}\boldsymbol{W}_{opt}$ is an eigenvalue equal matrix. The optimal weight vector is

$$\boldsymbol{W}_{opt}^{*}=\boldsymbol{C}^{-1} \quad (9.24)$$

Equation (9.24) indicates that the measurement vector Y is whitened, and to convert colored noise into white noise. After obtaining polarization whitening filtering, the image intensity is

$$y=|S_{hh}|^2+\left|S_{hv}/\sqrt{\varepsilon}\right|^2+\left|\left(S_{vv}-\rho^{*}\sqrt{\gamma}S_{hh}\right)\Big/\sqrt{\gamma(1-|\rho|^2)}\right|^2 \quad (9.25)$$

9.2.2 Polarization Whitening Filter Parameter Estimation Method

It can be seen from (9.25) that the key to the polarization whitening filtering method is to estimate the parameter ρ, ε, γ when the target scattering matrix is known. At first, scholars applied the ρ, ε, γ value of representative regional statistics data in the image to the whole image. This method is non-adaptive. The algorithms developed and improved on this basis include parameter estimation method for rectangular sliding window, parameter estimation method for Prewitt operator, improved Prewitt operator parameter estimation method, and fine Prewitt operator parameter estimation method.

9.2.2.1 Rectangular Sliding Window Parameter Estimation

The parameter estimation method of the rectangular sliding window first adopts a rectangular window (such as 5×5, 7×7, etc.) to sequentially slide over the polarization SAR measurement matrix, and estimates the scattering features of the pixel value of the window center pixel by using the statistical average value of the elements in the sliding window. Features calculate the corresponding polarization covariance matrix parameters and perform polarization whitening filtering processing. This method uses the statistical average of all the sample points in the window to calculate the polarization covariance matrix parameters, which is an approximation algorithm with low estimation accuracy.

The parameter estimation method of the rectangular sliding window first estimates the polarization SAR measurement matrix data by using a rectangular window (such as $5 \times 5, 7 \times 7$, etc.), and then uses the statistical average of the elements in the sliding window to estimate the scattering of the pixel values in the center of the window. The feature is used to calculate the corresponding polarization covariance matrix parameters and perform polarization whitening filtering processing. This method uses the statistical average of all the sample points in the window to calculate the polarization covariance matrix parameters, which is an approximation algorithm whose estimation accuracy is not satisfactory.

9.2.2.2 Parameter Estimation of Non-rectangular Window of Prewitt Operator

The parameter estimation method of the non-rectangular window of Prewitt operator first determines which area pixels belong to the same scattering mechanism as the central pixel in the sliding window and then selects those data that belong to the same scattering mechanism as the central pixel to estimate the ρ, ε, γ of the central pixel.

The detection template of this method is shown in Fig. 9.1 (taking 9×9 template as an example). Figure 9.1a–d gives a set of edge detection templates and the irmask structures. The right of each group is the mask, which is the same size as the detection template matrix. "Space," "+," and "−," respectively, indicate that the corresponding element values of the mask matrix are 0, +1, and −1.

When the topographical structure of the treated area is complex, the above detection template cannot meet the accuracy requirements of parameter estimation. The literature [11] improves the Prewitt operator parameter estimation method.

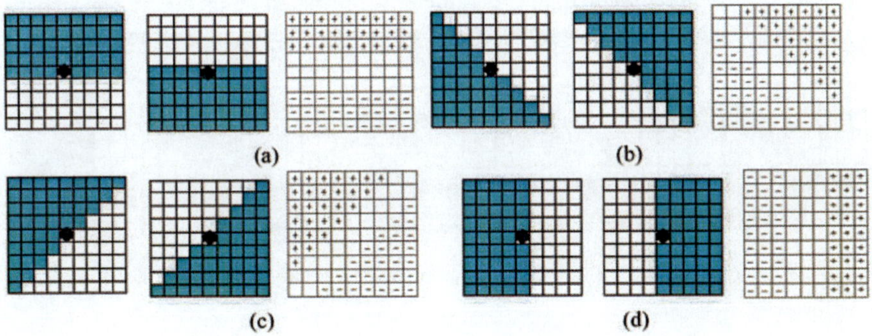

Fig. 9.1 Detection template and mask of non-rectangular window of Prewitt operator

9.2.2.3 Improved Prewitt Operator Parameter Estimation [9–11]

Based on the Fig. 9.1 detection template, the improved Prewitt operator parameter estimation method adds the edge detection type of the detection template, makes the edge detection type more in detail, and improves the discrimination method of the texture region to which the central pixel is to be processed. The added detection template is shown in Fig. 9.2. Only the gradual process of Fig. 9.1c–d is given, and the mask matrix and other templates can be analogized.

The method of Sect. 9.2.2.2 is a good way to determine the texture direction of the data image in the window to be processed. However, the discrimination of the area to which the central pixel belongs is not very accurate. Here, it is proposed to determine the region to which the center point belongs by the variation coefficient f. This method is more scientific and accurate for the identification of the region to which the central pixel is located. The coefficient of variation is defined as

$$variation\,coefficient(f) = standard\,deviation(d) / mean(m) \tag{9.26}$$

The parameter estimation process of this method is as follows:

(1) Determine the texture direction of the region where the center pixel of the template is located: multiply the various masks by the amplitude values of the image data in the window, respectively (the result is represented by the matrix M), and then find the absolute value of the matrix M sum. The direction of the detection template corresponding to the mask with the largest absolute value is the texture direction of the region to which the central pixel point belongs.
(2) If the edge is judged to be the structure shown in Fig. 9.1d, the shaded portions f_1,f_2 of the left two figures in the template of Fig. 9.2 are calculated using (9.26), respectively.
(3) The shaded area corresponding to the smaller one of f_1 and f_2 is the same scattering mechanism area as the central pixel.
(4) The central pixel point $\rho, \varepsilon, \gamma$ parameter is estimated by using the region data, and can be applied to the polarization whitening filtering algorithm.

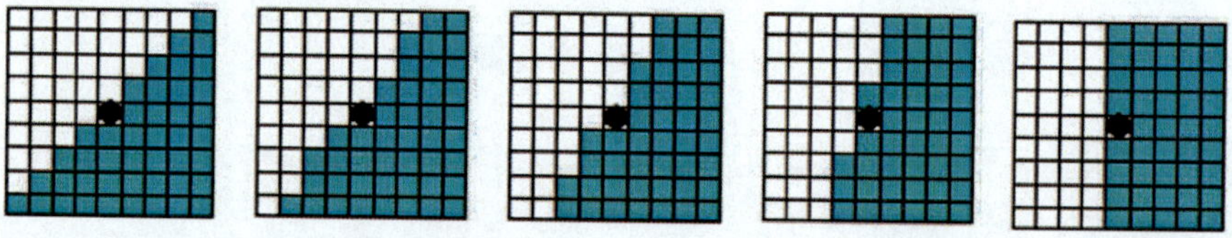

Fig. 9.2 Example of an added template for the improved Prewitt operator parameter estimation method

9.2.2.4 Fine Prewitt Operator Parameter Estimation

When there is a strong point target in the window, or there is a line and the center pixel belongs to this line, the above detection method cannot detect and judge it. The parameter estimation method of the fine Prewitt operator is further refined based on the structure detection model from Lopes et al. [4, 5] Gamma map algorithm is the improved detection method of window image texture structure.

The single-polarized synthetic aperture radar Gamma map algorithm uses the coefficient of variation and the intensity contrast coefficient to detect the texture structure of the data image in the window. As shown in Fig. 9.1, the left- and right-shaded portions, respectively, represent a texture, and I1 and I2, respectively, represent the average intensity of each texel. The intensity contrast factor is defined as

$$r = \max[(I_1/I_2), (I_2/I_1)] \tag{9.27}$$

The template added by the fine Prewitt operator method is shown in Fig. 9.3 (only one of the intermediate gradient examples is shown in the figure). A detection template for the fine Prewitt operator parameter estimation method consisting of the detection templates is introduced in Figs. 9.1, 9.2, and 9.3 and the class is obtained. Figures 9.1 and 9.2 show the detection of the existence and direction of the edge; Fig. 9.3 shows the existence and direction of the first five templates to detect the line structure, and the last template to detect the strong point target. When performing point target detection, the point target template should not be too large. Considering the side lobe effect, the point target side lobe data can be appropriately increased when estimating the parameters, as the same as is the line target.

The intensity of the full-polarization synthetic aperture radar scattering vector element is uniformly weighted to obtain the total power (Span) map of the fully polarized synthetic aperture radar. As shown in formula (9.28), it has a certain speckle suppression effect. Window structure detection using total power image data further increases the accuracy of texture detection.

$$P_{Span} = |S_{hh}|^2 + 2|S_{hv}|^2 + |S_{vv}|^2 \tag{9.28}$$

The steps of the fine Prewitt operator parameter estimation method are as follows:

(1) Calculate the total power image pixel value of the fully polarized synthetic aperture radar using Eq. (9.28).

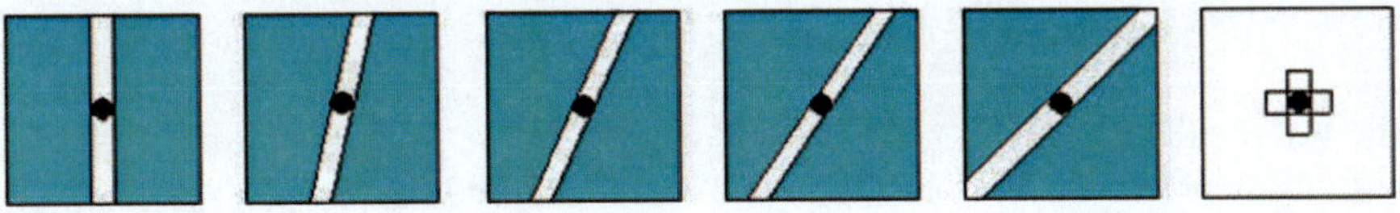

Fig. 9.3 Example of an added template for the parameter estimation method of the fine Prewitt operator

(2) Calculate the coefficient of variation f of the total window power image using Eq. (9.26). When $f < f_{\max}$, ($f_{\max}$ is the threshold), the data in the window belongs to a non-textured structure. Take all the data in the window to estimate the $\rho, \varepsilon, \gamma$ parameters of the central pixel; otherwise, the data in the window belongs to the texture structure, and continue to detect the image texture in the window.
(3) For a single-polarized synthetic aperture radar image $f_{\max} = 1/\sqrt{2 \times ENL + 1}$. Let the full-polarized total power image have a speckle reduction factor of $(s/m)_{Span}$ and an equivalent look number of $(ENL)_{Span} = 1/\left((s/m)_{Span}\right)^2$. Then, the detection threshold of the full-polarization total power image variation coefficient is $f_{\max} = 1/\sqrt{2 \times ENL \times (ENL)_{Span} + 1}$.
(4) According to the line detection template in Fig. 9.3, calculate the intensity contrast coefficient r of each direction line in the span image by using Eq. (9.27), take the maximum value as r line, and record the line direction corresponding to the maximum value; Template, calculate the point target intensity contrast coefficient r point.
(5) If r line > rl, it is determined that the center pixel of the window belongs to the line structure corresponding to r line. The white pixel data of the line is used to perform $\rho, \varepsilon, \gamma$ parameter estimation on the central pixel. If r point > r p, it is determined that the center pixel of the window belongs to a strong point structure. The central pixel point parameter $\rho, \varepsilon, \gamma$ is estimated using the center pixel and its side lobe region data, where rl and rp are detection thresholds of line and point.
(6) If none of the items in (5) is true, the window belongs to the edge texture structure, and the edge detection and parameter estimation are performed by the improved Prewitt operator method in Sect. 9.2.2.3.

This method increases the detection of points and line targets, and reduces the influence of strong points and line structures on the edge detection in the window. After the point and line structures have been detected, they can be shielded in the subsequent structure detection, and the influence of the point and line structure on the image texture structure detection is avoided as much as possible. When the structure of the area to be processed is relatively simple, the detection template can be appropriately simplified to reduce calculation time and unnecessary trouble.

9.2.3 Experiment and Result Analysis of Polarization Whitening Filtering Algorithm

The test data is the SIR-C L-band fully polarized SAR image data of NASA JPL somewhere near Lake Berga, Russia. The data is subjected to full-polarization SAR polarization whitening filtering, and the speckle suppression image is obtained.

The speckle suppression performance of several parameter estimation methods is analyzed in detail below.

9.2.3.1 Comparison of Amplitude After Polarization Whitening Filtering

Figure 9.4 shows the amplitude of the image before and after polarization whitening filtering. Figure 9.4a is the amplitude of the original data HH polarization. Figure 9.4b is the speckle reduction image obtained by polarization whitening filtering from the parameters obtained by the parameter estimation method of the fine Prewitt operator. From the overall and partial view of Fig. 9.4b, the speckle suppression effect is better: the left part is smoother and the details are clearer; the right part details and strong targets are obvious, and the image texture is clearer. The strong targets at D, E, and F in the figure are clearer and more obvious, and the image is very good. The texture of the middle and right images is more complicated. There are many point and line structures. The detailed Prewitt operator parameter estimation method can detect complex point and line structures, with small parameter estimation error and good polarization whitening filtering effect.

(a) The amplitude of the HH Image of Raw Data

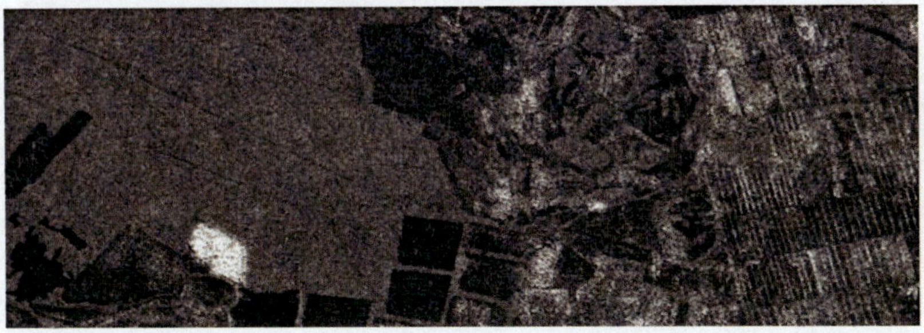

(b) PWF Amplitude of Specimen Prewitt Operator Parameter Estimation Method

Fig. 9.4 Polarized whitening filtered image

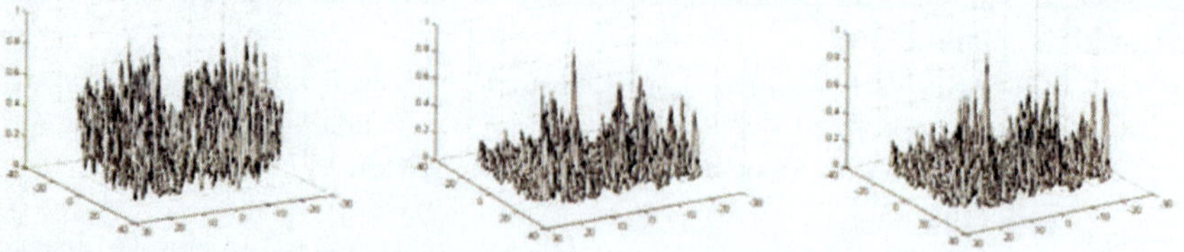

(a) HH Amplitude of Raw Data (b) Improved Prewitt Operator Method (c) Fine Prewitt operator method

Fig. 9.5 Amplitude of the A region before and after filtering

The filter is usually judged by the speckle suppression index and the edge enhancement index. The speckle reduction index is used to identify the suppression performance of the filter on the image speckle. The edge enhancement index is used to identify the enhancement of the edge of the filter while suppressing the speckle. A good filter should have both good speckle suppression and edge enhancement.

Figure 9.5 shows the amplitude before and after polarization whitening filtering. The effect of the speckle reduction before and after the polarization whitening filtering process can be more clearly seen in the figure. The image in Fig. 9.5a is very messy. The noise in Fig. 9.5b, c is suppressed. The strong target is more obvious, and the middle and right half edges are more obvious. The border texture in the lower right corner of Fig. 9.5c is clearer than Fig. 9.5b.

9.2.3.2 Filter Performance Evaluation of Polarization Whitening Filter

Before performing the polarization whitening filtering process, the original data is processed by two-look averaging. According to the relationship between the speckle factor (s/m) and the equivalent look number ENL, the (s/m) theoretical value of the HH polarized image is

$$(s/m)_{HH} = 1\Big/\sqrt{2} = 0.7071 \tag{9.29}$$

It can be seen from the prior that the polarization whitening filtering algorithm can achieve the equivalent speckle suppression effect equivalent to three-look averaging. Therefore, the (s/m) theoretical value of the image obtained after the polarization whitening filtering process is

$$(s/m)_{PWF} = \left(1\Big/\sqrt{3}\right) \times (s/m)_{HH} = 0.4082 \tag{9.30}$$

Typically, the (s/m) value is calculated from a single, stationary scattering region. Generally, choose a gentle desert and calm sea surface, lake surface, and other areas where the echo signal is weak and constant. The B region in the graph is selected to

Table 9.1 Statistical table of speckle factor in Region B

	HH image	Rect window	Prewitt	Improved prewitt	Fine prewitt
(s/m)	0.7083	0.4140	0.4116	0.4102	0.4089

evaluate the filtering performance of the polarization whitening filter. The speckle factor for this region is shown in Table 9.1.

It can be seen from Table 9.1 that several parameter estimation methods for polarization whitening filtering have obtained good speckle suppression effects. The speckle suppression factor (s/m) is close to the theoretical value of 0.4082. The results of the last two methods are closer to the theoretical values, indicating that the two parameter estimation methods are more accurate and the speckle suppression effect is better.

9.2.3.3 Comparison of Image Resolution After Polarization Whitening Filtering

Figure 9.6 is the amplitude of the polarization whitening filter speckle reduction image for the different parameter estimation methods in the C region in Fig. 9.4a. Figure 9.6a is the HH image amplitude, Figure (b) is the Prewitt operator filtered image amplitude, Figure (c) is the improved Prewitt operator filtered image amplitude, and Figure (d) is the fine Prewitt operator filtered image amplitude. Comparing the four images, it can be found that the noise suppression effect of the filtered data is obvious, the image strong target is clearer, and the characteristics of the point target are well maintained. In order to specifically study the image resolution before and after filtering, the resolution of the intermediate strong point target in the region is calculated below.

The spatial resolution is defined as the main response 3 dB bandwidth of the impulse response function. Manually designed point targets are typically used to calculate spatial resolution. When the artificial point target cannot be obtained, it can be calculated by finding an appropriate area with the point target characteristic in the image. In this experiment, the point C target in Fig. 9.4a is selected to calculate the image resolution before and after filtering. The raw data range resolution is 15.3445 m and the azimuth resolution is 20.21128 m.

Table 9.2 is the table of the 3 dB width of the point C of the C zone. As can be seen from the table, the target azimuth dimension is close to the image resolution and has a point target characteristic. The results show that the polarization whitening filter does not reduce the resolution of the polarimetric SAR image when the speckle reduction is performed.

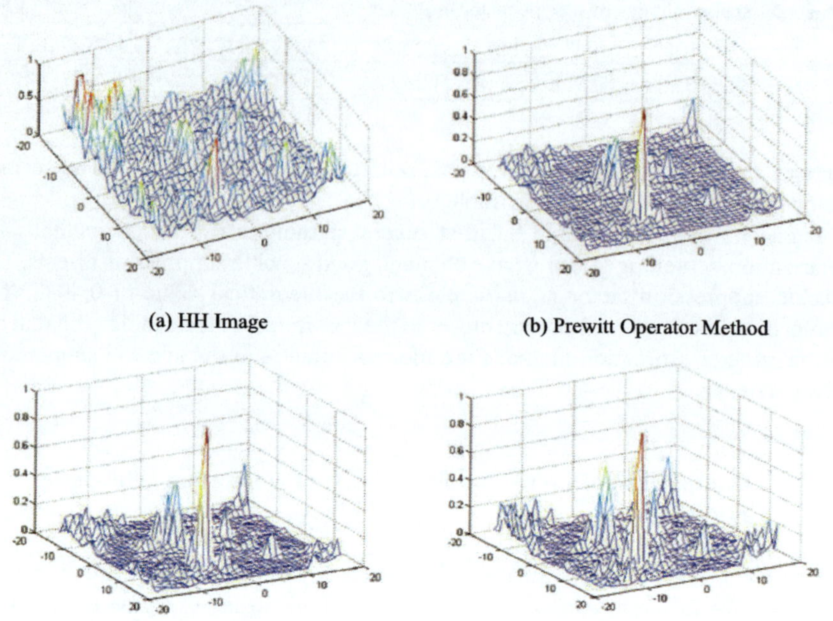

(a) HH Image

(b) Prewitt Operator Method

(c) Improved Prewitt Operator Method

(d) Fine Prewitt Method

Fig. 9.6 Image amplitude of polarization whitening filtering speckle reduction of different parameter estimation methods in C region

Table 9.2 Point target azimuth and distance to 3 dB width in Zone C

	HH Image	Prewitt	Improved prewitt	Fine prewitt
Range (m)	14.424	14.432	14.422	14.270
Azimuth (m)	33.146	29.386	27.487	27.926

9.3 Vector Lee Filtering for Polarized SAR Data

In order to analyze the target polarization scattering mechanism well, it is often desirable to reduce the data noise level and maintain the data polarization information before the polarization information processing. The polarization whitening filter introduced in the previous section is a weighting algorithm, which inevitably brings the loss of polarization information. The polarization whitening filter can only obtain the speckle suppression image, and the filtered data loses the polarization information of the data. The multi-look average filtering method can reduce the SAR image noise while maintaining the polarization information of the data, but it equivalently reduces the resolution of the image. Therefore, the number of multi-look processing is usually not high [2–6].

Lee [1, 2] believes that any ideal polarimetric SAR image speckle filtering algorithm should be developed around the following three basic criteria:

(1) Try to avoid interference between different polarization channels. It is required that each element in the polarization measurement vector must be independently spatially filtered. Any filtering algorithm that attempts to exploit the correlation of polarization data will inevitably introduce crosstalk between channels.
(2) After filtering, the resolution of the image and other characteristics such as edge sharpness and strong target diffusion degree should be kept as much as possible. It is required that the selected filtering algorithm should be adaptive and try to select isotropic adjacent pixels for processing.
(3) Keep the polarization characteristics of the data after filtering as much as possible, which can be verified by comparing the polarization characteristics of the data before and after filtering.

The local statistical filtering algorithm is derived based on the mathematical principle of minimum mean square error. It is a typical adaptive spatial filtering algorithm. Maintaining the filtered image resolution and the sharpness of the image edge and the degree of target diffusion are as important as possible, which is an important index for the quality evaluation of the filtered image.

9.3.1 Lee Filtering Algorithm Based on Minimum Root Mean Square Error

The minimum rms error local statistical vector filter (Lee filter) is a filtering method based on local mean of minimum mean square error, which avoids mutual interference between different polarization channels. It is an adaptive statistical filtering method, which is very good. The polarization information of the original data is maintained without significantly reducing the image resolution.

The ideal noise-free model of a fully polarized SAR scattering vector can be expressed as

$$X = \left[x_{hh} \; x_{hv} \; x_{vv} \right]^T \tag{9.31}$$

The subscript h indicates horizontal polarization and v indicates vertical polarization. Then the scattering vector of the multiplicative speckle noise is expressed as

$$Z = VX \tag{9.32}$$

where

$$V = \begin{bmatrix} v_{hh} & 0 & 0 \\ 0 & v_{hv} & 0 \\ 0 & 0 & v_{vv} \end{bmatrix}$$

is a multiplicative speckle noise with a mean of I and a variance of σ_v^2.

Assume that X and V are statistically independent. Let $\hat{X}$ and $\overline{X}$ be the estimated and average values of pixel X, respectively, then $\hat{X}$ can be expressed as

$$\hat{X} = a\overline{X} + bZ \tag{9.33}$$

Since the prior statistical average of each pixel X cannot be obtained, it is usually approximated by local statistical average. The data samples used for statistics can be selected by the valid data selection criteria in Sect. 9.2.2.3. By (9.32) there are

$$\overline{Z} = E[Z] = \overline{X} \tag{9.34}$$

Here $\overline{Z}$ can also be estimated by a local statistical window. Parameters a and b can be found using the minimum root mean square error criterion. The minimum root mean square error criterion is as follows:

$$J = E\left[\left(\hat{X} - X\right)^2\right] \rightarrow \min \tag{9.35}$$

Substituting (9.34) into (9.35), the optimal a and b values should satisfy the following:

$$\begin{aligned} \frac{\partial J}{\partial a} &= 0 \Rightarrow E\left[\overline{X}\left(a\overline{X} + bZ - X\right)\right] = 0 \\ \frac{\partial J}{\partial b} &= 0 \Rightarrow E\left[Z\left(a\overline{X} + bZ - X\right)\right] = 0 \end{aligned} \tag{9.36}$$

By (9.36):

$$a = 1 - b \tag{9.37}$$

$$E\left[Z\left(X - \overline{X}\right) + b\left(\overline{X} - Z\right)Z\right] = 0 \tag{9.38}$$

The relationship between X and V is statistically independent.
Since $E\left[\overline{X}\left(X - \overline{X}\right)\right] = \overline{X}E\left[\left(X - \overline{X}\right)\right] = 0$, the first item is

$$\begin{aligned} &E\left[Z\left(X - \overline{X}\right)\right] = E\left[XV\left(X - \overline{X}\right)\right] = E\left[X\left(X - \overline{X}\right)\right] \\ &= E\left[X\left(X - \overline{X}\right) - \overline{X}\left(X - \overline{X}\right)\right] = E\left[\left(X - \overline{X}\right)^2\right] \end{aligned} \tag{9.39}$$

Since $\overline{Z} = E[Z] = \overline{X}$, the second term of (9.38) is

$$E\left[b\left(\overline{X}-Z\right)Z\right]=bE\left[\left(\overline{Z}-Z\right)Z-\overline{Z}\left(\overline{Z}-Z\right)\right]=-bE\left[\left(Z-\overline{Z}\right)^2\right] \quad (9.40)$$

Substituting equations (9.40), (9.38) into Eq. (9.37), for each polarization, we get

$$b=\frac{\mathrm{var}(X)}{\mathrm{var}(Z)} \quad (9.41)$$

Thus, under the minimum root mean square error criterion, the estimated value $\hat{X}$ of the vector X is

$$\hat{X}=\overline{X}+MP^{-1}(Z-\overline{X}) \quad (9.42)$$

where $M=\left(P-\overline{Z}^2\sigma_v^2\right)/\left(1+\sigma_v^2\right)$.

$\overline{X}$ is the a priori average of each pixel, and $\overline{Z}$ is the average of the measurement vectors, thus t.

$$\overline{X}=\overline{Z}$$

where $M=\mathbf{Cov}(X)$,$P=\mathbf{Cov}(Z)$, and **Cov** represent covariance matrices.

For an isotropic homogeneous medium,$\mathbf{Cov}(X)\approx\mathbf{Cov}(Z)$, and $\hat{X}=\overline{X}$, the estimated value, is equal to the local statistical average. In areas where the ground objects differ greatly, $\mathbf{Cov}(X)\approx\mathbf{Cov}(Z)$, and $\hat{X}=Z$. That is, the filtered pixel is substantially equal to the original pixel, and the filtering smoothing effect is very weak. It can be seen that the local statistical Lee filtering method of the minimum root mean square error is an adaptive algorithm.

According to formula (9.42), the data filtering algorithm in single polarization or multi-polarization mode can be obtained. When independently filtering each polarization mode in a fully polarized vector, the following relationship is used:

$$\hat{x}=\overline{y}+b(y-\overline{y}) \quad (9.43)$$

$$b=\frac{\mathrm{var}(x)}{\mathrm{var}(y)},\quad \mathrm{var}(x)=\frac{\mathrm{var}(y)-\overline{y}^2\sigma_v^2}{1+\sigma_v^2} \quad (9.44)$$

where $\mathrm{var}(y)$ is the variance of the image data and can be obtained by local statistics of the image data. By means of (9.43), (9.44), it can filter each element of the covariance matrix of the fully polarized SAR image, which has the advantage of avoiding mutual interference between different polarization channels as much as possible. This is critical to eliminating crosstalk between different channels in the filtering algorithm.

9.3.2 Refined Lee Vector Filtering Algorithm

The Refined Lee filter algorithm is based on the data selection criterion of the Prewitt operator parameter estimation method, and uses the local statistical minimum root mean square error criterion to perform vector filtering on the PolSAR data [7, 8]. The PolSAR total power (Span) image is used as the edge detection map. The Span image is the sum of HH, HV + VH, VV, and its noise level is lower than that of a single polarization image. In addition, the scattering characteristics of HH, HV, and VV images are different, and the scattering characteristics displayed in different polarization images may be included in the Span image. After selecting the window edge based on the Span image, the pixels at the edge of the window are used to calculate the mean of each element in the covariance matrix, and the filter weighting coefficients calculated by the Span image can be equally used for the filtering of each element of the covariance matrix. The use of Span images avoids the effects of channel crosstalk when using a polarized image and maintains the scattering characteristics of the homogeneous scattering region. The steps of the refined Lee filtering algorithm are as follows:

(1) Data selection in the window: For each pixel, use the Span image to select a non-rectangular window that matches the edge. See Sects. 9.2.2.2 and 9.2.2.3 for the selection method. All elements in the covariance matrix C are filtered by means of the data within the selected window edge.
(2) Calculating the filter weighting coefficient: Calculating the filter weighting coefficient b of the Span image by the local statistical method according to Eq. (9.44).
(3) Filtering the covariance matrix: each element of the covariance matrix C is independently filtered using the same window data and the same weighting coefficient b. The filtered covariance matrix is

$$\hat{C} = \overline{C} + b(C - \overline{C}) \tag{9.45}$$

where in each element in $\overline{C}$ is calculated from the local average of the covariance matrix of the pixels in the same edge detection window.

9.3.3 Refined Lee Filtering Algorithm Based on Fine Prewitt Operator

The refined Lee filtering algorithm based on the fine Prewitt operator according to the data selection criterion of the fine Prewitt operator parameter estimation method, and the refined Lee filtering algorithm is used to perform vector filtering on the PolSAR data [10]. The algorithm uses the PolSAR total power (Span) image as the edge detection input image, and uses the texture detection template effective

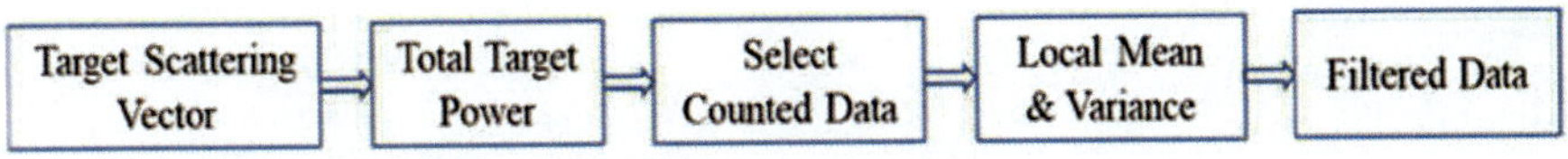

Fig. 9.7 PolSAR image data vector filtering algorithm process

data selection template of the fine Prewitt operator parameter estimation method in Sect. 9.2.2.4. The data is filtered using the refined Lee filtering method of Sect. 9.3.2.

Valid data is selected by the adaptive structure detection window in Sect. 9.2.2.4. This improves the accuracy of valid data selection in the window, maintaining image resolution, edge sharpness, and avoiding the spread of strong target. The process of Lee filtering algorithm for fully polarized SAR data is shown in Fig. 9.7. The specific steps are as follows:

(1) Obtaining a polarization measurement vector Z of the fully polarized SAR data.
(2) Calculate the total power map according to formula (9.18).
(3) Using the effective data selection criteria of Sect. 9.2.2.3, preprocessing the polarization measurement vector complex data to obtain the local statistical average vector $\overline{Z}$ of the polarization measurement vector.
(4) Find the covariance value of the local statistics of each image.
(5) Filtering each element in the SAR image covariance matrix according to (9.33) (9.34), or performing vector filtering on each pixel in each polarization mode of the scattering matrix.

9.3.4 *Experiment and Result Analysis*

The test data is shown in Fig. 9.8, where (a) the HH polarization map is the original data grayscale map and (b) the graph is the speckle patch filtered grayscale image. Comparing (a) and (b), it can be clearly seen that the speckle vector filtering method

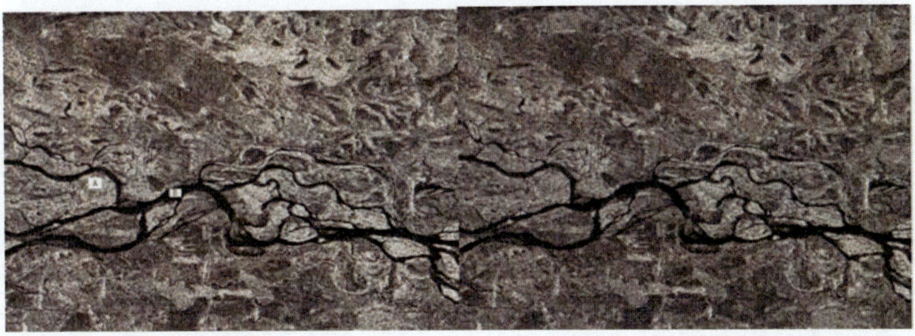

(a) Amplitude image of HH raw data (b) Amplitude image of filtered HH

Fig. 9.8 HH amplitude image before and after filtering

Table 9.3 Mean, variance, and variance mean ratio of raw data in Region B

	Mean	Variance	Variance mean ratio
HH	0.2845	0.1484	0.5215
HV	0.0931	0.0584	0.6270
VV	0.2179	0.1155	0.5303

Table 9.4 Mean, variance, and variance mean ratio of filtered data in Region B

	Mean	Variance	Variance mean ratio
HH	0.2447	0.0826	0.3376
HV	0.0758	0.0321	0.4235
VV	0.1698	0.0605	0.3563

effectively suppresses the speckle of the HH channel image without causing severe loss of image detail information and resolution.

The data in region B is selected as the sample, and the mean, variance, and variance to mean ratio of the polarized image data before and after statistical filtering are shown in Tables 9.3 and 9.4. The magnitude of the variance mean ratio reflects the magnitude of the noise level of the image. It can be concluded from Tables 9.3 and 9.4 that the noise level of the filtered data is significantly lower than the noise level of the original data.

In order to further compare the polarization information characteristics of the data before and after filtering, the data of the A rectangular area in Fig. 9.8a is selected as the sample, and the polarization signature and contour of the image before and after filtering are calculated. The results are shown in Figs. 9.9 and 9.10. Figure 9.9 shows the co-polarization signature and contour of the sample data before and after filtering. Figure 9.10 shows the cross-polarization signature and contour of the sample data before and after filtering. It can be seen that the co-polarization and cross-polarization signatures and contour of the sample data before and after filtering maintain good consistency. This shows that the filtering algorithm can well maintain the characteristics of data polarization information.

The concept of polarization decomposition of partially polarized electromagnetic waves shows that a partially polarized wave can be decomposed into the sum of a fully polarized wave and a completely unpolarized wave. Polarization wave decomposition is one of the methods to reflect the characteristics of polarimetric information of polar SAR data. The B region data is selected as a sample for polarization decomposition, and the obtained fully polarized and fully non-polarized signatures are shown in Figs. 9.11, 9.12, 9.13 and 9.14. Figure 9.11 shows the polarization signature of the region. Figures 9.12 and 9.13 show the fully polarized feature and fully non-polarized signature before and after filtering. Figure 9.14 shows the full-polarization and complete non-polarization ratio before and after filtering.

It can be found that the data polarization signature before and after filtering the fully polarized signature and the fully non-polarized signature all maintain good

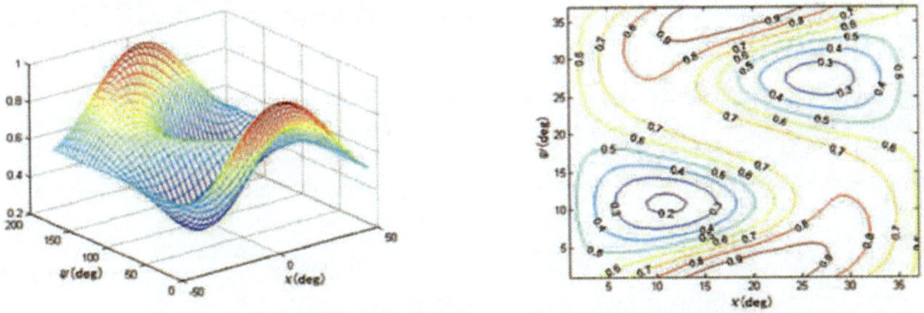

(a) Co-polar Signature and Contour of Raw Data

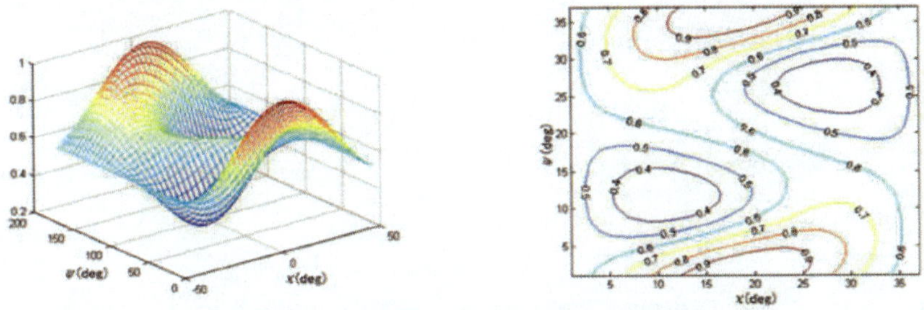

(b) Co-polar Signature and Contour of Filtered Data

Fig. 9.9 Co-polar signatures and contours of data before and after filtering in Region A

consistency. This shows that the minimum rms error local statistical filtering algorithm can well maintain the data polarization information characteristics. And the ratio of fully polarized to fully unpolarized data after filtering is much higher than the ratio of the original data. This shows that after filtering, the data noise is reduced and the depolarization component is reduced.

9.4 Summary

This chapter mainly introduces the PWF method and several traditional parameter estimation methods such as rectangular sliding window method, Prewitt operator method, improved Prewitt operator parameter estimation method, and fine Prewitt operator parameter estimation method. Experimental results and theoretical analysis show that the improved Prewitt operator method and the fine Prewitt operator method have higher accuracy of parameter estimation than the traditional method. The parameter estimation method of the fine Prewitt operator has strong adaptability and robustness, and the estimation accuracy is high.

This chapter also focuses on the full-polarization SAR data vector Lee filtering. The Lee filtering algorithm based on the minimum root mean square error, the refined Lee polarization vector filtering algorithm, and the refined Lee filtering algorithm

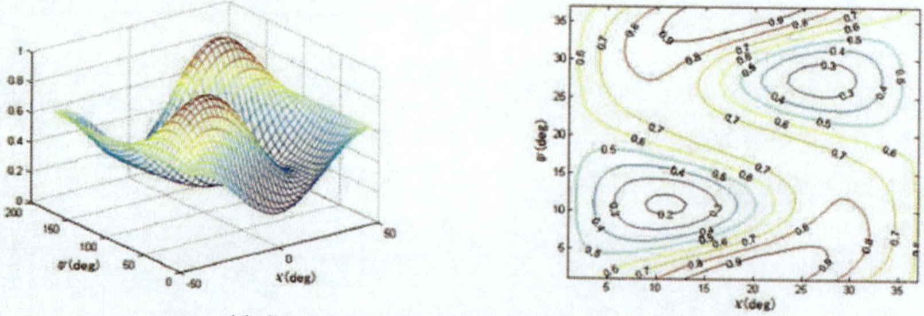

(a) Cross-polar Signature and Contour of Raw Data

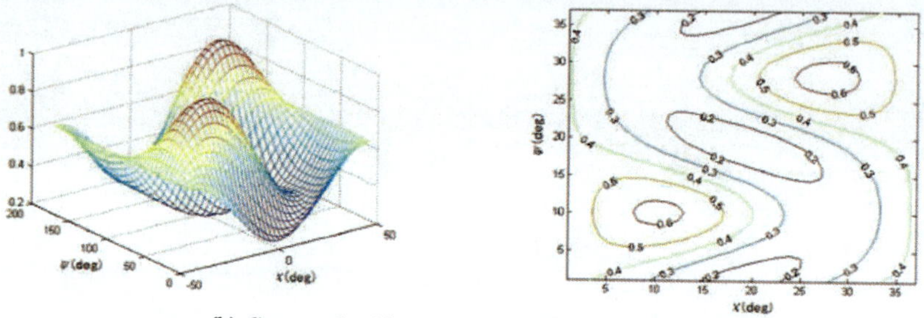

(b) Cross-polar Signature and Contour of Filtered Data

Fig. 9.10 Cross-polar signatures and contours of data before and after filtering in Region A

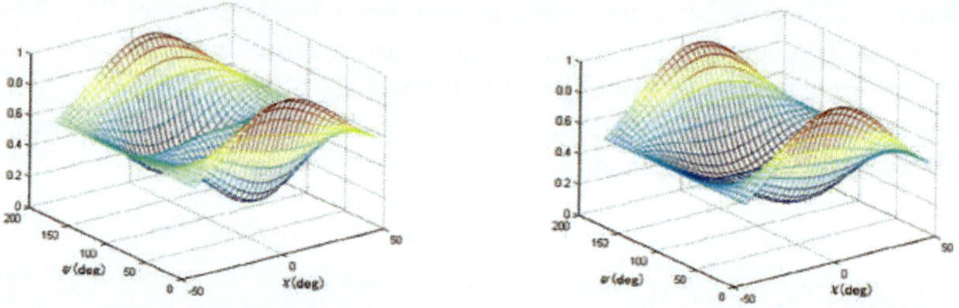

(a) Co-polar Signature of Raw Data (b) Co-polar Signature of Filtered Data

Fig. 9.11 Co-polar signatures of data before and after filtering in Region B

based on the fine Prewitt operator are mainly studied. The experimental results show that the speckles of the data after Lee filtering are greatly suppressed, and the image maintains the characteristics of the edge and point targets of the features. The filtered data polarization information is in good agreement with the original data. The noise level of the filtered data is reduced, and the quality of the polarization information is improved, which can be used in the field of polarization information processing such as polarization target decomposition, image classification, and target recognition.

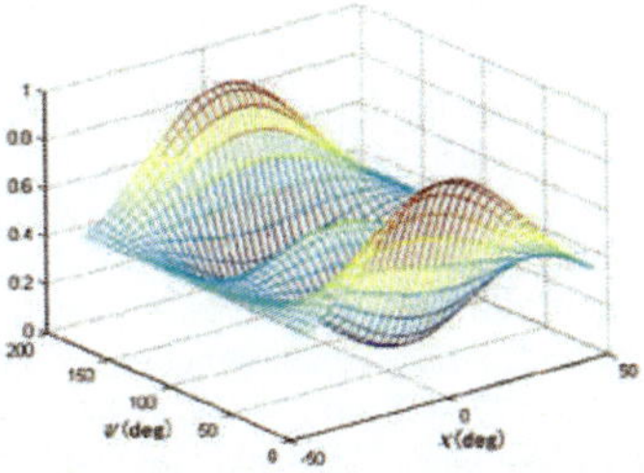

(a) Fully Polarized Co-polar Signature of Raw Data

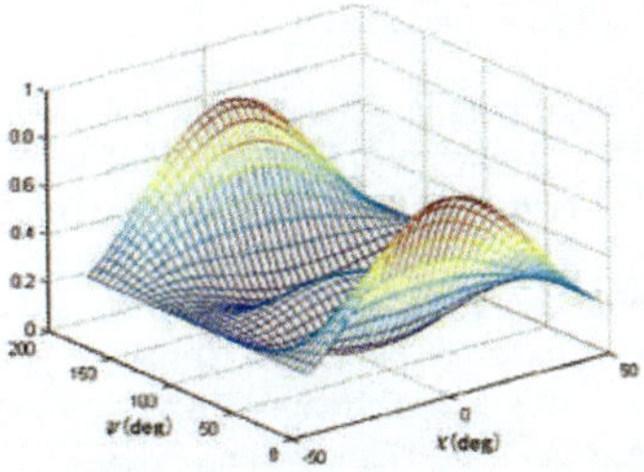

(b) Fully Polarized Co-polarSignature of Filtered Data

Fig. 9.12 The fully polarized co-polarized signature of the raw data and the filtered data of the B region

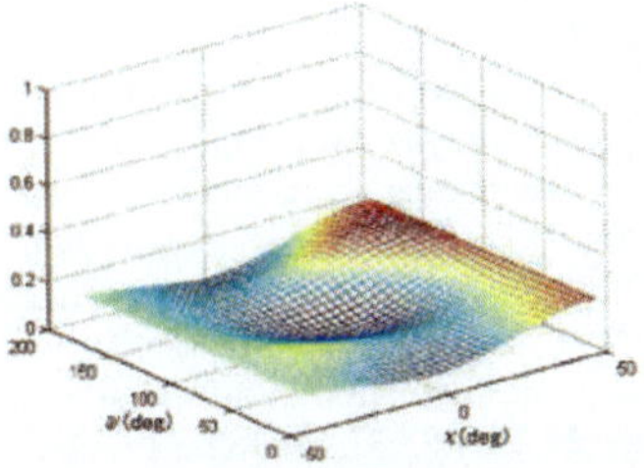

(a) Fully Unpolarized Co-polar Signature of Raw Data

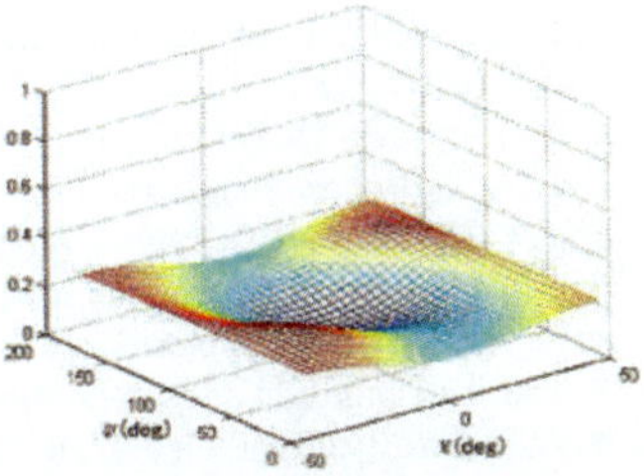

(b) Fully Unpolarized Co-polar Signature of Filtered Data

Fig. 9.13 The fully unpolarized co-polarized signature of the raw data and the filtered data of the B region

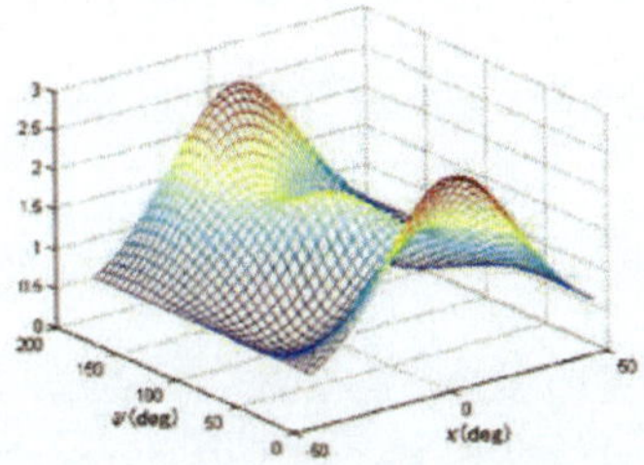

(a) Ratio of Fully Polarized and Fully Unpolarized Co-polarized Signature of the Raw Data

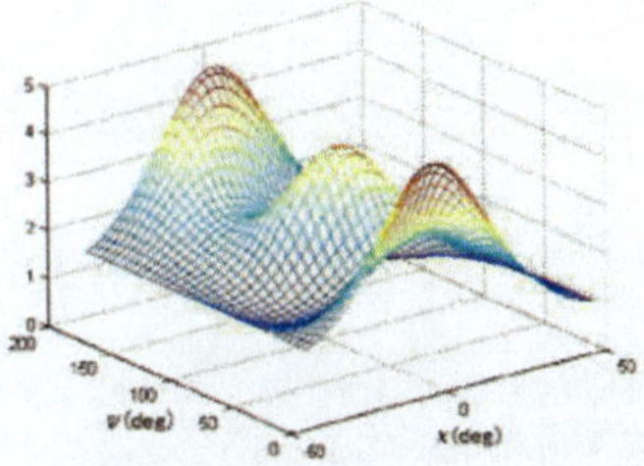

(b) Ratio of Fully Polarized and Fully Unpolarized Co-polarized Signature of the Filtered Data

Fig. 9.14 Ratio of fully polarized and fully unpolarized co-polarized signature of the raw data and the filtered data of the B region

References

1. J.S. Lee, M.R. Grunes, G. De Grandi, Polarimetric SAR speckle filtering and its implication for classification. IEEE Trans. GRS **5**(37), 2363–2373 (1999)
2. J.S. Lee, E. Pottier, *Polarimetric Radar Imaging From Basics to Applications*
3. J.S. Lee, M.R. Grunes, W.M. Boerner, Polarimetric property preservation in SAR speckle filtering, in *Proceedings of SPIE*, vol. 3120 (San Diego, California, USA, 1997), pp. 236–242
4. A. Lopes, E. Nezry, R. Touzi, H. Laur, Structure detection and statistical adaptive speckle filtering in SAR images. Int. J. Remote Sens. **14**(9), 735–1758 (1993)
5. R. Touzi, A. Lopes, The principle of speckle filtering in polarimetric SAR imagery. IEEE. Trans. GRS **32**(5), 1110–1114 (1994)
6. J.S. Lee, G. De Grandi, M.R. Grunes, E. Nezry, Polarimetric signature preservation in SAR speckle filtering. Geosci. Remote. Sens. Symp. IGARSS '96, **3**, 1574–1576 (1996)
7. G. Liu, S. Huang, Bayesian classification of multi-look polarimetric SAR images with a generalized multiplicative speckle model and adaptive a priori probabilities. Int. J. Remote Sens. **19**(1), 161–170 (1998)
8. J.-S. Lee, M.R. Grunes, G. Grandi, Polarimetric SAR speckle filtering and its implication for classification. IEEE Trans. GRS **37**(5), 2363–2373 (1999)
9. B. Dai, *Studies on Information Processing Technique of Fully-Polarimetric Synthetic Aperture Radar* (Research on Multipolarization SAR System and Polarization Information Processing Technology. Ph.D., Institute of Electronics, Chinese Academy of Sciences, 2000)
10. X. Liu, *Studies on Information Processing Technique of Fully-Polarimetric Synthetic Aperture Radar* (University of Chinese Academy of Sciences, Ph.D., 2004)
11. X. Liu, Z. Yang, R. Yang, Improvement research of parameter estimation in polarimetricwhitering filter of full-polarization SAR image. Acta Electron. Sin. **12**, 37–41 (2003)
12. X. Liu, Z. Yang, R. Yang, Improvement research on texture-detection in full-polarization SAR image filter. IEEE Int. Geosci. Remote. Sens. Symp. 21–25 (2003), Toulouse France

Chapter 10
Enhanced Processing of Polarimetric Synthetic Aperture Radar Images

Polarization synthesis technology plays an important role in polar SAR image processing. The polarization synthesis technique can be used to calculate the echo power of any combination of the transmitting antenna and the receiving antenna from the polarization scattering matrix of the target. This chapter will focus on two typical applications: contrast enhancement of two types of targets and suppression of clutter in polarimetric SAR images. Here, another important theory is used: the co-null theory of received polarization power.

Since the sensitivity of the human eye to the color image is much higher than that of the gray image, the gray image is converted into a color image, which can enhance the recognition effect of the image and achieve the purpose of highlighting the image information. This image processing method is called pseudo-color synthesis processing. This technique is described in Sect. 10.4 of this chapter.

10.1 Polarization Data Preprocessing Method

The fully polarized synthetic aperture radar needs to record the data of four different polarization channels. The amount of calculation is very large in the imaging processing, and the amount of polarization data after imaging processing is also large. In order to reduce the amount of polarization data, preprocessing is generally performed before storing data, mainly for polarization data compression processing. Polarized scattering matrices, Mueller matrices, and Stokes matrices are in fact equivalent, and polarized data can be represented in three ways. In order to facilitate the processing of images, the polarimetric SAR data of the scene is usually stored in a Stokes matrix. The Stokes matrix compression processing algorithm is based on two conditions: (1) the Stokes matrix is a symmetric matrix and (2) the dynamic range of the polarization data is less than 30 dB.

The Stokes matrix is a 4×4 real array. The most important prerequisite for compressing stored polarization data is based on the theory of reciprocity of the

R. Yang et al., *Polarimetric Microwave Imaging*,
https://doi.org/10.1007/978-981-15-8897-6_10

target: the polarization scattering matrix is a symmetric matrix, i.e., $h_{hv} = h_{vh}$. Thus, the Stokes matrix is also a symmetric matrix. The relationship between the Stokes matrix and the polarization scattering matrix is as follows:

$$M_{11} = \frac{1}{2}\left(h_{vv}h_{vv}^* + 2h_{vh}h_{vh}^* + h_{hh}h_{hh}^*\right) \tag{10.1}$$

$$M_{22} = \frac{1}{2}\left(h_{vv}h_{vv}^* - 2h_{vh}h_{vh}^* + h_{hh}h_{hh}^*\right) \tag{10.2}$$

$$M_{33} = \frac{1}{2}\left(h_{vv}h_{hh}^* + 2h_{vh}h_{vh}^* + h_{hh}h_{vv}^*\right) \tag{10.3}$$

$$\begin{aligned} M_{44} &= -\frac{1}{2}\left(h_{vv}h_{hh}^* - 2h_{vh}h_{vh}^* + h_{hh}h_{vv}^*\right) \\ &= M_{11} - M_{22} - M_{33} \end{aligned} \tag{10.4}$$

$$M_{12} = \frac{1}{2}\left(h_{vv}h_{vv}^* - h_{hh}h_{hh}^*\right) \tag{10.5}$$

$$M_{13} = \mathrm{Re}\left(h_{vv}h_{vh}^* + h_{vh}h_{hh}^*\right) \tag{10.6}$$

$$M_{14} = \mathrm{Im}\left(h_{vv}h_{vh}^* + h_{vh}h_{hh}^*\right) \tag{10.7}$$

$$M_{23} = \mathrm{Re}\left(h_{vv}h_{vh}^* - h_{vh}h_{hh}^*\right) \tag{10.8}$$

$$M_{24} = \mathrm{Im}\left(h_{vv}h_{vh}^* - h_{vh}h_{hh}^*\right) \tag{10.9}$$

$$M_{34} = \frac{1}{2}j\left(h_{vv}h_{hh}^* - h_{hh}h_{vv}^*\right) \tag{10.10}$$

Other elements can be derived from a matrix symmetric relationship $M_{ji} = M_{ij}$. According to the above relationship, the following results can be derived:

$$M_{11} = M_{22} + M_{33} + M_{44} \tag{10.11}$$

$$M_{13}M_{23} + M_{14}M_{24} = M_{12}(M_{11} - M_{22}) \tag{10.12}$$

$$M_{13}M_{14} + M_{23}M_{24} = M_{34}(M_{33} + M_{44}) \tag{10.13}$$

$$M_{13}^2 + M_{14}^2 + M_{23}^2 + M_{24}^2 = M_{11}^2 - M_{22}^2 \tag{10.14}$$

$$M_{13}^2 - M_{14}^2 - M_{23}^2 + M_{24}^2 = M_{33}^2 - M_{44}^2 \tag{10.15}$$

When merging the Stokes matrices of adjacent points, only the first formula holds in the above formula. This means that after multi-look processing, the Stokes matrix has a maximum of only nine independent components. Therefore, only nine quantities need to be considered when storing the Stokes matrix.

The first step in the process of compressing and storing polarization data by polarization synthetic aperture radar is to average the h_{hv} and h_{vh} components of the scattering matrix. The average formula is as follows:

$$h_{hv} = \frac{1}{2}\left(h_{hv} + h_{vh}e^{i\theta_{vh-hv}}\right) \tag{10.16}$$

Among them, θ_{vh-hv} is the phase difference between the h_{hv} component and the h_{vh} component of the scattering matrix. The components of the symmetric Stokes matrix can be solved according to the scattering matrix after the imaging process and the calibration process.

Since the dynamic range of the radar image is generally less than 30 dB, the components of the Stokes matrix are divided by an appropriate scale factor, and the distribution of the absolute values of the components can be considered in 2^{-128}–2^{-127}. The M_{11} component of the Stokes matrix can be stored in only two bytes, one byte representing the coefficient term and the other byte representing the exponent term. The remaining Stokes matrix components can be normalized to M_{11}, and their normalization coefficients are represented by 1 byte. Thus, the contents of the Stokes matrix can be fully represented by 10 bytes.

The following is the formula for the compressed Stokes matrix:

$$byte(1) = Fix\left(\frac{\log_2 M_{11}}{\log_2 2}\right) \tag{10.17}$$

where Fix is the rounding of the value.

$$byte(2) = Fix\left(254 \cdot \left(\frac{M_{11}}{2^{byte(1)}} - 1.5\right)\right) \tag{10.18}$$

Then, the scale factor x is calculated to approximate M_{11}. The purpose of using the scaling factor is to reduce the error caused by the direct use of M_{11}

$$x = \left(\frac{byte(2)}{254} + 1.5\right) \cdot 2^{byte(1)} \cdot gen\_fac \tag{10.19}$$

where gen_fac is the average of M_{11} of all Stokes matrices in the image:

$$byte(3) = 127 \cdot M_{12}/x \tag{10.20}$$

$$byte(4) = 127 \cdot sign(M_{13}/x) \cdot \sqrt{M_{13}/x} \tag{10.21}$$

$$byte(5) = 127 \cdot sign(M_{14}/x) \cdot \sqrt{M_{14}/x} \tag{10.22}$$

$$byte(6) = 127 \cdot sign(M_{23}/x) \cdot \sqrt{M_{23}/x} \tag{10.23}$$

$$byte(7) = 127 \cdot sign(M_{24}/x) \cdot \sqrt{M_{24}/x} \tag{10.24}$$

$$byte(8) = 127 \cdot M_{33}/x \tag{10.25}$$

$$byte(9) = 127 \cdot M_{34}/x \tag{10.26}$$

$$byte(10) = 127 \cdot M_{44}/x \tag{10.27}$$

where $sign$ is a sign function, corresponding to a positive or negative value of ± 1. In order to find the original Stokes matrix, the corresponding relationship is as follows:

$$M_{11} = \left(\frac{byte(2)}{254} + 1.5\right) \cdot 2^{byte(1)} \cdot gen\_fac \tag{10.28}$$

$$M_{12} = byte(3) \cdot \frac{M_{11}}{127} \tag{10.29}$$

$$M_{13} = sign(byte(4)) \cdot \left(\frac{byte(4)}{127}\right)^2 \cdot M_{11} \tag{10.30}$$

$$M_{14} = sign(byte(5)) \cdot \left(\frac{byte(5)}{127}\right)^2 \cdot M_{11} \tag{10.31}$$

$$M_{23} = sign(byte(6)) \cdot \left(\frac{byte(6)}{127}\right)^2 \cdot M_{11} \tag{10.32}$$

$$M_{24} = sign(byte(7)) \cdot \left(\frac{byte(7)}{127}\right)^2 \cdot M_{11} \tag{10.33}$$

$$M_{33} = byte(8) \cdot \frac{M_{11}}{127} \tag{10.34}$$

$$M_{34} = byte(9) \cdot \frac{M_{11}}{127} \tag{10.35}$$

$$M_{44} = byte(10) \cdot \frac{M_{11}}{127} \tag{10.36}$$

$$M_{22} = M_{11} - M_{33} - M_{44} \tag{10.37}$$

10.2 Contrast Enhancement of Polarimetric SAR Images

Image contrast methods are often used to enhance the image contrast of two types of targets in SAR images to better discriminate between two different types of objects, which is called contrast enhancement of SAR images. One of the typical applications of polarization synthesis techniques is the Optimal Polarization Contrast Enhancement (OPCE) of polarimetric SAR images. A typical method is to obtain their maximum contrast image by using polarization synthesis techniques from selecting the optimal polarization modes of the transmitting and receiving antennas that have maximized the ratio of the average received power of the two types of targets.

The OPCE problem includes three typical radar polarization channels (co-polarization, cross-polarization, and matching polarization). If the polarization state of the receiving antenna depends on the polarization state of the transmitting antenna, only the transmitting antenna polarization state is an independent variable. If the transmit polarization state and the receive polarization state are independent, that is, there are two independently changeable polarization states, a larger receive energy ratio may be obtained. Ioannidis et al. [1] studied the related OPCE problem by Lagrangian operator method. Swartz et al. [2] gave a simple procedure to solve this problem.

10.2.1 Contrast Enhancement Based on Stokes Matrix [3]

The Stokes matrix of the target can fully represent its polarization scattering characteristics. Therefore, M_{target} and $M_{clutter}$ are set to represent the scattering characteristics of specific targets and ground clutter, respectively, where the Stokes matrix is the Stokes matrix average of adjacent resolving units. The goal of polarization image contrast enhancement is to choose the appropriate polarization combination to maximize the echo of targets:

$$CR\max = \frac{S'_{rec} M_{target} S_{tr}}{S'_{rec} M_{clutter} S_{tr}} \tag{10.38}$$

where S_{tr} is the Stokes vector of the polarization mode of the transmitting antenna and S_{rec} is the Stokes vector of the polarization mode of the receiving antenna.

Let the receiving antenna be fully polarized, and the transmitting electromagnetic wave is also a fully polarized wave, so that

$$S_{trc0}^2 = S_{tr1}^2 + S_{tr2}^2 + S_{tr3}^2, S_{tr0} > 0 \tag{10.39}$$

$$S_{rec0}^2 = S_{rec1}^2 + S_{rec2}^2 + S_{rec3}^2, S_{rec0} > 0 \tag{10.40}$$

Set a matrix R_0:

$$R_0 = \begin{bmatrix} 1 & 0 & 0 & 0 \\ 0 & -1 & 0 & 0 \\ 0 & 0 & -1 & 0 \\ 0 & 0 & 0 & -1 \end{bmatrix} \tag{10.41}$$

Then Eqs. (10.39) and (10.40) can be expressed as follows:

$$S_{tr}' R_0 S_{tr} = 0 \tag{10.42}$$

$$S_{rec}' R_0 S_{rec} = 0, \quad S_{tr0} > 0, S_{rec0} > 0 \tag{10.43}$$

To avoid the presence of $S_{rec}' M_{clutter} S_{tr} = 0$ under a certain polarization combination (S_{tr}, S_{rec}), and $S_{rec}' M_{clutter} S_{tr} \neq 0$ is set in the following analysis.

The echo power calculated by the Stokes matrix is a positive real number, so the maximum value of the ratio of the echo power according to Eq. (10.38) can be equivalent to

$$\text{Solve } P_{\max} = S_{rec}' M_{target} S_{tr} \tag{10.44}$$

The restrictions are

$$S_{rec}' M_{clutter} S_{tr} = C \tag{10.45}$$

C is a constant.

$$S_{tr}' R_0 S_{tr} = 0 \text{ and } S_{rec}' R_0 S_{rec} = 0, \quad S_{tr0} > 0, \quad S_{rec0} > 0 \tag{10.46}$$

The Laplace extreme method can be used to solve the above equation. First, set a function:

$$\begin{aligned} \Phi(S_{tr}, S_{rec}) = S_{rec}' M_{target} S_{tr} - a\left(S_{rec}' M_{clutter} S_{tr} - C\right) \\ - b S_{tr}' R_0 S_{tr} - c S_{rec}' R_0 S_{rec} \end{aligned} \tag{10.47}$$

Thus, deriving the variables of Eq. (10.43) yields

$$\frac{\partial \Phi}{\partial S_{rec}'} = M_{target} S_{tr} - a M_{clutter} S_{tr} - c R_0 S_{rec} = 0 \tag{10.48}$$

$$\frac{\partial \Phi}{\partial S_{tr}} = S'_{rec} M_{target} - a S'_{rec} M_{clutter} - b S'_{tr} R_0 = 0$$

$$\text{which is } M_{target} S_{rec} - a M_{clutter} S_{rec} - b R_0 S_{tr} = 0 \tag{10.49}$$

From the above two formulas, the following two relations can be obtained separately:

$$S_{rec} = \frac{1}{c} R_0 \left(M_{target} - a M_{clutter} \right) S_{tr} \tag{10.50}$$

$$\left[\left(R_0 M_{target} - a M_{clutter} \right)^2 - bcI \right] S_{tr} = 0 \tag{10.51}$$

where I is a fourth-order unit matrix. In the derivation of the two formulas used in the relationship $R_0 = R_0^{-1} = R'_0$.

According to formulas (10.42), (10.43), (10.48), and (10.49), the following relationship can be obtained:

$$S'_{rec} \left(M_{target} - a M_{clutter} \right) S_{tr} = 0 \tag{10.52}$$

As can be seen from the above formula, $a = CR_{\max}$.

Define the matrix:

$$M_{lp} = \left(R M_{target} - a R M_{clutter} \right)^2 \tag{10.53}$$

And $d^2 = bc$. Since both b and c are real numbers, d is required to be a positive real number or a pure imaginary number.

According to Eq. (10.51), if S_{tr} has a non-zero solution, the determinant of the matrix on the left side of the relation is required to have a value of zero. The following relationship can be obtained:

$$\begin{aligned} D(p, bc) &= \left| \left(R_0 M_{target} - a R_0 M_{clutter} \right)^2 - bcI \right| \\ &= \left| M_{lp} - d^2 I \right| = 0 \end{aligned} \tag{10.54}$$

Define the following relationship:

$$S_{tr} = \begin{bmatrix} S_{tr0} \\ S_{tr1} \\ S_{tr2} \\ S_{tr3} \end{bmatrix} = \begin{bmatrix} S_{tr0} \\ S_{trs} \end{bmatrix} \text{ where, } S_{trs} = \begin{bmatrix} S_{tr1} \\ S_{tr2} \\ S_{tr3} \end{bmatrix} \tag{10.55}$$

$$M_{target} = \begin{bmatrix} \alpha_t & m'_t \\ m_t & M_t \end{bmatrix} \tag{10.56}$$

$$M_{cluttert} = \begin{bmatrix} \beta_c & m_c' \\ m_c & M_c \end{bmatrix} \tag{10.57}$$

$$M_{lp} = \begin{bmatrix} \gamma_l & m_l' \\ -m_l & -M_l \end{bmatrix} \tag{10.58}$$

Among them, m_t, m_c, and m_l are 1 × 3 column vectors, and M_{target}, $M_{clutter}$, and M_{lp} are 3 × 3 matrices.

According to formula (10.52), we can find the relationship between each element of M_{lp} and the elements of S_{tr}, M_{target}, and $M_{clutter}$:

$$\begin{aligned} \gamma_l &= (\alpha_t - a\beta_c)^2 - (m_t - am_c)'(m_t - am_c) \\ m_l &= (\alpha_t - a\beta_c)(m_t - am_c) - (M_t - aM_c)(m_t - am_c) \\ M_l &= (m_t - am_c)(m_t - am_c)' - (M_t - aM_c)^2 \end{aligned} \tag{10.59}$$

Substituting Eqs. (10.52) and (10.59) into (10.51), the following two relations can be obtained:

$$(\gamma_l - d^2)S_{tr0} + m_l' S_{trs} = 0 \tag{10.60}$$

$$S_{tr0} m_l + (M_l + d^2 I_3) S_{trs} = 0 \tag{10.61}$$

Among them, I_3 is a 3 × 3 unit matrix.

Assume $d^2 = v$, satisfy

$$|M_l + vI_3| \neq 0 \tag{10.62}$$

Solve the S_{trs} substitutions (10.39) and (10.59) and find the following two formulas:

$$\gamma_l - v - m_l'(M_l + vI_3)^{-1} m_l = 0 \tag{10.63}$$

$$m_l'(M_l + vI_3)^{-2} m_l = 1 \tag{10.64}$$

The above formula can be expanded into variables a and v polynomials. Roots M_{target},$M_{clutter}$, and a and v can be obtained by numerical methods. Because of $d^2 = v$, so request $v > 0$, and a is the receiving polarization power ratio. Therefore, the transmit polarization and receive polarization corresponding to the maximum a value are the optimal polarization combinations.

The Laplace extremum method is used in the solution of Eq. (10.38). The method of finding the extremum involves solving a large number of complex nonlinear equations. In order to reduce the difficulty of solving nonlinear equations in analytical methods, Dubois and van Zyl proposed a simple method for optimal polarization of echo power ratio in 1989. This method firstly uses the analytical method to find the optimal receiving polarization mode in the fixed transmit polarization mode, and then uses the numerical method to find the ratio of the echo power obtained by the optimal receive polarization antenna under different transmit polarization modes. From which, the optimal transmit polarization mode and its corresponding receiving polarization mode are found.

According to the Laplace extreme value method, the ratio of the optimal echo power can be obtained under the fixed transmit polarization mode:

$$CR_{\max}(\psi_t, \chi_t) = \frac{S_{01}S_{02} - s_1 \cdot s_2}{S_{02}^2 - s_2 \cdot s_2} + \sqrt{\left(\frac{S_{01}S_{02} - s_1 \cdot s_2}{S_{02}^2 - s_2 \cdot s_2}\right)^2 - \frac{S_{01}^2 - s_1 \cdot s_2}{S_{02}^2 - s_2 \cdot s_2}} \tag{10.65}$$

$$CR_{\min}(\psi_t, \chi_t) = \frac{S_{01}S_{02} - s_1 \cdot s_2}{S_{02}^2 - s_2 \cdot s_2} - \sqrt{\left(\frac{S_{01}S_{02} - s_1 \cdot s_2}{S_{02}^2 - s_2 \cdot s_2}\right)^2 - \frac{S_{01}^2 - s_1 \cdot s_2}{S_{02}^2 - s_2 \cdot s_2}} \tag{10.66}$$

where $\begin{bmatrix} S_{01} \\ s_1 \end{bmatrix} = M_{target} S(\psi_t, \chi_t)$, $\begin{bmatrix} S_{02} \\ s_2 \end{bmatrix} = M_{clutter} S(\psi_t, \chi_t)$.

Since $CR_{\max}$ is a function of (ψ_t, χ_t), CR_{opt} can be found within the range of (ψ_t, χ_t). According to the corresponding (ψ_t, χ_t) of CR_{opt}, the optimal transmit polarization mode can be obtained. The receiving polarization mode is

$$S_{reco} = \begin{bmatrix} 1 \\ s_{rec} \end{bmatrix}, \text{ here, } s_{rec} = \frac{s_1 - CR_{opt}s_2}{\sqrt{(s_1 - CR_{opt}s_2) \cdot (s_1 - CR_{opt}s_2)}} \tag{10.67}$$

The purpose of polarization contrast enhancement is to increase the echo power difference between the two types of targets. The specific processing steps are as follows:

(1) Find two targets for contrast enhancement on the scene, and use the average method to find the Stokes matrix of the two types of targets.
(2) Calculate the ratio of the target echo power corresponding to different transmit polarizations under optimal receive polarization according to Eq. (10.66), and find the maximum point from which the corresponding transmit polarization state and the usage formula (10.67) are obtained. The calculated polarization state is the optimal polarization combination.
(3) According to the optimal polarization combination, the echo power method is used to calculate the echo power value of the scene at this time and normalize into a grayscale image.

The test data is the fully polarized SAR data for the San Francisco Area. The target terrain used for contrast enhancement is the ocean and the city. Two regions S (ocean) and C (city) with the same area are selected, and their sizes are about

1200 m (azimuth direction) × 1000 m (range), and the incident angle is 47.40. Add the Stokes matrices of the two regions and average them:

$$M_{sea} = \begin{bmatrix} 0.9789 & -0.7148 & -0.0276 & 0.0737 \\ -0.7148 & 0.9279 & 0.0226 & -0.1420 \\ -0.0276 & 0.0226 & 0.5303 & 0.0471 \\ 0.0737 & -0.1420 & 0.0471 & -0.4793 \end{bmatrix}$$

$$M_{city} = \begin{bmatrix} 29.2889 & 2.6604 & 2.9416 & 1.3156 \\ 2.6604 & 23.1374 & 8.8277 & -1.6098 \\ 2.9416 & 8.8277 & -5.2439 & -0.6808 \\ 1.3156 & -1.6098 & -0.6808 & 11.3954 \end{bmatrix}$$

It is worth noting that the optimal echo power ratio under the co-polarization and cross-polarization can be determined from the polarization signature of the target. According to the Stokes matrix of the ocean and city, the polarization signature is obtained, as shown in Fig. 10.1. It can be seen from the figure that the echo power of the city under arbitrary polarization combination is always greater than the echo power of the ocean, so the urban–ocean echo power ratio is always greater than 1; at the co-polarization, near the HH-polarization, the ratio between the city and ocean echo power is relatively large; at the cross-polarization, near linear polarization with

Fig. 10.1 Polarized signature of the San Francisco area, the left is Co-Polarization, the right is X-Polarization

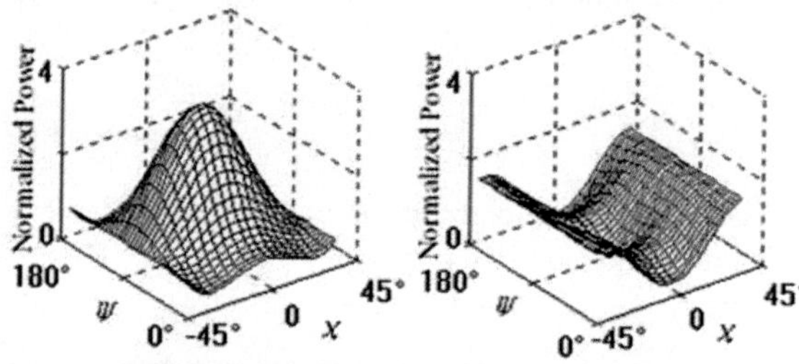

(a) Polarization Signature of the ocean

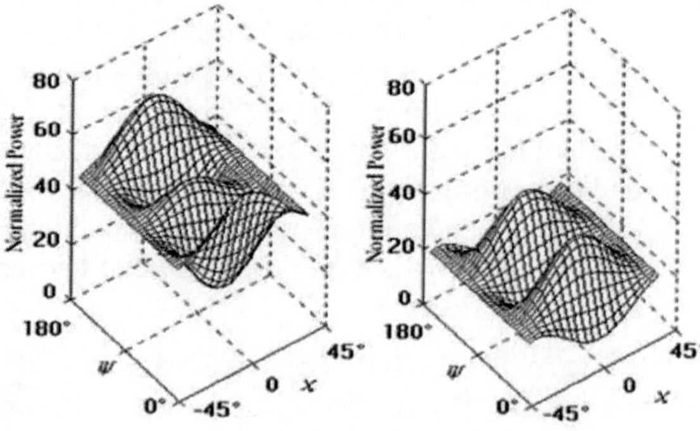

(b) Polarization Signature of the city

direction angles of 50° and 130°, the ratio of urban-to-ocean echo power is relatively large.

When solving the optimal polarization method by using Eqs. (10.63) and (10.64), it involves the simplification and solution of the binary higher order equations. This processing step is very complicated, it requires professional mathematical knowledge, and the formula for solving is also very cumbersome. Equation (10.66) is very simple, and its solution accuracy is determined by the interval between the direction angle and the ellipticity angle. According to Fig. 10.1, when the 5° interval is used in the value domain of the direction angle and the ellipticity angle, the optimal power ratio is 483.4, and the corresponding transmit and receive polarization ellipse parameters are (direction angle 0°, ellipticity angle 50°), (direction angle 165°, ellipticity angle −5°), see Fig. 10.2. It is noted that the transmit and receive polarizations are approximately cross-polarized. At this time, the transmit polarization mode is substantially the same as the optimal power ratio determined according to the cross-polarization in the polarization signature, so it is possible to use the polarization signature. Directly draw a better power ratio.

Table 10.1 shows the ratio of urban-to-ocean echo power for several typical polarization combinations. The theoretical power ratio is calculated from the average Stokes matrix of the city and the ocean under different polarization conditions. The image power ratio is obtained by using the polarization synthesis method to generate the echo power image under different polarization conditions, and then the average ratio of the city and ocean power is obtained. Due to the difference in the average

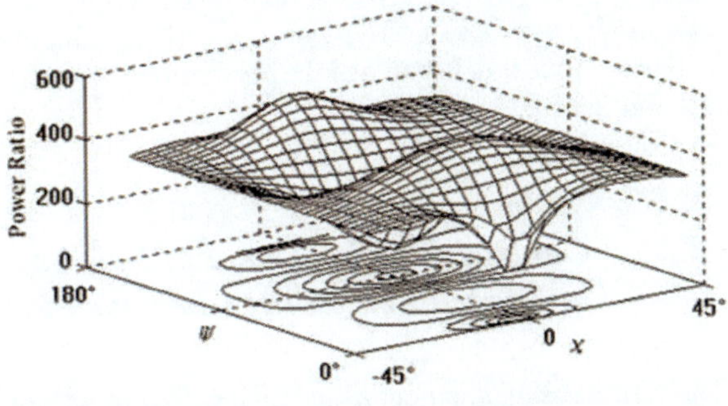

Fig. 10.2 Maximum receiving polarization power ratio of adjusted transmitting polarization

Table 10.1 Ratio of urban to ocean echo power

Polarization mode	HH	HV	VV	Optimal polarization
Polarized elliptic Parameter of the transmitted signal ψ_t, χ_t	(0°, 0°)	(0°, 0°)	(90°, 0°)	(0°, 50°)
Polarized elliptic parameter of the received signal ψ_r, χ_r	(0°, 0°)	(0°, 0°)	(0°, 0°)	(165°, −5°)
Theoretical power ratio	121.3	120.2	14.3	483.4
Image power ratio	101.5	157.9	11.1	437.0

(a) HH (b) Optimal polarization

Fig. 10.3 Total power image for HH and optimal polarizations

object, the mean values of the two are different. At the same time, the size and position of the data used in the average of the two values will affect both values, but the difference between the values compares the power ratio will be less than 20%. The theoretical power ratio can express the influence of the Stokes matrix on the echo power ratio when solving the optimal polarization mode, and the image power ratio can more intuitively reflect the target power phase difference.

From Table 10.1, it can see that at HH and HV polarizations, the power ratio of city and ocean is always higher than 20 dB. However, at VV-polarization, the power ratio of city and ocean only is near 10 dB, and the optimal polarization power ratio increase about 6 dB than at HH polarization. Figure 10.3 is total power image for HH and optimal polarizations

10.2.2 Contrast Enhancement of Lagrangian Operator Based on Maximum Average Received Power Ratio [4, 5]

The contrast enhancement based on the maximum average received power ratio is to use the measured full polarization information to enhance the contrast of two different types of scattering targets, that is, to select the optimal combination of reflection and reception polarization to maximize the ratio of the average received power of the two types of targets.

There are three independent complex elements in the polarization scattering matrix of the reciprocal medium, and the polarization measurement vector **X** is

$$X = \begin{bmatrix} s_{hh} & s_{hv} & s_{vv} \end{bmatrix}^T \tag{10.68}$$

It is assumed that there are two types of ground objects, each of which can be represented by its statistical covariance matrix:

$$C_j = E[\mathbf{X} \cdot \mathbf{X}^H] \tag{10.69}$$

Here $E[\cdot]$ represents the statistical average, the superscript H represents the conjugate transpose, and $j = a, b$ represents the two types of scattering targets A and B, respectively. Let's look for the best linear weight vector (or polarization matching filter) and preprocess the polarization eigenvector to get a linear combination:

$$Y = \mathbf{W}^H \cdot \mathbf{X} \tag{10.70}$$

Here, $\mathbf{W} = \begin{bmatrix} W_{hh} & W_{hv} & W_{vv} \end{bmatrix}^T$ is a linear weight vector. The optimal contrast r of the two types of targets A and B is defined as

$$r = \max\left\{\max_{\mathbf{W}}(r_{ab}), \max_{\mathbf{W}}(r_{ba})\right\} \tag{10.71}$$

Where

$$r_{ab} = \frac{\mathbf{W}^H C_a \mathbf{W}}{\mathbf{W}^H C_b \mathbf{W}} \quad r_{ba} = \frac{\mathbf{W}^H C_b \mathbf{W}}{\mathbf{W}^H C_a \mathbf{W}} \tag{10.72}$$

The elements in the polarization filter vector W are linear weighting coefficients that change the amplitude and phase of the polarimetric radar measured data.

To reveal the relationship between the polarization matched filter vector W and the transmitted and received polarized electromagnetic waves, Eq. (10.70) is expressed as the form of the scattering matrix:

$$Y = \begin{bmatrix} E_h^r & E_v^r \end{bmatrix} \cdot \begin{bmatrix} s_{hh} & s_{hv} \\ s_{vh} & s_{vv} \end{bmatrix} \cdot \begin{bmatrix} E_h^t \\ E_v^t \end{bmatrix} \tag{10.73}$$

Here, E_h^t, E_h^r are the electric field components transmitted and received under H-polarization, respectively, and E_v^t, E_v^r are the electric field components transmitted and received under V-polarization, respectively. In general, set

$$\left|E_h^t\right|^2 + \left|E_v^t\right|^2 = 1 \quad \left|E_h^r\right|^2 + \left|E_v^r\right|^2 = 1 \tag{10.74}$$

From Eqs. (10.70) and (10.72):

$$E_h^t E_h^r = W_{hh}^* \quad E_h^t E_v^r + E_v^t E_h^r = W_{hv}^* \quad E_v^t E_v^r = W_{vv}^* \tag{10.75}$$

For a given linear weight vector W, the corresponding polarization state elements E_h^t, E_v^t, E_h^r, E_v^r can be fully represented by the above equations.

It should be noted that the transmitting and receiving polarization vectors in Eq. (10.72) are reciprocal. That is, the electric field components E_h^t, E_v^t, E_h^r, E_v^r can be interchanged, and Y can remain unchanged. Therefore, in reciprocal backscattering, if the transmit and receive vectors are interchanged, the same co-polarization ratio can still be obtained.

Next, the Lagrangian operator method is used to find the optimal polarization matching filter vector W_{opt} that maximizes the contrast r in Eqs. (10.71) and (10.72). In order to get the maximum value of the objective function r_{ab}, the following formula:

$$\max\left\{\mathbf{W}^H C_a \mathbf{W}\right\} \tag{10.76}$$

It is obtained under the following constraints:

$$\mathbf{W}^H C_b \mathbf{W} = 1 \tag{10.77}$$

The Lagrangian function is

$$f(\mathbf{W}) = \mathbf{W}^H C_a \mathbf{W} + \lambda\left[1 - \mathbf{W}^H C_b \mathbf{W}\right] \tag{10.78}$$

Here, λ is the Lagrange multiplier. The auxiliary function is represented by the real part and the imaginary part, respectively, and the gradient is zero, and the following simultaneous equations are obtained:

$$\begin{cases} \partial f(\mathbf{W}) / \partial \mathbf{W}_R = 0 \\ \partial f(\mathbf{W}) / \partial \mathbf{W}_I = 0 \end{cases} \tag{10.79}$$

Here subscripts R and I represent the real and imaginary parts of the complex number, respectively. This gives the eigenvalue equation:

$$C_a \mathbf{W} = \lambda C_b \mathbf{W} \tag{10.80}$$

The λ n formula (10.79) is the required contrast r_{ab}. The maximum value of this eigenvalue corresponds to the optimal polarization matching filter vector W_{opt}. After obtaining the optimal polarization matching filter vector, the polarization states of the corresponding transmitting and receiving electromagnetic waves can be obtained. Generally, it can be obtained by (10.74) and (10.75), when W_{hh} is not zero:

$$E_h^t = \left[\left|\frac{W_{hv}^* \pm \left[\left(W_{hv}^*\right)^2 - 4(W_{hh}W_{vv})^*\right]^{1/2}}{2W_{hh}^*}\right|^2 + 1\right]^{-1/2} \tag{10.81a}$$

$$E_v^t = E_h^t \left[\frac{W_{hv}^* \pm \left[\left(W_{hv}^*\right)^2 - 4(W_{hh}W_{vv})^* \right]^{1/2}}{2W_{hh}^*} \right] \quad (10.81b)$$

$$E_v^r = \frac{\tau}{E_h^t} \left[\frac{W_{hv}^* \mp \left[\left(W_{hv}^*\right)^2 - 4(W_{hh}W_{vv})^* \right]^{1/2}}{2} \right] \quad (10.81c)$$

where

$$\tau = \frac{1}{(\tau_1 \cdot \tau_2 \cdot \tau_3)^{1/2}} \quad (10.82a)$$

$$\tau_1 = \left|W_{hh}^*\right|^2 \quad (10.82b)$$

$$\tau_2 = 1 + \left| \frac{W_{hv}^* \mp \left[\left(W_{hv}^*\right)^2 - 4(W_{hh}W_{vv})^* \right]^{1/2}}{2W_{hh}^*} \right|^2 \quad (10.82c)$$

$$\tau_3 = 1 + \left| \frac{W_{hv}^* \mp \left[\left(W_{hv}^*\right)^2 - 4(W_{hh}W_{vv})^* \right]^{1/2}}{2W_{hh}^*} \right| \quad (10.82d)$$

The values of the received power (image intensity) I obtained from the above equations are

$$I = \left|\tau \mathbf{W}^H \mathbf{X}\right|^2 \quad (10.83)$$

Generally, the cross-polarization and the co-polarization signals received by the radar are statistically independent from each other, and the statistical covariance matrix is

$$C = \sigma \cdot \begin{bmatrix} 1 & 0 & \rho\sqrt{r} \\ 0 & f & 0 \\ \rho^*\sqrt{r} & 0 & r \end{bmatrix} \quad (10.84a)$$

Where

$$\sigma = \left\langle |s_{hh}|^2 \right\rangle = \sigma_{hh} \quad (10.84b)$$

$$\gamma = \frac{\left\langle |s_{vv}|^2 \right\rangle}{\left\langle |s_{hh}|^2 \right\rangle} = \frac{\sigma_{vv}}{\sigma} \quad (10.84c)$$

$$f = \frac{\langle |s_{hv}|^2 \rangle}{\langle |s_{hh}|^2 \rangle} = \frac{\sigma_{hv}}{\sigma} \tag{10.84d}$$

$$\rho = \frac{\langle s_{hh} s_{vv}^* \rangle}{\sqrt{\langle |s_{hh}|^2 \rangle \langle |s_{vv}|^2 \rangle}} = |\rho| \exp[j\phi_\rho] \tag{10.84e}$$

Formula (10.84a–10.84e) is a few commonly polarization characteristic parameters. σ_{hh}, σ_{hv}, and σ_{vv} are the average scattering cross sections under HH, HV, and VV polarizations, respectively. The eigenvalues in Eq. (10.80) can be obtained from the statistical covariance matrix. The eigenvalues of the matrix $C_b^{-1} C_a$ are determined and can be expressed as

$$\lambda_1 = f_a \sigma_a / f_b \sigma_b \tag{10.85a}$$

$$\begin{aligned} \lambda_{2,3} = & \frac{\sigma_a}{2\sigma_b \gamma_b (1 - |\rho_b|^2)} \Big\{ \gamma_a + \gamma_b - 2(\gamma_b \gamma_a)^{1/2} |\rho_a||\rho_b| \cos(\phi_{\rho_a} - \phi_{\rho_b}) \\ & \pm \Big[\big(\gamma_a + \gamma_b - 2(\gamma_a \gamma_b)^{1/2} |\rho_a||\rho_b| \cos(\phi_{\rho_a} - \phi_{\rho_b}) \big)^2 - 4\gamma_a \gamma_b (1 - |\rho_b|^2)(1 - |\rho_a|^2) \Big]^{1/2} \Big\} \end{aligned} \tag{10.85b}$$

The eigenvector is

$$\lambda_1 \to W_1 = \begin{bmatrix} 0 \\ 1 \\ 0 \end{bmatrix} \tag{10.86a}$$

$$\lambda_{2,3} \to W_{2,3} = \begin{bmatrix} 1 \\ 0 \\ \Gamma_{2,3} \end{bmatrix} \tag{10.86b}$$

Where

$$\begin{aligned} \Gamma_{2,3} = & \Big\{ \gamma_a - \gamma_b - i2(\gamma_b \gamma_a)^{1/2} |\rho_a||\rho_b| \sin(\phi_{\rho_a} - \phi_{\rho_b}) \\ & \pm \Big[\big(\gamma_b + \gamma_a - 2(\gamma_a \gamma_b)^{1/2} |\rho_a||\rho_b| \cos(\phi_{\rho_a} - \phi_{\rho_b}) \big)^2 - 4\gamma_a \gamma_b (1 - |\rho_b|^2) \Big]^{1/2} \Big\} \\ & \cdot \big[2\big(\gamma_b (\gamma_a)^{1/2} \rho_a - \gamma_a (\gamma_b)^{1/2} \rho_b \big) \big]^{-1} \end{aligned} \tag{10.86c}$$

The optimal polarization matching filter vector can be determined by the eigenvector corresponding to the maximum eigenvalue, and the maximum contrast of any two types of targets is obtained.

Taking the full-polarization SAR data of the San Francisco Bay Area as an example, we determine the optimal polarization matching filter vector for the park and ocean regions, and obtain the image of the maximum contrast of the two types of

(a) Span (b) Training Area

Fig. 10.4 Polarimetric SAR image of the San Francisco Bay area

targets, as shown in Fig. 10.4. Polarization contrast enhancement essentially selects the appropriate transmit and receive polarization states through optimized signal processing, resulting in the largest ratio of the scattered power of the two types of ground objects. Any polarization state can be represented by its corresponding polarization ellipse, that is, it can be described by its elliptical azimuth and ellipticity angle. Therefore, the optimal combination of transmit and receive polarization states is also represented by its corresponding polarization elliptical azimuth angle ψ and elliptic rate angle χ.

According to formula (10.84a–10.84e), 20 × 80 pixels of the park and urban area are selected from the full polarization data of the San Francisco Bay Area as test samples, and the statistical covariance matrix of the two types of ground objects is calculated, as shown in Fig. 10.4b. The statistical covariance matrix elements for the two types of ground objects in the park and city are listed in Table 10.2.

The maximum contrast of the two types of ground objects is calculated below to determine the corresponding polarization state. When an H-V polarizer is used, several typical polarizations are combined with a polarization matched filter vector W, and the values are listed in Table 10.3.

The values of the polarization contrast ratios r_{ab} and r_{ba} for the above typical transmit and receive polarization combinations are given in Table 10.4, and the combinations of polarization states determined by r_{ab} and r_{ba} max are given, respectively. "a" represents a park area and "b" represents a city area. Consistent with

Table 10.2 Statistical covariance matrix elements for San Francisco Bay Area Parks and urban areas

	σ (dB)	f	γ	|ρ|	ϕρ
Parks	−41.3	0.479	1.531	0.225	−0.503
Urban areas	−33.8	0.172	0.887	0.398	3.012

Table 10.3 Corresponding polarization matching filter vectors in a typical transceiver polarization combination state

Transmit polarization	Receiving polarization	Optimal polarization matching filter vector W
H	H	[1.0 0.0 0.0]T
H	V	[0.0 1.0 0.0]T
V	H	[0.0 1.0 0.0]T
V	V	[0.0 0.0 1.0]T
L	L	[0.5 i −0.5]T
L	R	[0.5 0.0 0.5]T
‘R	L	[0.5 0.0 0.5]T
R	R	[0.5 −i −0.5]T

Table 10.4 *Contrasts of San Francisco Bay Area Parks and city targets

Transmit polarization	Receiving polarization	r_{ab} (dB)	r_{ba} (dB)
H	H	−7.30	7.30
H	V	−2.58	2.58
V	H	−2.58	2.58
V	V	−5.35	5.35
L	L	−6.94	6.94
L	R	−3.29	3.29
R	L	−3.29	3.29
R	R	−6.73	6.73
($\psi = 1.82°$, $\chi = 3.72°$)	($\psi = 107.0°$, $\chi = -1.64°$)	2.37	−2.37
($\psi = 48.7°$, $\chi = -6.44°$)	($\psi = 150.3°$, $\chi = 3.51°$)	−9.38	9.38

* “a” represents the park area, “b” represents the urban area, Max (r_{ab}) = 2.37 dB, Max (r_{ba}) = 9.38 dB, and optimal contrast $r = 9.38$ dB

the previous analysis, for reciprocal media, the interchange of transmit and receive polarization states does not change the value of the target contrast. This conclusion can be drawn from the fact that the results obtained by the combination of HV and VH, RL and LR are the same. The polarization ratios r_{ab} and r_{ba} can also be represented by their corresponding polarization elliptical azimuth angles ψ and elliptic rate angle χ. After calculation, the maximum value of r_{ab} is 2.37 dB, and the elliptic angle of transmission and receive polarization is ($\psi = 1.82°$, $\chi = 3.72°$) and ($\psi = 107.0°$, $\chi = -1.64°$). The maximum value of r_{ba} is 9.38 dB, and the corresponding transmit and transmit polarization ellipse angles are ($\psi = 48.7°$, $\chi = -6.44°$) and ($\psi = 150.3°$, $\chi = 3.51°$). According to Eq. (10.71), the optimal contrast is the maximum value of both $r = 9.38$ dB.

The contrast enhancement in the San Francisco Bay Area calculated from the above is shown in Fig. 10.5. In Fig. 10.5a is an image obtained in an optimal polarization matching state between a park and a city with a contrast ratio of $r = 9.38$ dB. Fig. 10.5b–d are images of three basic linear polarization combinations, namely, HH, VV, and HV, with contrasts of $r^{(HH)} = 7.30$ dB, $r^{(VV)} = 5.35$, and $r^{(HV)} = 2.58$, respectively. It can be clearly seen that in the image with higher contrast, the two types of land objects, park and city, are easier to distinguish. In an HV-polarized combined image with a contrast ratio of only 2.58 dB, it is difficult to distinguish between the two types of objects.

(a) Optimal matched polarization (r=9.38 dB)

(b) HH Image ($r^{(HH)}$=7.30 dB)

(c) VV Image ($r^{(VV)}$=5.35 dB)

(d) HV Image ($r^{(HV)}$=2.58 dB)

Fig. 10.5 Contrast enhancement image in the San Francisco Bay area

10.2.3 Numerical Solution of Contrast Enhancement Based on SUMT

This section describes another solution method to the OPCE problem—a numerical method based on SUMT. The method first converts the objective function of the optimal problem into a bilinear form, and then solves it by using numerical method by means of SUMT (Sequential Unconstrained Minimization Technique) [4]. This method is convergent and easy to program.

In this mathematical model, the polarization states of the transmitting and receiving antennas are independent variables, which may result in higher contrast. The target echo power ratio obtained by two independent polarization states may be better than that of the co-polarization, cross-polarization, and matched polarization channels.

A target corresponds to a scattering matrix, where $[S_1]$, $[S_2]$ represent the scattering matrix of the desired (target) and undesired (clutter) targets, respectively, a, b representing the transmit and receive polarization states, respectively. Then the received energy from the two targets is $\|[S_1]a \cdot b\|^2$ and $\|[S_2]a \cdot b\|^2$. Then according to the target-zero theory, there are countless pairs (a, b) such that the received clutter power is equal to 0 [6]. The mathematical model for such an OPCE problem is

$$\text{maximize:} \quad \|[S_1]a \cdot b\|^2 \tag{10.87a}$$

$$\text{Restrictions:} \quad \|[S_2]a \cdot b\|^2 = 0 \tag{10.87b}$$

$$\|a\|^2 = 1 \tag{10.87c}$$

$$\|b\|^2 = 1 \tag{10.87d}$$

Suppose $[K_1]$ and $[K_2]$ are the Stokes matrices of $[S_1]$ and $[S_2]$, respectively, and g and h are Stokes vectors of a and b, respectively. According to the polarization synthesis technique, (10.87a–10.87d) can be converted into the following form:

$$\text{minimize:} \quad -[K_1]g \cdot h \tag{10.88a}$$

$$\text{Restrictions:} \ [K_2]g \cdot h = 0 \tag{10.88b}$$

$$g_1^2 + g_2^2 + g_3^2 = 1 \tag{10.88c}$$

$$h_1^2 + h_2^2 + h_3^2 = 1 \tag{10.88d}$$

For any pair of g and h, $[K_2]g \cdot h$ represents the received energy of the clutter and is therefore always non-negative. If k is a large constant value, (10.87d) or (10.88a–10.88d) is also equivalent to the following form:

$$\text{minimize: } (-[K_1]g \cdot h + k[K_2]g \cdot h) \tag{10.89a}$$

$$\text{Restrictions: } g_1^2 + g_2^2 + g_3^2 = 1 \tag{10.89b}$$

$$h_1^2 + h_2^2 + h_3^2 = 1 \tag{10.89c}$$

The Continuous Unconstrained Minimization Technique (SUMT) is a numerical method that uses some sequences to transform constrained optimization problems into unconstrained optimization problems. Based on this, the following numerical method can be used to solve the solution of (10.89a–10.89c):

Step (1): Given an initial vector g_0 and an initial value k_0.
Step (2): Calculating the Stokes matrix $[K_1]$, $[K_2]$.
Step (3): Let $g_n = g_n^0 = (1, g_1^{n,0}, g_2^{n,0}, g_3^{n,0})^t$ and calculate the matrix:

$$[Q_n] = -[K_1] + k_n[K_2] \equiv \left(q_{ij}^n\right)_{4\times 4} \tag{10.90}$$

The superscript t indicates transposition. After getting $g_n^l = (1, g_1^{n,l}, g_2^{n,l}, g_3^{n,l})^t$, then

$$\begin{aligned} &-[K_1]g_n^l \cdot h + k_n[K_2]g_n^l \cdot h = [Q_n]g_n^l \cdot h = \\ &\left(\sum_{j=2}^{4} q_{1j}g_j^{n,l} + q_{11}\right) + \left(\sum_{j=2}^{4} q_{2j}g_j^{n,l} + q_{21}\right)h_1 \\ &+\left(\sum_{j=2}^{4} q_{3j}g_j^{n,l} + q_{31}\right)h2 + \left(\sum_{j=2}^{4} q_{4j}g_j^{n,l} + q_{41}\right)h_3 \end{aligned} \tag{10.91}$$

Obviously, if and only if:

$$h_i = -w_i/w \equiv h_i^{n,l}, \quad (i = 1, 2, 3) \tag{10.92}$$

The above expression (10.91) takes the smallest. Among them:

$$w_i = \left(\sum_{j=2}^{4} q_{(i+1)j}g_j^{n,l} + q_{(i+1)1}\right), \quad (i = 1, 2, 3) \tag{10.93a}$$

$$w = \sqrt{w_1^2 + w_3^2 + w_3^2} \tag{10.93b}$$

Step (4): After getting $h_n^l = (1, h_1^{n,l}, h_2^{n,l}, h_3^{n,l})^t$, $g_n^{l+1} = (1, g_1^{n,l+1}, g_2^{n,l+1}, g_3^{n,l+1})^t$ can be obtained by a similar method, as follows:

$$g_i^{n,l+1} = -v_i/v, \quad (i = 1, 2, 3) \tag{10.94}$$

Where

$$v_i = \left(\sum_{j=2}^{4} q_{j(i+1)} g_j^{n,l} + q_{1(i+1)}\right), \quad (i = 1, 2, 3) \tag{10.95a}$$

$$v = \sqrt{v_1^2 + v_3^2 + v_3^2} \tag{10.95b}$$

Step (5): If $\sum_{i=1}^{3} \left| g_i^{n,l+1} - g_i^{n,l} \right| \le \varepsilon_n$, and $\sum_{i=1}^{3} \left| h_i^{n,l+1} - h_i^{n,l} \right| \le \varepsilon_n$, then let $g_{n+1} \equiv g_n^{l+1}$, $h_{n+1} \equiv h_n^{l+1}$, and $k_{n+l} = 2k_n$.

Step (6): If n is large (for example, $n \ge 10$), then g_{n+1} and h_{n+1} can be considered to be the optimal solutions of (10.88a–10.88d).

The following points need to be noted:

(1) if $[S_1] = \begin{bmatrix} s1_1 & s1_2 \\ s1_3 & s1_4 \end{bmatrix}$, $[S_2] = \begin{bmatrix} s2_1 & s2_2 \\ s2_3 & s2_4 \end{bmatrix}$. From experience, it is recommended to use the following formula to choose k_0:

$$k_0 = \frac{10 \sum_{i=1}^{4} |s1_i|^2}{\sum_{i=1}^{4} |s2_i|^2} \tag{10.96}$$

(2) In order to reduce the number of iterations, in the above algorithm, it will be better if $\varepsilon_n = \varepsilon_0 / 2^n$.

(3) In calculation, if the initial vector that is very close to the optimal solution can be selected, the amount of calculation will be saved, and the optimal solution will be easily obtained. But in fact this is impossible because the optimal solution is unknown. In order to avoid losing some unknown optimal solutions, it is recommended to select several initial vectors and obtain the optimal polarization state by comparing the calculation results. For example, select the initial value of x as follows:

$$(1 \pm 1\, 0\, 0)^t, \ (1\, 0 \pm 1\, 0)^t, \ (1\, 0 \pm 1\, 0)^t \tag{10.97}$$

The effectiveness of the method is tested below using measured full-polarization SAR data from a region near Lake Baikal as shown in Fig. 10.6. The selected sample data is the two-region data labeled 1 and 2 in Fig. 10.6a. Among them, the scattering matrices of target 1 and target 2 (clutter) are

(a) Span

(b) Optimal contrast enhanced of region 1,2

Fig. 10.6 Image of optimal contrast enhanced

$$[S_1] = \begin{bmatrix} 0.1000 - 0.025 * j & -0.0069 + 0.0022 * j \\ -0.0069 + 0.0022 * j & -0.1173 - 0.0341 * j \end{bmatrix}$$

$$[S_2] = \begin{bmatrix} 0.0895 + 0.0831 * j & 0.0043 + 0.0054 * j \\ 0.0043 + 0.0054 * j & 0.0967 - 0.0545 * j \end{bmatrix}$$

Selecting $\varepsilon_n = 0.1/2^n, (n = 1, 2, \ldots, 10)$, after iterating 10 times, the six initials g_0 given by (10.97) above are selected to obtain the same g_n, h_n. The results are as follows:

$$g_n = \begin{pmatrix} 1 & -0.9422 & 0.1195 & 0.3130 \end{pmatrix}^t$$

$$h_n = \begin{pmatrix} 1 & 0.9460 & -0.2306 & -0.2277 \end{pmatrix}^t$$

In this optimal polarization state, the received energies from the target and clutter are

$$P_{r1} = 0.00151, \quad P_{r2} = 2.0015\text{e}{-16}$$

The number of iterations of the six starting g_0 is 212, 204, 210, 196, 224, 188 times. Note that each iteration of the method requires only 30 multiplications, 6 divisions, and 2 square root calculations. Thus, the efficiency of this method is very high.

Figure 10.6 shows the images of optimal contrast obtained by this method. Figure 10.6a is the total power image of a place near Lake Baikal, and Fig. 10.6b is the image of the optimal contrast enhancement of the areas 1 and 2. The echoes similar to the region 2 in Fig. 10.6b are greatly reduced. The echo in the region 1 becomes stronger. The advantage of this method is that the echo of the clutter (target b) can be suppressed to the utmost extent of the image.

10.3 Clutter Suppression of Polarization SAR Image Based on Polarization Synthesis and Target Null [6, 7]

The SUMT-based polarization contrast enhancement method is to maximize the echo energy of the other target under the constraint of ensuring that the echo power of one target is zero, so that the maximum contrast of the two targets is obtained. That is, in the case where the echo of one target (clutter) is suppressed, the echo energy of the other target is maximized. A method for simultaneously suppressing two types of clutter based on a polarization synthesis technique and a target-zero theory is described below.

10.3.1 Polarization Target Null Theory

In the 1950s, Kennaugh first introduced the concept of characteristic polarization. Huynen [8] then studied Kennaugh's work again and introduced the famous polarization fork concept. A polarization fork means that a characteristic polarization forms a fork shape on a Poincaré sphere in a co-polarized channel. Since the 1980s, researchers have proposed optimization procedures for characteristic polarization in various situations. In particular, Boerner and his colleagues [9], Van Zyl [10] and Lüneburg [11], have finished many projects in this area. In a co-polarized or cross-polarized channel, the polarization of the transmit and receive antennas is the same or orthogonal. The characteristic polarization state is defined as the polarization state of the transmitting and receiving antennas that maximize the received echo energy. Target zero (i.e., co-polarization (CO-POL) zero or cross-polarization (X-POL) zero) is defined as the polarization state of the transmitting and receiving antennas that receive zero energy. The co-polarization zero and cross-polarization zero are very important and useful. For example, Yamaguchi et al. [6] used the co-polarized zero and cross-polarization zeros of clutter to detect targets buried under snow and sand by means of the method of polarization. Yang Jian et al. [12, 13] extended the target-zero theory to two independent polarization states. It has been proved that for two arbitrary mutually independent symmetric scattering matrices, there is only one pair of transmitting and receiving antenna polarization states that cause the received energy of both scattering matrices to be zero. This polarization state is the co-null for two targets. Two independent targets mean that their scattering matrices are linearly independent.

10.3.1.1 CO-POL Null and X-POL Null

Rewrite the target scattering matrix as

$$[S] = \begin{bmatrix} s_{hh} & s_{hv} \\ s_{vh} & s_{vv} \end{bmatrix} \tag{10.98}$$

For the symmetric scattering case $s_{hv} = s_{vh}$. Let vector a and b indicate the polarization states of the transmitting and receiving antennas, respectively, then the received energy is expressed as

$$P = |[S]a \cdot b|^2 = \left|b^t[S]a\right|^2 \tag{10.99}$$

which represents the vector product, and the superscript t represents the transpose. In general, as summing $\|a\|^2 = \|b\|^2 = 1$. For the case of the co-polarization channel $b = a$, the co-polarization null is defined such that

$$P_c = |[S]a \cdot a|^2 = \left|a^t[S]a\right|^2 = 0 \tag{10.100}$$

The transmit antenna polarization state at (10.100). Typically, there are two co-polarized zeros. For the case of symmetric scattering matrices, the co-polarization zero is determined by the following equation [6, 12]:

$$[S]a = \lambda \begin{bmatrix} 0 & -1 \\ 1 & 0 \end{bmatrix} a \tag{10.101}$$

For cross-polarized channels:

$$b = \begin{bmatrix} 0 & -1 \\ 1 & 0 \end{bmatrix} a^* \tag{10.102}$$

The superscript * indicates complex conjugate. Cross-polarization zero is defined as the polarization state of the transmitting and receiving antennas when the following equation is established:

$$P_x = \left|[S]a \cdot \begin{bmatrix} 0 & -1 \\ 1 & 0 \end{bmatrix} a^*\right|^2 = 0 \tag{10.103}$$

For the case of symmetric scattering matrices, there are usually two X-POL zeros, which are determined by

$$[S]a = \lambda a^* \tag{10.104}$$

Let a_{x1} and a_{x2} denote two X-POL zeros, so it is easy to prove

$$a_{x2} = \begin{bmatrix} 0 & -1 \\ 1 & 0 \end{bmatrix} a_{x1}^* \tag{10.105}$$

10.3.1.2 Two Targets CO-Null of Symmetric Scattering Matrix

The target co-null is defined as [14]: $[S_1]$, $[S_2]$ represent the scattering matrix of target 1 and target 2, respectively. If a and b satisfy both $[S_1]a \cdot b = 0$, $[S_2]a \cdot b = 0$, we call (a, b) the same zero pair of the target 1, 2, or the co-null of $[S_1]$ and $[S_2]$. If a and b also satisfy $[S_3]a \cdot b = 0$, we call (a, b) the co-null of $[S_1]$, $[S_2]$, and $[S_3]$.

(a,b) is the co-null of $[S_1]$ and $[S_2]$, which can be obtained by $[S_1]a \cdot b = 0$, $[S_2]a \cdot b = 0$:

$$[S_1]a = \lambda_1 \begin{bmatrix} 0 & -1 \\ 1 & 0 \end{bmatrix} b \quad [S_2]a = \lambda_2 \begin{bmatrix} 0 & -1 \\ 1 & 0 \end{bmatrix} b \tag{10.106}$$

Then, $[S_1]a = \lambda[S_2]a$ is established. Similarly, since the two scattering matrices are symmetric, there is $[S_1]b = \lambda[S_2]b$. Therefore, when $[S_1]$ and $[S_2]$ are linearly independent, if and only if (a_1, a_2) is the eigenvector corresponding to the eigenvalue of the formula g, (a_1, a_2) is the co-null of (a_1, a_2).

This not only provides a way to obtain the co-null, but also gives a new physical meaning of the eigenvalue of equation $[S_1]a = \lambda[S_2]a$. In particular, when $[S] = [S_1]$ and $[S_2] = \begin{bmatrix} 1 & 0 \\ 0 & 1 \end{bmatrix}$, the feature vector eigenvector of $[S]a = \lambda a$ becomes the co-null of the following two targets: the target whose scattering matrix is $[S]$ and the sphere target (whose scattering matrix is a unit scattering matrix). This is a new interpretation of $[S] = [S_1]$ vectors eigenvector of equation $[S]a = \lambda a$.

According to the reciprocity theorem, for the symmetric scattering case, the order of the transmitting and receiving antennas in the equation is irrelevant. From this, it is concluded that if the amplitude of the polarization state is not considered, any two independent symmetric scattering matrices have one and only one co-null pair (a, b).

From the above results, it is easy to get the co-null pair of the two targets. Table 10.5 shows the co-null for some typical targets.

10.3.2 Clutter Suppression Based on Polarimetric SAR Images with CO-Null Target

It can be known from the target-zero theory that when the scattering vectors of the two types of targets are $[S_1]$ and $[S_2]$, as long as $[S_1]$ and $[S_2]$ are linearly independent, the polarization states (a_1, a_2) of the transmitting and receiving antennas

Table 10.5 The co-null matrix pair of several typical targets

Target1 / Target2	Wire $\begin{bmatrix}1 & 0\\0 & 0\end{bmatrix}$	Sphere $\begin{bmatrix}1 & 0\\0 & 1\end{bmatrix}$	Dihedral $\begin{bmatrix}1 & 0\\0 & -1\end{bmatrix}$	Left Helix $\begin{bmatrix}1 & i\\i & -1\end{bmatrix}$	Right Helix $\begin{bmatrix}1 & -i\\-i & -1\end{bmatrix}$
Wire $\begin{bmatrix}1 & 0\\0 & 0\end{bmatrix}$		$\left(\begin{bmatrix}1\\0\end{bmatrix},\begin{bmatrix}0\\1\end{bmatrix}\right)$	$\left(\begin{bmatrix}1\\0\end{bmatrix},\begin{bmatrix}0\\1\end{bmatrix}\right)$	$\left(\begin{bmatrix}\frac{1}{\sqrt{2}}\\\frac{i}{\sqrt{2}}\end{bmatrix},\begin{bmatrix}0\\1\end{bmatrix}\right)$	$\left(\begin{bmatrix}\frac{1}{\sqrt{2}}\\\frac{-i}{\sqrt{2}}\end{bmatrix},\begin{bmatrix}0\\1\end{bmatrix}\right)$
Sphere $\begin{bmatrix}1 & 0\\0 & 1\end{bmatrix}$	$\left(\begin{bmatrix}1\\0\end{bmatrix},\begin{bmatrix}0\\1\end{bmatrix}\right)$		$\left(\begin{bmatrix}1\\0\end{bmatrix},\begin{bmatrix}0\\1\end{bmatrix}\right)$	$\left(\begin{bmatrix}\frac{1}{\sqrt{2}}\\\frac{i}{\sqrt{2}}\end{bmatrix},\begin{bmatrix}\frac{1}{\sqrt{2}}\\\frac{i}{\sqrt{2}}\end{bmatrix}\right)$	$\left(\begin{bmatrix}\frac{1}{\sqrt{2}}\\\frac{-i}{\sqrt{2}}\end{bmatrix},\begin{bmatrix}\frac{1}{\sqrt{2}}\\\frac{-i}{\sqrt{2}}\end{bmatrix}\right)$
Dihedral $\begin{bmatrix}1 & 0\\0 & -1\end{bmatrix}$	$\left(\begin{bmatrix}1\\0\end{bmatrix},\begin{bmatrix}0\\1\end{bmatrix}\right)$	$\left(\begin{bmatrix}1\\0\end{bmatrix},\begin{bmatrix}0\\1\end{bmatrix}\right)$		$\left(\begin{bmatrix}\frac{1}{\sqrt{2}}\\\frac{i}{\sqrt{2}}\end{bmatrix},\begin{bmatrix}\frac{1}{\sqrt{2}}\\\frac{-i}{\sqrt{2}}\end{bmatrix}\right)$	$\left(\begin{bmatrix}\frac{1}{\sqrt{2}}\\\frac{i}{\sqrt{2}}\end{bmatrix},\begin{bmatrix}\frac{1}{\sqrt{2}}\\\frac{-i}{\sqrt{2}}\end{bmatrix}\right)$
Left Helix $\begin{bmatrix}1 & i\\i & -1\end{bmatrix}$	$\left(\begin{bmatrix}\frac{1}{\sqrt{2}}\\\frac{i}{\sqrt{2}}\end{bmatrix},\begin{bmatrix}0\\1\end{bmatrix}\right)$	$\left(\begin{bmatrix}\frac{1}{\sqrt{2}}\\\frac{i}{\sqrt{2}}\end{bmatrix},\begin{bmatrix}\frac{1}{\sqrt{2}}\\\frac{i}{\sqrt{2}}\end{bmatrix}\right)$	$\left(\begin{bmatrix}\frac{1}{\sqrt{2}}\\\frac{i}{\sqrt{2}}\end{bmatrix},\begin{bmatrix}\frac{1}{\sqrt{2}}\\\frac{-i}{\sqrt{2}}\end{bmatrix}\right)$		$\left(\begin{bmatrix}\frac{1}{\sqrt{2}}\\\frac{i}{\sqrt{2}}\end{bmatrix},\begin{bmatrix}\frac{1}{\sqrt{2}}\\\frac{-i}{\sqrt{2}}\end{bmatrix}\right)$
Right Helix $\begin{bmatrix}1 & -i\\-i & -1\end{bmatrix}$	$\left(\begin{bmatrix}\frac{1}{\sqrt{2}}\\\frac{-i}{\sqrt{2}}\end{bmatrix},\begin{bmatrix}0\\1\end{bmatrix}\right)$	$\left(\begin{bmatrix}\frac{1}{\sqrt{2}}\\\frac{-i}{\sqrt{2}}\end{bmatrix},\begin{bmatrix}\frac{1}{\sqrt{2}}\\\frac{-i}{\sqrt{2}}\end{bmatrix}\right)$	$\left(\begin{bmatrix}\frac{1}{\sqrt{2}}\\\frac{i}{\sqrt{2}}\end{bmatrix},\begin{bmatrix}\frac{1}{\sqrt{2}}\\\frac{-i}{\sqrt{2}}\end{bmatrix}\right)$	$\left(\begin{bmatrix}\frac{1}{\sqrt{2}}\\\frac{i}{\sqrt{2}}\end{bmatrix},\begin{bmatrix}\frac{1}{\sqrt{2}}\\\frac{-i}{\sqrt{2}}\end{bmatrix}\right)$	

that simultaneously make the echo energy of the two types of targets zero are obtained. Taking a_1, a_2 as the polarization state of the transmitting and receiving antennas, the SAR image of the transmitting and receiving antennas, respectively, a_1, a_2 can be synthesized by using polarization synthesis technology. The average received energy of the two types ($[S_1]$ and $[S_2]$) of clutter echoes will be suppressed to zero at the same time. This is very important in practical applications.

Take the regions 3 and 4 in Fig. 10.6a as training samples and calculate the clutter suppression of the target co-null. Figure 10.7 is the image after target co-null of regions 3 and 4. The scattering type which is the same as regions 3 and 4 is suppressed in the figure, and many targets under those clutter are revealed. Several goals are highlighted in Fig. 10.7. At this time, the polarization states of the transmitting antenna and the receiving antenna are, respectively,

$$\begin{bmatrix}0.8588\\0.4366-i*0.2681\end{bmatrix}\text{and}\begin{bmatrix}0.3193+i*0.3431\\0.8834\end{bmatrix}$$

Fig. 10.7 Clutter suppression image after co-null of regions 3 and 4

The average received powers of the regions 3 and 4 are P3 = 7.4476e–016 and P4 = 4.0030e–016, respectively. It follows that the average received energy of the regions 3 and 4 is close to zero.

10.4 Polarization Pseudo-Color Synthesis [3]

In general, the human eye's ability to distinguish color differences is much higher than that of black and white images. Thus, in order to achieve the purpose of highlighting image information, different colors can be used in place of the gray-level variation of the image during the processing of the SAR image. After image color enhancement processing, such as color density segmentation and pseudo-color enhancement, the image recognition effect of the image is greatly improved.

In a single-polarized SAR, since the data processed when the image is normalized is the echo power of the ground target under single polarization conditions, the grayscale image can only be generated first, and then the color enhancement method is used to generate the pseudo-color image. The color enhancement techniques used are often color density segmentation. In the single-polarized SAR, the amplitude can only be generated first, because the data processed when the image is normalized is the echo power of the ground target, then the color enhancement is produced using a color enhancement method, and the color enhancement techniques used are usually color density segmentation. The color density segmentation method first considers an amplitude image as a two-dimensional intensity function, and uses the color segmentation plane to crosscut the intensity function. Then, different colors are selected for the divided layers, so that the original amplitude image is transformed into a pseudo-color image. In PolSAR, the echo power can be generated by a polarization synthesis method after determining the Stokes matrix of the target. Therefore, it is possible to select a three-polarization combination mode echo power to synthesize a polarization pseudo-color image by a certain normalization method. Since the echo power of the ground targets from different polarization combinations is different, the difference information can be comprehensively reflected by the color synthesis method. In this way, the difference in target information on the image is greatly increased, and the recognition effect of the target is improved.

The following is a comparison between single-polarized color density segmentation and polarization pseudo-color synthesis methods:

(1) The purpose of the two is consistent, that is, to enhance the ability of the human eye to recognize image information.
(2) The result of the final processing is a pseudo-color image.
(3) The objects of the processed data are different, the object of color density segmentation is the single-polarized echo power, and the object of the polarization pseudo-color synthesis process is the echo power synthesized by polarization.
(4) In both methods, if the echo power is used in the same normalization method, then the maximum number of color types in the image synthesized by the polarization pseudo-color is the cube of the largest number of color species in the pseudo-color image generated by the color density segmentation. In the image synthesized by polarized color, the values of the three primary colors at any point determine the color of the point, and the values of the three primary colors are determined by the three-polarization modes selected and the normalization method.

Since the echo powers of the targets under different polarizations have large differences, for example, HH is 10 dB larger than HV, if the same normalization factor is used for the echo powers of the three different polarizations, the distribution of the three primary color values is uneven. The synthesized image is dominated by only one or several colors, and the recognition ability of the target cannot be significantly improved. If the three primary color channels are processed by their respective normalization factors, the dynamic range of the color can be increased, and the color difference of different targets can be increased.

The polarization combination mode is also a factor that affects the color of the target image. The echo power of different targets in different polarization combination modes is quite different. So, under the condition of a fixed normalization factor, the colors in the same resolution unit also change under different polarization combination modes. The simplest method of polarization pseudo-color synthesis processing is that the three primary colors correspond to the echo power of the HH, HV, and VV of the target, respectively. In this case, the target that can be accurately distinguished is the target with a large difference in echo power of HH, HV, and VV.

In order to increase the color difference between different targets, the polarization combination mode when the target echo power is the largest difference is obtained according to the Stokes matrix of the two types of targets. Then, the normalized echo power is used as a primary color. It is worth noting that there are three cases in the relationship between the magnitudes of the echo powers of the two types of targets at different polarizations:

(1) At any polarization, the echo power of target A is always greater than the echo power of target B; (2) at any polarization, the echo power of target B is always greater than the echo power of target A; (3) at some polarizations, the echo power of target A is greater than the echo power of target B. At the other polarizations, the echo power of target B is greater than the echo power of target A.

In the case of (1) and (2), the selected polarization combination is the point at which the power difference between the two types of targets is the largest. There is only one such point. The polarization combination of the remaining primary colors can be selected from two of HH, HV, and VV depending on the actual imaging effect. In the case of (3), the polarization combination points with the largest difference in echo power between the two types of targets are still selected. There are two such points. The polarization corresponding to the other primary colors may be selected from one of HH, HV, and VV depending on the actual imaging effect. In the above processing method, the solution to the polarization combination mode is actually consistent with the optimal polarization mode solution. Only let the echo power of the optimal polarization synthesis with HH, HV, and VV to produce a pseudo-color image.

Since the above solution process to the maximum point of the echo power difference is based on all polarization combinations, the solution process is more complicated. In the following examples, the polarization signature of two types of targets is utilized, and the optimal polarization is obtained by numerical methods. That is, the polarization combination modes corresponding to the three primary colors are selected from the co-polarization signature and the cross-polarization signature.

In the polarization pseudo-color synthesis, one or more of the primary colors may also be represented by polarization characteristic values other except the echo power, for example, phase difference of some polarizations, correlation coefficient as well. The polarization signatures of the target are important factors that distinguish the target category. For many types of targets, although the total power is different, the shape of the polarization signature is the same.

The shape of the polarization signature reflects the relationship between M_{11} (the total power of the target) and other elements in the Stokes matrix. Thus, the elements in the Stokes matrix can be normalized by M_{11} to produce a new matrix Mc. Based on this, the relative power of the echo is generated by HH, HV, and VV. At this time, the information reflected by the pseudo-color synthesis is the ratio of the echo power to the total power of the three polarized targets. This method is useful in dealing with the identification of certain targets whose echo power is too strong or too weak.

The polarization pseudo-color synthesis normalizes the echo power of the targets under three polarizations into three amplitude images, and uses color synthesis to represent the information of the three images. Since the introduction of a color representation, the ability to distinguish between scene targets is enhanced. At the same time, the optimal polarization under the co-polarization and cross-polarization can be obtained by means of the method of polarization signature, and the target echo power under the optimal polarization combination is synthesized into a pseudo-color image. As the polarization synthesis processing, it can use Stokes matrix, and also can use Stokes matrix normalized by M_{11}. The power image obtained by the latter can reflect the shape of the target polarization signature to a certain extent, which is beneficial to the recognition of the target.

The experimental data for polarization pseudo-color synthesis is the fully polarized data obtained by the AIRSAR system in Death Valley of the United States in 1992. Figure 10.8 shows the total power image of the region. The flight path of the aircraft is parallel to the top of the image, and the left orthogonal look is used. The area of the region is 12.2 km (in azimuth direction) × 8.5 km (in range direction).

The terrain of the Death Valley is very complex, and the area shown in Fig. 10.8 is called the Cottonball Basin. Above the figure is an alluvial area, and the upper right corner is an undulating mountain range. In the middle of the figure is a large saline-alkali land, where the brighter areas are coarse-grained saline-alkali land, while the darker areas are fairly saline-alkali land. On the left edge of the figure, one-third of the brighter part from the bottom is the local forestry station, including several houses and a small runway for take-off and landing. It is surrounded by forest farms.

Figure 10.9 is a color segmentation diagram generated using the total power map of Death Valley. The color conversion table used is the HSV in the software Matlab. Because the conversion color used is mainly red, yellow, green, and blue, the difference in total power of different targets is clearly expressed.

Figure 10.10 shows a polarization pseudo-color image synthesized using the echo power maps of HH, HV, and VV. The three primary colors are red = HH, green = HV, and blue = VV. The relationship between the HH, HV, and VV echo powers

Fig. 10.8 Total power of Death Valley

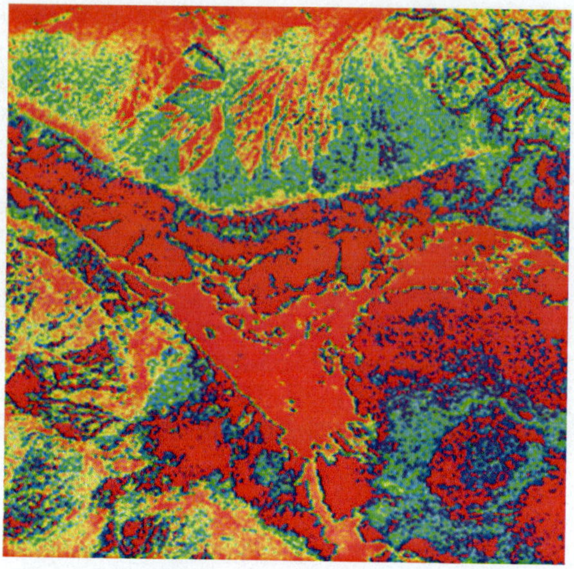

Fig. 10.9 Color segmentation image of Death Valley

Fig. 10.10 Polarized pseudo-color image of Death Valley (HH, HV, VV)

of different targets can be visually understood from the figure. For example, the HV echo intensity of the mountain in the upper right corner is much higher than other surrounding targets. The goal of higher HV echo intensity also includes looking at the forest around the forest station. The three echoes of coarse saline-alkali soils are relatively high, while the smooth saline-alkali grounds have low echo intensity.

Fig. 10.11 Polarized pseudo-color image after normalized by M11 in Death Valley (HH, HV, VV)

The alluvial area in the figure is purple, which indicates that the HH and VV echo intensities of this area are relatively high.

Figure 10.11 shows the polarization pseudo-color composite image obtained by normalizing M11 at HH, HV, and VV. The three primary colors are red = HH, green = HV, and blue = VV. The Stokes matrix used in the polarization synthesis in the image is Mc normalized by M11. So that, in a wide range, its color can represent a target having the same shape in the polarization signature. For example, the echo intensity of the coarse saline-alkali ground and the smooth saline-alkali ground in the total power diagram differs greatly.

However, in Fig. 10.11, the colors in both regions are blue. It is stated that in the polarization signature, the two salts have the same shape. Since the total power information is removed, the brightest and darkest regions of the total power image can be analyzed. For example, the forest farm is divided into orange and yellow in Fig. 10.11. The orange region is most likely an airstrip and the three yellow triangles in the alluvial region show only the bottommost piece in the total power image.

From the perspective of polarization synthesis, the relationship between the above three images is

$$P_{uc} = P_{TP} \cdot P_c \tag{10.107}$$

where P_{uc} is a polarization pseudo-color image, P_{TP} is a total power image, and P_c is a normalized polarization pseudo-color image.

Since the echo power changes of different targets are different under different polarizations, better polarization combination can be selected to increase the color difference between the targets after polarization pseudo-color. In the following, the difference in power of the target polarization signature is used as a criterion for determining the optimal polarization. The target is a high-voltage line that is erected in the forest. According to the Stokes matrix obtained by averaging about 50 points of the high-voltage line and the mountain, the difference between the echo power of the co-polarization and the cross-polarization and the echo power normalized by M_{11} is calculated, respectively. As shown in Figs. 10.12, 10.13, the polarization combination corresponding to the maximum and minimum values can be obtained from the graph, as shown in Table 10.6.

As can be seen from Fig. 10.14, the high-voltage line can be clearly displayed in the three pseudo-color maps. The more obvious the high-voltage line in the image, the

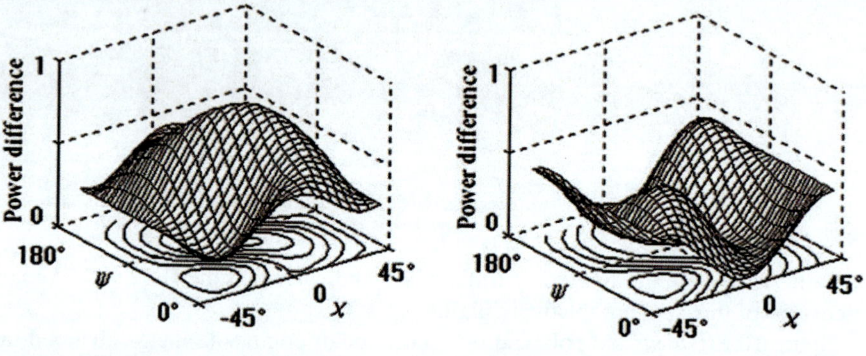

(a) Power Difference of co-polarization (b) Power Difference of cross-polarization

Fig. 10.12 Difference between polarization signature image of high-voltage line and mountain forest

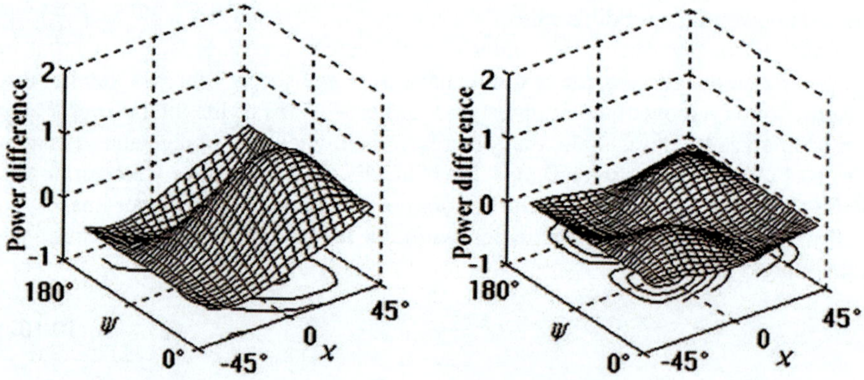

(a) Power Difference of co-polarization (b) Power Difference of cross-polarization

Fig. 10.13 Difference between polarization signature image of high-voltage line and mountain forest after normalization with M11

Table 10.6 Polarization combination method

		Co-polarization P_{max}	Co-polarization P_{min}	Cross-polarization P_{max}	Cross-polarization P_{min}
M	ψ_r	5	25	−30	0
	χ_r	85	140	45	90
	Difference	0.83*	0.10*	0.59*	0.05
M_c	ψ_r	15	5	20	−25
	χ_r	60	145	135	135
	Difference	1.02*	−0.82*	0.36*	−0.22

The * in the table indicates the polarization used in the polarization pseudo-color synthesis

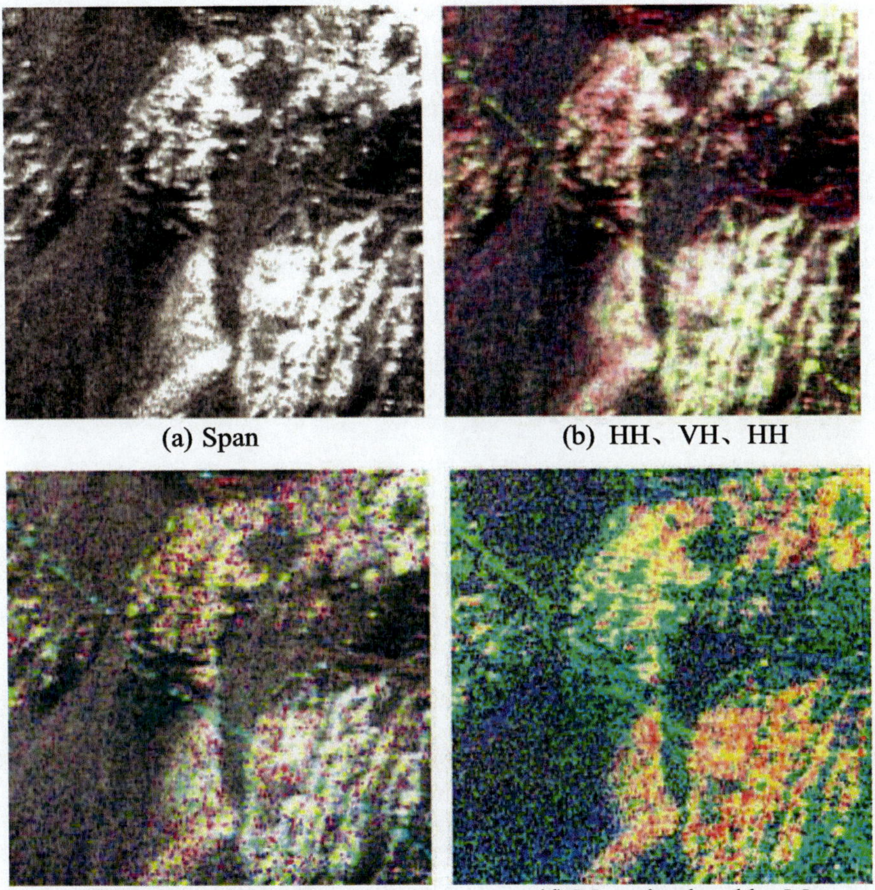

(a) Span (b) HH、VH、HH (c) Optimal polarization (d) M_c normalized by M_{11}

Fig. 10.14 Pseudo-color images of high-voltage line and mountain forest greater the fluctuation of the image. In Fig. 10.14a, the forest in the lower right part has strong echo power. So, in Fig. 10.14b, the high-voltage line is very blurred in this area and is barely visible. In Fig. 10.14c, d, high-voltage lines can still be distinguished from them. However, the brightness of the image in Fig. 10.14d fluctuates greatly, and there are many targets around it that are similar to the color of the high-voltage line. Therefore, it can be considered that the Stokes matrix is used first to find the optimal polarization, and then the pseudo-color synthesis is performed, so that the effect of distinguishing the target is better.

After analyzing the results of the previous three methods, it is proved that all three methods can effectively use the polarization SAR data to enhance the contrast of the image. These three methods are performed from three different perspectives: contrast enhancement is performed from the perspective of increasing the power ratio of the two types of targets. Because the two types of targets have different echo powers in different polarization combination modes, the power ratio of the two types of targets is maximized by selecting the optimal polarization combination mode, and speckle filtering is performed from the perspective of reducing the speckle in the image. A polarization whitening filter is obtained by calculating a polarization covariance matrix of the target, and then regenerating an image that effectively eliminates the speckle by optimally weighting the components of the polarization measurement vector. The color synthesis enhancement method is performed from the perspective of the representation of the image. Several suitable polarization modes are selected by calculating the polarization signatures of the two types of targets. The echo power is then used to generate a pseudo-color image, so that the color difference between a certain type of target and another type of target is larger.

10.5 Summary

This chapter mainly introduces the contrast enhancement of polarimetric SAR images, and the polarization clutter suppression and polarization pseudo-color synthesis based on polarization synthesis, target-zero of polarization SAR data. The basic idea of solving the optimal contrast problem based on Lagrangian operator is to obtain the optimal weight vector W_{opt} which makes the A and B target strengths the largest compared to the scattering covariance matrix of the targets A and B. Then, W_{opt} is used to weigh the scattering vector of the image data to obtain the intensity image with the largest contrast between the A and B targets. It assumes that the polarization states of the receiving and transmitting antennas are the same. The basic idea of the numerically enhanced numerical solution based on SUMT is the target-zero theory. According to the target-zero theory, the polarization state of the transmitting and receiving antennas with the largest target V (or target A) echo is obtained under the condition that the echo of the target A (or the target B) is zero. Then, the SAR image with the highest contrast of the target A and B is obtained by the polarization synthesis method. It eliminates a clutter when maximizing the target echo.

The theory of co-polarization zero and cross-polarization zero is derived in the co-polarization and cross-polarization channels. In both cases, the polarization state of the receiving antenna is related to the polarization state of the transmitting antenna. In order to make full use of the advantages of polarization, it is necessary to extend the theory of co-polarization zero and cross-zero to the polarization state of two independent antennas, which is the co-null of the two targets. These concepts are very important and useful. Suppose there are two different undesired targets (clutters) in the scene, and their co-null pair is (a, b), then (a, b) is used as the polarization state of the transmitting and receiving antennas. It is easy to make both undesired target echoes zero at the same time. In theory, this method can be used to suppress echoes in all spaces that have the same pair of zeros (a, b). Obviously, if a and b are neither identical nor orthogonal, then it is impossible to simultaneously suppress all echoes of the two undesired targets in the co-polarized or cross-polarized channels.

In fact, the contrast enhancement of the two targets based on the target-zero theory eliminates a clutter. The co-null of the two targets is to eliminate two different clutters at the same time.

According to the principle of three primary colors, color synthesis usually requires three components for three primary color synthesis. The resulting image is rich in color and easy to analyze and understand. By analyzing the color of the pixels in the image, it is also easier to analyze the difference in polarization scattering characteristics between each of the resolution unit features. The polarization pseudo-color composite image enhances the information recognition effect.

References

1. G.A. Ioannidis, D.E. Hammers, Optimum antenna polarizations for target discrimination in clutter. IEEE Trans. Antennas Propag. **27**(3), 357–362 (1979)
2. J.-S. Lee, K. Hoppel, Principal components transformation of multifrequency polarimetric SAR imagery. IEEE Trans. GRS **30**(4), 686–696 (1992)
3. L. Hongchun, data processing and analysis of polarized synthetic aperture radar. Ms, Institute of Electronics, Chinese Academy of Sciences, 2000
4. B. Dai, Studies on information processing technique of fully-polarimetric synthetic aperture radar, research on multipolarization SAR system and polarization information processing technology. Ph.D., Institute of Electronics, Chinese Academy of Sciences, 2000
5. S.R. Cloude, K.P. Papathanassiou, Polarimetric SAR interferometry. IEEE Trans. GRS **36**(5), 1551–1565 (1998)
6. J. Yang, Y. Yamaguchi, H. Yamada, Development of target null theory. IEEE Trans. GRS **39**(2), 330–338 (2001)
7. X. Liu, studies on information processing technique of fully-polarimetric synthetic aperture radar. Ph.D., University of Chinese Academy of Sciences, 2004
8. J.R. Huynen, Phenomenological theory of radar targets. Ph. D. Dissertation, 1970
9. W.M. Boerner, M.B. El-Arini, Polarization dependence in electromagnetic inverse problem. IEEE Trans. Antennas Propag. **29**(2), 262–271 (1981)
10. H. Zebker, J.J. Zyl et al., Calibrated imaging radar polarimetry: technique, examples and applications. IEEE Trans. GRS **29**(5), 942–961 (1991)

11. E. Luneburg, Polarimetric target matrix decompositions and the Karhunen-Loeve expansion, in *Geoscience and Remote Sensing Symposium, IGARSS '99 Proceedings*, vol. 5 (1999), pp. 2658–2660
12. J. Yang, Y. Yamaguchi, W.M. Boerner, Numerical methods for solving the optimal problem of contrast enhancement. IEEE Trans. GRS **38**(2), 965–971 (2000)
13. J. Yang, Y. Yamaguchi, H. Yamada, The characteristic polarization states and the equi-power curves. IEEE Trans. GRS **40**(2), 305–313 (2002)
14. Y.-Q. jin, S.R. Cloude, Numerical eigenanalysis of the coherency matrix for a lager of random nonspherical scatterers. IEEE Trans GRS **32**(6), 1179–1185 (1994)
15. S.N. Madsen, E.L. Christensen et al., The Danish SAR system: design and initial tests. IEEE Trans. GRS **29**(5), 417–426 (1991)
16. R. Horn, DLR airborne SAR project, objectives and status, in *Proceedings of the First International Airborne Remote Sensing Conference* (Strasbourg, France, 11–15 September, 1994)
17. F. Stuhr, R. Jordan, M. Werner, SIR-C/X-SAR: a multifaceted radar, in *IEEE International Radar Conference* (1995), pp. 53–61
18. R.L. Jordan, E.R. Caro et al., Design of Shuttle Radar Topography Mapper (SRTM). Proc. SPIE **2747**, 108–117 (1996)
19. P. Pasquali et al., Study of the ENVISAT ASAR alternate polarization mode: preliminary results, in *Proceedings of EUSAR Symposium* (1998), pp. 335–338
20. A.A. Swartz, H.A. Yueh, J.A. Kong, Optimal polarizations for achieving maximum contrast in radar images. J. Geophys. Res. **93**(B12), 252–260 (1988)
21. L. Brule, H. Baeggli, RADARSAT-2 program update, in *Proceedings of EUSAR Symposium* (2002)
22. [I, VIII] D.A. Garren, A.C. Odom, M.K. Osborn, Full-Polarization Matched-illumination for target detection and Identification. IEEE Trans. Aerosp. Electron. Syst. **38**(3), 824–837 (2000)
23. Y.-Q. Jin, N.-X. Zhang, M. Chang, The Muller and coherency matrices solution for Polarimetric Scattering simulation of tree canopy in SAR imaging at C band. J. Quant. Spectrosc. Radiat. Transfer, 599–611 (1999)

Chapter 11
Target Polarimetric Decompositions of PolSAR Image

Target polarimetric decomposition is one of the most important aspects of polarimetric microwave imaging. The objective of polarimetric decomposition is to express the polarimetric information of the target as a combination of many known objects, such as a plate, a dihedral, a helix, a line, and a sphere, so we can understand and analyze the scattering components and scattering mechanisms of the targets. Moreover, the data analysis of polarimetric data can be simplified. Polarimetric decomposition provides a method to extract useful information needed in application, which are the scattering characteristics. The characteristics parameters provide basis and useful information for image classification, target detection, target identification, parameters retrieval, etc.

There are two classes of polarimetric decompositions, which are coherent decomposition based on target scattering matrix and incoherent decomposition based on covariance matrix and coherency matrix.

The purpose of coherent decomposition is to express the measured scattering matrix as a combination of the scattering responses of simpler objects. Commonly used coherent decomposition includes Pauli decomposition, Krogager decomposition, [1] Cameron decomposition [2], and target similarity parameter decomposition [3]. Coherent decomposition is only applied in those cases in which both, the incident and the scattered waves are completely polarized waves. For a specific scattering matrix, the decompositions are not unique.

The purpose of incoherent decomposition is to separate the covariance matrix and coherency matrix as the combination of second-order descriptors which are easier to interpret. Most commonly used incoherent decompositions include polarization characteristic based decomposition method and model-based multi-look data decomposition.

Polarization characteristic-based decomposition models the covariance matrix as a contribution of three components, and eigenvalues and eigenvectors are used to characterize polarization information. The main polarization characteristic-based decomposition methods include Huynen decomposition, Cloude–Pottier decomposition, Holm decomposition, Van Zyl decomposition, etc.

R. Yang et al., *Polarimetric Microwave Imaging*,
https://doi.org/10.1007/978-981-15-8897-6_11

Model-based multi-look data incoherent decomposition models the target as the contribution of some simple objects corresponding to some known scattering models. These classes of decompositions include Freeman–Durden three components decomposition, Freeman two components decomposition, Yamaguchi four components decomposition, multi-components scattering model decomposition, etc.

In this chapter, coherent decomposition based on scattering matrices and incoherent decomposition based on covariance and coherency matrices are discussed and analyzed. The discussion and analysis will provide basis for the applications of polarimetric classification, polarimetric target detection, etc.

11.1 Coherent Decompositions

11.1.1 The Pauli Decomposition

The Pauli decomposition expresses the target as a combination of plate, dihedral oriented at 0° and dihedral oriented at 45°. Pauli decomposition takes Pauli basis as its basic matrix and expresses the scattering matrix **S** as the complex sum of Pauli basis matrices. If we consider the orthogonal linear (H, V) basis, the scattering matrix is given by

$$[\mathbf{S}] = \begin{bmatrix} S_{HH} & S_{HV} \\ S_{VH} & S_{VV} \end{bmatrix} = \frac{a}{\sqrt{2}}\begin{bmatrix} 1 & 0 \\ 0 & 1 \end{bmatrix} + \frac{b}{\sqrt{2}}\begin{bmatrix} 1 & 0 \\ 0 & -1 \end{bmatrix} + \frac{c}{\sqrt{2}}\begin{bmatrix} 0 & 1 \\ 1 & 0 \end{bmatrix} + \frac{d}{\sqrt{2}}\begin{bmatrix} 0 & -j \\ j & 0 \end{bmatrix} \tag{11.1}$$

Whereas, b, c, and d are all complex, and

$$\begin{aligned} a &= \frac{S_{HH} + S_{VV}}{\sqrt{2}}, b = \frac{S_{HH} - S_{VV}}{\sqrt{2}} \\ c &= \frac{S_{HV} + S_{VH}}{\sqrt{2}}, d = j\frac{S_{HV} - S_{VH}}{\sqrt{2}} \end{aligned} \tag{11.2}$$

Applying Pauli decomposition to a specific object can be interpreted as a combination of four scattering mechanisms. The first matrix corresponds to single-bounce scattering or odd-bounce scattering, like the scattering matrix of a plate. The second matrix represents the scattering mechanism of a dihedral oriented at 0° (double-bounce or even-bounce scattering). The third matrix corresponds to the scattering mechanism of a diplane oriented at 45° (double-bounce or even-bounce scattering). The fourth matrix represents the anti-symmetry component of [**S**].

Under the assumption of reciprocity condition, $S_{HV} = S_{VH}$, the Pauli basis can be reduced to three matrices, $d = 0$, and the final contribution of the fourth matrix to [**S**] is 0.

Fig. 11.1 Illustration of Pauli basis decomposition, R: $|S_{\mathrm{HH}} - S_{\mathrm{VV}}|$, G: $|S_{\mathrm{HV}}|$, B: $|S_{\mathrm{HH}} - S_{\mathrm{VV}}|$

Pauli decomposition is very simple, and the main drawbacks of Pauli decomposition are that it can only distinguish two scattering mechanisms, which are odd-bounce scattering and even-bounce scattering. The interpretation of targets is not integrated. Figure 11.1 shows the Pauli basis combination in an RGB image of San Francisco Bay acquired by AirSAR system. In Fig. 11.1, red channel represents $|S_{\mathrm{HH}} - S_{\mathrm{VV}}|$, green channel represents $|S_{\mathrm{HV}}|$, and blue channel represents $|S_{\mathrm{HH}} - S_{\mathrm{VV}}|$.

11.1.2 The Krogager Decomposition

In the Krogager decomposition, the symmetric scattering matrix [S] is decomposed as the combination of three scattering components which responses of a sphere, a diplane rotated θ degrees, and a helix. Krogager decomposition is called SDH decomposition too. In the linear orthogonal basis (H, V), the Krogager decomposition can be expressed as

$$\begin{aligned}\mathbf{S}_{(\mathrm{H,V})} &= \mathrm{e}^{j\phi}\left\{e^{j\phi_s}k_S\mathbf{S}_{\mathrm{sphere}} + k_D\mathbf{S}_{\mathrm{diplane}}(\theta) + k_{\mathrm{H}}\mathbf{S}_{\mathrm{helix}}(\theta)\right\} \\ &= \mathrm{e}^{j\phi}\left\{e^{j\phi_s}k_S\begin{bmatrix}1 & 0\\ 0 & 1\end{bmatrix} + k_D\begin{bmatrix}\cos 2\theta & \sin 2\theta\\ \sin 2\theta & -\cos 2\theta\end{bmatrix} + k_{\mathrm{H}}e^{\mp j2\theta}\begin{bmatrix}1 & \pm j\\ \pm j & -1\end{bmatrix}\right\}\end{aligned} \tag{11.3}$$

where k_S, k_D, and k_H correspond to the contributions of sphere, diplane, and helix, respectively. θ is the orientation angle, ϕ is the absolute phase, and ϕ_s is the displacement of the sphere respect to the diplane and helix components.

The displacement of the diplane and helix components cannot be measured, because only two phase values and three amplitude values can be extracted from the scattering matrix. It is noted that in a specific resolution cell, the helix component can be obtained by the sum of two or multiple diplane components, depending on their relative orientation angle and displacement.

In the circular polarization basis (r, l), the Krogager decomposition can be expressed as

$$\begin{aligned}\mathbf{S}_{(R,L)} &= \begin{bmatrix} S_{RR} & S_{RL} \\ S_{LR} & S_{LL} \end{bmatrix} \\ &= e^{j\phi}\left\{ e^{j\phi_s} k_S \begin{bmatrix} 0 & j \\ j & 0 \end{bmatrix} + k_D \begin{bmatrix} e^{j2\theta} & 0 \\ 0 & -e^{-j2\theta} \end{bmatrix} + k_H \begin{bmatrix} e^{j2\theta} & 0 \\ 0 & 0 \end{bmatrix} \right\}\end{aligned} \tag{11.4}$$

From Eq. (11.4), we can deduce the different Krogager decomposition parameters, that is

$$\begin{cases} k_S = |S_{RL}|, \phi = \frac{1}{2}(\phi_{RR} + \phi_{LL} - \pi) \\ \theta = \frac{1}{4}(\phi_{RR} - \phi_{LL} + \pi), \phi_S = \phi_{RL} - \frac{1}{2}(\phi_{RR} + \phi_{LL}) \end{cases} \tag{11.5}$$

It can be seen that the terms S_{RR} and S_{LL} can be represented, directly, the diplane component, but the two cases of analysis must be considered according to value of $|S_{RR}|$ and $|S_{LL}|$.

If $|S_{RR}| > |S_{LL}|$, $|S_{LL}|$ represents diplane component,

$$\begin{cases} k_D = |S_{LL}| \\ k_H = |S_{RR}| - |S_{LL}| \end{cases} \tag{11.6}$$

On the contrary, if $|S_{RR}| < |S_{LL}|$, $|S_{RR}|$ represents diplane component

$$\begin{cases} k_D = |S_{RR}| \\ k_H = |S_{LL}| - |S_{RR}| \end{cases} \tag{11.7}$$

In order to relate the formulations of the Krogager decomposition presented in Eqs. (11.3) and (11.4), the following relations can be referenced

$$\begin{cases} S_{RR} = iS_{HV} + 1/2(S_{HH} - S_{VV}) \\ S_{LL} = iS_{HV} - 1/2(S_{HH} - S_{VV}) \\ S_{RL} = \frac{i}{2}(S_{HH} + S_{VV}) \end{cases} \tag{11.8}$$

Figure 11.2 gives the RGB image of Krogager decomposition.

Fig. 11.2 Illustration of Krogager decomposition, R: k_D (dB), G: k_H (dB), and B: k_S (dB)

11.1.3 The Cameron Decomposition

The Cameron decomposition performs a factorization coherent decomposition method of the measured scattering matrix based on two basic properties of radar targets, reciprocity, and symmetry. The Cameron decomposition extracts the maximum symmetrical component using the rotation of the orientation angle.

The first step of the Cameron decomposition is factorizing the target scattering matrix **S** as the sum of reciprocal and nonreciprocal components. The second step is to further factorize the reciprocal component. The Cameron decomposition can be expressed as

$$\mathbf{S} = \mathrm{a}\left\{\cos\theta_{\mathrm{rec}}\left\{\cos\tau_{\mathrm{sym}}\mathbf{S}_{\mathrm{sym}}^{\max} + \sin\tau_{\mathrm{sym}}\mathbf{S}_{\mathrm{sym}}^{\min}\right\} + \sin\theta_{\mathrm{rec}}\mathbf{S}_{\mathrm{nonrec}}\right\}. \qquad (11.9)$$

where scalar $\mathrm{a} = \left\|\mathbf{S}_2^2\right\| = \mathrm{span}(\mathbf{S})$; θ_{rec} represents the degree to which scattering matrix **S** satisfies the reciprocal theorem; τ_{sym} represents the degree to which scattering matrix deviates from the corresponding symmetric component; $\mathbf{S}_{\mathrm{nonrec}}$ represents the normalized nonreciprocal component; $\mathbf{S}_{\mathrm{sym}}^{\max}$ represents the normalized maximum symmetrical component, while $\mathbf{S}_{\mathrm{sym}}^{\min}$ is the normalized minimum symmetrical component. Figure 11.3 shows the scheme of the Cameron decomposition.

From the Pauli decomposition, we can get

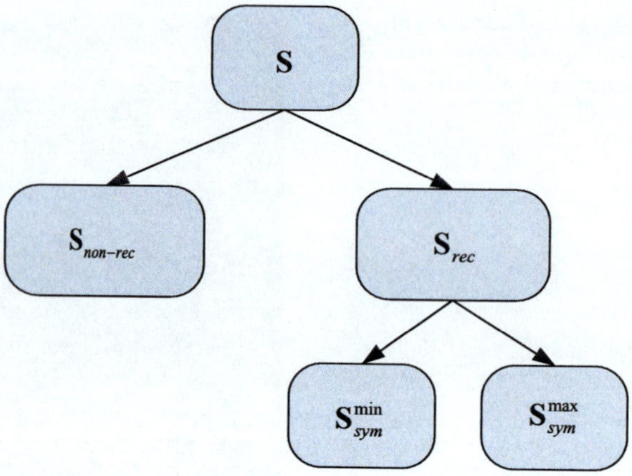

Fig. 11.3 Scheme of the Cameron decomposition

$$[\mathbf{S}] = \frac{a}{\sqrt{2}}\begin{bmatrix} 1 & 0 \\ 0 & 1 \end{bmatrix} + \frac{b}{\sqrt{2}}\begin{bmatrix} 1 & 0 \\ 0 & -1 \end{bmatrix} + \frac{c}{\sqrt{2}}\begin{bmatrix} 0 & 1 \\ 1 & 0 \end{bmatrix} + \frac{d}{\sqrt{2}}\begin{bmatrix} 0 & -j \\ j & 0 \end{bmatrix} \tag{11.10}$$

where a, b, c and d are all complex.

Expressing the scattering matrix **S** as the vector form

$$\mathbf{S} = \mathrm{V}(\mathbf{S}) = \frac{1}{2}\mathrm{tr}(\mathbf{S}\{\boldsymbol{\Psi}\}) = \begin{bmatrix} s_{\mathrm{hh}} \\ s_{\mathrm{hv}} \\ s_{\mathrm{vh}} \\ s_{\mathrm{vv}} \end{bmatrix} \tag{11.11}$$

where tr(·) represents the trace of a matrix, and $\{\boldsymbol{\Psi}\}$ is a set of lexicography matrices.

$$\boldsymbol{\Psi} = \left\{ \begin{bmatrix} 2 & 0 \\ 0 & 0 \end{bmatrix}, \begin{bmatrix} 0 & 2 \\ 0 & 0 \end{bmatrix}, \begin{bmatrix} 0 & 0 \\ 2 & 0 \end{bmatrix}, \begin{bmatrix} 0 & 0 \\ 0 & 2 \end{bmatrix} \right\} \tag{11.12}$$

The vector **S** can be expressed as

$$\mathbf{S} = \alpha\hat{\mathbf{S}}_{\mathrm{A}} + \beta\hat{\mathbf{S}}_{\mathrm{B}} + \gamma\hat{\mathbf{S}}_{\mathrm{C}} + \delta\hat{\mathbf{S}}_{\mathrm{D}} = \frac{\alpha}{\sqrt{2}}\begin{bmatrix} 1 \\ 0 \\ 0 \\ 1 \end{bmatrix} + \frac{\beta}{\sqrt{2}}\begin{bmatrix} 1 \\ 0 \\ 0 \\ -1 \end{bmatrix} + \frac{\gamma}{\sqrt{2}}\begin{bmatrix} 0 \\ 1 \\ 1 \\ 0 \end{bmatrix} + \frac{\delta}{\sqrt{2}}\begin{bmatrix} 0 \\ -1 \\ 1 \\ 0 \end{bmatrix} \tag{11.13}$$

When the scattering matrix is depicted in vector form, the scattering matrix corresponding to reciprocal targets is in the subspace $\mathbf{W}_{\mathrm{rec}} \subset \boldsymbol{C}^4$, and $\boldsymbol{C}^4$ represents four-dimensional complex space. The projector P_{rec} in $\boldsymbol{C}^4$ can be defined as

$$P_{\text{rec}} = \frac{1}{2}\begin{bmatrix} 2000 \\ 0110 \\ 0110 \\ 0002 \end{bmatrix} \tag{11.14}$$

Any scattering matrix can be uniquely decomposed to two orthogonal parts in complex domain, which are $\mathbf{S}_{\text{rec}}$ and $\mathbf{S}_{\perp}$. Among them, $\mathbf{S}_{\text{rec}} \in \mathbf{W}_{\text{rec}}$, $\mathbf{S}_{\perp} \in \boldsymbol{W}_{\text{rec}}^{\perp}$. $\boldsymbol{W}_{\text{rec}}^{\perp}$ is a subspace of $\boldsymbol{C}^4$ and it's orthogonal to $\mathbf{W}_{\text{rec}}$. Consequently, the decomposition can be expressed as

$$\begin{cases} \hat{\mathbf{S}} = \hat{\mathbf{S}}_{\text{rec}} + \hat{\mathbf{S}}_{\perp} \\ \hat{\mathbf{S}}_{\text{rec}} = P_{\text{rec}}\hat{\mathbf{S}} \\ \hat{\mathbf{S}}_{\perp} = \left(\hat{I} - P_{\text{rec}}\right)\hat{\mathbf{S}} \end{cases}. \tag{11.15}$$

In Eq. (11.15), I is the unit matrix in $\boldsymbol{C}^4$ space.

The degree to which scattering matrix satisfies the reciprocal theorem θ_{rec} can be expressed by the angle between the scattering matrix and the reciprocal subspace $\boldsymbol{W}_{\text{rec}}$

$$\theta_{\text{rec}} = \cos^{-1} P_{\text{rec}}\boldsymbol{S}, 0 \le \theta_{\text{rec}} \le \pi/2 \tag{11.16}$$

When $\theta_{\text{rec}} = 0$, the scattering matrix corresponds to the scattering volume strictly fulfill the reciprocal principle. When $\theta_{\text{rec}} = \pi/2$, the scattering matrix is completely in the orthogonal space of $\boldsymbol{W}_{\text{rec}}$, and it does not satisfy the reciprocal principle.

Define $\boldsymbol{DX}$ as

$$\boldsymbol{DX} = \left(\boldsymbol{X}, \hat{\mathbf{S}}_{\text{A}}\right)\hat{\mathbf{S}}_{\text{A}} + \left(\boldsymbol{X}, \hat{\mathbf{S}}'\right)\hat{\mathbf{S}}' \tag{11.17}$$

where

$$\begin{cases} \hat{\mathbf{S}}' = \cos\chi\hat{\mathbf{S}}_{\text{B}} + \sin\chi\hat{\mathbf{S}}_{\text{C}} \\ \tan 2\chi = \frac{\beta\gamma^* + \gamma\beta^*}{|\beta|^2 + |\gamma|^2} \end{cases}$$

The reciprocal component in the scattering matrix $\mathbf{S}$ can be further factorized as the maximum symmetric and minimum symmetric components

$$\hat{\mathbf{S}}_{\text{sym}}^{\max} = \frac{\boldsymbol{DS}}{\|\boldsymbol{DS}\|}\hat{\mathbf{S}}_{\text{sym}}^{\min} = \frac{(\mathbf{I} - D)P_{\text{rec}}\mathbf{S}}{\|(\mathbf{I} - D)P_{\text{rec}}\mathbf{S}\|} \tag{11.18}$$

Moreover, the degree to which $\hat{\mathbf{S}}_{\text{rec}}$ deviates from symmetric subspace $\boldsymbol{W}_{\text{rec}}$ can be expressed by τ_{sym}

$$\tau_{\mathrm{sym}} = \cos^{-1}\left(\frac{\left(\hat{\mathbf{S}}_{\mathrm{rec}}, D\hat{\mathbf{S}}_{\mathrm{rec}}\right)}{\left\|\hat{\mathbf{S}}_{\mathrm{rec}}\right\| \cdot \left\|D\hat{\mathbf{S}}_{\mathrm{rec}}\right\|}\right), 0 \leq \tau_{\mathrm{sym}} \leq \frac{\pi}{4} \tag{11.19}$$

When $\tau_{\mathrm{sym}} = 0$, $\hat{\mathbf{S}}_{\mathrm{rec}}$ is the scattering matrix of a symmetric scatter. When τ_{sym} reaches its maximum $\pi/4$, $\hat{\mathbf{S}}_{\mathrm{rec}}$ corresponds to non-symmetric left-handed or right-handed helix.

The Cameron decomposition states that a reciprocal target can be decomposed as two symmetric targets. One of them is the maximum symmetric component, and the other one is the minimum symmetric component. The process of getting the maximum symmetric component is to find an orientation angle which gets the maximum polarization energy.

11.1.4 Similarity Parameter Decomposition

Similarity parameter is a polarization characterization parameter used to represent the intrinsic characterization proposed by Yang Jian, etc. The target can be decomposed by extracting the similarity parameter.

If the reciprocity condition holds, the scattering matrix of a target is

$$\mathbf{S} = \begin{bmatrix} S_{\mathrm{HH}} & S_{\mathrm{HV}} \\ S_{\mathrm{VH}} & S_{\mathrm{VV}} \end{bmatrix}, S_{\mathrm{HV}} = S_{\mathrm{VH}} \tag{11.20}$$

Let ψ denote the orientation angle of the target, then the following scattering matrix can be defined

$$\mathbf{S}^0 = [\mathrm{J}(-\psi)][\mathbf{S}][\mathrm{J}(\psi)] = \begin{bmatrix} S^0_{\mathrm{HH}} & S^0_{\mathrm{HV}} \\ S^0_{\mathrm{VH}} & S^0_{\mathrm{VV}} \end{bmatrix} \tag{11.21}$$

where

$$[\mathrm{J}(\psi)] = \begin{bmatrix} \cos\psi & -\sin\psi \\ \sin\psi & \cos\psi \end{bmatrix} \tag{11.22}$$

According to Huynen's theory, [4] we can know that $\left[\mathbf{S}^0\right]$ denotes the scattering matrix of the above target in a special position. If the scattering matrix $\mathbf{S}$ can be known, $\left[\mathbf{S}^0\right]$ can be easily obtained. Let

$$\boldsymbol{k} = \frac{1}{\sqrt{2}}\left[S^0_{\mathrm{HH}} + S^0_{\mathrm{VV}} \; S^0_{\mathrm{HH}} - S^0_{\mathrm{VV}} \; 2S^0_{\mathrm{HV}}\right]^{\mathrm{T}} \tag{11.23}$$

where the vector $\boldsymbol{k}$ is called the modified Pauli-scattering vector. The similarity parameter between two scattering matrices $\mathbf{S}_1$ and $\mathbf{S}_2$ is defined as

$$r(\mathbf{S}_1, \mathbf{S}_2) = \frac{|\boldsymbol{k}_1^{\mathrm{H}} \boldsymbol{k}_2|}{\|\boldsymbol{k}_{12}^2 \boldsymbol{k}_{22}^2\|} \tag{11.24}$$

where $\boldsymbol{k}_1$ and $\boldsymbol{k}_2$ correspond to $\mathbf{S}_1$ and $\mathbf{S}_2$, $\| \|^2$ denote the square sum of the absolute values of the components of the vector.

The similarity parameter has the following properties:

(1) $r(J(\theta_1)\mathbf{S}_1 J(-\theta_1), J(\theta_2)\mathbf{S}_2 J(-\theta_2)) = r(\mathbf{S}_1, \mathbf{S}_2)$, where θ_1 and θ_2 are two arbitrary angles.
(2) $r(a_1\mathbf{S}_1, a_2\mathbf{S}_2) = r(\mathbf{S}_1, \mathbf{S}_2)$, where a_1 and a_2 are two arbitrary complex numbers.
(3) $0 \le r(\mathbf{S}_1, \mathbf{S}_2) \le 1$, where $r(\mathbf{S}_1, \mathbf{S}_2) = 1$
if and only if $\mathbf{S}_2 = aJ(\theta)\mathbf{S}_1 J(-\theta)$, and $r(\mathbf{S}_1, \mathbf{S}_2) = 1$
(4) If $\mathbf{S}_1$, $\mathbf{S}_2$, and $\mathbf{S}_3$ are three scattering matrices which satisfy

$$r(\mathbf{S}_1, \mathbf{S}_2) = r(\mathbf{S}_1, \mathbf{S}_3) = r(\mathbf{S}_2, \mathbf{S}_3) = 0 \tag{11.25}$$

then for an arbitrary scattering matrix $\mathbf{S}$, we have

$$r(\mathbf{S}, \mathbf{S}_1) = r(\mathbf{S}, \mathbf{S}_2) = r(\mathbf{S}, \mathbf{S}_3) = 1. \tag{11.26}$$

The similarity between arbitrary target and a plate is

$$r_1 = k\begin{bmatrix}1\\0\\0\end{bmatrix} = \frac{|S_{\mathrm{HH}}|^2 + |S_{\mathrm{VV}}|^2}{2(|S_{\mathrm{HH}}|^2 + |S_{\mathrm{VV}}|^2 + 2|S_{\mathrm{HV}}|^2)} = \frac{|S_{\mathrm{HH}} + S_{\mathrm{VV}}^0|^2}{2P_{\mathrm{span}}} = \frac{|S_{\mathrm{HH}} + S_{\mathrm{VV}}|^2}{2P_{\mathrm{span}}} \tag{11.27}$$

In Eq. (11.27), the superscript 0 denotes the element in the scattering matrices of targets with orientation angle of 0°. P_{span} is the polarization span.

The similarity between arbitrary Pauli vector $\boldsymbol{k}$ and a diplane is

$$r_2 = k\begin{bmatrix}0\\1\\0\end{bmatrix} = \frac{|s_{\mathrm{HH}} - s_{\mathrm{VV}}^0|^2}{2P_{\mathrm{span}}} \tag{11.28}$$

The similarity parameter of arbitrary target to the line is

$$r_2 = k\begin{bmatrix}1\\1\\0\end{bmatrix} = \frac{|s_{\mathrm{HH}}|^2}{2P_{\mathrm{span}}} \tag{11.29}$$

The parameters of similarities to the scattering matrices of a left and a right helix is

$$r_L = r\left(k\begin{bmatrix}0\\1\\f\end{bmatrix}\right) = \frac{|(s_{HH} - s_{VV}) - 2js_{HV}|^2}{4P_{span}}$$

$$r_R = r\left(k\begin{bmatrix}0\\1\\-f\end{bmatrix}\right) = \frac{|(s_{HH} - s_{VV}) + 2js_{HV}|^2}{4P_{span}} \quad (11.30)$$

The parameters of similarities can be used to get the similarities parameters between observed targets and typical targets, and consequently the scattering characteristics description of the targets can be obtained. Typical targets include the plate, diplane, helix, line, etc. The parameters of similarities have real physical meaning, and it is in accordance with existing target decomposition theories. Making use of the parameters of similarities to analyze the radar target scattering can get more parameters than the other common decompositions of the scattering matrices. Moreover, it can make us understand the geometry scattering characteristics of the targets better more comprehensive.

11.2 Eigenvector-Based Decompositions

Target decompositions based on eigenvectors include Cloude–Pottier decomposition, Huynen decomposition, Holm decomposition, Van Zyl decomposition, etc.

11.2.1 Cloude–Pottier Decomposition

Cloude–Pottier decomposition is a kind of target decomposition based on the incoherency matrix. This decomposition can extract the polarimetric information effectively and it's widely used in the classification of polarimetric SAR images. This method is based on the eigen decomposition of the coherency matrix. For a n-look polarimetric coherency matrix $\langle \boldsymbol{T}_3 \rangle$, it is a positive semi-definite 3×3 Hermitian matrix, and its eigenvalue decomposition can be described as

$$\langle \boldsymbol{T}_3 \rangle = \boldsymbol{U}_3 \boldsymbol{\Lambda} \boldsymbol{U}_3^{-1} \quad (11.32)$$

where $\boldsymbol{U}_3$ is a unitary eigenvector matrix with the orthogonal eigenvectors of $\langle \boldsymbol{T}_3 \rangle \mathbf{e}_i, i = 1, 2, 3$ as its column vectors. $\boldsymbol{\Lambda}$ is a diagonal matrix contains the eigenvalues of $\langle \boldsymbol{T}_3 \rangle$.

$$U_3 = \begin{bmatrix} e_1 & e_2 & e_3 \end{bmatrix}$$
$$e_i = e^{j\phi_i}\begin{bmatrix} \cos\alpha_i & \sin\alpha_i\cos\beta_i e^{j\delta_i} & \sin\alpha_i\sin\beta_i e^{j\gamma_i} \end{bmatrix}^{\mathrm{T}}, i = 1, 2, 3$$
$$\Lambda = \begin{bmatrix} \lambda_1 & 0 & 0 \\ 0 & \lambda_2 & 0 \\ 0 & 0 & \lambda_3 \end{bmatrix}, \lambda_1 \geq \lambda_2 \geq \lambda_3 \tag{11.33}$$

In Eq. (11.33), e_i can be taken as a specific scattering mechanism, and α_i denotes the type of the scattering mechanism. The value of α_i is in $[0^\circ, 90^\circ]$. β_i is the orientation angle of the target to the sight of radar. ϕ_i, β_i, and γ_i are the scattering angle of the target.

As we can see from Eqs. (11.32) and (11.33), the coherency matrix can be denoted as the sum of three independent components, that is

$$T_3 = \sum_{i=1}^{3} \lambda_i e_i e_i^{\mathrm{H}} = \sum_{i=1}^{3} \lambda_i T_{3i} \tag{11.34}$$

where $e_i e_i^{\mathrm{H}}, i = 1, 2, 3$ are independent rank 1 matrices, which can be related to a scattering mechanism, respectively. The corresponding $\lambda_i, i = 1, 2, 3 (\lambda_1 \geq \lambda_2 \geq \lambda_3)$ denotes the power intensity of scattering component.

According to the above decomposition, a parameter called polarimetric entropy is obtained to describe the randomness of the scattering process. The expression of polarimetric entropy is

$$H = \sum_{i=1}^{3} -P_i \lg_3 P_i \tag{11.35}$$

where $P_i = \frac{\lambda_i}{\sum_{j=1}^{3} \lambda_j}$. The value range of P_i is $0 \leq H \leq 1$.

If $H = 0$, there is only one eigenvalue is not 0, and it means the scattering process corresponds to a pure target. The scattering wave is completely polarized. If $H = 1$, the values of three eigenvalues are the same, and it means the contribution of the three scattering mechanisms in the scattering process is the same. The scattering wave is completely depolarized, and it degenerates to a random noise process.

According to the expression of e_i, the parameter α_i is extracted, and the polarimetric scattering angle α of the target is defined as

$$\alpha = \sum_{i=1}^{3} P_i \alpha_i \tag{11.36}$$

The scattering angle α interprets the average scattering mechanism of the target, and it's continuously varied from 0° to 90°. It represents the scattering mechanism of the target to some extent. When $\alpha = 0$, the scattering mechanism is isotropy surface scattering. When $\alpha = 45^{\circ}$, the scattering mechanism is double-bounce scattering. When $\alpha = 90^{\circ}$, the scattering mechanism is isotropy dihedral scattering.

The polarimetric anisotropy A is a parameter complementary to the H and α parameter, the definition of A is

$$A = \frac{\lambda_2 - \lambda_3}{\lambda_2 + \lambda_3} \tag{11.37}$$

If the scattering entropy H is high, the anisotropy A can measure the relative importance of the second and scattering mechanisms to the scattering results. Higher A value implies that the second scattering mechanism has an effect on the scattering process, while lower A value implies that the third scattering mechanism can't be ignored either.

A two-dimensional plane combining polariemetric entropy H and scattering angle α called H/α plane is obtained. The plane is divided into 9 zones, and each zone represents a specific scattering type, which is shown in Fig. 11.4. It is noticed that the scattering angle is an mean value as it is influenced by the probability P_i of scattering mechanism.

According to the definitions of polarimetric entropy H and scattering angle α, the value scope of α is decreasing as H increases. It can be induced that the value range of α is restricted between curve 1 and curve 2 shown in Fig. 11.4 when the value of H changes. Curve 1 and curve 2 can be defined by the following coherency matrices:

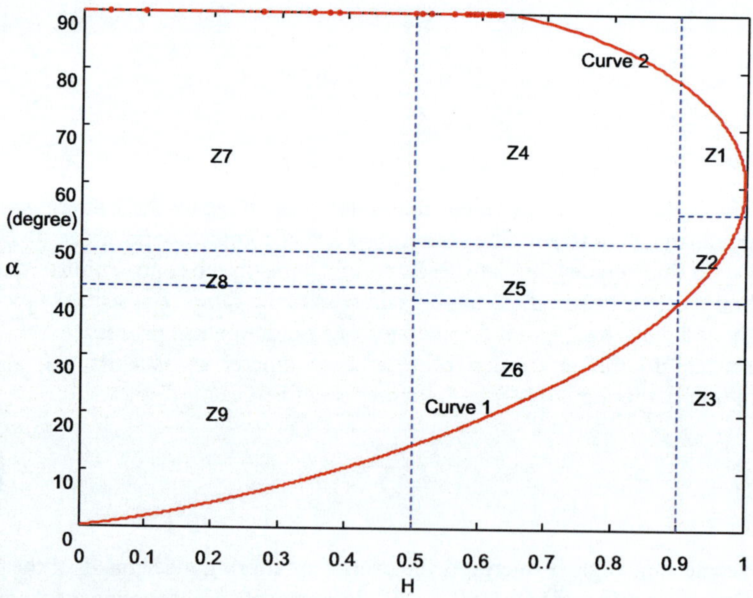

Fig. 11.4 H/α plane

$$\mathrm{T_I} = \begin{bmatrix} 1 & 0 & 0 \\ 0 & m & 0 \\ 0 & 0 & m \end{bmatrix}, 0 \le m \le 1$$

$$\begin{cases} \mathrm{T_I} = \begin{bmatrix} 1 & 0 & 0 \\ 0 & m & 0 \\ 0 & 0 & m \end{bmatrix}, 0 \le m \le 0.5 \\ \mathrm{T_{II}} = \begin{bmatrix} 2m-1 & 0 & 0 \\ 0 & 1 & 0 \\ 0 & 0 & 1 \end{bmatrix}, 0.5 \le m \le 1 \end{cases} \tag{11.38}$$

In Fig. 11.4, nine zones are specified related to specific scattering characteristics:

Z9: Low entropy surface scatter ($0 < H \le 0.5, 0 < \alpha \le 42.5°$).

In this zone, the polarimetric entropy is quite low, and it means there is only one dominant scattering mechanism. The scattering angle α is less than 42.5°. These include geometric optics and physical optics surface scattering—Bragg surface scattering and specular scattering phenomena which do not involve 180° phase inversions between HH and VV channels. Physical surfaces such as water at L- and P-bands, sea-ice at L-band, as well as very smooth land surfaces, all fall into this category.

Z8: Low entropy dipole scattering ($0 < H \le 0.5, 42.5° < \alpha \le 47.5°$).

In this zone, the polarimetric entropy is quite low, and the dominant scattering mechanism is dipole scattering. In this zone occur strongly correlated mechanisms which have a large imbalance between HH and VV in amplitude. Scattering from vegetation with strongly correlated orientation of anisotropic scattering elements would appear this scattering mechanism.

Z7: Low entropy multiple scattering events ($0 < H \le 0.5, 47.5° < \alpha \le 90°$).

This zone corresponds to low entropy double, or even, bounce scattering events, such as those provided by isolated dielectric and metallic dihedral scatters.

Z6: Medium entropy surface scatte ($0.5 < H \le 0.9, 0° < \alpha \le 40°$).

This zone reflects the increase in entropy compared to Z9 due to changes in surface roughness and due to canopy propagation effects. Leaves and scatters covered by plate-like surfaces fall into this category.

Z5: Medium entropy surface scatter ($0.5 < H \le 0.9, 40° < \alpha \le 50°$).

This zone has moderate entropy but with a dominant dipole type scattering mechanism. The increased entropy H is due to a central statistical distribution of orientation angle. Such a zone would include scattering from vegetated surfaces with anisotropic scatters.

Z4: Medium entropy multiple scattering ($0.5 < H \le 0.9, 50° < \alpha \le 90°$).

This zone accounts for dihedral scattering with moderate entropy. It corresponds to the scattering of urban areas and the scattering between the ground and the trunks when the radar wave penetrates the canopy.

Z3: High entropy surface scatter ($0.9 < H \le 1, 0° < \alpha \le 40°$).

This class is a non-feasible region because it is impossible to distinguish surface scattering with entropy $H > 0.9$. The value range of scattering angle α is decreasing as the high entropy will constrain the classification capability of SAR.

Z2: High entropy vegetation scattering ($0.9 < H \leq 1, 40° < \alpha \leq 55°$).

This zone has high entropy and moderate scattering angle. This can arise for single scattering from a cloud of anisotropic needle-like particles. As the entropy is very high, the radar signal can be noises without any polarimetric information.

Z1: High entropy multiple scattering ($0.9 < H \leq 1, 55° < \alpha \leq 90°$).

The scattering angle $\alpha > 55°$, and it reflects diplane scattering. Polarimetric entropy $H > 0.9$ means the degree of randomness is very high. Such mechanisms can be observed in forestry applications or in scattering from vegetation which has a well-developed branch and crown structure.

H/α and $H/\alpha/A$ classification are valid polarimetric SAR target decomposition method. The classification result can be connected with realistic scattering mechanisms, and it is widely applied.

Figure 11.5 shows the Cloude–Pottier decomposition result of L-band PolSAR data acquired by AirSAR system of San Francisco Bay. As we can see from Fig. 11.5, different terrain characteristics can be distinguished from the classification.

The definition of the other scattering angle is

$$\beta = \sum_{i=1}^{3} P_i \beta_i \tag{11.39}$$

It is a representation of the orientation of the target.

Two important parameters, which are polarimetric degree P and scattering anisotropy A, can be obtained from the first and second eigenvalue, and the second and third eigenvalues, respectively. The expressions of the two parameters are

$$\mathrm{P} = \frac{P_1 - P_3}{P_1 + P_2} \tag{11.40}$$

$$\mathrm{A} = \frac{P_2 - P_3}{P_2 + P_3} \tag{11.41}$$

Scattering anisotropy A represents the degree of the anisotropy of the targets. Considering the influence of noise, A is a quite important to the analysis of the target decomposition when the scattering entropy is medium. The applicability of A has to be considered due to the influence of noise.

The value of α and β is a combination of eigenvalues and eigenvectors, while the value of scattering entropy H, scattering anisotropy A and polarimetric degree P can be calculated from eigenvalues only.

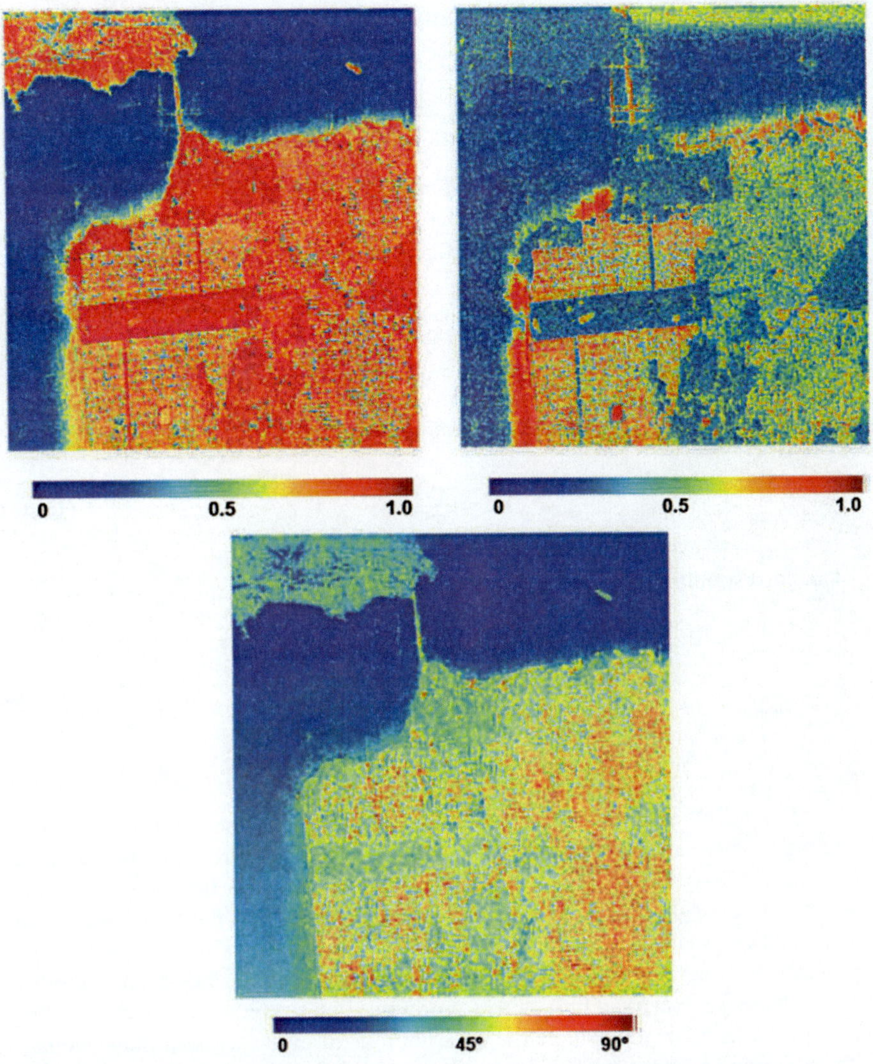

Fig. 11.5 Example of Cloude–Pottier decomposition **a** Entropy (H), **b** Anisotropy (A), and alpha (α) images

11.2.2 Information Enhancement of Cloude–Pottier Decomposition

Coherency matrix and covariance matrix are both Hermitian semi-definite, and they have the same eigenvalues. Therefore, covariance matrix can also be used to perform Cloude–Pottier decomposition. The covariance matrix can be decomposed as a sum of three components after unitary similarity transform

$$\mathbf{C} = \sum_{i=1}^{3} \lambda_i \boldsymbol{e}_i \boldsymbol{e}_i^{*T} \tag{11.42}$$

where λ_i and $\boldsymbol{e}_i$ are the eigenvalues and corresponding eigenvectors of covariance matrix $\mathbf{C}$.

Since covariance matrix $\mathbf{C}$ is semi-definite, λ_i are non-negative real numbers. We can get

$$\lambda_{1,2} = \sigma_{HH} \frac{\tau \pm \sqrt{\eta}}{2}, \lambda_3 = \sigma_{HH} \in \tag{11.43}$$

The corresponding normalized eigenvalues are

$$p_i = \frac{\lambda_i}{\sum_{i=1}^{3} \lambda_i} \tag{11.44}$$

The corresponding eigenvectors are

$$\boldsymbol{e}_{1,2} = \sqrt{\frac{\sqrt{\eta} \pm \delta}{2\sqrt{\eta}}} \begin{bmatrix} \frac{2\rho\sqrt{\gamma}}{\delta \pm \sqrt{\eta}} \\ 0 \\ 1 \end{bmatrix}, \boldsymbol{e}_3 = \begin{bmatrix} 0 \\ 1 \\ 0 \end{bmatrix} \tag{11.45}$$

where $\tau = \gamma + 1, \delta = \gamma - 1, \eta = \delta^2 + 4|\rho|^2\gamma$.

If the eigenvalues of the Hermitian matrix are different, the eigenvectors corresponding to different eigenvalues are orthogonal. Based on reciprocity and some other assumptions, $\langle S_{\text{HH}}^* S_{\text{HV}} \rangle$ and $\langle S_{\text{HV}}^* S_{\text{VV}} \rangle$ equal 0, and the first and the second eigenvector ($\boldsymbol{e}_1, \boldsymbol{e}_2$) contain co-polarization parts only, while the third eigenvector contains cross-polarization parts only. When $\text{Re}(\rho) > 0$, maximum eigenvalue λ_1 represents odd scattering, and the second eigenvalue λ_2 represents even scattering. On the contrary, if $\text{Re}(\rho) < 0$, λ_1 and λ_2 represent even scattering and odd scattering, respectively. In order to evaluate the contributions of every scattering mechanisms to the scattering process, the eigenvalues are normalized. The following eigenvalues refer to normalized eigenvalues.

Eigenvalues and eigenvectors can be obtained by Eqs. (11.44) and (11.45). Three kinds of new vectors can be obtained by the eigenvalues and eigenvectors, and they can be represented as three-dimensional form in the space. The construction of new vectors is as follows. Two new vectors are formed by the first and second eigenvalue and their corresponding eigenvectors. Only real parts are composed in the vectors. That is, the real part of the co-polarization coherency matrix $\text{Re}(\rho)$ is used to replace ρ in Eq. (11.45). The third new vector is formed by the third eigenvector plus the phase difference of the co-polarization channels. Three new vectors weighted by the eigenvectors are as follows:

$$\boldsymbol{\nu}_1 = \begin{bmatrix} \nu_1^1 \\ \nu_1^2 \\ \nu_1^3 \end{bmatrix} = p_1\sqrt{\frac{\sqrt{\eta}+S}{2\sqrt{\eta}}}\begin{bmatrix} \frac{2Re(\rho)\sqrt{\gamma}}{S+\sqrt{\eta}} \\ 0 \\ 1 \end{bmatrix} \tag{11.46}$$

$$\boldsymbol{\nu}_2 = \begin{bmatrix} \nu_2^1 \\ \nu_2^2 \\ \nu_2^3 \end{bmatrix} = p_2\sqrt{\frac{\sqrt{\eta}-S}{2\sqrt{\eta}}}\begin{bmatrix} \frac{2Re(\rho)\sqrt{\gamma}}{S-\sqrt{\eta}} \\ 0 \\ 1 \end{bmatrix} \tag{11.47}$$

$$\boldsymbol{\nu}_3 = \begin{bmatrix} \nu_3^1 \\ \nu_3^2 \\ \nu_3^3 \end{bmatrix} = p_3\begin{bmatrix} 0 \\ 1 \\ \Delta\phi \end{bmatrix} \tag{11.48}$$

where ν_i^j are the elements of the new vectors, i represents the number of the vector, and j represents the number of elements. Re(ρ) represents the real part of the co-polarization coherence coefficient. Figure 11.6 shows the spatial triangle representation of the new vectors. In Fig. 11.6, the vector from the original point O to $A(\nu_1^1, \nu_1^2, \nu_1^3)$ is $\boldsymbol{\nu}_1$, to $B(\nu_2^1, \nu_2^2, \nu_2^3)$ is $\boldsymbol{\nu}_2$, and to $C(\nu_3^1, \nu_3^2, \nu_3^3)$ is $\boldsymbol{\nu}_3$. The three vertices of every pixel can be uniquely determined in the three-dimensional space, that is $A(\nu_1^1, \nu_1^2, \nu_1^3)$, $B(\nu_2^1, \nu_2^2, \nu_2^3)$, and $C(\nu_3^1, \nu_3^2, \nu_3^3)$. A feature triangle can be obtained by connecting the three points, as shown in the dashed line of Fig. 11.6. The sides of the triangle can be represented by the following equations:

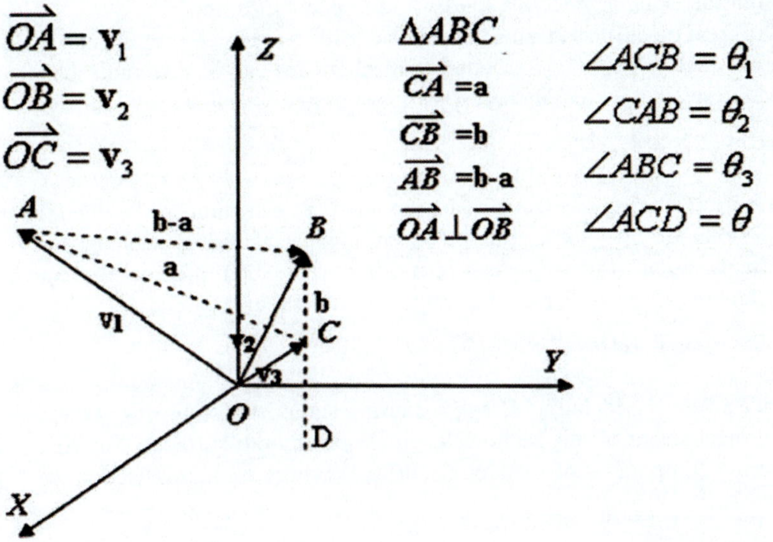

Fig. 11.6 The triangle representation of the new vectors, and the inner angle and external angle of the triangle

$$\overrightarrow{OA} = \boldsymbol{v}_1, \overrightarrow{OB} = \boldsymbol{v}_2, \overrightarrow{OC} = \boldsymbol{v}_3$$

$$a = \begin{bmatrix} v_1^1 - v_3^1 \\ v_1^2 - v_3^2 \\ v_1^3 - v_3^3 \end{bmatrix} b = \begin{bmatrix} v_2^1 - v_3^1 \\ v_2^2 - v_3^2 \\ v_2^3 - v_3^3 \end{bmatrix} b - a = \begin{bmatrix} v_2^1 - v_1^1 \\ v_2^2 - v_1^2 \\ v_2^3 - v_1^3 \end{bmatrix} \tag{11.49}$$

According to Eqs. (11.46)–(11.48), $v_1^1 = v_2^1 = v_3^1 = 0$. Let θ_1 be the inner angle of vector a and b, and so is θ_2 and θ_3. a, b, and $b - a$ are the length of vectors a, b, and $b - a$. The included angle between the vectors can be defined as

$$\begin{cases} \cos\theta_1 = \frac{a \cdot b}{ab} \\ \cos\theta_2 = \frac{(-a)\cdot(b-a)}{-ab-a} \\ \cos\theta_3 = \frac{(a-b)\cdot(-b)}{a-b-b} \end{cases} \tag{11.50}$$

In Eq. (11.50), $0 \leq \theta_i \leq \pi$, and three inter angles can be calculated by the arccosine. Moreover, $\theta_1 + \theta_2 + \theta_3 = 180°$. The three inter angles have the potential of interpreting scattering mechanisms. The external angles of the triangle can be calculated by two interior angles. Assuming γ is the external angle of θ_1, we can get

$$\gamma = \theta_2 + \theta_3 = 180° - \theta_1, 0 \leq \theta_i \leq \pi \tag{11.51}$$

The value scope of the external angle γ is 0°–180°, and the scattering characteristics can be described by γ. It provides stronger scattering information. This is the information enhancement of Cloude–Pottier decomposition.

Different types of scattering mechanism (different targets) correspond to different external angles. Three main scattering mechanisms can be distinguished clearly by the value scope of external angles. γ, it is summed as follows:

(1) $0 \leq \gamma \leq 90°$

Two main scattering mechanisms are included. This section can be divided into two subsections. $0 \leq \gamma \leq 60°$ represents even-bounce scattering, while $60° < \gamma \leq 90°$ represents dipole scattering. As a result, $0 \leq \gamma \leq 90°$ represents dispersive forest areas and urban areas. L-band represents residential areas.

(2) $90° \leq \gamma \leq 100°$

Surface scattering including Bragg scattering and mirror scattering is the main scattering mechanism of this section. It represents smooth surfaces. Moreover, mirror scattering components of very rough surface scatters are in this section too.

(3) $100° < \gamma \leq 180°$

This section can be divided into two subsections, that is $100° < \gamma \leq 115°$ and $115° < \gamma \leq 180°$. Multiple scattering are the main scattering mechanism in the $100° < \gamma \leq 115°$ section, and there's also some even-bounce scattering under some conditions. So, the forests with flourishing branches and leaves fall into this section. Volume scattering and diffuse scattering are the main scattering mechanisms in the $115^\circ < \gamma \leq 180^\circ$ section. Scattering from branches and canopy surfaces fall into this zone, and some random scattering may fall into this zone too. As a result, it interprets canopy area and some random rough surface scattering. The pixels in the $100^\circ < \gamma \leq 115^\circ$ zone are less, while much more pixels appear in $115^\circ < \gamma \leq 180^\circ$ zone. It is considered that volume scattering is the main scattering mechanism in this section.

The characteristics of scattering types can be implicated by the other inner angles too. The value of three inner angles vary from 0° to 180°, that is $0^\circ \leq \theta_i \leq \pi$. Their value scope is decreased in reality. The value scope is $0^\circ \leq \theta_1 \leq 140^\circ, 0^\circ \leq \theta_2 \leq 70^\circ$, $0^\circ \leq \theta_3 \leq 110^\circ$. Among them, $0^\circ \leq \theta_1 < 90^\circ$ represents volume scattering, $90^\circ \leq \theta_1 < 100^\circ$ denotes surface scattering, and $100^\circ \leq \theta_1 < 140^\circ$ denotes even-bounce scattering. $0^\circ \leq \theta_2 \leq 10^\circ$ denotes surface scattering, $\theta_2 > 10^\circ$ represents even-bounce scattering, and $\theta_2 \gg 10^\circ$ denotes volume scattering. The boundary of the latter two scattering zones is not very clear. $0^\circ \leq \theta_3 \leq 50^\circ$ denotes even-bounce scattering, $50^\circ \leq \theta_3 < 80^\circ$ represents volume scattering, and $80^\circ \leq \theta_3 \leq 110^\circ$ denotes surface scattering.

Figure 11.7 shows the $\mathrm{H} - \gamma$ plane obtained by the information enhancement method. The solid lines show the boundaries of all the zones, and the dashed lines show the boundaries which can be further divided. Figure 11.7 shows the $\mathrm{H} - \gamma$ plane. The scattering characteristics represented by the 9 zones can be interpreted as low entropy even-bounce scattering or dipole scattering, low entropy surface scattering, low entropy volume scattering, medium entropy double-bounce scattering or

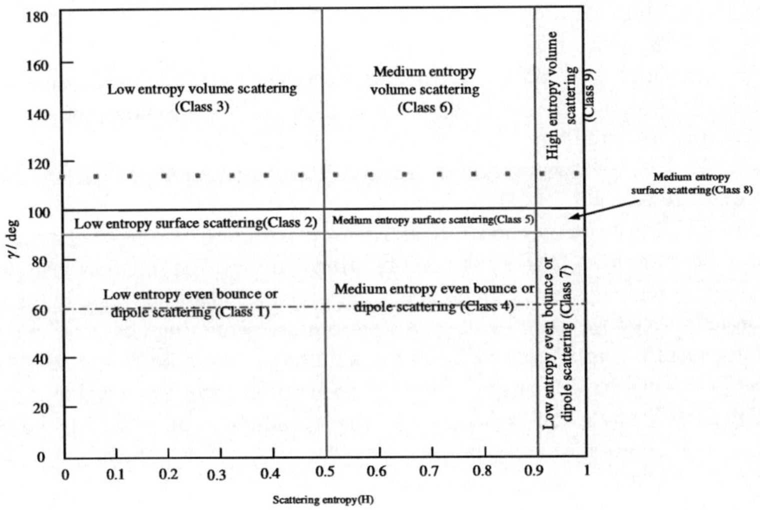

Fig. 11.7 $\mathrm{H} - \gamma$ plane

Fig. 11.8 The span of L-band SAR data of Tian Mountain

dipole scattering, medium entropy surface scattering, high entropy even-bounce scattering or dipole scattering, high entropy surface scattering, and high entropy volume scattering.

Media of high entropy surface scattering is not existed in natural world. However, the other scattering type in the zone $100° < \gamma \leq 115°$ is the mirror scattering component of very rough surface scatterers. Therefore, the zone is very rough surface scattering with high entropy.

L-band SAR data of Tian Mountain area acquired by NASA JPL is applied to investigate the information enhancement of Cloude–Pottier decomposition. Figure 11.8 shows the span of Tian Mountain. Figure 11.9a–c shows the three inner angles of this area, and Fig. 11.9d is the image of external angle γ.

Figure 11.10a, b shows the image of α and β angle, where Fig. 11.10a is the gray image of scattering entropy H, and Fig. 11.10b is the gray image of anisotropy A. Figure 11.11a, b is the $H-\gamma$ and $H-\alpha$ distribution of this area, respectively. As we can see from the γ angle and the α angle, more information can be found in γ angle. The rectangle area in the center of the image is very typical. The gray value in the image of γ angle is very low and it represents even-bounce scattering characteristic while in the image of α angle, gray value of this area is very similar to the neighboring area, and it's very hard to distinguish. Consequently, it brings convenience to the explanation of γ angle and the division of the boundary. Figure 11.12 is the $H-\gamma$ and $H-\alpha$ distribution of the data.

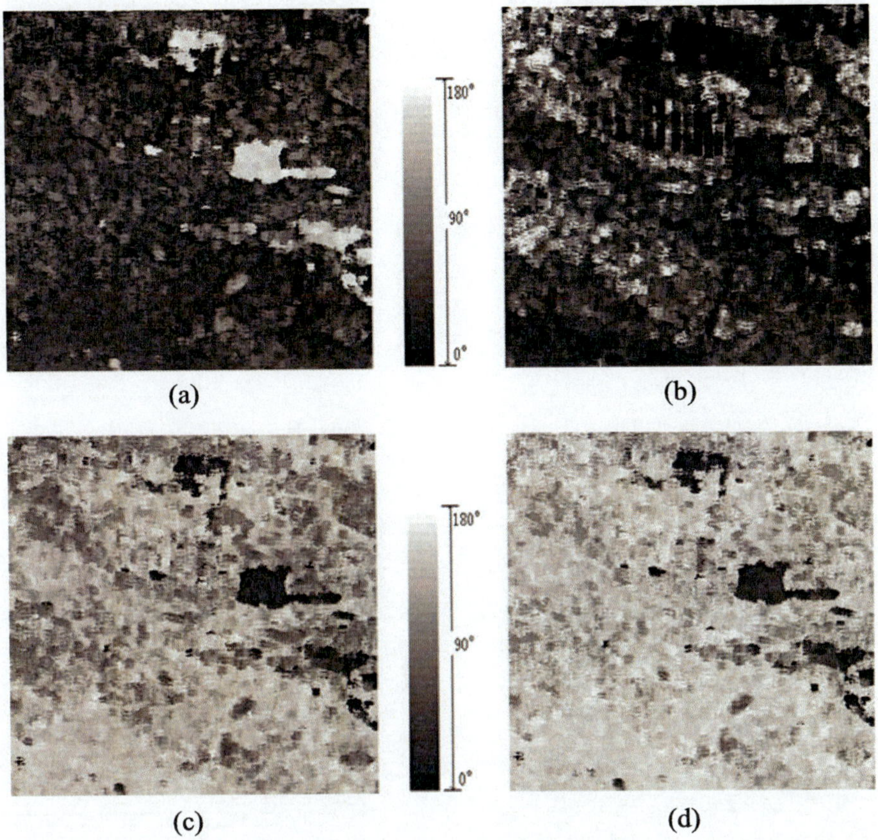

Fig. 11.9 Image of three inner angle θ_1, θ_2, θ_3 and external angle γ

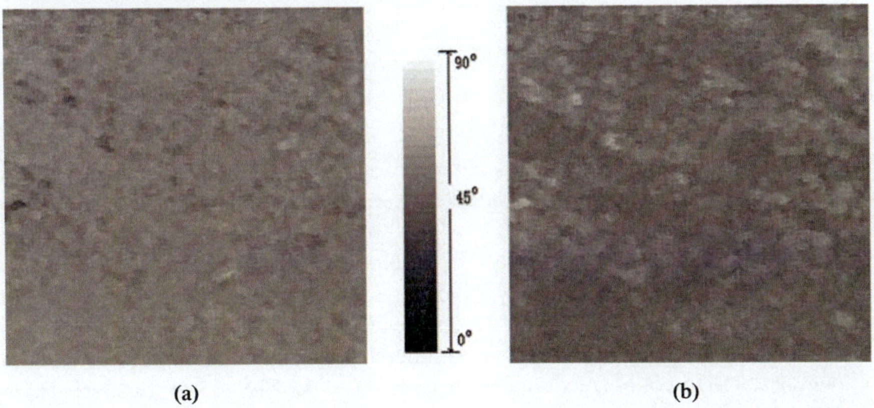

Fig. 11.10 α and β angle obtained from polarimetric decomposition **a** α angle image, **b** β angle image

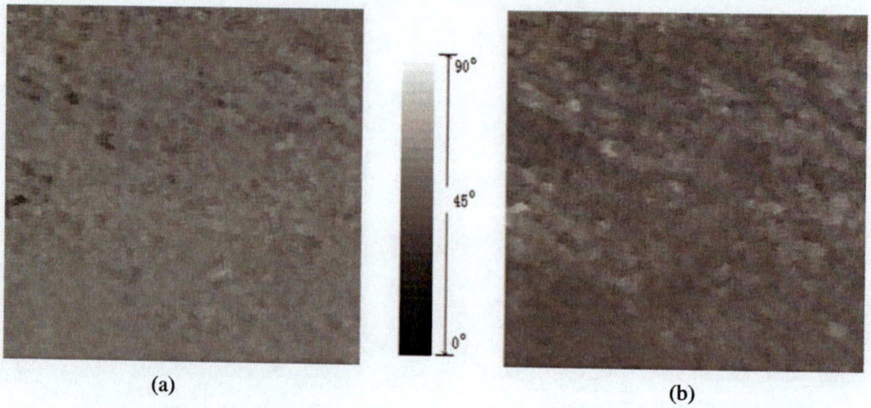

Fig. 11.11 Scattering entropy and anisotropy obtained from polarimetric decomposition. **a** Scattering entropy H, **b** Scattering anisotropy A

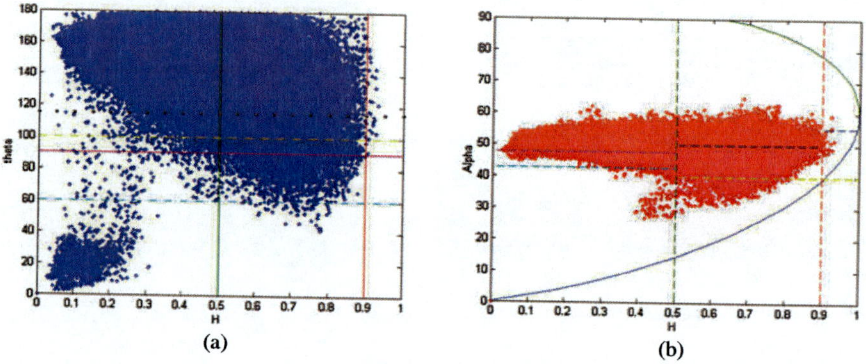

Fig. 11.12 H$-\gamma$ and H$-\alpha$ distribution of L-band polarimetric SAR data of Tian Mountain **a** H$-\gamma$ distribution, **b** H $-$ α distribution

11.2.3 Huynen Decomposition

The basic principle of the Huynen decomposition method is to decompose the echo data into two parts, one is the identifiable deterministic target, and the other is the residual or noise component.

Distributed targets can be expressed by Muller matrix $\boldsymbol{M}$, which represents time-variant average target determined by 9 independent parameters. Muller matrix $\boldsymbol{M}$ is incoherent averaging, and it can be decomposed to the sum of an equivalent deterministic target M_0 (determined by five parameters) and a residual target N (with the remaining four degrees of freedom parameters): $\boldsymbol{M} = M_0 + N$. These two targets are independent. The residual target N represents the target non-symmetric parameter, and it is the rotational invariant target. This is the basis of Huynen decomposition method.

The polarization vecto $\boldsymbol{k}$ of the target can be expressed by the scattering elements of the target.

$$\boldsymbol{k} = \frac{1}{\sqrt{2}}\left[S_{HH} + S_{VV} \; S_{VV} - S_{HH} \; 2S_{\mathrm{HV}} \right]^{\mathrm{T}} \tag{11.52}$$

According to Huynen decomposition, the polarimetric coherency matrix has the following form:

$$\boldsymbol{T} = \boldsymbol{k} \cdot \boldsymbol{k}^{*\mathrm{T}} = \begin{bmatrix} 2A_0 & C - jD & H + jG \\ C + jD & B_0 + B & E + jF \\ H - jG & E - jF & B_0 - B \end{bmatrix} \tag{11.53}$$

The equivalent form of Huynen polarimetric target decomposition is to decompose the coherency matrix $\boldsymbol{T}$ of distributed target into the sum of determinant target coherency matrix $\boldsymbol{T}_0$ and residual target coherency matrix $\boldsymbol{T}_N$.

$$\boldsymbol{T} = \boldsymbol{T}_0 + \boldsymbol{T}_N$$

$$\boldsymbol{T}_0 = \begin{bmatrix} 2A_0 & C - jD & H + jG \\ C + jD & B_{0\mathrm{T}} + B_{\mathrm{T}} & E_{\mathrm{T}} + jF_{\mathrm{T}} \\ H - jG & E_{\mathrm{T}} - jF_{\mathrm{T}} & B_{0\mathrm{T}} - B_{\mathrm{T}} \end{bmatrix}$$

$$\boldsymbol{T}_{\mathrm{N}} = \begin{bmatrix} 0 & 0 & 0 \\ 0 & B_{0N} + B_N & E_N + jF_N \\ 0 & E_N - jF_N & B_{0N} - B_N \end{bmatrix}$$

where A_0, B_0, C, D, E, F, G, and H are known under the name of "Huynen parameters." They are the functions of target scattering matrix parameters.

These nine parameters all contain some polarimetric information, so the nine parameters are useful for general target analysis without reference to any model. In Huynen's paper, the meaning [4] of these parameters is as follows (Table 11.1):

These parameters are obtained from the scattering matrix **S**, or the target scattering vector $\boldsymbol{k}$, so they are related.

These parameters are obtained from the knowledge of the Scattering Matrix **S** or the target scattering vector $\boldsymbol{k}$, which contains five independent parameters. As a result, there are only five independent parameters in these nine parameters, and there are four dependency relationships.

Table 11.1 The meaning of Huynen parameters

A_0	Total scattering power of the regular part of the target	B_0	Total scattering power of the non-regular part of the target
$B_0 - B$	Generator factor of Target Symmetry	$B_0 + B$	Generator Factor of target irregularity
C	Shape factor	D	Measure of local curvature difference
E	Surface torsion	F	Target helicity
G	Coupling between symmetric and non-symmetric parts	H	Target orientation

$$\begin{cases} 2A_0(B_0 + B) = C^2 + D^2 \\ 2A_0(B_0 - B) = G^2 + H^2 \\ 2A_0E = CH - DG \\ 2A_0F = CG + DH \end{cases} \tag{11.54}$$

The diagonal elements of the coherency matrix are called the "target factors." We can see at the non-diagonal Hunyen parameter pairs like (C, D), (E, F), and (G, H) can be generated from these parameters. More importantly, they are related to the physical characteristics of the target such as symmetry, non-irregularity, and non-uniformity. Usual scattering is composed of three simple scattering mechanisms, which are odd-bounce scattering, double even-bounce scattering, and volume scattering. The "target factor" of these three scattering mechanisms has the following characteristics:

Odd-bounce scattering: $A_0 \gg B_0 + B,\ B_0 - B$
Even-bounce scattering: $B_0 + B \gg A_0,\ B_0 - B$
Volume scattering: $B_0 - B \gg A_0,\ B_0 + B$.

11.2.4 Holm Decomposition

Holm decomposition is another explanation of the target characteristics spectrum. It decomposes the target into the sum of scattering matrix (coherency matrix with rank of 1) and two noises components or residual components. Holm decomposition is a mix method, which combines eigenvalues analysis (providing unitary invariance) and the single target and noise model. The eigenvalue matrix can be expressed as

$$\boldsymbol{\Sigma} = \begin{bmatrix} \lambda_1 & 0 & 0 \\ 0 & \lambda_2 & 0 \\ 0 & 0 & \lambda_3 \end{bmatrix}_{\lambda_1 \ge \lambda_2 \ge \lambda_3}$$

$$
\begin{aligned}
&= \begin{bmatrix} \lambda_1 - \lambda_2 & 0 & 0 \\ 0 & \lambda_2 & 0 \\ 0 & 0 & \lambda_3 \end{bmatrix} + \begin{bmatrix} \lambda_2 - \lambda_3 & 0 & 0 \\ 0 & \lambda_2 - \lambda_3 & 0 \\ 0 & 0 & \lambda_3 \end{bmatrix} + \begin{bmatrix} \lambda_3 & 0 & 0 \\ 0 & \lambda_3 & 0 \\ 0 & 0 & \lambda_3 \end{bmatrix} \\
&= \mathbf{\Sigma}_1 + \mathbf{\Sigma}_2 + \mathbf{\Sigma}_3
\end{aligned} \tag{11.55}
$$

The corresponding Holm decomposition is

$$
\mathbf{T} = \boldsymbol{U}_3 \mathbf{\Sigma} \boldsymbol{U}_3^{-1} = \boldsymbol{U}_3 \mathbf{\Sigma}_1 \boldsymbol{U}_3^{-1} + \boldsymbol{U}_3 \mathbf{\Sigma}_2 \boldsymbol{U}_3^{-1} + \boldsymbol{U}_3 \mathbf{\Sigma}_3 \boldsymbol{U}_3^{-1} = \boldsymbol{T}_1 + \boldsymbol{T}_2 + \boldsymbol{T}_3 \tag{11.56}
$$

The 3 × 3 coherency matrix $\boldsymbol{T}_1$ denotes the state of a pure target and provides the general expression of the target. The 3 × 3 coherency matrix $\boldsymbol{T}_2$ denotes the mix target state and provides the variance of the target. Finally, the matrix $\boldsymbol{T}_3$ denotes the non-polarimetric mix state of the noise term.

11.2.5 Van Zyl Decomposition

Van Zyl decomposition was first proposed to describe the 3 × 3 covariance matrix $\boldsymbol{C}_3$ of the single-static azimuth symmetric natural terrains. Van Zyl decomposition is based on the assumption that the coherence coefficient of co-polarization and cross-polarization polarization is 0. The corresponding average covariance matrix is

$$
\mathbf{C}_3 = \begin{bmatrix} |s_{HH}|^2 & 0 & s_{HH} S_{VV}^* \\ 0 & 2|s_{HV}|^2 & 0 \\ s_{VV} S_{HH}^* & 0 & |s_{VV}|^2 \end{bmatrix} = \alpha \begin{bmatrix} 1 & 0 & \rho \\ 0 & \eta & 0 \\ \rho^* & 0 & \mu \end{bmatrix} \tag{11.57}
$$

where

$$
\begin{aligned}
\alpha &= s_{HH} S_{HH}^*, \ \rho = s_{HH} S_{VV}^* / s_{HH} S_{HH}^* \\
\eta &= 2 s_{HV} S_{HV}^* / s_{HH} S_{HH}^*, \ \mu = s_{VV} S_{VV}^* / s_{HH} S_{HH}^*
\end{aligned} \tag{11.58}
$$

The parameters α, ρ, η, and μ are related to the size, shape, media characteristics, and statistical angles. The analytical expression of the corresponding eigenvalues are

$$
\begin{cases}
\lambda_1 = \frac{\alpha}{2}\left\{1 + \mu + \sqrt{(1-\mu)^2 + 4|\rho|^2}\right\} \\
\lambda_2 = \frac{\alpha}{2}\left\{1 + \mu - \sqrt{(1-\mu)^2 + 4|\rho|^2}\right\} \\
\lambda_3 = \alpha\eta
\end{cases} \tag{11.59}
$$

And the corresponding eigenvectors are

$$\boldsymbol{u}_1 = \sqrt{\frac{\mu - 1 + \sqrt{\Delta}}{\left(\mu - 1 + \sqrt{\Delta}\right)^2 + 4|\rho|^2}} \begin{bmatrix} \frac{2\rho}{\mu-1+\sqrt{\Delta}} \\ 0 \\ 1 \end{bmatrix} \tag{11.60}$$

$$\boldsymbol{u}_2 = \sqrt{\frac{\mu - 1 - \sqrt{\Delta}}{\left(\mu - 1 - \sqrt{\Delta}\right)^2 + 4|\rho|^2}} \begin{bmatrix} \frac{2\rho}{\mu-1-\sqrt{\Delta}} \\ 0 \\ 1 \end{bmatrix} \tag{11.61}$$

$$\boldsymbol{u}_3 = \begin{bmatrix} 0 \\ 1 \\ 0 \end{bmatrix} \tag{11.62}$$

In Eqs. (11.60)–(11.62), $\Delta = \sqrt{(1-\mu)^2 + 4|\rho|^2}$.
Therefore, the 3×3 covariance matrix $\mathbf{C}_3$ can be expressed as

$$\mathbf{C}_3 = \sum_{i=1}^{3} \lambda_i \boldsymbol{u}_i \boldsymbol{u}_i^H = \boldsymbol{\Lambda}_1 \begin{bmatrix} |\alpha|^2 & 0 & \alpha \\ 0 & 0 & 0 \\ \alpha^* & 0 & 1 \end{bmatrix} + \boldsymbol{\Lambda}_2 \begin{bmatrix} |\beta|^2 & 0 & \beta \\ 0 & 0 & 0 \\ \beta^* & 0 & 1 \end{bmatrix} + \boldsymbol{\Lambda}_3 \begin{bmatrix} 0 & 0 & 0 \\ 0 & 1 & 0 \\ 0 & 0 & 0 \end{bmatrix} \tag{11.63}$$

where

$$\boldsymbol{\Lambda}_1 = \lambda_1 \left[\frac{\left(\mu - 1 + \sqrt{\Delta}\right)^2}{\left(\mu - 1 + \sqrt{\Delta}\right)^2 + 4|\rho|^2} \right], \alpha = \frac{2\rho}{\mu - 1 + \sqrt{\Delta}} \tag{11.64}$$

$$\boldsymbol{\Lambda}_2 = \lambda_1 \left[\frac{\left(\mu - 1 - \sqrt{\Delta}\right)^2}{\left(\mu - 1 - \sqrt{\Delta}\right)^2 + 4|\rho|^2} \right], \beta = \frac{2\rho}{\mu - 1 + \sqrt{\Delta}} \tag{11.65}$$

$$\boldsymbol{\Lambda}_2 = \lambda_3 \tag{11.66}$$

The first two eigenvectors of Van Zyl decomposition can be interpreted as the scattering matrix of odd-bounce scattering and even-bounce scattering.

11.3 Model-Based Polarimetric Decomposition

The incoherence decomposition of multi-look data based on model decomposes the target into the sum of several simple objects, and the simple objects correspond to the physical scattering model of some simple objects. These kinds of target decompositions include Freeman–Durden three-component decomposition, Freeman two-component decomposition, Yamaguchi four-component decomposition, multi-component decomposition, etc.

11.3.1 Freeman–Durden Decomposition

The Freeman–Durden decomposition is an incoherent target decomposition method based on three-component scattering mechanism model. The three scattering mechanism components included in the model are volume scatter from randomly oriented dipoles, first-order Bragg surface scatter and a dihedral scattering mechanism for two surfaces with different dielectric constants.

For volume scattering, it is assumed that the radar return is from randomly oriented, very thin cylinder-like scatters. A typical terrain of this scattering mechanism is a vegetation area with many branches and leaves. Volume scattering can be represented by scattering from scatters with the following scattering matrix when in standard orientation.

$$\boldsymbol{S} = \begin{bmatrix} S_{\mathrm{H}} & 0 \\ 0 & S_{\mathrm{V}} \end{bmatrix} \tag{11.67}$$

It is assumed that the scatters are randomly oriented about the radar look direction with an angle ϕ from the vertical polarization direction. The scattering matrix of the scatters can be obtained by means of the coordinate basis transformation.

$$\begin{bmatrix} S_{\mathrm{HH}} & S_{\mathrm{HV}} \\ S_{\mathrm{VH}} & S_{\mathrm{VV}} \end{bmatrix} = \begin{bmatrix} \cos\phi & \sin\phi \\ -\sin\phi & \cos\phi \end{bmatrix} \boldsymbol{S} \begin{bmatrix} \cos\phi & -\sin\phi \\ \sin\phi & \cos\phi \end{bmatrix} \tag{11.68}$$

Equation (11.68) can be organized as

$$\begin{bmatrix} S_{\mathrm{HH}} & S_{\mathrm{HV}} \\ S_{\mathrm{VH}} & S_{\mathrm{VV}} \end{bmatrix} = \begin{bmatrix} S_{\mathrm{H}}\cos^2\phi + S_{\mathrm{v}}\sin^2\phi & (S_{\mathrm{V}} - S_{\mathrm{H}})\cos\phi\sin\phi \\ (S_{\mathrm{V}} - S_{\mathrm{H}})\cos\phi\sin\phi & S_{\mathrm{H}}\sin^2\phi + S_{\mathrm{v}}\cos^2\phi \end{bmatrix} \tag{11.69}$$

Assuming the probability density function for scatter orientation is $p(\phi)$, and the expected value of any function $f(\phi)$ is

$$f = \int_0^{2\pi} f(\phi)p(\phi)d\phi \tag{11.70}$$

The second-order scattering matrix statistics can be obtained from Eqs. (11.69) to (11.70). It can be further simplified by assuming thin cylindrical scatters so that $S_{\mathrm{H}} = 0$, $S_{\mathrm{V}} = 1$. Furthermore, assuming a uniform orientation distribution $p(\phi)$, the polarimetric covariance matrix of volume scattering is

$$\boldsymbol{C}_{3\mathrm{v}} = f_{\mathrm{v}} \begin{bmatrix} 1 & 0 & 1/3 \\ 0 & 2/3 & 0 \\ 1/3 & 0 & 1 \end{bmatrix} \tag{11.71}$$

where f_{v} is the volume scattering contribution to the $|S_{VV}|^2$ component.

Double-bounce scattering is modeled by scattering from a dihedral corner reflector, such as ground-tree trunk back scattering forest and in urban ground-wall backscatter where the reflector surfaces can be made of different dielectric materials. The vertical scattering surface has the reflection coefficient and $R_{\perp V}$ for horizontal and vertical polarizations, respectively. The horizontal ground scattering surface has the reflection coefficients $R_{\parallel H}$ and $R_{\parallel V}$. The model can be made more general by incorporating propagation factors $e^{j2\gamma_v}$ and $e^{j2\gamma_H}$, where the complex coefficients γ_v and γ_H represent any propagation attenuation and phase change effects. The scattering matrix of double-bounce scattering is

$$S = \begin{bmatrix} e^{j2\gamma_H} R_{\parallel H} R_{\perp H} & 0 \\ 0 & e^{j2\gamma_H} R_{\parallel V} R_{\perp V} \end{bmatrix} \tag{11.72}$$

The normalized covariance matrix double-bounce scattering is

$$\boldsymbol{C}_{3D} = f_D \begin{bmatrix} |\alpha|^2 & 0 & \alpha \\ 0 & 0 & 0 \\ \alpha^* & 0 & 1 \end{bmatrix} \tag{11.73}$$

where

$$\alpha = e^{j2(\gamma_H - \gamma_V)} \frac{R_{\parallel H} R_{\perp H}}{R_{\parallel V} R_{\perp V}} \tag{11.74}$$

f_D corresponds to the contribution of the double-bounce scattering to the $|S_{VV}|^2$ component. There are two properties in double-bounce scattering model. One is that the amplitude of HH and VV scattering are not necessarily equal, the other is that the phase difference between them is not necessarily $\pm\pi$.

Surface scattering can be modeled as first-order Bragg scattering. Typical terrains with surface scattering include rough ground surface with slight disturbance and water surface with waves. The covariance matrix of surface scattering model is

$$\boldsymbol{C}_{3S} = f_S \begin{bmatrix} |\beta|^2 & 0 & \beta \\ 0 & 0 & 0 \\ \beta^* & 0 & 1 \end{bmatrix} \tag{11.75}$$

where f_S corresponds to the contribution of the single-bounce scattering to the $|S_{VV}|^2$ component.

For all of these backscatter components, it is assumed that the co-polarized and cross-polarized returns are uncorrelated and that the backscatter is reciprocal. Further, assuming that the volume, double-bounce, and surface scatter components are uncorrelated, the total second-order statistics are the sum of the above statistics for the individual mechanisms. Thus the model for the total backscatter is

$$\begin{cases} |S_{\mathrm{HH}}|^2 = f_{\mathrm{S}}|\beta|^2 + f_{\mathrm{D}}|\alpha|^2 + f_{\mathrm{V}} \\ |S_{\mathrm{VV}}|^2 = f_{\mathrm{S}} + f_{\mathrm{D}} + f_{\mathrm{V}} \\ S_{\mathrm{HH}}S_{\mathrm{VV}}^* = f_{\mathrm{S}}\beta + f_{\mathrm{D}}\alpha + f_{\mathrm{V}}/3 \\ |S_{\mathrm{HV}}|^2 = f_{\mathrm{V}}/3 \end{cases} \tag{11.76}$$

and

$$S_{\mathrm{HH}}S_{\mathrm{HV}}^* = S_{\mathrm{HV}}S_{\mathrm{VV}}^* = 0 \tag{11.77}$$

In Eq. (11.76), f_{S}, f_{D}, and f_{V} correspond to the contribution of the surface scattering, double-bounce scattering, and volume scattering to the $|S_{VV}|^2$ component, respectively. This model gives four equations in five unknowns. In general, a solution can be found if one of the unknowns is fixed. Since neither the surface nor the double-bounce mechanisms contribute to the HV term in the model, we can use this to estimate the volume scatter contribution directly. The volume contribution, f_{V} or $f_{\mathrm{V}}/3$, can then be subtracted off the $|S_{\mathrm{HH}}|^2$, $|S_{\mathrm{VV}}|^2$, and $S_{\mathrm{HH}}S_{\mathrm{VV}}^*$ terms, leaving three equations in four unknowns.

$$\begin{cases} |S_{\mathrm{HH}}|^2 = f_{\mathrm{S}}|\beta|^2 + f_{\mathrm{D}}|\alpha|^2 \\ |S_{\mathrm{VV}}|^2 = f_{\mathrm{S}} + f_{\mathrm{D}} \\ S_{\mathrm{HH}}S_{\mathrm{VV}}^* = f_{\mathrm{S}}\beta + f_{\mathrm{D}}\alpha \end{cases} \tag{11.78}$$

Based on the sign of $\mathrm{Re}\left(S_{\mathrm{HH}}S_{\mathrm{HV}}^*\right)$, double-bounce and surface scatter can be considered as the dominant contribution in the residual. If $\mathrm{Re}\left(S_{\mathrm{HH}}S_{\mathrm{HV}}^*\right) > 0$, the surface scatter is considered as dominant and the parameter α is fixed with $\alpha = -1$. If $\mathrm{Re}\left(S_{\mathrm{HH}}S_{\mathrm{HV}}^*\right) < 0$, the double-bounce scatter is considered as dominant and the parameter β is fixed with $\beta = 1$. Then other parameters can be estimated from the residual radar measurements. Finally, the contribution of each scattering mechanism can be estimated to the span, following

$$\begin{cases} P_{\mathrm{S}} = f_{\mathrm{S}}\left(1 + |\beta|^2\right) \\ P_{\mathrm{D}} = f_{\mathrm{D}}\left(1 + |\alpha|^2\right) \\ P_{\mathrm{V}} = 8f_{\mathrm{V}}/3 \end{cases} \tag{11.79}$$

According to the estimated P_{S}, P_{D}, and P_{V}, the contributions of three scattering mechanisms to the span can be compared. The scattering mechanism with the maximum power is taken as the current scattering mechanism of the target. Therefore, targets can be classified into three classes volume scattering, double-bounce scattering, and surface scattering.

L-band PolInSAR data of Oberpfaffenhofen area, Germany, acquired by ESAR system is used for experiment. The size of the data is 800×900, and this area contains some typical terrain types, such as runways, buildings, and forests. Figure 11.13 shows the RGB composition of the experimental data with Freeman decomposition, where surface scattering component P_{S} is represented by blue channel (B),

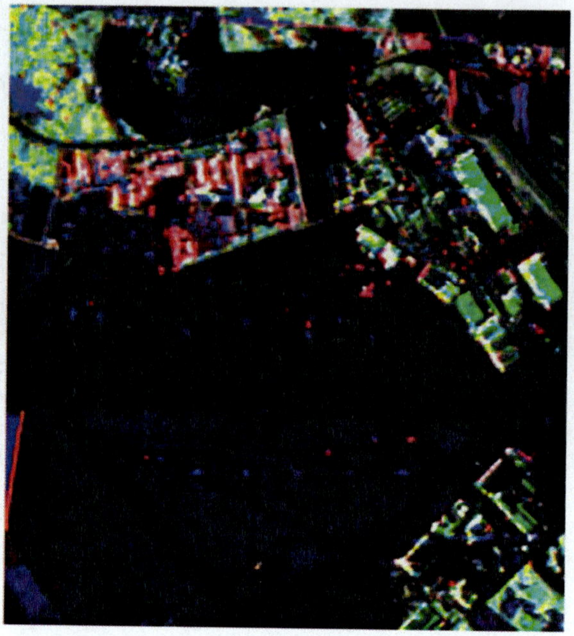

Fig. 11.13 RGB composition of the experimental data with Freeman decomposition

double-bounce scattering component P_D is represented by red channel (R), and the volume scattering component is represented by green channel (G). It can be seen from Fig. 11.13 that the forest area in the upper left corner of the image is green, indicating that volume scattering is dominant. The area shown in red in the middle of the image is the urban area dominated by double-bounce scattering components. The runway area in the lower part of the image is surface scattering based on blue.

11.3.2 Freeman Two-Component Decomposition

In 2007, Freeman proposed a new simplified target decomposition theory for forest target decomposition, the two-component target decomposition theory. This decomposition method models scattering polarization covariance matrix of the forest as the sum of two scattering mechanisms. Unlike the three-component decomposition, in addition to the volume scattering component, the decomposition uniformly expresses of odd-bounce scattering and direct ground scattering [5].

The assumption of volume scattering based on that the radar returns of forest canopy are formed from the reflection sofa cloud of randomly oriented cylinder-like scatters. The covariance matrix can be expressed as

$$C_{3v} = f_v \begin{bmatrix} 1 & 0 & \rho \\ 0 & 1-\rho & 0 \\ \rho^* & 0 & 1 \end{bmatrix} \tag{11.80}$$

where f_v and ρ represent the contribution of volume scattering component.

The second scattering mechanism represents double-bounce scattering or direct ground scattering, and its' covariance matrix can be expressed as

$$C_{3G} = f_G \begin{bmatrix} 1 & 0 & \alpha \\ 0 & 0 & 0 \\ \alpha^* & 0 & |\alpha|^2 \end{bmatrix} \tag{11.81}$$

Where f_G and α represent the contribution of double-bounce scattering component or direct ground scattering component.

When double-bounce scattering is dominant, $|\alpha| \leq 1$, $\arg(\alpha) = \pm\pi$; while odd-bounce scattering is dominant, $|\alpha| \geq 1$, $\arg(\alpha) \approx 2\phi$, where ϕ is the phase difference between HH and VV component.

Assuming that the volume, double-bounce, and surface scatter components are uncorrelated, the total target covariance matrix is the sum of the above statistics for the individual mechanisms. That is

$$C_3 = C_{3G} + C_{3V} = \begin{bmatrix} f_G + f_V & 0 & f_G\alpha + f_V\rho \\ 0 & f_V(1-\rho) & 0 \\ f_G\alpha^* + f_V\rho^* & 0 & f_G|\alpha|^2 + f_V \end{bmatrix} \tag{11.82}$$

Compared with Freeman–Duren three-component decomposition, the number of input and output parameters is equal in Freeman two-component decomposition (four equations in four unknowns). Therefore, the unknowns can be estimated without assumptions.

The contributions of every scattering mechanism to the span are

$$S_{span} = |S_{HH}|^2 + 2|S_{HV}|^2 + |S_{VV}|^2 = P_G + P_V \tag{11.83}$$

where

$$P_G = f_G\left(1 + |\alpha|^2\right),\ P_V = f_V(3-\rho) \tag{11.84}$$

The procedure of two-component decomposition is shown in Fig. 11.14.

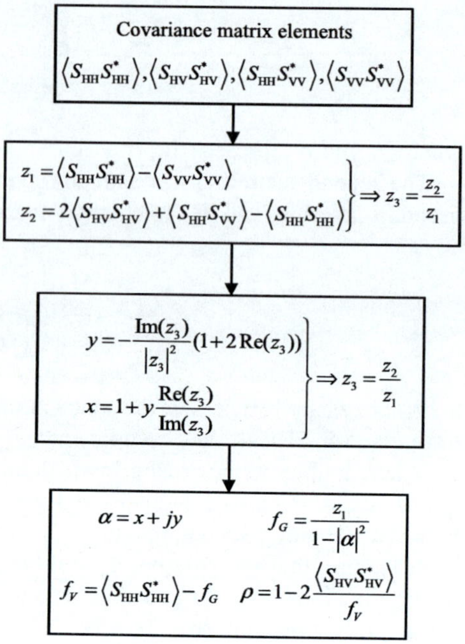

Fig. 11.14 The procedure of Freeman two components decomposition

11.3.3 Yamaguchi Four-Component Decomposition

Freeman decomposition can be successfully applied to decompose SAR observations under the reflection symmetry condition, which satisfies [6–10]

$$\langle S_{HH}S_{HV}^*\rangle = \langle S_{VV}S_{HV}^*\rangle = 0 \tag{11.85}$$

However, it can be possible to find some areas in an SAR image for which the reflection symmetry condition does not hold. The condition that the difference between $\langle S_{HH}S_{HV}^*\rangle$ and $\langle S_{VV}S_{HV}^*\rangle$ is obvious has to be considered. Based on the three-component scattering model approach, Yamaguchi et al. extended the Freeman decomposition and proposed a four-component scattering model [11, 12], which is called Yamaguchi decomposition. The model is as follows:

$$\langle \boldsymbol{C}\rangle = f_S\langle \boldsymbol{C}\rangle_S + f_D\langle \boldsymbol{C}\rangle_D + f_V\langle \boldsymbol{C}\rangle_V + f_H\langle \boldsymbol{C}\rangle_H \tag{11.86}$$

where f_S, f_D, f_V, and f_H represent the contributions of surface scattering, double-bounce scattering, volume scattering, and helix scattering.

The expression of surface scattering is

$$\langle \boldsymbol{C} \rangle_{\mathrm{S}} = \begin{bmatrix} |\beta|^2 & 0 & \beta \\ 0 & 0 & 0 \\ \beta^* & 0 & 1 \end{bmatrix}, |\alpha| < 1 \tag{11.87}$$

The model of double-bounce scattering is

$$\langle \boldsymbol{C} \rangle_{\mathrm{D}} = \begin{bmatrix} |\alpha|^2 & 0 & \alpha \\ 0 & 0 & 0 \\ \alpha^* & 0 & 1 \end{bmatrix}, |\alpha| < 1 \tag{11.88}$$

where α and β are unknowns to be determined.

These two terms are identical to Freeman–Durden decomposition.

Different from Freeman–Durden decomposition, volume scattering model is expanded by Yamaguchi. Considering the distribution of trunks and branches, the randomly oriented dipole model adopted in Freeman decomposition is modeled by Yamaguchi. In the theoretical modeling of volume scattering, a cloud of randomly oriented dipoles is implemented with a uniform probability function for the orientation angles. In Yamaguchi's modification, the scattering from tree trunks and branches displays a nonuniform angle distribution. According to the modified model, the expression of volume scattering model is as follows:

If $10\ \lg\left(\langle |S_{\mathrm{VV}}|\rangle^2 / \langle |S_{\mathrm{HH}}|\rangle^2\right) > 2\,\mathrm{dB}$,

$$\langle \boldsymbol{C} \rangle_{\mathrm{V}} = \frac{1}{15} \begin{bmatrix} 8 & 0 & 2 \\ 0 & 4 & 0 \\ 2 & 0 & 3 \end{bmatrix} \tag{11.90}$$

If $10\ \lg\left(\langle |S_{\mathrm{VV}}|\rangle^2 / \langle |S_{\mathrm{HH}}|\rangle^2\right) < 2\,\mathrm{dB}$,

$$\langle \boldsymbol{C} \rangle_{\mathrm{V}} = \frac{1}{8} \begin{bmatrix} 3 & 0 & 2 \\ 0 & 4 & 0 \\ 2 & 0 & 8 \end{bmatrix} \tag{11.91}$$

The helix scattering model is

$$\langle \boldsymbol{C} \rangle_{\mathrm{L-H}} = \frac{1}{4} \begin{bmatrix} 1 & -j\sqrt{2} & -1 \\ j\sqrt{2} & 2 & -j\sqrt{2} \\ -1 & j\sqrt{2} & 1 \end{bmatrix} \tag{11.92}$$

$$\langle \boldsymbol{C} \rangle_{\mathrm{R-H}} = \frac{1}{4} \begin{bmatrix} 1 & j\sqrt{2} & -1 \\ -j\sqrt{2} & 2 & j\sqrt{2} \\ -1 & -j\sqrt{2} & 1 \end{bmatrix} \tag{11.93}$$

Equation (11.92) denotes left helix, and Eq. (11.93) denotes right helix.

The measured covariance matrix can be expressed as follows if we get the $\langle \boldsymbol{C} \rangle_{\mathrm{V}}$ of condition $10\,\lg\big(\langle |S_{\mathrm{VV}}|\rangle^2 / \langle |S_{\mathrm{HH}}|\rangle^2\big) > 2\,\mathrm{dB}$

$$\langle \boldsymbol{C} \rangle = f_{\mathrm{S}} \begin{bmatrix} |\beta|^2 & 0 & \beta \\ 0 & 0 & 0 \\ \beta^* & 0 & 1 \end{bmatrix} + f_{\mathrm{D}} \begin{bmatrix} |\alpha|^2 & 0 & \alpha \\ 0 & 0 & 0 \\ \alpha^* & 0 & 1 \end{bmatrix} + \frac{f_{\mathrm{V}}}{15} \begin{bmatrix} 8 & 0 & 2 \\ 0 & 4 & 0 \\ 2 & 0 & 3 \end{bmatrix} + \frac{f_{\mathrm{C}}}{4} \begin{bmatrix} 1 & \pm j\sqrt{2} & -1 \\ \mp j\sqrt{2} & 2 & \pm j\sqrt{2} \\ -1 & \mp j\sqrt{2} & 1 \end{bmatrix} \tag{11.94}$$

Comparing the covariance matrix elements at both ends of (11.90), we get:

$$\langle |S_{\mathrm{HH}}|\rangle^2 = f_{\mathrm{S}}|\beta|^2 + f_{\mathrm{D}}|\alpha|^2 + \frac{8}{15} f_{\mathrm{V}} + \frac{f_{\mathrm{H}}}{4} \tag{11.95}$$

$$\langle |S_{\mathrm{HV}}|\rangle^2 = \frac{2}{15} f_{\mathrm{V}} + \frac{f_{\mathrm{H}}}{4} \tag{11.96}$$

$$\langle |S_{\mathrm{VV}}|\rangle^2 = f_{\mathrm{S}} + f_{\mathrm{D}} + \frac{3}{15} f_{\mathrm{V}} + \frac{f_{\mathrm{H}}}{4} \tag{11.97}$$

$$\langle S_{\mathrm{HH}} S_{\mathrm{VV}}^* \rangle = f_{\mathrm{S}} \beta \tag{11.98}$$

$$\frac{1}{2} \mathrm{Im}\{\langle S_{\mathrm{HH}} S_{\mathrm{VV}}^* \rangle + \langle S_{\mathrm{HV}} S_{\mathrm{VV}}^* \rangle\} = \frac{f_{\mathrm{H}}}{4} \tag{11.99}$$

As the left ends of Eq. (11.95)–(11.99) are measurable, the value of f_{H} can be determined

$$f_{\mathrm{H}} = 2\mathrm{Im}\{\langle S_{\mathrm{HV}}^* (S_{\mathrm{HH}} - S_{\mathrm{VV}}) \rangle\} \tag{11.100}$$

And we can get

$$f_{\mathrm{V}} = \frac{15}{2} \left(|S_{\mathrm{HV}}|^2 - \frac{f_{\mathrm{H}}}{4} \right) \tag{11.101}$$

The left four unknowns can be obtained by the procedures of Freeman–Durden decomposition. Moreover, the scattering power of surface scattering P_{S}, double-bounce scattering P_{S}, volume scattering P_{V}, and helix scattering P_{H} are, respectively,

$$P_{\mathrm{S}} = f_{\mathrm{S}} \big(1 + |\beta|^2\big) \tag{11.102}$$

$$P_{\mathrm{D}} = f_{\mathrm{D}} \big(1 + |\alpha|^2\big) \tag{11.103}$$

$$P_{\mathrm{V}} = f_{\mathrm{V}} \tag{11.104}$$

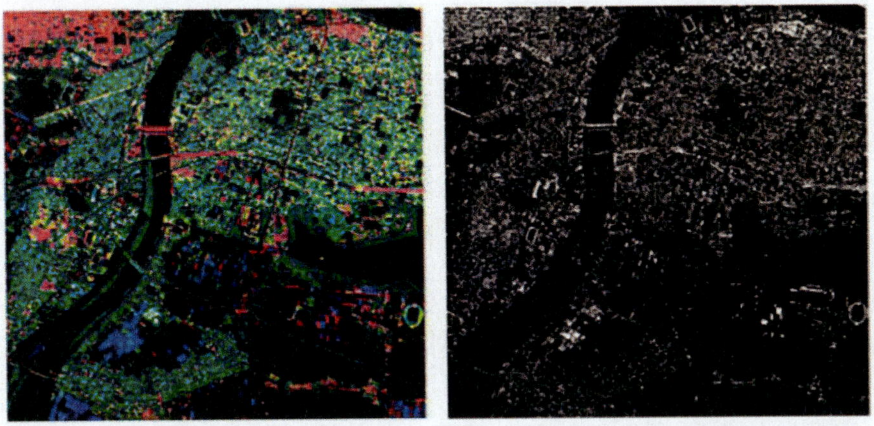

Fig. 11.15 The example of Yamaguchi four-component decomposition **a** RGB composition, R: P_D, G: P_V, B: P_S; **b** Helix scattering power P_H

$$P_H = f_H \tag{11.105}$$

$$P_T = P_S + P_D + P_V + P_H = \langle |S_{HH}| \rangle^2 + 2\langle |S_{HV}| \rangle^2 + \langle |S_{VV}| \rangle^2 \tag{11.106}$$

When applying Yamaguchi decomposition to PolSAR data directly, there may occur the situation that $f_S < 0$ or $f_D < 0$, and it is not consistent with the real situation. The typical characteristics of this area is that $\langle |S_{HV}|^2 \rangle \gg \langle |S_{HH}|^2 \rangle$ or $\langle |S_{HV}|^2 \rangle \gg \langle |S_{VV}|^2 \rangle$, and it is generally found in artificial scatterers and forest areas with complex geometries. This inconsistency can be avoided with the following energy ratios:

$$2\langle |S_{HV}| \rangle^2 : |S_{HH}|^2 \; or \; |S_{VV}|^2 \tag{11.107}$$

Theoretical research shows that the ratio of the power of the co-polarized channel and the power of the cross-polarized channel is statistically 2: 1

After the above analysis, Fig. 11.15 shows an example of Yamaguchi four-component decomposition. Figure 11.16 shows the flowchart of Yamaguchi decomposition.

11.3.4 Multi-component Scattering Model

Multi-component scattering model is an extension of Freeman–Durdan decomposition proposed by Zou Bin, Zhang Lamei, etc. This decomposition expands three-component models into surface, double-bounce, volume, helix, wire scattering, etc.

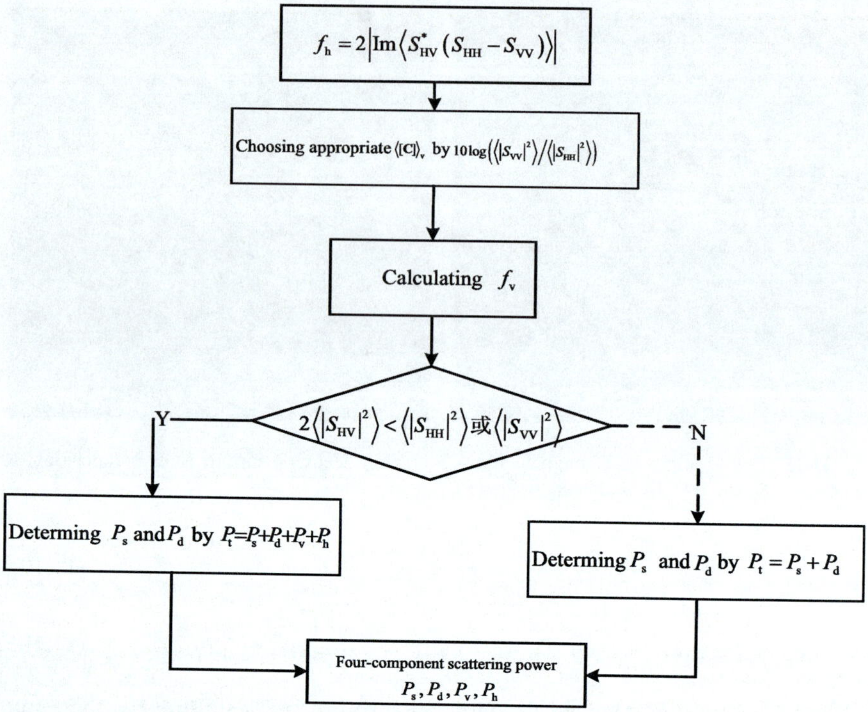

Fig. 11.6 Flowchart of Yamaguchi four-component decomposition

Accordingly, the covariance matrix can be expressed as follows:

$$\langle \boldsymbol{C} \rangle = f_S \langle \boldsymbol{C} \rangle_S + f_D \langle \boldsymbol{C} \rangle_D + f_V \langle \boldsymbol{C} \rangle_V + f_H \langle \boldsymbol{C} \rangle_H + f_W \langle \boldsymbol{C} \rangle_W \tag{11.108}$$

where f_S, f_D, f_V, f_H, and f_W represent the contributions of surface scattering, double-bounce scattering, volume scattering, helix scattering, and wire scattering.

Among them, the terms of surface scattering, double-bounce scattering, volume scattering are identical to Freeman–Durden decomposition. The helix scattering model is identical to that introduced in Yamaguchi decomposition. Wire scattering is a special scattering mechanism exist in urban areas, and the scattering matrix can be expressed as

$$\boldsymbol{S}_W = \begin{bmatrix} \gamma & \rho \\ \rho & 1 \end{bmatrix} \tag{11.109}$$

In Eq. (11.109), $\gamma = \frac{S_{HH}}{S_{VV}}$, $\rho = \frac{S_{HV}}{S_{VV}}$, and the corresponding covariance matrix is

$$\langle \boldsymbol{C} \rangle_{\mathrm{W}} = \begin{bmatrix} |\gamma|^2 & \sqrt{2}\gamma\rho^* & \gamma \\ \sqrt{2}\gamma\rho^* & 2|\rho|^2 & \sqrt{2}\rho \\ \gamma^* & \sqrt{2}\rho^* & 1 \end{bmatrix} \tag{11.110}$$

From Eqs. (11.88), (11.99), (11.108), and (11.110), we can get

$$\left\langle |S_{\mathrm{HH}}|^2 \right\rangle = f_{\mathrm{S}}|\beta|^2 + f_{\mathrm{D}}|\alpha|^2 + f_{\mathrm{V}} + \frac{f_{\mathrm{H}}}{4} + f_{\mathrm{W}}|\gamma|^2 \tag{11.111}$$

$$\left\langle |S_{\mathrm{VV}}|^2 \right\rangle = f_{\mathrm{S}} + f_{\mathrm{D}} + f_{\mathrm{V}} + \frac{f_{\mathrm{H}}}{4} + f_{\mathrm{W}} \tag{11.112}$$

$$\left\langle S_{\mathrm{HH}} S_{\mathrm{VV}}^* \right\rangle = f_{\mathrm{S}}\beta + f_{\mathrm{D}}\alpha + \frac{1}{3} f_{\mathrm{V}} - \frac{f_{\mathrm{H}}}{4} + f_{\mathrm{W}}\gamma \tag{11.113}$$

$$\left\langle |S_{\mathrm{HV}}|^2 \right\rangle = \frac{f_{\mathrm{V}}}{3} + \frac{f_{\mathrm{H}}}{4} + f_{\mathrm{W}}|\rho|^2 \tag{11.114}$$

$$\left\langle \left| S_{\mathrm{HH}} S_{\mathrm{HV}}^* \right|^2 \right\rangle = \pm j \frac{1}{4} f_{\mathrm{H}} + f_{\mathrm{W}}\gamma\rho^* \tag{11.115}$$

$$\left\langle \left| S_{\mathrm{HH}} S_{\mathrm{VV}}^* \right|^2 \right\rangle = \pm j \frac{1}{4} f_{\mathrm{H}} + f_{\mathrm{W}}\rho \tag{11.116}$$

Combining Eqs. (11.111)–(11.116), we can get

$$f_{\mathrm{W}} = \frac{\left\langle S_{\mathrm{HH}} S_{\mathrm{HV}}^* \right\rangle - \left\langle S_{\mathrm{HV}} S_{\mathrm{VV}}^* \right\rangle}{\gamma\rho^* - \rho} \tag{11.117}$$

$$f_{\mathrm{H}} = 2\,\mathrm{Im}\left\{ \left\langle S_{HH} S_{HV}^* \right\rangle + \left\langle S_{HH} S_{HV}^* \right\rangle - f_{\mathrm{W}}\left(\gamma\rho^* + \rho\right) \right\} \tag{11.118}$$

$$f_{\mathrm{V}} = 3\left\{ \left\langle |S_{\mathrm{HV}}|^2 \right\rangle - \frac{1}{4} f_H - f_{\mathrm{W}}|\rho|^2 \right\} \tag{11.119}$$

The scattering power of surface scattering, volume scattering, helix scattering, and wire scattering is

$$P_{\mathrm{S}} = f_{\mathrm{S}}\left(1 + |\beta|^2\right) \tag{11.120}$$

$$P_{\mathrm{D}} = f_{\mathrm{D}}\left(1 + |\alpha|^2\right) \tag{11.121}$$

$$P_{\mathrm{V}} = 8 f_{\mathrm{V}}/3 \tag{11.122}$$

$$P_{\mathrm{H}} = f_{\mathrm{H}} \tag{11.123}$$

$$P_W = f_W\left(1 + |\gamma|^2 + 2|\rho|^2\right) \tag{11.124}$$

$$P = P_S + P_D + P_V + P_H + P_W \tag{11.125}$$

It is shown by the experiment that the multi-component scattering model decomposition method has certain advantages for analyzing urban area data.

11.4 Data Decomposition Based on PolInSAR Data

A scattering model decomposition technique with Pol-InSAR data was proposed by Japanese scholar Yamada, etc. [13]. The target covariance matrix is modeled as the sum of surface scattering, double-bounce scattering, and volume scattering based on three-component model. Compared with the scattering model based on PolSAR data, the number of negative scattering power pixels is obviously reduced.

When the set of PolInSAR data is available, we can assume that the covariance matrices of corresponding pair would be highly correlated. They can be modeled as

$$\langle \boldsymbol{C}\rangle^{(1)} = f_S\langle \boldsymbol{C}\rangle_S + f_D\langle \boldsymbol{C}\rangle_D + f_V\langle \boldsymbol{C}\rangle_V \tag{11.126}$$

$$\langle \boldsymbol{C}\rangle^{(2)} = f_S\langle \boldsymbol{C}\rangle_S e^{j\phi_S} + f_D\langle \boldsymbol{C}\rangle_D e^{j\phi_D} + f_V\langle \boldsymbol{C}\rangle_V e^{j\phi_V} \tag{11.127}$$

where the superscripts 1 and 2 indicate the master and slave channels, respectively. ϕ_S, ϕ_D, and ϕ_V are the interferometric phase to the corresponding component.

As we can see from Freeman decomposition, the HV-related components in $\langle \boldsymbol{C}\rangle_S$ and $\langle \boldsymbol{C}\rangle_D$ are 0, and ϕ_V can be easily estimated.

Suppose if the precise $\langle \boldsymbol{C}\rangle_V$ is known, f_V can be derived. This means that we can easily remove the volume scattering component from the original covariance matrices.

$$\left\langle \tilde{\boldsymbol{C}}\right\rangle^{(1)} = \langle \boldsymbol{C}\rangle^{(1)} - f_V\langle \boldsymbol{C}\rangle_V \tag{11.128}$$

$$\left\langle \tilde{\boldsymbol{C}}\right\rangle^{(2)} = \langle \boldsymbol{C}\rangle^{(2)} - f_V\langle \boldsymbol{C}\rangle_V e^{j\phi_V} \tag{11.129}$$

Obviously, the rank of matrix $\left\langle \tilde{\boldsymbol{C}}\right\rangle^{(1)}$ and $\left\langle \tilde{\boldsymbol{C}}\right\rangle^{(2)}$ is 2. In addition, when the proper $\langle \boldsymbol{C}\rangle_V$ is selected, magnitude of estimated the complex eigenvalues corresponding to complex coherence of local scatters which by means of the ESPRIT analysis for Eqs. (11.128) and (11.129) becomes 1 approximately. When potential candidates for $\langle \boldsymbol{C}\rangle_V$ can be prepared in advance, we can easily estimate the most proper $\langle \boldsymbol{C}\rangle_V$ among the candidates by this property. We can obtain more candidates with Pol-InSAR data than those in the conventional decomposition technique with PolSAR data can be obtained.

When the power of volume scattering component can be correctly estimated, estimation of the power of remaining components (P_S and P_D) is the next problem. As the rank of $\left\langle \tilde{C} \right\rangle^{(1)}$ is 2, the eigenvalues decomposition of $\left\langle \tilde{C} \right\rangle^{(1)}$ is

$$\left\langle [\tilde{\mathbf{C}}] \right\rangle^{(1)} = \left[\mathbf{e}_1 \ \mathbf{e}_2 \right] \begin{bmatrix} \lambda_1 & 0 \\ 0 & \lambda_2 \end{bmatrix} \begin{bmatrix} \mathbf{e}_1^{\mathrm{H}} \\ \mathbf{e}_2^{\mathrm{H}} \end{bmatrix} = \mathbf{E}\Lambda\mathbf{E}^{\mathrm{H}} \tag{11.30}$$

where λ_i and $\mathbf{e}_i$ ($i = 1, 2$) are the eigenvalues and corresponding eigenvectors of the ith component. It is clear that $\lambda_1 + \lambda_2$ is the sum of the surface scattering power and double-bounce scattering power. In addition, according to the formula of Freeman decomposition giving $\langle [\mathbf{C}] \rangle_s$ and $\langle [\mathbf{C}] \rangle_d$, $\left\langle [\tilde{\mathbf{C}}] \right\rangle^{(1)}$ can be written as

$$\left\langle [\tilde{\mathbf{C}}] \right\rangle^{(1)} = \begin{bmatrix} \beta & \alpha \\ 0 & 0 \\ 1 & 1 \end{bmatrix} \begin{bmatrix} f_s & 0 \\ 0 & f_d \end{bmatrix} \begin{bmatrix} \beta^* & 0 & 1 \\ \alpha^* & 0 & 1 \end{bmatrix} \tag{11.131}$$

From Eqs. (11.130) to (11.131), we can get

$$\begin{bmatrix} \beta & \alpha \\ 0 & 0 \\ 1 & 1 \end{bmatrix} \begin{bmatrix} \sqrt{f_S} & 0 \\ 0 & \sqrt{f_D} \end{bmatrix} = \boldsymbol{E}\boldsymbol{\Lambda}^{1/2}\boldsymbol{U} \tag{11.132}$$

where $\boldsymbol{U}$ is a unitary matrix defined by

$$\boldsymbol{U} = \begin{bmatrix} e^{j\psi} \cos\varsigma & e^{j\delta} \cos\varsigma \\ e^{-j\delta} \sin\varsigma & e^{-j\psi} \cos\varsigma \end{bmatrix} \tag{11.133}$$

Dominant component can be easily distinguished by the (3, 1)-element of $\tilde{\boldsymbol{C}}^{(1)}$. If the real part of the element is positive, contribution from double-bounce component is dominant, and vice versa, the contribution from surface scattering component is dominant,

when the surface scattering component is dominant and we can define that $\lambda_1 > \lambda_2$, the following relation holds:

$$\begin{cases} \frac{\lambda_2}{\lambda_1} \le \frac{P_D}{P_S} < 1 \\ P_S + P_D = \lambda_1 + \lambda_2 \end{cases} \tag{11.134}$$

This means that P_D/P_S value has large ambiguity of power ratio for relative small λ_2/λ_1, however we can narrow the possible range of P_d/P_s by using the equation derived from Eq. (11.134) as

$$\begin{cases} \frac{P_D}{P_S} = \frac{k^2+\lambda_r}{1+\lambda_r k^2} \\ k = \sin\varsigma / \cos\varsigma \\ \lambda_r = \frac{\lambda_2}{\lambda_1} \end{cases} \tag{11.135}$$

And the parameters in surface scattering and double-bounce scattering are

$$\beta = \frac{(e_1)_1 - \sqrt{\lambda_r} k (e_1) e^{-j\xi}}{(e_2)_1 - \sqrt{\lambda_r} k (e_2) e^{-j\xi}} \tag{11.136}$$

$$\alpha = \frac{(e_1)_1 + \sqrt{\lambda_r} k^{-1} (e_1) e^{-j\xi}}{(e_2)_1 - \sqrt{\lambda_r} k^{-1} (e_2) e^{-j\xi}} \tag{11.137}$$

In this case for dominant surface scattering contribution, $|\alpha| > 1$ and $|\beta| > 1$ can be assumed. Possible power ratio $\frac{P_D}{P_S}$ can be obtained by estimating proper combination of k $(= \sin\varsigma / \cos\varsigma)$ and ξ $(= \psi + \delta)$.

11.5 Conclusion

Basic polarimetric radar target decomposition method is introduced in this chapter. Polarimetric target decomposition is the basis of polarization data analysis, understanding, and application. It is widely applied in image classification, target detection, target recognition, and parameter inversion.

References

1. E. Krogager, New decomposition of the radar target scattering matrix. Electron. Lett. **26**(18), 1525–1527 (1990)
2. W.L. Cameron, L.K. Leung, Feature motivated polarization scattering matrix decomposition, in *Proceedings of IEEE International Radar Conference 1990* (Arlington, VA)
3. J. Yang, Y. Peng, S. Lin, Similarity between two scattering matrices. Electron. Lett. **37**(3), 193–194 (2001)
4. J.R. Huynen, Phenomenological theory of radar targets. Ph. D. Dissertation, 1970
5. A. Freeman, S.L. Durden, A three-component scattering model for polarimetric SAR data. IEEE Trans. Geosci. Remote Sens. **36**(3), 963–973 (1998)
6. S.R. Cloude, E. Pottier, A review of target decomposition theorems in radar polarimetry. IEEE Trans. Geosci. Remote Sens. **34**(2), 498–518 (1996)
7. J.-S. Lee, E. Pottier, *Polarimetric Radar Imaging: from Basics to Applications*. New York: CRC Press (2008)
8. A. Freeman, Fitting a two-component scattering model to polarimetric SAR data from forests. IEEE Trans. Geosci. Remote Sens. **45**(8), 10 (2007)
9. Y. Yamaguchi et al., Four-component scattering model for polarimetric SAR image decomposition. IEEE Trans. Geosci. Remote Sens. **43**(8), 8 (2005)
10. W. An, The polarimetric decomposition and scattering characteristics extraction of polarimetric SAR, Tsinghua University, 2010

11. Y. Yamaguchi et al., Four-component scattering model for polarimetric SAR image decomposition. IEEE Trans. Geosci. Remote Sens. **43**(8), 1699–1706 (2005)
12. Y. Yajima et al., POLSAR image analysis of wetlands using a modified four-component scattering power decomposition. IEEE Trans. Geosci. Remote Sens. **46**(6), 1667–1673 (2008)
13. H. Yamada, M. Yamazaki, Y. Yamaguchi, On scattering model decomposition of POLSAR images and its application to the ESPRIT-based pol-In SAR, in *Proceedings of EUSAR' 2006* (2006)

Chapter 12
Polarimetric Synthetic Aperture Radar Target Detection Technology

Target detection of synthetic aperture radar images is an important issue in military applications. Since the rise of SAR image target detection in the 1990s, it has been widely used due to its huge application value and the urgent needs of image interpretation applications.

Polarimetric SAR can simultaneously provide target information for multiple polarimetric channels. How to effectively use multi-dimensional information to suppress clutter interference, find potential targets, and enhance target detection performance has been the focus of research. The target detection research of polarimetric SAR mainly focuses on two aspects, including the extraction as well as use of effective features of the target and the research of statistical detection algorithms.

Polarimetric target detection is a binary probability hypothesis detection, that is,

H_0: The resolution unit is clutter

H_1: The resolution unit is the target

At present, the data forms used for polarization detection can be divided into two types. One is directly using the measured three-dimensional scattering vector. The other is converting the measured data into a Stokes vector, and then using the Stokes vector for detection.

This chapter summarizes the research situation of SAR target detection, focusing on several classic polarimetric target detection methods.

12.1 SAR Target Detection Overview

The essence of SAR image target detection is to complete the target detection based on the characteristics differences between the target and clutter. SAR image target detection methods can be divided into the following categories [1–3], including contrast-based target detection algorithms, other-features-based target detection algorithms, and complex image target detection algorithms.

R. Yang et al., *Polarimetric Microwave Imaging*,
https://doi.org/10.1007/978-981-15-8897-6_12

(a) Contrast-based target detection algorithm

Contrast-based target detection algorithms include CFAR detection algorithms, Generalized Likelihood Ratio Detection Algorithms (GLRT), and energy loop detection algorithms.

CFAR detection algorithm is a pixel-level target detection method, which is on the premise of a strong contrast between the target and the background. The specific implementation process of the CFAR method is as follows. Firstly, according to the classic statistical detection theory, the detection threshold is adaptively calculated based on the statistical characteristics of the background clutter around the target, under the given false alarm rate. Secondly, the pixels to be detected and the adaptive threshold are compared to determine whether it is the target point. By sliding the reference window, automatic detection of all pixels is achieved. The statistical characteristics of the background clutter around the target are usually determined by the pixels in the reference window around the target pixel. CFAR detection algorithm does not consider the probability distribution function of the target, which is a suboptimal statistical detection method. CFAR detection has three input factors, including false alarm probability, clutter distribution model, and CFAR detector. Clutter distribution model describes the statistical characteristics of clutter background. CFAR detector is a complete CFAR algorithm, which should be an algorithm that uses a CFAR detector based on a certain distribution. So far, in order to improve the speed and accuracy of the detection algorithm, scholars have developed a variety of CFAR detection algorithms from the perspective of different background statistical distribution models and different CFAR detectors, such as Mean Class (ML) CFAR detection and Ordered Statistics (OS) CFAR detection.

Generalized Likelihood Ratio Detection Technology (GLRT) takes into account the probability distribution function of the target. In theory, the theoretical Bayesian optimal solution can be obtained using the GLRT algorithm. However, the GLRT detection algorithm needs to know the statistical distribution of the target and the background. The actual statistical distribution of the target cannot be known. And because the shape, size, and direction of the actual target are different, it is difficult to build a unified statistical model of the target. The GLRT algorithm has not been widely used.

The energy loop (PR) detection algorithm is based on the fact that the ratio of the mean value μ_T within a certain size neighborhood of the target pixel to the clutter mean value μ_c in the surrounding background ring area should be above a certain threshold λ. μ_T/μ_c reflects the local signal-to-noise ratio of the image. In theory, the PR detection algorithm uses the local contrast information of the image. If the parameters are selected properly, it can achieve better performance. But λ cannot be adaptively selected, and how to properly select the target area size restricts the application and promotion of the algorithm.

In addition, people have also proposed a variety of polarimetric SAR image contrast enhancement techniques to improve the contrast between targets and thereby improve the performance of the detector.

(b) Target detection algorithm based on other image features

The construction of various target detection algorithms is basically based on the difference between the target and clutter electromagnetic scattering characteristics. Targets and clutter false alarms show differences in features on images that can be used to complete target detection, such as size, shape, texture, etc. Target detection algorithms that use other image features include clustering-based target detection algorithms, context-based target detection algorithms, and multi-feature joint detection methods.

(c) Target detection algorithm based on complex image features

The target detection algorithm based on complex image features is a more general SAR image target detection algorithm. The above two types of target detection algorithms detect the entire image area after the image is generated. In fact, the difference between the target and the clutter on the real image is essentially caused by the different echo characteristics, while the real image contains only the amplitude information, and the phase information is not used for target detection. Therefore, it is feasible to develop a more accurate target detection algorithm through in-depth research on the echo characteristics and imaging mechanism of the two-dimensional SAR. Typical methods in this respect include sub-aperture coherence, coherent spatial filtering, and so on.

12.2 Classic Polarimetric Target Detector

12.2.1 Optimal Polarimetric Detector (OPD)

In ideal case, if it is assumed that the statistical distribution of the target and clutter is known, the statistical signal processing theory shows that the optimal detection problem can be reduced to the form of a likelihood ratio test as follows:

$$\frac{f(\mathbf{X}|\omega_{t+c})}{f(\mathbf{X}|\omega_c)} > T_D \tag{12.1}$$

In the above equation, ω_{t+c} represents the target and clutter hypothesis, ω_c represents the clutter hypothesis, and T_D is the detection threshold. The complex Gaussian target/clutter polarization scattering statistical model is brought into Eq. (12.1). The simplified form can be obtained as follows:

$$\mathbf{X}^H\left(\sum\nolimits_c^{-1} - \sum\nolimits_{c+t}^{-1}\right)\mathbf{X} + \ln\frac{|\Sigma_c|}{|\Sigma_{t+c}|} \lessgtr \ln T_D \tag{12.2}$$

Σ_c and Σ_{t+c} represent the covariance matrix of the target and the clutter, respectively.

Let

$$d_c(\mathbf{X}) = \mathbf{X}^H\left(\sum\nolimits_c^{-1}\right)\mathbf{X} + \ln|\Sigma_c| \tag{12.3}$$

$$d_c(\mathbf{X}) = \mathbf{X}^H\left(\sum\nolimits_{t+c}^{-1}\right)\mathbf{X} + \ln|\Sigma_{t+c}| \tag{12.4}$$

Equation (12.2) can be simplified as

$$d_c(\mathbf{X}) - d_{t+c}(\mathbf{X}) \lessgtr \ln T_D \tag{12.5}$$

The optimal polarimetric detector is optimal in the sense of maximum likelihood, which can obtain the best detection probability for a given false alarm probability. However, the optimal polarimetric detector requires to know statistical information of targets. In practice, due to the lack of such prior information, optimal polarimetric detection is often difficult to achieve. Thus, it is only used as the upper limit of the performance in conventional polarimetric detection algorithms to measure the performance of other polarimetric detection algorithms.

12.2.2 Polarimetric Whitening Filter (PWF)

The polarimetric whitening filter detector was originally explained from the perspective of coherent speckle suppression. In SAR images, the ratio of the standard deviation of the pixel intensity to the average value is usually used as the measure of speckle. It is expressed as

$$\frac{s}{m} = \frac{\sqrt{\mathrm{var}(y)}}{mean(y)} \tag{12.6}$$

The polarimetric whitening filter reduces the speckle of a fully polarimetric SAR image by optimally weighting the three complex elements s_{hh}, s_{hv} and s_{vv} in the polarimetric scattering vector.

For isotropic scattering medium satisfied the reciprocity theorem, the target's backscattering vector can be expressed as

$$\mathbf{X} = \left[s_{hh}\ \sqrt{2}s_{hv}\ s_{vv}\right]^T \tag{12.7}$$

The corresponding target covariance matrix is

$$\mathbf{C} = E\left[\mathbf{X}\mathbf{X}^H\right] = \sigma_{hh}\begin{bmatrix} 1 & 0 & \rho\sqrt{\gamma} \\ 0 & 2\varepsilon & 0 \\ \rho^*\sqrt{\gamma} & 0 & \gamma \end{bmatrix} \tag{12.8}$$

where $E[\bullet]$ is statistical averaging. The definition of the polarization parameter in Eq. (12.8) is as follows:

$$\sigma_{hh} = E\left[|s_{hh}|^2\right], \gamma = \frac{E\left[|s_{vv}|^2\right]}{E\left[|s_{hh}|^2\right]},$$
$$\varepsilon = \frac{E\left[|s_{hv}|^2\right]}{E\left[|s_{hh}|^2\right]}, \rho = \frac{E\left[s_{hh}s_{vv}^*\right]}{\left(E\left[|s_{hh}|^2\right]E\left[|s_{vv}|^2\right]\right)^{1/2}} \tag{12.9}$$

Generally, speckle noise is considered to be multiplicative, and its model is

$$\mathbf{Y} = \sqrt{g}\mathbf{X} \tag{12.10}$$

g is a spatial texture variable, which represents the radiation characteristics of the ground surface. It is related to the natural fluctuations and uniformity of the scene, and obeys the Gamma distribution. The core of polarimetric whitening speckle reduction is to construct a new image through the optimal combination of each polarization channel to get the lowest speckle noise.

Assuming **A** is the weighted matrix, which is a semi-definite Hermitian matrix. The weighted combined intensity image of each polarization channel can be constructed as follows:

$$y = \mathbf{Y}^H\mathbf{A}\mathbf{Y} = g\mathbf{X}^H\mathbf{A}\mathbf{X} \tag{12.11}$$

$$\text{var}\left[\mathbf{X}^H\mathbf{A}\mathbf{X}\right] = tr(\mathbf{C}\cdot\mathbf{A})^2 = \sum_{i=1}^{3}\lambda_i^2 \tag{12.12}$$

We can obtain Eq. (12.13).

$$E\left[\mathbf{X}^H\mathbf{A}\mathbf{X}\right] = tr(\mathbf{C}\cdot\mathbf{A}) = \sum_{i=1}^{3}\lambda_i \tag{12.13}$$

where $\mathbf{C} = \mathbf{X}^H\mathbf{X}$, λ is the eigenvalues of matrix $\mathbf{C}\cdot\mathbf{A}$, and $tr(\cdot)$ is the trace of the matrix.

Therefore, finding the minimum of Eq. (12.6) is equivalent to finding the minimum of below equation.

$$\sum_{i=1}^{3}\lambda_i^2 \Big/ \left(\sum_{i=1}^{3}\lambda_i\right)^2 \tag{12.14}$$

Obviously, only when $\lambda_1 = \lambda_2 = \lambda_3$, the above equation obtains the minimum value of 1/3, then

$$\mathbf{A}_{opt} = \mathbf{C}^{-1} \tag{12.15}$$

It can be obtained that the image intensity after polarimetric whitening filtering is

$$y = |s_{hh}|^2 + \left|\frac{s_{hv}}{\sqrt{\varepsilon}}\right|^2 + \left|\frac{s_{vv} - \rho^* \sqrt{\gamma} s_{hh}}{\sqrt{\gamma(1-|\rho|^2)}}\right| \tag{12.16}$$

It can be seen from the above equation that the polarimetric whitening filtering algorithm is a weighting algorithm. The filtered intensity image is finally obtained by weighting different polarization channels, but the processed data loses the original polarization information. The polarimetric whitening filtering process is also a fusion process of different polarization channels.

The commonly used polarization CFAR detector is actually an adaptive polarimetric whitening filter implementation. The research results show that the performance of the polarimetric whitening filter channel in Constant False Alarm Rate Detection (CFAR) is better than any single polarimetric channel.

The PWF detector needs to know the coherent matrix of clutter, but does not need to know the statistical information of the target. So the applicable range of PWF is wider than OPD. The main disadvantage of PWF is that it requires the inverse of a coherent matrix, that is, a feature decomposition operation is performed for each pixel, which has a large amount of calculation and is difficult to apply in real time.

12.2.3 Polarimetric Match Detector (PMD)

The polarimetric match filter is a linear combination of the received scattering vectors. By selecting an optimal linear weighting vector, that is, the polarimetric matching filter, the ratio of the received power of the signal to the noise is maximized.

$$\max_{\mathbf{w}}(T/C)_{out} = \frac{E[\mathbf{y}_t\mathbf{y}_t^H]}{E[\mathbf{y}_c\mathbf{y}_c^H]} = \frac{\mathbf{w}^H E[\mathbf{x}_t\mathbf{x}_t^H]\mathbf{w}}{\mathbf{w}^H E[\mathbf{x}_c\mathbf{x}_c^H]\mathbf{w}} = \frac{\mathbf{w}^H \Sigma_t \mathbf{w}}{\mathbf{w}^H \Sigma_c \mathbf{w}} \tag{12.17}$$

The above equation is equivalent to

$$\begin{cases} \max\limits_{\mathbf{w}} \mathbf{w}^H \Sigma_t \mathbf{w} \\ s.t. \mathbf{w}^H \Sigma_c \mathbf{w} = 1 \end{cases} \tag{12.18}$$

Apply Lagrangian multiplier method.

$$\begin{cases} \max\limits_{\mathbf{w}} f(\mathbf{w}) \\ f(\mathbf{w}) = \mathbf{w}^H \Sigma_t \mathbf{w} - \lambda(\mathbf{w}^H \Sigma_c \mathbf{w} - 1) \end{cases} \tag{12.19}$$

Let $\frac{\partial f(\mathbf{w})}{\partial \mathbf{w}} = 0$ to obtain

$$\Sigma_t \mathbf{w} = \lambda \Sigma_c \mathbf{w} \tag{12.20}$$

Thus, the optimal weight vector $\mathbf{w}$ corresponds to the generalized eigenvalue problem of Σ_t and Σ_c. Multiply Σ_c^{-1} both sides of Eq. (12.20) left by

$$\Sigma_c^{-1} \Sigma_t \mathbf{w} = \lambda \mathbf{w} \tag{12.21}$$

The optimal weight vector $\mathbf{w}_{opt}$ is the eigenvector corresponding to the maximum eigenvalue $\lambda_{\max}$ of $\Sigma_c^{-1}\Sigma_t$.

12.2.4 Span Detector (SD)

Span detector makes the non-coherent weighted sum of the received scattered power from each received polarization channel, and the expression is as follows:

$$Span = |s_{hh}|^2 + 2|s_{hv}|^2 + |s_{vv}|^2 \lessgtr Th \tag{12.22}$$

The total polarization span detector does not require statistical information of the target and clutter, and has good detection performance in the case of high signal-to-noise ratio

12.2.5 Power Maximization Synthesis (PMS)

PMS is an improved version of SD, and its expression is

$$Y_{PMS} = \frac{Span}{2} + \sqrt{\left(|S_{hh}|^2 - |S_{vv}|^2\right) + 4\left|s_{hh}^* s_{hv} + s_{vv} s_{hv}^*\right|^2} \lessgtr Th \tag{12.23}$$

For single-look data, PMS is equivalent to the maximum power that can be obtained when the scattered electromagnetic waves of a target are received using a matched polarization channel. Multi-look data can be considered as the average of several single-view targets. Therefore, Y_{PMS} can be considered as the sum of the maximum power, which can be received by its respective matched polarization channel from each single-look target that composes the multi-look data. The PMS detector is equivalent to enhancing the received power through a polarization matched filter before detecting the target.

12.3 Polarimetric Target Detection Method Based on Target Characteristic Parameters

In addition to the classic polarimetric target detection method introduced in the previous section, another more commonly used polarimetric target detection method is on the basis of target characteristic parameters, which can usually select different target decomposition methods according to different application purposes to obtain different characteristic parameters [4–6]. For example, the even scattering and spiral scattering components obtained by Krogager decomposition can well characterize the scattering characteristics of artificial targets, and are usually used for the detection of artificial targets.

12.3.1 Polarimetric Target Detection Method Based on Polarization Similarity Parameters

12.3.1.1 Polarization Similarity Parameters Based on Scattering Matrix

The polarization similarity parameter is an effective method to analyze the characteristics of the target. Professor Yang Jian of Tsinghua University proposed the concept of similarity parameters for single-look polarimetric SAR data.

Let the polarimetric Pauli vectors of two target from single-look SAR data be $\mathbf{k}_1$ and $\mathbf{k}_2$. In order to eliminate the influence of the respective orientation angles, firstly, a de-orientation operation is used to rotate both targets to a 0° orientation angle, and the modified Pauli-scattering vectors $\mathbf{k}_1^0$ and $\mathbf{k}_2^0$ can be obtained. The similarity parameters between two single-look targets are

$$r(\mathbf{k}_1, \mathbf{k}_2) = \frac{\left|(\mathbf{k}_1^0)^{\mathrm{H}}\mathbf{k}_2^0\right|}{\left\|\mathbf{k}_1^0\right\|^2\left\|\mathbf{k}_2^0\right\|^2} \tag{12.24}$$

where $\|\bullet\|$ represents 2-norm of vectors. Equation (12.24) can also be transformed into the following form:

$$r(\mathbf{k}_1, \mathbf{k}_2) = \left|\left(\frac{\mathbf{k}_1^0}{\left\|\mathbf{k}_1^0\right\|}\right)^{\mathrm{H}}\left(\frac{\mathbf{k}_2^0}{\left\|\mathbf{k}_2^0\right\|}\right)\right| = \left|\mathbf{v}_1^H\mathbf{v}_2\right| \tag{12.25}$$

The similarity parameter can be considered as the inner product square of the two targets polarization vectors after the respective their first curved and normalized.

The similarity parameter has the following properties:

(1) $r(\mathbf{k}_1, \mathbf{k}_2) = r(\mathbf{k}_2, \mathbf{k}_1)$, which means that the similarity parameter has nothing to do with the parameter order.
(2) Target rotation invariance:

$$r([\mathbf{Q}(2\theta_1)]\mathbf{k}_1, [\mathbf{Q}(2\theta_2)]\mathbf{k}_2) = r(\mathbf{k}_1, \mathbf{k}_2) \tag{12.26}$$

where $\mathbf{Q}$ is the orientation angle rotation matrix, θ_1 and θ_2 are any two rotation angles.
(3) Target power and phase independence:

$$r(a_1\mathbf{k}_1, a_2\mathbf{k}_2) = r(\mathbf{k}_1, \mathbf{k}_2) \tag{12.27}$$

where a_1 and a_2 is any two non-zero complex numbers.
(4) $0 \leq r(\mathbf{k}_1, \mathbf{k}_2) \leq 1$, and only when $\mathbf{k}_2 = a[\mathbf{Q}(2\theta)]\mathbf{k}_1$, $r(\mathbf{k}_1, \mathbf{k}_2) = 1$, where a is any non-zero complex numbers. θ is any rotation angle, where $\mathbf{Q}$ is rotation matrix for orientation angle.
(5) If $\mathbf{k}_1, \mathbf{k}_2, \mathbf{k}_3$ are three non-zero and non-similar Puali vectors with each other, i.e.,

$$r(\mathbf{k}_1, \mathbf{k}_2) = r(\mathbf{k}_1, \mathbf{k}_3) = r(\mathbf{k}_2, \mathbf{k}_3) = 0 \tag{12.28}$$

there is an Eq. (12.29) for any non-zero Pauli vector $\mathbf{k}$.

$$r(\mathbf{k}, \mathbf{k}_1) + r(\mathbf{k}, \mathbf{k}_2) + r(\mathbf{k}, \mathbf{k}_3) = 1 \tag{12.29}$$

A typical example that satisfies (12.29) is three basis vectors under a circularly polarized basis, including plane, left helix, and right helix.

12.3.1.2 Polarization Similarity Parameters Based on Stokes Matrix

For some complex targets, the scattering characteristics are jointly determined by different independent sub-scatterers and their interactions. Therefore, in the radar measurement process, the scattering characteristics of complex targets have certain randomness and depolarization characteristics, which should be descript by statistical methods. In addition, for multi-look polarimetric SAR images, the scattering matrices of different ground targets are usually described by Muller matrix or Stokes matrix.

Let the Stokes matrix of the two targets be $[\mathbf{K}_1]$ and $[\mathbf{K}_2]$, respectively. Then the similarity parameters of the two targets based on the Stokes matrix are defined as

$$r([\mathbf{K}_1], [\mathbf{K}_2]) = \frac{|\langle[\mathbf{K}_1], [\mathbf{K}_2]\rangle|}{\|[\mathbf{K}_1]\|_F \|[\mathbf{K}_2]\|_F} \tag{12.30}$$

where $\langle\bullet, \bullet\rangle$ is inner-product of matrix, and $\|\bullet\|_F$ is F-norm of matrix.

Muller matrix $[\mathbf{M}]$ has relationship with scatter matrix $[\mathbf{S}]$ as follows:

$$[\mathbf{M}] = [\mathbf{R}]\big([\mathbf{S}] \otimes [\mathbf{S}]^{*}\big)[\mathbf{R}]^{-1} \tag{12.31}$$

$\otimes$ is Kronecker product. $[\mathbf{R}]$ and $[\mathbf{R}]^{-1}$ are

$$[\mathbf{R}] = \begin{bmatrix} 1 & 0 & 0 & 1 \\ 1 & 0 & 0 & -1 \\ 0 & 1 & 1 & 0 \\ 0 & j & -j & 0 \end{bmatrix}, [\mathbf{R}]^{-1} = \frac{1}{2}[\mathbf{R}]^{\mathrm{H}} = \frac{1}{2}\begin{bmatrix} 1 & 1 & 0 & 0 \\ 0 & 0 & 1 & -j \\ 0 & 0 & 1 & j \\ 1 & -1 & 0 & 0 \end{bmatrix} \tag{12.32}$$

Muller after matrix rotation transforms to

$$[\mathbf{M}(\phi)] = [\mathbf{R}]\big([\mathbf{S}(\phi)] \otimes [\mathbf{S}(\phi)]^{-1}\big)[\mathbf{R}]^{-1} \tag{12.33}$$

We can derive

$$[\mathbf{M}(\phi)] = [\mathbf{U}(\phi)]^{-1}[\mathbf{M}][\mathbf{U}(\phi)] \tag{12.34}$$

$[\mathbf{U}(\phi)]$ is the unitary matrix, i.e., $[\mathbf{M}(\phi)]$ is the unitary transform for $[\mathbf{M}]$. According to the nature of the unitary transformation, $\|[\mathbf{M}(\phi)]\|_F = \|[\mathbf{M}]\|_F$ and $\langle[\mathbf{M}_1], [\mathbf{M}_2]\rangle = \langle[\mathbf{M}_1(\phi_1)], [\mathbf{M}_2(\phi_2)]\rangle$.

The relationship between the Stokes matrix $[\mathbf{K}]$ and the Muller scattering matrix $[\mathbf{M}]$ is known as

$$[\mathbf{K}] = [\mathbf{U}_4][\mathbf{M}] \tag{12.35}$$

where $[\mathbf{U}_4] = diag\{1, 1, 1, -1\}$.

Thus,

$$\langle[\mathbf{K}_1(\phi_1)], [\mathbf{K}_2(\phi_2)]\rangle = \langle[\mathbf{M}_1(\phi_1)], [\mathbf{M}_2(\phi_2)]\rangle = \langle[\mathbf{M}_1], [\mathbf{M}_2]\rangle = \langle[\mathbf{K}_1], [\mathbf{K}_2]\rangle \tag{12.36}$$

$$\|[\mathbf{K}(\phi)]\|_F = \|[\mathbf{M}(\phi)]\|_F = \|[\mathbf{M}]\|_F = \|[\mathbf{K}]\|_F \tag{12.37}$$

$[\mathbf{K}(\phi)]$ is a unitary transform of $[\mathbf{K}]$. So, $r([\mathbf{K}_1(\phi_1)], [\mathbf{K}_2(\phi_2)]) = r(\mathbf{K}_1, \mathbf{K}_2)$, which means that the similarity parameters defined based on the Stokes matrix have rotation invariance.

According to the relationship between the Stokes matrix and the scattering matrix, it can be deduced that the similarity parameters of the two are mathematically equivalent, but each has a different application.

12.3.1.3 Target Detection Based on Polarization Similarity Parameters

According to the definition of polarization similarity, the average scattering characteristics of radar targets can be described by calculating the similarity parameters between radar targets and typical targets. Thus, targets can be detected based on similarity parameters between targets and typical targets.

Assuming that any target has a Stokes matrix of $[\mathbf{K}]$, the similarity parameters of the target with the known typical targets can be calculated to determine the scattering characteristics of the target, and then determine the type of the target.

(1) Similarity parameter between radar target and plane

The Stokes matrix of the plane is $[\mathbf{K}_s] = diag\{1, 1, 1, -1\}$. Then the similarity parameter of the target and the plane is

$$r_s([\mathbf{K}], [\mathbf{K}_s]) = \frac{|\langle[\mathbf{K}], [\mathbf{K}_s]\rangle|}{\|[\mathbf{K}]\|_F \|[\mathbf{K}_s]\|_F} = \frac{|K_{11} + K_{22} + K_{33} - K_{44}|}{2\|[\mathbf{K}]\|_F} \tag{12.38}$$

(2) Similarity parameter between radar target and dihedral

The Stokes matrix of the dihedral is $[\mathbf{K}_d] = diag\{1, 1, -1, 1\}$. The similarity parameter between the target and the dihedral is

$$r_d([\mathbf{K}], [\mathbf{K}_d]) = \frac{|\langle[\mathbf{K}], [\mathbf{K}_d]\rangle|}{\|[\mathbf{K}]\|_F \|[\mathbf{K}_d]\|_F} = \frac{|K_{11} + K_{22} - K_{33} + K_{44}|}{2\|[\mathbf{K}]\|_F} \tag{12.39}$$

(3) Similarity parameter between radar target and helix

The Stokes matrices of the left and right helix are $[\mathbf{K}_{lh}] = \begin{bmatrix} 1 & 0 & 0 & -1 \\ 0 & 0 & 0 & 0 \\ 0 & 0 & 0 & 0 \\ -1 & 0 & 0 & 1 \end{bmatrix}$ and $[\mathbf{K}_{rh}] = \begin{bmatrix} 1 & 0 & 0 & 1 \\ 0 & 0 & 0 & 0 \\ 0 & 0 & 0 & 0 \\ 1 & 0 & 0 & 1 \end{bmatrix}$, respectively. The similarity parameters between the target and the left or right helix are

$$r_{lh}([\mathbf{K}], [\mathbf{K}_{lh}]) = \frac{|\langle[\mathbf{K}], [\mathbf{K}_{lh}]\rangle|}{\|[\mathbf{K}]\|_F \|[\mathbf{K}_{lh}]\|_F} = \frac{|K_{11} - K_{14} - K_{41} + K_{44}|}{2\|[\mathbf{K}]\|_F} \tag{12.40}$$

$$r_{rh}([\mathbf{K}], [\mathbf{K}_{rh}]) = \frac{|\langle[\mathbf{K}], [\mathbf{K}_{rh}]\rangle|}{\|[\mathbf{K}]\|_F \|[\mathbf{K}_{rh}]\|_F} = \frac{|K_{11} + K_{14} + K_{41} + K_{44}|}{2\|[\mathbf{K}]\|_F} \tag{12.41}$$

(4) Similarity parameter between radar target and line target

The Stokes matrix of the line target is $[\mathbf{K}_w] = \begin{bmatrix} 1\,1\,0\,0 \\ 1\,1\,0\,0 \\ 0\,0\,0\,0 \\ 0\,0\,0\,0 \end{bmatrix}$. The similarity parameter between the target and the line target is

$$r_w([\mathbf{K}], [\mathbf{K}_w]) = \frac{|\langle[\mathbf{K}], [\mathbf{K}_w]\rangle|}{\|[\mathbf{K}]\|_F \|[\mathbf{K}_w]\|_F} = \frac{|K_{11} + K_{12} + K_{21} + K_{22}|}{2\|[\mathbf{K}]\|_F} \tag{12.42}$$

The specific process of target detection based on similarity parameters is as follows:

(1) Preprocess the polarimetric SAR data, and then set the type of target to be detected according to the purpose of target detection. The main scattering types of buildings are dihedral scattering, helix scattering, and line scattering.
(2) Calculate the similarity parameters of the image and the target to be detected. Use the detector to judge the similarity parameter results, and obtain the detection results.

12.3.1.4 Cross-Entropy-Based Polarimetric Target Detection Method

Polarization cross-entropy is another polarization characteristic parameter introduced by the similarity parameter. The method of detecting polarimetric targets based on cross-entropy is introduced below.

Entropy is a measure to describe the state of nature. S. R. Cloude, Jin Yaqiu, etc. introduced the concept of "entropy" into polarization theory to analyze the scattering characteristics of targets.

The polarization entropy is defined as follows:

$$H = -\sum_{i=1}^{3} P_i \log_3 P_i \tag{12.43}$$

$P_i = \frac{\lambda_i}{\sum_{j=1}^{3} \lambda_j}$ and λ_i is the eigenvalues of the polarization coherence matrix.

In order to use polarization cross entropy, we introduce r_1, r_2 and r_3. r_1 is the similarity parameters between the target and the plane. r_2 is the similarity parameters between the target and the dihedral. $r_3 = 1 - r_1 - r_2$. Then, the polarization cross entropy is introduced on this basis.

$$I(H_1; H_2; r) = \sum_{i=1}^{3} r_i(1) \log \frac{r_i(1)}{r_i(2)} \tag{12.44}$$

$r_i(1)$ and $r_i(2)$ correspond to H_1 and H_2, respectively. $I(H_1; H_2; r)$ is the difference between the average scattering mechanisms of the two scatterers.

If power is added to the above definition of cross entropy, there is

$$I(H_1; H_2; r, P) = \sum_{i=1}^{3} [P(1)r_i(1)] \left[\log \frac{P(1)r_i(1)}{P(2)r_i(2)} \right] \tag{12.45}$$

Then, we can obtain

$$I(H_1; H_2; r, P) = P(1) \sum_{i=1}^{3} [r_i(1)] \left[\log \frac{r_i(1)}{r_i(2)} + \log \frac{P(1)}{P(2)} \right] \tag{12.46}$$

The characteristic parameter $I(H_1; H_2; r, P)$ obtained by Eq. (12.46) is not only related to the average scattering mechanism, but also depends on the power difference of the targets. For areas where the power of the target is significantly better than the background, this characteristic parameter has better detection performance.

12.3.2 *Object Detection Method Based on Polarimetric Target Decomposition*

Extraction of polarimetirc targets based on the scattering mechanism classification is the main strategy for target extraction, especially for artificial targets. The main idea can be summarized as follows. Classify the scattering mechanism of polarimetric SAR image pixels based on polarization decomposition methods. Because Even-scattering mechanism have a large proportion in the region of artificial targets, the even-scattering point set of the classification result is used as the artificial target point to realize the extraction of the artificial target.

Polarization decomposition is the most effective method to analyze the target scattering mechanism of polarimetric SAR images, and has been widely used. Except for a few deterministic scattering structures on a few artificial targets, most of the targets in the SAR data are more suitable for analysis by incoherent decomposition. The H/α decomposition classification and the Freeman decomposition classification described in the Polarimetric target Decomposition section of this book are the two most representative methods. Through the polarimetric target decomposition, the target points dominated by even-order scattering are selected as the artificial target.

12.4 Target Detection Method Based on Optimal Polarization Contrast Enhancement

In order to better distinguish different targets in the image, the signal processing method is usually used to enhance the contrast between the targets and improve the performance of the polarimetric target detector. According to the theory of receiving polarimetric electromagnetic waves, changing the polarization state of the transmitting antenna and the receiving antenna can change the target's echo power. Thus, the contrast between the target and the clutter can be enhanced by selecting the appropriate polarization state of the transmitting and receiving antenna. Chap. 10 of this paper has described in detail the classical methods of polarimetric target enhancement, including Optimal Polarization Contrast Enhancement (OPCE) and the Target CO-Null Method.

In 2004, Professor Yang Jian extended the Optimal Polarization Contrast Enhancement (OPCE) method and proposed the Generalized Relative Optimal Polarization Method (GOPCE). The OPCE method uses only the power information of the target, while doesn't use reflection scattering. The polarization parameter fusion information representing the scattering mechanism is added in GOPCE method to further increase the contrast between the target and the background clutter.

GOPCE model is shown below.

$$GP_x = \left[\sum_{i=1}^{3} x_i r_i\right]^2 \times \vec{h}^T[\mathbf{K}_x]\vec{g} \tag{12.47}$$

$\vec{h}^T[\mathbf{K}_x]\vec{g}$ is the traditional OPCE model. $\vec{g} = [1\ g_1\ g_2\ g_3]$ and $\vec{h} = [1\ h_1\ h_2\ h_3]$ are the Stokes vectors of the radar transmitting antenna and the receiving antenna, respectively. $[\mathbf{K}_x]$ are the Kennaugh matrix of the pixels in the image. $\vec{r} = \left[r_1\ r_2\ H\right]^T$ is the polarization parameter vector. $\vec{x} = \left[x_1\ x_2\ x_3\right]^T$ is the fusion coefficient vector that needs to be optimized, where $\|\vec{x}\| = 1$. The three polarization parameters in $\vec{r}$ reflect the different scattering characteristics of the target. r_1 and r_2 represent the similarity parameters between the target and plane scattering or dihedral scattering, respectively. H represents the polarization entropy obtained from the eigenvalue decomposition. The GOPCE problem solves the following optimization problems:

$$\begin{aligned} &\max\left\{\frac{(\vec{x}^T\vec{r}_A)^2}{(\vec{x}^T\vec{r}_B)^2} \times \frac{\vec{h}^T\mathbf{K}_A\vec{g}}{\vec{h}^T\mathbf{K}_B\vec{g}}\right\} \\ &s.t.\ \ g_1^2 + g_2^2 + g_3^2 = 1;\ h_1^2 + h_2^2 + h_3^2 = 1;\ x_1^2 + x_2^2 + x_3^2 = 1 \end{aligned} \tag{12.48}$$

By using GOPCE contrast enhancement technology, road detection and maritime target detection can be applied. For different target detection application purposes, different criterion functions can be constructed to optimize the fusion coefficients in the GOPCE model and improve the accuracy of detection.

12.5 Target Detection Method Based on Polarimetric Interferometric SAR

With the gradual deepening research on polarimetric interferometric SAR in recent years [7], scholars have also carried out research on target detection methods based on P-band polarimetric interferometric SAR data. This section mainly introduces the target detection technology under ground.

In 2001, L. Sagues et al. carried out research using polarimetric interferometric SAR to detect land-covered landmines. Using the unique advantages of polarimetric interferometric SAR, the phase centers of landmines and clutter were distinguished, and their Relative altitude can be estimated to obtain effective mine detection. This section focuses on this target detection method.

Figure 12.1 shows the interference imaging geometry of the target scene under the ground. The key step to detect the target under the ground is to distinguish the scattering mechanism of the target and obtain their relative height. Different scattering centers (targets and ground clutter) in the resolution unit cannot be distinguished by pure interference measurement, which can be achieved using polarization techniques. L. Sagues et al. proposed a method to distinguish the scattering mechanism by using the Polarization Subspace Coherent Optimal Method (PSM).

This method can obtain the optimal scattering mechanism that maximizes the coherence coefficient, and can obtain the height information of the target. If the

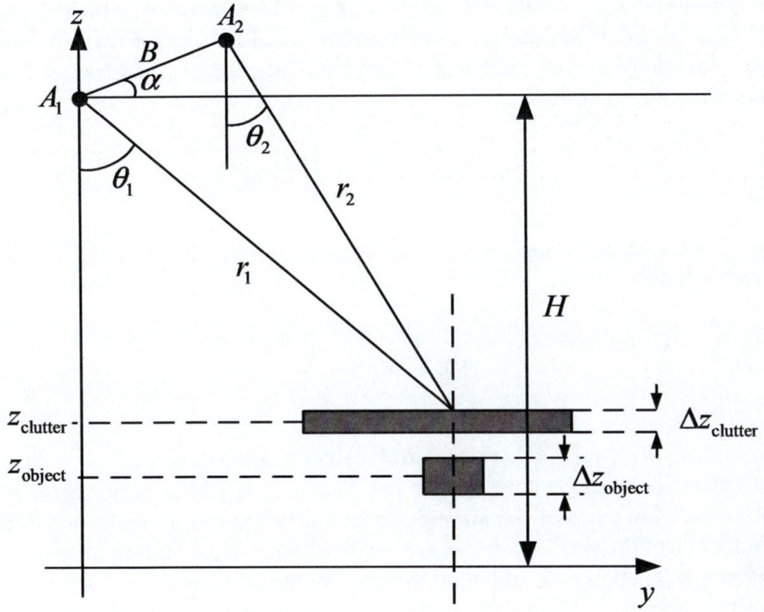

Fig. 12.1 Interferometric geometry of the target scene under ground

co-polarization and cross-polarization coherence functions in different polarization modes are represented graphically, different independent scattering mechanisms in the same resolution unit can be distinguished. For the case where the single scattering mechanism is dominant, both the co-polarization and cross-polarization coherence functions show only an absolute maximum. However, for the case of a multilayer vertical structure as shown in Fig. 12.1, there are different local coherence maximums in different polarization modes. These optimal polarization states correspond to different independent scattering mechanisms, and can obtain local optimal height estimates. For a target covered by the ground surface, two maximum coherence coefficients can be obtained, and each corresponds to a scattering mechanism at different heights, namely the target and the surface clutter. After generating two interferograms corresponding to the optimal polarization state, the height difference between them can be extracted. The location where the target is buried corresponds to the peak position of the height difference, and the area where the height difference is close to zero corresponds to the area where only ground scattering exists.

12.6 Target Detection Based on Eigenvalue Decomposition of Polarization Interference Coherent Matrix

Zhang Lamei and Zou Bin proposed a target detection method based on eigenvalue decomposition of polarization interference coherent matrix [8].

The two sets of fully polarimetric scattering matrices obtained by the Polarimetric Interferometric SAR System are vectorized by Pauli basis and formed into a 6-dimensional complex vector **k**. 6×6 polarization interference coherent matrix from **k** can be obtained as follows:

$$[\mathbf{T}_6] = \langle \mathbf{k}\mathbf{k}^{\mathrm{H}} \rangle = \left\langle \begin{bmatrix} \mathbf{k}_1 \\ \mathbf{k}_1 \end{bmatrix} \begin{bmatrix} \mathbf{k}_1^{\mathrm{H}} & \mathbf{k}_2^{\mathrm{H}} \end{bmatrix} \right\rangle \tag{12.49}$$

The eigenvalue decomposition of $[\mathbf{T}_6]$ can be written in the weighted sum of eigenvector bases.

$$[\mathbf{T}_6] = [\mathbf{E}][\mathbf{\Lambda}][\mathbf{E}]^{\mathrm{H}} = \sum_{i=1}^{6} \lambda_i \mathbf{e}_i \mathbf{e}_i^H \tag{12.50}$$

$[\mathbf{\Lambda}] = diag\{\lambda_1, \lambda_2, \cdots, \lambda_6\}$ is eigenvalue matrix, and $\lambda_1 \geq \lambda_2 \geq \cdots \geq \lambda_6 \geq 0.\mathbf{e}_i$ is the i th feature column vector. $[\mathbf{E}] = \begin{bmatrix} \mathbf{e}_1, & \mathbf{e}_2, & \cdots, & \mathbf{e}_6 \end{bmatrix}$ is the eigenvector matrix.

Through the analysis of the six eigenvalues distribution, it can be seen that the value intervals of the six eigenvalues are very different. Thus, the eigenvalues can be used for target detection. The dihedral structure of the building reflects strong energy, and most of the energy is concentrated in the first eigenvalue. So the building presents a very strong energy in the first eigenvalue compared to the forest and farmland, which expresses more distinctly in the image. Due to the randomness of forest distribution, the forest area has its energy scattered among all the eigenvalues, and the proportion of forests in λ_5 is relatively large. Because each eigenvalue has different sensitivity to different types of features, the target can be effectively detected from the image according to the eigenvalue decomposition.

12.7 Summary

Polarization detectors can be expressed in a unified form as follows:

$$Y > Z \tag{12.51}$$

Y is test statistics, and Z is decision thresholds. Different detectors correspond to different thresholds Y. The thresholds Z are usually determined by the statistical distribution Y of background clutter and the artificially set false alarm probability P_0.

Based on the above-mentioned classic polarimetric target detection algorithms, scholars have proposed a series of methods to detect and identify target information such as roads, bridges, and ships in polarimetric radar images. These methods are thinking about how to effectively use multidimensional information suppression clutter interference, find potential targets, and enhance target detection performance. For different background characteristics, different information dimensions should be selected to improve the accuracy of target detection.

References

1. L.M. Novak, M.B. Sechtin etc., Studies of target detection algorithms that use polarimetric radar data [J]. IEEE Trans. Aerosp. Electron. Syst. **25**(2), 150–165 (1989)
2. L.M. Novak, M.C. Burl, Optimal polarimetric processing for enhanced target detection [J]. IEEE Trans. Geosci. Remote Sens. **29**(1), 234–243 (1993)
3. C. Burl, On the performance of polarimetric target decectíon algorithms. IEEE Trans. Aerosp. Electron. Syst. **29**(1), 234–244 (1993)
4. X. Shunping, W. Xuesong, D. Haida et al., *Polarization Radar Imaging Processing and Application [M]* (Science Press, Beijing, 2013)
5. J. Yang, Y.N. Peng, S.M. Lin, Similarity between two scattering matrices [J]. Electron. Lett. **37**(3), 193–194 (2001)
6. W. An, The polarimetric decomposition and scattering characteristic extraction of polarimetric SAR [D], Ph.D. thesis, Tsinghua University, 2010
7. S.R. Cloude, K.P. Papathanassiou, Polarimetric SAR interferometry [J]. IEEE Trans. Geosci. Remote Sens. **36**(5), 1551–1565 (1998)
8. L. Zhang, Research on feature extraction and detection of Man-made target using polarimetric SAR image [D], Ph.D. Dissertation, Harbin Institute of Technology, 2010

Chapter 13
Classification of Polarimetric SAR Images

Polarized synthetic aperture image classification is an important application of image analysis and processing. Polarimetric SAR image classification can be divided into single-band, dual band, multi-band and interferometric fully polarized synthetic aperture radar image classification. From the processing method, the polarization SAR image classification method can be divided into two categories: supervised classification based on prior knowledge and unsupervised classification without prior knowledge. Among them, the supervised classification method includes supervised classification of multi-polarized synthetic aperture radar image based on statistical characteristics, supervised classification of polarization synthetic aperture radar image based on artificial neural network and supervised classification of Maximum Likelihood (ML) estimation based on Wishart distribution; unsupervised classification methods include unsupervised classification algorithms based on polarization scattering mechanism, unsupervised classification methods based on Cloude–Pottier decomposition for target scattering entropy H/scattering angle α, and H/γ unsupervised classification methods based on Cloude–Pottier decomposition information enhancement, and classification method Based on Cloude–Pottier decomposition and Wishart's unsupervised.

This chapter introduces the supervised method, unsupervised method, and dual band full-polarization SAR image classification method. Polarization interference SAR image classification method and polarization interference of multi-band SAR images unsupervised Wishart ML classification method.

R. Yang et al., *Polarimetric Microwave Imaging*,
https://doi.org/10.1007/978-981-15-8897-6_13

13.1 Supervised Classification of Polarimetric SAR Images

13.1.1 Classification of Polarimetric SAR Images Based on Artificial Neural Networks [1]

Artificial neural network methods are more and more used in the field of remote sensing image processing. Studies have shown that [2–4], under the premise of no prior knowledge of data, neural network methods have more obvious advantages than statistical classification.

There are many neural network classification methods for polarimetric SAR images, such as algorithms based on Backward Propagation (BP) neural networks [5], algorithms based on self-organizing competitive neural networks [2, 6], classification algorithms based on dynamic learning neural networks [3, 4], etc. All of these algorithms select specific polarization parameters to train the samples and make full use of the rich information volume specific for the polarization synthetic aperture radar data to establish classification criteria, which has better classification accuracy [2–9].

Due to the limitation of spatial resolution, there may be many features belong to different categories in the same resolution unit in the SAR image, and they can only be represented as one pixel in the SAR image. Thus, the SAR image pixels themselves have a certain ambiguity. The traditional neural network classification method uses pixel points as a classification reference, which sometimes cannot meet the requirements of high-precision classification. Therefore, people have proposed a classification algorithm based on fuzzy neural network [10–12]. The method of combining fuzzy logic with neural network makes the classification of microwave remote sensing images more efficient.

13.1.2 Supervised Classification Based on Wishart Maximum Likelihood Estimation (ML)

A simple statistical supervised classification method is to preprocess the speckle reduction of multi-polarized SAR data, and then use Bayesian methods (such as maximum likelihood estimation MLE method) to classify images in a specific polarization state [13, 14]. This classification method can classify two types of targets as well, and its classification accuracy can be higher than 90% [15, 16], but the classification effect for multiple types of targets is not good [17–19]. The iterative MLE method proposed by H. Derin et al. has better classification accuracy. However, this method is only suitable for classification of three types of targets. In general, the statistical classification method assumes that each type of SAR image can be uniquely classified by using local statistical information unique to each type of target. Therefore, each image must be locally trained and tested, so it is not suitable for a wide range of applications. A classification method based on a priori knowledge that incorporates the interaction between SAR and ground objects is likely to construct a classification algorithm that is more widely used and independent of time and place. Once the training of this algorithm is completed, a better classification result can be obtained for each image [1].

The supervised classification based on Wishart's Maximum Likelihood Estimation (ML) is a typical representative of the multi-polarization SAR image supervised classification method based on statistical properties. The method is described in detail below.

The target scattering vector of the uniform region approximates the 0-means complex Gaussian distribution $N_C(0, C)$, where C is the scattering covariance matrix. The scattering covariance matrix C obeys the complex Wishart distribution $W_C(n, C)$ with a degree of freedom n. The scattering correlation matrix T also obeys a complex Wishart distribution with a degree of freedom n, where n is a multi-look number. The n-look correlation matrix is expressed as

$$\langle T \rangle = \frac{1}{n} \sum\nolimits_{i=1}^{n} z_i * z_i^H \tag{13.1}$$

where z_i is the Pauli-based scattering vector. The uniform surface scattering coherence matrix $\langle T \rangle$ also obeys the complex Wishart distribution, and its probability density function is [20]

$$p_T^{(n)}(\langle T \rangle) = \frac{n^{qn} |\langle T \rangle|^{n-q} \exp\left[-n \cdot Tr\left(V^{-1}\langle T \rangle\right)\right]}{K(n,q)|V|^n} \tag{13.2}$$

$$K(n,q) = \pi^{\frac{1}{2}q(q-1)} \Gamma(n) \ldots \Gamma(n-q+1), \Gamma(n) = (n-1)! \tag{13.3}$$

where n is the visual number; for the reciprocal medium, in the case of backscattering, there is $q = 3$; K is the normalization factor, which is a constant; $V = E[\langle T \rangle]$, $E[]$ is the average.

According to the definition of the maximum likelihood (Maximum Likelihood, ML) classifier, the decision criteria are as follows:

If for all a, there is b, Then the target belongs to class c

$$\begin{aligned} &\text{If for all } j \neq m,\ P(\omega_m|\langle T \rangle) \geq P\left(\omega_j\middle|\langle T \rangle\right) \text{is true.} \\ &\text{Then the target belongs to class} \end{aligned} \tag{13.4}$$

where ω_m is a collection of pixels belonging to the class m.

Apply the Bayesian formula to the above decision criteria:

$$P(\omega_m|\langle T \rangle) = \frac{p(\langle T \rangle|\omega_m)P(\omega_m)}{p(\langle T \rangle)} \tag{13.5}$$

The following decision criteria can be obtained:

$$\begin{array}{l}\text{If for all } j \neq m \text{, the following is true} \\ p(\langle T\rangle|\omega_m)P(\omega_m) > p\big(\langle T\rangle|\omega_j\big)P\big(\omega_j\big), \\ \text{Then the target belongs to class}\end{array} \quad (13.6)$$

where $P(\omega_m)$ is the prior probability of class ω_m.

V_m is defined as the average scattering correlation matrix of class $m(\omega_m)$. The c training samples of class ω_m can be used to estimate the average scattering correlation matrix V_m of class ω_m:

$$V_m = E[\langle T\rangle|\langle T\rangle \in \omega_m] \quad (13.7)$$

Then the distance factor of any one of the target $\langle T\rangle$ to the class $m(\omega_m)$ is [20]:

$$d_2(\langle T\rangle, \omega_m) = n\big[\ln|V_m| + Tr\big(V_m^{-1}\langle T\rangle\big)\big] - \ln P(\omega_m) \quad (13.8)$$

In general, it is difficult to get the prior probability of class ω_m. It is therefore generally assumed that the prior probabilities $P(\omega_m)$ of all classes are equal. At this point, the distance factor can be equivalently simplified to

$$d_2(\langle T\rangle, \omega_m) = \ln|V_m| + Tr\big(V_m^{-1}\langle T\rangle\big) \quad (13.9)$$

It can be seen that the distance factor is independent of the look number. Thus, the decision criteria for the Bayesian Maximum Likelihood (ML) classifier expressed by the distance factor are obtained as follows:

$$\begin{array}{l}\text{If for all } j \neq m\text{, the following is true} \\ d_2(\langle T\rangle, \omega_m) \leq d_2\big(\langle T\rangle, \omega_j\big), \\ \text{Then the target belongs to class}\end{array} \quad (13.10)$$

This is the Maximum Likelihood estimation (ML) supervised classification method based on the correlation matrix Wishart distribution. It computes d_2 using the average scattering correlation matrix V_m of the class m.

That is to say, this method utilizes the statistical prior knowledge of the target and belongs to the statistical classifier. It has the advantages of clear physical meaning and simple design. Moreover, since the method processes the pixels in the image one by one, the detailed information of the scene can be better maintained.

The coherence matrix has a linear relationship with the covariance matrix, so the classification result using the correlation matrix is the same as the classification result using the covariance matrix.

Supervised classification has high classification accuracy, but it requires enough training samples. It requires training samples to be of high quality and to be representative.

Supervised classification has high classification accuracy, but it requires enough training samples. It requires good training sample quality, and the training sample must be representative. In order to improve the classification accuracy, it is better to select multiple sample groups for training in different time and space, which greatly increases the actual workload of the supervised classification method. In contrast, unsupervised classification methods are fast and simple, with a certain degree of precision. And it helps to better understand the scattering characteristics of the target [20–27].

13.2 Unsupervised Classification of Polarimetric SAR Images

The unsupervised classification of polarimetric SAR images is based on extensive research on target scattering matrix decomposition and target scattering component analysis. The scattering characteristic information of the target can be obtained by decomposing the target scattering matrix and the scattering correlation matrix.

13.2.1 Image Classification Based on Polarization Scattering Mechanism

In 2001, Van Zyl et al. proposed an unsupervised classification algorithm based on polarization scattering mechanism [21]. Later, Moghaddam and Freeman et al. improved and extended the method [22]. Van Zyl pointed out that the polarization SAR data can be used to identify the main scattering mechanism in the average Mueller matrix. And a set of top-down judgment criteria can be used to distinguish some basic feature types.

According to the scattering matrix of a typical target, the transmitted wave is scattered once in an infinite plane, twice scattered on the dihedral reflector, and three times scattered on the trihedral reflector, so that the simple target can be divided into odd times scattering and even scattering. In the scattering matrix of the infinitely large plane scattered echoes, the phases of h_{hh} and h_{vv} are the same, and the phase difference of h_{hh} and h_{vv} changes by 180° after each scattering. Therefore, for the target of odd scattering, the phase difference between h_{hh} and h_{vv} is 0°; for the target of even scattering, the phase difference between h_{hh} and h_{vv} is 180°. Many targets in nature are composed of many independent scatterers. There may be odd and even scattering at the same time. The h_{hh} and h_{vv} phase changes of such targets are irregular, called diffuse scattering. Therefore, the target can be divided into three scattering mechanisms according to the relationship between the phase difference of h_{hh} and h_{vv} and the number of scattering: odd scattering, even scattering, and diffuse scattering.

According to the principle of reciprocity, the Mueller matrix of a symmetric target can be expressed as follows:

$$[L] = \begin{bmatrix} A + B_0 & B & 0 & 0 \\ B & A - B_0 & 0 & 0 \\ 0 & 0 & C + B_0 & D \\ 0 & 0 & -D & C - B_0 \end{bmatrix} \tag{13.11}$$

where

$$\begin{aligned} A &= \frac{1}{2}\left\langle |h_{hh}|^2 + |h_{vv}|^2 \right\rangle \\ B &= \frac{1}{2}\left\langle |h_{hh}|^2 - |h_{vv}|^2 \right\rangle \\ B_0 &= \left\langle |h_{hv}|^2 \right\rangle \\ C &= \left\langle R\left(h_{hh}^* h_{vv}\right)\right\rangle \\ D &= \left\langle I\left(h_{hh}^* h_{vv}\right)\right\rangle \end{aligned}$$

Suppose there are two polarization modes of incident waves, and their Stokes vectors are

$$S_l = \begin{bmatrix} S_0 \\ S_1 \\ S_2 \\ S_3 \end{bmatrix} = S_0 \begin{bmatrix} 1 \\ \cos 2\chi \\ \sin 2\chi \\ 0 \end{bmatrix} \tag{13.12}$$

$$S_c = \begin{bmatrix} S_0 \\ S_1 \\ S_2 \\ S_3 \end{bmatrix} = S_0 \begin{bmatrix} 1 \\ 0 \\ 0 \\ \sin 2\psi \end{bmatrix} \tag{13.13}$$

where S_l is the Stokes vector of the linearly polarized wave, S_c is the Stokes vector of the circularly polarized wave, ψ is the polarization ellipse angle, and χ is the polarization.

According to the relationship between the Stokes vector of the incident wave and the Mueller matrix of the target, the Stokes vector of the scattered wave can be calculated:

$$S_{rl} = \begin{bmatrix} A + B_0 + B\cos(2\chi) \\ B + A\cos(2\chi) - B_0\cos(2\chi) \\ (C + B_0)\sin(2\chi) \\ -D\sin(2\chi) \end{bmatrix} \tag{13.14}$$

$$S_{rc} = \begin{bmatrix} A + B_0 \\ B \\ D\sin(2\psi) \\ (C - B_0)\sin(2\psi) \end{bmatrix} \tag{13.15}$$

When the target scattering mechanism is odd scattering, the phase difference between h_{hh} and h_{vv} is 0°, and C in the target Mueller matrix is positive. At the same time, the HV echo power value of the common target is much smaller than the HH and VV echo power values, thus $C+B_0 > 0$, $C - B_0 > 0$. When the direction angle of the transmitted linearly polarized wave increases, the direction of the scattered wave also increases, and the direction of rotation of the scattered wave is opposite to the direction of rotation of the emitted circularly polarized wave.

When the scattering of the target is even scattering, the phase difference between h_{hh} and h_{vv} is 180°, and C in the target Mueller matrix is negative. Similarly, the echo power value of the ordinary target HV is much smaller than the echo power value of HH and VV, thus $C+B_0 < 0$, $C - B_0 < 0$. Therefore, when the direction angle of the transmitted linearly polarized wave increases, the direction angle of the scattered wave decreases, and the direction of rotation of the scattered wave is the same as the direction of rotation of the emitted circularly polarized wave.

The classification criteria based on the polarization scattering characteristics of the above scattering targets are shown in Table 13.1.

Next, polarization SAR image classification is performed from the polarization scattering mechanism. The experimental data is the fully polarized SAR data in the San Francisco area. The processing steps are as follows:

Table 13.1 Classification criteria for polarization scattering mechanisms

Class type	The relationship between the increase of the direction angle of the transmitted line-polarized wave and the direction angle of the scattered wave		The relationship between the circle-polarized wave and the direction of rotation of the scattered wave	
	$0° < \psi_t < 90°$	$90° < \psi_t < 180°$	Clockwise	Counter clockwise
Odd scattering	Increase	Increase	Counter clockwise	Clockwise
Even scattering	Decrease	Decrease	Clockwise	Counter clockwise
Bragg scattering	Increase	Increase	Clockwise	Counter clockwise

(1) transmitting a linearly polarized electromagnetic wave, and then analyzing the relationship of the direction angle of the polarization ellipse between the target scattering wave and the emitted linearly polarized electromagnetic wave;
(2) transmitting a circularly polarized electromagnetic wave, and then analyzing the relationship between the rotational direction of the polarization of the target scattered wave and the rotational direction of the transmitted circularly polarized wave.

The above-mentioned transmitted electromagnetic waves are realized in the processing by Stokes vector and target Stokes matrix operations of different polarization modes.

The classification criteria are processed according to Table 13.1, with blue, green, and red representing odd-order scattering, even-order scattering, and bragge scattering, respectively. The processing results are shown in Fig. 13.1. As can be seen from the figure, the terrain of the odd scattering is mainly the ocean. The terrain of even scattering is mainly the streets of the city. The scattering mechanism at the concentration of vegetation in the city is bragge scattering. The results of these classifications are reasonable. The scattering characteristics of the ocean are the same as the surface scattering model with slight roughness, and the scattering mechanism of this model is odd; since the electromagnetic waves are scattered twice from the streets of the city, the scattering characteristics are similar to dihedral scattering. Forest vegetation is a multiple scattering mechanism that exhibits diffuse scattering characteristics. Therefore, the results of classifying targets using the polarization scattering mechanism can be used to analyze and determine local terrain to a large extent.

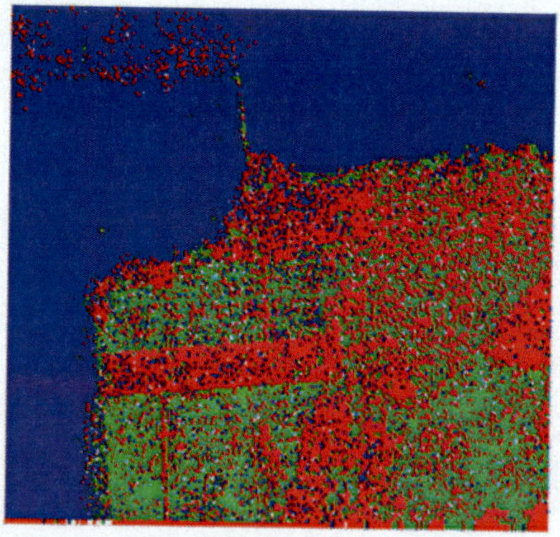

Fig. 13.1 Classification of polarization scattering in the San Francisco area. Blue–odd scattering, green–even scattering, red–diffuse scattering

13.2.2 H/α Unsupervised Classification Based on Cloude–Pottier Decomposition

In 1997, British scholar Cloude [28] proposed a non-supervised classification method for target scattering entropy H/scattering angle α(H/α) based on Cloude–Pottier decomposition. This provides a new way to fully utilize and extract polarized radar remote sensing information without prior knowledge.

The H/α unsupervised classification method combines the nine regions of the H-α plane division with the H-α curve envelope to classify the ground object targets into eight representative scattering categories, namely, Surface scattering with low scattering entropy, dipole scattering with low scattering entropy, multiple scattering with low scattering entropy, surface scattering with medium scattering entropy, vegetation scattering with medium scattering entropy, multiple scattering with medium scattering entropy, vegetation scattering with high scattering entropy, and multiple scattering with high scattering entropy [1, 29].

Taking the L-band fully polarized SAR image data of the Tianshan Mountains as an example, Fig. 13.2 gives the H-α classification results for this region. Figure 13.2a, b are the grayscale and pseudo-color images of the data H-α classification results for the region, respectively. The H/α image classification method understands the intrinsic physical properties of natural media by estimating the polarization scattering mechanism of the target. It is a better unsupervised classification method. The classification results are consistent with the actual scene and can basically reflect the scattering characteristics of the actual objects. However, since the boundaries between the categories are linearly divided, the judgment of the category is too arbitrary, and the details of the image are lost a lot.

(a) The grayscale of H-α Classification (b) The Pseudo-color image of H-α Classification

Fig. 13.2 H-α classification results of Tianshan Mountains of L-band data

13.2.3 *H/γ Unsupervised Classification Method Based on Information Enhancement of Cloude–Pottier Decomposition*

The H/γ unsupervised classification method based on information enhancement of Cloude–Pottier decomposition divides H-γ planes into nine regions for image classification. The scattering characteristics of the features represented by the nine regions of the H-γ plane can be understood as follows: Even scattering or dipole scattering with low scattering entropy, surface scattering with low scattering entropy, volume scattering with low scattering entropy, even or dipole scattering of medium scattering entropy, surface scattering with medium scattering entropy, Volume scattering with medium scattering entropy, even or dipole scattering with high scattering entropy, surface scattering with high scattering entropy, and volume scattering with high scattering entropy.

Taking the L-band fully polarized SAR image data of the Tianshan Mountains as an example, the results of the H-γ classification in this area are given. Figure 13.3a, b are the grayscale and pseudo-color image of the H-γ classification for the region, respectively.

It can be seen from Figs. 13.2 and 13.3 that the classification results of the H/α and H/γ methods are highly consistent. At the same time, there are differences between them. This can be more clearly identified from the pseudo-color map of the classification results. The most obvious is that the H/γ classification result separates the scattering type represented by red from the scattering type represented by blue in the H/α classification result. As can be seen from the Span image of the area, the scattering characteristics of this area are indeed different from the blue area. This also shows that the H/γ classification method using γ angle makes full use of the ground polarization information, and the classification results are significantly improved.

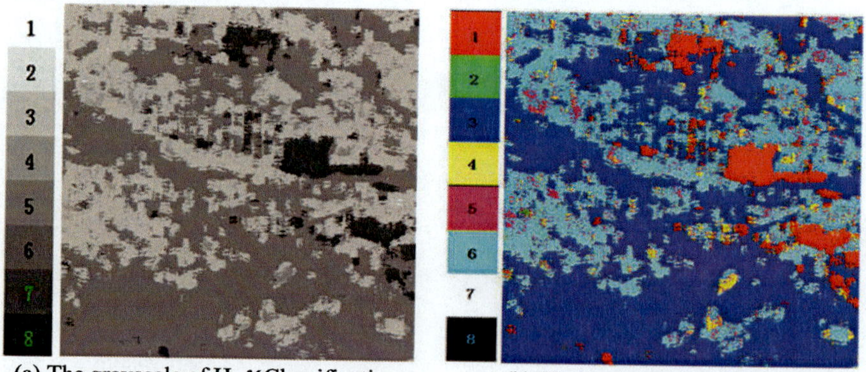

(a) The grayscale of H-γ Classification

(b) The Pseudo-color image of H-γ Classification

Fig. 13.3 H-γ classification results of Tianshan Mountains of L-band data

13.2.4 Unsupervised Classification of Polarimetric SAR Based on Cloude–Pottier Decomposition and Wishart

Supervised classification methods have higher classification accuracy, but it requires some prior knowledge. The H/α classification method of Cloude–Pottier decomposition can obtain classification results with certain classification accuracy without prior knowledge. And it can explain the scattering mechanism of the classification type; However, the H-α classification criteria are empirical. It cannot adapt to the complex changes of the earth's surface. In 1998, Lee [30] improved the unsupervised classification of Cloude–Pottier. It uses the classification result obtained by H/α decomposition as the basis for the initial training sample selection of the ML method and uses the ML classification method of Wishart to reclassify the pixels. This improves the accuracy of the classification.

The average correlation matrix for each class can be calculated as follows V_m:

$$V_m = \frac{1}{n_m}\sum_{i=1}^{n_m} \langle T\rangle_i, \forall \langle T\rangle_i \in \omega_m \tag{13.16}$$

m is the class obtained by H/α classification method, $m = 1, 2, 3 \ldots 8$, n_m is the total number of m th pixels. $\langle T\rangle_i$ is the set of all pixels belonging to the class in the H/α classification method. The distance factor of each pixel to each class is calculated according to the formula (13.9), and the pixel is reclassified according to the decision criterion of (13.10).

The process of the full-polarization SAR based on the H/α decomposition of the Wishart ML classification algorithm is shown in Fig. 13.4. The steps are as follows:

(1) Performed speckle filtering of fully polarized SAR data correlation matrices;
(2) Perform additive noise filtering and H/α decomposition classification;
(3) Select the initial training sample of the ML classification method according to formula (13.16);
(4) Reclassified by the ML method the pixels according to (13.10);
(5) Terminate if the termination condition is satisfied; otherwise, return to Step (4), and continue the iteration.

The termination condition may select that the number of transitions of pixels between categories between two iterations is less than a certain value, or the number of iterations reaches a pre-specified value.

Taking the L-band fully polarized SAR image data of the Tianshan Mountains as an example, the H/α classification result is used as the first iteration input of the ML classification, and the specified iteration number (eight iterations is selected) is used as the termination condition. The resulting classification results are shown in Fig. 13.5. Figure 13.5a, b is pseudo-color representations of the results of the second iteration and the eight iterations, respectively.

Compared with the H/α classification results in Fig. 13.2, the texture information of the features after ML classification (Fig. 13.5) has been greatly improved, typically

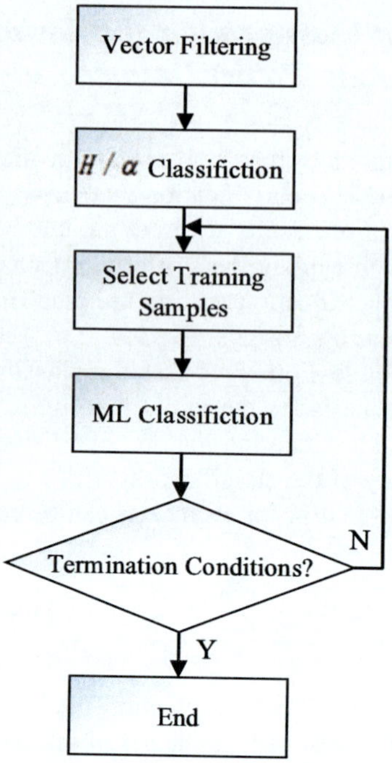

Fig. 13.4 Flowchart of Wishart ML classification based on H/α decomposition

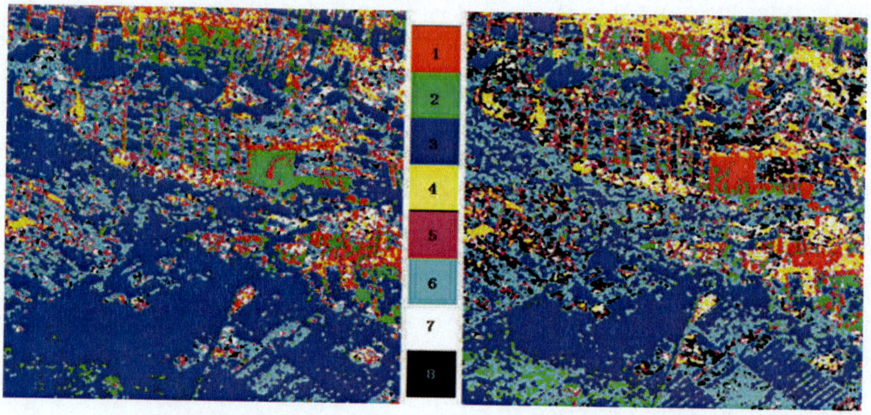

(a) Results after two iterations (b) Results after eight iterations

Fig. 13.5 Pseudo-color images of L-band unsupervised classification results the strip-shaped region in the middle and upper parts. The texture pattern of the area is hardly seen in the H/α class, and the texture details are clear after ML classification. It can be seen that the ML supervised classification improves the classification accuracy of the H/α class. It can be further seen from Fig. 13.5 that as the number of iterations increases, some pixels shift between different classes. This is because each iteration recalculates the average correlation matrix of the class based on the last classification result. The average correlation matrix of the class changes after each iteration, and the distance factor of the same pixel to each type changes every iteration. Each time the ML classification further adjusts the previous classification results to improve the precise and accuracy of the classification.

As can be seen from the triangle area in the lower right corner of the image, the details of the area from (a) to (b) in the graph are increasing, which reflects that the ML classification effectively adjusts the class to which the pixel belongs.

13.3 H/α/Wishart Unsupervised Classification Method for Dual Band Full-Polarization SAR Images [29, 31]

The backscattering properties of natural media vary with the physical properties of the observed frequency, such as the structure, the geometry of the relative radar wavelength, and the like. The electromagnetic waves received by the radar are related not only to the polarization of the radar electromagnetic waves, but also to the frequency of the electromagnetic waves (i.e., the wavelength). The L- and C-band electromagnetic waves have little difference to the scattering mechanism of the bare surface, but the P- and C-band electromagnetic waves have great difference to the echo signals of the forest. Figure 13.6a, b are the L-band and C-band HH polarization grayscale maps of the same area (somewhere near the foothills of the Tianshan Mountains). The L-band electromagnetic wave has a long wavelength and a strong penetrating power and can penetrate the branches and leaves of the vegetation to reach the branches. The echo signals received by the L-band radar are scattered waves formed by multiple scattering of branches, leaves, and rods of vegetation; the C-band electromagnetic waves have shorter wavelengths and weaker penetrating power. The echo signals received by the C-band radar are mainly scattered waves formed by vegetation branches and the like. As can be seen from Fig. 13.6, there are significant differences in SAR images in different bands in the same region.

13.3.1 Classification Method for Dual Band Full-Polarization SAR [12, 14, 32]

Using the polarization information of two bands for image classification will greatly improve the classification accuracy. Recombining a 6×6 correlation matrix with

(a) L band

(b) C band

Fig. 13.6 L-band and C-band HH polarization amplitude images of the Tianshan Mountains

dual band complex data, including the correlation matrix of the two images, and their cross-correlation terms [8, 9]. The matrix still obeys the Wishart distribution and meets the criteria for maximum likelihood estimation criteria. The basic idea of dual band full-polarization SAR classification is firstly, H/α classification and ML classification processing are used for the two images, respectively; then the classification results of the two images are combined and can be divided up to 64 categories; finally, the type reduction polymerization technique is used to reduce the number of classification types and improve the degree of characteristic aggregation.

The scattering vector of a dual band full-polarization SAR is defined as

$$W = \begin{bmatrix} z_{b1} \\ z_{b2} \end{bmatrix} \tag{13.17}$$

z_{b1}, z_{b2} are the Pauli-scattering vectors of the respective polarization SAR data of the bands b_1 and b_2, respectively. Assuming that z_{b1}, z_{b2} are both q-dimensional vectors, then W is a vector with a dimension of $\mathrm{p} = 2\mathrm{q}$. Thus, the p-p-dimensional n-looks correlation matrix A is

$$A = \frac{1}{n}\sum\nolimits_{j=1}^{n} W_{\mathrm{v}} W_{\mathrm{j}}^{\mathrm{H}} = \begin{bmatrix} A_{11} & A_{12} \\ A_{21} & A_{22} \end{bmatrix} \tag{13.18}$$

where, $A_{rs} = \frac{1}{n}\sum\nolimits_{j=1}^{n} \mathrm{z_{rj}} \mathrm{z_{sj}^{H}}$.

Target scattering vector W obeys complex normal distribution N_c (0, Σ_W) [10]. Here, Σ_W is a p $\times$ p-dimensional covariance matrix. The p $\times$ p-dimensional correlation matrix A obeys the complex Wishart distribution with degree of freedom n:

$$p(A) = \frac{n^{pn}|A|^{n-p}\exp\left(-n\mathrm{tr}\left(\Sigma_W^{-1}A\right)\right)}{\mathrm{K}(n,p)|\Sigma_W|^n} \tag{13.19}$$

where $\mathrm{K}(n, p)$ is as defined in (13.3).

The matrices $\boldsymbol{A}_{11}$ and $\boldsymbol{A}_{22}$ are q × q-dimensional correlation matrices of n-looks of the b_1 and b_2 band SAR images, respectively. $\boldsymbol{A}_{12}(= \boldsymbol{A}_{21}^H)$ is a q × q-dimensional cross-correlation matrix containing z_{b1} and z_{b2} polarization cross-correlation information. $\boldsymbol{A}$ is a p × p-dimensional n-looks correlation matrix containing joint information of two-band full-polarization SAR images of b_1 and b_2. A dual-frequency data distance factor similar to that of Eq. (13.9) is obtained using the method of Sect. 13.1.2:

$$d_2(\langle \boldsymbol{A}\rangle, \boldsymbol{T}_m) = \ln|\boldsymbol{T}_m| + \mathrm{tr}\left(\boldsymbol{T}_m^{-1}\langle \boldsymbol{A}\rangle\right) \tag{13.20}$$

T_m is the p × p-dimensional average scattering correlation matrix of the m-th class (or the class-like feature matrix). Estimating the average scattering correlation matrix of the mth class by training samples of class ω_m:

$$T_m = E[\langle A\rangle | \langle A\rangle \in \omega_m] \tag{13.21}$$

The classification decision criterion for the maximum likelihood (ML) classifier of the dual band fully polarized SAR is obtained in the same way as in Sect. 13.1.2:

$$\begin{aligned}&\text{If for all } j \neq m\text{, there is } d_2(\langle \boldsymbol{A}\rangle, \boldsymbol{T}_m) \le d_2\left(\langle \boldsymbol{A}\rangle, \boldsymbol{T}_j\right) \text{ established}\\ &\text{Then, the pixel belongs to the } m\text{th class}\end{aligned} \tag{13.22}$$

The advantage of using the p × p-dimensional correlation matrix is that, according to the formula (13.19), by using the distance factor (13.20), the two-band data can be simultaneously applied to the classification, and no assumption is made about their independent relationship.

13.3.2 Methods for Classifying Aggregates

Combine the classification results of the two-band images separately and divide them up to 64 categories. In fact, too many types of images are not consistent with the actual situation, and the classification results are not conducive to the understanding and interpretation of the images. In order to be able to interpret the characteristics of various classes, and to visualize the geographical location of different kinds of features, the number of classification categories should not be too large. Therefore, it is necessary to perform a reduction aggregation of the number of classification results. Lee et al. proposed a class aggregation technique based on the degree of separability between species [32].

The basic principle of type compact aggregation is that for all n classification categories, the two categories that are less separable are combined into one category. If the class is compact internally and the average distance between elements is high, then the two classes are separable. Therefore, the degree of separability between the two types of X_i and X_j classes is defined as the ratio of their intra-class dispersion to the distance between classes. For a given class X_i, the intra-class dispersion S_i is defined as the average distance of its elements to the p × p dimension class center (or average) feature matrix $\boldsymbol{T}_i$:

$$S_i = \frac{1}{N_i} \sum {}_{N_i} d_2[(\langle \boldsymbol{A} \rangle \in X_i), X_i] \tag{13.23}$$

Substituting Eq. (13.20) into Eq. (13.23) is

$$S_i = \ln|\boldsymbol{T}_i| + \mathrm{tr}[\boldsymbol{I}] = \ln|\boldsymbol{T}_i| + u \tag{13.24}$$

tr represents the trace of the matrix, and the trace of the p × p unit matrix is a constant p. u is a constant term, and the equivalent expression for the internal dispersion of the kind is

$$S_i = \ln|\boldsymbol{T}_i| \tag{13.25}$$

S_i represents the internal tightness of the i-th class. Therefore, the sum of all S_i of the mth class implies the degree of compactness within the class. The distance B_{ij} between the X_i and X_j classes is defined as the average distance from each class of elements to another type of central feature matrix. The expression is shown in (13.26).

$$B_{ij} = \left\{(1/N_i) \sum {}_{N_i} d_2\left[(\langle \boldsymbol{A} \rangle \in X_i), X_j\right] + (1/N_j) \sum {}_{N_j} d_2\left[(\langle \boldsymbol{A} \rangle \in X_j), X_i\right]\right\} \Big/ 2 \tag{13.26}$$

According to the definition of the distance factor, the average distance between the two classes is reduced to:

$$B_{ij} = \frac{\left[S_i + S_j + \mathrm{tr}\left(\boldsymbol{T}_i^{-1}\boldsymbol{T}_j + \boldsymbol{T}_j^{-1}\boldsymbol{T}_i\right)\right]}{2} \tag{13.27}$$

A larger B_{ij} indicates a better separability between the two classes.

If the intra-class dispersions (S_i,S_j) of the two classes (X_i, X_j) are high and the distance factor (B_{ij}) between the classes is small, then the two classes should be combined into one class. The degree of separability of the species of natural features, $\mathrm{Sp}(X_i, X_j)$, is defined as

$$\mathrm{Sp}(X_i, X_j) = \frac{(S_i + S_j)}{B_{ij}} \tag{13.28}$$

The smaller $\mathrm{Sp}(X_i, X_j)$, the better the separability of the two types of features. Thus the two classes with the largest $\mathrm{Sp}(X_i, X_j)$ can be combined into one class.

Thus, the type of polymerization termination condition is derived: for each class i, the maximum value for all $j \neq i$, $\mathrm{Sp}(X_i, X_j)$ is an indicator that measures its degree of dispersion from all other classes. Calculated as

$$J_i = \mathop{\mathrm{Max}}_{j}(\mathrm{Sp}(X_i, X_j)),\ j \neq i. \tag{13.29}$$

The most appropriate class number N_C should be the number of classes that minimize the sum of the discreteness indicators N_C of all classes. That makes J:

$$J = (1/N_C)\sum_{i=1}^{N_C} J_i \tag{13.30}$$

Minimum number of classes N_C.

Therefore, according to Eqs. (13.24) to (13.30), it is considered that the two classes with the lowest degree of resolution are combined into one class, so that the number of classes will be gradually reduced until the termination condition is satisfied.

The termination criterion (13.30) can also be selected when the number of classification categories reaches a predetermined value.

13.3.3 *Processing Flow and Result Analysis of Dual-Frequency Band Full-Polarization H/α/Wishart Classification*

The flow of the dual band full-polarization SAR classification algorithm is shown in Fig. 13.7. Specific steps are as follows:

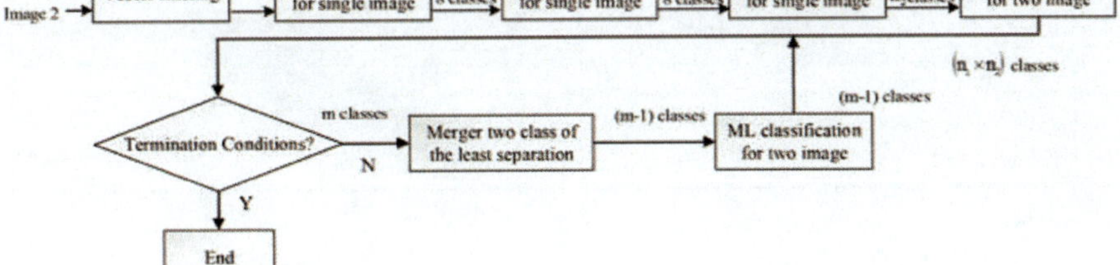

Fig. 13.7 Flowchart of a dual band full-polarization SAR image classification

(1) Firstly, vector filtering method is used to perform coherent spackle vector filtering on b_1 and b_2 band polarization SAR data;
(2) Classification processing of H/α classification and maximum likelihood estimation ML of the filtered two images;
(3) The types of the two images are separately aggregated according to the situation. That is, in a single image the two categories that have lower degree of separacity can be combined into one category. The two images are divided into n_1 and n_2 categories;
(4) All X_{1i} classes in the first image and the types of pixels belonging to the X_{2j} class in the second image are numbered $X_{i+n_1 j}$, and there are $m = n_1 \times n_2$ classes;
(5) Calculate the average feature correlation matrix $\boldsymbol{T}_i$ of $p \times p$ dimensions of each type of X_i;
(6) Calculate the distance factor $d_2(\langle \boldsymbol{A} \rangle, \boldsymbol{T}_m)$ of the pixel to all m classes according to the formula (13.20), and classify the pixel into the class with the smallest distance factor according to (13.22);
(7) Using the type aggregation technology, the two categories with the lowest degree of separability are combined into one class;
(8) ML classification processing of two images;
(9) terminate if the termination condition is satisfied; otherwise, return to Step (5).

The test data is L-band and C-band fully polarized SAR image data as shown in Fig. 13.6. Figure 13.8a is the H-α plane distribution and division of the L-band data, and Fig. 13.8b is the H-α plane distribution and division of the C-band data. The L-band data mainly shows the even-scattering and multi-scattering characteristics of low entropy and medium entropy, and the C-band data mainly shows the surface scattering characteristics of low entropy and medium entropy.

Figure 13.9a and b is pseudo-color maps of the second iteration classification results of L- and C-bands, respectively. There are rivers passing through the area.

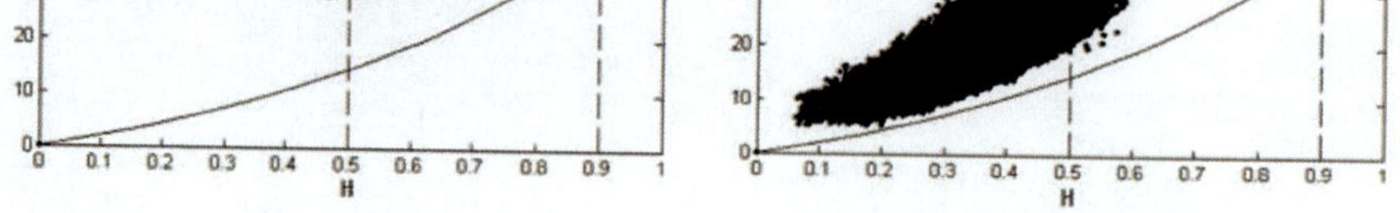

(a) L Band (b) C Band

Fig. 13.8 L, C-band H/α plane distribution and division diagram

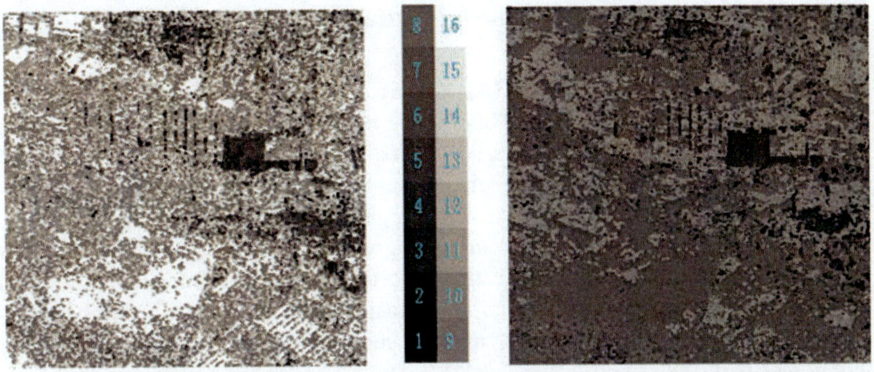

(a) classification results of 16 classes (b) Classification result of minimum J (10 classes)

Fig. 13.9 ML classification results of L, C dual band image

In the L-band image, the wet and vegetation areas near the river to the lower left corner are divided into multiple scattering classes with low scattering entropy. The middle and upper parts are dominated by medium scattering entropy vegetation scattering and multiple scattering; the entire region of the C-band image is dominated by surface scattering with low scattering entropy and medium scattering entropy. Figure 13.9a, b shows the classification results of the polarization data of the L- and C-bands. Figure 13.9a is the classification result using the number of pre-specified final classifications (16 classes) as the termination condition; Fig. 13.9b is the classification result using the estimation of classification accuracy by (13.30) as the termination condition.

The 6 × 6 correlation matrix recombined with dual band complex data contains cross-correlation terms of two images, which further improves the classification accuracy. The final classification number is 16, which more accurately refines the species of the area. In Fig. 13.9a, b, the river part in the lower right corner can be clearly identified, which is divided into the same class as the surrounding wet area.

Computational efficiency is an important factor in the practicality of a method. The dual band full-polarization SAR classification method first classifies the single-band full-polarization SAR, combines the results into the $m = n_1 \times n_2$ class, and then performs the type aggregation. The calculation amount of this method is very large. Therefore, the aggregation of single-band SAR classification results (Step (3)) greatly improves the computational efficiency of the algorithm. If Step (3) is not used, the number of types of the two single-band SAR classification results is $8 \times 8 = 64$. Taking the final result of classifying the species into 16 classes as an example, a 48-number type polymerization process is required. If only two images are aggregated into seven types using Step (3), the number of types after combination is $7 \times 7 = 49$, and the type is aggregated into 16 types, which only requires 33 times of type polymerization process. This saves nearly a third of the amount of calculations. It can be seen that the type aggregation of a single image greatly improves the efficiency of the algorithm.

13.4 Classification of PolInSAR Images

The classification method of PolInSAR mainly analyzes the correlation coefficients of two image pairs. The correlation coefficient between pairs of interference images is related to polarization. The optimal coherent decomposition of PolInSAR is helpful to understand the scattering mechanism of ground objects. Therefore, using the polarization interference technique to classify the ground scene can get a good classification result. Therefore, the classification of ground scenes by using the PolInSAR technique can obtain good classification results. The specific implementation and experimental results of unsupervised classification of SAR images using polarization interference information are given below based on the unsupervised Wishart ML classification of polarimetric SAR.

13.4.1 H_{Int}/A_{Int} Unsupervised Classification of PolInSAR

The eigenvalues and eigenvectors of the polarization interference correlation matrix can be obtained by the optimal correlation decomposition of the polarization interference SAR. From these parameters, two important parameters H_{Int} and A_{Int} in the PolInSAR process can be defined. H_{Int} is the optimal entropy of polarization interference, and A_{Int} is the optimal coherence anisotropy of polarization interference. H_{Int} and A_{Int} describe the distribution characteristics of interference correlation coefficients in different optimized polarimetric SAR channels [33, 34].

$$H_{Int} = \sum_{i=1}^{3} -P_{opt_i} \log_3 P_{opt_i}, 0 \le H_{Int} \le 1 \tag{13.31}$$

where $P_{opt_i} = \frac{\gamma_{opt_i}}{\sum_{j=1}^{3} \gamma_{opt_j}}$.

$$A_{Int} = \frac{P_{opt_2}}{P_{opt_1}}, \quad 0 \le A_{Int} \le 1 \tag{13.32}$$

Here, γ_{opt_i} is the three optimal correlation coefficients obtained by the optimal correlation decomposition.

H_{Int} shows the complexity of the potential correlation scattering mechanism in the scene. $H_{Int} = 0$ shows that there is only one single correlation scattering mechanism in the resolution unit, and three equal correlation coefficient values result in $H_{Int} = 1$. A_{Int} is mainly used to indicate the relative amplitude variation between different optimal polarization channels when there is a medium H_{Int} value. The smaller A_{Int} indicates that there is only one pair of major correlation scattering mechanisms in the resolution unit. The larger A_{Int} indicates that there are at least two pairs of major correlation scattering mechanisms.

Figure 13.10 shows the H_{Int}–A_{Int} spatial distribution. It can be seen from the figure that the range of values of A_{Int} gradually increases as the value of H_{Int} increases. This shows that A_{Int} helps to understand the relevant scattering process at medium and high H_{Int} values. According to the values of H_{Int} and A_{Int}, the H_{Int}–A_{Int} plane can be divided into four regions, and the physical scattering characteristics of each region and its representative are shown in Fig. 13.11. The boundaries of the regions can be appropriately adjusted based on the distribution of actual data H_{Int} and A_{Int} values.

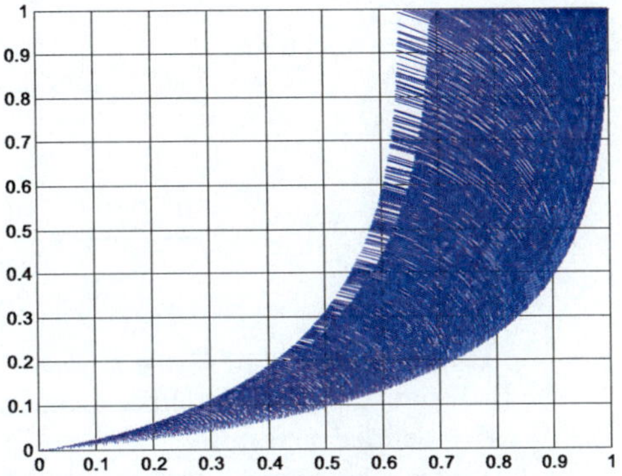

Fig.13.10 Spatial distribution of H_{Int}–A_{Int}

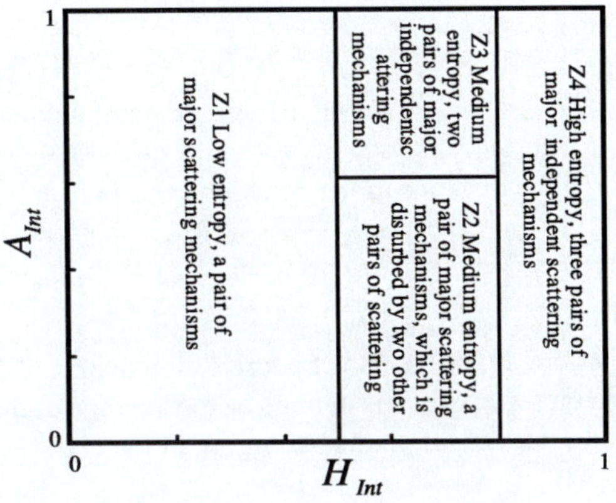

Fig. 13.11 Plane of H_{Int}–A_{Int}

13.4.2 Unsupervised Wishart ML Classification of PolInSAR Images [35, 36]

The correlation matrix of (6 × 6) polarized interferometric SAR after multi-look processing of two images can be expressed as

$$\langle T_6 \rangle = \frac{1}{n}\sum_{i=1}^{n} \vec{w}_i \vec{w}_i^{*T} = \begin{bmatrix} \langle T_{11} \rangle & \langle \Omega_{12} \rangle \\ \langle \Omega_{12} \rangle^{*T} & \langle T_{22} \rangle \end{bmatrix} \tag{13.33}$$

where $\vec{w} = \begin{bmatrix} \vec{k}_1 \\ \vec{k}_2 \end{bmatrix}$, n is the number of looks. $\vec{k}_1$ and $\vec{k}_2$ are the scattering vectors of the two images, respectively.

The matrix $\langle T_6 \rangle$ also obeys the complex Wishart distribution, whose probability density function is defined as

$$p(\langle T_6 \rangle) = \frac{|\langle T_6 \rangle|^{n-6} \exp\left(-n \cdot Tr\left(\Sigma_w^{-1} \langle T_6 \rangle\right)\right)}{K(n, 6)|\Sigma_w|^n} \tag{13.34}$$

where $K(n, 6)$ is a constant, $\Sigma_w = E\left(\vec{w}_i \vec{w}_i^{*T}\right)$ is (6 × 6) correlation matrix.

The Bayesian maximum likelihood classifier of the correlation matrix $\langle T_6 \rangle$ of the PolInSAR is defined as follows:

For pixels in a polarized interferometric SAR image pair:

$$\begin{aligned} &\text{If for all } j \neq m \\ &d(\langle T_6 \rangle, X_m) \leq d\left(\langle T_6 \rangle, X_j\right) \\ &\text{Then the pixel belongs to class } X_m \end{aligned} \tag{13.35}$$

The distance in the decision criteria is defined as

$$d(\langle T_6 \rangle, X_m) = \ln|\Sigma_m| + Tr\left(\Sigma_m^{-1} \langle T_6 \rangle\right) \tag{13.36}$$

where, Σ_m is the (6 × 6) correlation matrix of the characteristic PolInSAR of the mth target X_m, which is defined as the average of the polarization interference correlation matrices of all pixels belonging to the class X_m, namely,

$$\Sigma_m = \frac{1}{n_m}\sum_{i=1}^{n_m} \langle T_6 \rangle_i, \forall \langle T_6 \rangle_i \in X_m \tag{13.37}$$

where n_m is the number of pixels in class X_m.

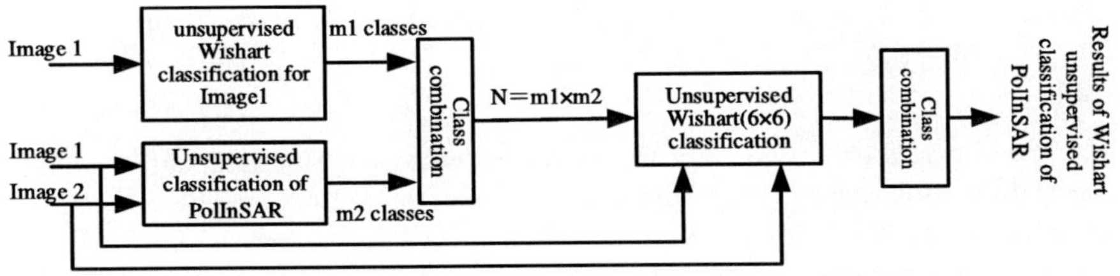

Fig. 13.12 Classification flowchart for unsupervised Wishart ML for PolInSAR

Species aggregation is also required in the unsupervised Wishart ML classification for PolInSAR. The intra-class dispersion W_i and the average distance B_{ij} between classes are

$$W_i = \frac{1}{n_i}\sum_{k=1}^{n_i} d\big[(\langle T_6\rangle_k \in X_i), X_i\big] = \ln|\Sigma_i| + 6 \tag{13.38}$$

$$\begin{aligned} B_{ij} &= \frac{\frac{1}{n_i}\sum_{k=1}^{n_i} d\big[(\langle T_6\rangle_k \in X_i), X_i\big] + \frac{1}{n_j}\sum_{k=1}^{n_j} d\big[(\langle T_6\rangle_k \in X_j), X_j\big]}{2} \\ &= \frac{W_i + W_j + Tr\left(\Sigma_i^{-1}\Sigma_j + \Sigma_j^{-1}\Sigma_i\right)}{2} \end{aligned} \tag{13.39}$$

The classification flowchart for unsupervised Wishart ML for PolInSAR is shown in Fig. 13.12. Proceed as follows:

(1) Unsupervised classification of $H/\overline{\alpha}$ Wishart ML for a single full-polarization SAR image is used, and m1 class is obtained; The optimal coherent decomposition is performed on the two interferometric SAR image pairs of full-polarization, and the H_{Int}–A_{Int} classification is used, and the m2 class is obtained by the H_{Int}–A_{Int} method.
(2) Pixels belonging to the class X_{1i} in the m1 class and belonging to the class X_{2j} in the m2 class are classified into the class $X_{i+m1\times(j-1)}$. Thus, a total of N = m1 × m2 is obtained.
(3) For each class X_k, calculate its (6 × 6) eigen covariance matrix:

$$\Sigma_k = \frac{1}{n_k}\sum_{i=1}^{n_k} \langle T_6\rangle_i, \forall \langle T_6\rangle_i \in X_k \tag{13.40}$$

(4) Classify each pixel into a class with the smallest distance $d(\langle T_6\rangle, X_m)$ among all N classes.
(5) Class aggregation until the specified standard is reached; otherwise, to Step (3).

13.4.3 Results of Unsupervised Wishart ML Classification Tests for PolInSAR Images

The experimental data used were I1 and I2 somewhere in Tianshan, and the image size of this region was 200 × 250 pixels. In the polarization SAR image classification, since the polarization information contained in the two images is very similar, only I1 is processed. The polarized pseudo-color images are shown in Fig. 13.13a, and red, green, and blue in the image represent HH, HV, and VV, respectively. Figure 13.14a–c gives the scattering entropy H and scattering angle $\overline{\alpha}$ of each pixel in image I1, The amplitude in the graph is proportional to the magnitude of the d and f values. The amplitude in the figure is proportional to the magnitude of the H and $\overline{\alpha}$ values. Black with an amplitude of 0 represents $H = 0$ and $\overline{\alpha} = 0$, and $H = 1$ and $\overline{\alpha} = 90°$ correspond to white with the amplitude of 255. Both H and $\overline{\alpha}$ are calculated using a 5 × 5 window to obtain an average estimate. It can be seen from the entropy H image of the region that the scattering randomness of the target does have strong and weak differences and has a certain relationship with the distribution of the features. The $\overline{\alpha}$ values of the pixels in this area are not very large, mainly concentrated around the medium value. Figure 13.14 shows the distribution of the pixels of the experimental data in the $H/\overline{\alpha}$ plane. Comparing the distribution in the $H/\overline{\alpha}$ plane, it can be seen that most of the area is medium entropy and low entropy body scattering (vegetation), and there are also some entropy surface scattering (rough Brag scattering) and medium entropy low entropy multiple scattering. This result is consistent with the features of the area. The existence of multiple scattering is mainly due to the strong penetrating ability of the electromagnetic wave in the L-band. The electromagnetic wave not only interacts with the top surface of the canopy, but also enters the canopy to scatter between the branches and even interacts with the trunk and the ground through the canopy. According to the value of y of the pixel, the experimental data is divided into seven categories according to the distribution in the z-plane (it can be seen from

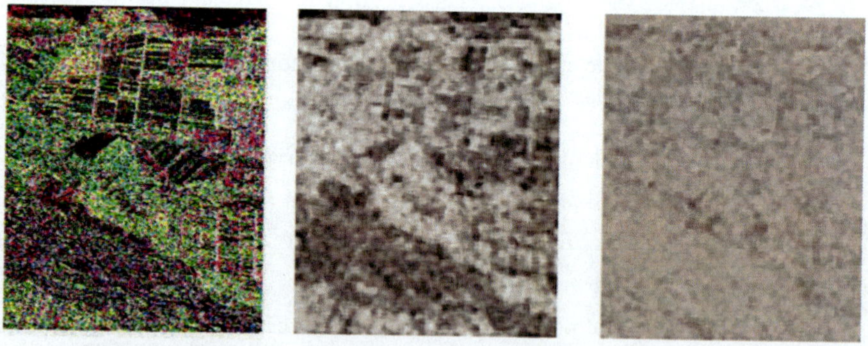

(a) Pseudo color image (b) Scattering Entropy H (c) Scattering Angle $\overline{\alpha}$

Fig. 13.13 Polarized pseudo-color image, the scattering entropy and the scattering Angle of data I1

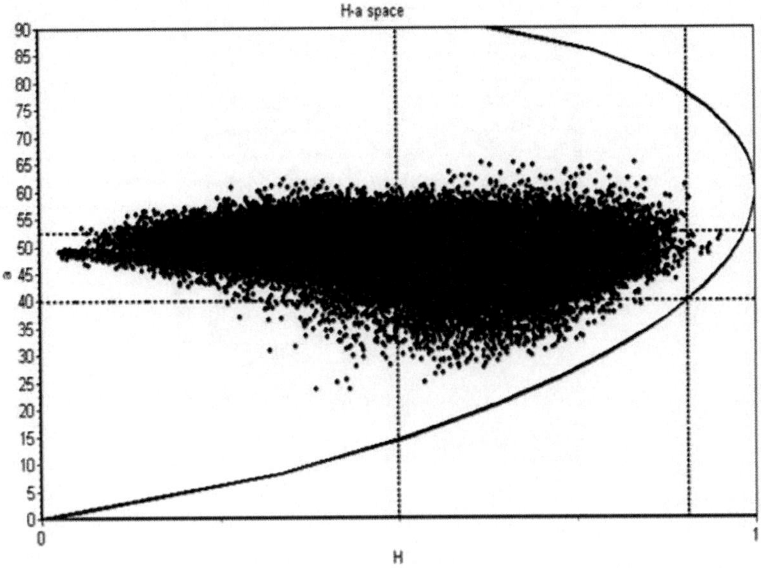

Fig. 13.14 $H/\overline{\alpha}$ plane

Fig. 13.14 that there is no pixel in the Z8 region, so the class is not considered), and the result is shown in Fig. 13.16. It can be seen that the results of the y classification are not very good, and the details of the features disappear.

The classification result of $H/\overline{\alpha}$ is used as a training set to initialize the Wishart ML classifier. The result obtained after first iteration is shown in Fig. 13.16a. Comparing Figs. 13.15 and 13.16, it can be clearly seen that after only first iteration of the Wishart ML classification, the results obtained have been considerably improved in terms of the preservation of feature details. Figure 13.16b and c shows the results obtained after second and fourth iterations, respectively. Comparing the results of different iterations, it can be seen that as the time of iterations increases, the result of the classification of features is better, and the details are displayed. In the iteration of the Wishart ML classification, there is no processing of the merged class, and the original seven-type results of $H/\overline{\alpha}$ are still maintained, and the corresponding representative colors of the various types are not changed.

The above mainly analyzes the results of unsupervised Wishart ML classification of polarimetric SAR images. The influence of polarization interference SAR on classification results is discussed below.

Figure 13.17 shows the correlation coefficient between the two images for HH polarization. The gray scale from 0 to 255 corresponds to a correlation coefficient from 0 to 1.

As can be seen from the image, there are several lines with a correlation coefficient of almost zero in the lower left part of the area, which can be judged as a river from

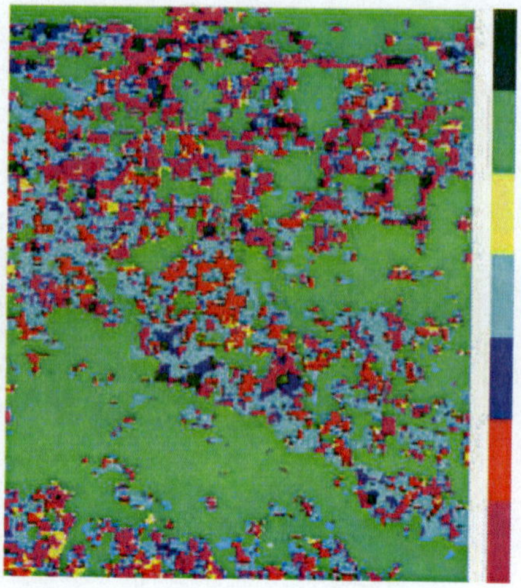

Fig. 13.15 Image of $H/\overline{\alpha}$ classification (7 classes)

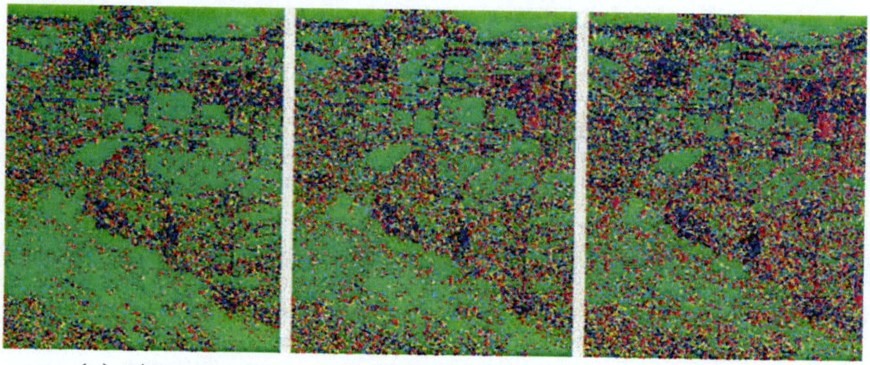

(a) 1st iteration (b) 2nd iteration (c) 4th iteration

Fig. 13.16 $H/\overline{\alpha}$-Wishart ML classification results for data I1

its shape. This river is not visible in the unsupervised classification of H-Wishart ML in Fig. 13.16.

Figure 13.18 shows a pseudo-color image formed by three optimal correlation coefficients obtained by polarization correlation analysis of two image pairs. The images are red, green, and blue, corresponding to γ_{opt_1}, γ_{opt_2}, and γ_{opt_3}, respectively.

Figure 13.19 shows the distribution of pixels in the H_{Int}/A_{Int} plane. Comparing Fig. 13.10, it is found that the pixels are mainly concentrated in the upper right corner of the H_{Int}/A_{Int} plane, the area of $H_{Int} > 0.6$, $A_{Int} > 0.3$. According to the pixel distribution characteristics of Fig. 13.19, the region boundaries in the H_{Int}/A_{Int} plane need to be adjusted:

Fig. 13.17 Interference correlation coefficient of HH between I1 and I2

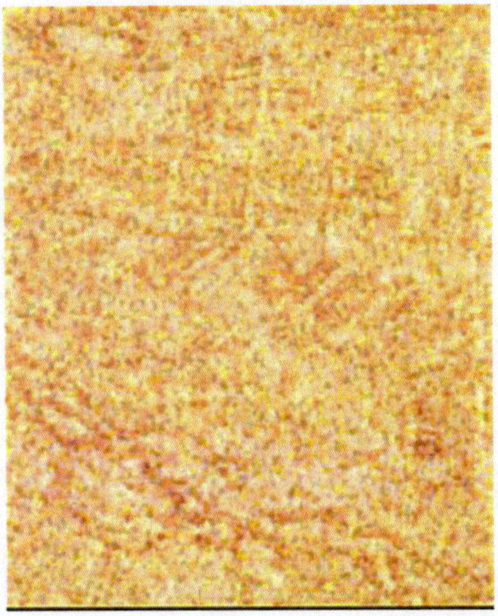

Fig. 13.18 Pseudo-color image of the optimal correlation coefficient

$H_{Int} < 0.85$ as the class of low entropy; $H_{Int} > 0.95$ as the class of high entropy; the middle is entropy between them, and the middle entropy part is subdivided into two parts, $A_{Int} < 0.75$ and $A_{Int} > 0.75$. Thus, four classes are obtained, and the results are shown in Fig. 13.21.

Among them, Z1 has a low H_{Int} value, which mainly includes single scattered ground objects such as water surface, land, and rock; Z2 has a medium H_{Int} equivalent and a low A_{Int} value, indicating that there is a pair of main scattering mechanisms, but the other two pairs also have some effects on it, such as farmland and low bushes.

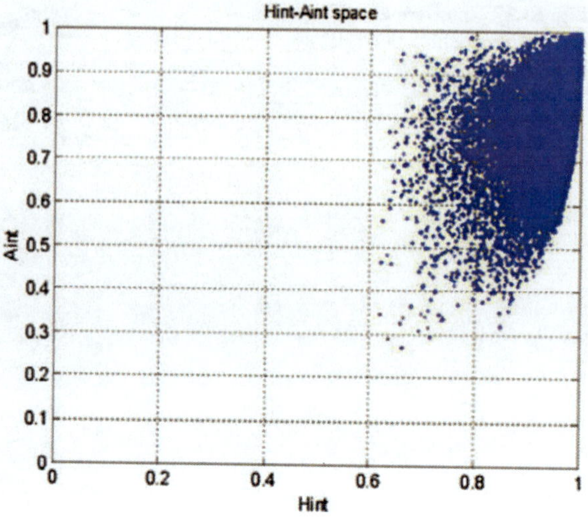

Fig. 13.19 Distribution of pixels in I1 and I2 the image pairs in H_{Int}/A_{Int} plane

Z3 also has a medium H_{Int} value, and its A_{Int} value is also high, indicating that there are two main pairs of scattering mechanisms, such as the middle and high and not very dense trees; Z4 has a higher H_{Int} value, indicating that the scattering is more random, such as a forest composed of tall and dense trees can be considered to belong to this class. As can be seen from the results in Fig. 13.20, the river in the lower left part of the figure is clearly separated, but similar to the result of $H/\overline{\alpha}$, the specific details of the object are not very clear.

The seven types of results obtained by the Wishart ML unsupervised polarimetric SAR classifier through four iterations are combined with the four types of results initially classified according to the H_{Int}/A_{Int} plane, and 28 classification results are

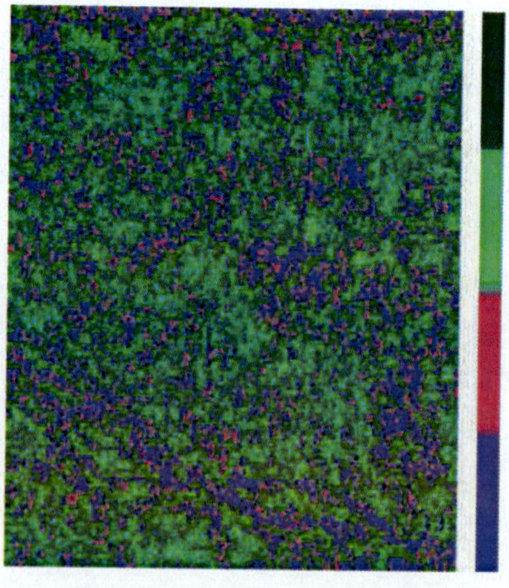

Fig. 13.20 Results of the H_{Int}/A_{Int} classification (four classes)

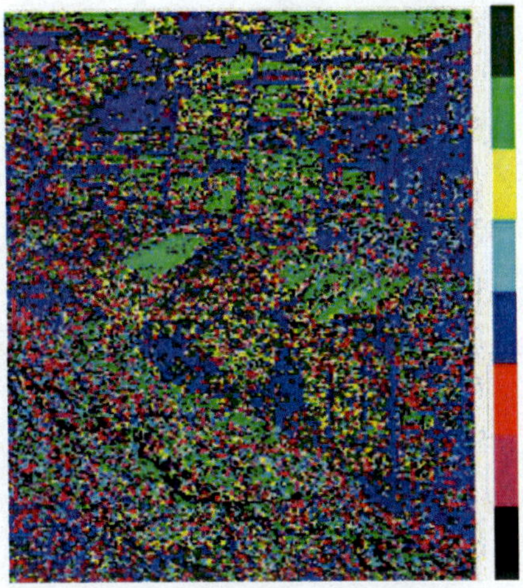

Fig. 13.21 Polarization interference Wishart ML unsupervised classification results of I1 and I2 (eight classes)

obtained. This is used as the initialization input for the Wishart ML unsupervised classifier for PolInSAR. The results of the unsupervised classification of the Wishart ML of the PolInSAR can be obtained by processing the experimental data according to the steps given in Sect. 13.4.2, as shown in Fig. 13.21. Figure 13.21 shows the aggregation of classes in the process, until there are only eight types of results. The river (black) in the lower left part of the image can be clearly seen from the figure, and the categories of the images can be clearly distinguished. The details of the features are significantly higher than the polarization unsupervised classification in Fig. 13.16 and the H_{Int}/A_{Int} classification results in Fig. 13.20. It can be seen that the unsupervised classification of the polarization interference of the SAR image pair Wishart ML can get very good results. When processing the actual data, the flat ground phase removal processing should be performed before calculating the polarization interference correlation matrix $\langle T_6 \rangle$.

13.4.4 Problems and Solutions for the Unsupervised Classification of the Wishart ML for PolInSAR [35,37]

The experiment found that although the flat ground phase removal process has been performed, there are still some stripes in the resulting image, as shown in Fig. 13.22. These stripes are mainly due to the fact that the terrain of this part has a certain slope and is not flat. The interference fringe image after removing the flat phase of I1 and I2 is shown in Fig. 13.23. Comparing Fig. 13.22 with Fig. 13.23, it can be seen that the stripes of the two images fit very well. Polarization interference correlation matrix $\langle T_6 \rangle$ is used in the Wishart ML unsupervised classification for PolInSAR. And this matrix contains the interference phase obtained by multiplying the conjugates of the

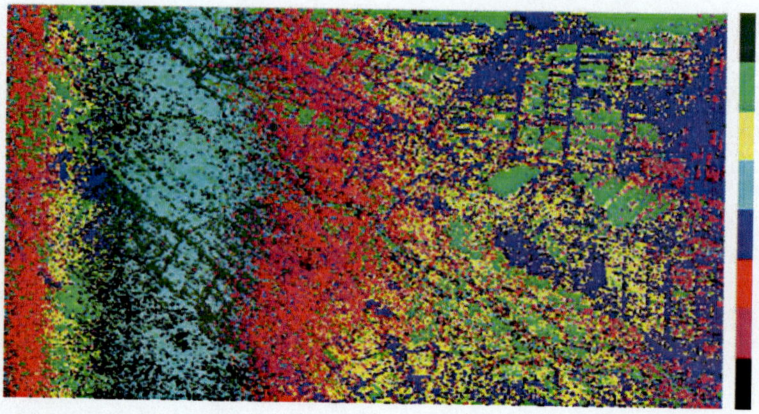

Fig. 13.22 Results of the Wishart ML unsupervised classification of polarization interference SAR for I1 and I2 (450 × 250 pixels) (eight classes)

Fig. 13.23 Interference fringe image after removing the flat phase of I1 and I2 (450 × 250 pixels) (HH polarization)

two image scattering vectors. Therefore, interference fringes will inevitably occur in slopes. The interference fringes appearing at the slope of Fig. 13.22 severely affect the classification results. Appropriate methods must be found to remove the effects of interference fringes. Otherwise, the method can only be used when the terrain is flat and cannot be applied to any terrain.

Based on the above discussion, this paper proposes the following solutions:

Method 1: Use real DEM data from the area under study.

Interference fringes can be reversed based on the real DEM and the orbital and imaging parameters of the two observations. In this way, the influence of the terrain in the original data can be eliminated as if the phase of the ground is removed, and then classified by the Wishart ML unsupervised method of PolInSAR.

The advantage of this method is that it can effectively eliminate the influence of terrain on the classification results and get the correct classification results. This method is more feasible after the "Digital Earth" task is actually implemented, but at present, it is generally difficult to obtain the actual DEM data of any region. Since the true DEM of the experimental data is not available, this method is not discussed further.

Method 2: Select two pairs of interference images with a sufficiently short baseline. The baseline is short enough to mean that the interference fringes disappear completely after removing the flat phase.

The data used previously is images I1 and I2 with a longer baseline. Shorter baselines are less sensitive to terrain changes and the resulting interference fringes are sparse. This facilitates the use of polarization interference in the unsupervised Wishart ML classification method. Therefore, I2 and I3 with shorter baselines were selected for experiments. Figure 13.24 shows the interference fringe image after removing the flat phase between I2 and I3. It can be seen that the fringe disappears after the flat phase is removed. This shows that the baseline length between I2 and I3 is short enough so that it can be classified using the unsupervised Wishart ML method of PolInSAR. Figure 13.25 shows the results of the final eight categories. It can be seen that the classification results are indeed unaffected by the stripes.

Although this method can be used effectively, the premise is that pairs of graphs with different baseline lengths are optional, and the baseline must be short enough. For the airborne or spaceborne dual-antenna single-track interferometric polarimetric SAR mode, the baseline is short, and the images obtained by this mode generally satisfy most of the terrain on the earth. However, for repeated track interference, it is not possible under any circumstances to select an image pair whose baseline length

Fig. 13.24 Interference fringe image after removal of the ground phase for I2 and I3 (450 × 250 pixels) (HH polarization)

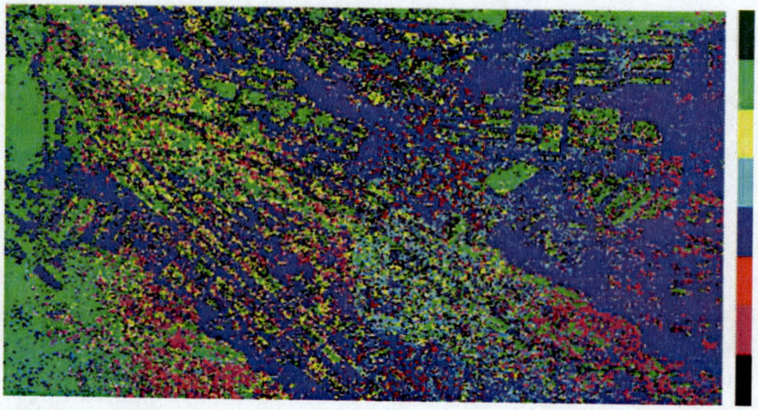

Fig. 13.25 Wishart ML Unsupervised Classification Results for I2 and I3 (450 × 250 pixels) of PolInSAR (eight classes)

meets the requirements. Especially when the local shape is steep, the requirement for the length of the baseline becomes very demanding.

Method 3: Use the interference phase between HH (or VV) polarization as a reference to eliminate terrain effects.

The two methods presented above have some special requirements that are sometimes not possible. A general scalar interference can only obtain an HH (or VV) polarization image, thereby calculating the interference phase and obtaining ground elevation information. Therefore, in the absence of real ground DEM data, in order to eliminate the influence of terrain on the classification results, the ground elevation information obtained by HH (or VV) polarization can be used as a terrain reference. In the actual data processing, only the interference phase of HH (or VV) is calculated, and the phase is replaced by the flat phase to remove the influence of the terrain. Figure 13.26 shows the classification results obtained for the I1 and I2 images by this method. Comparing Fig. 13.22, it can be seen that the streaks disappear, showing that the method does effectively eliminate the terrain effects.

It must be pointed out here that using the HH (or VV) interference phase as a reference will inevitably affect the final classification result. Because the HH (or VV) interference phase contains not only terrain information, but also certain feature information. While eliminating the topographical information of HH (or VV) interference, it also inevitably loses the feature information unique to this part of HH (or VV) polarization. By carefully comparing the results of Fig. 13.26 with Figs. 13.22 and 13.25, it can be seen that the rivers flowing through the upper left to the lower right corners, which are clearly visible in Figs. 13.22 and 13.25, are difficult to distinguish in Fig. 13.26. And some other places in the image are not consistent with the results of the previous classification. However, the benefits of this method are also obvious. It does not require external auxiliary data and can be used under any circumstances. It is an effective means and is of great help in understanding the information of the terrain.

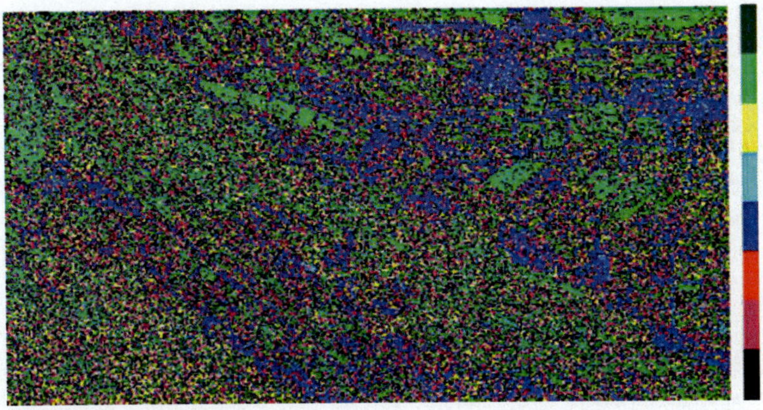

Fig. 13.26 Classification results of I1 and I2 (450 × 250 pixels) after eliminating HH interference phase (eight types)

13.5 Unsupervised Wishart ML Classification of Multi-band PolInSAR Images

The scattering characteristics of the target depend on the relationship between the target size and the radar wavelength. Incident waves of different wavelengths interact with different parts of a complex target, and the target exhibits different scattering characteristics. For example, for poor land without vegetation cover, the scattering mechanisms under L- and C-band illumination are almost identical; However, for dense forests with a variety of different vegetation, exposures in the L- and C-bands may show different results. It can be seen that the comprehensive application of multi-band polarimetric SAR data will increase the understanding of the target scattering characteristics, thus providing assistance for feature analysis and target classification of features [35].

It has been pointed out in [3] that the distance measurement formula in the Wishart ML classifier can be extended to multi-band data. If the frequencies between the radar bands are significantly different, the data between the different bands is statistically independent. Since the joint probability density function of statistical independent data is the product of the probability density function of each band of data, the following distance formula can be obtained:

$$d(\langle T\rangle, \omega_m) = \sum_{j=1}^{J} n_j \left\{ \ln|V_m(j)| + Tr\left[V_m(j)^{-1}\langle T\rangle_j\right]\right\} - \ln[P(\omega_m)] \quad (13.41)$$

where $V_m(j)$ is the characteristic scattering correlation matrix of the mth class in the j-band. $\langle T\rangle_j$ and n_j are, respectively, the scattering correlation matrix and the look number of the j-band pixels, and J is the total number of bands.

Based on the above analysis, the unsupervised classification method of Wishart ML for polarization interference of multi-band SAR images is described in detail below.

Assuming that the data of the L- and C-bands have the same look number n, the (6×6) polarization interference correlation matrix of the two images of the L-band is

$$\langle T_6 \rangle = \frac{1}{n} \sum_{i=1}^{n} \vec{w}_i \vec{w}_i^{*T} = \begin{bmatrix} \langle T_{11} \rangle & \langle \Omega_{12} \rangle \\ \langle \Omega_{12} \rangle^{*T} & \langle T_{22} \rangle \end{bmatrix} \tag{13.42}$$

where $\vec{w} = \begin{bmatrix} \vec{k}_1 \\ \vec{k}_2 \end{bmatrix}$, and the (3×3) scattering correlation matrix of the C-band image is

$$\langle T_3 \rangle = \frac{1}{n} \sum_{i=1}^{n} \vec{k}_i \vec{k}_i^{*T} \tag{13.43}$$

The distance in the multi-band polarization interference Wishart ML classification is defined as follows:

$$\begin{aligned} d(\langle A \rangle, X_m) &= d_1 + d_2 \\ &= \left[\ln\left|V_{m_{6\times6}}\right| + Tr\left(V_{m_{6\times6}}^{-1}\langle T_6 \rangle\right)\right] + \left[\ln\left|V_{m_{3\times3}}\right| + Tr\left(V_{m_{3\times3}}^{-1}\langle T_3 \rangle\right)\right] \end{aligned} \tag{13.44}$$

where $V_{m_{6\times6}}$ and $V_{m_{3\times3}}$ are the (6×6) polarization interference correlation matrix and the (3×3) scattering correlation matrix of the m-th class target X_m:

$$V_{m_{6\times6}} = \frac{1}{n_m} \sum_{i=1}^{n_m} \langle T_6 \rangle_i, \forall \langle T_6 \rangle_i \in X_m \tag{13.45}$$

$$V_{m_{3\times3}} = \frac{1}{n_m} \sum_{i=1}^{n_m} \langle T_3 \rangle_i, \forall \langle T_3 \rangle_i \in X_m \tag{13.46}$$

n_m is the number of pixels in class X_m.

The decision criterion for the Bayesian ML (Maximum Likelihood) classifier is

$$\begin{aligned} &\text{For pixels in the image, if for all } j \neq m, \\ &d(\langle A \rangle, X_m) \le d\big(\langle A \rangle, X_j\big) \text{ is true, the pixel belongs to class } X_m \end{aligned} \tag{13.47}$$

The dispersion within the class and the average distance between classes are

$$W_i = \frac{1}{n_i}\sum_{k=1}^{n_i} d\big[(\langle A\rangle_k \in X_i), X_i\big] = \ln\big|V_{i_{6\times6}}\big| + \ln\big|V_{i_{3\times3}}\big| + 9 \tag{13.48}$$

$$\begin{aligned} B_{ij} &= \frac{\frac{1}{n_i}\sum_{k=1}^{n_i} d\big[(\langle A\rangle_k \in X_i), X_i\big] + \frac{1}{n_j}\sum_{k=1}^{n_j} d\big[(\langle A\rangle_k \in X_j), X_j\big]}{2} \\ &= \frac{W_i + W_j + Tr\Big(V_{i_{6\times6}}^{-1} V_{j_{6\times6}} + V_{j_{6\times6}}^{-1} V_{i_{6\times6}}\Big) + Tr\Big(V_{i_{3\times3}}^{-1} V_{j_{3\times3}} + V_{j_{3\times3}}^{-1} V_{i_{3\times3}}\Big)}{2} \end{aligned} \tag{13.49}$$

The steps of data processing for multi-band PolInSAR classification are as follows:

(1) Performing $H/\overline{\alpha}$ classification on the full-polarization SAR images of the L- and C-bands, and obtaining m1 × m2 classes; we perform the optimal coherent decomposition of the two interferometric images of the L-band full-polarization SAR and classify them by the H_{Int}–A_{Int} method to obtain the m3 classes.
(2) Pixels belonging to the X_{1i} class in the m1 class, the X_{2j} class belonging to the m2 class, and the X_{3k} class belonging to the m3 class are classified into the class $X_{i+m1\times[(j-1)+m2\times(k-1)]}$. Thus, a total of N = m1 × m2 × m3 class numbers is obtained.
(3) For each class X_k, calculate its (6 × 6) polarization interference correlation matrix and (3 × 3) scattering correlation matrix: $V_{k_{6\times6}} = \frac{1}{n_k}\sum_{i=1}^{n_k}\langle T_6\rangle_i, \forall\langle T_6\rangle_i \in X_k$, $V_{k_{3\times3}} = \frac{1}{n_k}\sum_{i=1}^{n_k}\langle T_3\rangle_i, \forall\langle T_3\rangle_i \in X_k$.$n_k$ is the number of pixels in class X_k.
(4) Calculate the distance from each pixel to each type according to the formula (13.44) and classify the pixel as one of the smallest distances among all N types.
(5) Perform class aggregation until the specified standard is reached; otherwise, return to Step (3).

The experimental data is I1 and I2 of the L-band, and the C-band data I4 corresponding to I1. Figure 13.27a shows the polarized pseudo-color image of the C-band image I4. Red, green, and blue represent HH, HV, and VV, respectively. Comparing with the polarization pseudo-color image of the L-band data I1 of Fig. 13.22, it can be found that there is a significant difference between the two images. The details of the L-band image are very obvious. This is because the L-band has a longer wavelength, which is more sensitive to the details of the ground changes. However, a large number of object targets are rough relative to the wavelength of the C-band. Figure 13.27b shows the distribution of I4 pixels in the H/$\overline{\alpha}$ plane. It can be seen that the distribution of the C-band image in the H/$\overline{\alpha}$-plane is quite different from that of the L-band image. The pixels of the C-band image are mostly distributed in the Z1 region, and some of the pixels are in the Z4 region, with only a few points in the Z2 and Z5 regions. This is because the electromagnetic wave penetration ability of the C-band is poor, and the electromagnetic wave mainly interacts with the target on the surface of the ground object, so the surface scattering with medium entropy and low

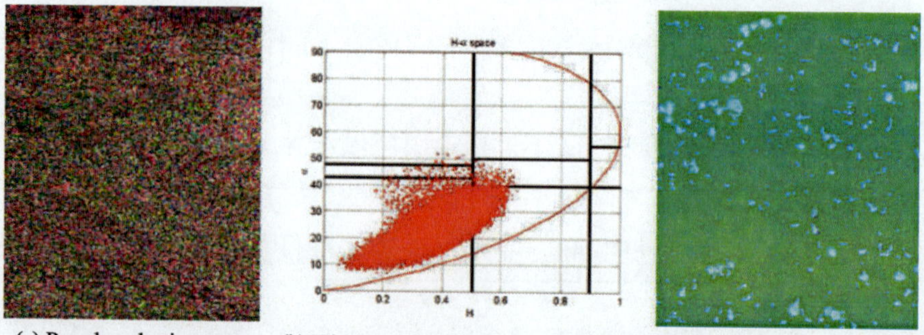

(a) Pseudo color image (b) H/$\overline{\alpha}$ plane distribution (c) H/$\overline{\alpha}$ classification result

Fig. 13.27 Images of the C-band image I4

entropy is dominant. According to the distribution of pixels in the H/$\overline{\alpha}$ plane, and also to reduce the amount of calculation for subsequent processing, only the C-band image is divided into two classes of $H < 0.5$ and $H > 0.5$, and the result is shown in Fig. 13.27c.

For the L-band image I1, it can be seen that only a few pixels are in the Z7 region from observing the distribution of its pixels in H/space (Fig. 13.23). In order to reduce the amount of calculation of subsequent data processing, the pixels of the area are classified into the Z5 area, so that the L-band image I1 is divided into six classes. Six types (m1) of the L-band I1, two types (m2) of the C-band I4, and four types (m3) of the L-band I1 and I2 are combined to obtain 48 types. Using these 48 classes as initialization inputs, the experimental data was processed according to the steps given above, and the classes were combined. The resulting 10 classes are given in Fig. 13.28a. For a convenient comparison, the results of the Weartart ML unsupervised classification of the PolInSAR of the L-band I1 and I2 are again given in

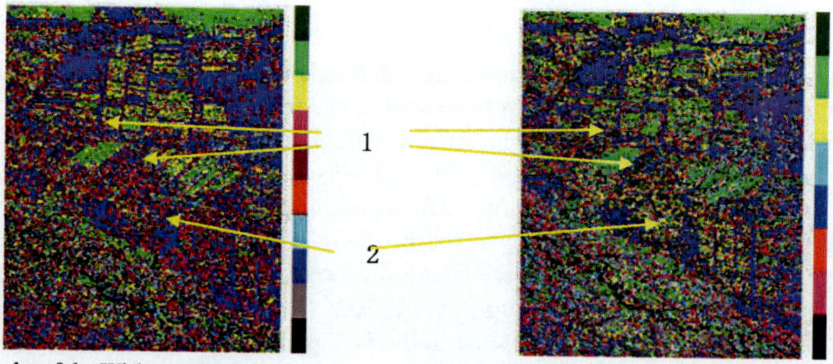

(a) Results of the Wishart ML unsupervised classification of L and C-band PolInSAR (10 classes)

(b) Results of the Wishart ML unsupervised classification of L-band PolInSAR (8classes)

Fig. 13.28 Unsupervised classification of the Wishart ML for polarized interferometric SAR

Fig. 13.28b (Fig. 13.21). The main differences between the two results are indicated by "1" and "2" in Fig. 13.28. It can be seen that the farmland area at "1" is mainly a type of crop on (b), and (a) it can be clearly seen that there are 2–3 different types of crops; In the area "2," the number of classes of (a) images is increased compared to (b), and the details of the features are also greatly improved. It should be noted that the influence of the ground phase must also be removed first when calculating $\langle T_6 \rangle$.

It can be seen from the above analysis that the comprehensive use of multi-band polarimetric SAR data is helpful to understand the characteristics of the ground object and improve the classification ability of the ground object.

13.6 Summary

The methods for supervising the target classification of polarimetric SAR images, unsupervised classification methods, and experimental results for various classification methods are presented in this chapter. The supervised classification of polarimetric SAR images is discussed, including image classification based on artificial neural network and supervised classification based on Wishart's Maximum Likelihood estimation (ML). The unsupervised classification of polarimetric SAR images is analyzed and discussed, including unsupervised classification algorithm based on polarization scattering mechanism, target scattering entropy H/scattering angle α(H/α) unsupervised classification method based on Cloude–Pottier decomposition, based on Cloude–Pottier decomposition information enhanced H/γ unsupervised classification method, Cloude–Pottier decomposition and Wishart-based polarization SAR unsupervised classification method and dual band full-polarization SAR image H/α/Wishart unsupervised classification method, etc.

The classification of polarized interferometric SAR images and the multi-band PolInSAR images of Wishart ML unsupervised classification techniques are also presented in this chapter.

References

1. B. Dai, Studies on multi polarization SAR system and polarization information processing technology. PhD, Institute of Electronics. Institute of Electronics, Chinese Academy of Sciences, PhD (2000)
2. L.E. Pierce, F.T. Ulaby, K. Sarabandi, M.C. Dobson, Knowledge based classification of polarimetric SAR images. IEEE Trans. GRS **32**(5), 1081–1086 (1994)
3. J.S. Lee, M.K. Grunes, Classification of multi-look polarimetric SAR imagery based on Complex Wishart distribution. Int. J. Remote Sensing **15**(11), 2299–2311 (1994)
4. P. Ferrazzoli, L. Guerriero, G. Schiavon, A vegetation classification scheme validated by model simulations, in *Geoscience and Remote Sensing, 1997. IGARSS '97. 'Remote Sensing—A Scientific Vision for Sustainable Development'*, vol. 4, pp. 1618–1620
5. M. Hosokawa, T. Hoshi, Polarimetric SAR data classification method using the self-organizing map, in *Geoscience and Remote Sensing Symposium, 2002. IGARSS '02*, vol. 6, pp. 3468–3470

6. B. Scheuchl, I. Hajnsek, I.G. Cumming, Model-based classification of polarimetric SAR sea ice data, in *Geoscience and Remote Sensing Symposium, 2002. IGARSS '02*, vol. 3, pp. 1521–1523 (2002)
7. J.S. Lee, M.R. Grunes, Classification of multi-look polarimetric SAR data based on complex Wishart distribution, in *Telesystems Conference, 1992. NTC-92*, pp. 7/21–7/24
8. J.J. Van Zyl, Unsupervised classification of scattering behavior using radar polarimetry data. IEEE Trans. GRS **27**(1), 36–45 (1989)
9. P. Lombardo, T. Macri Pellizzeri, A. Tomasuolo, Classification of Polarimetric SAR images of suburban areas using joint annealed segmentation and "H/A/α" decomposition, in *Remote Sensing and Data Fusion over Urban Areas, IEEE/ISPRS Joint Workshop 2001*, pp. 117–121
10. P. Lombardo, T, Macri Pellizzeri, A, Tomasuolo, A joint classification technique based on multivariate annealing segmentation and "H/A/α" decomposition for fully polarimetric SAR images of suburban areas, in *Geoscience and Remote Sensing Symposium, 2001. IGARSS '01*, vol. 6, pp. 2922–2924
11. H. McNairn, V. Decker, K. Murnaghan, The sensitivity of C-band polarimetric SAR to crop condition, in *Geoscience and Remote Sensing Symposium, 2002. IGARSS '02*, vol. 3, pp. 1471–1473
12. F. Mattia, T. Toan, J.-C. Souyris, The effect of surface roughness on multifrequency polarimetric SAR data. IEEE Trans. GRS **35**(4), 954–965 (1997)
13. J.S. Lee, M.R. Grunes, G. De Grandi, Polarimetric SAR speckle filtering and its impact on classification, in *Geoscience and Remote Sensing, 1997. IGARSS '97*, vol. 2, pp. 1038–1040
14. S. Baronti, G. Macelloni, S. Paloscia, P. Pampaloni, Analysis of an automatic method for surface classification using multifrequency multipolarization SAR data, in *Geoscience and Remote Sensing Symposium, 1995. IGARSS '95*, vol. 2, pp. 973–975
15. L. Du, J.S. Lee, S.A. Mango, Fuzzy classification of earth terrain covers using multi-look polarimetric SAR image data, in *Geoscience and Remote Sensing Symposium, 1993. IGARSS '93*, vol. 4, pp. 1602–1604
16. L.-J. Du, M.R. Grunes, Unsupervised segmentation of multi-polarization SAR images based on amplitude and texture characteristics, in *Geoscience and Remote Sensing Symposium, 2000. Proceedings. IGARSS 2000*, vol. 3, pp. 1122–1124
17. Z. Belhadj, A. Saad, S. El Assad, J. Saillard, D. Barba, Comparative study of some algorithms for terrain classification using SAR images, in *1994 IEEE International Conference on Acoustics, Speech, and Signal Processing, 1994. ICASSP-94*, vol. 5, pp. 165–168
18. J.S. Lee, M.R. Grunes, T.L. Ainsworth, E. Pottier, E. Krogager, W.-M. Boerner, Quantitative comparison of classification capability: fully-polarimetric versus partially polarimetric SAR, in *Geoscience and Remote Sensing Symposium, 2000. Proceedings. IGARSS 2000*, vol. 3, pp. 1101–1103
19. J.-Y. Xu, J. Yang, Y.-N. Peng, C. Wang, Using cross-entropy for polarimetric SAR image classification, in *Geoscience and Remote Sensing Symposium, 2002. IGARSS '02*, vol. 3, pp. 1917–1919
20. J.-S. Lee, M.R. Grunes, E. Pottier, Quantitative comparison of classification capability: fully polarimetric versus dual and single-polarization SAR. IEEE. Trans. GRS **39**(11), 2343–2351 (2001)
21. K.Y. Lee, S.C. Liew, L. Kwoh, M. Nakayama, Land cover classification Using NASA/JPL polarimetric synthetic aperture rader (Pol SAR) data. Paper presented at the 22nd Asian Conference on Remote Sensing, November 2001, Singapore, pp. 5–9
22. Y. Dong, A.K. Milne, B.C. Forster, Segmentation and classification of vegetated areas using polarimetric SAR image data. IEEE. Trans. GRS **39**(2), 321–329 (2001)
23. D.B. Trizna, C. Bachmann, M. Sletten, Projection pursuit classification of multiband polarimetric SAR land images. IEEE. Trans. GRS **39**(11), 2380–2385 (2001)
24. R. Touzi, F. Charbonneay, Characterization of target symmetric scattering using polarimcetric SARs. IEEE Trans. GRS **40**(11), 2507–2516 (2002)
25. Y. Dong, B.C. Forster, C. Ticehurst, A new decomposition of radar polarization signatures. IEEE Trans. GRS **36**(3), 933–939 (1998)

26. Y.-Q. Jin, S.R. Cloude, Numerical eigenanalysis of the coherency matrix for a layer of random nonspherical scatterers. IEEE Trans. GRS **32**(6), 1179–1185 (1994)
27. T.L. Ainsworth, S.R. Cloude, J.S. Lee, Eigenvector analysis of polarimetric SAR data, in *Geoscience and Remote Sensing Symposium, 2002. IGARSS '02*, vol. 1, pp. 626–628
28. W.M. Boerner, M.B. El-Arini, Polarization dependence in electromagnetic inverse problem. IEEE Trans. Ant. Propagat. **29**(2), 262–271 (1981)
29. X. Liu, Studies on information processing technique of fully-polarimetric synthetic aperture radar. Institute of Electronics, Chinese Academy of Sciences, PhD (2004)
30. J.S. Lee, M.R. Grunes, T.L. Ainsworth, L. Du, D.L. Schuler, S.R. Cloude, Unsupervised classification using polarimetric decomposition and complex Wishart classifier, in *Geoscience and Remote Sensing Symposium Proceedings, 1998. IGARSS '98*, vol. 4 (1998), pp. 2178–2180
31. X. Liu, R. Yang, Z. Yang, Unsupervised classification methods and experimental research of dual frequency fully polarimetric SAR images. J. Electron. Inf. Technol. **26**(11), 1738–1745 (2004)
32. J.-S. Lee, K. Hoppel, Principal components transformation of multifrequency polarimetric SAR imagery. IEEE Trans. GRS **30**(4), 686–696 (1992)
33. L. Ferro-Famil, E. Pottier, J.S. Lee, Unsupervised classification and analysis of natural scenes from polarimetric interferometric SAR data, in *IEEE Proceedings of IGARSS'01* (2001), pp. 2715–2717
34. L. Ferro-Famil, E. Pottier, J.S. Lee, Classification and interpretation of polarimetric interferometric SAR data, in *IEEE Proceedings of IGARSS'02* (2002), pp. 635–637
35. Z. Yang, Studies on synthetic aperture radar interferometric and polarimetric interferometric techniques. Institute of Electronics, Chinese Academy of Sciences, PhD (2003)
36. Z. Yang, R. Yang, X. Liu, Study of Polarimetric Interferometric Unsupervised Wishart Classification of SAR Images. J. Electron. Inf. Technol. **26**(5), 752–759 (2004)
37. Z. Yang, R. Yang, Application of polarimetric synthetic aperture radar interferometry for land cover classification, in *Proceedings of the IEEE Radar Conference* (2002), pp. 459–463

Chapter 14
Surface Parameter Inversion of Polarimetric Synthetic Aperture Radar Data

14.1 Introduction

Soil moisture inversion using polarimetric SAR data is an important contribution of polarized microwave imaging radars in hydrogeography, meteorology, and agricultural process economic optimization. The main difficulty in quantitatively estimating soil moisture or surface roughness is how to separate their effects on radar backscatter. In order to accurately estimate soil moisture, surface roughness information is often required. It is not possible to estimate the surface roughness independently using only one frequency or one polarization SAR data, and roughness is considered as an interference factor for soil moisture estimation. This chapter will describe the methods for estimating surface roughness and soil moisture independently from fully polarized SAR data [1–19].

The surface characteristics of the scattering media are primarily described by their material (dielectric) and geometrical properties. On the one hand, soil moisture content per unit volume is an important material parameter. It affects the dielectric constant of the medium and thus affects the radar backscatter. On the other hand, the surface Root Mean Square (RMS) height s and surface correlation length l, respectively, describe the effects of horizontal and vertical spatial roughness on radar geometry. On the other hand, the surface Root Mean Square (RMS) height s and the surface correlation length l, respectively, describe the effects of spatial roughness in the horizontal and vertical directions on the geometrical characteristics of the radar. In the backscattered signal, both the horizontal and vertical surface roughness parameters are normalized to ks and kl based on the actual wavelength, where $k = 2\pi / \lambda$. There is no any simple relationship between the backscattered signal and the surface parameters. The polarization information can directly or indirectly distinguish the effect of surface roughness and soil moisture on radar backscattering. Therefore, polarization information plays an important role in surface parameter inversion.

Electromagnetic wave scattering from random rough surfaces has been a hot topic in practical research, and there is no exact closed-loop solution. However, for many practical applications, an approximate solution is sufficient. The most commonly

R. Yang et al., *Polarimetric Microwave Imaging*,
https://doi.org/10.1007/978-981-15-8897-6_14

used approximation in the field of microwave imaging radars is the Small Perturbation Model (SPM). In the Bragg scattering region, roughness can be considered as perturbation of a smooth surface. The SPM performs a Fourier spectral component decomposition on a random surface. Each surface is an ideal sinusoidal surface. Scattering is primarily caused by the surface that resonates with the incident wavelength and the angle of incidence. The backscatter coefficient can be obtained by SPM. SPM assumes that the surface height changes less than the wavelength, and it is more suitable for longer wavelength bands, such as S, P, L, and other bands. SPM is one of the classic methods and is widely used in practice.

In SPM, the co-polar scattering ratio of a slightly rough ($ks << 0.3$) surface is sensitive to the dielectric constant, and the more rough ($ks > 0.3$) the surface, it becomes the less sensitive. For exposed natural surfaces in the L-band, the typical range of ks values is $0.3 << ks << 1$ ($1 << s\,[\text{cm}] << 4$), which is far beyond the required range of SPM. In order to extend the application of SPM to a large range of surface roughness, people have obtained empirical or semi-empirical deterministic relationships of the amplitude ratios of co-polarization and cross-polarization. This technology is simple and fast, and its main limitation is the limitation on the application of natural surface roughness.

Another promising method for surface parameter extraction is the polarized surface scattering model. This model is based on the analysis of the eigenvalues and eigenvectors of the polarization correlation matrix. It starts from the study of first-order scattering statistics of surface scatters. The correlation coefficient between different polarizations is related to the dielectric constant or/and the measured rms height. This indicates that relevant information may separate soil moisture from surface roughness. Some experimental results also show the relationship between the surface roughness and the eigenvalues of the polarization correlation matrix.

The SPM, the second-order scattering distribution of surface scatters, and the polarization surface scattering model are discussed in this chapter. And the applications of polarimetric SAR data in surface roughness inversion and soil moisture inversion are analyzed.

14.2 SPM (Small Perturbation Model) [20, 21]

At oblique incidence, the backscatter of a completely smooth surface is zero. In the Bragg scattering region, the change in surface height is much smaller than the wavelength, which is $ks << 0.3$. Surface roughness can be viewed as a perturbation problem with smooth surfaces. In this case, the backscattering coefficient can be obtained by SPM. The polarization scattering matrix $[S]$ describes the backscattering of the Bragg surface polarization and can be written as

$$[S] = \begin{bmatrix} S_{HH} & S_{HV} \\ S_{VH} & S_{VV} \end{bmatrix} = m_s \begin{bmatrix} R_S(\theta, \varepsilon) & 0 \\ 0 & R_P(\theta, \varepsilon) \end{bmatrix} \tag{14.1}$$

where m_s is the backscatter amplitude and it contains surface roughness information. R_S and R_P are Bragg scattering coefficients that are perpendicular and parallel to the incident plane, respectively. They are both functions of the complex dielectric constant ε and the local incident angle θ.

$$R_S = \frac{\cos\theta - \sqrt{\varepsilon - \sin^2\theta}}{\cos\theta + \sqrt{\varepsilon - \sin^2\theta}}$$
$$R_P = \frac{(\varepsilon - 1)(\sin^2\theta - \varepsilon(1 + \sin^2\theta))}{\left(\varepsilon\cos\theta + \sqrt{\varepsilon - \sin^2\theta}\right)^2} \tag{14.2}$$

Equation (14.2) leads to one of the important arguments of SPM. That is, the co-polarization ratio R_S/R_P depends only on the complex dielectric constant ε and the local incident angle θ and has nothing to do with the surface roughness. From the Eqs. (14.1) and (14.2), the relationship between the co-polarization ratio and the complex permittivity ε (soil moisture) can be obtained. The relationship between soil moisture and complex permittivity ε can be found in the literature [22]. Figure 14.1 shows this relationship when the incident angle changes from 25° to 60°. The co-polarization ratio of dry surface is close to 1, and the co-polarization ratio decreases with increasing humidity. The co-polarization ratio changes quickly at $0 < m_v < 20$ [vol%] for all angles of incidence. But when $m_v > 20$, the co-polarization ratio is saturated. This indicates that the SPM model is not sensitive to very wet surfaces. When the soil moisture reaches saturation, there is great uncertainty in the inversion results. Some experimental studies have corrected the relationship between co-polarization ratio and soil moisture when the soil moisture on slightly rough surfaces is greater than saturation ($m_v > 20$ [vol%]). However, the surface roughness of SPM is too severe for most natural surfaces. In any case, even if the surface

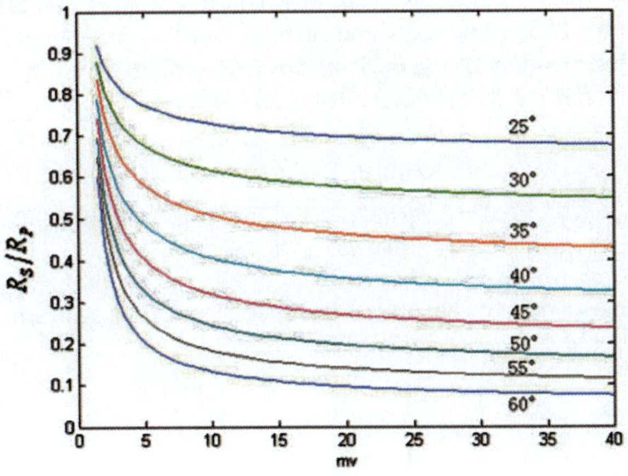

Fig. 14.1 Relationship between co-polarization ratio and soil moisture

roughness of the measured area exceeds the limit of the effective range of the SPM, the estimated results will still have a general understanding of the surface parameters.

In general, due to the effect of roughness, for the corresponding m_v value, the co-polarization ratio R_S/R_P of natural surface experimental data is much higher than the predicted value of SPM. Because of this underestimation, the SPM inversion results are usually less than the soil moisture. SPM can perform m_v inversion directly. Nonlinear formula is obtained directly from R_S/R_P. For a given angle of incidence, this ratio depends only on the dielectric constant ε'. Solve ε' and convert it to m_v to get an estimate of soil moisture.

14.3 Second-Order Scattering Distribution of Surface Scatterers

In order to describe the correlation of natural media, the scattering correlation matrix $[T]$ needs to be introduced. Let us first review three main physical property parameters related to scattering characteristics: polarization scattering entropy H, polarization scattering angle α, and scattering non-uniformity A.

The two parameters H and A vary between 0 and 1. The scattering entropy H is interpreted as a measure of the randomness of the scattering process. The scattering inhomogeneity A defines the relationship between the second and third eigenvalues. For smooth surfaces, H is 0, which indicates the scattering process described by a single scattering matrix, without depolarization effects. Non-zero H characterize scattering from depolarized surfaces. For medium H surfaces, a large A indicates that there is only one secondary scattering process; a small A indicates that there are two scattering processes of equal intensity. For the azimuthally symmetric surface, $\lambda_2 = \lambda_3$, A is 0. For rough surfaces, A can also be zero. Thus, A is complementary to H and can help explain surface scattering. The advantage of these two parameters is that the eigenvalues are invariant in a single transformation: the same eigenvalues are obtained on the same surface, and the same H and A are obtained, and they are independent of the basis of the scattering matrix.

For the SPM, the Pauli-based scattering vector is

$$\vec{k}_P = 1/\sqrt{2}\begin{bmatrix} S_{HH} + S_{VV} \\ S_{HH} - S_{VV} \\ 2S_{HV} \end{bmatrix} = m_S \begin{bmatrix} R_S + R_P \\ R_S - R_P \\ 0 \end{bmatrix} = m \begin{bmatrix} \cos\alpha \exp(i\phi_1) \\ \sin\alpha \exp(i\phi_2) \\ 0 \end{bmatrix} \tag{14.3}$$

where m is the absolute amplitude. The Bragg scattering correlation matrix is shown in (14.4):

$$[T] = \left\langle \vec{k}_P \vec{k}_P^{+} \right\rangle$$

$$= m_s^2 \begin{bmatrix} \langle |R_s + R_P|^2 \rangle & \langle (R_s + R_P)(R_s - R_P)^* \rangle & 0 \\ \langle (R_s - R_P)(R_s + R_P)^* \rangle & \langle |R_s - R_P|^2 \rangle & 0 \\ 0 & 0 & 0 \end{bmatrix} \quad (14.4)$$

Since the cross-polarization is zero, two of the three off-diagonal terms are also zero. The third term is only related to the surface permittivity and radar incident angle. The corresponding normalized correlation coefficient is

$$\begin{aligned} \gamma_{(HH+VV)(HH-VV)} &= \frac{|\langle (HH+VV)(HH-VV)^* \rangle|}{\sqrt{\langle |HH+VV|^2 \rangle \langle |HH+VV|^2 \rangle}} \\ &= \frac{|\langle (R_s + R_P)(R_s - R_P)^* \rangle|}{\sqrt{\langle |R_s + R_P|^2 \rangle \langle |R_s - R_P|^2 \rangle}} \\ &= 1 \end{aligned} \quad (14.5)$$

This result indicates that SPM cannot describe the depolarization effect.

The scattering angle α indicates the type of the average scattering process. Similar to the same polarization ratio R_S / R_P, the angle α from Eqs. (14.3) and (14.6) is independent of the surface roughness. Therefore, if the local angle of incidence has been determined, the α angle can be used to estimate the dielectric constant.

$$\alpha = \arccos\left(\frac{|R_s + R_P|}{\sqrt{\langle |R_s + R_P|^2 \rangle \langle |R_s - R_P|^2 \rangle}} \right) \quad (14.6)$$

The relationship between the angle α and m_v is shown in Fig. 14.2. The dry surface has a small α angle, which increases with increasing soil moisture. The maximum

Fig. 14.2 Relationship between α and soil moisture

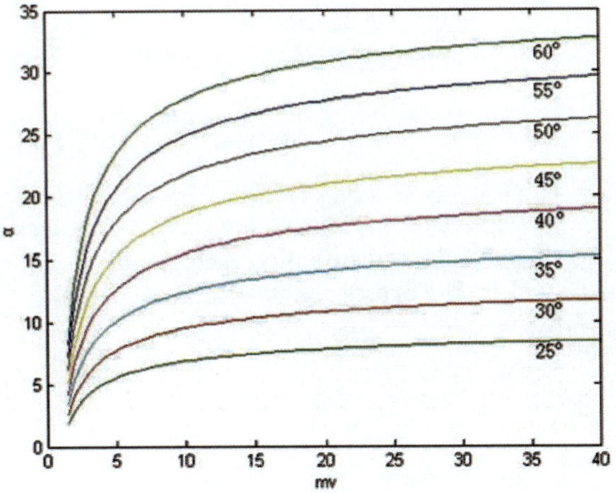

sensitivity at angle α is between $0–20m_v$, and then as m_v increases, the angle α saturates.

The main limitations of the SPM model are

(1) The applicable range of the surface roughness is very small (less than $0.3k_s$);
(2) When soil moisture $m_v > 20$ [vol%], the measurement sensitivity is saturated;
(3) It cannot describe the effect of depolarization, which limits its applicability to the inversion of natural surface measured data and interpretation of inversion results.

Because SPM is very robust in its applicable range, scholars use it as a starting point for extending models.

14.4 Polarized Surface Scattering Model

In order to extend the Bragg scattering model to a wider range of roughness, two main influences must be introduced: non-zero cross-polarized backscatter and depolarization. In fact, the traditional two-scale or multi-scale expansion introduces the cross-polarization term, but cannot explain the depolarization effect. Another way to introduce roughness is to model the surface as a reflective symmetrical depolarizer. The Bragg correlation matrix $[T]$ is described by rotating β angle from the vertical plane to the scattering plane, such as (14.7)

$$[T(\beta)] = \boldsymbol{V}\begin{bmatrix} \langle |R_S + R_P|^2 \rangle & \langle (R_S + R_P)(R_S - R_P)^* \rangle & 0 \\ \langle (R_S - R_P)(R_S + R_P)^* \rangle & \langle |R_S - R_P|^2 \rangle & 0 \\ 0 & 0 & 0 \end{bmatrix}\boldsymbol{V}^T \quad (14.7)$$

where $\boldsymbol{V} = \begin{bmatrix} 1 & 0 & 0 \\ 0 & \cos 2\beta & \sin 2\beta \\ 0 & -\sin 2\beta & \cos 2\beta \end{bmatrix}$.

For terrain averages of the β angles of any given distribution $P(\beta)$, we have

$$[T] = \int_0^{2\pi} [T(\beta)]P(\beta)d\beta \quad (14.8)$$

The angle β corresponds to the surface roughness of the model. Assume that the angle β is a uniform distribution with a width of 2 β_1 and a mean value of 0, as shown in Fig. 14.3. The expression is as follows:

$$P(\beta) = \begin{cases} \frac{1}{2\beta_1} & |\beta| \le \beta_1 \\ 0 & \beta_1 < |\beta| \le 2\pi \end{cases} \quad (14.9)$$

The correlation matrix $[T]$ of the rough surface is obtained, such as (14.10):

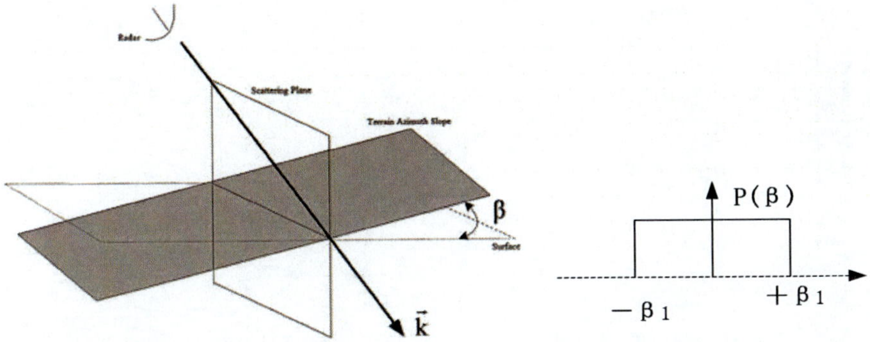

Fig. 14.3 Model of β and the distribution width

$$[T] = \begin{bmatrix} T_{11} & T_{12} & T_{13} \\ T_{21} & T_{22} & T_{23} \\ T_{31} & T_{32} & T_{33} \end{bmatrix} = \begin{bmatrix} C_1 & C_2 \sin c(2\beta_1) & 0 \\ C_2 \sin c(2\beta_1) & C_3(1 + \sin c(4\beta_1)) & 0 \\ 0 & 0 & C_3(1 - \sin c(4\beta_1)) \end{bmatrix} \tag{14.10}$$

where $\sin c(x) = \sin(x)/x$, the correlation coefficient C_1, C_2, C_3 describes the Bragg component of the surface.

$$C_1 = |R_S + R_P|^2$$

$$C_2 = (R_S + R_P)\left(R_S^* - R_P^*\right)$$

$$C_3 = \frac{1}{2}|R_S - R_P|^2 \tag{14.11}$$

On the one hand, Eq. (14.10) represents an extended Bragg scattering correlation matrix that introduces cross-polarization energy. Its polarization correlation coefficient is smaller than 1:

$$\gamma_{(\mathrm{HH+VV})(\mathrm{HH-VV})} = \frac{T_{12}}{\sqrt{T_{11}T_{22}}} = \frac{C_2}{\sqrt{C_1 C_3}} \frac{\sin c(2\beta_1)}{\sqrt{\frac{(1+\sin c(4\beta_1))}{2}}} \le 1 \tag{14.12}$$

The distribution width β_1 describes the degree of surface roughness, which is related to the correlation coefficient between the cross-polarization energy and (HH + VV)(HH − VV). Figure 14.4a shows the change of the correlation coefficient (HH + VV)(HH − VV), and Fig. 14.4b shows the normalized cross-polarization energy. For a completely smooth surface, $\beta_1 = 0$, the correlation coefficient

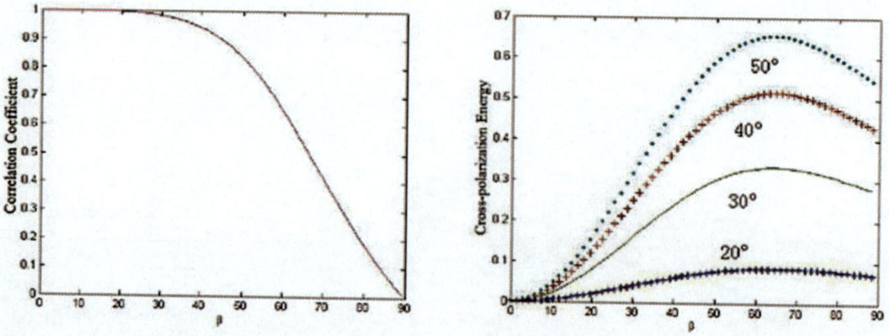

(a) The correlation coefficient (HH + VV)(HH − VV)

(b) Normalized cross-polarization energy

Fig. 14.4 The correlation coefficient (HH + VV)(HH − VV) and the normalized cross-polarization energy

(HH + VV)(HH − VV) is 1, the HV backscattering energy is 0, and Eq. (14.4) is expressed as a complete Bragg correlation matrix. The HV energy increases with increasing roughness β_1, the (HH + VV)(HH − VV) correlation coefficient decreases monotonically from 1 to 0. When the roughness is strictly limited, the surface becomes azimuthally symmetric. In addition, although the cross-polarization energy increases with increasing roughness and is related to the surface dielectric constant (and angle of incidence), but also the decrease in the correlation coefficient (HH + VV)(HH − VV) is not related to either.

On the other hand, according to (14.10), the ratio is

$$\frac{T_{22} + T_{33}}{T_{11}} = \frac{|R_S - R_P|^2}{|R_S + R_P|^2} \tag{14.13}$$

It is not related to surface roughness, it is only related to the surface's dielectric constant and incidence angle, as shown in Fig. 14.5. Thus, the extended Bragg scattering model can essentially separate the roughness from soil moisture effect obtained from the correlation matrix element ratio. This gives independent estimates of surface roughness and soil moisture.

14.5 Surface Parameter Inversion

14.5.1 Surface Roughness Inversion

From Eq. (14.10), the correlation coefficients of left-left handedness (L-L) and right-right handedness (R-R) circular polarization can be obtained as follows:

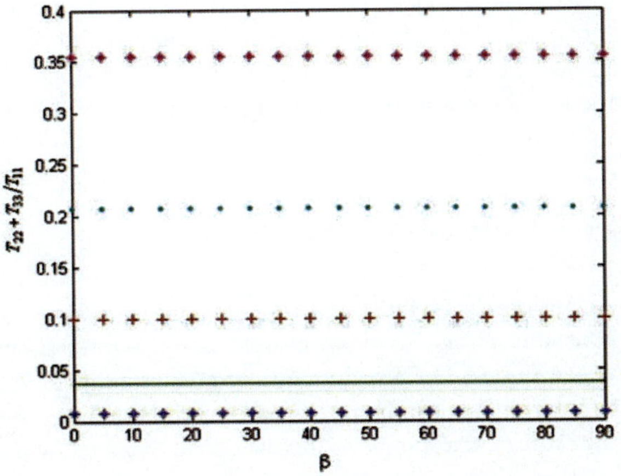

Fig. 14.5 Relationship between $T_{22} + T_{33} / T_{11}$ and β_1 when the dielectric constants are different

$$R_{\mathrm{LLRR}} = \frac{T_{22} - T_{33}}{T_{22} + T_{33}} = \sin c(4\beta_1) \tag{14.14}$$

It is only related to surface roughness. The scattering inhomogeneity A can be interpreted as a generalized rotation-invariant expression of R_{LLRR}, so A is also considered to be independent of the surface dielectric constant.

Indeed, as β_1 increases, A decreases monotonically from 1 to 0, regardless of the dielectric constant (and angle of incidence), as shown in Fig. 14.6. When the value of ks is increased to 1, that is, β_1 is increased to 90°, A and ks are close to a linear relationship, which has nothing to do with the surface dielectric constant, that is, it is not related to soil moisture. A and ks have almost a linear relationship, which has nothing to do with the surface dielectric constant, that is, it is not related to soil moisture. Above $ks = 1$, A is not sensitive to the increase in surface roughness. This

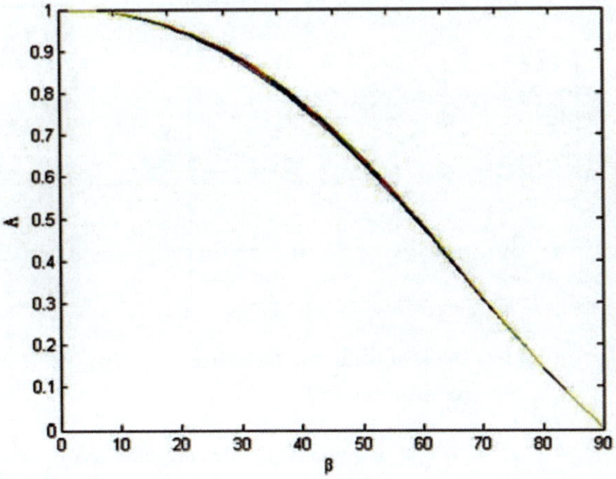

Fig. 14.6 Relationship between A and β_1 of different dielectric constants

directly separates the estimation of surface roughness from the estimation of soil moisture. This is one of the main advantages of this model. In addition, this result has nothing to do with the distribution of the slope β, the form of $P(\beta)$ only affects the mathematical expression of parameter A.

14.5.2 Soil Moisture Retrieval

According to formula (14.10), H-α track at various points of different dielectric constants ε' and inclination distribution width β_1 can be plotted at any incident angle. Figure 14.7a is the track of H-α at an incidence angle of 45°. This track can be interpreted as polar coordinates with H = 0 and $\alpha = 0$ as the origin. In this sense, the radius corresponds to the dielectric constant, and the azimuth angle represents a change in roughness. For an absolutely smooth surface, the entropy H is 0, and the angle α corresponds directly to the dielectric constant through (14.6). However, as H increases in the figure with increasing roughness, and α decreases, which leads to underestimation of the dielectric constant. In order to correctly estimate soil moisture, both H and α values are used. Figure 14.7b shows the effect of the local incident angle on the track of H-α. Both the scattering angle α and the entropy H value increase as the incident angle increases from 20° to 50°.

Figure 14.8 shows the effect of A-α track of different dielectric constants and the effect of local incident angle. Figure 14.8a is the A-α track of different dielectric constants and β_1. Figure 14.8b is the effect of local incident angle. Figure 14.9 shows A-H track of different dielectric constants and β_1, and the effect of local incident angle. Figure 14.9a is A-H track of different dielectric constants and β_1. Figure 14.9b is effect of local incident angle. These four figures show again that the A is independence with vary of soil moisture and local incident angle. The scattering inhomogeneity parameter A is measured parameter of surface roughness, and it is no

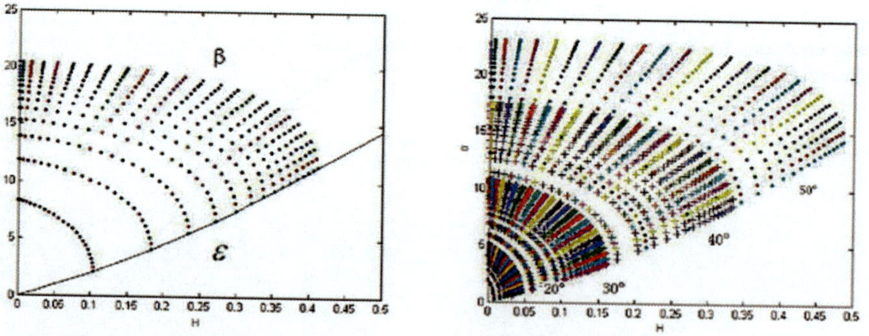

(a) H-α track of different dielectric constants and β_1

(b) Effect of local incident angle

Fig. 14.7 H-α track of different dielectric constants and β_1, and the effect of local incident angle

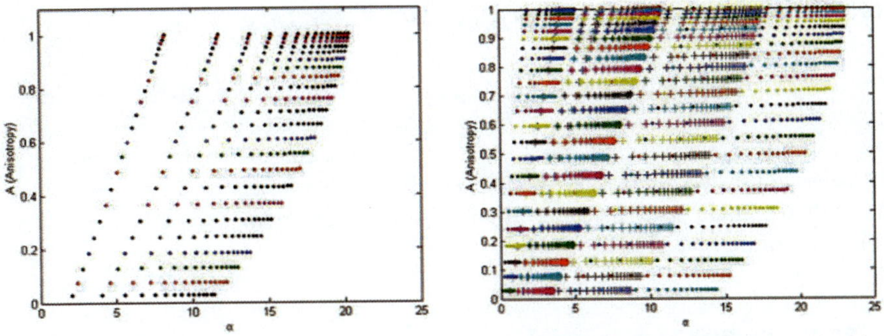

(a) A-α track of different dielectric constants and β_1

(b) Effect of local incident angle

Fig. 14.8 A-α track of different dielectric constants and β_1, and the effect of local incident angle

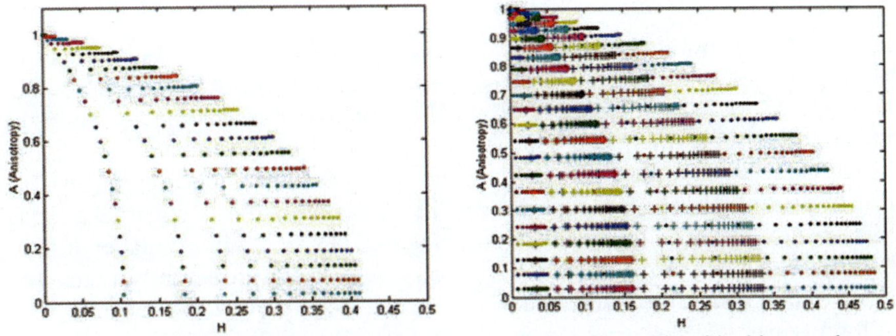

(a) A-H track of different dielectric constants and β_1

(b) Effect of local incident angle

Fig. 14.9 A-H track of different dielectric constants and β_1, and the effect of local incident angle

change with vary of the dielectric constant and local incident angle. It will provide a basis on the surface roughness inversion from the effect of soil moisture.

In summary, the three parameters of the scattering entropy H, the scattering non-uniformity parameter A, and the scattering angle α can independently estimate the surface roughness and the surface dielectric constant. The surface roughness is directly estimated from the scattering non-uniformity parameter A, and the dielectric constant is estimated from the scattering entropy H and the scattering angle α.

14.5.3 Experiment and Surface Parameter Inversion

In order to reduce the influence of multiplicative coherent speckle noise and additive noise, the polarization SAR data is filtered first, and then the polarization entropy H, the scattering non-uniformity parameter A, and the scattering angle α are calculated by polarization decomposition method. In order to determine that the processed area has only one main surface scattering mechanism, those areas $H > 0.5, \alpha > 45°$ are shielded before inversion [23, 24].

14.5.3.1 Surface Roughness Inversion

For other regions except the region $H > 0.5, \alpha > 45°$, the linear relationship between A and ks is approximated by the following formula:

$$ks = 1 - A \quad (14.15)$$

Since there is no obvious analytical relationship between A and ks, the optimal estimation result cannot be obtained. In any case, the inversion formula of Eq. (14.15) has obtained a good estimation result within the range of rough conditions on the surface for most natural media.

Surface scattering characteristics are indicated by the dominant scattering mechanism (represented by the first eigenvalue). The magnitude of the secondary scattering is represented by the second and third eigenvalues. They are 25 dB or smaller than the first eigenvalue. They are close to the system noise level and are severely affected by additive noise. Therefore, compared with the scattering entropy H, the scattering non-uniformity parameter A is a parameter with noise. Especially in the low scattering entropy H region, noise causes an underestimation of A, that is, an overestimation of surface roughness. Therefore, it is necessary to perform additive noise filtering on the data very much. The experimental data is P-band fully polarized SAR image data acquired by NASA JPL in the FLEVOLAN area of the United States. Figure 14.10 is an image of the total polarization power in the region. The result of surface roughness estimation after additive noise filtering is shown in Fig. 14.11. Figure 14.11a is the amplitude of the surface roughness inversion result, and Fig. 14.11b is the pseudo-color image of the surface roughness inversion result. The black area in the figure indicates the area where the scattering entropy and the scattering angle are beyond the suitable applicable range of the model.

14.5.3.2 Soil Moisture Retrieval

Before inversion of soil moisture, the dielectric constant was estimated using H, α. The correlation matrices of different dielectric constants and incidence angles are modeled by Eq. (14.10). The complex dielectric constant and the corresponding H

Fig. 14.10 Total polarization power in P-band of FLEVOLAN

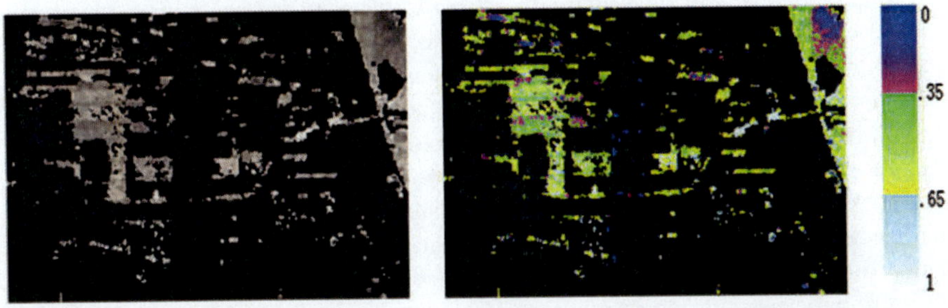

(a) Amplitude of Surface roughness (b) Pseudo-color Image of Surface Roughness

Fig. 14.11 Inversion results of surface roughness

and α are stored in a look-up table in advance. According to H and α, the dielectric constant value can be directly queried from the look-up table. The polarization surface scattering model is an extension of the SPM model. The sensitivity of the two models to soil moisture estimation is similar. For slightly rough surfaces, when the soil moisture is higher than $m_v = 20$ [vol%]. The SPM model reaches saturation, the polarization surface scattering model improves the applicable range. When $m_v =$ 35 [vol%], accurate estimation can still be made; When $m_v > 35$ [vol%], estimation performance decreased greatly.

Figure 14.12 shows the results of soil moisture inversion from the FLEVOLAN data shown in Fig. 14.10. Figure 14.12a is the amplitude image, and Fig. 14.12b is the pseudo-color image. Similarly, the black area in the figure indicates the area where the scattering entropy and scattering angle are beyond the applicable range of the model. The existence of vegetation greatly limits the applicable area of the surface parameter model.

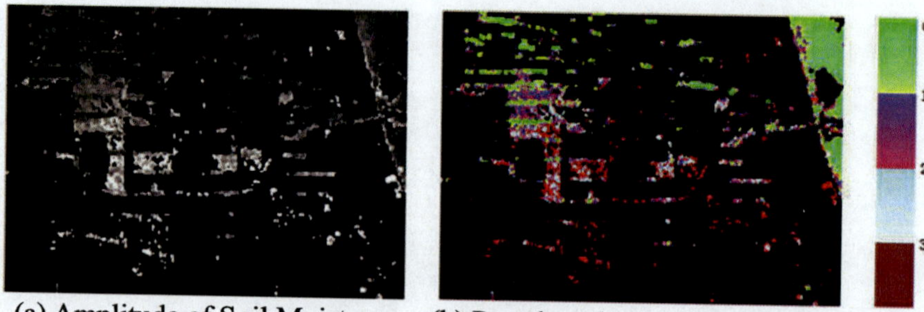

(a) Amplitude of Soil Moisture (b) Pseudo-color Image of Soil Moisture

Fig. 14.12 Inversion results of soil moisture

14.5.3.3 Discussion

The polarization surface scattering model is an extension of the SPM model. It can independently and quantitatively estimate the bare natural surface roughness and dielectric constant in a large range and solves the problems of non-zero cross-polarized backscattering and depolarization. Its main advantages are that it separates the estimation of the roughness from the estimation of the dielectric constant (that is, the soil moisture), and it can complete the parameter inversion without any data iteration. It does not have strict requirements on the absolute calibration of the scattering matrix. When estimating ks, only crosstalk and relative calibration of the channel are needed. When estimating ε', phase calibration relative to polarization is critical. The estimation of ε' has a great relationship with the angle of α. The co-polarized phase imbalance angle $\Delta\varphi$ directly affects the α angle estimation ($\Delta\alpha \sim \Delta\varphi/2$), so a high accuracy of phase calibration is required. The effect of noise can lead to overestimation of surface roughness in small eigenvalue regions. Multiplicative noise and additive noise filtering can reduce the effects of noise. In addition, a low correlation length will cause multiple scattering effects, leading to an increase in S_{HH}/S_{VV} value and introducing an error in α angle estimation. However, since the surface-dependent length term is not reflected in the model, its impact is unknown.

The main limitation of surface parameter estimation using PolSAR data is the presence of vegetation. On the one hand, vegetation increases the scattering entropy H and decreases A, resulting in overestimation of surface roughness. On the other hand, the presence of vegetation will increase the angle α, leading to overestimation of the complex dielectric constant. In fact, most natural surfaces are temporarily or permanently covered by vegetation, which greatly limits the widespread application of radar remote sensing in geophysics and environment. However, with the development of radar technology and techniques, it is believed that this limitation will be gradually overcome. It has been proposed that a longer wavelength (lower frequency) can be used to invert surface parameters with vegetation coverage. For example, the P-band has the potential to have a fairly high ability to penetrate and pass through vegetation layers. However, in the case of dense vegetation, it is difficult to invert surface parameters to effectively separate vegetation scattering from surface scattering components. In addition, the effective roughness is normalized by wavelength. The medium-roughness bare surface appears very smooth in the P-band, which implies that the low scattering coefficient is often close to the system noise level. Therefore, the effect of additive noise becomes very important. A more promising approach is to use dual band (e.g., L- and P-band) full-polarization SAR. The combination of two or more frequencies may relax the restrictions on natural surfaces on the one hand, and the parameter inversion will be more robust on the other hand. However, a change in wavelength may also imply a change in the scattering process, which affects the performance of the inversion algorithm [25].

Another challenging method is polarization interference technology [23, 26]. The sensitivity of the interference phase, the correlation of the effective scattering center position in the resolution unit, etc., combined with the strong influence of ground scattering on the scattering center position. These allow polarization interference techniques to be used to invert surface parameters with even ground scattering from vegetation cover. In addition, interference-related changes are a function of the baseline, which can calculate the vegetation layer on the ground. Polarization is important for surface scattering inversion. Polarized interference, that is, the combination of polarization and interference, has very promising application prospects for information extraction under vegetation layers.

14.6 Summary

This chapter introduces a polarized surface scattering model for inversion of surface roughness and soil moisture from PolSAR data. The model is based on the eigenvalue and eigenvector analysis of the polarization correlation matrix. It shows that in the inversion of soil surface parameters, the three polarization parameters, scattering entropy H, scattering non-uniformity parameter A, and scattering angle α, can remove the coupling between soil moisture and surface roughness. The algorithm was experimentally studied using the measured data of P-band full-polarization SAR in FLEVOLAN area by NASA JPL, and the inversion results of surface roughness and soil moisture in this area were obtained independently.

References

1. L.-S. Lee, D.L. Schuler, Polarimetric SAR data compensation for terrain azimuth slope variation. IEEE Trans. GRS **38**(5), 2153–2163 (2000)
2. L.H. Ulander, Radiometric slope correction of synthetic aperture radar images. IEEE. Trans. GRS **34**(5), 1115–1122 (1996)

3. D.L. Schuler, J.-S. Lee, T.L. Ainsworth, W.M. Boerner, Correction for terrain azimuthal slope effects in polarimetric SAR images, in *Geoscience and Remote Sensing Symposium, 2000. Proceedings. IGARSS 2000*, vol. 3 (2000), pp. 1304–1306
4. P.E. O'Neill, A.Y. Hsu, M.S. Seyfried, SAR terrain correction for improved soil moisture estimation in a mountain watershed, in *Geoscience and Remote Sensing Symposium, 1994. IGARSS '94*, vol. 3 (1994), pp. 1429–1431
5. G. Riegler, W. Mauser, Geometric and radiometric terrain correction of ERS SAR data for applications in hydrologic modelling, in *Geoscience and Remote Sensing Symposium Proceedings, 1998. IGARSS '98*, vol. 5 (1998), pp. 2603–2605
6. I. Hajnsek, E. Pottier, Terrain correction for quantitative moisture and roughness retrieval using polarimetric SAR data, in *Geoscience and Remote Sensing Symposium, 2000. Proceedings. IGARSS 2000*, vol. 3 (2000), pp. 1307–1309
7. J.T. Parr, Geometric transformation and terrain correction of remotely-sensed imagery, in *Geoscience and Remote Sensing Symposium, 1989. IGARSS'89*, vol. 2 (1989), pp. 440–441
8. H.A. Zebker, J.J. Van Zyl, S.L. Durden, L. Norikane, Calibrated imaging radar polarimetry: technique, examples, and applications. IEEE Trans. GRS **29**(6), 942–961 (1991)
9. M.S. Dawson, J. Olvera, A.K. Fung, M.T. Manry, Inversion of surface parameters using fast learning neural networks, in *Geoscience and Remote Sensing Symposium, 1992. IGARSS '92*, vol. 2 (1992), pp. 910–912
10. I. Hajnsek, S.R. Cloude, J.-S. Lee, E. Pottier, Inversion of surface parameters from polarimetric SAR data, in *Geoscience and Remote Sensing Symposium, 2000. Proceedings. IGARSS 2000*, vol. 3 (2000), pp. 1095–1097
11. M. Moghaddam, S. Saatchi, An inversion algorithm applied to SAR data to retrieve surface parameters, in *Geoscience and Remote Sensing Symposium, 1993. IGARSS '93*, vol. 2 (1993), pp. 587–589
12. K.S. Chen, Y.C. Tzeng, W.L. Kao, Retrieval of surface parameters using dynamic learning neural network, in *Geoscience and Remote Sensing Symposium, 1993. IGARSS '93. 'Better Understanding of Earth Environment'*, vol. 2 (1993), pp. 505–507
13. R. Hoeben, P.A. Troch, Z. Su, M. Mancini, K.-S. Chen, Sensitivity of radar backscattering to soil surface parameters: a comparison between theoretical analysis and experimental evidence, in *Geoscience and Remote Sensing, 1997. IGARSS'97*, vol. 3 (1997), pp. 1368–1370
14. E.G. Njoku, A. Chehbouni, F. Cabot, B. Rague, K. Fleming, Y.H. Kerr, An approach to estimating surface parameters and fluxes using modeling and multispectral remote sensing, in *Geoscience and Remote Sensing Symposium, 1994. IGARSS '94*, vol. 1 (1994), pp. 187–189
15. S.Foucher, F. Zagolski, C. Gaillard, Y. Voirin, M.N. Nguyen, Influence of speckle filtering on the assessment of agricultural surface parameters, in *Geoscience and Remote Sensing Symposium, 1999. IGARSS '99 Proceedings*, vol. 4 (1999), pp. 2137–2139
16. E.D. Hernandez-Baquero, J.R. Schott, Atmospheric and surface parameter retrievals from multispectral thermal imagery via reduced-rank multivariate regression, in *Geoscience and Remote Sensing Symposium, 2000. Proceedings. IGARSS 2000*, vol. 4 (2000), pp. 1525–1527
17. A. Ishimaru, J.-N. Hwang, K. Yoshitomi, J.-S. Chen, Application of artificial neural network (ANN) technique to rough surface inverse scattering problems, in *Antennas and Propagation Society International Symposium, 1992. AP-S. 1992 Digest*, vol. 3 (1992), pp. 1397
18. Y.-A. Liou, Y.C. Tzeng, A.W. England, Retrieval of soil moisture using a dynamic learning neural network trained with a 1-dimensional hydrology/radiobrightness model, in *Geoscience and Remote Sensing, 1997. IGARSS '97*, vol. 3 (1997), pp. 1096–1098
19. Y. Oh, K. Sarabandi, F.T. Ulaby, An inversion algorithm for retrieving soil moisture and surface roughness from polarimetric radar observation, in *Geoscience and Remote Sensing Symposium, 1994. IGARSS '94*, vol. 3 (1994), pp. 1582–1584
20. I. Hajnsek, E. Pottier, S.R. Cloude, Inversion of surface parameters from polarimetric SAR. IEEE Trans. GRS **41**(4), 727–744
21. D.L. Schuler, J.S. Lee, D. Kasilingam, G. Nesti, Remote sensing of surface roughness using polarimetric SAR data, in *Geoscience and Remote Sensing Symposium, 2001. IGARSS '01*, vol. 2 (2001), pp. 916–918

22. Z. Li, K. Feng, Inversion of soil moisture and complex dielectric constant from radar backward scattering coefficient. Sci. China Ser. E Inf. Sci. **27**(3), 223–228 (1997)
23. L. Sagues, X. Fabregas, A. Broquetas, Extraction of vegetation structure and surface parameters using polarimetric and interferometric SAR techniques, in *Geoscience and Remote Sensing Symposium, 2000. Proceedings. IGARSS 2000*, vol. 1 (2000), pp. 141–143
24. X. Liu, *Studies on Information Processing Technique of Fully-Polarimetric Synthetic Aperture Radar*. Institute of Electronics, Chinese Academy of Sciences, Ph.D. 2004.01
25. I. Hajnsek, K.P. Papathanassiou, S.R. Cloude, L- and P-band for surface parameter estimation, in *Geoscience and Remote Sensing Symposium, 2001. IGARSS '01*, vol. 6 (2001), pp. 2775–2777
26. I. Hajnsek, K.P. Papathanassiou, A. Moreira, S.R. Cloude, Surface parameter estimation using interferometric and polarimetric SAR, in *Geoscience and Remote Sensing Symposium, 2002. IGARSS '02*, vol. 1 (2002), pp. 420–422

Chapter 15
Information Processing of Polarimetric SAR Interferometry

Polarimetric SAR interferometry (PolInSAR) plays an important role in microwave remote sensing, which is a frontier topic in microwave remote sensing. There are two classes of applications of the information processing of polarimetric SAR interferometry. One of them is performing quantitative analysis using polarimetric SAR interferometry observations. The application fields include forest parameters retrieval [1–3], agricultural vegetation parameters retrieval [4], surface parameters retrieval [5, 6], glacier parameter retrieval [7–9], DEM retrieval [10, 11], etc. The other class is performing qualitative analysis with the information provided by polarimetric interferometric SAR. They can be applied in terrain classification [12–14], forest biomass extraction [15, 16], target indication [17, 18], change detection [19, 20] urban area analysis [21], etc. In this chapter, application of polarimetric SAR interferometry such as canopy parameters retrieval, DEM retrieval, and polarimetric coherence tomography is introduced.

15.1 Canopy Parameters Retrieval of Polarimetric SAR Interferometry

Full polarimetric SAR interferometry has great potential in obtaining vertical structure information. Among them, tree height retrieval is one of the most important applications of PolInSAR. Tree height is a very important biophysics parameter [22], which can be used to estimate the population, volume, density, and biomass of the forest.

R. Yang et al., *Polarimetric Microwave Imaging*,
https://doi.org/10.1007/978-981-15-8897-6_15

15.1.1 Multi-layers Vegetation Structure Model

In order to extract parameters from PolInSAR data, scattering model which combines observations values with retrieved parameters should be established. A good model should contain enough complex physical structure to describe the characteristics of the real object, while it can be simple enough for the retrieval. Among many forest canopy scattering models, Random Volume over Ground (RVoG) model proposed by Treuhaft, etc. is widely accepted.

15.1.1.1 The Random Volume over Ground (RVoG) Model

The Random Volume over Ground (RVoG) model is very close to real scattering situation. In this model, canopy is simulated as a homogeneous randomly oriented particle layer, and the backscattering of the canopy is irrelevant to the polarization. The ground is horizontal oriented and is very sensitive to the electromagnetic polarization.

In the Random Volume over Ground (RVoG) model, there are two cases in which the ground scattering influence SAR signals, which are with interactional ground scattering model and with direct ground scattering model. If the two scattering components are considered at same time, the scattering model will become very complex. In order to simplify scattering model, supposing the one scattering component occupies main contribution and the other scattering component can be ignored. These two situations will be analysed, respectively, as following:

(1) **Randomly oriented volume with interactional ground scattering model**

Figure 15.1 is the scheme of the randomly oriented volume with interactional ground scattering model, where the gray filled part represents ground surface while the particles represent vegetation. As we can see from Fig. 15.1, the interaction between the ground surface and the volume is the main scattering component. If the polarization states of the transmitting and receiving antennas both are $\hat{t}$, the voltages of both receiving antennas are $\boldsymbol{E}_{\hat{\mathrm{t}}}(\boldsymbol{R}_1)$ and $\boldsymbol{E}_{\hat{\mathrm{t}}}(\boldsymbol{R}_2)$. The complex coherence of this model is

$$\frac{\hat{t}\cdot\boldsymbol{E}_{\hat{\mathrm{t}}}(\boldsymbol{R}_1)\cdot\hat{t}^{*}\cdot\boldsymbol{E}_{\hat{t}^{*}}(\boldsymbol{R}_2)}{\sqrt{\left\langle\left|\hat{t}\cdot\boldsymbol{E}_{\hat{\mathrm{t}}}(\boldsymbol{R}_1)\right|^2\right\rangle\left\langle\left|\hat{t}^{*}\cdot\boldsymbol{E}_{\hat{t}^{*}}(\boldsymbol{R}_2)\right|^2\right\rangle}}
= Ae^{j\phi_0}\frac{\int_0^{h_v}e^{i\alpha_z z'}e^{2\sigma z'/\cos\theta_0}dz' + 4\Delta_{\hat{t}}^{s}h_v\sin(k_z h_v)/(k_z h_v)}{\frac{\cos\theta_0\left(e^{2\sigma h_v/\cos\theta_0 - 1}\right)}{2\sigma} + 4\Delta_{\hat{t}}^{s}h_v} \tag{15.1}$$

where $\boldsymbol{R}_1$ and $\boldsymbol{R}_2$ are the position vectors of the radar receiver at both ends of the baseline; A is the normalized amplitude; θ_0 is the incidence angle; z_0 is the topography

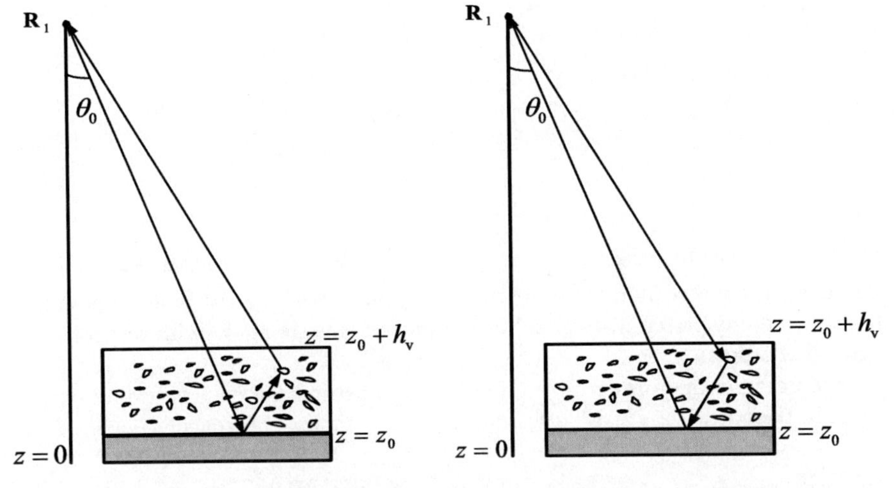

Fig. 15.1 The scheme of the ground scattering model of random volume interacted with scatterers

height; k_z is the partial differential of the interferometric phase with respect to the vertical coordinate z; $\Delta_{\hat{t}}^{s}$ is a parameter that relates the interaction between the ground and the scatterers. $\Delta_{\hat{t}}^{s}$ is the ratio of surface scattering to volume scattering in the echo, which is called surface–volume ratio. σ is the vegetation extinction coefficient.

(2) **Randomly oriented volume with direct ground scattering model**

Figure 15.2 shows the scattering model of randomly oriented volume with direct ground scattering model. It can be seen that direct surface scattering is the main contribution, and the interaction between ground surface and the vegetation is ignored. The complex coherence of this model is

Fig. 15.2 The scheme of randomly oriented volume and direct ground scattering

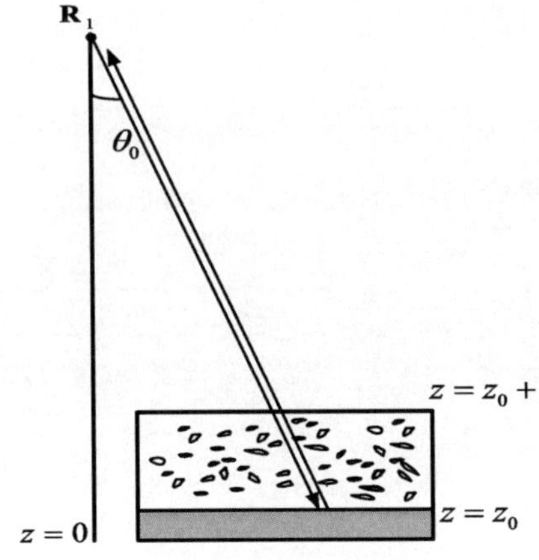

$$\frac{\hat{t} \cdot \boldsymbol{E}_{\hat{t}}(\boldsymbol{R}_1) \cdot \hat{t}^* \cdot \boldsymbol{E}_{\hat{t}^*}(\boldsymbol{R}_2)}{\sqrt{\left\langle\left|\hat{t} \cdot \boldsymbol{E}_{\hat{t}}(\boldsymbol{R}_1)\right|^2\right\rangle\left\langle\left|\hat{t}^* \cdot \boldsymbol{E}_{\hat{t}^*}(\boldsymbol{R}_2)\right|^2\right\rangle}}$$

$$= Ae^{j\phi_0} \frac{\int_0^{h_v} e^{i\alpha_z z'} e^{2\sigma z'/\cos\theta_0} dz' + 4\Delta_{\hat{t}}^D)}{\frac{\cos\theta_0\left(e^{2\sigma h_v/\cos\theta_0} - 1\right)}{2\sigma} + 4\Delta_{\hat{t}}^D} \tag{15.2}$$

where $\Delta_{\hat{t}}^D$ is the ratio of ground scattering to the volume scattering, and it represents the direct ground scattering compared to volume scattering. z_0 is the topography height; k_z is the partial differential of the interferometric phase with respect to the vertical coordinate$z$.

According to the former model, the interferometric coherence is further derived by S. R. Cloude, K. P. Papathanassiou, etc. The complex coherence is

$$\gamma(\boldsymbol{\omega}) = e^{j\phi_0} \frac{\gamma_v + m(\boldsymbol{\omega})}{1 + m(\boldsymbol{\omega})} \tag{15.3}$$

where $\boldsymbol{\omega}$ is the projection vector relating the polarization;ϕ_0 is the interferometric phase relating the topographic height; $m(\boldsymbol{\omega})$ is the ground-to-volume scattering ratio, which represents power ratio between the ground scattering and upper volume scattering. $m(\boldsymbol{\omega})$ is the only variable that relates to the polarization state. γ_v is the polarization relevant volume decorrelation, which relates to extinction coefficient σ and volume height h_v. The expression of γ_v is

$$\gamma_v = \frac{I}{I_0} = \frac{\int_0^{h_v} e^{jk_z z'} e^{(2\sigma z')/\cos\theta_0} dz'}{\int_0^{h_v} e^{(2\sigma z')/\cos\theta_0} dz'} \tag{15.4}$$

The relationship between the coherence matrix and vegetation parameters in RVoG model is derived by Cloude, etc. The expression of coherence matrix is

$$R = \begin{bmatrix} \boldsymbol{T} & \boldsymbol{X} \\ \boldsymbol{X}^H & \boldsymbol{T} \end{bmatrix} \tag{15.5}$$

where

$$\boldsymbol{T} = \boldsymbol{I}_1 \boldsymbol{T}_v + a\boldsymbol{T}_g \tag{15.6}$$

$$\boldsymbol{X} = e^{j\phi}\left(\boldsymbol{I}_1 \boldsymbol{T}_v + a\boldsymbol{T}_g\right) \tag{15.7}$$

$$a = e^{-2\sigma h_v/\cos\theta_0} \tag{15.8}$$

$$\boldsymbol{I}_1 = a \int_0^{h_v} e^{(2\sigma z')/\cos\theta_0} dz' \tag{15.9}$$

$$\boldsymbol{I}_2 = a \int_0^{h_v} e^{(2\sigma z')/\cos\theta_0} e^{jk_z z'} dz' \tag{15.10}$$

In Eqs. (15.6)–(15.10), $\boldsymbol{T}_v$ is the coherency matrix for the volume scattering and $\boldsymbol{T}_g$ the ground scattering coherency matrix, and z is volume scattering decorrelation. The expression of z is shown in Eq. (15.4).

$\boldsymbol{T}_v$ and $\boldsymbol{T}_g$ is defined as

$$\boldsymbol{T}_v = m_v \begin{bmatrix} 1 & 0 & 0 \\ 0 & \rho & 0 \\ 0 & 0 & \rho \end{bmatrix}, 0 \le \rho \le 0.5 \tag{15.11}$$

$$\boldsymbol{T}_g = \begin{bmatrix} t_{11} & t_{12} & 0 \\ t_{12}^H & t_{22} & 0 \\ 0 & 0 & t_{33} \end{bmatrix} \tag{15.12}$$

15.1.1.2 Independent Scattering Centers Model

Independent scattering center model is shown in Fig. 15.3. According to the similarity of the scatters in different layers, the backscattering is equivalent to the echo of a point-like scattering center. Considering the difference of the shape, orientation of the scatters in different layers, the scattering mechanisms of different layers can be assumed to be independent. Ignoring the influence of double and multiple scattering echo, the full polarized observation can be expressed as the sum of multiple scattering

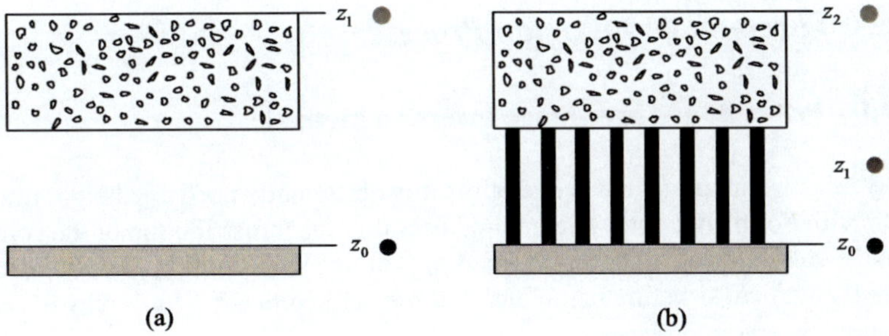

Fig. 15.3 The scheme of independent scattering centers. **a** Two independent scattering centers model; **b** three independent scattering centers model

centers.

$$k_B = \sum_{m=1}^{d} \sigma^m e^{j\kappa r^m} \boldsymbol{s}^m + \boldsymbol{n} \tag{15.13}$$

where d is the number of independent scattering centers; σ^m is the polarization span of the mth independent scattering centers; r^m is the distance from the mth scattering center to the receiving antenna; κ is the wavenumber of the radar signal; $e^{j\kappa r^m}\boldsymbol{s}^m$ is the unitary scattering vector of the mth scattering center; $\boldsymbol{s}^m$ is the unitary vector related to the scattering mechanism; $\boldsymbol{n}$ is the noise in the observations, and it is the three-dimensional vector composed of each polarization channel.

Usually, the noise of each polarization channel is zero-mean Gaussian white noise.

According to independent scattering center model, the observed PolInSAR data $\boldsymbol{k}_{1B}$ and $\boldsymbol{k}_{2B}$ from two different locations can be expressed as

$$k_{1B} = \sum_{m=1}^{d} \sigma^m e^{j\kappa r_1^m} \boldsymbol{s}^m + \boldsymbol{n}_1 \tag{15.14}$$

$$k_{2B} = \sum_{m=1}^{d} \sigma'^m e^{j\kappa r_2^m} \boldsymbol{s}'^m + \boldsymbol{n}_2 \approx \sum_{m=1}^{d} \sigma^m e^{j\kappa r_1^m} e^{j\phi^m} \boldsymbol{s}^m + \boldsymbol{n}_2 \tag{15.15}$$

In Eqs. (15.14)–(15.15), r_1^m is the distance from the mth scattering center to the receiving antenna. σ' and $\boldsymbol{s}'$ are the observed signal in location 2. r_2^m is the distance from the mth scattering center to the receiving antenna. $\boldsymbol{n}_1$ and $\boldsymbol{n}_2$ denote the noise in both observations.

During single-pass PolInSAR data acquire, or during short-interval repeated-pass PolInSAR data acquire, the location difference of the two observations is very little. It is assumed that the scattering centers are the same, that is $\sigma' \approx \sigma$, $s' \approx s$. As a result, the interferometric phase ϕ^m is relevant to the difference of r_2^m and r_1^m.

15.1.2 Three-Stage Inversion Process

15.1.2.1 Principle of Three-Stage Inversion Method

Three-stage inversion method is one of the most frequently used tree height retrieval method with PolInSAR data. The method uses the linear distribution of the complex coherence coefficients and bases on Random Volume over Ground (RVoG) model to retrieve the vertical structure parameter in three-stage process. That's why it's called three-stage inversion process.

The expression in Fig. (15.3) can be expressed as

$$\gamma(\omega) = e^{j\phi_0}\frac{\gamma_v + m(\omega)}{1 + m(\omega)} = e^{j\phi_0}\left(\gamma_v + \frac{m(\omega)}{1 + m(\omega)}(1 - \gamma_v)\right)$$
$$= e^{j\phi_0}(\gamma_v + L(\omega)(1 - \gamma_v)) \tag{15.16}$$

where $L(\omega) = \frac{m(\omega)}{1+m(\omega)}$, $0 \leq L(\omega) \leq 1$, and it denotes the ground-to-volume scattering ratio. It is a variable relevant to the polarization state. When $L(\omega) = 0$, complex coherence coefficient $\gamma(\omega) = e^{j\phi_0}\gamma_v$, it only relevant to volume scattering component, and it is volume decorrelation. When $L(\omega) = 1$, $\gamma(\omega) = e^{j\phi_0}$, and it is only relevant to ground surface scattering component. As we can see from Eq. (15.16), $\gamma(\omega)$ changing with ω is a straight line inside the unit circle of interferometric coherence plane. It is shown in Fig. 15.4.

As $L(\omega)$ changes from 0 to 1, the complex coherence coefficient (shown as point P in Fig. 15.4) varies from $e^{j\phi_0}\gamma_v$ to $e^{j\phi_0}$. This line has three important features.

(1) The line intersects the circle at two points. One of these is the underlying topography related phase point (shown as point Q in Fig. 15.4).
(2) The volume coherence γ_v lies at one end of the line ($\mu = 0$). It is shown as the point P in Fig. 15.4. This point corresponds to $L(\omega) = 0$, that is surface scattering does not exist.
(3) Distribution line of complex coherency $\gamma(\omega)$ for different polarization channel is called the visible line. The visible length of the line in the data may only be a fraction of PQ and neither P nor Q may be directly observed. The visible length depends on baseline, operating frequency, and vegetation density. However, the line can be extrapolated to estimate the parameters.

According to the aforementioned linear model of the coherence coefficients, the three-stage inversion process is as follows:

Stage 1: Least squares line fit and topographic phase estimation.

Fig. 15.4 Line model for the interferometric coherence

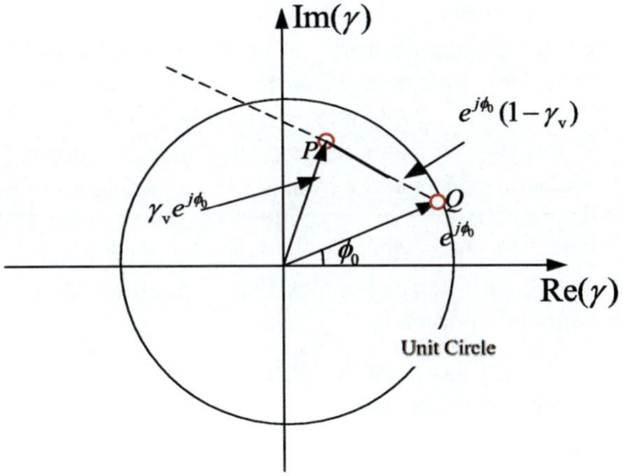

The first stage is to find the best-fit straight line inside the unit circle of interferometric coherence. A usually method is to estimate N complex coherence coefficients $\gamma_i (i = 1, 2, \cdots, N)$ of N polarization states of every pixel, and then make a least squares line fit from the N coherence coefficients. Let the slope of the line is m, and the intercept of the line is c, [32] then

$$\begin{cases} m = \frac{SS_{xy} - S_x S_y}{SS_{xx} - S_x^2} \\ c = \frac{S_{xx} S_y - S_x S_{xy}}{SS_{xx} - S_x^2} \end{cases} \tag{15.17}$$

$$\begin{cases} S = \sum_{i=1}^{N} w_i \\ S_x = \sum_{i=1}^{N} w_i \mathrm{Re}(\gamma_i) \\ S_{xx} = \sum_{i=1}^{N} w_i \mathrm{Re}(\gamma_i) \mathrm{Re}(\gamma_i) \end{cases} \tag{15.18}$$

$$\begin{cases} S_{xy} = \sum_{i=1}^{N} w_i \mathrm{Re}(\gamma_i) \mathrm{Im}(\gamma_i) \\ S_x = \sum_{i=1}^{N} w_i \mathrm{Im}(\gamma_i) \end{cases} \tag{15.19}$$

In Eqs. (15.17)–(15.19), w_i denotes the weight of specific points. If we adopt the uniform weighting least square fit, the weight is set to be $1/N$.

The fitted line has two intersects with the unit circle, which are $e^{j\phi_1}$, $e^{j\phi_2}$

$$e^{j\phi_1} = x_1 + j(mx_1 + c) \tag{15.20}$$

$$e^{j\phi_2} = x_2 + j(mx_2 + c) \tag{15.21}$$

where

$$x_1 = \frac{-mc + \sqrt{m^2 - c^2 - 1}}{1 + m^2}, x_2 = \frac{-mc - \sqrt{m^2 - c^2 - 1}}{1 + m^2} \tag{15.22}$$

After getting two intersects, we have to choose one of the points as the underlying ground topographic phase point. According to the characteristics of the scattering mechanism, we can see that the least ground-to-volume ratio will arise in the HV channel. That is, volume scattering is dominant in the HV channel. While HH + VV and HH-VV can change relative position depending on whether direct ground or dihedral effects are dominant, it is seen that the strongest ground-to-volume ratio will arise. That is surface scattering is dominant in HH + VV and HH-VV channel. Figure 15.5 shows the relative locations of different polarization channels coherence coefficients in the coherence line. Therefore, there are two ways of selecting the topographic phases.

(1) Choose the point which is further to γ_{HV} than γ_{HH+VV}(orγ_{HH-VV}) as topographic phase point.

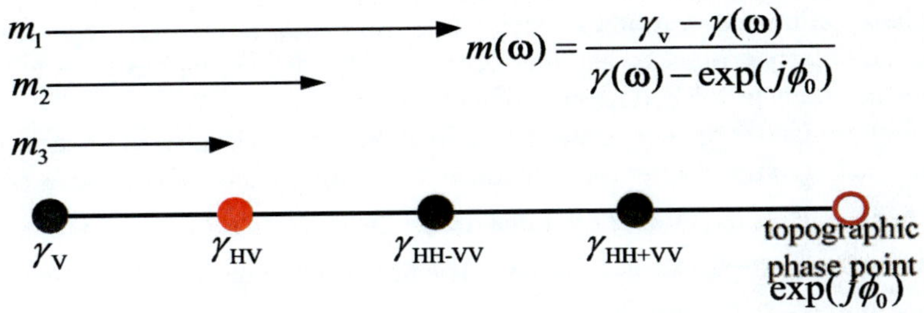

Fig. 15.5 The relative location of the coherence coefficients of different polarization channels

$$e^{j\phi_0} = \begin{cases} e^{j\phi_1} |\gamma_{HV} - \exp(j\phi_1)| > |\gamma_{HH+VV} - \exp(j\phi_1)| \\ e^{j\phi_2} |\gamma_{HV} - \exp(j\phi_2)| > |\gamma_{HH+VV} - \exp(j\phi_2)| \end{cases} \tag{15.23}$$

(2) Choose the point which has a larger amplitude angle difference of γ_{HV} than γ_{HH+VV}(or γ_{HH-VV}) as topographic phase point [33].

$$e^{j\phi_0} = \begin{cases} e^{j\phi_1} \left|\arg(\gamma_{HV}) - \phi_1\right| \geq |\arg(\gamma_{HH+VV}) - \phi_1| \\ e^{j\phi_2} \left|\arg(\gamma_{HV}) - \phi_2\right| \geq |\arg(\gamma_{HH+VV}) - \phi_2| \end{cases} \tag{15.24}$$

In Eqs. (15.23)–(15.24), $e^{j\phi_0}$ is the topographic phase point.

Stage 2: Volume decorrelation estimation.

Theoretically, the coherence coefficient of point corresponding to $m(\omega) = 0$ can be taken as the estimation of volume decorrelation. However, the polarization state which makes $m(\omega) = 0$ is hard to confirm. Usually, it is assumed that in HV channel, $m(\omega) = 0$. That is, there is no ground scattering in HV channel, and the estimation of volume correlation is $\hat{\gamma}_v = \gamma_{HV}\exp(-j\phi_0)$.

Stage 3: Vegetation parameter estimation.

According to Eq. (15.4), the volume decorrelation $\hat{\gamma}_v$ depends on the vegetation height h_v and extinction coefficient σ. Consequently, a look-up table of γ_v with h_v and σ can be established. Comparing estimated $\tilde{\gamma}_v$ with γ_v, we can get the estimation of vegetation height h_v by look-up table.

15.1.2.2 Error Analysis of Three-Stage Inversion Process

The error source of three-stage inversion process is as follows.

(1) Ground surface phase estimation error. Usually, the inaccuracy of the estimation of coherence coefficient, the error of straight line fit, the incompliance of the forest construction to the linear model, and the ground phase estimation induced by the misjudge of the topographic phase are the main error sources of the ground

phase estimation. Ground surface phase estimation is the first stage of three-stage inversion process, and the accuracy of the estimation has great effect on the accuracy of the inversion of other parameters.

(2) Volume scattering decorrelation estimation error. Usually, it is caused by residual ground scattering components in a hypothetical volume scattering channel. The estimation of volume decorrelation $\widetilde{\gamma}_v$ is a key step in three-stage inversion process. $\widetilde{\gamma}_v$ is of great relevance to the tree height and extinction coefficients. Therefore, the performance of the estimation of $\widetilde{\gamma}_v$ affects the estimation accuracy of vegetation height and extinction coefficient. Usually, only if the surface-to-volume scattering amplitude ratio $m(\boldsymbol{\omega})$ for the specified volume scattering polarization channel is $m(\boldsymbol{\omega}) \leq -10\text{dB}$, the accuracy of the estimation of $\widetilde{\gamma}_v$ can be guaranteed and the estimation error of the vertical structure parameter be controlled under 10% [35, 36].

(3) The influence of temporal decorrelation and noise decorrelation, etc. The RVoG model describes the quantitative relationship between PolInSAR observations and vertical structure parameters while ignoring decorrelation sources such as noise decorrelation and temporal decorrelation. Decorrelation factors in actual data will affect the performance of vertical structure parameter estimation.

15.1.2.3 Improved Three-Stage Inversion Process

As we can see from the aforementioned three-stage inversion process, the estimation of vegetation height relies on the assumption that the ground-to-volume scattering ratio $m(\boldsymbol{\omega}) = 0$ in the HV channel. It is considered that there is only vegetation scattering component in the HV channel. In fact, there's still some ground scattering component in HV channel. Therefore, it is not accurate to take $\gamma_{\text{HV}}\exp(-j\phi_0)$ as volume decorrelation. The inaccuracy of volume scattering decorrelation estimation will directly affect the accuracy of subsequent inversions.

Coherence region (CR) refers to the distribution area of all complex interference coherence coefficients in the complex plane. According to the properties of volume decorrelation, we can see that it is the most distinct point from the topography phase lying in the correlation region. The volume decorrelation coefficient is located at the furthest point away from the ground surface phase point on the coherent line. There must be a pair of polarization vectors meeting the condition. Literature [39] proposed a method of coherent region boundary extraction technology combined with polarimetric SAR interferometry to estimate the volume decorrelation coefficient. By seeking the boundary point of the coherence region and taking the most distinct point from the topography phase as the estimation of volume decorrelation, the ambiguity estimation of volume decorrelation is resolved, and the more accurate and reliable estimation of the volume decorrelation coefficient can be obtained (Fig. 15.6).

If the polarization state of master and slave images are the same, the interferometric coherence coefficients of arbitrary scattering vectors are

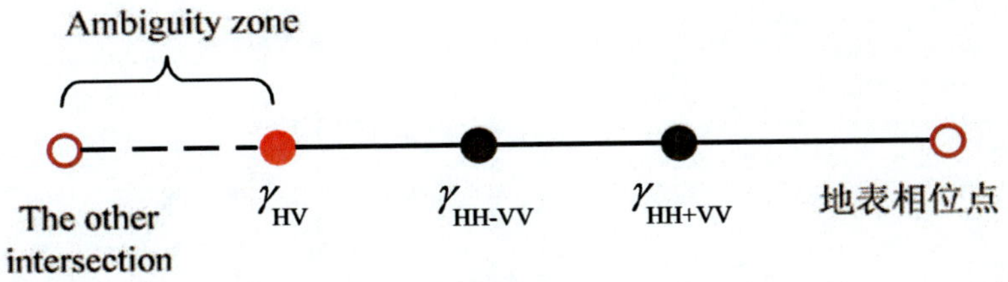

Fig. 15.6 The scheme of solving volume decorrelation

$$\gamma = \frac{\boldsymbol{\omega}^H \boldsymbol{\Omega}_{12} \boldsymbol{\omega}}{\sqrt{\boldsymbol{\omega}^H \boldsymbol{T}_{11} \boldsymbol{\omega} \cdot \boldsymbol{\omega}^H \boldsymbol{T}_{22} \boldsymbol{\omega}}} \tag{15.25}$$

$\boldsymbol{\omega}$ is the corresponding scattering mechanism in the master and slave images. $\boldsymbol{T}_{11}$ and $\boldsymbol{T}_{22}$ are, respectively, the covariance matrix of the master and slave images. $\boldsymbol{\Omega}_{12}$ is the polarization mutual covariance matrix. Infact, $\boldsymbol{T}_{11}$ and $\boldsymbol{T}_{22}$ are similar. Therefore, in order to simplify the calculation, the correlation coefficient can be modified as

$$\tilde{\boldsymbol{\gamma}} = \frac{\boldsymbol{\omega}^H \boldsymbol{\Omega}_{12} \boldsymbol{\omega}}{\boldsymbol{\omega}^H \boldsymbol{T} \boldsymbol{\omega}} \tag{15.26}$$

where $T = (T_{11} + T_{22})/2$. As

$$2\sqrt{\boldsymbol{\omega}^H \boldsymbol{T}_{11} \boldsymbol{\omega}} \cdot \sqrt{\boldsymbol{\omega}^H \boldsymbol{T}_{22} \boldsymbol{\omega}} \le \boldsymbol{\omega}^H \boldsymbol{T}_{11} \boldsymbol{\omega} + \boldsymbol{\omega}^H \boldsymbol{T}_{22} \boldsymbol{\omega} \tag{15.27}$$

We can get $\left|\tilde{\gamma}\right| \le |\gamma|$, and $\arg(\gamma) = \arg(\tilde{\gamma})$. From Eq. (15.26),

$$Re(\tilde{\gamma}) = \frac{\boldsymbol{\omega}^H \left((\boldsymbol{\Omega}_{12} + \boldsymbol{\Omega}_{12}^H)/2\right) \boldsymbol{\omega}}{\boldsymbol{\omega}^H \boldsymbol{T} \boldsymbol{\omega}} \tag{15.28}$$

As we can see from the former section, topographic phase can be estimated and all the points in the coherence region are biased by $\exp(j\phi_0)$. The points in the coherence region are $\tilde{\gamma}\exp(j\phi_0)$. According to RVoG model, the coherence region approximately is a straight line, and the boundary in the coherence line is exactly the coherence value associated with the extremum of $\mathrm{Re}(\tilde{\gamma}\exp(j\phi_0))$, as shown in Fig. 15.7.

$$\mathrm{Re}(\tilde{\gamma}\exp(j\phi_0)) = \frac{\boldsymbol{\omega}^H \left((e^{j\phi_0} \boldsymbol{\Omega}_{12} + e^{-j\phi_0} \boldsymbol{\Omega}_{12}^H)/2\right) \boldsymbol{\omega}}{\boldsymbol{\omega}^H \boldsymbol{T} \boldsymbol{\omega}} \tag{15.29}$$

Let $\boldsymbol{A} = (e^{j\phi_0} \boldsymbol{\Omega}_{12} + e^{-j\phi_0} \boldsymbol{\Omega}_{12}^H)/2$, Eq. (15.29) can be written as [38]

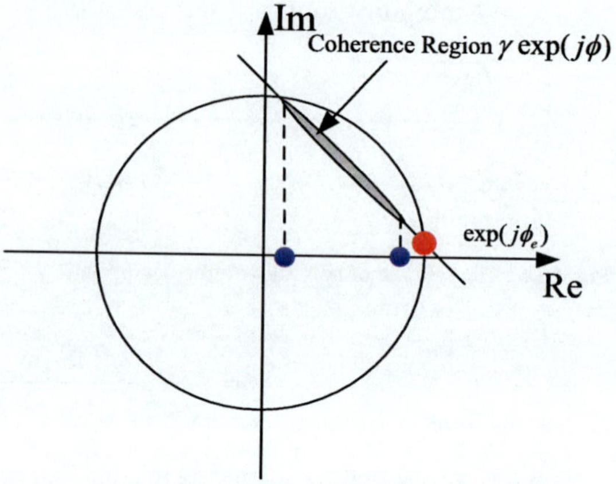

Figure 15.7 The solution to the boundary of coherence region

$$\mathrm{Re}(\tilde{\gamma}\exp(j\phi_0)) = \frac{\boldsymbol{\omega}^H \boldsymbol{A}\boldsymbol{\omega}}{\boldsymbol{\omega}^H \boldsymbol{T}\boldsymbol{\omega}} \tag{15.30}$$

We can obtain the extremum of Eq. (15.30) by maximizing the complex Lagrangian defined as

$$L = \boldsymbol{\omega}^H \boldsymbol{A}\boldsymbol{\omega} + \lambda\left(\boldsymbol{\omega}^H \boldsymbol{T}\boldsymbol{\omega} - C_1\right) \tag{15.31}$$

Let $\frac{\partial L}{\partial \boldsymbol{\omega}^H} = 0$, we can get

$$\boldsymbol{A}\boldsymbol{\omega} = -\lambda \boldsymbol{T}\boldsymbol{\omega} \tag{15.32}$$

That is,

$$-\boldsymbol{T}^{-1}\boldsymbol{A}\boldsymbol{\omega} = \lambda\boldsymbol{\omega} \tag{15.33}$$

Then the solution of (15.33) can be converted to a generalized eigenvalue problem. Performing eigenvalue decomposition of matrix $-\boldsymbol{T}^{-1}\boldsymbol{A}$, and the eigenvectors corresponding to the maximum and minimum eigenvalue are, respectively, $\boldsymbol{\omega}_1$ and $\boldsymbol{\omega}_2$. The boundary points of the coherence region are

$$\gamma_1 = \frac{\boldsymbol{\omega}_1^H \boldsymbol{\Omega}_{12}\boldsymbol{\omega}_1}{\boldsymbol{\omega}_1^H \boldsymbol{T}\boldsymbol{\omega}_1}, \gamma_2 = \frac{\boldsymbol{\omega}_2^H \boldsymbol{\Omega}_{12}\boldsymbol{\omega}_2}{\boldsymbol{\omega}_2^H \boldsymbol{T}\boldsymbol{\omega}_2} \tag{15.34}$$

The estimated volume decorrelation is the point which is most distinct from the ground phase point. The judge criterion is as following:

$$\gamma_{\mathrm{v,est}} = \begin{cases} \gamma_1\exp(-j\phi_e), |\gamma_1 - \exp(j\phi_e)| > |\gamma_2 - \exp(j\phi_e)| \\ \gamma_2\exp(-j\phi_e), |\gamma_2 - \exp(j\phi_e)| > |\gamma_1 - \exp(j\phi_e)| \end{cases} \tag{15.35}$$

where $\gamma_{v,\text{est}}$ is the estimated volume decorrelation; ϕ_e is the known ground topographic phase.

After the estimation of volume decorrelation, the height of vegetation can be obtained by the look-up table.

15.1.3 Maximum Likelihood Inversion Method

The maximum likelihood tree height inversion method [41, 42] is based on the sample coherency matrix of RVoG model. The logarithm likelihood function of the sample coherency matrix can be obtained. Since the vegetation height, extinction coefficient, and other information are included in the sample coherency matrix, the maximum likelihood estimation of the matrix can be used to derive parameters such as vegetation height.

15.1.3.1 The Principle of Maximum Likelihood

Let $\widehat{\mathbf{R}}$ be the coherency matrix measured by full PolInSAR system

$$\widehat{\mathbf{R}} =< \begin{bmatrix} \boldsymbol{k}_1 \\ \boldsymbol{k}_2 \end{bmatrix} [\boldsymbol{k}_1^H \ \boldsymbol{k}_2^H] > \tag{15.36}$$

where $\boldsymbol{k}_1$ and $\boldsymbol{k}_2$ are the scattering vectors of the master and slave PolSAR images.

The probability density of $\widehat{\mathbf{R}}$ is the complex Wishart distribution,

$$p(\widehat{\mathbf{R}}; \mathbf{R}, n) = c\left(\widehat{\mathbf{R}}\right)\exp\left(-n\text{tr}\left(\mathbf{R}^{-1}\widehat{\mathbf{R}}\right)\right)\left(|\mathbf{R}^{-1}|\right)^n \tag{15.37}$$

where $\mathbf{R}$ is the coherency matrix containing scattering model of vegetation parameters to be retrieved; $\widehat{\mathbf{R}}$ is the coherency matrix estimated from observations; tr(·) is the trace of the matrix; |·| is the determinant of matrix, and $c\left(\widehat{\mathbf{R}}\right)$ a normalization term which is independent of $\mathbf{R}$.

The corresponding likelihood function is

$$f(\mathbf{R}) = -\log|\mathbf{R}| - \text{tr}\left(\mathbf{R}^{-1}\widehat{\mathbf{R}}\right) \tag{15.38}$$

The maximum parameters of the likelihood function can be used to estimate the vegetation parameters, such as the height, surface phase, and extinction coefficient, contained in the coherence matrix.

15.1.3.2 Maximum Likelihood Inversion Process

According to the coherency matrix expression of RVoG model given in Eqs. (15.3)–(15.12), the vegetation parameters inversion process based on maximum likelihood method can be derived. Let

$$\boldsymbol{U}\boldsymbol{\Lambda}\boldsymbol{U}^H = aI_1^{-1}\boldsymbol{T}_v^{-1/2}\boldsymbol{T}_G\boldsymbol{T}_v^{-1/2} \tag{15.39}$$

The definitions of a, I_1, $\boldsymbol{T}_G$ and $\boldsymbol{T}_v$ are shown in Eqs. (15.3)~(15.12). $\boldsymbol{\Lambda} = \mathbf{diag}\{\lambda_i\}$, $i = 1, 2, 3$; and λ_i are the eigenvalues of $aI_1^{-1}\boldsymbol{T}_v^{-1/2}\boldsymbol{T}_G\boldsymbol{T}_v^{-1/2}$. So,

$$\boldsymbol{T} = I_1\boldsymbol{T}_v + a\boldsymbol{T}_G = (I_1\boldsymbol{T}_v)^{1/2}(I_1\boldsymbol{T}_v)^{1/2} + (I_1\boldsymbol{T}_v)^{1/2}\boldsymbol{U}\boldsymbol{\Lambda}\boldsymbol{U}^H(I_1\boldsymbol{T}_v)^{1/2} \tag{15.40}$$

$$\begin{aligned}\boldsymbol{X} = e^{j\phi_0}(I_1\boldsymbol{T}_v + a\boldsymbol{T}_G) = e^{j\phi_0}((I_1\boldsymbol{T}_v)^{\frac{1}{2}}(z\boldsymbol{I})(I_1\boldsymbol{T}_v)^{\frac{1}{2}} \\ + (I_1\boldsymbol{T}_v)^{1/2}\boldsymbol{U}\boldsymbol{\Lambda}\boldsymbol{U}^H(I_1\boldsymbol{T}_v)^{1/2}\end{aligned} \tag{15.41}$$

Let $\widehat{\boldsymbol{T}}_v = I_1T_v$,$\Lambda_0 = \mathrm{diag}\{1 + \lambda_i\}$, $\Lambda_1 = \mathrm{diag}\{e^{j\phi_0}(z + \lambda_i)\}$, and Eqs. (15.40)–(15.41) can be rewritten as

$$\boldsymbol{T} = \widehat{\boldsymbol{T}}_v^{\frac{1}{2}}\boldsymbol{U}(1 + \boldsymbol{\Lambda})\boldsymbol{U}^H\widehat{\boldsymbol{T}}_v^{\frac{1}{2}} = \widehat{\boldsymbol{T}}_v^{\frac{1}{2}}\boldsymbol{U}\boldsymbol{\Lambda}_0\boldsymbol{U}^H\widehat{\boldsymbol{T}}_v^{\frac{1}{2}} \tag{15.42}$$

$$X = e^{j\phi_0}\left(\widehat{\boldsymbol{T}}_v^{\frac{1}{2}}\boldsymbol{U}(z\boldsymbol{I} + \boldsymbol{\Lambda})\boldsymbol{U}^H\widehat{\boldsymbol{T}}_v^{\frac{1}{2}}\right) = e^{j\phi_0}\widehat{\boldsymbol{T}}_v^{\frac{1}{2}}\boldsymbol{U}\boldsymbol{\Lambda}_1\boldsymbol{U}^H\widehat{\boldsymbol{T}}_v^{\frac{1}{2}} \tag{15.43}$$

Then the expression of coherency matrix is

$$\boldsymbol{R} = \begin{bmatrix} \boldsymbol{T} & \boldsymbol{X} \\ \boldsymbol{X}^H & \boldsymbol{T} \end{bmatrix} = \begin{bmatrix} \widehat{\boldsymbol{T}}_v^{\frac{1}{2}} & 0 \\ 0 & \widehat{\boldsymbol{T}}_v^{\frac{1}{2}} \end{bmatrix}\begin{bmatrix} \boldsymbol{U}^H & 0 \\ 0 & \boldsymbol{U}^H \end{bmatrix}\begin{bmatrix} \widehat{\boldsymbol{T}}_v^{\frac{1}{2}} & 0 \\ 0 & \widehat{\boldsymbol{T}}_v^{\frac{1}{2}} \end{bmatrix} \tag{15.44}$$

In order to facilitate the derivation of the following formula, a coordinate base transformation is performed on Eq. (15.44). Let

$$m = \frac{z - e^{j\theta}}{e^{j\theta} - 1} \tag{15.45}$$

and

$$z = e^{j\theta}(1 + m) - m \tag{15.46}$$

$$z + \lambda_i = e^{j\theta}(1 + m) + \lambda_i - m \tag{15.47}$$

Let $z + \lambda_i = p(e^{j\theta} + \mu_i)$, and we can get

$$p = 1 + m, \mu_i = \frac{\lambda_i - m}{p} \tag{15.48}$$

The transform expression from basis (λ_i, z) to basis (μ_i, p, θ) is

$$1 + \lambda_i = p(1 + \mu_i) \tag{15.49}$$

$$z = p\left(e^{j\theta} - 1\right) + 1 \tag{15.50}$$

After coordinate basis transformation, the coherency matrix $\boldsymbol{R}$ can be expressed as

$$\boldsymbol{R} = \begin{bmatrix} \boldsymbol{T} & \boldsymbol{X} \\ \boldsymbol{X}^H & \boldsymbol{T} \end{bmatrix} = \begin{bmatrix} \widehat{\boldsymbol{T}}_v^{\frac{1}{2}} & 0 \\ 0 & \widehat{\boldsymbol{T}}_v^{\frac{1}{2}} \end{bmatrix} \begin{bmatrix} \boldsymbol{U}^H & 0 \\ 0 & \boldsymbol{U}^H \end{bmatrix} \begin{bmatrix} \widehat{\boldsymbol{T}}_v^{\frac{1}{2}} & 0 \\ 0 & \widehat{\boldsymbol{T}}_v^{\frac{1}{2}} \end{bmatrix} \tag{15.51}$$

where

$$\boldsymbol{\Lambda}_0 = \text{diag}\{1 + \mu_i\} \tag{15.52}$$

$$\boldsymbol{\Lambda}_1 = \text{diag}\left\{e^{j\phi_0}\left(e^{j\theta} + \mu_i\right)\right\} \tag{15.53}$$

$$\boldsymbol{T}_v = pI_1\widehat{\boldsymbol{T}}_v \tag{15.54}$$

Then

$$|\boldsymbol{R}| = \left|\widehat{\boldsymbol{T}}_v\right|^2 (2 - 2\cos\theta)^3 \prod_i \mu_i \tag{15.55}$$

According to the matrix inverse formula, we can get [46]

$$\boldsymbol{R}^{-1} = 2^{-1}(1 - \cos\theta)^{-1} \begin{bmatrix} \widehat{\boldsymbol{T}}_v^{-1/2} & 0 \\ 0 & \widehat{\boldsymbol{T}}_v^{-1/2} \end{bmatrix} \begin{bmatrix} \boldsymbol{U} & 0 \\ 0 & \boldsymbol{U} \end{bmatrix} \begin{bmatrix} \boldsymbol{\Lambda}_2 & -\boldsymbol{\Lambda}_3 \\ -\boldsymbol{\Lambda}_3^H & \boldsymbol{\Lambda}_2 \end{bmatrix} \begin{bmatrix} \widehat{\boldsymbol{T}}_v^{-1/2} & 0 \\ 0 & \widehat{\boldsymbol{T}}_v^{-1/2} \end{bmatrix} \tag{15.56}$$

where

$$\boldsymbol{\Lambda}_2 = \text{diag}(1 + \mu_i^{-1}) \tag{15.57}$$

$$\mathbf{\Lambda}_3 = \text{diag}(1 + e^{j\theta}\mu_i^{-1}) \tag{15.58}$$

Therefore, we can get

$$\begin{aligned}
\boldsymbol{tr}(\boldsymbol{R}^{-1}\boldsymbol{R}) &= (1-\cos\theta)^{-1}\left[\text{tr}\left(\widehat{\boldsymbol{T}}_{\text{v}}^{-\frac{1}{2}}\boldsymbol{U}\mathbf{\Lambda}_2\boldsymbol{U}^H\widehat{\boldsymbol{T}}_{\text{v}}^{-\frac{1}{2}}\bar{\boldsymbol{T}}\right) - \frac{1}{2}\text{tr}\left(\widehat{\boldsymbol{T}}_{\text{v}}^{-\frac{1}{2}}\boldsymbol{U}\mathbf{\Lambda}_3\boldsymbol{U}^H\widehat{\boldsymbol{T}}_{\text{v}}^{-\frac{1}{2}}\boldsymbol{O}_{12}^H\right)\right] \\
&= (1-\cos\theta)^{-1}\left[\text{tr}\left(\boldsymbol{U}^H\widehat{\boldsymbol{T}}_{\text{v}}^{-\frac{1}{2}}\boldsymbol{A}_\phi\widehat{\boldsymbol{T}}_{\text{v}}^{-\frac{1}{2}}\boldsymbol{O}_{12}^H + \boldsymbol{tr}\boldsymbol{U}^H\widehat{\boldsymbol{T}}_{\text{v}}^{-\frac{1}{2}}\boldsymbol{A}_{\phi+\theta}\widehat{\boldsymbol{T}}_{\text{v}}^{-\frac{1}{2}}\boldsymbol{U}\mathbf{\Lambda}_{\mu^{-1}}\right)\right] \\
&= (1-\cos\theta)^{-1}\left[\text{tr}\left(\widehat{\boldsymbol{T}}_{\text{v}}^{-\frac{1}{2}}\boldsymbol{A}_\phi\right) + \text{tr}\left(\boldsymbol{U}^H\boldsymbol{V}\mathbf{\Lambda}_{a2}\boldsymbol{V}^H\boldsymbol{U}\mathbf{\Lambda}_{\mu^{-1}}\right)\right]
\end{aligned} \tag{15.59}$$

where

$$\bar{\boldsymbol{T}} = \frac{\boldsymbol{T}_{11} + \boldsymbol{T}_{22}}{2} \tag{15.60}$$

$$\mathbf{A}_\alpha = \bar{\boldsymbol{T}} - \bar{\boldsymbol{O}}_\alpha \tag{15.61}$$

$$\bar{\boldsymbol{O}}_\alpha = \frac{1}{2}e^{j\alpha}\mathbf{\Omega}_{12}^H + \frac{1}{2}e^{-j\alpha}\mathbf{\Omega}_{12} \tag{15.62}$$

$$\mathbf{\Lambda}_{\mu_i^{-1}} = \textbf{diag}\{\mu_i^{-1}\} \tag{15.63}$$

$$\boldsymbol{V}\mathbf{\Lambda}_{a2}\boldsymbol{V}^H = \widehat{\boldsymbol{T}}_v^{-\frac{1}{2}}\boldsymbol{A}_{\phi+\theta}\widehat{\boldsymbol{T}}_v^{-\frac{1}{2}} \tag{15.64}$$

The likelihood function is

$$\begin{aligned}
f(\boldsymbol{R}) &= -\lg|\boldsymbol{R}| - \text{tr}\left(\boldsymbol{R}^{-1}\widehat{\boldsymbol{R}}\right) \\
&= -2\lg\left|\widehat{\mathbf{T}}_{\text{v}}\right| - 3\lg(1-\cos\theta) - \sum\lg\mu_{\text{i}} \\
&\quad - (1-\cos\theta)^{-1}\left[\text{tr}\widehat{\boldsymbol{T}}_{\text{v}}^{-1}\boldsymbol{A}_\phi + \text{tr}\left(\boldsymbol{U}^H\boldsymbol{V}\mathbf{\Lambda}_{a2}\boldsymbol{V}^H\boldsymbol{U}\mathbf{\Lambda}_{\mu^{-1}}\right)\right]
\end{aligned} \tag{15.65}$$

Let $\frac{\partial f(\boldsymbol{R})}{\partial \mu_i} = 0$, and we can get

$$\frac{\partial f(\boldsymbol{R})}{\partial \mu_i} = -\frac{1}{\mu_i} - \frac{1}{1-\cos\theta}\left(\boldsymbol{U}^H\widehat{\boldsymbol{T}}_v^{-\frac{1}{2}}\boldsymbol{A}_{\phi+\theta}\widehat{\boldsymbol{T}}_{\text{v}}^{-\frac{1}{2}}\right)_{ii} - (1-\cos\theta)^{-1}\left[\text{tr}\widehat{\boldsymbol{T}}_v^{-1}\boldsymbol{A}_\phi\right] = 0 \tag{15.66}$$

$$\mu_i = (1-\cos\theta)^{-1}\left(\boldsymbol{U}^H\boldsymbol{V}\mathbf{\Lambda}_{a2}\boldsymbol{V}^H\boldsymbol{U}\right)_{ii} \tag{15.67}$$

In Eqs. (15.66)–(15.67), $(\cdot)_{ii}$ means the i-th diagonal element of the matrix. Substituting Eq. (15.67) into Eq. (15.65),

$$f(\boldsymbol{R}) = -2\lg\left|\widehat{\mathbf{T}}_v\right| - \lg\prod\left(\boldsymbol{U}^H\boldsymbol{V}\boldsymbol{\Lambda}_{a2}\boldsymbol{V}^H\boldsymbol{U}\right)_{ii} - (1-\cos\theta)^{-1}\mathrm{tr}(\widehat{\boldsymbol{T}}_v^{-1}\boldsymbol{A}_\phi) \quad (15.68)$$

When $\boldsymbol{V} = \boldsymbol{U}$, the second term of $f(\boldsymbol{R})$ takes the maximum, and

$$\begin{aligned} f(\boldsymbol{R}) &= -2\lg\left|\widehat{\mathbf{T}}_{\mathrm{v}}\right| - \lg|\boldsymbol{\Lambda}_{a2}| - (1-\cos\theta)^{-1}\mathrm{tr}(\widehat{\boldsymbol{T}}_{\mathrm{v}}^{-1}\boldsymbol{A}_\phi) \\ &= -\lg\left|\widehat{\mathbf{T}}_{\mathrm{v}}\right| - \lg\left|\boldsymbol{A}_{\phi+\theta}\right| - (1-\cos\theta)^{-1}\mathrm{tr}\left(\widehat{\boldsymbol{T}}_{\mathrm{v}}^{-1}\boldsymbol{A}_\phi\right) \end{aligned} \quad (15.69)$$

From Eq. (15.11), we can get

$$\boldsymbol{T}_{\mathrm{v}} = f_v\begin{bmatrix} k & 0 & 0 \\ 0 & 1 & 0 \\ 0 & 0 & 1 \end{bmatrix} \quad (15.70)$$

and

$$f(\boldsymbol{R}) = -3\lg f_v - \lg k - \lg\left|\boldsymbol{A}_{\phi+\theta}\right| - (1-\cos\theta)^{-1}f_v^{-1}(k^{-1}a_{\phi11} + a_{\phi22} + a_{\phi33}) \quad (15.71)$$

where $a_{\phi ii}, i = 1, 2, 3$ is the ith diagonal element of the matrix $\boldsymbol{A}_\phi$.

Differentiate $f(\boldsymbol{R})$ for f_v and k, respectively, and we can get

$$\frac{\partial f(\boldsymbol{R})}{\partial f_v} = -\frac{3}{f_v} + \frac{1}{1-\cos\theta}\frac{1}{f_v^2}(k^{-1}a_{\phi11} + a_{\phi22} + a_{\phi33}) \quad (15.72)$$

$$\frac{\partial f(\boldsymbol{R})}{\partial k} = -\frac{1}{k} + \frac{1}{f_v(1-\cos\theta)}\frac{1}{k^2}a_{\phi11} = 0 \quad (15.73)$$

From Eqs. (15.72) and (15.73),we can get

$$f_{\mathrm{v}} = \frac{1}{3}(1-\cos\theta)^{-1} + (k^{-1}a_{\phi11} + a_{\phi22} + a_{\phi33}) \quad (15.74)$$

$$k = \frac{2a_{\phi11}}{a_{\phi22} + a_{\phi33}} \quad (15.75)$$

Substituting Eqs. (15.74) and (15.75) into Eq. (15.71), we can get

$$f(\boldsymbol{R}) = 3\lg(1 - cos\theta) - \lg\left|\boldsymbol{A}_{\phi+\theta}\right| - \lg a_{\phi11} - 2\lg(a_{\phi22} + a_{\phi33}) \quad (15.76)$$

Differentiate $f(\boldsymbol{R})$ for θ and ϕ, respectively, and we can get

$$\frac{\partial f(\boldsymbol{R})}{\partial\theta} = \frac{3\sin\theta}{1-\cos\theta} - \frac{\partial\lg\left|\boldsymbol{A}_{\phi+\theta}\right|}{\partial\theta} \quad (15.77)$$

$$\frac{\partial f(\boldsymbol{R})}{\partial \phi} = -\frac{\partial \lg\left|\boldsymbol{A}_{\phi+\theta}\right|}{\partial \phi} - \frac{1}{a_{\phi 11}}\frac{\partial a_{\phi 11}}{\partial \phi} - \frac{2}{a_{\phi 22} + a_{\phi 33}}\left(\frac{\partial a_{\phi 22}}{\partial \phi} + \frac{\partial a_{\phi 33}}{\partial \phi}\right) \tag{15.78}$$

Since $\frac{\partial \lg|\boldsymbol{A}_{\phi+\theta}|}{\partial \theta} = \frac{\partial \lg|\boldsymbol{A}_{\phi+\theta}|}{\partial \phi}$, subtracting Eq. (15.78) from Eq. (15.77),

$$\frac{3\sin\theta}{1-\cos\theta} = -\frac{1}{a_{\phi 11}}\frac{\partial a_{\phi 11}}{\partial \phi} - \frac{2}{a_{\phi 22} + a_{\phi 33}}(\frac{\partial a_{\phi 22}}{\partial \phi} + \frac{\partial a_{\phi 33}}{\partial \phi}) \tag{15.79}$$

where

$$\frac{\partial a_{\phi 11}}{\partial \phi} = -\frac{j}{2}e^{j\phi}\left(\boldsymbol{\Omega}_{12}^{H}\right)_{11} + \frac{j}{2}e^{j\phi}(\boldsymbol{\Omega}_{12})_{11} \tag{15.80}$$

Similarly,

$$\frac{\partial a_{\phi 22}}{\partial \phi} = -\frac{j}{2}e^{j\phi}\left(\boldsymbol{\Omega}_{12}^{H}\right)_{22} + \frac{j}{2}e^{j\phi}(\boldsymbol{\Omega}_{12})_{22} \tag{15.81}$$

$$\frac{\partial a_{\phi 22}}{\partial \phi} = -\frac{j}{2}e^{j\phi}\left(\boldsymbol{\Omega}_{12}^{H}\right)_{33} + \frac{j}{2}e^{j\phi}(\boldsymbol{\Omega}_{12})_{33} \tag{15.82}$$

Let $(\boldsymbol{\Omega}_{12})_{ii} = \overline{O}_{ii} + j\hat{O}_{ii}, i = 1, 2, 3$, and $\bar{O}_{ii}$ and $\widehat{O}_{ii}$ are the real part and imaginary part of $(\boldsymbol{\Omega}_{12})_{ii}$, respectively. So,

$$\begin{aligned} a_{\phi 11} =& \bar{t}_{11} - \frac{j}{2}e^{j\phi}\left(\boldsymbol{\Omega}_{12}^{H}\right)_{11} - \frac{1}{2}e^{-j\phi}(\boldsymbol{\Omega}_{12})_{11} \\ =& \bar{t}_{11} - \frac{1}{2}e^{j\phi}\left(\overline{O}_{11} - j\hat{O}_{11}\right) \\ & - \frac{1}{2}e^{-j\phi}\left(\overline{O}_{11} + j\hat{O}_{11}\right) \\ =& \bar{t}_{11} - \cos\phi \cdot \overline{O}_{11} - \sin\phi \cdot \hat{O}_{11} \\ =& d_1 \end{aligned} \tag{15.83}$$

$$\frac{\partial a_{\phi 22}}{\partial \phi} = \sin\phi \cdot \overline{O}_{11} - \cos\phi \cdot \widehat{O}_{11} = n_1 \tag{15.84}$$

Similarly,

$$a_{\phi 22} + a_{\phi 33} = \bar{t}_{22} + \bar{t}_{33} - \left(\overline{O}_{22} + \overline{O}_{33}\right)\cos\phi - \left(\widehat{O}_{22} + \widehat{O}_{33}\right)\sin\phi = d_2 \tag{15.85}$$

$$\frac{\partial\left(a_{\phi 22} + a_{\phi 33}\right)}{\partial \phi} = \sin\phi \cdot \left(\overline{O}_{22} + \overline{O}_{33}\right) - \cos\phi \cdot \left(\widehat{O}_{22} + \widehat{O}_{33}\right) = n_2 \tag{15.86}$$

In Eqs. (15.83)–(15.85), $\overline{t}_{ii}$ is the i-th diagonal element of the matrix $\overline{\boldsymbol{T}}$. So Eq. (15.79) can be rewritten as

$$\frac{3}{\tan(\frac{\theta}{2})} = -\frac{n_1}{d_1} - \frac{2n_2}{d_2} = -\frac{n_1 d_2 + 2n_2 d_1}{d_1 d_2} \tag{15.87}$$

So,

$$\theta = \pi - 2\arctan\left[\frac{n_1 d_2 + 2n_2 d_1}{d_1 d_2}\right] \tag{15.88}$$

According to the aforementioned derivations, the maximum likelihood inversion process can be obtained as follows:

(1) Solving ϕ that maximize $\frac{(1-\cos\theta)^3}{a_{\phi 11}(a_{\phi 22}+a_{\phi 33})^2 |A_{\phi+\theta}|}$, where θ is given by Eq. (15.88).
(2) Get the estimation of k and f_v from Eqs. (15.75) and (15.74), and then get the estimation of $\widehat{\mathbf{T}}_v$.
(3) Performing eigenvalue decomposition to matrix $\widehat{\boldsymbol{T}}_v^{-\frac{1}{2}} \boldsymbol{A}_{\phi+\theta} \widehat{\boldsymbol{T}}_v^{-\frac{1}{2}}$ from Eq. (15.64), and then get the eigenvalues of $\boldsymbol{\Lambda}_{a2}$, which are $\lambda_{a2}, i = 1, 2, 3$.
(4) Let $p = \frac{1+\lambda_3}{1+(1-\cos\theta)^{-1}\lambda_{a2,3}}$, and λ_3 is generally set to 0. Substituting p into Eq. (15.50), we can get the estimation of z.
(5) z is the estimation of volume decorrelation coefficient. The estimation of vegetation height h_v and extinction coefficient σ can be obtained by the look-up table similar to three-stage inversion process.

15.1.4 Scattering Centers Resolution Method

Scattering centers resolution method is based on independent scattering centers model. Multi-layer structure can be taken as some stable, distributed independent scattering models. According to the assumption of independent scattering centers, the full polarimetric observation of multi-layer structure can be modeled by the sum of some independent scattering signals with different scattering centers. According to the aforementioned analysis, the observation of different locations can be modeled as

$$E_1^{(kl)} = \sum_{i=1}^{d} \sigma_i s_i^{(kl)} e^{j\frac{4\pi}{\lambda}\rho} + n_1^{(kl)} \tag{15.89}$$

$$E_2^{(kl)} = \sum_{i=1}^{d} \sigma_i' s_i'^{(kl)} e^{j\frac{4\pi}{\lambda}(\rho+\Delta\rho_i)} + n_2^{(kl)} \tag{15.90}$$

where k and l denote polarization combinations (like HH, HV, VH, VV); d represents the number of dominant scattering mechanisms in the resolution cell; $s_i^{(kl)}$ and $s_i^{\prime(kl)}$ represents the backscattering coefficients observed by two antennas of kl polarization; $\sigma_i(\sigma_i^{\prime})$ denotes the scattering power of the i-th scattering center measured by antenna 1(2); ρ is the slant range of antenna 1 to the range cell. $\Delta\rho_i$ in Eq. (15.90) is the distance difference between the position of the first scatterer to antenna 2 and the position of the first scatter to antenna 1.$n_1^{(kl)}$ and $n_2^{(kl)}$ are the additive Gaussian white noise of both antennas in kl polarization.

Expressing Eqs. (15.89) and (15.90) in matrix form, and we can get

$$\boldsymbol{E}_1 = \left[E_1^{(HH)} \; E_1^{(HV)} \; E_1^{(VH)} \; E_1^{(VV)} \right]^{\mathrm{T}} = \boldsymbol{S\sigma} + \boldsymbol{n}_1 \tag{15.91}$$

$$\boldsymbol{E}_2 = \left[E_2^{(HH)} \; E_2^{(HV)} \; E_2^{(VH)} \; E_2^{(VV)} \right]^{\mathrm{T}} = \boldsymbol{S}^{\prime}\boldsymbol{\sigma}^{\prime} + \boldsymbol{n}_2 \tag{15.92}$$

where

$$\boldsymbol{S} = [s_1, \quad s_2, \ldots, \quad s_d] = \begin{bmatrix} s_1^{(\mathrm{HH})} & \cdots & s_d^{(\mathrm{HH})} \\ s_1^{(\mathrm{HV})} & \cdots & s_d^{(\mathrm{HV})} \\ \vdots & & \vdots \\ s_1^{(\mathrm{VV})} & \cdots & s_d^{(\mathrm{VV})} \end{bmatrix} \tag{15.93}$$

$$\boldsymbol{S}^{\prime} = \left[s_1^{\prime}, \quad s_2^{\prime}, \ldots, \quad s_d^{\prime}\right] = \begin{bmatrix} s_1^{\prime(\mathrm{HH})} & \cdots & s_d^{\prime(\mathrm{HH})} \\ s_1^{\prime(\mathrm{HV})} & \cdots & s_d^{\prime(\mathrm{HV})} \\ \vdots & & \vdots \\ s_1^{\prime(\mathrm{VV})} & \cdots & s_d^{\prime(\mathrm{VV})} \end{bmatrix} \tag{15.94}$$

$$\boldsymbol{\sigma} = \left[\sigma_1 e^{j\frac{4\pi}{\lambda}\rho}, \ldots, \sigma_d e^{j\frac{4\pi}{\lambda}\rho}\right] \tag{15.95}$$

$$\boldsymbol{\sigma}^{\prime} = \left[\sigma_1^{\prime} e^{j\frac{4\pi}{\lambda}(\rho+\Delta\rho_1)}, \ldots, \sigma_d^{\prime} e^{j\frac{4\pi}{\lambda}(\rho+\Delta\rho_d)}\right] \tag{15.96}$$

$$\boldsymbol{n}_j = \left[n_j^{(HH)}, n_j^{(HV)}, n_j^{(VH)}, n_j^{(VV)}\right]^{\mathrm{T}}, j = 1, 2 \tag{15.97}$$

Assuming that all the targets are in the far-field region of radar, and both antennas have the same looking angle, so the polarization state of the scattering centers of the master and slave receiving channels are the same, that is

$$s_i^{(kl)} \approx s_i^{\prime(kl)} \sigma_i \approx \sigma_i^{\prime} \quad i = 1, \ldots, d \tag{15.98}$$

According to Eq. (15.98), Eq. (15.92) can be simplified as

$$\boldsymbol{E}_2 = \boldsymbol{S}^{\prime}\boldsymbol{\sigma}^{\prime} + \boldsymbol{n}_2 \approx \boldsymbol{SD\sigma} + \boldsymbol{n}_2 \tag{15.99}$$

$$D = \mathrm{diag}\{e^{j\phi_1}, e^{j\phi_2}, \cdots, e^{j\phi_d}\} \tag{15.100}$$

$$\phi_i = \frac{4\pi}{\lambda}\Delta\rho_i, i = 1, \cdots, d \tag{15.101}$$

The estimation of $\boldsymbol{D}$ can be obtained with Eqs. (15.91) and (15.92), and then the estimation of the interferometric phases of the scatters.

15.1.4.1 ESPRIT Method

According to the aforementioned model, ESPRIT algorithm can be applied to PolInSAR data for tree height inversion. Applying the ESPRIT algorithm to PolInSAR data is based on the following assumptions:

(1) The noise of each polarization channel is independent of each other and is Gaussian white noise.
(2) The noise of each polarization channel is independent of the backscattered signals of the scatterers.
(3) The coupling among the polarization channels is negligible.
(4) The polarization states of the scattering waves of the master and slave channels are the same.
(5) The number of scatters in the range cell is smaller than the number of polarization channels.
(6) The backscattered signals can be taken as far-field signals since the antennas are distant away from the scatters.
(7) The influence of temporal decorrelation can be ignored for repeated-pass interferometry system.

The realization of ESPRIT algorithm is introduced. Total least square ESPRIT algorithm is applied here.

From Eqs. (15.91) and (15.99), the data vector can be defined as

$$x = \begin{bmatrix} E_1 \\ E_2 \end{bmatrix} = \begin{bmatrix} \boldsymbol{S} \\ \boldsymbol{SD} \end{bmatrix} \sigma + \begin{bmatrix} \boldsymbol{n}_1 \\ \boldsymbol{n}_2 \end{bmatrix} = \bar{\boldsymbol{S}}\,\sigma + \bar{\boldsymbol{n}} \tag{15.102}$$

where $\bar{\boldsymbol{S}} = \left[\boldsymbol{S}^{\mathrm{T}}\ (\boldsymbol{SD})^{\mathrm{T}}\right]^{\mathrm{T}}$.

For a data set consisting M channels, $\boldsymbol{x}$ is a vector with 2 M dimensions. The coherency matrix of $\boldsymbol{x}$ can be defined as

$$\boldsymbol{R}_{xx} = <\boldsymbol{x}\boldsymbol{x}^H> = \bar{\boldsymbol{S}}\,\boldsymbol{\Sigma}_S \bar{\boldsymbol{S}}^{\mathrm{H}} + \sigma_N^2 \boldsymbol{\Sigma}_N \tag{15.103}$$

$$\boldsymbol{\Sigma}_S = <\sigma\sigma^{\mathrm{H}}> \tag{15.104}$$

$$\boldsymbol{\Sigma}_S = <\bar{\boldsymbol{n}}\,\bar{\boldsymbol{n}}^{\mathrm{H}}> \tag{15.105}$$

where $\langle\cdot\rangle$ denotes spatial averaging; H denotes conjugate transpose operation; σ_N^2 is the average noise power; matrix $\boldsymbol{\Sigma}_S$ and $\boldsymbol{\Sigma}_N$ denote the coherency matrix of scattering wave and noise, respectively, and their ranks are

$$\begin{cases} rank(\boldsymbol{\Sigma}_S) = d \\ rank(\boldsymbol{\Sigma}_N) = 2M \end{cases} \tag{15.106}$$

If the polarization states of every local scattering wave are linearly independent, that is

$$s_i \neq ks_j, i \neq j, k \text{ is a complex constant} \tag{15.107}$$

Then rank$(\boldsymbol{S}) = d$. Further, there is

$$\text{rank}\left(\bar{\boldsymbol{S}}\,\boldsymbol{\Sigma}_S\bar{\boldsymbol{S}}^{H}\right) = d \tag{15.108}$$

Performing eigenvalue decomposition on $\boldsymbol{R}_{xx}$, we can get

$$|\boldsymbol{R}_{xx} - \lambda\boldsymbol{\Sigma}_N| = 0 \tag{15.109}$$

Assuming that the eigenvalues of $\boldsymbol{R}_{xx}$ are $\lambda_1, \lambda_2, \cdots, \lambda_{2M}$, and their amplitude are

$$\lambda_1 \geq \lambda_2 \geq \cdots \geq \lambda_d > \lambda_{d+1} = \cdots = \lambda_{2M}(=\sigma_N^2) \tag{15.110}$$

When $d < M$, the amplitude of 2M $- d$ the least eigenvalue is the power of noise. The number of scattering centers can be estimated by this property. The eigenvectors corresponding to d maximum eigenvalues can be expressed as

$$[e_1, e_2, \cdots, e_d] = \begin{bmatrix} \boldsymbol{F}_1 \\ \boldsymbol{F}_2 \end{bmatrix} = \begin{bmatrix} \boldsymbol{ST} \\ \boldsymbol{SDT} \end{bmatrix} \tag{15.111}$$

where $\boldsymbol{T}$ is a $d \times d$ non-sigular matrix.

Defining

$$\boldsymbol{F}_{12} = \left[\boldsymbol{F}_1\ \boldsymbol{F}_2\right] \tag{15.112}$$

Since the rank of $\boldsymbol{F}_{12}$ is d, there is a unique $2d \times d$ matrix $\boldsymbol{G}$ with rank d. Let

$$0 = \boldsymbol{F}_{12}\boldsymbol{G} = \boldsymbol{F}_1\boldsymbol{G}_1 + \boldsymbol{F}_2\boldsymbol{G}_2 = \boldsymbol{ST}\boldsymbol{G}_1 + \boldsymbol{SDT}\boldsymbol{G}_2 \tag{15.113}$$

$$\boldsymbol{G} = \begin{bmatrix} \boldsymbol{G}_1 \\ \boldsymbol{G}_2 \end{bmatrix} \tag{15.114}$$

G is a matrix which is made up of the eigenvectors corresponding to the 0 eigenvalues of matrix $\boldsymbol{F}_{12}^H\boldsymbol{F}_{12}$.

Defining matrix $\boldsymbol{\Psi}$ as

$$\boldsymbol{\Psi} = -\boldsymbol{G}_1\boldsymbol{G}_2^{-1} \tag{15.115}$$

Substituting Eq. (15.115) into Eq. (15.113), we can get

$$\boldsymbol{ST\Psi} = \boldsymbol{SDT} \tag{15.116}$$

So,

$$\boldsymbol{D} = \boldsymbol{T\Psi T}^{-1} \tag{15.117}$$

As we can see from Eq. (15.117), $\boldsymbol{D}$ and $\boldsymbol{\Psi}$ are similar transformations to each other, so they have the same eigenvalues. Therefore, the eigenvalues of $\boldsymbol{\Psi}$ $\lambda_1, \lambda_2, \ldots, \lambda_d$ form the diagonal elements of matrix $\boldsymbol{D}$. The interferometric phases of the each scattering centers are.

$$\phi_i = \arg\left(\lambda_i'\right)\text{ch} \tag{15.118}$$

15.1.4.2 CAPON Method [47]

CAPON method is a Minimum Variance Method (MVM). As the aforementioned signal model, the observations of PolInSAR system can be expressed by Eq. (15.102). It is rewritten as

$$\boldsymbol{x} = \begin{bmatrix} E_1 \\ E_2 \end{bmatrix} = \begin{bmatrix} \boldsymbol{S} \\ \boldsymbol{SD} \end{bmatrix}\sigma + \begin{bmatrix} \boldsymbol{n}_1 \\ \boldsymbol{n}_2 \end{bmatrix} = \bar{\boldsymbol{S}}\,\sigma + \bar{\boldsymbol{n}} \tag{15.119}$$

where $\bar{\boldsymbol{S}}$ consists the polarimetric and interferometric information, and $\bar{\boldsymbol{n}}$ reflects the influence of noise.

According to the estimation principle of direction of arrival in array signal processing, The CAPON method obtains the interferometric information of each scattering center from the PolInSAR observations by finding the vector $\mathbf{a}(\phi)$ that minimizes noise interference while maintaining the invariant signal power. The mathematical description of the solution process is as follows:

$$\begin{gathered}\arg\min\left\{\mathbf{a}(\phi)^{\mathrm{H}}\boldsymbol{C}_6^{-1}\mathbf{a}(\phi)\right\}\\ \mathbf{a}(\phi)=\left[a_1a_2a_3e^{j\phi}a_1e^{j\phi}a_2e^{j\phi}a_3\right],\ \mathbf{a}(\phi)^{\mathrm{H}}\mathbf{a}(\phi)=2\end{gathered} \tag{15.120}$$

where the vector $[a_1a_2a_3]$ is the estimation of independent scattering centers. $\boldsymbol{C}_6$ is the estimation of polarimetric covariance matrix. ϕ is the interferometric phase estimation for the scattering center.

The Lagrangian multiplier method is used to solve the conditional minimization problem, and a spectral function $P(\phi)$ can be introduced.

$$P(\phi)=\frac{1}{\left|\mathbf{a}(\phi)^{\mathrm{H}}\boldsymbol{C}_6^{-1}\mathbf{a}(\phi)\right|} \tag{15.121}$$

The peak of $P(\phi)$ is searched for the estimation of independent scattering centers. That is, by searching the sampling angle ϕ_{m} which makes $P(\phi_{\mathrm{m}})$ maximum, and then the vector $\mathbf{a}(\phi_m)$ is obtained. The local maximum value of the spectrum function can be obtained by comparing the $P(\phi_{\mathrm{m}})$ of all sampling angles. Then the phase corresponding to local maximum value is the interferometric phase of the scattering center.

15.1.4.3 MUSIC Method [48]

MUSIC method is multiple signal classification method. As we can see from Eq. (15.103), the covariance matrix of PolInSAR data can be decomposed as to the signal part $\boldsymbol{\Sigma}_s$ relevant to the scattering centers and the noise part $\boldsymbol{\Sigma}_n$ relevant to noise. Based on the assumption that signal and noise are independent, the signal subspace and noise subspace are orthogonal. Therefore, in MUSIC method, eigenvalue decomposition of covariance matrix is applied to extract the noise subspace and noise. Spectrum function $P(\phi)$ is introduced to describe the degree of irrelevance to the noise part.

$$P(\phi)=\frac{\mathbf{a}(\phi)^{\mathrm{H}}\mathbf{a}(\phi)}{\left|\mathbf{a}(\phi)^{\mathrm{H}}\boldsymbol{\Sigma}_N\mathbf{a}(\phi)\right|} \tag{15.122}$$

$P(\phi)$ can be used to estimate the signal of scattering centers. The expression of $\mathbf{a}(\phi)$ is shown in Eq. (15.120). $\boldsymbol{\Sigma}_N$ is obtained by calculating the eigenvectors $\boldsymbol{e}_m$ corresponding to small eigenvalues λ_m. When the estimation is independent to noise, the denominator of $P(\phi)$ is zero, and the spectrum function is the local maximum value. According to the assumption that the signal and noise are not correlated, the MUSIC method obtains the information of each independent scattering center by solving the peak of the spectral function $P(\phi)$.

15.2 Agricultural Parameter Inversion Technique Based on PolInSAR

Agricultural parameters inversion is one of the important applications of PolInSAR [4, 49]. Unlike forest vegetation, the height of agricultural crop vegetation is much lower than the forest height, and the effect of ground scattering effects on PolInSAR data is greatly enhanced. Valid inversion model has to be established in order to inverse the ground surface parameters effectively.

15.2.1 Agricultural Crop Vegetation Scattering Model

Due to the obvious orientation of crops, the most commonly used scattering model for agricultural parameter inversion is the oriented Volume over ground Model (OVoG) [24].

In oriented volume model, the extinction coefficients are highly polarization dependent. Due to the obvious directivity of crops, for inversion of farmland parameters is the directional scatter model containing ground scattering components. Therefore, the vertical distribution of the vegetation is polarization dependent. The model is illustrated in Fig. 15.8.

The expression of complex coherence coefficient is

$$\gamma = \frac{\boldsymbol{\omega}^{\mathrm{H}} \boldsymbol{\Omega}_{12} \boldsymbol{\omega}}{\sqrt{\boldsymbol{\omega}^{\mathrm{H}} \boldsymbol{T}_{11} \boldsymbol{\omega} \boldsymbol{\omega}^{\mathrm{H}} \boldsymbol{T}_{22} \boldsymbol{\omega}}} \tag{15.123}$$

Assuming that $\boldsymbol{T}_{11} = \boldsymbol{T}_{22}$, Eq. (15.123) is expressed as

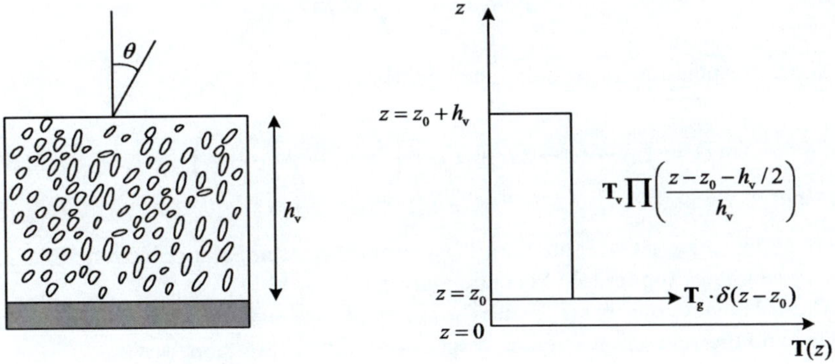

Fig. 15.8 Illustration of OVoG model

$$\gamma = \frac{\boldsymbol{\omega}^{\mathrm{H}}\boldsymbol{\Omega}_{12}\boldsymbol{\omega}}{\sqrt{\boldsymbol{\omega}^{\mathrm{H}}\boldsymbol{T}_{11}\boldsymbol{\omega}}} \tag{15.124}$$

The matrices in Eq. (15.124) can be expressed as

$$\boldsymbol{T}_{11} = e^{-\frac{\sigma_{T}(\boldsymbol{\omega})h_V}{\cos\theta_0}} e^{-\frac{\sigma_{R}(\boldsymbol{\omega})h_V}{\cos\theta_0}} \int\limits_{z} e^{\frac{\sigma_T'(\boldsymbol{\omega})+\sigma_R(\boldsymbol{\omega})}{\cos\theta_0} z} \boldsymbol{T}(z)dz \tag{15.125}$$

$$\boldsymbol{\Omega}_{12} = e^{j\phi_0} e^{-\frac{\sigma_{T}(\boldsymbol{\omega})h_V}{\cos\theta_0}} e^{-\frac{\sigma_{R}(\boldsymbol{\omega})h_V}{\cos\theta_0}} \int\limits_{z} e^{\left(\frac{\sigma_T'(\boldsymbol{\omega})+\sigma_R(\boldsymbol{\omega})}{\cos\theta_0}+jk_z\right) z} \boldsymbol{T}(z)dz \tag{15.126}$$

where z denotes the vertical coordinate, h_V is the vegetation height, θ_0 is the mean incidence, k_z is the mean vertical wavenumber, $\boldsymbol{T}(z)$ is the target coherency matrix as a function of height, $\sigma_{\boldsymbol{T}}(\boldsymbol{\omega})$ and $\sigma_{\boldsymbol{R}}(\boldsymbol{\omega})$ are the extinction coefficients of transmitting and receiving polarizations.

Expressions in Eqs. (15.125) and (15.126) are formally correct only for a random volume because in that case the polarization $\boldsymbol{\omega}$ remains unchanged as the wave propagates through the volume. In the case of an oriented volume, the polarization of the incidence wave changes with the variance of the polarization state z, i.e., $\boldsymbol{\omega} = \boldsymbol{\omega}(z)$. Only the eigenpolarizations [24] states are not be affected by this phenomenon, i.e., $\boldsymbol{\omega} = \boldsymbol{\omega}(z)$, which are two orthogonal polarizations that correspond to the maximum and minimum values of the extinction coefficient. For agricultural crops, eigenpolarizations are normally horizontal and vertical polarization. Accordingly, in the following formulation, a constant effective polarization $\boldsymbol{\omega}$ will be assumed for simplicity.

Defining an equivalent extinction coefficient as

$$\sigma_{\boldsymbol{\omega}} = \frac{\sigma_{\boldsymbol{T}}(\boldsymbol{\omega}) + \sigma_{\boldsymbol{R}}(\boldsymbol{\omega})}{2} \tag{15.127}$$

The corresponding target coherency matrix is

$$\boldsymbol{T}(z) = \boldsymbol{T}_G \cdot \delta(z - z_0) + \boldsymbol{T}_V \prod\left(\frac{z - z_0 - h_V/2}{h_V}\right) \tag{15.128}$$

where $\delta(\cdot)$ is Dirac's delta function, $\prod(\cdot)$ denotes a rectangular pulse function. The ground height (or topography) is defined by z_0.

The coherency matrices $\boldsymbol{T}_G$ and $\boldsymbol{T}_V$ are associated with the contributions of the ground and the vegetation volume, respectively. The expressions are

$$\boldsymbol{T}_{\mathrm{G}} = m_G \cdot \begin{bmatrix} 1 & t_{G12} & 0 \\ t_{G12}^{*} & t_{G22} & 0 \\ 0 & 0 & t_{G33} \end{bmatrix} \tag{15.129}$$

$$\boldsymbol{T}_{\mathrm{V}} = m_G \cdot \begin{bmatrix} 1 & t_{V12} & 0 \\ t_{V12}^* & t_{V22} & 0 \\ 0 & 0 & t_{V33} \end{bmatrix} \tag{15.130}$$

Without loss of generality, it can be assumed that $z_0 = 0$. Then, Eqs. (15.129) and 15.130) can be rewritten as

$$\boldsymbol{T}_{11} = e^{-\frac{\sigma_T(\omega)h_V}{\cos\theta_0}} \left[\boldsymbol{T}_{\mathrm{G}} + \boldsymbol{T}_{\mathrm{V}} \int\limits_0^{h_V} e^{\frac{2\sigma_\omega}{\cos\theta_0}} dz \right] = e^{-\frac{\sigma_T(\omega)h_V}{\cos\theta_0}} [\boldsymbol{T}_{\mathrm{G}} + \boldsymbol{I}_1^V] \tag{15.131}$$

$$\begin{aligned} \boldsymbol{\Omega}_{12} &= e^{j\phi_0} e^{-\frac{2\sigma_T(\omega)h_V}{\cos\theta_0}} \left[\boldsymbol{T}_{\mathrm{G}} + \boldsymbol{T}_{\mathrm{V}} \int\limits_0^{h_V} e^{\left(\frac{2\sigma_\omega}{\cos\theta_0} + jk_z\right)z} dz \right] \\ &= e^{j\phi_0} e^{-\frac{2\sigma_T(\omega)h_V}{\cos\theta_0}} [\boldsymbol{T}_{\mathrm{G}} + \boldsymbol{I}_2^V] \end{aligned} \tag{15.132}$$

Assuming that the extinctions coefficients are constant along the volume, the integrals in Eqs. (15.131) and (15.132) can be written as

$$\boldsymbol{I}_1^V = \frac{\cos\theta_0}{2\sigma_\omega} \left(e^{\frac{2\sigma_\omega h_V}{\cos\theta_0}} - 1 \right) \boldsymbol{T}_{\mathrm{V}} \tag{15.133}$$

$$\boldsymbol{I}_2^V = \frac{1}{\frac{2\sigma_\omega}{\cos\theta_0} + jk_z} \left(e^{\frac{2\sigma_\omega h_V}{\cos\theta_0}} e^{jk_z z} - 1 \right) \boldsymbol{T}_{\mathrm{V}} \tag{15.134}$$

The resulting complex interferometric coherence is

$$\gamma = e^{j\phi_0} \frac{\boldsymbol{\omega}^{\mathrm{H}} \left[e^{-\frac{2\sigma_\omega h_V}{\cos\theta_0}} \boldsymbol{T}_{\mathrm{G}} + e^{-\frac{2\sigma_T(\omega)h_V}{\cos\theta_0}} \boldsymbol{I}_2^V \right] \boldsymbol{\omega}}{\boldsymbol{\omega}^{\mathrm{H}} \left[e^{-\frac{2\sigma_\omega h_V}{\cos\theta_0}} \boldsymbol{T}_{\mathrm{G}} + e^{-\frac{2\sigma_T(\omega)h_V}{\cos\theta_0}} \boldsymbol{I}_1^V \right] \boldsymbol{\omega}} \tag{15.135}$$

Equation (15.135) can be transformed into

$$\gamma = e^{j\phi_0} \frac{\mu(\sigma_\omega, \boldsymbol{\omega}) + \gamma_{\mathrm{V}}(\sigma_\omega, \boldsymbol{\omega})}{1 + \mu(\sigma_\omega, \boldsymbol{\omega})} \tag{15.136}$$

where the volume coherence and the ground-to-volume ratio can be defined as

$$\gamma_{\mathrm{V}}(\sigma_\omega, \boldsymbol{\omega}) = \frac{\boldsymbol{\omega}^{\mathrm{H}} e^{-\frac{2\sigma_\omega h_V}{\cos\theta_0}} \boldsymbol{I}_2^V \boldsymbol{\omega}}{\boldsymbol{\omega}^{\mathrm{H}} e^{-\frac{2\sigma_\omega h_V}{\cos\theta_0}} \boldsymbol{I}_1^V \boldsymbol{\omega}} \tag{15.137}$$

$$\mu(\sigma_\omega, \omega) = \frac{\omega^{\mathrm{H}} e^{-\frac{2\sigma_\omega h_V}{\cos\theta_0}} T_G \omega}{\omega^{\mathrm{H}} e^{-\frac{2\sigma_\omega h_V}{\cos\theta_0}} I_1^V \omega} \tag{15.138}$$

When a particular polarization is selected, volume decorrelation coefficient can be simplified as

$$\gamma_V(\sigma_\omega, \omega) = \frac{2\sigma_\omega}{\cos\theta_0\left(e^{-\frac{2\sigma_\omega h_V}{\cos\theta_0}} - 1\right)} \frac{e^{\left(\frac{2\sigma_\omega}{\cos\theta_0} + jk_z\right)h_v} - 1}{\frac{2\sigma_\omega}{\cos\theta_0} + jk_z} \tag{15.139}$$

Consequently, the complex interferometric coherence coefficient for the model of an oriented volume over the ground can be finally expressed as

$$\begin{aligned}
\gamma &= e^{j\phi_0} \frac{\mu(\sigma_\omega, \omega) + \gamma_V(\sigma_\omega, \omega)}{1 + \mu(\sigma_\omega, \omega)} \\
&= e^{j\phi_0}\left[\gamma_V(\sigma_\omega, \omega) + \frac{\mu(\sigma_\omega, \omega)}{1 + \mu(\sigma_\omega, \omega)}(1 - \gamma_V(\sigma_\omega))\right] \\
&= e^{j\phi_0}\left[\gamma_V(\sigma_\omega, \omega) + L(\sigma_\omega, \omega)(1 - \gamma_V(\sigma_\omega))\right], 0 \le L(\sigma_\omega, \omega) \le 1
\end{aligned} \tag{15.140}$$

In contrast with the random volume case, the expression in Eq. (15.140) does not result in a straight line on the complex plane because the volume de-coherence coefficient γ_v is dependent to the polarization state.

In order to identify the position of the coherence values inside the unit circle, the following scenario is assumed. A thickness of the vegetation layer $h_V = 2\mathrm{m}$ and a ground topographic phase $\phi_0 = 20°$. The remote sensing platform is assumed to be the German Aerospace Center (DLR) E-SAR airborne system working at L-band with a 50-m baseline, which corresponds to a vertical wavenumber $k_z = 0.5$. It is also assumed that the extinction coefficients will range from 0~3 dB/m, and the whole range of the ground-to-volume ratios μ (from 0 to ∞) is sampled and plotted in the complex plane. The resulting loci of coherence values are shown in Fig. 15.9.

As we can see from Fig. 15.9, the shape of the region is basically a narrow triangle. The width of the feasible region depends on $k_z \cdot h_V$, as well as the magnitude of the differential extinction coefficient. If the range of $k_z \cdot h_V$ is restricted, the width of the triangle can be narrow enough.

15.2.2 *Agricultural Crop Parameters Inversion Process*

According to the aforementioned agricultural crop model, the process of agricultural crop parameters inversion is shown in Fig. 15.10.

Specific steps are as follows:

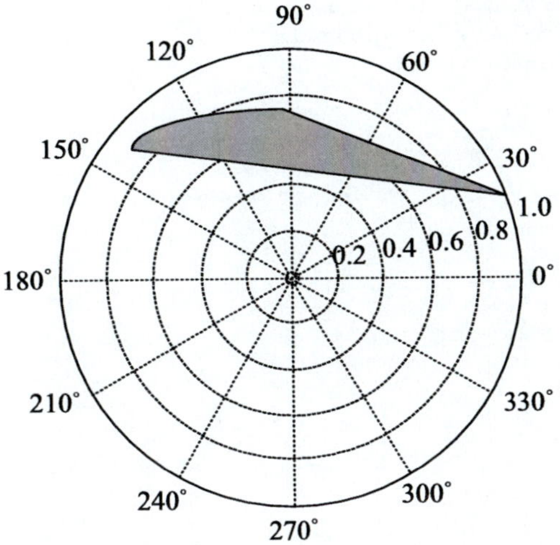

Fig. 15.9 The possible loci of coherence coefficient values in the unit circle

(a) Similar to the RVoG model inversion method, the estimated coherences coefficient of different polarization channels is used to find the fitted total least square line. Then, the intersection between the line and the unit circle defines the phase of the ground position ϕ_0, which is directly converted to topographic height by the expression: $z_0 = \phi_0/k$, shown in Fig. 15.10a.
(b) Once the topographic phase ϕ_0 is determined, all possible lines with the end points at $\exp(j\phi_0)$ and each measured coherence point are plotted. The set of lines intersecting the estimated topographic phase is enclosed in the triangular feasible region shown in Fig. 15.10b. the points at the right end are those with minimum extinction coefficient $\sigma_{\min}$, whereas the points at the left end show the maximum extinction coefficient $\sigma_{\max}$. All these straight lines correspond to a unique vegetation height h_v, which is still unknown.
(c) In order to estimate the vegetation height h_v, As set of volume coherence coefficient functions γ_V, departing from the unit circle ground topographic point, is plotted as a function of two variables: h_v and σ. The used range of extinctions coefficient goes from 0 dB/mup to 2 dB/m. Finally, the functions $\gamma_V(h_v, \sigma_{\min})$ and $\gamma_V(h_v, \sigma_{\max})$ on the unit circle are also plotted as a function of height h_v, as shown in Fig. 15.10c. From this set of possible solutions, an average height estimation h_v can be obtained.

15.3 DEM Extraction from PolInSAR

The principle of DEM extraction from InSAR is introduced in Chap. 4. The performance of InSAR measurement depends on the correlation of the signals between the corresponding pixels of the two images forming the interferogram. Phase estimation error increases rapidly as coherence coefficient decreases. In addition, the

Fig. 15.10 The scheme of agricultural crop parameters inversion process

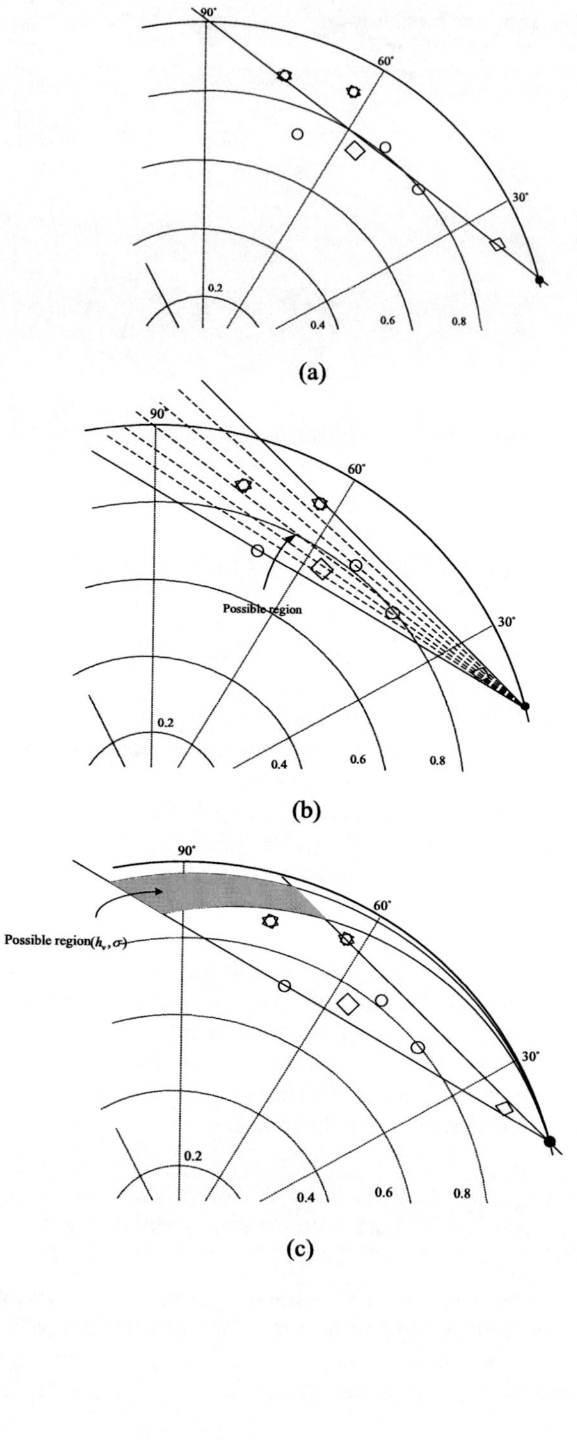

SAR interferometry technique implies the assumption that the signal echo of each pixel in the image is backscattered from a scattering center above a reference plane at a fixed height. The measured phase difference is therefore proportional to the height of this reference plane. However, due to the existence of factors such as ground slope and roughness, as well as the effects of surface vegetation and volume scattering, the actual process of ground objects scattering electromagnetic waves is extremely complicated. There are often multiple scattering mechanisms in the resolution unit, and phase centers of various scattering mechanisms may be at different heights. Then the phase difference between the two interferograms cannot reflect the actual height of any scattering center but reflects the average height of all scatters. This effect is particularly serious in vegetation-covered areas. Combing polarimetric techniques with interferometric techniques can effectively improve the performance of interferometry [50].

Figure 15.11 shows the DEM extraction process of PolInSAR. First, full PolInSAR data is coregistered, and then the six coregistered HH, HV, and VV polarization images are used for the optimization coherence coefficient calculation. Therefore, three optimized coherence coefficients and the corresponding scattering mechanisms are obtained. After obtaining the optimal interferogram from Eq. (4.46), similar to the conventional interferometric SAR registration procedure, the steps of flattening the interferogram, phase filtering, phase unwrapping, and DEM inversion are performed.

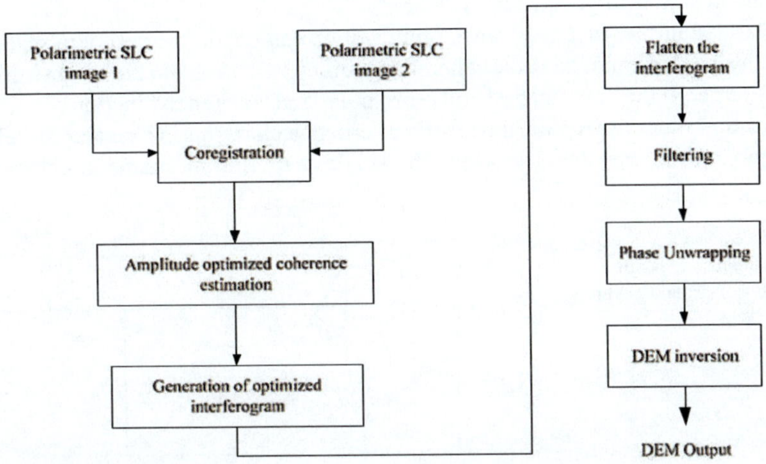

Fig. 15.11 Flowchart of DEM extraction from PolInSAR

15.4 Target Detection from PolInSAR

Target detection is an important research area in remote sensing. With the in-depth study of PolInSAR information extraction technology, the study of artificial target detection with PolInSAR data has begun.

15.4.1 Target Detection Beneath Ground Cover

In 2001, Sagues et al. proposed an investigation on detection and identify buried mines with PolInSARdata [17, 52]. The unique advantage of PolInSAR is used for separating the phase center corresponding to the mine from that related to the clutter and estimating their relative height difference or depth. The detection method is introduced.

Figure 15.12 shows the interferometric imaging geometry of the target scene under the ground cover. The key step to detect the target under the ground cover is to distinguish the scattering mechanism of the targets to obtain their relative height. Different scattering centers (target and ground clutter) in the resolution cell cannot be distinguished by pure interferometry, and this can be achieved using polarization techniques. Polarization Subspace Method (PSM) is applied by L. Sagues, etc. to distinguish scattering mechanisms.

Polarization subspace coherence optimization can obtain the optimum scattering mechanism which maximize the coherence coefficient, and obtain the height information of target. If the co-polarized and cross-polarized coherence function of different polarization states in graphic, different scattering mechanisms in a resolution cell can be distinguished. For the case where the single scattering mechanism is dominant,

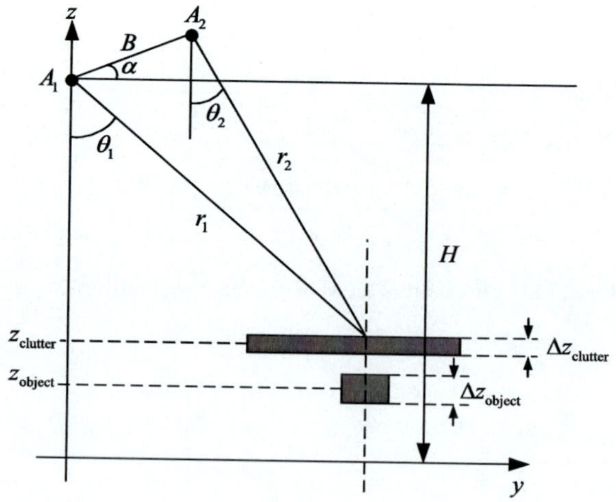

Fig. 15.12 The interferometric measure geometry of ground covered target scene

both the co-polarized and cross-polarized coherence functions show only an absolute maximum. However, in those cases where the target exhibits a multilayer vertical structure which is shown in Fig. 15.12, different maxim local coherent values can be obtained under different polarization states. These optimum polarization states will be related to different independent scattering mechanisms that lead to the locally optimum height estimation. For targets covered by ground, it is expected to find two coherence maxima that corresponding to the different scattering mechanisms: the target and the surface clutter. After generating the two interferograms corresponding to these optimum polarization states, we can extract the height difference between them. The resulting height difference map will exhibit some peaks at those positions where the targets are covered, whereas it will remain close to zero for the zones of the image where only natural surface scattering process can be found.

15.4.2 Target Detection Technique Beneath Foliage

Detection of artificial targets obscured by foliage clutter is one of the applications of PolInSAR information extraction. Target detection which takes interferometric phases as observations was investigated by S. R. Cloude. The theoretical fundamental of detection method is that the interferometric coherence coefficient of random volume scattering is polarization independent, and the interferometric coherence coefficient is polarization dependent when there is a target, and the change is linear in the complex coherence plane. The target detection method is described as following.

For random vegetation, the expression volume decorrelation is

$$
\begin{aligned}
\gamma_V = & \frac{2\sigma_\omega}{\cos\theta_0\left(e^{-\frac{2\sigma_\omega h_V}{\cos\theta_0}} - 1\right)} \int_0^{h_V} e^{ik_z z'} \exp\left(\frac{2\sigma z'}{\cos\theta_0}\right) dz' \\
= & f(h_v, \sigma) = \frac{p_1\left(e^{p_2 h_v} - 1\right)}{p_2\left(e^{p_1 h_v} - 1\right)} \begin{cases} p_1 = \frac{2\sigma}{\cos\theta_0} \\ p_2 = \frac{2\sigma}{\cos\theta_0} + ik_z \end{cases}
\end{aligned} \tag{15.141}
$$

where θ_0 is the mean average incidence angle, σ is the extinction coefficient, and k_z is the vertical wave number. $\mathbf{T_v}$ is coherency matrix of the vegetation scattering.

If we extend the analysis to a two-layer problem consisting of a random vegetation layer above a localized scattering center, which acts as a model for a vehicle or other targets beneath the canopy. Figure 15.13 shows the geometry to be considered. In the figure, $d(z)$ is the one-dimensional vertical profile function. The target is located at some unknown depth z_0 and has an apparent magnitude m_g. The coherence coefficient of this model is

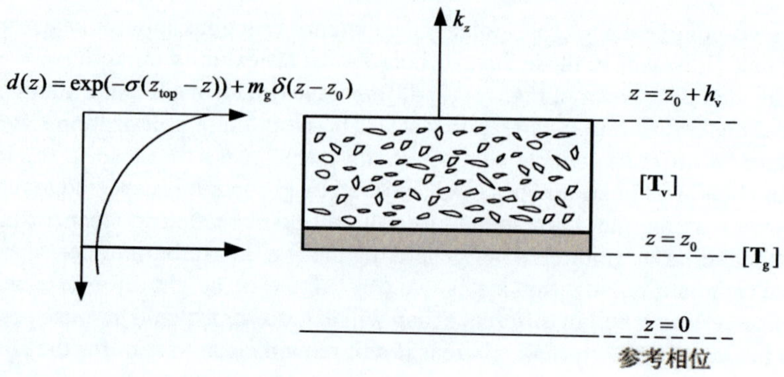

Fig. 15.13 Two-layer random volume with object model

$$
\begin{cases}
\gamma = e^{j\phi_0} \dfrac{\boldsymbol{\omega}^{\mathrm{H}}\left(e^{i\phi_2} \boldsymbol{I}_2^{\mathrm{V}} + \exp\left(-\frac{2\sigma h_{\mathrm{V}}}{\cos\theta_0}\right) \boldsymbol{T}_{\mathrm{G}} e^{i\phi_1}\right)\boldsymbol{\omega}}{\boldsymbol{\omega}^{\mathrm{H}}\left[\boldsymbol{I}_1^{\mathrm{V}} + exp\left(-\frac{2\sigma h_{\mathrm{V}}}{\cos\theta_0}\right)\boldsymbol{T}_{\mathrm{G}}\right]\boldsymbol{\omega}} \\
\quad = \dfrac{\boldsymbol{\omega}^{\mathrm{H}}(e^{i\phi_2} \nu^{-1} \boldsymbol{I}_2^{\mathrm{V}} + e^{i\phi_1} \nu^{-1} \boldsymbol{T}_{\mathrm{G}}))\boldsymbol{\omega}}{1 + \boldsymbol{\omega}^{\mathrm{H}} \nu^{-1} \boldsymbol{T}_{\mathrm{G}} \boldsymbol{\omega}} \\
\nu^{-1} = \dfrac{1}{\boldsymbol{\omega}^{\mathrm{H}} \boldsymbol{I}_1^{\mathrm{V}} \boldsymbol{\omega}} \\
\widehat{\gamma}(\boldsymbol{\omega}) = e^{i\phi_1} \dfrac{\widehat{\gamma}_{\mathrm{V}} + \mu(\boldsymbol{\omega})}{1 + \mu(\boldsymbol{\omega})} = e^{i\phi_1}\left(\widehat{\gamma}_{\mathrm{V}} + \dfrac{\mu(\boldsymbol{\omega})}{1 + \mu(\boldsymbol{\omega})}\left(1 - \widehat{\gamma}_{\mathrm{V}}\right)\right) \\
\quad = e^{i\phi_1}\left(\widehat{\gamma}_{\mathrm{V}} + L(\boldsymbol{\omega})\left(1 - \widehat{\gamma}_{\mathrm{V}}\right)\right), 0 \le L(\boldsymbol{\omega}) \le 1
\end{cases}
\tag{15.142}
$$

In Eq. (15.142), $\boldsymbol{\omega}$ is the polarization state. The angles ϕ_1 and ϕ_2 are the phase centers of the target and bottom of vegetation layer, respectively. The parameter μ is the target-to-volume scattering ratio.

$$
\mu(\boldsymbol{\omega}) = \frac{2\sigma}{\cos\theta_0\left(\exp\left(\frac{2\sigma h_{\mathrm{V}}}{\cos\theta_0}\right) - 1\right)} \frac{\boldsymbol{\omega}^H \boldsymbol{T}_G \boldsymbol{\omega}}{\boldsymbol{\omega}^H \boldsymbol{T}_V \boldsymbol{\omega}} \ge 0 \tag{15.143}
$$

The matrices $\boldsymbol{T}_G$ and $\boldsymbol{T}_V$ are the coherency matrices of target and scatterers, and they are Hermitian. We can see from Eqs. (15.142)–(15.143) that the presence of a target beneath the canopy influences the observed coherence. If the target scattering changes with polarization then this will be reflected in changes in the observed coherence. Hence by using polarimetric interferometry we can detect the presence of targets by looking at the variation of coherence with polarization. For vegetation only this will yield zero or small changes. However, when a target is present the changes will be greater.

As we can see from Eq. (15.142), similar to RVoG model, the coherence is a line in the complex plane inside the unit circle as shown in Fig. 15.14. The left point is the $\mu = 0$ or volume coherence point alone. It can be obtained by means of measuring the polarization coherence coefficient in a polarization channel with less volume scattering component, such as the cross-polarized HV channel. The right point is the bare earth topography and it can be estimated as the stage 1 of three-stage

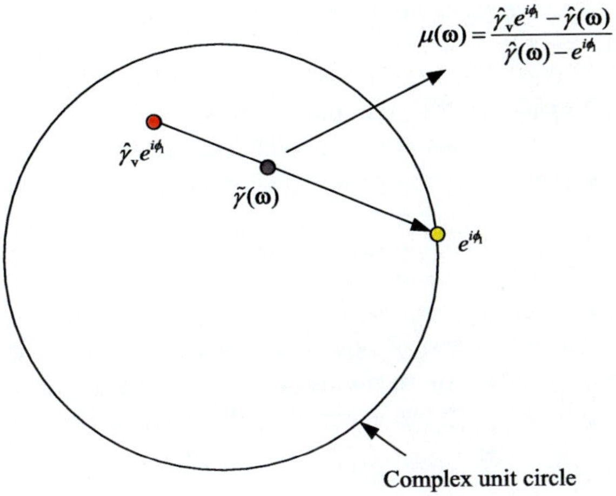

Fig. 15.14 The line coherence model for target + random volume scattering

inversion process. By measuring the coherence coefficient of specific polarization state, and comparing the coherence coefficient with both ends of the line, the target-to-volume scattering ratio of this polarization state can be estimated. Polarization state corresponds to bigger μ is more benefit for detecting the target.

According to the former analysis, the introduced algorithm is based on estimation of a scalar radar parameter μ, obtained from complex interferometric coherence estimates in different polarization channels $\boldsymbol{\omega}$. The parameter μ is the ratio of all scattered power contributions with an apparent phase center located on the surface, to the total volume scattering. The former includes the direct ground echo, the target backscattering, and the effect of the dihedral scattering caused by second interaction between ground scattering and vegetation. The volume scattering echo includes the sum of scattering for all phase centers deviate to ground surface. The target exists the position corresponding local maximum μ. It follows from this that the presence of a target corresponds to a local maximum in μ. We can obtainan estimate of the sum of these two contributions $s(\boldsymbol{\omega})$ as shown in equation Eq. (15.144).

$$s(\boldsymbol{\omega}) = \frac{\cos\theta_0\left(1 - \exp\left(-\frac{2\sigma h_V}{\cos\theta_0}\right)\right)}{2\sigma}\boldsymbol{\omega}^H \boldsymbol{T}_V \boldsymbol{\omega} + \exp\left(-\frac{2\sigma h_V}{\cos\theta_0}\right)\boldsymbol{\omega}^H \boldsymbol{T}_G \boldsymbol{\omega} \quad (15.144)$$

Hence we can solve Eqs. (15.143) and (15.144) to obtain an estimate of the surface scattering components.

$$\begin{aligned}\boldsymbol{\omega}^H \boldsymbol{T}_G \boldsymbol{\omega} &= \exp\left(\frac{2\sigma h_V}{\cos\theta_0}\right)\frac{\mu(\boldsymbol{\omega})}{1+\mu(\boldsymbol{\omega})}s(\boldsymbol{\omega}) = \exp\left(\frac{2\sigma h_V}{\cos\theta_0}\right)L(\boldsymbol{\omega})s(\boldsymbol{\omega}) \\ &= \exp\left(\frac{2\sigma h_V}{\cos\theta_0}\right)F(\boldsymbol{\omega}), 0 \le L(\boldsymbol{\omega}) \le 1 \end{aligned} \quad (15.145)$$

The function $L(\boldsymbol{\omega})$ acts as a filter on the intensity channel to produce an image domain with better signal to clutter ratio.

As the former analysis, the target detection algorithm with PolInSAR data is as follows:

Stage 1: coherence estimation. The first stage is to generate complex coherence estimates for different polarization states.

Stage 2: least squares line fit. The next stage is to find the best-fit straight line to the polarimetric coherence values inside the unit circle.

Stage 3: select the reference volume coherence channel. The next is to select the reference volume coherence channel, and usually HV channel is selected.

Stage 4: estimate ground scattering components and filtered intensity. Having estimated the extreme points on the coherence line, we can now estimate the μ value in any desired polarization channel. From μ we can then obtain $L(\boldsymbol{\omega})$ using Eq. (15.142). We then estimate the mean backscatter intensity in the $\boldsymbol{\omega}$ channel and finally multiply this intensity by the function $L(\boldsymbol{\omega})$ to obtain $F(\boldsymbol{\omega})$.

15.5 Polarimetric Coherence Tomography

Polarimetric coherence tomography is a new radar imaging technique, which employs variation of the interferometric coherence with polarization to reconstruct a vertical profile function in penetrable volume scattering [52, 53]. This is a very effective tomography method and can be applied in forest classification and target detection beneath foliage.

The fundamental principle of polarimetric coherence tomography is performing Fourier-Legendre series expansion to an unknown vertical profile of a limited range. The variation of the interferometric coherence with polarization and baseline can be used to retrieve and determine the expansion coefficients.

The coherence coefficient of random volume scattered can be expressed as

$$\widehat{\gamma}_{\mathrm{V}} = e^{ik_z z_0} \frac{\int_0^{h_{\mathrm{V}}} f(z) e^{ik_z z} dz}{\int_0^{h_{\mathrm{V}}} f(z) dz} = e^{i\phi_0} \frac{\int_0^{h_{\mathrm{V}}} f(z) e^{ik_z z} dz}{\int_0^{h_{\mathrm{V}}} f(z) dz} \tag{15.146}$$

where z_0 is the position of the bottom of the scattering layer, ϕ_0 is the topographic phase, and $f(z)$ is the vertical structure function, as shown in Fig. 15.15.

The estimation of h_V and ϕ_0 is obtained from PolInSAR data, and then the reconstruction of vertical structure function $f(z)$ is considered. Removing ground phase term and performing normalization to the integral, we can get

$$\int_0^{h_{\mathrm{V}}} f(z) e^{ik_z z} dz \xrightarrow{z' = \frac{2z}{h_{\mathrm{V}}} - 1} \int_{-1}^{1} f(z') e^{ikz'} dz' \tag{15.147}$$

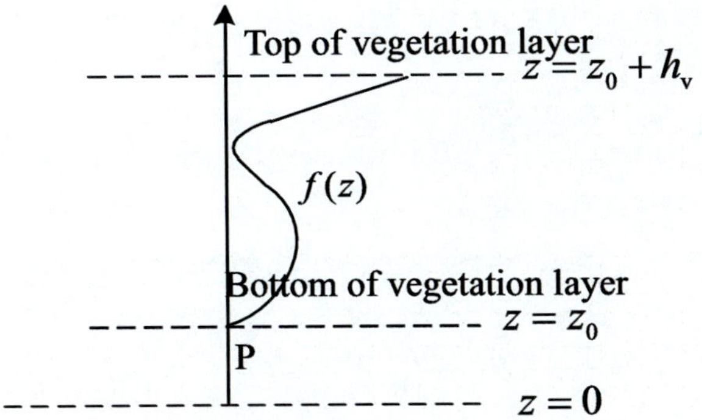

Fig. 15.15 Vertical structure function $f(z)$ at a point P on the surface

We also rescale variation of the real non-negative function $f(z)$ so that if $0 \leq f(z) \leq \infty$ then $-1 \leq f(z') \leq \infty$. We can develop the Fourier-Legendre series on [−1,1] as

$$\begin{cases} f(z') = \sum_n a_n P_n(z') \\ a_n = \frac{2n+1}{2} \int_{-1}^{1} f(z') P_n(z') dz' \end{cases} \tag{15.148}$$

where the first few Legendre polynomials are given as

$$\begin{cases} P_0(z) = 1 \\ P_1(z) = z \\ P_2(z) = \frac{1}{2}(3z^2 - 1) \\ P_3(z) = \frac{1}{2}(5z^3 - 3z) \\ P_4(z) = \frac{1}{8}(35z^4 - 30z^2 + 3) \end{cases} \tag{15.149}$$

The numerator and denominator for Eq. (15.146) are

$$\begin{cases} \int_0^{h_V} f(z) e^{ik_z z} dz = \frac{h_V}{2} e^{i\frac{k_z h_V}{2}} \int_{-1}^{1} (1 + f(z')) e^{i\frac{k_z h_V}{2} z'} dz' \\ \int_0^{h_V} f(z) dz = \frac{h_V}{2} \int_{-1}^{1} (1 + f(z')) dz' \end{cases} \tag{15.150}$$

from which that the coherence coefficient can be written as follows:

$$\tilde{\gamma} = e^{i\frac{k_z h_V}{2}} \frac{\int_{-1}^{1} (1 + f(z')) e^{i\frac{k_z h_V}{2} z'} dz'}{\int_{-1}^{1} (1 + f(z')) dz'}$$

$$= e^{ik_V} \frac{\int_{-1}^{1}\left(1+\sum_n a_n P_n(z')\right)e^{ik_V z'}dz'}{\int_{-1}^{1}\left(1+\sum_n a_n P_n(z')\right)dz'} \tag{15.151}$$

where $k_V = \frac{k_z h_V}{2}$. By expanding the series and collecting terms, it can be rewritten as

$$\begin{aligned}\tilde{\gamma} &= e^{ik_V} \frac{(1+a_0)\int_{-1}^{1} e^{ik_V z'}dz' + a_1\int_{-1}^{1} P_1(z')e^{ik_V z'}dz' + a_2\int_{-1}^{1} P_2(z')e^{ik_V z'}dz' + a_3\int_{-1}^{1} P_3(z')e^{ik_V z'}dz' + \cdots}{(1+a_0)\int_{-1}^{1} dz' + a_1\int_{-1}^{1} P_1(z')dz' + a_2\int_{-1}^{1} P_2(z')dz' + a_3\int_{-1}^{1} P_3(z')dz' + \cdots} \\ &= e^{ik_V} \frac{(1+a_0)f_0 + a_1 f_1 + a_2 f_2 + \cdots a_n f_n}{(1+a_0)}\end{aligned} \tag{15.152}$$

The function f_n can be analytically expressed as

$$\begin{cases} f_0 = \frac{\sin k_V}{k_V} \\ f_1 = i\left(\frac{\sin k_V}{k_V^2} - \frac{\cos k_V}{k_V}\right) \\ f_2 = \frac{3\cos k_V}{k_V^2} - \left(\frac{6-3k_V^2}{2k_V^3} + \frac{1}{2k_V}\right)\sin k_V \\ f_3 = i\left(\left(\frac{30-5k_V^2}{2k_V^3} + \frac{3}{2k_V}\right)\cos k_V - \left(\frac{30-15k_V^2}{2k_V^4} + \frac{3}{2k_V^2}\right)\sin k_V\right) \\ f_4 = \left(\frac{35(k_V^2-6)}{2k_V^4} - \frac{15}{2k_V^2}\right)\cos k_V + \left(\frac{35(k_V^4-12k_V^2+24)}{8k_V^5} + \frac{30(2-k_V^2)}{8k_V^3} + \frac{3}{8k_V}\right)\sin k_V \end{cases} \tag{15.153}$$

As the former analysis, the coherence for an arbitrary vertical structure function $f(z)$ can be written in the form

$$\tilde{\gamma}\, e^{-ik_V} = \tilde{\gamma}_k = f_0 + a_{10} f_1 + a_{20} f_2 + \cdots + a_{n0} f_n \tag{15.154}$$

Let $a_{n0} = \frac{a_n}{1+a_0}$, and if we can estimate the set of unknown coefficients a_{n0}, then we can use them to generate an estimate of the unknown vertical structure profile $\hat{f}(z')$

$$\hat{f}\left(z'\right) = 1 + \hat{a}_{10} P_1\left(z'\right) + \hat{a}_{20} P_2\left(z'\right) + \cdots \hat{a}_{n0} P_n\left(z'\right) \tag{15.155}$$

If the height of vegetation h_V and topographic phase is known, the realization process of vertical structure function is as follows:

(1) For single-baseline PolInSAR system, we can only achieve a single complex coherence $\tilde{\gamma}_k$ can be achieved. Hence only two unknown parameters can be obtained in the Fourier-Legendre series, namely, $\hat{a}_{10}$ and $\hat{a}_{20}$.

$$\begin{bmatrix} 1 & 0 & 0 \\ 0 & f_1 & 0 \\ 0 & 0 & f_2 \end{bmatrix} \begin{bmatrix} a_{00} \\ a_{10} \\ a_{20} \end{bmatrix} = \begin{bmatrix} 1 \\ \mathrm{Im}\left(\tilde{\gamma}_k\right) \\ \mathrm{Re}\left(\tilde{\gamma}_k\right) - f_0 \end{bmatrix} \Rightarrow [\boldsymbol{F}]\boldsymbol{a} = \boldsymbol{b} \Rightarrow \hat{\boldsymbol{a}} = [\boldsymbol{F}]^{-1}\boldsymbol{b} \tag{15.156}$$

(2) For double-baseline PolInSAR system we can achieve the estimation of four unknown parameters

$$\begin{bmatrix} f_1^x & 0 & f_3^x & 0 \\ 0 & f_2^x & 0 & f_4^x \\ f_1^y & 0 & f_3^y & 0 \\ 0 & f_2^y & 0 & f_4^y \end{bmatrix} \begin{bmatrix} a_{10} \\ a_{20} \\ a_{30} \\ a_{40} \end{bmatrix} = \begin{bmatrix} \mathrm{Im}(\tilde{\gamma}_k^x) \\ \mathrm{Re}(\tilde{\gamma}_k^x) - f_0^x \\ \mathrm{Im}(\tilde{\gamma}_k^y) \\ \mathrm{Re}(\tilde{\gamma}_k^y) - f_0^y \end{bmatrix}$$
$$\Rightarrow [\boldsymbol{F}]\boldsymbol{a} = \boldsymbol{b} \Rightarrow \hat{\boldsymbol{a}} = [\boldsymbol{F}]^{-1}\boldsymbol{b} \tag{15.157}$$

In Eq. (15.157), subscript x and y represent the measurement of two baselines, respectively.

15.6 Conclusion

In this chapter, the information process techniques with PolInSAR is introduced. The applications of PolInSAR in forest parameter inversion, agricultural parameter inversion, and target detection are introduced. As the acquire of more high-precision PolInSAR data, the application of PolInSAR can be more and more extensive.

References

1. S.R. Cloude, K.P. Papathanassiou, Three-stage inversion process for polarimetric SAR interferometry. IEEE Proc.-Radar Sonar Navig. **150**(3), 125–134 (2003)
2. S.R. Cloude, K.P. Papathanassiou, A. Reigber, *Polarimetric SAR Interferometry at P-Band for Vegetation Structure Extraction* (Pro. of EUSAR Munich, Germany, 2000)
3. I. Hajnsek, et al., Tropical-forest-parameter estimation by means of pol-InSAR: the INDREX-II campaign. IEEE Trans. Geosci. Remote Sens **47**(2), 481–493 (2009)
4. J.D. Ballester-Berman, J.M. Lopez-Sanchez, J. Fortuny-Guasch, Retrieval of biophysical parameters of agricultural crops using polarimetric SAR interferometry. IEEE Trans. Geosci. Remote Sens. **43**(4), 683–694 (2005)
5. I. Hajnsek, K.P. Papathanassiou, S.R. Cloude, *A Hybrid Scattering Model for Surface Parameter Estimation Using Polarimetric SAR Interferometry* (Proc. of IGARSS '03, Toulouse, 2003), pp. 693–695
6. I. Hajnsek, et al., Surface Parameter Estimation Using Interferometric and Polarimetric SAR (Proc. of IGARSS'02, Toronto, 2002), pp. 420–422
7. O. Stebler et al., Pol-InSAR observations from an alpine glacier in the cold infiltration zone at L- and P-band. IEEE Geosci. Remote Sens. Lett. **2**(3), 357–361 (2005)

8. T. Busche, *Potential of Airborne Pol-InSAR Measurements to Derive Vertical Structure Parameters of Alpine Glaciers* (Proc. of EUSAR, Friedrichshafen, Germany, 2008)
9. J.J. Sharma, I. Hajnsek, K.P. Papathanassiou, *Extinction Estimation Over Land Ice Using Long-Wavelength Pol-InSAR* (Proc. of EUSAR, Friedrichshafen, Germany, 2008)
10. M. Brandfass, *Generation of Bald Earth Digital Elevation Models as Applied to Polarimetric SAR Interferometry* (Proc. of IGARSS, Canada, Toronto, 2002), pp. 1014–1016
11. Xiong Tao etc, Approach to reversion of DEM based on polarimetric SAR interferometry. J. Tsinghua Univ. (Sci. Technol.) **47**(7), 1170–1173 (2007)
12. L. Ferro-Famil, E. Pottier, J.S. Lee., *Unsupervised Classification and Analysis of Natural Scenes from Polarimetric Interferometric SAR Data* (Proc. of IGARSS'01, 2001), pp. 2715–2717
13. M. Hellmann, S.R. Cloude, K.P. Papathanassiou, *Classification Using Polarimetric and Interferometric SAR Data* (Proc. of IGARSS '97, 1997), pp. 1411–1413
14. J.S. Lee, et al., *Forest Classification Based on L-Band Polarimetric and Interferometric SAR Data* (PolInSAR 2005 Workshop, 2005)
15. T. Mette, K. Papathanassiou, I. Hajnsek, *Biomass estimation from polarimetric SAR interferometry over heterogeneous forest terrain* (Proc. of IGARSS, Anchorage, USA, 2004), pp. 511–514
16. T. Mette, et al., *Forest Biomass Estimation using Polarimetric SAR Interferometry* (Proc. of IGARSS'02, 2002), pp. 817–819
17. L. Sagues et al., Polarimetric radar interferometry for improved mine detection and surface clutter rejection. IEEE Trans. Geosci. Remote Sens. **39**(6), 1271–1278 (2001)
18. S.R. Cloude, D.G. Corr, M.L. Williams, Target detection beneath foliage using polarimetric synthetic aperture radar interferometry. Waves Random Media **14**, S393–S414 (2004)
19. M. Qong, Polarization state conformation and its application to change detection in polarimetric SAR data. IEEE Geosci. Remote Sens. Lett. **1**(4), 304–308 (2004)
20. R.Z. Schneider, et al., *Polarimetric Interferometry over Urban Areas: Information Extraction Using Coherent Scatterers* (Proc. of IGARSS 2005, Seoul, Korea, 2005)
21. S. Guillaso et al., Building characterization using L-band polarimetric interferometric SAR data. IEEE Geosci. Remote Sens. Lett. **2**(3), 347–351 (2005)
22. C. Erxue et al., Polarimetric synthetic aperture radar interferometry based mean tree height extraction technique. Scientia Silvae Sinicae **43**(4), 66–70 (2007)
23. R.N. Treuhaft, P.R. Siqueira, Vertical structure of vegetated land surfaces from interferometric and polarimetric radar. Radio Sci. **35**(1), 141–177 (2000)
24. R.N. Treuhaft, S.R. Cloude, The structure of oriented vegetation from polarimetricinterferometry. IEEE Trans. Geosci. Remote Sens. **37**(5), pp. 2620–2624 (1999)
25. J.D. Ballester-Berman, J.M. Lopez-Sanchez, Coherence loci for a homogeneous volume over a double-bounce ground return. IEEE Geosci. Remote Sens. Lett. **4**(2), 317–321 (2007)
26. Li. Xinwu, *Information Extraction Methods and Application Study of Polarimetric SAR Interferometry* (Institute of Remote Sensing Applications, Chinese Academy of Sciences, 2002)
27. K.P. Papathanassiou, Single-Baseline Polarimetric SAR Interferometry. IEEE Trans. Geosci. Remote Sens. **39**(11), 2352–2363 (2001)
28. K.P. Papathanassiou, et al., *Airborne Polarimetric SAR Interferometry* (Proc IGARSS'98, 1998), pp. 1901–1903.
29. C. Bing, *Investigation on Interferometry and Polarimetric interferometry SAR Singal Processing Techniques* (Institute of Electronics, Chinese Academy of Sciences, 2008)
30. H. Yamada, Y. Yamaguchi, W.M. Boerner, *Forest Height Feature Extraction in Polarimetric SAR Interferometry by using Rotational Invariance Property* (Proc. IGARSS'03, 2003), pp. 1426–1428
31. B. Lu, *Multiple-Layer Model Based PolInSAR Methods to Estimate Vertical Structure Parameters and Their Applications in Forest Areas* (Institute of Electronics, Chinese Academy of Sciences, 2009)
32. M. Isola, S.R. Cloude, *Forest Height Mapping Using Space-Borne Polarimetric SAR Interferometry* (Proc. IGARSS'01, 2001), pp. 1095–1097

33. F. Garestier, P.C. Dubois-Fernandez, I. Champion, Forest Height Inversion Using High-Resolution P-Band Pol-InSARData[J]. IEEE Trans. on Geoscience and Remote Sensing, 2008. Vol. 46(No. 11): p. 3544–3559.
34. T. Mette, et al., *Forest and the Random Volume over Ground-Nature and Effect of 3 Possible Error Types* (Proc. of EUSAR 2006, Dresden, Germany)
35. S.R. Cloude,*Robust Parameter Estimation Using Dual Baseline Polarimetric SAR Interferometry* (Proc. of IGARSS'02, 2002), pp. 838–840
36. F. Koudogbo, et al., *Dual-Baseline Approach for the Retrieval of Vegetation Parameters* (Proc. EUSAR, 2006, Dresden, Germany)
37. K.P. Papathanassiou, S.R. Cloude., *The Effect of Temporal Decorrelation on the Inversion of Forest Parameters from Pol-InSAR Data* (Proc of IGARSS 2003)
38. T. Flynn, M. Tabb, R. Carande., *Coherence Region Shape Extraction for Vegetation Parameter Estimation in Polarimetric SAR Interferometry* (Proc. of IGARSS'02, 2002) pp. 2596–2598
39. L. Tan, R. Yang., *Investigation on Tree Height Retrieval with Polarimetric SAR Interferometry* (Proc. of IGARSS '08, Boston, USA, 2008)
40. M. Tabb, et al., Phase Diversity: A Decomposition for Vegetation Parameter Estimation Using Polarimetric SAR Interferometry (Proc. of EUSAR, Cologne, Germany, 2002)
41. T. Flynn, M. Tabb, R. Carande, *Direct Estimation of Vegetation Parameters from Convariance Data in Polarimetric SAR Interferometry* (Proc. of IGARSS'02, 2002, Toronto, Canada) pp. 1908–1910
42. M. Tabb, T. Flynn, R. Carande, *Full Maximum Likelihood Inversion of PolInSAR Scattering Models* (Proceeding of IGARSS'04, 2004), pp. 1232–1235
43. Z. Bin et al., Information extraction using polarimetric interferometric SAR data: present and future. J. Electron. Inform. Technol. **28**(10), 1979–1984 (2006)
44. N.R. Goodman, Statistical analysis based on a certain multivariate complex Gaussian distribution (an introduction). Ann. Math. Stat. **34**(1), 152–177 (1963)
45. J.-S. Lee et al., Intensity and phase statistics of multilook polarimetric and interferometric SAR imagery. IEEE Trans. Geosci. Remote Sens. **32**(5), 1017–1028 (1994)
46. Z. Xianda, *Matrix Analysis and Applications* (Tsinghua University Press, Beijing, 2004)
47. D. Kasilingam, M. Nomula, S. Cloude, *A Technique for Removing Vegetation Bias from Polarimetric SAR Interferometry* (Proc. of IGARSS 2002, Toronto, Canada) pp. 1017-1019
48. S. Guillaso, et al., *PolarimetricInterferometric SAR Data Analysis Based on ESPRIT/MUSIC Methods* (Proc. of PolInSAR, 2003, Roma)
49. J.M. Lopez-Sanchez, J.D. Ballester-Berman, *Analysis of Approaches for a Complete Estimation of Parameters in Agricultural Crops Applications Using Polarimetric SAR Interferometry* (URSI Commision F Symposium on Microwave Remote Sensing, 2005)
50. Y. Zhen, *Studies on Synthetic Aperture Radar Interferometry and Polarimetric Interferometry Techniques* (Institute of Electronics, Chinese Academy of Sciences, 2003)
51. T. Lulu, *Study on Information Processing of Polarimetric and Compact Polarimtric SAR Interferometry* (Institute of Electronics, Chinese Academy of Sciences, 2010)
52. L. Sagues, et al., *Buried Mine Detection by Polarimetric Radar Interferometry* (Ultra-Wideband, Short-Pulse Electromagnetics 5, 2002)
53. S.R. Cloude, Polarization coherence tomography. Radio Sci. **41**, RS 4017 (2006)
54. S. Cloude, *Polarisation Coherence Tomography (PCT)* (Proc. of EUSAR 2006, Dresden, Germany)
55. S.R. Cloude, *Application of Polarization Coherence Tomography (PCT)* (International Radar Symposium, Krakow, Poland, 2006)

Chapter 16
Information Processing of Compact Polarimetric SAR Interferometry

Compact polarimetric SAR can preserve most of the information of full polarimetric SAR. The investigation of the information processing of polarimetric SAR interferometry is the hot spots in recent years. In this chapter, tree height retrieval and terrain classification application of compact PolInSAR are discussed more deeply.

16.1 Tree Height Retrieval of Compact Polarimetric SAR Interferometry

Like full polarimetric SAR interferometry, compact polarimetric SAR interferometry can effectively extract the vertical structure information of vegetation. Inversion of ground parameters beneath foliage is one of the important research directions of compact polarimetric SAR interferometry information extraction.

16.1.1 Three-Stage Inversion Process of Compact PolInSAR Data

There are several vegetation height inversion methods with compact PolInSAR data: (1) pseudo full PolInSAR data reconstruction method. Pseudo full PolInSAR data reconstructed from compact PolInSAR data is used for the inversion. However, the reconstruction algorithm is based on the symmetry assumption and some error can be induced by the assumption, so there's some bias in this algorithm [11]. Moreover, the reconstruction does not increase the amount of information contained in the data and does not improve the inversion results. (2) To traversely solving the coherence region of compact PolInSAR data. The transmit polarization is fixed, and the compact PolInSAR interferometric coherence coefficients of different receiving polarization states are calculated. The distribution of the coherence coefficients in the unit circle

R. Yang et al., *Polarimetric Microwave Imaging*,
https://doi.org/10.1007/978-981-15-8897-6_16

is used in three-stage inversion process for vegetation height retrieval. Although this method can obtain high inversion accuracy, it needs to traverse and calculate the complex interference coherence coefficients in all receiving polarization states, which requires a large amount of calculation.

16.1.1.1 Inversion Method of Solving the Coherence Region of Compact PolInSAR Data

The interferometric coherence coefficients of all the transmitting and receiving polarization states can be achieved in full PolInSAR data. However, for compact PolInSAR data, the transmitting polarization is fixed, only the interferometric coherence coefficients of arbitrary receive polarization states can be obtained. In reference [5], a vegetation height inversion method based on traversally solving the coherence region is proposed by Dubois-Fernandez, which is referred to as Dubois method for short.

The coherence coefficient of arbitrary receive channel is

$$\gamma(\psi,\eta)=\frac{\langle k_1(\psi,\eta)k_2^*(\psi,\eta)\rangle}{\sqrt{k_1(\psi,\eta)k_1^*(\psi,\eta)}\sqrt{k_1(\psi,\eta)k_2^*(\psi,\eta)}} \tag{16.1}$$

where $k_i(\psi,\eta)=\begin{bmatrix}\cos\psi\\ e^{j\eta}\sin\psi\end{bmatrix}^{\mathrm{T}}\boldsymbol{k}_j,\boldsymbol{k}_j,\ j=1,2$ are the compact polarimetric target scattering vector of the master and slave channel, respectively, $\psi\in[0,\pi],\eta\in[0,\pi]$.

By traverse all the ψ and η, the corresponding coherence region of compact PolInSAR data is obtained. The region is a subspace of the coherence region of full PolInSAR. Three-stage inversion process introduced in the previous sections can then be applied to the coherence region for vegetation height inversion. This method needs to calculate a large number of coherence coefficients in the receiving polarization state and has a large amount of calculation.

16.1.1.2 Inversion Method Based on Compact PolInSAR Coherence Optimization and Coherence Region Extraction

Compact PolInSAR coherence optimization is applied and the boundary of coherence region is extracted for the three-stage inversion process is introduced in this section. This method is referred as a compact PolInSAR three-stage inversion process for short, which overcomes the drawbacks of large calculation amount of aforementioned method.

(1) Topographic phase estimation based on coherence optimization of compact PolInSAR

The optimized coherence coefficient of compact PolInSAR can be achieved by the coherence optimization algorithm of compact PolInSAR data.

From the data set of compact PolInSAR, the interferometric coherence coefficients of N polarization states can be obtained. If $N = 7$, the interferometric coherence coefficients set is

$$(\gamma_{\text{opt1}}, \gamma_{\text{opt2}}, \gamma_{\text{div1}}, \gamma_{\text{div2}}, \gamma_{\text{HH+VV}}, \gamma_{\text{HH+HV}}, \gamma_{\text{HH-VV}}). \tag{16.2}$$

where $\gamma_{\text{opt},i}$ and $\gamma_{\text{div},i}, i = 1, 2$ are the amplitude optimization and phase optimization coherence coefficients.

Coherence line can be fit by the N polarization coherence coefficients. The line intersects the unit circle at two points, which are $e^{j\phi_1}$ and $e^{j\phi_2}$. One of these is the underlying topography related phase. It is considered that volume scattering is dominant in the HV channel while surface scattering is dominant in HH + VV channel. In the vegetation height retrieval of full PolInSAR, the point with larger amplitude angle difference of γ_{HV} than $\gamma_{\text{HH+VV}}$ is taken as ground topographic phase point. However, the estimation of γ_{HV} is not available in compact PolInSAR data. From the physical meaning of optimized interferometric coherence coefficient,γ_{div1} and γ_{div2} can be obtained as the coherence coefficients of volume scattering dominant and surface scattering dominant channel. If $e^{j\phi_i}, i = 1, 2$ is the ground topographic phase, and the estimation of volume scattering is

$$\widehat{\gamma}_{V,i} = \begin{cases} \gamma_{\text{div1}} e^{-j\phi_i}, & |\gamma_{\text{div1}} - \exp(j\phi_i)| > |\gamma_{\text{div2}} - \exp(j\phi_i)| \\ \gamma_{\text{div2}} e^{-j\phi_i}, & |\gamma_{\text{div1}} - \exp(j\phi_i)| < |\gamma_{\text{div2}} - \exp(j\phi_i)| \end{cases} \tag{16.3}$$

Two candidates of ground topographic phase and volume decorrelation $\left(\gamma_{\text{V1}}, e^{j\phi_1}\right)$ and $\left(\gamma_{\text{V2}}, e^{j\phi_2}\right)$ can be obtained.

The two-candidate estimations and the look-up table of γ_{v} with h_{v} and σ are applied to obtain two groups of vegetation heights and extinction coefficients, which are $(h_{\text{v1}}, \sigma_1)$ and $(h_{\text{v2}}, \sigma_2)$. As more reasonable height estimation value can fit RVoG model better, the ground topographic phase corresponding to the solution which makes the following equation minimum can be taken as the estimation of ground topographic phase.

$$\left\| \begin{bmatrix} \gamma_1 \\ \gamma_2 \\ \gamma_3 \end{bmatrix} - \begin{bmatrix} e^{i\phi_i} \frac{\widehat{\gamma}_{V,i} + \mu_1}{1+\mu_1} \\ e^{i\phi_i} \frac{\widehat{\gamma}_{V,i} + \mu_2}{1+\mu_2} \\ e^{i\phi_i} \frac{\widehat{\gamma}_{V,i} + \mu_3}{1+\mu_3} \end{bmatrix} \right\| \tag{16.4}$$

In Eq. (16.4), $\widehat{\gamma}_{\text{V}}$ can be given as

$$\widehat{\gamma}_V = \frac{2\sigma}{\cos\theta_0(e^{\frac{(2\sigma h_V)}{\cos\theta_0}} - 1)} \int_0^{h_V} e^{ik_z z'} e^{\frac{(2\sigma h_V)}{\cos\theta_0}} dz' \quad (16.5)$$

where σ is the mean extinction coefficient, k_z is the mean wavenumber of the interferometry, θ_0 is the mean incidence, h_V is the vegetation height.

In Eq. (16.4), γ_i, $(i = 1, 2, 3)$ is the interferometric coherence coefficient of different polarization channels. The ground-to-volume scattering ratio of the three polarization channels is

$$\mu_i = \frac{\widehat{\gamma}_{V,i} - \gamma_i}{\gamma_i - \exp(j\phi_i)} \quad (16.6)$$

(2) **Volume decorrelation estimation based on coherence region extraction**

Based on the initial estimation of vegetation height, boundary extraction of coherence region can be used to modify volume decorrelation estimation. Moreover, the inversion precision of vegetation height estimation can be improved.

According to the properties of volume decorrelation, we can see that it is the most distinct point from the topography phase lying in the correlation region. By seeking the boundary point of the coherence region and taking the most distinct point from the topography phase as the estimation of volume decorrelation, the ambiguity estimation of volume decorrelation is resolved.

The simplified definition of complex coherence coefficient is

$$\tilde{\gamma} = \frac{\boldsymbol{\omega}^H \boldsymbol{J}_{12} \boldsymbol{\omega}}{\boldsymbol{\omega}^H \boldsymbol{J} \boldsymbol{\omega}}, \ \boldsymbol{J} = \frac{\boldsymbol{J}_{11} + \boldsymbol{J}_{22}}{2} \quad (16.7)$$

As we can see from the former section, topographic phase can be estimated and all the points in the coherence region are biased by $\exp(i\phi_e)$. The points in the coherence region are $\gamma(\boldsymbol{\omega})\exp(i\phi_e)$. According to RVoG model, the coherence region is a straight line, and the boundary in the coherence line is exactly the coherence value associated with the extremum of the real part of $\gamma(\boldsymbol{\omega})\exp(i\phi_e)$. The expression of the real part of $\gamma(\boldsymbol{\omega})\exp(i\phi_e)$ is

$$\text{Re}(\gamma(\boldsymbol{\omega})\exp(i\phi_e)) = \frac{\boldsymbol{\omega}^{\text{H}}\left((e^{j\phi_e}\boldsymbol{J}_{12} + e^{-j\phi_e}\boldsymbol{J}_{12}^{\text{H}})/2\right)\boldsymbol{\omega}}{\boldsymbol{\omega}^H \boldsymbol{J} \boldsymbol{\omega}} \quad (16.8)$$

The scattering vectors $\boldsymbol{\omega}_i$, $i = 1, 2$ which gets the extremum of $\text{Re}(\gamma(\boldsymbol{\omega})\exp(i\phi_e))$ can be obtained by performing eigenvalue decomposition to the following matrix:

$$\frac{1}{2}\boldsymbol{J}^{-1}(\boldsymbol{J}_{12} + \boldsymbol{J}_{12}^H)\boldsymbol{\omega} = \lambda\boldsymbol{\omega} \quad (16.9)$$

Therefore, boundary points of the corresponding coherence line are

$$\gamma_1 = \frac{\boldsymbol{\omega}_1^H \boldsymbol{J}_{12}\boldsymbol{\omega}_1}{\boldsymbol{\omega}_1^H \boldsymbol{J}\boldsymbol{\omega}_1}, \gamma_2 = \frac{\boldsymbol{\omega}_2^H \boldsymbol{J}_{12}\boldsymbol{\omega}_1}{\boldsymbol{\omega}_2^H \boldsymbol{J}\boldsymbol{\omega}_2} \tag{16.10}$$

The estimated volume decorrelation is the point which is most distant from the ground phase point, that is

$$\gamma_{\text{v,est}} = \begin{cases} \gamma_1 \exp(-j\phi_e), \, |\gamma_1 - \exp(j\phi_e)| \geq |\gamma_2 - \exp(j\phi_e)| \\ \gamma_2 \exp(-j\phi_e), \, |\gamma_2 - \exp(j\phi_e)| > |\gamma_1 - \exp(j\phi_e)| \end{cases} \tag{16.11}$$

(3) **Vegetation Height estimation**

According to Eq. (16.5), the volume decorrelation γ_v depends on the vegetation height h_v and extinction coefficient σ. Consequently, a look-up table of γ_v with h_v and σ can be established. Comparing estimated $\gamma_{V,\text{est}}$ with γ_v, we can get the estimation of vegetation height h_v by look-up table.

16.1.1.3 Experiment Result and Analysis

In order to validate the effectiveness of the proposed tree height inversion method, simulated PolInSAR data is used for experiment. The HH and HV channel data of the simulated PolInSAR data set is summed to obtain the HH + HV and HV + VV channel data, so the $\pi/4$ mode compact PolInSAR data set is generated. Figure 16.1 shows the amplitude image of the two channels of $\pi/4$ mode, which on the left is HH + HV channel while on the right is HV + VV channel.

Figure 16.2 shows the vegetation height inversion result of compact PolInSAR three-stage inversion method and Dubois method. Figure 16.3 is the inversion result of the marked line in Fig. 16.1. As we can see from Figs. 16.2 and 16.3 that the inversion results of two methods are consistent and are all quite accurate.

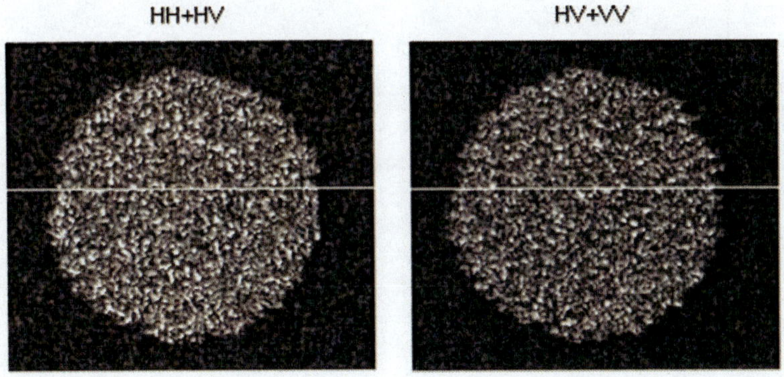

Fig. 16.1 Amplitude image of $\pi/4$ mode compact PolInSAR data

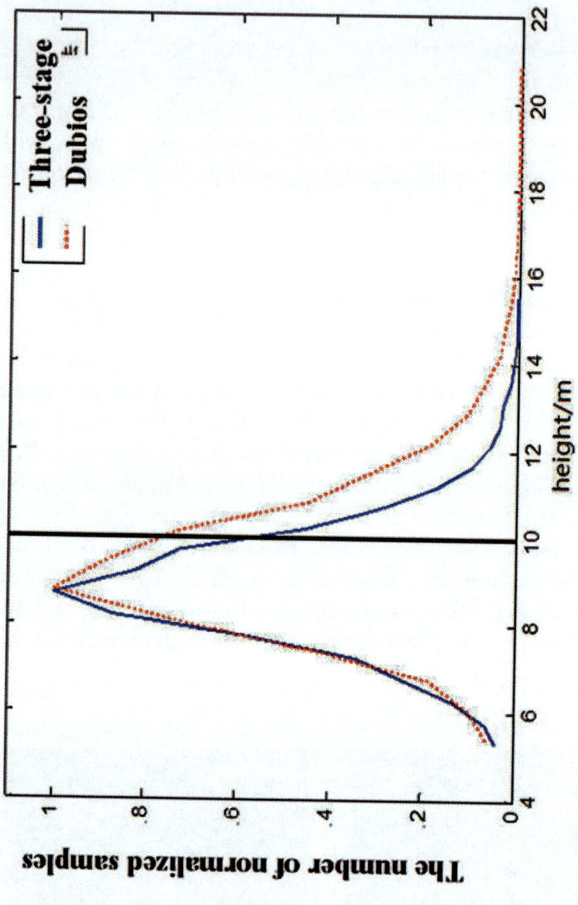

Fig. 16.2 Histogram of the retrieval result of two methods

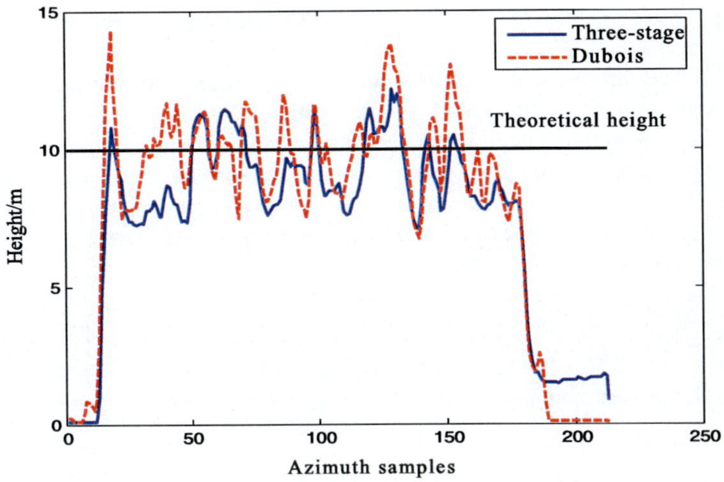

Fig. 16.3 Inversion result of the line marked in Fig. 16.1

Table 16.1 Quantitative comparison of inversion methods

	Mean Error/m	RMSE/m
Compact POLInSAR three-stage process	1.3927	1.6648
Dubois method	1.3443	1.6469
Full POLINSAR three-stage process	1.1948	1.4396

Tabel 16.2 gives the inversion result of both methods, and the mean errors and root mean square errors (RMSE). As we can see from Table 16.1 that both methods give good estimation that approximates theoretical value.

16.1.2 Performance Comparison of Different Compact PolInSAR Modes in Tree Height Retrieval

According to former analysis, vegetation height information can be extracted from compact PolInSAR data. However, the tree height inversion capabilities of different compact PolInSAR modes are slightly different. In this chapter, the maximum and minimum phase difference resolved by different modes of data is used to compare the capabilities in vegetation height retrieval of different compact PolInSAR data.

16.1.2.1 Phase Difference Based on Phase Optimized Coherence

The optimized phase coherence can obtain the maximum and minimum interferometric phase in case that the master and slave image have the same scattering mechanism. If phase optimized coherence algorithm is applied to the compact PolInSAR data of vegetation area, the obtained maximum, and minimum phase ϕ_{max} and ϕ_{min} can be taken as the phase of canopy and ground topography phase. Therefore, tree height retrieval based on the optimized phase difference can be expressed as

$$h_{V,est} = \frac{\phi_{max} - \phi_{min}}{k_z} \tag{16.12}$$

where k_z is the vertical wavenumber.

16.1.2.2 Phase Difference Based on ESPRIT Method

The phase centers of the canopy and the covered ground surface can be distinguished by ESPRIT method. It can not only be applied to full PolInSAR data, but also to dual PolInSAR data and compact PolInSAR data as well as.

ESPRIT method introduced in Chapter 15 is applied to compact PolInSAR data, and the phases of ground surface and canopy are obtained. If the estimated ground phase and canopy phase are ϕ_g and ϕ_e, respectively, then the estimated tree height is

$$h_{V,est} = \frac{\phi_e - \phi_g}{k_z} \tag{16.13}$$

16.1.2.3 Experiment Results and Analysis

In order to compare the tree height retrieval capabilities of different mode compact PolInSAR data, the $\pi/4$, DCP and CTLR mode compact PolInSAR data are generated from simulated full PolInSAR data of 10 m tree height. The phase optimized coherence and ESPRIT algorithm on the compact PolInSAR data. Figure 16.4 shows the histogram of the tree heights retrieved by phase optimized coherence algorithm. Figure 16.5 shows the histogram of tree heights retrieved by ESPRIT algorithm. Figure 16.6 shows the histogram of tree heights retrieved by three-stage inversion method.

As we can see from Figs. 16.4 to 16.5, the performance of $\pi/4$ mode is better than DCP mode and CTLR mode.

In order to quantitatively analyze the performance of different compact PolInSAR modes, the mean error and Root Mean Square Error (RMSE) of former retrieval results are listed in Table 16.2.

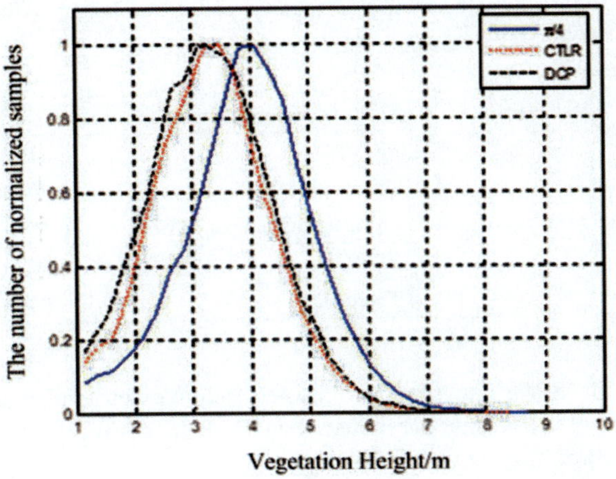

Fig. 16.4 Histogram of tree heights retrieved by phase optimized coherence algorithm

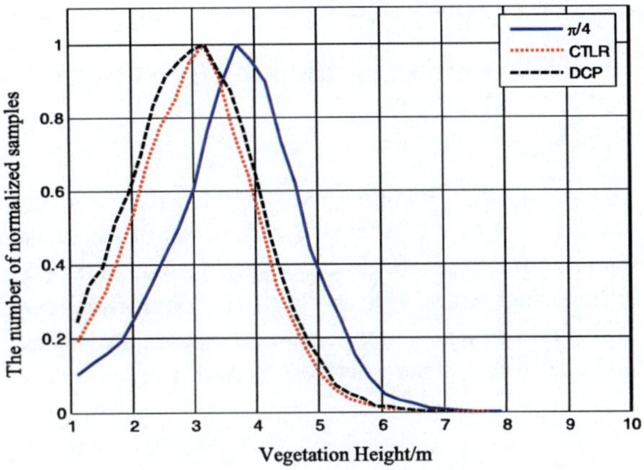

Fig. 16.5 Histogram of tree heights retrieved by ESPRIT algorithm

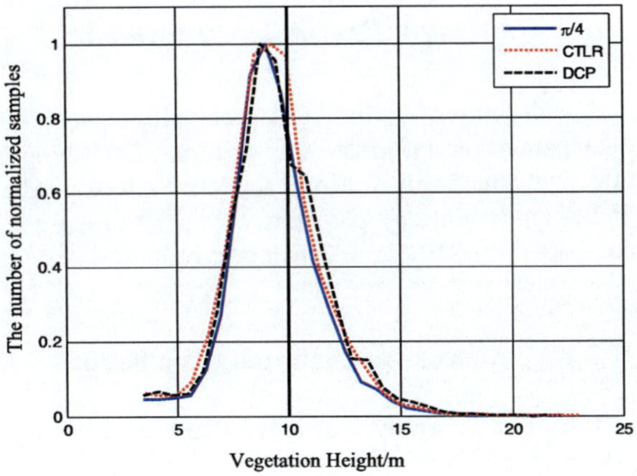

Fig. 16.6 Histogram of tree heights retrieved by three-stage algorithm

Table 16.2 Quantitative comparison of the tree height retrieval capabilities of three compact PolInSAR modes

Mode		π/ 4	DCP	CTLR
Mean error/m	Phase optimized coherence	6.0585	6.6958	6.6699
	ESPRIT	6.3167	6.9724	6.9538
	Three-stage	1.3443	1.3913	1.3790
RMSE/m	Phase optimized coherence	6.1548	6.7751	6.7451
	ESPRIT	6.4019	7.0406	7.0179
	Three-stage	1.6469	1.7040	1.6861

As we can see from Table 16.2, the retrieval results of π/ 4 mode compact PolInSAR data are better than DCP mode and CTLR mode if the same retrieval algorithms are performed.

16.2 Terrain Classification Technique Based on Compact PolInSAR

The key point of terrain classification with compact PolInSAR data is to extract the parameters benefit for classification. An unsupervised terrain classification method with compact PolInSAR data of π/ 4 mode is proposed in reference [8]. The data is initially segmented with an alternative conformity coefficient extracted from compact PolInSAR data and then the initial classes are segmented into more subclasses with Shannon entropy parameter extracted from compact PolInSAR data.

16.2.1 The Alternative Conformity Coefficient and Shannon Entropy Parameter of Compact PolInSAR

Although compact PolInSAR data contains most of the information of full PolInSAR data, part of the information is still loss. Extracting useful parameter is the key to effectively making use of compact PolInSAR data. Alternative conformity coefficient and Shannon entropy parameter are used to extract the information contained in compact PolInSAR data, which is introduced later.

16.2.1.1 Alternative Conformity Coefficient

In [7], the conformity coefficient extracted from CTLR compact PolSAR data is proposed, which can effectively distinguish different scattering mechanisms of the dominant area. With considering Faraday rotation, the formulation of the conformity coefficient is

$$\mu = \frac{2\mathrm{Im}\langle S_{\mathrm{CTLR},1} S^*_{\mathrm{CTLR},2}\rangle}{\langle S_{\mathrm{CTLR},1} S^*_{\mathrm{CTLR},1}\rangle + \langle S_{\mathrm{CTLR},2} S^*_{\mathrm{CTLR},2}\rangle} \tag{16.4}$$

where $S_{\mathrm{CTLR},1} = S_{\mathrm{HH}} - iS_{\mathrm{HV}}$,$S_{\mathrm{CTLR},2} = -iS_{\mathrm{VV}} + S_{\mathrm{HV}}$. $\mathrm{Im}(\cdot)$ means taking the imaginary part, and $\langle\cdot\rangle$ denotes spatial averaging.

The method of extracting conformity parameter from $\pi/4$ mode compact PolSAR data is introduced. The expression is

$$\mu_a = \frac{2\mathrm{Re}\left\langle S_{\pi/4,1} S^*_{\pi/4,2}\right\rangle}{\left\langle S_{\pi/4,1} S^*_{\pi/4,1}\right\rangle + \left\langle S_{\pi/4,2} S^*_{\pi/4,2}\right\rangle} \tag{16.15}$$

where $S_{\pi/4,1} = S_{\mathrm{HH}} + S_{\mathrm{HV}}$, $S_{\pi/4,2} = S_{\mathrm{HV}} + S_{\mathrm{VV}}$. $\mathrm{Re}(\cdot)$ means taking the real part.

Under reflection symmetry assumption, $\left\langle S_{\mathrm{HH}} S^*_{\mathrm{HV}}\right\rangle = \left\langle S_{\mathrm{HV}} S^*_{\mathrm{VV}}\right\rangle \approx 0$, and alternative conformity coefficient can be simplified as

$$\mu_a \approx \frac{2\left(\mathrm{Re}\left\langle S_{\mathrm{HH}} S^*_{\mathrm{VV}}\right\rangle + |S_{\mathrm{HV}}|^2\right)}{|S_{\mathrm{HH}}|^2 + 2|S_{\mathrm{HV}}|^2 + |S_{\mathrm{VV}}|^2} \tag{16.16}$$

According to the scattering characteristics of different targets, the alternative conformity coefficient can be a discriminator of different scattering mechanism dominated areas.

For surface scattering dominated area, $|S_{\mathrm{HV}}|^2$ is quite small, and $\mathrm{Re}\langle S_{\mathrm{HH}} S^*_{\mathrm{VV}}\rangle > 0$. Consequently, μ_a is quite large. For double-bounce scattering dominated area, $\mathrm{Re}\langle S_{\mathrm{HH}} S^*_{\mathrm{VV}}\rangle < 0$ and $|S_{\mathrm{HV}}|^2$ is quite small, so μ_a is quite small. For volume scattering dominated area, $|S_{\mathrm{HV}}|^2$ is quite large, and $\mathrm{Re}\langle S_{\mathrm{HH}} S^*_{\mathrm{VV}}\rangle \approx 0$. As a result, the value of μ_a of volume scattering dominated area is between the value of surface scattering dominated area and double bounce scattering dominated area [7, 10]. Obviously, conformity coefficient can help discriminate different scattering mechanism dominated areas.

16.2.1.2 Shannon Entropy Parameter of Compact PolInSAR Data

Shannon entropy is a measurement of the statistical disorder of the medium illuminated by the radar [9]. Shannon entropy parameter reflects the influence of scattering intensity, polarimetry, and interferometry and can embody the information contained in compact PolInSAR data. It is reasonable to use Shannon entropy parameter for the classification of compact PolInSAR data.

The scattering vector of $\pi/4$ mode compact PolInSAR data $\boldsymbol{k}_{\pi/4}$ follows a four-dimensional Gaussian circular probability density function (pdf) with zero mean, and the formulation of pdf is

$$p_J(\boldsymbol{k}_{\pi/4}) = \frac{1}{\pi^4 \det(\boldsymbol{J})} \exp(-\boldsymbol{k}_{\pi/4}^{\mathrm{H}} \boldsymbol{J}^{-1} \boldsymbol{k}_{\pi/4}) \tag{16.17}$$

where $\boldsymbol{J}$ is the 4×4 covariance matrix defined by

$$\boldsymbol{J} = \begin{bmatrix} J_{11} & J_{12} \\ J_{21} & J_{22} \end{bmatrix} \tag{16.18}$$

In Eq. (16.17), det(·) is determinant of the matrix. The Shannon entropy of the four-dimensional complex vector can be expressed as

$$S = -\int p_T(\boldsymbol{k}_{\pi/4}) \lg[p_T(\boldsymbol{k}_{\pi/4})] d\boldsymbol{k}_{\pi/4} \tag{16.19}$$

where $\int \cdot d\boldsymbol{k}_{\pi/4}$ standing for complex four-dimensional integration. The entropy expressed in Eq. (16.19) can be written as

$$S = \lg\{\pi^4 e^4 \det[\boldsymbol{J}_4]\} \tag{16.20}$$

According to [9], Eq. (16.20) can be expressed as

$$S[\boldsymbol{J}_4] = S_I[\boldsymbol{J}_4] + S_p[\boldsymbol{J}_4] + S_\mu[\boldsymbol{J}_4] \tag{16.21}$$

where

$$S_I[\boldsymbol{J}_4] = 2\lg\left(\frac{e\pi I_1}{2}\right) + 2\lg\left(\frac{e\pi I_2}{2}\right) \tag{16.22}$$

$$I_i = |S_{i\mathrm{HH}} + S_{i\mathrm{HV}}|^2 + |S_{i\mathrm{HV}} + S_{i\mathrm{VV}}|^2$$

In Eq. (16.22), I_1 and I_2 represent the span of compact PolInSAR image pairs. $S_I[\boldsymbol{J}_4]$ represents the intensity contribution to the Shannon entropy.

$$S_\mu[\boldsymbol{J}_4] = \lg\{(1-\gamma_1^2)(1-\gamma_2^2)\} \tag{16.23}$$

where γ_1 and γ_2 are the coherence optimization coefficient obtained from compact PolInSAR data [4]. $S_\mu[\boldsymbol{J}_4](-\infty < S_\mu[\boldsymbol{J}_4] \le 0)$ denotes the interferometry contribution to entropy. Lower value corresponds to coherent scattering area while higher value corresponds to non-coherent scattering area.

$$S_p[\boldsymbol{J}_4] = \sum_{i=1}^{2} \lg(2\tilde{\lambda}_i) + \sum_{i=1}^{2} \lg(2\tilde{\rho}_i), i = 1, 2 \tag{16.24}$$

where $\widetilde{\lambda}_i$ and $\widetilde{\rho}_i$ are the eigenvalues of $\boldsymbol{J}_{11}$ and $\boldsymbol{J}_{22}$, and $S_p[\boldsymbol{J}_4]$ represents the polarimetry contribution to the Shannon entropy.

16.2.2 The Classification Scheme of Compact PolInSAR Data

According to the aforementioned analysis, alternative conformity coefficient can embody the polarimetric information contained in the compact PolInSAR data. Moreover, interferometric partial entropy and polarimetric partial entropy can reflect the effect of scattering intensity and interferometry. As a result, these three parameters are used to perform classification of compact PolInSAR data. Figure 16.7 shows the flowchart of the classification of $\pi/4$ mode compact PolInSAR data.

Specific steps are as following [8]:

(1) Using compact polarization interferometric SAR data, the alternative conformity coefficient, interferometric partial entropy and polarimetric partial entropy of each pixel are estimated according to Eqs. (16.14), (16.22), and (16.23).
(2) The compact PolInSAR data is initially divided into six classes depending on the alternative conformity coefficient.
(3) Wishart maximum likelihood clustering is performed on the six classes of data, so we can get the initial segmentation of the data.

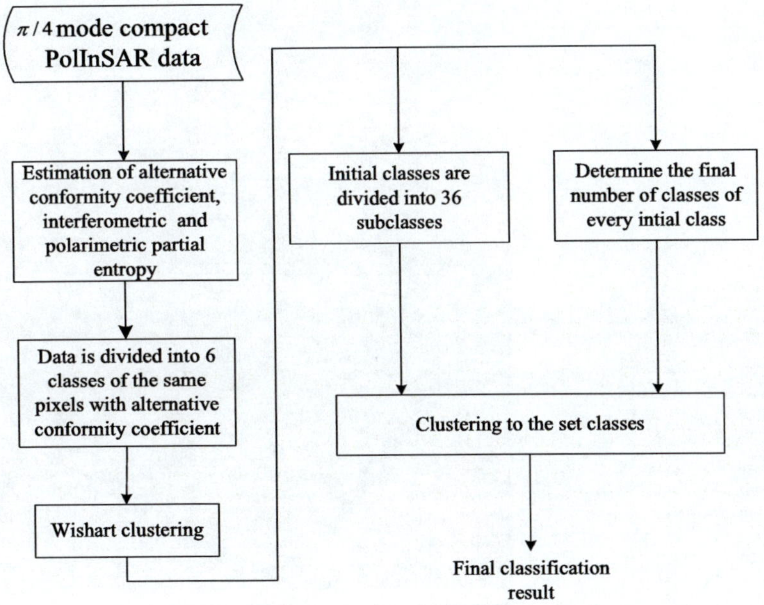

Fig. 16.7 Flowchart of the unsupervised classification of $\pi/4$ mode compact PolInSAR data

(4) Initial segmentation is further divided into thirty-six subclasses depending on the interferometric partial entropy and polarimetric partial entropy parameters. Specific method is diving every class into twelve subclasses of the same sample number depending on interferometric partial entropy, and every dataset of these twelve subclasses is further divided into three subclasses according to polarimetric partial entropy. At the same time, the number of classes to be divided into each initial class is set.

(5) In every initial segmented class, clustering and merging method are applied to merge the subclasses into the classes whose number is set. Consequently, the final classification result is obtained.

16.2.3 Experiment Results and Analysis

$\pi/4$ mode compact PolInSAR data generated from full PolInSAR data of Oberpfaffenhofen area acquired by ESAR is used. The classification result of compact PolInSAR data is compared with the classification result of full PolInSAR data of maximum likelihood classification method making use of Shannon entropy parameters. Figure 16.8 illustrates the result of presented method applied to compact PolInSAR data.

Fig. 16.8 Classification result of compact PolInSAR data

Table 16.3 Classification accuracy comparison between compact PolInSAR and full PolInSAR

Terrain type	Compact PolInSAR data	Full PolInSAR data
Forest	0.9422	0.9545
Cropland	0.9866	0.9995
City	0.5786	0.7128
Runway	1.000	1.000
Grassland	0.98657	0.9920
Overall accuracy	0.9589	0.9739

In order to quantitatively evaluate the performance of the classification result, several testing areas representing different typical terrain types are selected to calculate the classification accuracy of the classification results. The classification accuracy of compact PolInSAR data and full PolInSAR data is shown in Table 16.3.

As we can see from Table 16.3, the classification result of the proposed classification method is quite satisfactory, whose classification accuracy is very close to the result of full PolInSAR data. Experiment results confirm the validity of the proposed method.

16.3 Conclusion

In this chapter, information processing technology of compact PolInSAR data is preliminarily studied. The tree height retrieval technique and the terrain classification of compact PolInSAR are studied. Three-stage inversion process combining phase optimized coherence and coherence region extraction is introduced to overcome the drawbacks of traversing calculating the coherent region method of compact PolInSAR data, which requires too much calculation. The terrain classification method based on alternative conformity parameters and Shannon entropy parameters extracted from compact PolInSAR data is introduced.

References

1. P.C. Dubois-Fernandez et al., The compact polarimetry alternative for spaceborne SAR at low frequency. IEEE Trans. Geosci. Remote Sens. **46**(10), 3208–3222 (2008)
2. T. Lulu, *Study on Information Processing of Polarimetric and Compact Polarimtric SAR Interferometry* (Institute of Electronics, Chinese Academy of Sciences, 2010)
3. M. Moghaddam, Effect of medium symmetries on parameter estimation with polarimetricinterferometry. J. Electromagn. Waves Appl. **14**(2), 173–184 (2000)
4. A. Reigber et al., Multi-baseline coherence optimisation in partial and compact polarimetric modes, in *Proceedings of IGARSS 2008* (Boston, USA, 2008)
5. P. Dubois-Fernandez et al., The specificity of P band PolInSAR data over vegetation, in *Proceedings of PolInSAR 2007* (2007)

6. T. Flynn, M. Tabb, R. Carande. Coherence Region Shape Extraction for Vegetation Parameter Estimation in Polarimetric SAR Interferometry, in *Proceedings of IGARSS'02*, pp. 2596–2598
7. M.-L. Truong-Loï et al., The conformity coefficient or how to explore the scattering behavior from compact polarimetry mode, in *2009 IEEE Radar Conference* (2009)
8. L. Tan, P. Huang, Improved investigation on unsupervised classification of compact PolInSAR data, in *Proceedings of EUSAR 2012* (2012)
9. J. Morio et al., A characterization of shannon entropy and bhattacharyya measure of contrast in polarimetric and interferometric SAR image. Proc. IEEE **97**(6), 1097–1108 (2009)
10. A. Freeman, S.L. Durden, A three-component scattering model for polarimetric SAR data. IEEE Trans. Geosci. Remote Sens. **36**(3), 963–973 (1998)
11. M. Lavalle, E. Pottier, Y.-L. Desnos, Investigation on the performance of compact polarimetry in SAR interferometry, in *Proceedings of EUSAR 2008* (Friedrichshafen, Germany, 2008)
12. S.R. Cloude, K.P. Papathanassiou, Three-stage inversion process for polarimetric SAR interferometry. IEEE Proc.-Radar Sonar Navig. **150**(3), 125–134 (2003)
13. R.Z. Schneider et al., Polarimetric interferometry over urban areas: information extraction using coherent scatterers, in *Proceedings of IGARSS 2005* (Seoul, Korea, 2005)
14. S. Guillaso et al., Building characterization using L-Band polarimetricinterferometric SAR data. IEEE Geosci. Remote Sens. Lett. **2**(3), 347–351 (2005)
15. C. Erxue et al., Polarimetric synthetic aperture radar interferometry based mean tree height extraction technique. ScientiaSilvaeSinicae **43**(4), 66–70 (2007)
16. R.N. Treuhaft, P.R. Siqueira, Vertical structure of vegetated land surfaces from interferometric and polarimetricRadar. Radio Sci. **35**(1), 141–177 (2000)
17. R.N. Treuhaft, S.R. Cloude, The structure of oriented vegetation from polarimetricinterferometry. IEEE Trans. Geosci. Remote Sens. **37**(5), 2620–2624 (1999)
18. J.D. Ballester-Berman, J.M. Lopez-Sanchez, Coherence loci for a homogeneous volume over a double-bounce ground return. IEEE Geosci. Remote Sens. Lett. **4**(2), 317–321 (2007)
19. Li. Xinwu, *Information Extraction Methods and Application Study of Polarimetric SAR Interferometry* (Institute of Remote Sensing Applications, Chinese Academy of Sciences, 2002)
20. K.P. Papathanassiou, Single-baseline polarimetric SAR interferometry. IEEE Trans. Geosci. Remote Sens. **39**(11), 2352–2363 (2001)
21. K.P. Papathanassiou et al., Airborne polarimetric SAR interferometry, in *Proceedings of IGARSS'98*, pp. 1901–1903 (1998)
22. B. Chen, Investigation on Interferometry and Polarimetric interferometry SAR Singal Processing Techniques. 2008, Institute of Electronics, Chinese Academy of Sciences (2008)
23. H. Yamada, Y. Yamaguchi, W.M. Boerner, Forest height feature extraction in polarimetric SAR interferometry by using rotational invariance property, in *Proceedings of IGARSS'03*, pp. 1426–1428 (2003)
24. B. Lu, Multiple-Layer Model Based PolInSAR Methods to Estimate Vertical Structure Parameters and Their Applications in Forest Areas. Institute of Electronics, Chinese Academy of Sciences (2009)
25. M. Isola, S.R. Cloude, Forest height mapping using space-borne polarimetric sar interferometry, in *Proceedings of IGARSS'01*, pp. 1095–1097 (2001)
26. F. Garestier, P.C. Dubois-Fernandez, I. Champion, Forest height inversion using high-resolution P-Band Pol-InSARData. IEEE Trans. Geosci. Remote Sens. **46**(11), 3544–3559 (2008)
27. T. Mette et al., Forest and the random volume over ground—nature and effect of 3 possible error types, in *Proceedings of EUSAR 2006* (Dresden, Germany, 2006)
28. S.R. Cloude, Robust parameter estimation using dual baseline polarimetric SAR interferometry, in *Proceedings of IGARSS'02*, pp. 838–840 (2002)
29. F. Koudogbo et al., Dual-baseline approach for the retrieval of vegetation parameters, in *Proceedings of EUSAR* (Dresden, Germany, 2006)
30. K.P. Papathanassiou, S.R. Cloude, The effect of temporal decorrelation on the inversion of forest parameters from Pol-InSAR data, in *Proceedings of IGARSS 2003* (2003)
31. T. Flynn, M. Tabb, R. Carande, Coherence region shape extraction for vegetation parameter estimation in polarimetric SAR interferometry, in *Proceedings of IGARSS'02*, pp. 2596–2598 (2002)

32. L. Tan, R. Yang, Investigation on tree height retrieval with polarimetric SAR interferometry, in *Proceedings of IGARSS '08* (Boston, USA, 2008)
33. M. Tabb et al., Phase diversity: a decomposition for vegetation parameter estimation using polarimetric SAR interferometry, in *Proceedigns of EUSAR 2002* (Cologne, Germany, 2002)
34. T. Flynn, M. Tabb, R. Carande, Direct estimation of vegetation parameters from convariance data in polarimetric SAR interferometry, in *Proceedings of IGARSS'02*, (Toronto, Canada, 2002), pp. 1908–1910
35. M. Tabb, T. Flynn, R. Carande, Full maximum likelihood inversion of PolInSAR scattering models, in *Proceeding of IGARSS'04*, pp. 1232–1235 (2004)
36. Z. Bin et al., Information extraction using polarimetricinterferometric SAR data: present and future. J. Electron. Inf. Technol. **28**(10), 1979–1984 (2006)
37. N.R. Goodman, Statistical analysis based on a certain multivariate complex gaussian distribution (An introduction). Ann. Math. Stat. **34**(1), 152–177 (1963)
38. J.-S. Lee et al., Intensity and phase statistics of multilookpolarimetric and interferometric SAR imagery. IEEE Trans. Geosci. Remote Sens. **32**(5), 1017–1028 (1994)
39. Z. Xianda, *Matrix Analysis and Applications* (Tsinghua University Press, Beijing, 2004)
40. D. Kasilingam, M. Nomula, S. Cloude, A technique for removing vegetation bias from polarimetric SAR interferometry, in *Proceedings of IGARSS 2002* (Toronto, Canada, 2002), pp. 1017–1019
41. S. Guillaso et al., Polarimetric interferometric SAR data analysis based on ESPRIT/MUSIC methods, in *Proceedings of PolInSAR 2003* (Roma, 2003)
42. J.M. Lopez-Sanchez, J.D. Ballester-Berman, Analysis of approaches for a complete estimation of parameters in agricultural crops applications using polarimetric SAR interferometry, in *URSI Commision F Symposium on Microwave Remote Sensing* (2005)
43. Z. Yang, Studies on synthetic aperture radar interferometry and polarimetric interferometry techniques (Institute of Electronics, Chinese Academy of Sciences, 2003)
44. L. Sagues et al., Buried mine detection by polarimetric radar interferometry. Ultra-Wideband, Short-Pulse Electromagn. **5** (2002)
45. S.R. Cloude, Polarization coherence tomography. Radio Sci. **41**, RS 4017 (2006)
46. S. Cloude, Polarisation coherence tomography (PCT), in *Proceedings of EUSAR 2006* (Dresden, Germany, 2006)
47. S.R. Cloude, Application of polarization coherence tomography (PCT), in *International Radar Symposium 2006*. (Krakow, Poland, 2006)

Chapter 17
Influence and Compensation of Terrain Azimuth Slopes on Polarimetric Synthetic Aperture Radar Data

The azimuth slopes of the terrain shift the ellipse slopes θ in the polarized ellipse, which will have a great impact on applications such as target decomposition and information extraction based on the polarized synthetic aperture radar data covariance matrix and correlation matrix (such as h/α decomposition, feature classification, crop density estimation, soil moisture estimation, and surface roughness). In order to reduce the impact of terrain azimuth tilt on the application, azimuth tilt estimation and data compensation can be performed using polarized SAR data. In this chapter, the influence of terrain slopes on electromagnetic wave polarization is mainly analyzed, the method of azimuth offset angle estimation using polarized SAR data is discussed, and the influence factors of estimation accuracy and data compensation methods are analyzed.

17.1 Influence of Terrain Slopes on Electromagnetic Wave Polarization

17.1.1 Influence of Azimuth Slopes of Terrain on Electromagnetic Wave Polarization

When polarized SAR is imaging uneven areas, the surface tilt has two main effects on SAR imaging: First, it affects the radar cross section of the imaging unit; Second, the azimuth tilt of the terrain causes the azimuth angle of the polarization state to change, thereby affecting the polarization state of the electromagnetic wave. When polarized SAR images a horizontal plane, the horizontally polarized electric field is parallel to this plane. For an inclined surface, the horizontal electric field is no longer parallel to this surface. As a result, the slopes of the ground surface will affect the polarization response of horizontal transmission and horizontal reception (HH). Similarly, the polarization response of VV, VH, and HV will also be affected. The

R. Yang et al., *Polarimetric Microwave Imaging*,
https://doi.org/10.1007/978-981-15-8897-6_17

azimuth offset θ introduced by the inclined surface can be regarded as the $v - h$ polarization base rotating an angle around the radar line of sight (Los), so that the horizontally polarized electric field is parallel to the plane [1–5].

17.1.2 Dependence of Polarization Azimuth Angle on Terra Slopes

Figure 17.1 illustrates the relationship between radar polarization and polarization azimuth angle. Figure 17.2 shows the geometric relationship of radar imaging. The geometric relationship between the change in polarization azimuth and terrain slopes and radar, LOS is analyzed below. The unit vector pair $(\hat{x}, \hat{y})$ defines a horizontal plane, $(\hat{y}, \hat{z})$ defines the radar incidence plane, and the radar line of sight is in the reverse direction of the axis $\hat{I}_1$. The angle ϕ between $\hat{I}_1$ and $\hat{z}$ is the radar look angle The axis $\hat{x}$ is in the azimuth direction, and $\hat{y}$ is in the ground range direction. The surface normal for a ground patch is denoted by $\boldsymbol{N}$. Assume that the polarimetric SAR is calibrated so that the horizontal polarization (H) is parallel to the horizontal plane $(\hat{x}, \hat{y})$, and the vertical polarization (V) is in the incidence plane. For a horizontal surface patch, its surface normal $\boldsymbol{N}$ is in the incidence plane, and no orientation angle shift is induced. However, for a surface patch with an azimuthal tilt, its surface normal $\boldsymbol{N}$ is no longer in the incidence plane. The induced polarization orientation angle shift θ is the angle that rotates the incidence plane $(\hat{y}, \hat{z})$ about the line of sight to the surface normal by the following equation [1]. The definition is as follows:

$$\tan\theta = \frac{\tan\omega}{-\tan\gamma\cos\phi + \sin\phi} \tag{17.1}$$

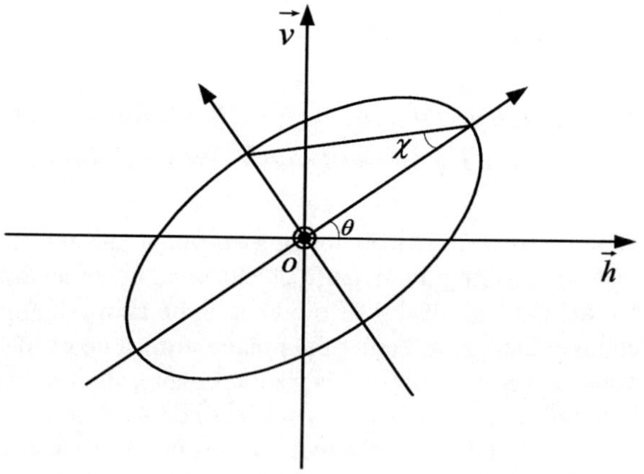

Fig. 17.1 The relationship between radar polarization and azimuth

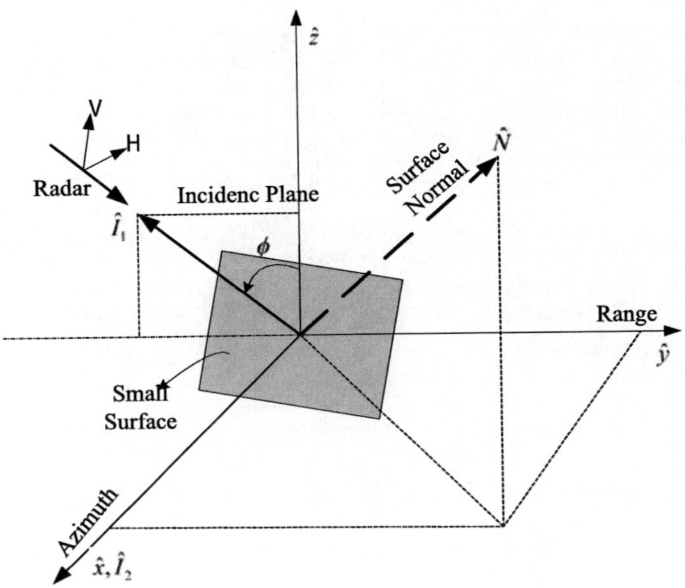

Fig. 17.2 Schematic diagram of the radar imaging geometry

θ is the orientation angle shift. ϕ is the radar look angle. $\tan\omega$ is the azimuth slope, and $\tan\gamma$ is the slope in the ground range direction. This equation shows that the orientation angle shift is mainly induced by azimuth slope, and it is a function of the range slope and the radar look angle.

For example, the relationship between polarization azimuth slop and terrain azimuth shift is shown in Fig. 17.3a–c for radar look angle of 30°, 45°, and 60°, respectively. The slope in the ground range direction −60° to 20°, −50° to –30°, −30° to 50°, respectively. It is customary to agree that a positive range inclination angle indicates that the surface is inclined toward the radar Los, and a negative value indicates that the surface is inclined toward the direction away from the radar Los. Those shows that when the radar Los is less than 60° and the range slopes are positive, the absolute value of the θ angle offset is always greater than the azimuth slope angle. That is, unless the radar Los is large and the range inclination is negative, the azimuth angle must be greater than the azimuth inclination.

17.2 Estimation of Azimuth Orientation Angle Shift Introduced by Terrain Azimuth Slopes

There are two types of estimation methods for azimuth orientation angle shift introduced by terrain azimuth slopes: The first category is the Digital Elevation Model (DEM). DEM can be obtained by interference SAR or other methods, and then use

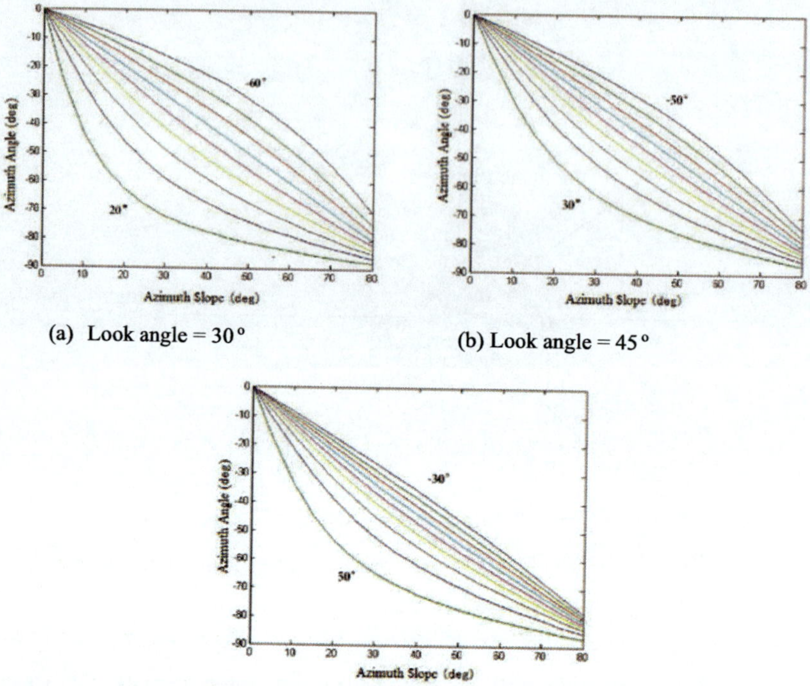

(a) Look angle = 30°

(b) Look angle = 45°

(c) Look angle = 60°

Fig. 17.3 Azimuth slope angle

formula (17.1) to estimate the azimuth orientation angle shift introduced by azimuth slopes (in the subsequent analysis, the θ angle is used to represent the azimuth orientation angle shift). The estimation accuracy of this method is very high, but the disadvantage is that DEM needs to be obtained in advance. In many cases, there is usually no DEM for the area of interest, and applications are limited.

The second type is to extract the θ angle directly from the polarized SAR data. The advantage is that no DEM model is needed, the disadvantage is that the estimation accuracy is not as high as the DEM model method. This section mainly introduces the second type of methods, including circular polarization estimation methods of azimuth orientation angle shift (including RR-LL circular polarization method, RR-RL circular Polarization method), scattering matrix estimation method, polarization signature method, and Cloude decomposition method. In the derivation and proof of the θ angle estimation algorithm, knowledge of the influence of the azimuth orientation angle shift on the circular polarization basis matrix, scattering symmetry, and its theoretical model will be used.

17.2.1 Estimation Method by Circular Polarization for Azimuth Orientation Angle

17.2.1.1 Theoretical Model of Circular Polarization Estimation Method for Azimuth Orientation Angle

(a) Relationship between circular polarization matrix and azimuth orientation angle

Any combination of receiving and transmitting polarization can be obtained from the h-v scattering matrix through polarization basis transformation. The components of circular polarization can be derived from the scattering matrix:

$$E^{rt} = \begin{bmatrix} E_h^r & E_v^r \end{bmatrix} S \begin{bmatrix} E_h^t \\ E_v^t \end{bmatrix} \tag{17.2}$$

where E_h^t and E_v^t are unit vectors of emission polarization in the $\left(\hat{h}, \hat{v}\right)$ coordinate system. The superscripts r and t indicate reception and transmission, respectively. The right-handed (R) and left-handed (L) circular polarization unit vectors are

$$\vec{P}_R = \frac{1}{2}\begin{bmatrix} 1 \\ -i \end{bmatrix}, \text{ and } \vec{P}_L = \frac{1}{2}\begin{bmatrix} -i \\ 1 \end{bmatrix} \tag{17.3}$$

Defining the unit vectors $\vec{P}_R$, $\vec{P}_L$ in this way guarantees their orthogonality, and at the same time guarantees that the absolute phases are consistent.

The transformation from h-v linear polarization basis to R-L circular polarization basis can be obtained by the polarization basis conversion. In this way, Eq. (17.2) is transformed into:

$$\begin{aligned} E^{rt} &= \frac{1}{2}\begin{bmatrix} E_R^r & E_L^r \end{bmatrix}\begin{bmatrix} 1 & i \\ i & 1 \end{bmatrix}\begin{bmatrix} S_{HH} & S_{HV} \\ S_{HV} & S_{VV} \end{bmatrix}\begin{bmatrix} 1 & i \\ i & 1 \end{bmatrix}\begin{bmatrix} E_R^t \\ E_L^t \end{bmatrix} \\ &= \frac{1}{2}\begin{bmatrix} E_R^r & E_L^r \end{bmatrix}\begin{bmatrix} S_{RR} & S_{RL} \\ S_{RL} & S_{LL} \end{bmatrix}\begin{bmatrix} E_R^t \\ E_L^t \end{bmatrix} \end{aligned} \tag{17.4}$$

This gives three circular polarization basis components:

$$\begin{aligned} S_{RR} &= (S_{HH} - S_{VV} + i * S_{HV})/2 \\ S_{LL} &= (S_{VV} - S_{HH} + i * S_{HV})/2 \\ S_{RL} &= i^*(S_{HH} + S_{VV})/2 \end{aligned} \tag{17.5}$$

The Pauli-based scattering vector of the circular polarization basis can be easily derived from (17.5):

$$k = \frac{1}{\sqrt{2}} \begin{bmatrix} -i2S_{RL} \\ (S_{RR} - S_{LL}) \\ -i(S_{RR} + S_{LL}) \end{bmatrix} \tag{17.6}$$

J. S. Lee and D. L. Schuler et al. Obtained the scattering vector components after introducing orientation angle as [2–7]

$$\begin{aligned} \tilde{S}_{RR} &= S_{RR} e^{-i2\theta} \\ \tilde{S}_{LL} &= S_{LL} e^{i2\theta} \\ \tilde{S}_{RL} &= S_{RL} \end{aligned} \tag{17.7}$$

The upper wavy line represents the vector after rotating around Los. (17.7) shows that this rotation only affects the phases of S_{RR} and S_{LL}, and S_{RL} is rotation-invariant. This is because S_{RL} depends only on $(S_{HH} + S_{VV})$. This characteristic of circular polarization is very useful for deriving and analyzing azimuth estimation algorithms.

The circular polarization scattering vector is defined as follows:

$$c = \begin{bmatrix} S_{RR} \\ \sqrt{2} S_{RL} \\ S_{LL} \end{bmatrix} \tag{17.8}$$

The covariance matrix G of the circular polarization is obtained from the scattering vector c:

$$G = \langle c c^{*T} \rangle = \begin{bmatrix} \langle |S_{RR}|^2 \rangle & \sqrt{2}\langle (S_{RR} S_{RL}^*) \rangle & \langle (S_{RR} S_{LL}^*) \rangle \\ \sqrt{2}\langle (S_{RL} S_{RR}^*) \rangle & 2\langle |S_{RL}|^2 \rangle & \sqrt{2}\langle (S_{RL} S_{LL}^*) \rangle \\ \langle (S_{LL} S_{RR}^*) \rangle & \sqrt{2}\langle (S_{LL} S_{RL}^*) \rangle & \langle |S_{LL}|^2 \rangle \end{bmatrix} \tag{17.9}$$

From the Eqs. (17.7) and (17.9), the circular polarization covariance matrix after the θ angle rotation is introduced as follows:

$$\tilde{G} = \begin{bmatrix} \langle |S_{RR}|^2 \rangle & \sqrt{2}\langle (S_{RR} S_{RL}^*) \rangle e^{-i2\theta} & \langle (S_{RR} S_{LL}^*) \rangle e^{-i4\theta} \\ \sqrt{2}\langle (S_{RL} S_{RR}^*) \rangle e^{i2\theta} & 2\langle |S_{RL}|^2 \rangle & \sqrt{2}\langle (S_{RL} S_{LL}^*) \rangle e^{-i2\theta} \\ \langle (S_{LL} S_{RR}^*) \rangle e^{i4\theta} & \sqrt{2}\langle (S_{LL} S_{RL}^*) \rangle e^{i2\theta} & \langle |S_{LL}|^2 \rangle \end{bmatrix} \tag{17.10}$$

It is clear that the orientation angle only affects the phase of the terms on off-diagonal lines. The method of finding the θ angle by the circular polarization is directly related to the (1, 3) term of this matrix.

(b) Scattering symmetry and its theoretical model

Scattering symmetry is important to understand the nature of azimuth orientation angle caused by azimuth slopes. If the scatterer is symmetrical about the radar incident plane in the resolution unit, the scattering medium has reflection symmetry. In other words, within a resolution unit, the scatterer responses with azimuth orientation angle of opposite signs are numerically equal. Averaging neighboring cell data results in better scattering symmetry of the scatterer. The average of neighboring unit pixels can reveal changes in the scattering mechanism of a given scattering medium. This change is related to reflection symmetry. Polarized SAR data can be averaged by means of an average covariance matrix, a Stocks matrix, a correlation matrix, or a circularly polarized covariance matrix.

Using the theoretical model of Cloude et al., the scattering matrix form of a symmetric target with Bragg scattering and even scattering characteristics is as follows:

$$S = \begin{bmatrix} S_{HH} & 0 \\ 0 & S_{VV} \end{bmatrix} \tag{17.11}$$

For bragg scattering

$$\begin{aligned} S_{HH} &= \frac{\cos\theta_i - \sqrt{\varepsilon_r - \sin^2\theta_i}}{\cos\theta_i + \sqrt{\varepsilon_r - \sin^2\theta_i}} \\ S_{VV} &= \frac{(\varepsilon_r - 1)\left(\sin^2\theta_i - \varepsilon_r\left(1 + \sin^2\theta_i\right)\right)}{\left(\varepsilon_r\cos\theta_i + \sqrt{\varepsilon_r - \sin^2\theta_i}\right)^2} \end{aligned} \tag{17.12}$$

where ε_r is the complex dielectric constant, and θ_i is the local incident angle. The single looks correlation matrix has the following form:

$$T = \frac{1}{2}\begin{bmatrix} S_{HH} + S_{VV} \\ S_{HH} - S_{VV} \\ 0 \end{bmatrix}\begin{bmatrix} S_{HH} + S_{VV} & S_{HH} - S_{VV} & 0 \end{bmatrix} \equiv \begin{bmatrix} A & B & 0 \\ B^* & 2C & 0 \\ 0 & 0 & 0 \end{bmatrix} \tag{17.13}$$

Here, A, B, and C are derived from (17.12).

Let the θ angle introduced by the azimuth slope change of each scatter have a $p(\varphi)$ distribution. If $p(\varphi)$ is symmetric about $\varphi = 0$, the average correlation matrix T of the distribution is

$$\int UTU^T p(\varphi)d\varphi = \begin{bmatrix} A & B\mu & 0 \\ B^*\mu & 2C\omega & 0 \\ 0 & 0 & 2C(1-\omega) \end{bmatrix} \tag{17.14}$$

where $U = \begin{bmatrix} 1 & 0 & 0 \\ 0 & \cos 2\theta & \sin 2\theta \\ 0 & -\sin 2\theta & \cos 2\theta \end{bmatrix}$, $\mu = \int \cos 2\phi p(\phi)d\phi$, $\omega = \int \cos^2 2\phi p(\phi)d\phi$.

Equation (17.14) represents symmetrical scattering characteristics, that is, the term S_{HV} with the exception of $\langle |S_{HV}|^2 \rangle$ is 0. The orientation angle shift for reflection symmetric media is zero. If the orientation angle distribution is not centered at zero, the coherency matrix will not display a reflection symmetric form. For example, if $p(\varphi)$ is symmetrical about $\varphi = \theta_s$, instead of $\varphi = 0$, the integration of the shifted $p(\varphi)$ produces the following coherency:

$$\tilde{T} = \begin{bmatrix} A & B\mu\cos 2\theta_s & -B\mu\sin 2\theta_s \\ B^*\mu\cos 2\theta_s & 2C\left(\sin^2 2\theta_s + \omega\cos 4\theta_s\right) & C(1-2\omega)\sin 4\theta_s \\ -B^*\mu\sin 2\theta_s & C(1-2\omega)\sin 4\theta_s & 2C\left(\cos^2 2\theta_s - \omega\cos 4\theta_s\right) \end{bmatrix} \tag{17.15}$$

17.2.1.2 Method of RR-LL Circular Polarization

The (1, 3) term in the circular polarization covariance matrix (17.10) can be used to estimate the azimuth orientation angle θ. If the θ angle is mainly caused by the terrain azimuth slope, then

$$\left\langle \tilde{S}_{RR}\tilde{S}_{LL}^* \right\rangle = \langle S_{RR}S_{LL}^* \rangle e^{-i4\theta} \tag{17.16}$$

For a reflection symmetrical medium, $\langle S_{RR}S_{LL}^* \rangle$ is required to be real in value so that it will not corrupt the orientation angle θ related to the phase term $e^{-i4\theta}$. We will prove that $\langle S_{RR}S_{LL}^* \rangle$ is real for a reflection symmetrical medium.

Substituting (17.5) into $\langle S_{RR}S_{LL}^* \rangle$ and setting terms that contain S_{HV} to zero (except the $\langle |S_{HV}|^2 \rangle$ term), we have

$$\langle S_{RR}S_{LL}^* \rangle = \left(-\langle |S_{HH} - S_{VV}|^2 \rangle + 4\langle |S_{HV}|^2 \rangle\right)/4 \tag{17.17}$$

This term is real, so the argument of $\langle S_{RR}S_{LL}^* \rangle$ is zero or π. Consequently, the phase difference between S_{HH} and S_{VV} does not cause error in the estimation of orientation angles θ. The factor of 4θ in (17.16) limits the range of θ to $[-\pi/4, \pi/4]$. Substituting (17.5) into $\left\langle \tilde{S}_{RR}\tilde{S}_{LL}^* \right\rangle$, we have

$$\left\langle \tilde{S}_{RR}\tilde{S}_{LL}^* \right\rangle = \frac{1}{4}\left\{\left\langle -\left|\tilde{S}_{HH} - \tilde{S}_{VV}\right|^2 + 4\left|\tilde{S}_{HV}\right|^2 \right\rangle - i4\mathrm{Re}\left(\left\langle \left(\tilde{S}_{HH} - \tilde{S}_{VV}\right)\tilde{S}_{HV}^* \right\rangle\right)\right\} \tag{17.18}$$

From the formulas (17.16) and (17.18):

$$-4\theta = \mathrm{Arg}\left(\left\langle \tilde{S}_{RR}\tilde{S}_{LL}^* \right\rangle\right) = \tan^{-1}\left(\frac{-4\mathrm{Re}\left(\left\langle \left(\tilde{S}_{HH} - \tilde{S}_{VV}\right)\tilde{S}_{HV}^* \right\rangle\right)}{-\left\langle \left|\tilde{S}_{HH} - \tilde{S}_{VV}\right|^2 \right\rangle + 4\left\langle \left|\tilde{S}_{HV}\right|^2 \right\rangle}\right) \tag{17.19}$$

If (17.19) is applied directly, it would introduce errors, because for an azimuth symmetrical medium, $\left\langle\left|\tilde{S}_{HH}-\tilde{S}_{VV}\right|^2\right\rangle$ is normally greater than $4\left\langle\left|\tilde{S}_{HV}\right|^2\right\rangle$. The denominator is negative. Consequently, when the numerator is near zero, the arctangent is near $\pm\pi$. The orientation angle θ would be rather than near zero, as it should be $\pm\pi/4$. To match the orientation angle θ corresponding to the azimuth slope angle, the errors must be removed by adding π. The circular polarization estimator is

$$\theta=\begin{cases}\eta, & \eta\le\pi/4\\ \eta-\pi/2, & \eta>\pi/4\end{cases} \tag{17.20}$$

$$\eta=\frac{1}{4}\left[\tan^{-1}\left(\frac{-4\mathrm{Re}\left(\left\langle\left(\tilde{S}_{HH}-\tilde{S}_{VV}\right)\tilde{S}_{HV}^{*}\right\rangle\right)}{-\left\langle\left|\tilde{S}_{HH}-\tilde{S}_{VV}\right|^2\right\rangle+4\left\langle\left|\tilde{S}_{HV}\right|^2\right\rangle}\right)+\pi\right] \tag{17.21}$$

where the arctangent is computed in the range $(-\pi,\pi)$.

The theoretical model of (17.15) correlation matrix can also verify this circular polarization algorithm. The denominator in (17.21) is the difference between the (2, 2) and (3, 3) terms of the coherency matrix, that is

$$\begin{aligned}&-\left\langle\left|\tilde{S}_{HH}-\tilde{S}_{VV}\right|^2\right\rangle+4\left\langle\left|\tilde{S}_{HV}\right|^2\right\rangle\\&=2C\left(-\sin^2 2\theta_s+\cos^2 2\theta_s-2\omega\cos 4\theta_s\right)\\&=2C(1-2\omega)\cos 4\theta_s\end{aligned} \tag{17.22}$$

The numerator is the (2, 3) term, multiplied by 2, or

$$4\mathrm{Re}\left(\left\langle\left(\tilde{S}_{HH}-\tilde{S}_{VV}\right)\tilde{S}_{HV}^{*}\right\rangle\right)=2C(1-2\omega)\sin 4\theta_s \tag{17.23}$$

From the previous two equations, we have

$$\tan 4\theta=\frac{4\mathrm{Re}\left(\left\langle\left(\tilde{S}_{HH}-\tilde{S}_{VV}\right)\tilde{S}_{HV}^{*}\right\rangle\right)}{-\left\langle\left|\tilde{S}_{HH}-\tilde{S}_{VV}\right|^2\right\rangle+4\left\langle\left|\tilde{S}_{HV}\right|^2\right\rangle} \tag{17.24}$$

Equation (17.24) has the same form as Eq. (17.19), indicating that the algorithm can successfully estimate the angle θ. Figure 17.4 shows the Span image of the P-band of BONANZAC and the θ image of RR-LL estimation. The data is the full-polarization SAR data collected by the AIRSAR system of NASA JPL in the United States on September 12, 1991, in the BONANZA (Montana, USA) area, and processed with four looks. Figure 17.4a is a P-band Span image, Fig. 17.4b is the estimation result of θ, and Fig. 17.4c is the distribution histogram of θ. As can be

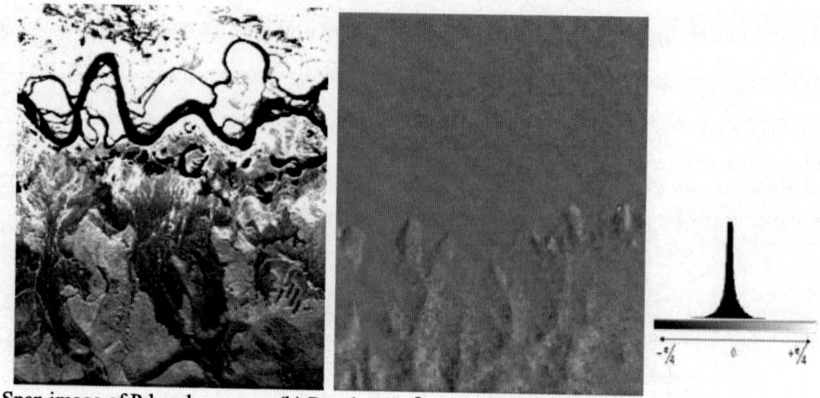

(a) Span image of P-band (b) Results of θ (c) Distribution histogram

Fig. 17.4 Polarization orientation angles results of P-band images

seen from Fig. 17.4b, the angle θ obtained from the P-band data is consistent with the Span image of Fig. 17.4a, and the noise is low. It can be seen from the histogram of θ that the mean value of the distribution is around zero.

17.2.1.3 Method of RR-RL Circular Polarization

Two other estimators can also be derived from the circular polarization matrix (17.17) using the (1, 2) term or the (2, 3) term. Assuming that the dominant orientation angle shift is induced by azimuth slopes, we have

$$\left\langle \tilde{S}_{RR}\tilde{S}_{RL}^{*} \right\rangle = \left\langle S_{RR}S_{RL}^{*} \right\rangle e^{-i2\theta} \tag{17.25}$$

If (17.25) is a valid estimator, $\left\langle S_{RR}S_{RL}^{*} \right\rangle$ must not be complex in value for reflection symmetrical media. However, the assumption is not valid, because for symmetrical media

$$\left\langle S_{RR}S_{RL}^{*} \right\rangle = \left[2\left\langle \mathrm{Im}\left(S_{HH}S_{VV}^{*} \right) \right\rangle - i\left(\left\langle |S_{HH}|^2 \right\rangle - \left\langle |S_{VV}|^2 \right\rangle \right) \right] / 4 \tag{17.26}$$

where Im() denotes the imaginary part of a complex number. The previous equation indicates that $\left\langle S_{RR}S_{RL}^{*} \right\rangle$ is complex in value, except when S_{HH} and S_{VV} are same in phase. For general terrain having vegetated or rough surfaces, the difference in the locations of scattering phase centers between HH and VV polarizations induces phase differences. Consequently, this estimator produces erroneous results in vegetated and rough surface areas.

Similar to the circular polarization algorithm of RR-LL, a offset must be taken into consideration, because, for dominant ground scatter, such as Bragg scattering, the real part of (17.26) is much smaller than the imaginary part. A offset of $\pi/2$ is introduced. For other media, the offset is difficult to determine. Consequently, for Bragg scattering, the second circular estimator is

$$\theta = \begin{cases} \xi, & \xi \le \pi/2 \\ \xi - \pi, & \xi > \pi/2 \end{cases} \tag{17.27}$$

$$\xi = \frac{1}{2}\left[\tan^{-1}\left(\frac{\mathrm{Re}\left(\tilde{S}_{RR}\tilde{S}_{RL}^{*}\right)}{\mathrm{Im}\left(\tilde{S}_{RR}\tilde{S}_{RL}^{*}\right)}\right) + \frac{\pi}{2}\right] \tag{17.28}$$

Since $\langle\tilde{S}_{RR}\tilde{S}_{RL}^{*}\rangle = \frac{-i}{4}[\langle(\tilde{S}_{HH} - \tilde{S}_{VV})(\tilde{S}_{HH} + \tilde{S}_{VV})^{*}\rangle + i2\langle\tilde{S}_{HV}(\tilde{S}_{HH} + \tilde{S}_{VV})^{*}\rangle]$. Then, the orientation angle θ estimated by this algorithm is dominated by the phase difference of $\langle(\tilde{S}_{HH} - \tilde{S}_{VV})(\tilde{S}_{HH} + \tilde{S}_{VV})^{*}\rangle$, and this domination produces inaccurate results.

17.2.2 *Method of Estimating Orientation Angle by Scattering Matrix*

17.2.2.1 The First Method of Estimating Orientation Angle by Scattering Matrix

Equation (17.7) shows that the phase of $\tilde{S}_{RR}$ is directly related to the azimuth rotation angle. Assuming that the S_{RR} of the symmetrical medium is real, substituting Eqs. (17.5) and (17.7) into $\tilde{S}_{RR}$ gives

$$|S_{RR}|\cos 2\theta + i|S_{RR}|\sin 2\theta = \left(\tilde{S}_{HH} - \tilde{S}_{VV} + i2\tilde{S}_{HV}\right)/2 \tag{17.29}$$

If $\tilde{S}_{HH}$, $\tilde{S}_{VV}$, $\tilde{S}_{HV}$ are real or their phases are real, a simple estimator can be obtained by using (17.29):

$$\tan 2\theta = \frac{2\tilde{S}_{HV}}{\tilde{S}_{VV} - \tilde{S}_{HH}} \tag{17.30}$$

This formula shows that the θ increases with the increase of $\tilde{S}_{HV}$. The measurement scope of formula (17.30) is $-\pi/2 \sim \pi/2$. The actual scope of the θ angle is $-\pi/4 \sim \pi/4$. This is because the sign of the denominator of (17.30) is not only affected by the azimuth rotation, but also by the influence of both the azimuth rotation and the media scattering characteristics. When radar scattering approaches specular scattering or when the θ is near 45°, $\tilde{S}_{HH} \approx \tilde{S}_{VV}$, the algorithm is unstable.

In addition, because the coherent speckle noise of SAR data is high, the actual result noise is very large.

17.2.2.2 The Second Method of Estimating Orientation Angle by Scattering Matrix

Another method is derived from the scattering theory model (17.15). Equation (17.15) indicates that the angle θ can be obtained from the ratio of terms (1, 3) to (1, 2), as follows:

$$\tan(2\theta_s) = \frac{-2\left\langle\left(\tilde{S}_{HH}+\tilde{S}_{VV}\right)\tilde{S}_{HV}^{*}\right\rangle}{\left\langle\left(\tilde{S}_{HH}+\tilde{S}_{VV}\right)\left(\tilde{S}_{HH}-\tilde{S}_{VV}\right)^{*}\right\rangle} \tag{17.31}$$

For the case of the single looks scattering matrix, (17.31) will be simplified to (17.26). Since B is a complex number, it has to be calculated using the real part of the terms (1, 3) and (1, 2). It can be found that (17.31) can be directly derived from the (1, 2) terms of (17.10). Equation (17.31) is also not stable. It can be seen that, except that the real part is used as the numerator and denominator, the estimator of (17.31) is similar to the estimator derived from $\langle S_{RR}S_{RL}^{*}\rangle$, and they are numerically complex. Equation (17.31) is transformed into the following form:

$$\tan(2\theta) = \frac{\mathrm{Re}\left[2\left\langle\left(\tilde{S}_{HH}+\tilde{S}_{VV}\right)\tilde{S}_{HV}^{*}\right\rangle\right]}{\mathrm{Re}\left[\left\langle\left(\tilde{S}_{HH}-\tilde{S}_{VV}\right)\left(\tilde{S}_{HH}+\tilde{S}_{VV}\right)^{*}\right\rangle\right]} \tag{17.32}$$

The denominator of (17.32) is affected by both the θ caused by the azimuth slope and the scattering characteristics of the medium. It will produce the phase wrapped and get a poor result. However, for media dominated by a single scattering mechanism (such as Bragg scattering from slightly rough seas), the HH and VV phases are close to the same phase, and this method is available.

17.2.3 Method of Estimation Orientation Angle by Polarization Signature

Schuler found that it is reasonable to estimate the azimuth orientation angle θ using the maximum position of the co-polarization signature for the bare mountain area where the Bragg scattering is dominant. The polarization signature method proposed by Van Zyl shows the characteristics of the scattering mechanism in a 3D map. The polarization response of each pixel is a function of the polarization orientation angle φ and the ellipticity angle χ. The eigenvalue peak of the ideal Bragg scattering horizontal surface lies at the VV polarization state ($\varphi = 90°$, $\chi = 0°$). The azimuthal surface slope will shift the peak from the VV polarization state. Similarly, for forests containing even-order scattering, this offset can be measured by the HH polarization ($\varphi = 0°$ or $180°$, $\chi = 0°$) state. Its value corresponds to the amount of change in the azimuth of the polarization state introduced by the azimuth slope (see formula (17.1)). The exact measurable range of the θ angle is limited to $-45°$ to $+45°$. Figures 17.5 and 17.6 show the simulation results of the polarization signature method. Figure 17.5a, b is the signatures and contours of standard Bragg scattering. Figure 17.6a, b is Bragg scattering signatures and contours after azimuth tilt of $+30°$. The maximum in Fig. 17.6 is shifted. The signatures method estimates the azimuth orientation angle θ by means of calculating this offset. The main characteristic of the signature estimation results are consistent with the terrain but are greatly affected by noise.

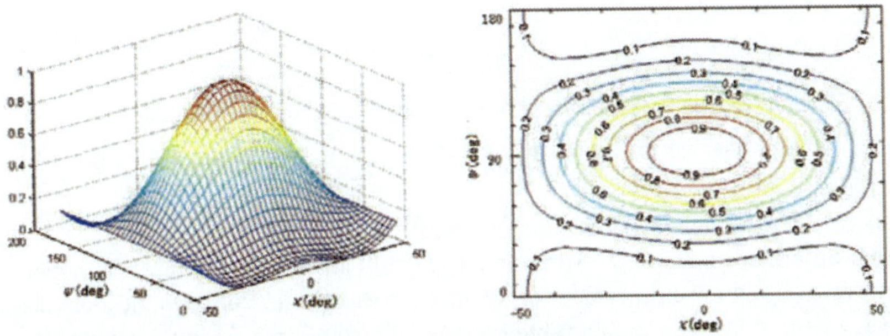

(a) Signature map (b) Contour map

Fig. 17.5 Signatures and contour maps of standard Bragg scattering

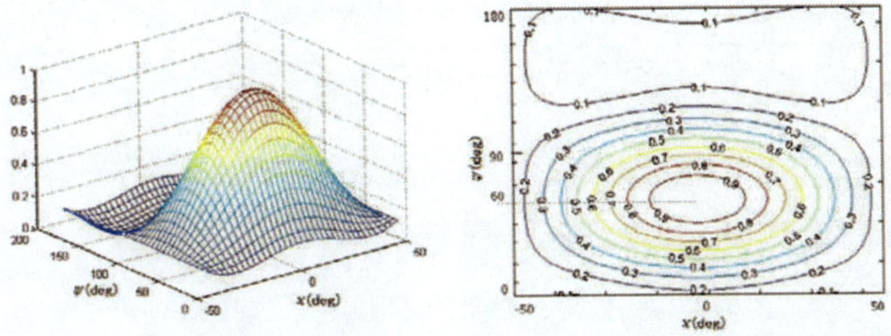

(a) Signature map (b) Contour map

Fig. 17.6 Signatures and contour maps of standard Bragg scattering with an azimuth tilt of 30°

17.2.4 Method of Estimating Orientation Angle by Target Decomposition

In the polarization target decomposition theory proposed by Cloude and Pottier et al., The angle α represents the characteristics of the scattering mechanism and represents the surface scattering from $\alpha = 0°$ to the dipole scattering $\alpha = 45°$, and to the even-order scattering $\alpha = 90°$. Cloude et al. verified that β is twice the angle of θ. Because the scattering mechanism is complex, intuitively, the dominant scattering vector should be more accurate in estimating the θ angle introduced by the azimuth slope. The result is shown in Fig. 17.7, which is heavy noisy. Figure 17.7a is the result of the θ angle obtained with the dominant eigenvector. Figure 17.7b is the result of the θ angle obtained by weighing three eigenvectors according to eigenvalues. The noise in Fig. 17.7b is smaller than that in Fig. 17.7a, but they are not as clear as the RR-LL method.

17.3 The Factors Affecting the Accuracy of Azimuth Orientation Angle Estimation

It can be seen from the above analysis that the estimation results of the RR-LL circular polarization azimuth estimation algorithm are relatively stable, and there are many restrictions on the application of other methods. In subsequent analysis, the RR-LL estimation method is adopted unless otherwise specified. In addition to the

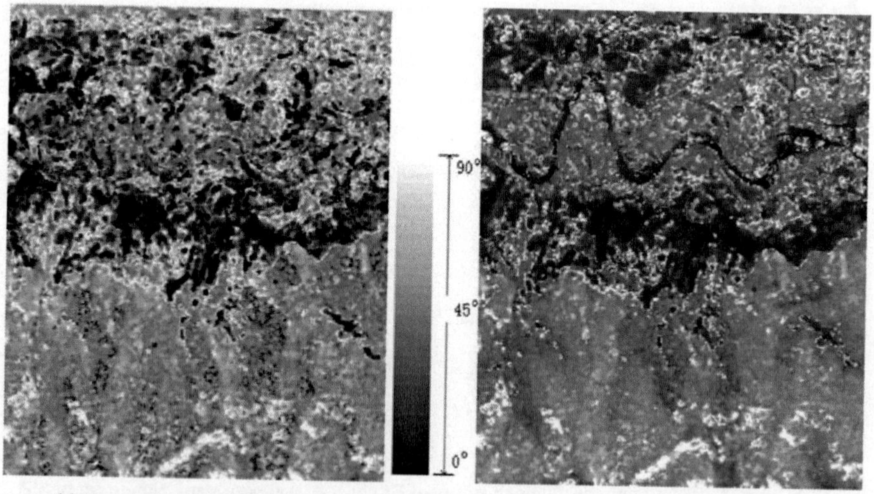

(a) Results of dominant feature vector (b) Results of three eigenvector weighted

Fig. 17.7 Results by target decomposition method

estimation accuracy of the algorithm itself, there are many other factors that affect the θ estimation.

17.3.1 Effect of Radar Frequency

A good azimuth orientation angle θ can be obtained from P- and L-band full-polarization SAR data, but the estimation effect of C or higher frequency bands is not good. The high-frequency electromagnetic wave has a short wavelength and weak penetrating ability and is sensitive to small scatters in a resolution unit, so its response is not sensitive to azimuth slope changes. The change in θ angle caused by smaller scatters overwhelms the change in angle caused by terrain tilt. It can be found that the θ angle noise of C-band data is very serious. On the other hand, the wavelength of P-band is longer, the penetrating ability is strong, and it is not sensitive to small scatterers. The θ angle obtained is better than that of the L-band. The following uses the multi-band (P-, L-, C-band) fully polarized SAR image data of NASA JPL in the BONANZA region as an example to analyze the θ angle results obtained from the data in different bands. The θ angles obtained from the data in different frequency bands are shown in Figs. 17.8, 17.9 and 17.10. Figures 17.8, 17.9, and 17.10a are a Span image of P-, L-, and C-band data; Fig. 17.10b is an θ result estimated from P-, L-, and C-band data; Fig. 17.10c is the histogram of θ. From the calculation results, the P-band data obtained an θ angle that matched the topography of the Span image well, and the noise was small. The L-band calculation results compare with the P-band results that the L-band specific noise increase significantly, but the θ angle map matching the terrain is still obtained. C-band data did not obtain a satisfactory θ angle. The θ angle is not mainly introduced by the azimuth slope,

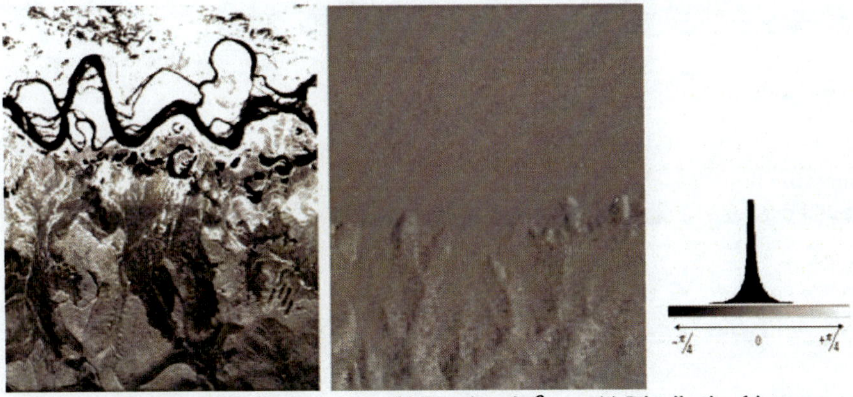

(a) Span image of P-band (b) Results of θ (c) Distribution histogram

Fig. 17.8 Azimuth orientation angle estimation of P-band

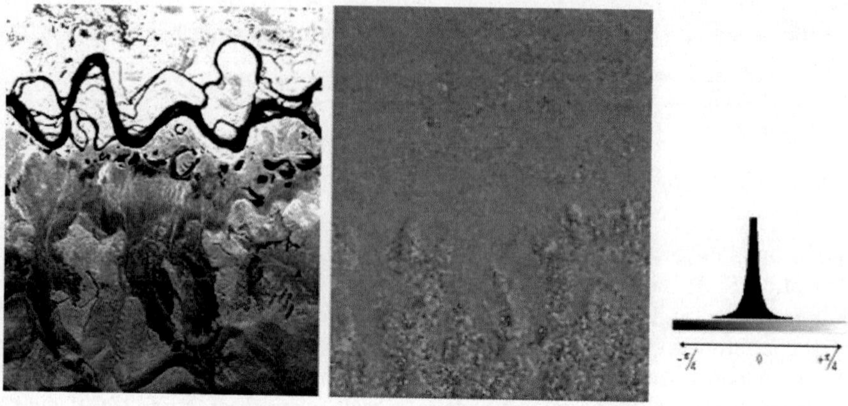

(a) Span image of P-band (b) Results of θ (c) Distribution histogram

Fig. 17.9 Azimuth orientation angle estimation of L-band

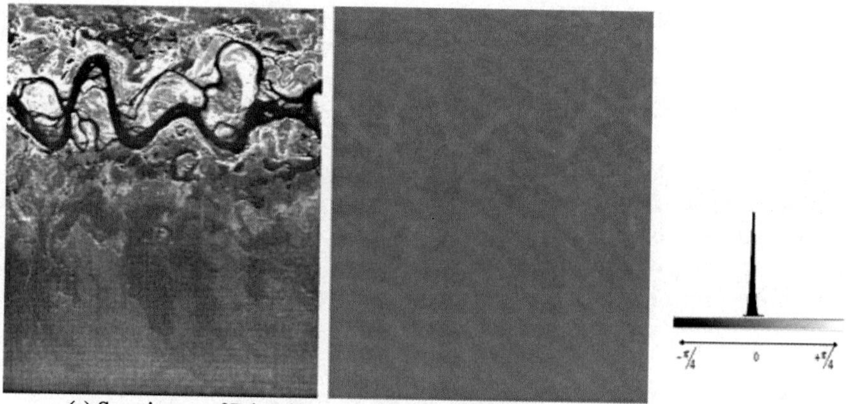

(a) Span image of P-band (b) Results of θ (c) Distribution histogram

Fig. 17.10 Azimuth orientation angle estimation of C-band

but by the branches and leaves of the vegetation. The above analysis shows that the P-band data can obtain the θ angle in dense forest areas.

17.3.2 Effects of Depolarization

The estimation accuracy of the θ angle depends on whether the azimuth offset is mainly caused by the azimuth slope. If the azimuth orientation angle of scatterers with different scattering mechanisms is caused by the azimuth slope, their confounding

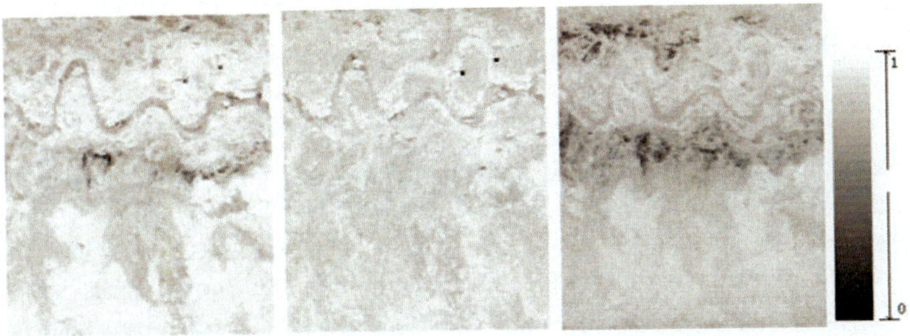

(a) P band (b) L band (c) C band

Fig. 17.11 H images of different band

has little effect on the estimation accuracy. For a fixed azimuth orientation angle θ, the matrix of the scattering (or correlation) sum is equal to the sum of the scattering (or correlation) matrix, the overall θ angle will be rotated, and the mixing of different scatterers will not introduce any errors. The previous analysis shows that the robustness of the circular polarization method stems from its independence from co-polarization and cross-polarization. As long as the θ angle is caused by the azimuth slope, the estimation accuracy of the circular polarization method is independent of the mixing of scatterers with different scattering mechanisms. The randomness of the scattering medium can be measured by entropy, which is also a measure of the depolarization effect. Figure 17.11a–c shows the scattering entropy of the data in different bands from Figs. 17.8, 17.9 and 17.10, respectively. The scattering entropy of the bright region in Fig. 17.11 is above 0.7. Comparing the estimation results of the scattering entropy and θ-angle in different bands, we cannot see any relationship between high scattering entropy and θ angle noise. However, if the azimuth orientation angle offset of large number of scatterers is introduced by tree branches or other influences, it is necessary to study how to separate the azimuth orientation angle caused by the azimuthal slope of the terrain from other components. This requires further research.

17.3.3 Importance of Polarization Calibration

The calibration of polarized SAR data is very important for the accuracy of θ angle estimation. The isolation of polarized channels and the accuracy of amplitude and phase calibration all affect the estimation of θ angle. Because the effects of $|S_{HV}|^2$ and co-polarization and cross-polarization phase differences are particularly significant, many polarization SAR calibration algorithms assume zero correlation between the co-polarized and cross-polarized terms. This assumption will introduce θ angle estimation errors. In addition, the tilt of the radar platform will also cause azimuth orientation angle. These should be compensated before the θ is estimated [2, 3, 8].

17.3.4 *Impact of Dynamic Range of SAR*

The dynamic range of the radar receiver is also very important for θ angle estimation. The estimation accuracy of the circular polarization method depends on the measurement accuracy of the $\left\langle\left(\tilde{S}_{HH} - \tilde{S}_{VV}\right)\tilde{S}^*_{HV}\right\rangle$ term. This term is usually much smaller than $|S_{HH}|^2$ or $|S_{VV}|^2$. The lack of dynamic range can make this term very noisy. For radar systems with small dynamic range and poor channel isolation, it is impossible to extract the θ angle.

17.4 Compensation of Topographic Azimuth Slopes for Fully Polarized SAR Data

The θ angle estimated by the previous method can be directly applied to polarized SAR data to compensate for uneven areas. Correct compensation of polarized SAR data is important to ensure the estimation accuracy of physical parameters such as soil moisture, surface roughness, snow thickness, and biological density. Lee et al. have given compensation studies for polarized SAR data. The compensation results for the terrain azimuth slope show that all the terms in the correlation matrix are adjusted except that the $\left\langle\left|\tilde{S}_{HH} + \tilde{S}_{VV}\right|^2\right\rangle$ is rotation-invariant. The biggest change is the real part of $\left\langle\left(\tilde{S}_{HH} - \tilde{S}_{VV}\right)\tilde{S}^*_{HV}\right\rangle$ [9].

Terrain azimuth slope affects the relative amplitude and phase of all terms except $\langle|S_{HH} + S_{VV}|^2\rangle$ in the polarization covariance matrix and correlation matrix. The angle of θ introduced by the surface slope can be regarded as the $v - h$ polarization base rotating around the radar sight angle Los by an angle θ. The basic principle of correction is to rotate the data in reverse around Los, a polarization azimuth orientation angle introduced by terrain slope θ. This is equivalent to rotating the antenna by an angle around the Los direction, so that the horizontal polarization component is parallel to the inclined surface. This compensation method is only applicable to fully polarized SAR data [10–18].

17.4.1 Azimuth Slope Compensation Method for Fully Polarized SAR Data

The methods of compensating the terrain azimuth slope in four formats of polarized SAR data are described below.

(a) **Compensation Single-look Complex Scattering Matrix**

The scattering matrix S is

$$S = \begin{bmatrix} S_{HH} & S_{HV} \\ S_{VH} & S_{VV} \end{bmatrix} \tag{17.33}$$

The new matrix after azimuth slope compensation is

$$S^{(new)} = \begin{bmatrix} \cos\theta & \sin\theta \\ -\sin\theta & \cos\theta \end{bmatrix} \begin{bmatrix} S_{HH} & S_{HV} \\ S_{VH} & S_{VV} \end{bmatrix} \begin{bmatrix} \cos\theta & -\sin\theta \\ \sin\theta & \cos\theta \end{bmatrix} \tag{17.34}$$

(b) **Compensation for single-look and multi-looks covariance matrices**

The covariance matrix C is

$$C = \begin{bmatrix} \langle |S_{HH}|^2 \rangle & \left\langle \sqrt{2} S_{HH} S_{HV}^* \right\rangle & \langle S_{HH} S_{VV}^* \rangle \\ \left\langle \sqrt{2} S_{HV} S_{HH}^* \right\rangle & \langle 2|S_{HV}|^2 \rangle & \left\langle \sqrt{2} S_{HV} S_{VV}^* \right\rangle \\ \langle S_{VV} S_{HH}^* \rangle & \left\langle \sqrt{2} S_{VV} S_{HV}^* \right\rangle & \langle |S_{VV}|^2 \rangle \end{bmatrix} \tag{17.35}$$

$\langle\rangle$ is the multi-looks average. A superscript * indicates a complex conjugate. The data compensation method is as follows:

$$C^{(new)} = VCV^T \tag{17.36}$$

where $V = \frac{1}{2} \begin{bmatrix} 1 + \cos 2\theta & \sqrt{2}\sin 2\theta & 1 - \cos 2\theta \\ -\sqrt{2}\sin 2\theta & 2\cos 2\theta & \sqrt{2}\sin 2\theta \\ 1 - \cos 2\theta & -\sqrt{2}\sin 2\theta & 1 + \cos 2\theta \end{bmatrix}$

(c) **Compensation of correlation matrix**

The Pauli-based scattering vector is

$$k = \frac{1}{\sqrt{2}} \begin{bmatrix} S_{HH} + S_{VV} \\ S_{HH} - S_{VV} \\ 2S_{HV} \end{bmatrix} \tag{17.37}$$

The correlation matrix is

$$T = \frac{1}{2}\begin{bmatrix} \langle |S_{HH}+S_{VV}|^2\rangle & \langle (S_{HH}+S_{VV})(S_{HH}-S_{VV})^*\rangle & 2\langle (S_{HH}+S_{VV})S_{HV}^*\rangle \\ \langle (S_{HH}-S_{VV})(S_{HH}+S_{VV})^*\rangle & \langle |S_{HH}-S_{VV}|^2\rangle & 2\langle (S_{HH}-S_{VV})S_{HV}^*\rangle \\ 2\langle S_{HV}(S_{HH}+S_{VV})^*\rangle & 2\langle S_{HV}(S_{HH}-S_{VV})^*\rangle & 4\langle |S_{HV}|^2\rangle \end{bmatrix} \tag{17.38}$$

The new matrix after azimuth slope compensation is

$$T^{(new)} = UTU^T \tag{17.39}$$

where:

$$U = \begin{bmatrix} 1 & 0 & 0 \\ 0 & \cos 2\theta & \sin 2\theta \\ 0 & -\sin 2\theta & \cos 2\theta \end{bmatrix} \tag{17.40}$$

(d) **Compensation of circularly polarized basis covariance matrix**

The compensated circularly polarized basis scattering vector elements are rewritten as follows:

$$\begin{aligned} \tilde{S}_{RR} &= S_{RR}e^{-i2\theta} \\ \tilde{S}_{LL} &= S_{LL}e^{i2\theta} \\ \tilde{S}_{RL} &= S_{RL} \end{aligned} \tag{17.41}$$

Equation (17.10) also gives the compensated circularly polarized covariance matrix. Rewrite it as follows:

$$\tilde{G} = \begin{bmatrix} \langle |S_{RR}|^2\rangle & \sqrt{2}\langle (S_{RR}S_{RL}^*)\rangle e^{-i2\theta} & \langle (S_{RR}S_{LL}^*)\rangle e^{-i4\theta} \\ \sqrt{2}\langle (S_{RL}S_{RR}^*)\rangle e^{i2\theta} & 2\langle |S_{RL}|^2\rangle & \sqrt{2}\langle (S_{RL}S_{LL}^*)\rangle e^{-i2\theta} \\ \langle (S_{LL}S_{RR}^*)\rangle e^{i4\theta} & \sqrt{2}\langle (S_{LL}S_{RL}^*)\rangle e^{i2\theta} & \langle |S_{LL}|^2\rangle \end{bmatrix} \tag{17.42}$$

17.4.2 Azimuth Slope Compensation Results and Analysis of Fully Polarized SAR Data

There is a corresponding relationship between the various representations of polarization data. For the compensation result analysis and explanation, the compensation of the correlation matrix as an example, only be taken below as following. It can be seen from Eqs. (17.38) and (17.39) that $\langle |S_{HH} + S_{VV}|^2\rangle$ is rotation-invariant. If the θ angle is caused entirely by the azimuthal tilt of the terrain, in theory, compensation will make the average correlation matrix transpose symmetrical. To achieve transposition symmetry, both $\langle (S_{HH} + S_{VV})S_{HV}^{*} \rangle$ and $\langle (S_{HH} - S_{VV})S_{HV}^{*} \rangle$ should be zero, because only one parameter (angle θ) can be adjusted. This is mathematically impossible. It can be found that the compensation has the largest impact on the real part of $\langle (S_{HH} - S_{VV})S_{HV}^{*} \rangle$, but has a smaller impact on the $\langle (S_{HH} + S_{VV})S_{HV}^{*} \rangle$. Because there are many other scatters in a resolution unit, it can also cause azimuth orientation angle. Therefore, the azimuth orientation angle component caused only by the azimuth of the terrain should be extracted to compensate the polarized SAR data.

Taking the P-band full-polarization SAR image data of NASA JPL in the BONANZA area as an example, the results of the experimental research on the compensation of the terrain azimuth slope are given below. Figure 17.12a–d is the comparisons of the amplitudes of the significant terms of the correlation matrix before and after compensation. The left column is the original data amplitude image and the right column is the compensated data amplitude image. $\langle |S_{HH} + S_{VV}|^2 \rangle$ is unchanged before and after compensation. The estimated θ angle shows that the θ angle changes greatly in the middle and lower regions of the figure, that is, the changes are more obvious in this part of the image before and after compensation.

However, the amplitude of the original data is weak in this part of the image. In order to obtain better visual effects, the image amplitude value is adjusted to make this part of the image brighter. Obviously, the $\langle (S_{HH} + S_{VV})S_{HV}^{*} \rangle$ changes most great in the amplitude image, followed by $\langle (S_{HH} + S_{VV})S_{HV}^{*} \rangle$ and $\langle (S_{HH} + S_{VV})S_{HV}^{*} \rangle$, and other terms also change to some extent.

Figure 17.13a is a Span image of the P-band full-polarization SAR data in BONANZA. Figure 17.13b is the estimated θ angle and Fig. 17.13c is the θ in bright line. Take the bright line in Fig. 17.13b as example and compare the amplitudes of the correlation matrix terms before and after the row compensation, as shown in Figs. 17.14a–h. The dotted line indicates the original data amplitude, and the solid line indicates the compensated data amplitude. Figure 17.14a, b are the amplitude before and after the compensation of $\langle |S_{HH} - S_{VV}|^2 \rangle$ and $\langle |S_{HV}|^2 \rangle$. There is no change in $\langle |S_{HH} + S_{VV}|^2 \rangle$ before and after compensation. Figure 17.14c–h are the real part and amplitude changes of $\langle (S_{HH} + S_{VV})S_{HV}^{*} \rangle$, $\langle (S_{HH} + S_{VV})S_{HV}^{*} \rangle$, and $\langle (S_{HH} + S_{VV})S_{HV}^{*} \rangle$ three terms before and after compensation. From the comparison of these images, it is found that the biggest change is $\langle (S_{HH} + S_{VV})S_{HV}^{*} \rangle$. Followed by $\langle (S_{HH} + S_{VV})S_{HV}^{*} \rangle$ and $\langle (S_{HH} + S_{VV})(S_{HH} - S_{VV})^{*} \rangle$. And other items also have some changes. And this change matches with the θ angle change, where the real part changes of $\langle (S_{HH} + S_{VV})S_{HV}^{*} \rangle$ and $\langle (S_{HH} + S_{VV})S_{HV}^{*} \rangle$ are particularly significant. This result is in good agreement with the comparison of amplitude images.

17.4.3 Compensation Example of Terrain Azimuth Slope

Terrain azimuth slope affects the amplitude and phase of polarized SAR data, and it will have a greater impact on the classification of polarized SAR data. The

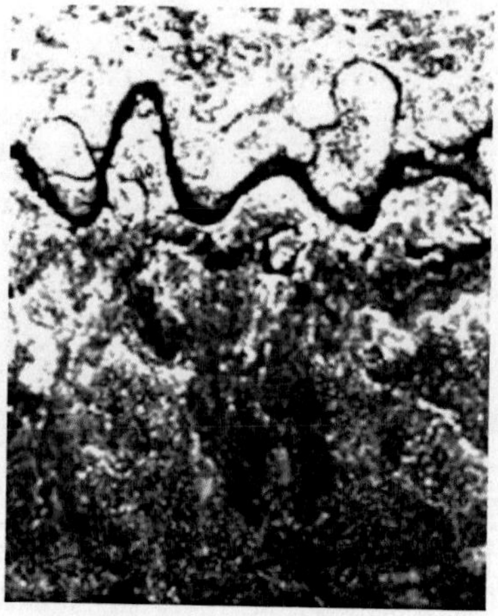

(a) Original image of $\langle (S_{HH} + S_{VV}) S_{HV}^{*} \rangle$

(b) Compensated image of $\langle (S_{HH} + S_{VV}) S_{HV}^{*} \rangle$

(c) Original image of $\langle (S_{HH} - S_{VV}) S_{HV}^{*} \rangle$

(b) Compensated image of $\langle (S_{HH} - S_{VV}) S_{HV}^{*} \rangle$

Fig. 17.12 Comparison of amplitude images before and after compensation of correlation matrix

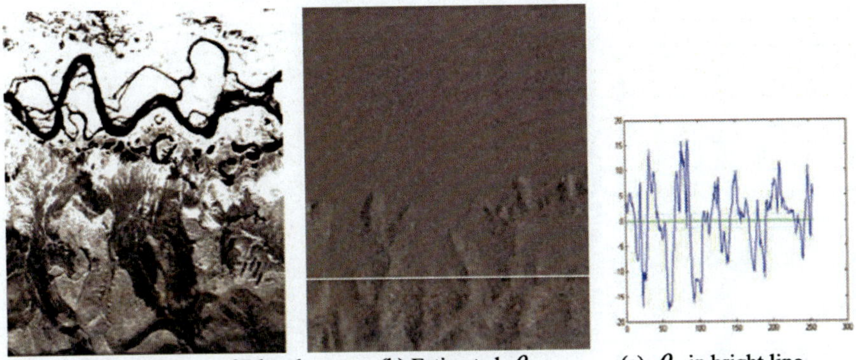

(a) Span image of P band (b) Estimated θ (c) θ in bright line

Fig. 17.13 Span image of P-band in BONANZAC and the estimated θ

following discusses unsupervised classification of polarized SAR data and compares the differences in classification results before and after compensation.

Because the θ angle accuracy obtained from P-band data is high, the noise is small, and the data compensation effect is well, so the P-band data in BONANZA is selected for analysis. Figure 17.15 is the θ angle of the P-band data. The θ angle estimation in the middle and lower areas of the image is effective and the data compensation is obvious. Therefore, it will pay attention to compare the classification results before and after compensation in this area. The eigenvalue decomposition is rotation-invariant. In theory, the terrain azimuth slope will not affect the H/α classification results. Therefore, the original data H/α classification results are used as input to analyze the changes of ML classification results before and after compensation. Figure 17.16 is the ML classification result of the original data, Fig. 17.16a is the ML classification result after second iterations, and Fig. 17.16b is the ML classification result after fourth iterations. Figure 17.17 is the ML classification result of the compensated data, Fig. 17.17a is the ML classification result after second iterations, and Fig. 17.17b is the ML classification result after fourth iterations. Combining the θ angle and comparing the classification results before and after compensation, It can be seen that the classification results before compensation are seriously affected by the ridges and piedmonts, and the classification results are strongly matched with the shape of the mountain terrain; After compensation, the influence of the mountain terrain is greatly reduced and obtained classification results which can much more reflect the features of the terrain. The comparison between the images in the elliptical area in Figs. 17.16, 17.17 and the elliptical area in Fig. 17.15 can illustrate this situation as well. It can be seen that the polarization SAR data is greatly affected by the classification of the azimuth slope of the terrain. And great impact on the classification of the polarization SAR data. In order to obtain a classification result which more fully reflects the characteristics of ground scattering, it is necessary to compensate the terrain azimuth slope.

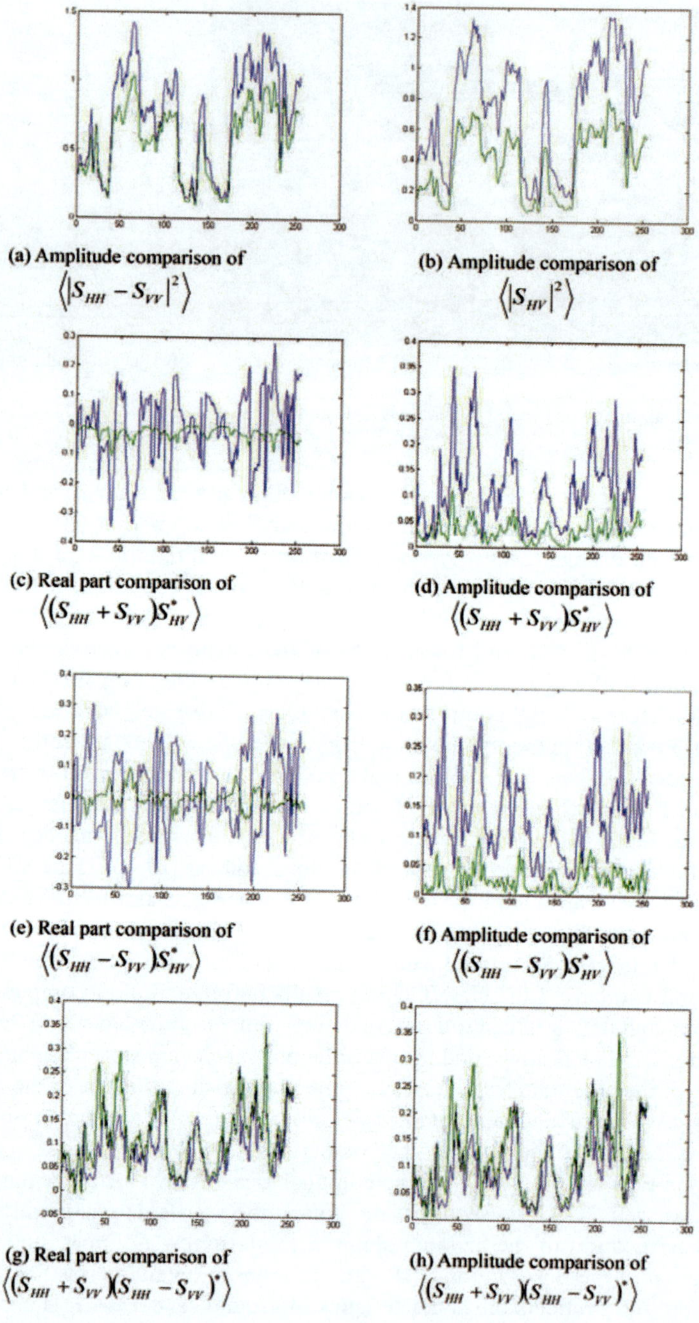

(a) Amplitude comparison of $\langle |S_{HH} - S_{VV}|^2 \rangle$

(b) Amplitude comparison of $\langle |S_{HV}|^2 \rangle$

(c) Real part comparison of $\langle (S_{HH} + S_{VV}) S_{HV}^{*} \rangle$

(d) Amplitude comparison of $\langle (S_{HH} + S_{VV}) S_{HV}^{*} \rangle$

(e) Real part comparison of $\langle (S_{HH} - S_{VV}) S_{HV}^{*} \rangle$

(f) Amplitude comparison of $\langle (S_{HH} - S_{VV}) S_{HV}^{*} \rangle$

(g) Real part comparison of $\langle (S_{HH} + S_{VV})(S_{HH} - S_{VV})^{*} \rangle$

(h) Amplitude comparison of $\langle (S_{HH} + S_{VV})(S_{HH} - S_{VV})^{*} \rangle$

Fig. 17.14 Comparison of the terms of the correlation matrix before and after compensation

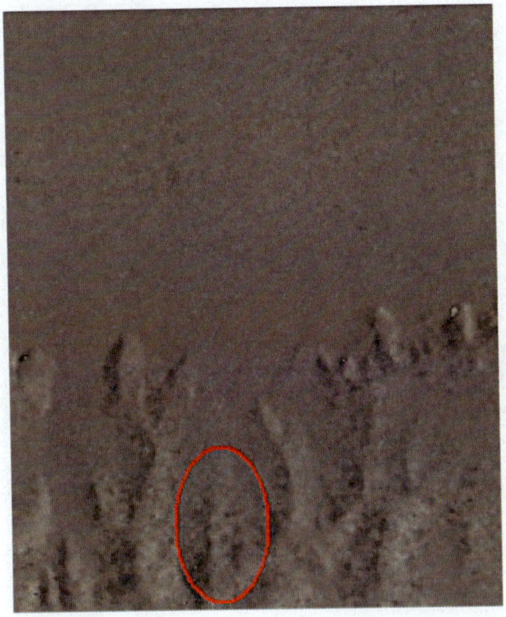

Fig. 17.15 θ angle of P-band

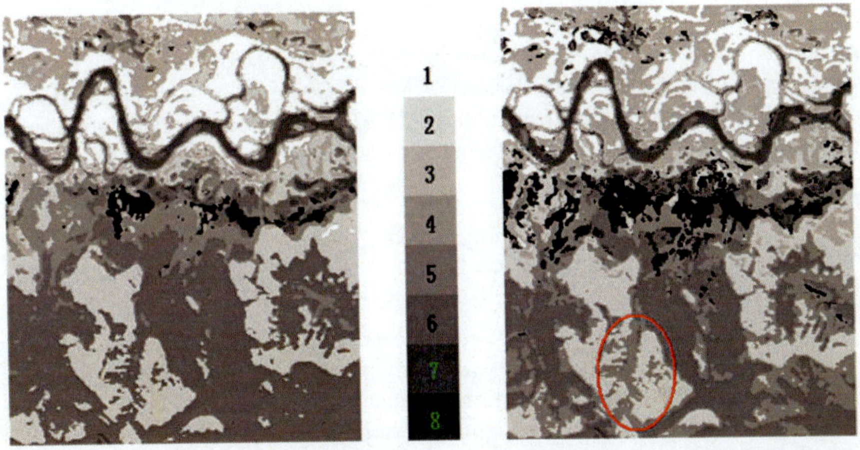

(a) ML classification after 2 iterations (b) ML classification after 4 iterations

Fig. 17.16 ML classification results of original data

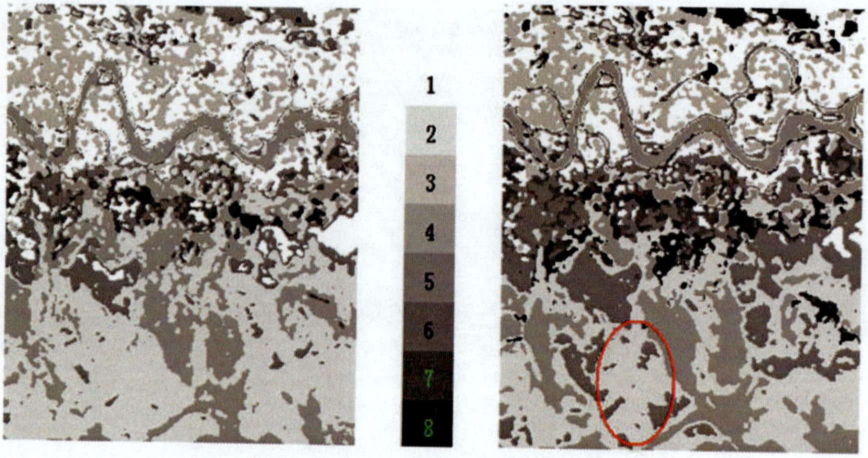

(a) ML classification after 2 iterations (b) ML classification after 4 iterations

Fig. 17.17 ML classification results of compensated data

17.5 Application of Azimuth Orientation Angle Introduced by Terrain Slope

On the one hand, the azimuth orientation angle introduced by the terrain azimuth slope affects the application of polarized SAR data; On the other hand, the estimation of azimuth orientation angle θ from polarized SAR data has certain application value in obtaining DEM of terrain and estimation of the tilt of the ocean surface.

17.5.1 Generating DEM Model

The extracted θ angles can be used to generate a topographic map. In order to extract the θ angle in the vertical direction, two orthogonal polarized SAR data are needed. The ground inclination in two directions can be calculated by Eq. (17.1). The slope data is then integrated to obtain a Digital Elevation Map (DEM) of the terrain. Due to the folding effect of the radar image, it is difficult to match the two images. At present, the accuracy of DEM results obtained by this method is lower than that obtained by SAR interference [9, 19–23].

17.5.2 Application on the Ocean Surface

Another application of the θ angle is to directly estimate the tilt of the ocean surface. The surface of the ocean can be considered homogeneous in most cases and is a Bragg scattering property. This provides excellent conditions for azimuth estimation. Because the range slope of the ocean surface is very small, the mediation term for range in Eq. (17.1) can be omitted. However, the radar look angle must be corrected. Lee et al. used angles to study the small surface slope angle of the Gulf Stream ocean surface, and obtained results with an accuracy of $<1°$. Schuler, Kasilingam et al. extended this study to estimating the slope spectrum of ocean waves [5–7, 9, 20, 21]. Ainsworth et al. extended this research to the study of ocean surface characteristics [4, 5].

17.6 Summary

This chapter describes the influence of terrain azimuth tilt on radar polarization, that is, terrain azimuth slope will cause the radar polarization azimuth to shift. The method of extracting θ angle from polarized SAR data is mainly introduced. Among them, the estimation result of RR-LL method is relatively stable, which is suitable for wide application. This method was used to conduct experimental research on the fully polarized SAR data collected in BONANZA by the AIRSAR system of NASA JPL, and the estimated results of the θ angle of the P-, L-, and C-band data were obtained.

This chapter focuses on the azimuth slope compensation algorithm for polarized SAR data scattering matrix, covariance matrix, correlation matrix, and circular polarization matrix format data. The classification of the data before and after the P-band compensation in BONANZA is studied. The experimental results show that the effect of the classification of the data after compensation is significantly smaller than the original data.

Compared with the scattering entropy H and the scattering α angle, it is found that there is no clear relationship between the noise model of the θ angle and the entropy H or α angle. But it is corresponding to the brighter areas in $\langle |S_{HV}|^2 \rangle$, $\langle |S_{HH} - S_{VV}|^2 \rangle$ and $\langle |S_{HH} + S_{VV}|^2 \rangle$. Strong radar echoes correspond to the high complexity of smaller radar incident angles and scattering mechanisms. And it is related to trees and very rough ground. For multiplicative coherent speckle noise, the higher the noise mean, the higher the standard deviation. The strong amplitude changes in the brighter areas of polarized SAR data cause noise in the θ estimates.

References

1. J.J. Van Zyl, The effect of topography on radar scattering from vegetated areas. IEEE Trans. GRS **31**(1), 153–160 (1993)
2. J.-S. Lee, D.L. Schuler, T.L. Ainsworth, On the estimation of radar polarization orientation shifts induced by terrain slopes. IEEE Trans. GRS **40**(1), 30–40 (2001)
3. D.L. Schuler, J.-S. Lee, D. Kasilingam, Surface roughness and slope measurements using polarimetric SAR data. IEEE Trans. GRS **40**(3), 687–698 (2002)
4. J.S. Lee, E. Krogager, D.L. Schuler, T.L. Ainsworth, On the estimation of polarization orientation angles induced form azimuth slopes using polarimetric SAR data, in *Proceedings Geoscience and Remote Sensing Symposium. IGARSS 2000*, vol. 3 (2000), pp 1310–1312
5. D.L. Schuler, J.S. Lee, T.L. Ainsworth, Terrain slope measurement accuracy using polarimetric SAR data, in *Geoscience and Remote Sensing Symposium. IGARSS '99 Proceedings*, vol. 5 (1999), pp. 2652–2654
6. D.L. Schuler, J.S. Lee, G. de Grandi, Measurement of topographic surface slopes using polarimetric SAR images, in *Geoscience and Remote Sensing Symposium, IGARSS '94. Surface and Atmospheric Remote Sensing: Technologies, Data Analysis and Interpretation*, vol. 3 (1994), pp. 1467–1469
7. D.L. Schuler, J.S. Lee, G. De Grandi, The remote sensing of topography using polarimetric SAR, in *Geoscience and Remote Sensing Symposium. IGARSS '95*, vol. 3, (1995), pp. 1789–1791
8. M.S. Dawson, J. Olvera, A.K. Fung, M.T. Manry, Inversion of surface parameters using fast learning neural networks, in *Geoscience and Remote Sensing Symposium. IGARSS '92*, vol. 2 (1992), pp. 910–912
9. D.L. Schuler, J.-S. Lee, Measurement of topography using polarimetric SAR images. IEEE Trans. GRS **34**(5), 1266–1276 (1996)
10. I. Hajnsek, E. Pottier, Terrain correction for quantitative moisture and roughness retrieval using polarimetric SAR data, in *Geoscience and Remote Sensing Symposium. Proceedings IGARSS 2000*, vol. 3 (2000), pp. 1307–1309
11. L. Iong-Sen, D.L. Schuler, Polarimetric SAR data compensation for terrain azimuth slope variation. IEEE Trans. GRS **38**(5), 2153–2163 (2000)
12. L.H. Ulander, Radiometric slope correction of synthetic aperture radar images. IEEE. Trans. GRS, **34**(5), 1115–1122 (1996)
13. D.L. Schuler, L. Jong-Sen, T.L. Ainsworth, W.M. Boerner, Correction for terrain azimuthal slope effects in polarimetric SAR images, in *Geoscience and Remote Sensing Symposium. Proceedings IGARSS 2000*, vol. 3 (2000), pp. 1304–1306
14. P.E. O'Neill, A.Y. Hsu, M.S. Seyfried, SAR terrain correction for improved soil moisture estimation in a mountain watershed, in *Geoscience and Remote Sensing Symposium. IGARSS '94*, vol. 3 (1994), pp. 1429–1431
15. G. Riegler, W. Mauser, Geometric and radiometric terrain correction of ERS SAR data for applications in hydrologic modelling, in *Geoscience and Remote Sensing Symposium Proceedings. IGARSS '98*, vol. 5 (1998), pp. 2603–2605
16. J.T. Parr, Geometric transformation and terrain correction of remotely-sensed imagery, in *Geoscience and Remote Sensing Symposium. IGARSS'89*, vol. 2 (1989), pp. 440–441
17. X. Liu, R. Yang, Experimental research on data compensation of terrain azimuth and its application in classification using polarimetric SAR data. J. Electron. Inf. Technol. **9**(26), 343–347 (2004)
18. X. Liu, *Studies on Information Processing Technique of Fully-Polarimetric Synthetic Aperture Radar* (Chinese Academy of Sciences, PhD, Institute of Electronics, 2004), p. 01
19. S. Li, M. Sturm, S. Kyriazis, T. Logan, Mapping snowmelt sequence at Ivotuk, Alaska, using RADARSAT SAR images acquired with different standard beams, in *Geoscience and Remote Sensing Symposium. Proceedings IGARSS 2000*, vol.2 (2000), pp. 481–483

20. D.L. Schuler, J.S. Lee, Terrain slope and surface roughness studies using circularly polarized SAR backscatter, in *Geoscience and Remote Sensing Symposium Proceedings. IGARSS '98*, vol. 4 (1998), pp. 2187–2189
21. D.L. Schuler, J.S. Lee, G. de Grandi, Measurement of topographic surface slopes using polarimetric SAR images, in *Geoscience and Remote Sensing Symposium. IGARSS '94*, vol. 3 (1994), pp. 1467–1469
22. H. Lee, J.G. Liu, Spatial decorrelation due to topography in the interferometric SAR coherence imagery, in *Geoscience and Remote Sensing Symposium. IGARSS '99 Proceedings*, vol. 1 (1999), pp. 485–487
23. R. Eisler, C. Lewis, Cooperative robotic map correlation from relative position and terrain slope measurements, in *World Automation Congress, 2002. Proceedings of the 5th Biannual*, vol. 14 (2000), pp. 329–333